Ethology and Behavioral Ecology of Marine Mammals

Series Editor

Bernd Würsig, Department of Marine Biology, Texas A&M University at Galveston, Galveston, TX, USA

The aim of this series is to provide the latest ethological information on the major groupings of marine mammals, in six separate books roughly organized in similar manner. These groupings are the 1) toothed whales and dolphins, 2) baleen whales, 3) eared seals and walrus, 4) true seals, 5) sea otter, marine otter and polar bear, and 6) manatees and dugong, the sirens. The scope shall present 1) general patterns of ethological ways of animals in their natural environments, with a strong bent towards modern behavioral ecology; and 2) examples of particularly well-studied species and species groups for which we have enough data. The scope shall be in the form of general and specific reviews for concepts and species, with an emphasis especially on data gathered in the past 15 years or so. The editors and authors are all established scientists in their fields, even though some of them are quite young.

More information about this series at http://www.springer.com/series/15983

Claudio Campagna • Robert Harcourt
Editors

Ethology and Behavioral Ecology of Otariids and the Odobenid

Springer

Editors
Claudio Campagna
Wildlife Conservation Society
Marine and Argentina Programs
Buenos Aires, Argentina

Robert Harcourt
Department of Biological Sciences
Macquarie University
Sydney, NSW, Australia

ISSN 2523-7500 ISSN 2523-7519 (electronic)
Ethology and Behavioral Ecology of Marine Mammals
ISBN 978-3-030-59183-0 ISBN 978-3-030-59184-7 (eBook)
https://doi.org/10.1007/978-3-030-59184-7

This Springer imprint is published by the registered company Springer Nature Switzerland AG.
The registered company address is: Gewerbestrasse 11, 6330 Cham, Switzerland

Fig. 1 Walrus. Photo: Paul Nicklen

Fig. 2 South American fur seal. Photo: Enric Sala

Fig. 3 Southern sea lion. Photo: Manu San Félix

You mustn't swim till you're six weeks old,
Or your head will be sunk by your heels;
And summer gales and Killer Whales
Are bad for baby seals.

Are bad for baby seals, dear rat,
As bad as bad can be;
But splash and grow strong,
And you can't be wrong.
Child of the Open Sea!

R. Kipling "The White Seal" 1893, a tale of the Pribilof fur seal

Introduction to the Series

We—multiple topic editors and authors—are pleased to provide a series on ethology and behavioral ecology of marine mammals. We define ethology as "the science of animal behavior," and behavioral ecology as "the science of the evolutionary basis for animal behavior due to ecological pressures." Those ecological pressures include us, the humans. We determine, somewhat arbitrarily but with some background, that "marine mammals" habitually feed in the sea, but also include several mammals that went from saltwater oceans back into rivers, as seen in the chapter by Sutaria et al., in the first book on *Odontocetes*. Polar bears represent a somewhat outlier "marine mammal," as they are quite at home in the sea, but can also feed on terrestrial mammals, birds, berries, lichens, and mosses.

In six books, we include toothed whales (the odontocetes); baleen whales (the mysticetes); sea lions and fur seals (the otariids) as well as the walrus; true seals (the phocids); the special cases of the sea otter and polar bear; and manatees and the dugong (the sirens). Each of our chosen editors and their chapter authors have their own schedules, so the series will not arrive in the order given above, but within the 5 years of 2019 through 2023, all six marine mammal books on *Ethology and Behavioral Ecology of Marine Mammals* will see the light of day, and you, the readers, will be able to ascertain their worth and their promise, as to current knowledge and to accumulating data while our fields of science advance.

Since the first book on odontocetes came out in 2019, we added a seventh final book, on *The Human Factor*, with chapters on past assaults on marine mammals, continuing assaults on the marine and other environments, dawning of awareness of assaults, and perhaps ways that we humans can and must do better. Several of us simply felt that to detail modern science of marine mammal ethology and behavioral ecology was not enough—we need to be aware of the amazingly destructive Anthropocene epoch in which we live, and try to improve, for all of nature (and therefore also for us). While topics of human influence run throughout each of the first six books, a concentration on human actions and potential solutions is needed.

Not all mammals that occur in marine waters are represented, nor all that have gone back to freshwater. Thus, there is nary a mention of marine-feeding bats, marine-feeding river otters, those aspects of beluga whales that foray way up into major rivers, seals living in land-locked lakes at times thousands of kilometers from the ocean, and other species that occasionally make the marine environment or—as generally accepted marine mammals—adjacent freshwater systems their home. Such are the ways of a summary, and we apologize that we do not fully encompass all.

As series editor, I have been a science partner to all major taxonomic entities of this series, but to this only because I have been in the marine mammal field for about 50 years now, with over 65 graduate students who—in aggregate—have conducted research on all seven continents. In no manner do I pretend to have kept up with all aspects of diverse fields of modern enquiry. It is a special privilege (and delight) to have multiple up-to-date editors and their fine authors involved in this modern compilation and am extremely grateful (and humbled) for this. Still learning, and ever so.

Each chapter is reviewed by the book editors, peer reviewed by other scientists as chosen by the editors, and perused and commented on by me. If you learned something new and imparted that to your colleagues, students, or your own mentors, then the series and sections of it shall have been worthwhile.

With respect and best wishes

Tortolita Desert, Arizona Bernd Würsig
December 2020

Preface

When we started this journey, we were told that editing a book is a labor of love. That advice proved sage as we discovered that it is also a source of much angst. The usual editing headaches were compounded by each of us facing evacuation from severe bushfires first in Australia (RGH) and then for CC's lovely home in Patagonia, followed by floods for one of us (RGH) and then, of course, COVID-19. However, we persevered with strong and continuing support from our beloved series editor Bernd Würsig and now we are in the fortunate position of presenting an edited collection from a wonderful group of authors almost as diverse as the delightful animals we study.

Sea lions, fur seals, and walrus are semi-aquatic mammals found across the cool, temperate seas of the Pacific, Indian, and South Atlantic Oceans, as well as the frigid waters of the Arctic. Their distribution overlaps that of many human populations, and they inhabit harbors and ports across the world, ensuring their familiarity to the populace at large. The intense motivation and high trainability of some (explored in Chap. 17) has seen California sea lions (*Zalophus californianus*) a popular feature of circuses and aquatic shows. For millennia, they have played a valuable role in human societies, primarily for food and/or skins, and have totemic importance for many indigenous cultures. They were also terribly overexploited in the eighteenth and nineteenth centuries, particularly fur seals, and many populations have only started recovering over the past one-half century. The latter recovery coincides with a flurry of scientific attention that we are pleased to present here.

In this volume, we have brought together a collection of authors from across the globe, representatives of all the continents on which fur seals, sea lions, and walrus (*Odobenus rosmarus*) occur. We are delighted to have in this mix, authors ranging from highly respected pioneers of ethological studies of otariid and odobenid behavior to dynamic young researchers with their fingers on the pulse of the latest advances in behavioral ecology.

We, the editors, started our careers with behavioral studies of South American otariids, CC in Argentina investigating mating strategies in southern sea lions (*Otaria byronia*) and RGH in Peru on maternal strategies and behavioral development in South American fur seals (*Arctocephalus australis*). We continue to be captivated by these delightful animals, and in this volume, we hope to showcase many of the developments that have refined our understanding of their behavior, from the nascent studies over half a century ago to the highly diverse and complex research of the present day.

Part 1 opens with a narrative from those early days, vividly recalled by Roger Gentry (Chap. 1). Many of these earliest behavior observations have stood the test of time, as has Roger's plea for field-based experimental manipulations in the pursuit of scientific excellence. His chapter also includes personal revelations about the foundation of the Society for Marine Mammalogy—fascinating snippets that provide insight into the etiology of many research directions in marine mammal behavior, as well as sage advice for those just starting out. His chapter sets the scene for the whole volume and is followed by a state-of-the-art overview of the intricate links between foraging ecology, energetics, and reproduction by Daniel Costa and Ana Valenzuela Toro (Chap. 2), building on many decades of exploration and synthetic thinking about these topics by Daniel and his colleagues. They explain the evolutionary context for the income breeding pattern that is common to all female fur seals, sea lions, and walrus and explore the energetic constraints that drive this pattern. This link between energetics, reproduction, and income breeding has important consequences for maternal behavior strategies and social interactions, reviewed by Maritza Sepulveda and Robert Harcourt (Chap. 3).

As semi-aquatic mammals, finding food at sea, diving to great depths to forage, efficiently dismantling large and potentially dangerous prey, and evading predation are all critical to their phenomenal success—and this is the focus of Part 2. Many of the breakthroughs that provide insights into their adaptations to a deep, dark, and dangerous watery world have come with developments in and miniaturization of technology that have opened a window into the underwater world. Tiphaine du Dot and Christophe Guinet (Chap. 4) provide a comprehensive account of the different dive strategies displayed by fur seals and sea lions, showing convincingly that there are distinct foraging strategies—benthic, epi-pelagic, and mesopelagic—with different representations across and within species. The implications of different strategies and their energetic costs are explored and related back to the success of different populations. Delving even deeper into feeding, David Hocking et al. (Chap. 5) provide a fascinating exploration of the evolutionary origin of their vastly different feeding strategies, describing in detail the exact mechanisms required for raptorial, semi-aquatic, and suction feeding in a visually rich essay.

Reproduction is the core of evolution and due to their accessibility during the breeding season as semi-aquatic mammals, and because they provide an excellent model for comparative studies, fur seals and sea lions (and to a lesser extent the walrus) have been intensively studied in relation to reproduction. Indeed, their behavior has become fundamental to our understanding of mammalian reproductive strategies. Part 3 opens with two chapters by Marcelo Cassini. Chapter 6 revisits the classic work on sexual selection and male success, reviews modern genetic evidence, and shows that behavioral estimates of male success greatly overestimate the degree of reproductive skew between males compared to genetic estimates. In Chap. 7,

Marcelo further refines our views on the evolution of mating systems, systematically dismantling many long-standing tenets and rightly promoting the role of females as active agents in mate selection. These two chapters set the scene for the remainder of this section. Masashi Kiyota (Chap. 8) reinforces some of the important ideas promoted in Cassini's chapters with an elegant study using meticulous observations of known animals over multiple years, revealing a number of subtle behavior strategies by both male and female northern fur seals (*Callorhinus ursinus*). He shows that male territorial strategies change as males mature, and how female behavior drives male strategies. In Chap. 9, Claudio Campagna et al. discuss some of the seminal work on raiding parties by male southern sea lions. This chapter highlights the plasticity of mating systems of otariids, with males adopting multiple diverse strategies in their attempts to successfully reproduce. However, in sexually dimorphic mammals such as these, the consequences of selfish strategies can be severe on others. The, often sad, consequences of the determination of individuals to succeed regardless of the cost to others are apparent in the overview of infanticide by Sergei Ryazanov (Chap. 10). He explores the different hypotheses promoted for infanticide and shows that intentional intra-specific infanticide, whether for food or other resources, is relatively rare in otariids, but that inter-specific infanticide or accidental killing is relatively more common.

Reproductive decisions, whether harassment of females by males, philopatry by females, or dispersal decisions by either sex, have population consequences beyond those of the individual that makes them, particularly for species that have gone through periods of intense exploitation such as fur seals during the fur trade era. Carolina Bonin (Chap. 11) explores the consequences of these decisions for fur seals and sea lions and elegantly marries the strong connections between behavior, ecology, and evolution. She outlines just how otariids escaped the worst consequences of potential population bottlenecks that should theoretically arise from severe exploitation, probably because more animals escaped than formerly realized. To cap off this section, Valentina Franco-Trecu (Chap. 12) presents empirical, i.e., genetic, evidence of a rare event, intergeneric pinniped hybridization between a southern sea lion and a South American fur seal with a resultant fertile hybrid successfully producing a live pup. But hybridization has costs, and selection should produce mechanisms such as reinforcement to select against its occurrence. Jennifer Sinclair and colleagues (Chap. 13) convincingly reveal one such mechanism in a fur seal hybridization zone on Crozet Island. Recovering populations of these two species of fur seals now breed sympatrically at three locations around the southern hemisphere, Crozet Is., Marion Is., and Macquarie Is. Whether this is due to a mix-up of habitats as a result of sealing, or a natural overlap, the consequent hybridization provides a powerful experiment in reproductive selection. By manipulating Subantarctic fur seal (*A. tropicalis*) males in the field, so that they imitate male Antarctic fur seals (*A. gazella*), Jennifer Sinclair and colleagues show that females not only actively choose mates but can choose the correct species to mate with using visual cues.

This elegant field experiment leads directly from a focus on reproduction into the world of communication, cognition, learning, and recognition (Part 4). On land, the acoustic medium is highly effective at conveying directional and complex information, and Isabelle Charrier (Chap. 14) provides a comprehensive review of acoustic calls by all potential protagonists: males, females, and pups. Income breeding results in repeated separations between mothers and their pups throughout lactation, and Isabelle details the use of calls in mother–pup recognition, including when recognition appears during development and its persistence, as well as the unique codas they carry. She shows that otariid and walrus males show greater vocal diversity than females, and produce calls both in air and under water, with sufficient individual vocal stereotypy to allow individual recognition among rivals and mate choice for females. In Chap. 15, Isabelle outlines the state of the art in field investigations of communication using vision and olfaction. Mothers can visually distinguish young by age class, and olfactory cues have an important role in mother–pup recognition.

The use of sensory organs for communication demands a detailed description of how semi-aquatic animals can effectively use their sensory organs in both air and water. We know that when we humans open our eyes underwater, vision is blurred, and when we listen, sounds come from all directions. This is because the difference in density between water and bone is so low that sound transmits to our ears through our skull (bone conduction), and we cannot detect where it is coming from. Frederike Hanke et al. (Chap. 16) provide a clear description of just how otariids overcome these difficulties and the extent of their remarkable capabilities. This is followed by the same authors, this time led by Peter Cook (Chap. 17). Peter and colleagues provide further testament to the late Ron Schusterman (earlier discussed by Gentry in Chap. 1) and his lead in experimental explorations of otariid cognition and language learning, particularly with his protégé Rocky, the California sea lion. In so doing, they also describe the latest research demonstrating the behavioral and cognitive flexibility of otariid species, their intense task focus and high motivation, and unique features of their neurobiology.

The application of modern theories of learning to current human–wildlife conflicts is the subject of Chap. 18 by Zachary Schakner and Daniel Blumstein, with a case history of the adaptable California sea lion. Given that many otariid populations are recovering from nineteenth-century exploitation and coming into conflict with fisheries, this is a particularly timely essay. They outline how fundamental associative and non-associative learning mechanisms in California sea lions could be used to manage wildlife–human conflicts involving otariids. Of course, as is clear from that chapter, not all animals are the same; some individuals are particularly problematic and aggressive or bold around humans; others shy away and avoid interactions with humans. The importance of these individual behavioral differences is now well recognized. Eugene DeRango and Jonas Schwartz (Chap. 19) provide a thoroughly modern exploration of the importance of personality in evolution, in behavioral development, and delve into its physiological basis. They explore its importance in individual foraging strategies and outline how it may have a crucial role to play in developing successful conservation efforts. An important aspect highlighted in the personality chapter is the boldness–shyness continuum, and DeRango and Schwartz discuss how this may lead to fundamental differences in risks faced during exploratory and play behavior by young pups. Fur seals, sea lions, and walrus are notoriously playful, and Clara Llamazares-Martín and Elisabetta Palagi (Chap. 20) review the theoretical underpinnings of play, and then present an up-to-date description of play behavior, a joyful read for anyone who has watched puppies chasing each other around the rookery, fighting over a rock or piece of seaweed, or lining up in the surf.

In Part 5, we dive deeper into our understanding of some of the best-known species, as well as some exciting comparative studies. It is not the intention of this book to cover all species individually, but instead to provide examples of interesting case studies that explore some of the cross-species aspects of biology treated in Part 1. The history of exploitation, recovery, or continued precarious situation for some species is described in detail. Our experts show how there are strong links to behavior in both vulnerability to exploitation and their ability to adapt to human-induced changes in the environment. We start with a fascinating investigation by Diego Rodriguez and colleagues (Chap. 21) of the ever plastic southern sea lion, looking at how males in particular now haul out in marinas and ports, habituating and adapting to humans to an extraordinary degree. This is followed by a detailed monograph on that most charismatic and most gregarious of pinnipeds: the walrus, by Ted Miller and Anatoly Kochnev (Chap. 22), who focus on the social functions of the characteristic walrus tusks. We have taken the opportunity with this chapter to present to the world many hitherto little known findings from Russian ethologists who have been studying walrus over many decades. Staying with the giants of this world, we turn to the largest and one of the most intensely studied otariids, the North Pacific Steller's sea lion (*Eumetopias jubatus*). Andrew Trites (Chap. 23) discusses the differing trajectories of the eastern (stable) and western (endangered) stocks with insights into how behavior has influenced conservation efforts and success. As we continue our journey around the sea lion world, Enrique (Kike) Crespo (Chap. 24) takes us south to Argentina, where he provides a history of modern exploitation and recovery for southern sea lions. He shows how their opportunistic and plastic behavior together with high juvenile survival has fortunately resulted in an increasing rate of population recovery in recent decades. Crossing the South Pacific, Louise Chilvers (Chap. 25) explores how behavior interacts with vulnerability in the deepest diving otariid, the New Zealand sea lion (*Phocarctos hookeri*). New Zealand sea lions, already greatly reduced in abundance and distribution by extirpation from the main New Zealand islands, have been pushed to the edge in recent years by disease outbreaks, and remain highly vulnerable to anthropogenic pressures. Fortunately, there are some positive signs, and Louise details how maternal behavior is driving the recolonization process as these now protected animals return to mainland New Zealand. Further west, Rebecca McIntosh and Benjamin Pitcher (Chap. 26) reveal their deep passion for the "labrador of the sea" the Australian sea lion (*Neophoca cinerea*) and provide some intriguing insights into why of all the otariids, only this species exhibits a unique, asynchronous, aseasonal reproductive cycle.

Still in Australia, Roger Kirkwood and Rebecca McIntosh (Chap. 27) describe the behavior of the largest of fur seals, the Australian fur seal (*A. pusillus doriferus*). This species is not unlike the present human inhabitants of this smallest continent, being larger and more boisterous than the other Arctocephalinae (except for its subspecific sister the Cape fur seal (*A. pusillus pusillus*)). A benthic feeder, it is in many ways the ecological equivalent of the sea lion, rather than the agile, epipelagic, fur seal norm.

The contrast in behavioral ecology between sea lions and fur seals is of increasing importance, given that many of the latter were nearly exterminated during the peak of the fur trade, and now are recovering and potentially competing with sympatric sea lions. Fernando Elorriaga-Verplancken et al. (Chap. 28) compare foraging habits of the only northern hemisphere fur seal, the Guadalupe fur seal (*A. townsendi*) in San Benito Archipelago, where it is sympatric with California sea lions. In an almost classic differentiation, stable isotope analyses from juveniles and adults of both species show that Guadalupe fur seals forage offshore mainly on squid across widely distributed foraging grounds, while California sea lions are more coastal and feed primarily on fish. Validating this approach, Valentina Franco-Trecu et al. (Chap. 29) compare stable isotope analysis with traditional scat analysis and tracking of animals at sea to show that female South American fur seals in Uruguay are typical fur seals, i.e., epi-pelagic divers, feeding on vertically migrating prey that rise to shallower depths during the night.

Fittingly for a book focused on behavioral ecology, Marjorie Riofrío-Lazo and Diego Páez-Rosas (Chap. 30) compare the Galapagos fur seal (*A. galapagoensis*) and Galapagos sea lions (*Zalophus wollebaeki*) that inhabit a cradle of evolutionary studies, the Galapagos Islands. Being exceptionally vulnerable to environmental variability with its location at the end of the highly productive, cold upwelling Humboldt Current, and with limited ability to disperse from the archipelago, both species are extremely vulnerable to environmental change. As they evolved in the Galapagos, we now see adaptation at work, both species showing reduced body size and prolonged lactation strategies to cope with low and variable productivity. Yet, reflecting the fundamental foraging dichotomy between the two groups, fur seals and sea lions also show divergent adaptations. Galapagos fur seal mothers are dependent upon midwater schooling prey and so adjust the length of foraging trips and the duration of visits to attend pups in response to fluctuations in food, while benthic Galapagos sea lions reduce competition among individuals by displaying different foraging strategies in relation to prey availability.

In writing this book, we had the good fortune to work with colleagues and friends from all over the world. We thank the many of you who gave freely of your time, your insights, and your exquisite photos to produce what we hope is an enthralling book on these most wonderful models for the behavioral disciplines. Many people, both authors and beyond, have contributed to peer review to ensure that the science presented here is correct and of interest. Our dedicated, patient, and insightful Series Editor, Bernd Würsig, has been with us every step of the way and we would never have completed this work without his support, his good-natured editing, and his endless patience. We extend our thanks to Springer International and the editors and advisors of Springer International, with special thanks to Éva Lörinczi and Bibhuti Sharma for their help, patience, and understanding as deadlines were necessarily extended. RGH thanks Patricia Majluf and John Croxall for introducing him to fur seals all those years ago, to his students and colleagues for helping maintain the passion, and to CC for inviting him to join the editing team. CC goes back in time to thank his doctorate advisor, Burney Le Boeuf, and the members of his graduate committee, Robert Trivers, Jim Estes, and Ron Schusterman, for providing, at the outset of his career, a guide of how to blend objectivity and creativity, indispensable to keeping alive his passion for stimulating science.

Buenos Aires, Argentina — Claudio Campagna
Sydney, NSW, Australia — Robert Harcourt

Contents

Part V Diving Deeper into Species

Part I
Behavior on Land and at Sea: Introduction

Chapter 1
Otariid Ethology: One Researcher's Historical Perspective

Roger L. Gentry

Abstract This chapter summarizes methods used to study otariids from the late 1700s to the early 1960s, when a scientific publication claimed that the California sea lion (*Zalophus californianus*) used echolocation. That paper triggered the first-ever conference on all diving mammals which, over time, became the Society for Marine Mammalogy. The chapter then briefly discusses otariid life cycles by sex, differences in otariid annual cycles by latitude, and nature of social behavior and social organization on land. It describes manipulative experiments used to test females for the duration of estrus, and the possible existence of female mate choice. Finally, it discusses the first deployments of multiple dive time-depth recorders to reveal diving behavior at sea, and the first deployments of those recorders that were coupled with isotope injections to explore diving physiology in marine mammals.

Keywords Social organization · Life cycle · Annual cycle · Estrus · Mate choice · Diving behavior · Diving physiology

1.1 Introduction

Field studies on otariid seals began in the 1700s and 1800s in different parts of the world. Researchers back then asked very basic questions because it was not known how many species existed, where they were found, population sizes, seasons for pupping and mating, and basic information about social behavior. Species studied in the earliest years included the Galapagos (*Arctocephalus galapagoensis*), northern (*Callorhinus ursinus*), subantarctic (*Arctocephalus tropicalis*), South African (*Arctocephalus pusillus pusillus*) and South American (*Arctocephalus australis*) fur seals; and the Steller (*Eumetopias jubatus*), California (*Zalophus californianus*), Galapagos (*Zalophus c. wollebaeki*) and South American (*Otaria flavescens*) sea

R. L. Gentry (✉)
Grass Valley, CA, USA

C. Campagna, R. Harcourt (eds.), *Ethology and Behavioral Ecology of Otariids and the Odobenid*, Ethology and Behavioral Ecology of Marine Mammals,
https://doi.org/10.1007/978-3-030-59184-7_1

lions. Early researchers differed somewhat in their research interests, and they usually published their results in the languages they spoke, usually French, English, Spanish, or German. The use of different languages somewhat limited the sharing of information among field researchers. Some researchers possibly knew of a few others that had similar research interests, but no regional or international communities of researchers had yet formed. Furthermore, this early research was conducted by established, working scientists. No college courses were available on marine mammal biology at that time, few graduate students were looking for projects, and the public then had no concern about marine mammal welfare.

This chapter traces how our present international community of researchers developed out of this background. This is my personal view of the field, based on my research, meetings I attended, and the literature published between 1963 and 1998. For another assessment of this field, see LeBoeuf and Würsig (1985).

1.2 Otariid Study Methods Over Time

The earliest research methods used to study pinnipeds were voyages of discovery to lands unknown or little known to westerners, and laboratory dissections of cadavers brought back from those voyages. Little was known in the 1700s about which species of mammals existed where, so voyages of discovery were made to see and collect specimens for anatomical and taxonomic purposes. Vitus Bering made two voyages into the North Pacific Ocean in that century. Georg W. Steller (1749) was the naturalist on the second trip, and he published the first scientific descriptions of the northern fur seal and Steller's sea lion. Karl Wilhelm Illiger (1811) first identified pinnipeds as a separate taxonomic group of mammals, based on anatomy. Elliott (1882) and later Scammon (1884) conducted voyages of discovery that involved new marine mammal sightings. These journeys continued into the mid 1960s with little change in methods. On one such voyage to Guadalupe Island, Mexico, Carl L. Hubbs (1956) discovered that the Guadalupe fur seal (*Arctocephalus townsendi*) had survived the era of fur sealing in the eastern Pacific islands.

The first modern otariid researcher specifically trained in animal behavior, and who used modern methods to study otariid social behavior in the field was Richard S. Peterson. He collected quantitative data on the social behavior of tagged northern fur seals on St. Paul Island, Alaska from 1961–1964 (Peterson 1965). His data collection methods were not described in his dissertation, in his book on the California sea lion (Peterson and Bartholomew 1967), nor to me when I was his graduate student. He started an elephant seal research project at Año Nuevo Island which Burney J. Le Boeuf and students later turned into a world-class program. Peterson died before any of his other otariid work was published, a scientific loss.

The modern field of pinniped research began to coalesce as a result of some research by Thomas C. Poulter (Fig. 1.1), a polymath and former munitions designer. While looking for a remote site from which to launch small rockets in

Fig. 1.1 Dr. Thomas C. Poulter, early 1960s (Photo by Richard Jennings)

1961, he landed on Año Nuevo Island in central California. The island was inhabited by four species of pinnipeds; harbor (*Phoca vitulina*) and northern elephant seals (*Mirounga angustirostris*), and Steller and California sea lions (Fig. 1.2). Poulter recognized the island and the animals thereon as valuable research resources. He contacted Robert T. Orr of the California Academy of Science, a pinniped specialist, to help him survey the island and publish papers describing the populations there.

Poulter established a pinniped research laboratory in an abandoned Nuclear missile base at the southern end of San Francisco Bay, near Fremont, California (Fig. 1.3). He built pens and water pools for keeping and feeding pinnipeds, and populated the facility with representatives of each of the species found on the island. He built an anechoic water tank in one of the buildings for making high quality underwater recordings of the sounds these animals made. He located the anatomical source of those sounds by recording from three directions. Based on his recordings of California sea lions (but not the other species) he published a paper in the journal *Science* claiming that this species uses echolocation (Poulter 1963). He somehow acquired a totally blind but robust northern fur seal female which he used as (apparent) evidence that other otariids used echolocation as well. He named this facility the Biological Sonar Laboratory.

While searching the literature for a topic to present in a 1963 graduate seminar, I first read Donald Griffin's recent book on bat echolocation *Listening in the Dark* (Griffin 1959), and was surprised to also find Poulter's paper claiming pinniped

Fig. 1.2 Año Nuevo Island, California in summer 1968. All animals shown are non-breeding *Zalophus* males

Fig. 1.3 The Biological Sonar Laboratory, Fremont California in the late 1960s. It is now Coyote Hills Regional Park, Fremont (Photo by Roger L. Gentry)

echolocation. I called him to ask for an interview, and we met a week later. He began the interview by describing how his idea of pinniped echolocation had originated. When he was in the Antarctic with Admiral Byrd in the 1930s, the long, eerie-sounding calls of Weddell seals often came up through the ice on which he stood and he wondered what function they served. Griffin's book suggested to him that they may have been used in echolocation.

Poulter showed me two other projects that were then under way in his laboratory. Charles E. Rice was studying the echolocation abilities of blind humans in an anechoic room, and Ronald J. Schusterman was studying perception, cognition, learning, and problem-solving in California sea lions in his own compound and test tank. He used that species because they are highly food-motivated and easily trainable. Schusterman was a comparative psychologist and used all the data collection methods common to others in his field: event recording, time-ruled check sheets, the longitudinal histories of known individuals, and others (Schusterman 1981). When the interview ended, Poulter invited me to join the laboratory and measure underwater auditory localization in California sea lions for my master's thesis (Gentry 1966). The ability to locate the source of a sound was critical to his claim of echolocation. I conducted the hearing study after learning from Schusterman how to train research animals and how to make the required measurements.

After finishing my research, I assisted Schusterman in testing California sea lion females for visual acuity, learning, cognition, and problem solving. We also studied social behavior between his females and a half-grown male. He later published a paper stating that this species could not use echolocation (Schusterman 1967). This was probably a correct assertion because the visual acuity of this species is sufficient to find food without using sound. However, the paper forever severed the working relations between him and Poulter. The Bio-Sonar Laboratory remained active until Poulter's death at the laboratory in 1978.

During the summer of 1965 the laboratory's photographer, Richard Jennings, asked me to help transport photographic equipment to Año Nuevo Island, where he was filming the social behavior of Steller sea lions. He worked from a blind on the edge of a breeding group, at times within arm's reach of the animals. He had a close view of all aspects of Steller sea lion behavior, including male territorial behavior, births, mating, pup-rearing, female aggression, movements to sea and back by females, and pup play. Some aspects of social behavior changed with time of day, wind speed, and tide height. This trip made it clear to me that the behavior of animals in the wild was much closer to my own interests than was underwater directional hearing.

After finishing the laboratory work, I studied social behavior of Steller sea lions at the island for my doctorate (Gentry 1970). The field work covered three breeding seasons and raised questions about the effect that climate and physical factors, like wind and tide, had on otariid social behavior. To answer those questions I then studied several other species of otariids. The first was social behavior of the New Zealand fur seal (*Arctocephalus forsteri*) on the South Neptune Islands of Australia. Following that I studied social behavior and diving in the northern fur seal in Alaska and Russia (from 1974 to 1998), then social and diving behavior of the

New Zealand sea lion (*Phocarctos hookeri*), and finally diving behavior of the South African fur seal. The results of all these studies were presented at marine mammal conferences but not all the data reached publication due to time constraints between projects. This background, as well as the published literature, contribute to the remainder of this chapter.

1.3 Otariid Life Cycles

The family Otariidae consists of six species of sea lions and nine of fur seals. Male and female otariids have very different life cycles. Juvenile males gather on traditional non-breeding sites where they develop fighting skills in play matches, alternating with feeding bouts at sea. After a number of years, which varies among species, males grow large enough to compete for territories on sites where females traditionally land and give birth. If successful in establishing a territory, they spend one or more breeding seasons defending the territory from other males, and mating when possible. Males risk losing their territory if they leave land to feed during the breeding season. Therefore, all otariid males fast while holding territory. They prepare for this fast by laying down extra fat (termed the "fatted male effect") before the breeding season. This effect even occurs in captive California sea lion males and is accompanied by an increase in male-male aggression (Schusterman and Gentry 1971). The breakdown of stored fat during fasting substitutes for both food and water for approximately 4 weeks. Territorial males drink seawater if it is available in their territory (Gentry 1980). When older males lose their territories to younger males, they usually move to landing areas[1] where juvenile males gather. They may return yearly until they become senescent. The rigors of territorial life limit the male's lifespan to about three quarters that of females.

Females first arrive on breeding areas[2] at about age 4 or 5 at the start or end of a breeding season, depending on species. They mate with an established male and undergo a delayed implantation of the blastocyst that results in birth about 12 months later. Females come into estrus and mate a few days after giving birth. In contrast to the territorial behavior of males, adult females of all species have some freedom to move about the breeding area, both before and after parturition. Females can live for 20 to 25 years.

[1]The places where non-breeding males landed were referred to as "Hauling Grounds" by early sealers. The term "Landing Area" avoids jargon and more clearly describes what animals do there.

[2]Similarly, early sealers called the places where mating occurred, "Rookeries," in reference to rooks, a bird species. The term "Breeding area" more clearly describes what animals do there.

1.4 Otariid Annual Cycles

Three types of annual cycles exist among otariids. The Galapagos fur seal and sea lion breed and feed in tropical waters and do not undergo an annual migration. Their reproductive behavior tends to be asynchronous without the formation of dense breeding colonies. Two species, the northern fur seal and Antarctic fur seal (*Arctocephalus gazella*), breed at high latitudes in a brief season (4 to 6 weeks) after which males depart while females remain nearby, foraging for 4 to 5 months until their offspring reach weaning age. Thereafter, females migrate alone to more temperate waters where they forage until the next breeding season begins. All other otariids breed and feed in temperate waters. Females and young may remain at the breeding site until the young are able to swim to more distant landing sites. These mothers often support their young until they are 1 year old, and a few females suckle both a newborn and the unweaned pup from the previous season. The sexes usually separate after a breeding season that lasts 4 to 6 weeks.

The movements that male otariids use in fighting, threat displays, and mating appear to be instinctual, not learned by young males that observe and mimic the behavior of adult males. Two sets of observations support that statement. Poulter raised orphaned Steller sea lion pups in his laboratory in the early 1960s, and I kept northern fur seal mothers and pups in the laboratory in the late 1970s. Pups in both cases used recognizable components of adult male fights and threat displays, and sexually mounted each other correctly, without ever having witnessed such behavior in adult males.

I helped Gerald (Gerry) L. Kooyman record diving behavior of South African fur seals in 1978 at Kleinzee, South Africa (Fig. 1.4). We did not observe their breeding behavior and social organization because the season had ended before we arrived. However, whenever we approached, captured, or physically restrained females for instrument attachment, they acted more like California sea lions than like other fur seals. During the approach phase of a capture if one female fled, then all others near her would also flee, and ultimately the beach would be cleared of animals. California sea lions flee in that way but northern fur seals do not. Also, when we caught these females with a noose pole they vigorously rolled sideways like noosed California sea lions do, and resisted being restrained for instrument attachment much more vigorously than did northern fur seals. Interestingly, Berta and Churchill (2012) consider the South African fur seal to be taxonomically closer to sea lions than to other fur seals. Our observations of female behavior at capture tend to support that statement.

1.5 Social Behavior and Organization

All otariids have similarities in their general form of social organization (McCann 1980). Adult males are larger than females and establish seasonal breeding territories. Females are gregarious, but do not appear to form stable social bonds except

Fig. 1.4 Gerry Kooyman (right) and the author in Africa

with their young. Breeding occurs seasonally at specific locations and times of year when individuals congregate in shoreline breeding colonies (with the exception of the Australian sea lion, *Neophoca cinerea,* McIntosh and Pitcher, Chap. 26). Social organization among otariids, as a rule, involves male territoriality during the breeding season Otariids have not developed the alternative forms of social organization that occur in group-living terrestrial mammals, such as social dominance hierarchies, or group territorial defense (Schein 1975). One factor that may explain this difference is that otariids tend to feed individually at sea and only form groups seasonally and briefly on land.

Male otariids usually establish territories before the first females arrive for the breeding season. The males' body conditions deteriorate during an extended fasting interval, and many begin to abandon their territories before all females have mated. Most species mate and females rear their young on sand or cobble beaches, or on rock substrate at the sea's edge. A few species have been seen climbing nearly vertical rock surfaces.

Behavioral differences are known among otariids. The New Zealand sea lion, for example, may breed on dirt or grass under shade trees instead of on beaches (The Snares Island, Crawley and Cameron 1972; Figure of Eight Island, New Zealand, Gentry, pers. obs.). Females that breed on Enderby Island mate on beaches as do the females of other otariids. But after mating, they move uphill and inland where they raise their young under dense brush (Gentry, unpublished data) unlike the other species I studied. Male Galapagos fur seals tend to hold territories where shade exists, and about 16% of male Juan Fernandez fur seals (*Arctocephalus philippii*) establish fully aquatic territories (Boness and Francis 1991). These are only minor deviations from the typical otariid pattern.

Otariid breeding sites can be quite stable over time. Northern fur seals have bred on Bering Island, Russia for 275 years. This stability results either from natal philopatry (females giving birth at the site of their own births), or site fidelity (repeat birthing at the same location when the mother's birth site is unknown). The best evidence for breeding near the natal site comes from tagging studies of northern fur seals. From 1956 to 1968, the U.S. government intentionally killed 315,000 females on terrestrial breeding colonies to reduce herd size and stimulate the population growth rate. Some 7940 of these animals had been tagged at birth, and about 78% of them were later killed on the breeding area of their own births. A few were killed within meters of the exact birth sites recorded during tagging (reported in Gentry 1998). The females in my study that had been marked as adults gave birth within about 8.3 m of the exact location in successive years. The dates of their arrival at the colony depended on whether the individual female was pregnant; those that were pregnant in successive years arrived within about 3 days of the same date each year. Those that missed a yearly pregnancy arrived about 7 days later than their usual date, but returned to their typical date the following year if they were again pregnant (Gentry 1998) This predictability implies that the female population is not the mass of anonymous animals it appears to be from casual observation. It is a collection of individuals, each having tendencies to use a specific birthing site at a specific time of season. While such fine-scale patterns of social organization are not well-known for other species, it is possible that long-term detailed behavioral observations of known individuals, and/or genetic studies, will reveal even more subtle aspects of social structure in otariid colonies.

Female otariids arriving at the start of a breeding season typically join existing female groups rather than resting alone among males. They are gregarious but not social; at least there is no evidence of female-female social bonds beyond kin. Such evidence might be found by observing known individuals repeatedly resting together, moving among female groups together, or leaving for and returning from foraging trips together. Females on shore typically move from one female resting group to another in unpredictable pattern, and in the absence of male interference. If female social bonds occur in any species, they would most likely develop in those that do not make an annual migration (see Wolf and Trillmich 2007, 2008; Wolf et al. 2007).

The unhindered movement of females from one resting group to another during the breeding season shows that female groups are not "harems" that are under the control of males. The opposite seems to be true. Female northern fur seals form groups on highly predictable breeding areas irrespective of the individual males that have established territories there in any given year. Females may give birth and mate at a given site for 15 or more seasons, whereas males hold a territory for from one to three seasons. This means that females choose birthing and mating sites, not the individual males that are present.

The tendency of male otariids to herd females is not uniform among species. Male New Zealand sea lions on territory never herded females during my study. Some half-grown males attempted to herd females that were arriving from sea, briefly and unsuccessfully. Steller sea lion males in California occasionally blocked the movements of females, but only briefly and early in the season, especially when females were on their way to sea. Northern fur seal males herded females only at the start of the season when the female group was small. Later in the season, when female group size had increased, some females escaped the group whenever the male left to capture a newly-arriving female. That is, male herding of females in the northern fur seal became counter-productive at some group size and most males eventually stopped all attempts to herd.

The male reproductive strategy in northern fur seals is not to increase their reproductive success by forcibly herding females into groups. It is to establish a well-defined territory on ground where females predictably gather, and to take advantage of female gregariousness to increase their reproductive success. The large size and aggressive demeanor of adult male otariids did not evolve in the context of dominating females, but as essential components of male-male competition for access to female gathering sites. The term "harem," used to describe collections of females associated with a single male, is misleading when applied to otariids, and should be abandoned.

The social behavior of New Zealand sea lions seems to differ from that of the other otariids I studied in terms of male territorial behavior, male-female interactions, and stability of the female population. In two seasons, breeding males established territorial sites only after the females had arrived, not before as in other temperate breeding otariids. Females at the east end of the breeding beach formed a single group, and thereafter moved westward. Every few days a group of 10 or more females that lacked pups would gather at the west end of the female group, huddle tightly, then rush about 10 m west and stop where there were no other females. A few days later all remaining females would join them and abandon the area they had formerly occupied. The males holding these just-abandoned territories would then swim west of the female group and establish a new territory into which the females would move later in the season. Some males defended three different territorial sites along the beach, apparently trying to keep in contact with the moving female group. The breeding aggregation moved on average about 3 m/day throughout the season.

Group movements of New Zealand sea lions suggest that adults have no apparent attachment to specific land sites as do other otariids. Perhaps breeding on featureless sand makes one site as good as any other for parturition and mating. However, just breeding on sand does not explain why females constantly moved along the beach. These movements seem to explain why males did not try to establish territories before the females arrived.

In my experience, New Zealand sea lion males do not herd females, even those that are obviously ready to mate. Several males showed no response when fully receptive females left their territories and entered the territory of an adjacent male. However, small juvenile males frequently herded pups into small groups, blocked their movements, and tried to mount them sexually just as males of some other species act toward females. This herding behavior by small juvenile males wanes with age and apparently does not serve as practice for some future adult need, as it seems to in other species.

1.6 Otariids and the Society for Marine Mammalogy

Otariid seals, especially the California sea lion, played a central role in focusing scientific attention on marine mammal biology, and on bringing marine mammal welfare to the public's attention. These trends started with a series of marine mammal conferences that Poulter held at his laboratory. He had spent decades in research and understood the importance to researchers of having contact with peers. He announced that a first-ever conference would be held on biological sonar and all diving mammals at his laboratory in the summer of 1964. The response was very positive. The 25 active marine mammal researchers at that time attended. The group included academics but also some researchers from a U.S. Navy laboratory who were using California sea lions in their research. The dolphin biologist John C. Lilly attended, as did Winthrop Kellogg and Victor B. Scheffer, whose book on pinnipeds (Scheffer 1958) had recently become available. Gerry Kooyman and I were the only two graduate students there.

From the formal way conference attendees first met and conversed it was clear that they did not know each other personally. By conference's end, however, they had clearly found common ground and were sharing research ideas and addresses, just as modern researchers do, and were looking forward to another conference. Poulter complied and held a few more conferences at the laboratory but moved it across the bay to the Stanford Research Institute in Palo Alto when the audience outgrew his laboratory space. Thereafter, the audience grew steadily until Poulter, when quite old, turned over future conferences to Kenneth S. Norris, then at the University of California in Santa Cruz. Norris expanded the meetings to include all marine mammal species, and all research topics that were of interest to the attendees. A vote held in 1981 formally created the Society for Marine Mammalogy with its first meeting in San Francisco. Today, over 50 years after Poulter's first meeting, few of the Society's well over 2000 international members know that the Society's original birthplace was an abandoned missile base in rural California, or that a paper wrongly claiming echolocation in the California sea lion was the catalyst that brought society members together. However, most members realize that marine

Fig. 1.5 Attendees of the Fur Seal Symposium, Cambridge, England 23–27 April 1984. Front row, left to right, G.I.H. Kerley, L.A. Fleischer, D. Torres. Second row: R.L. Gentry, R.L. DeLong, P.D. Shaughnessy, G.L. Shaughnessy, M.E. Goebel, A.E. York, A. Trites, P. Majluf, T. S. McCann, A. Ponce-de-Leon. Third row: B. Tollu, C.W. Fowler, D.F. Costa, M.N. Bester G.L. Kooyman, I. Stirling, R.W. Davis, D.W. Doidge, J.L. Bengtson, F. Trillmich, R. Vaz-Ferreira. Back row: P. Jouventin, J.-P. Roux, J.P. Croxall, T.G. Smith, D.M. Lavigne, R.H. Mattlin, J.H.M. David, R.M. Laws, W.N. Bonner, and M.O. Pierson (Photographer unknown)

mammal research questions being asked today are far more complex than attendees of the 1964 conference could have imagined.

The two best-known otariid species in the 1960s were the northern fur seal and California sea lion. Research on northern fur seals was funded by the U.S. government due to its treaty obligation to three other nations[3] to kill northern fur seals for pelts for the fashion industry. Most of the funded research was on population dynamics, but one study involved social behavior of known individuals (Peterson 1965). Peterson later worked with George Bartholomew and published a major work on the California sea lion (Peterson and Bartholomew 1967). These were two of the earliest papers on otariids that focused on social behavior.

As scientific interest in pinnipeds grew, a major symposium on all pinniped species was held in 1977 at Guelph University, Canada, that attracted more than 100 attendees. By 1981 some data were available on most of the otariid species, especially on fur seals, that were more numerous and widespread than sea lions. Researchers were still working in isolation from each other. To further improve communication, a symposium on fur seals was held in 1984 at the British Antarctic Survey in Cambridge, England (Croxall and Gentry 1984). Papers were presented by

[3]Japan, Russia, and England (acting through Canada).

27 first authors representing 13 nations, reporting on 11 species of fur seals (Fig. 1.5).

Most studies in the 1980s focused on observing social behavior and describing it in print. Field conditions and the individual interests of the researchers probably determined the data being collected. Some of the topics of interest were means by which males won and defended territories, duration of territorial tenure, and males' mating success. The topics of interest for females were their arrival dates, dates and locations of parturition and mating, and timing and duration of feeding trips to sea. Together, these seemed to be the key factors that affected the next generation of adults.

1.7 Manipulative Behavioral Experiments

U.S. managers that were responsible for northern fur seal populations were concerned that killing young males for pelts had reduced the overall pregnancy rate, resulting in fewer pups. When too few males were available for mating, the critical factor in the overall pregnancy rate was the duration of estrus. To measure it, we fitted females with a harness that prevented coitus and paired them with a captive adult male for 15 minute test sessions several times daily while scoring the fine details of their behavioral interactions (Gentry 1998). The females had ready access to tanks of fresh seawater where they were fed fish and squid for the duration of their tests, about 30 days. Test females not in estrus were consistently aggressive toward the male. When estrus began, their threats toward the male stopped and then resumed when estrus ended, as indicated by their recorded behavior toward the test male. All females in our tests followed this pattern. These tests may have been the first use of behavioral indicators to estimate the duration of a physiological process in otariids.

We used the same methods to test whether the concept of female mate choice applied to northern fur seals. Females that our tests had shown to be in estrus were fitted with the harness mentioned above and transported to a landing area where males of varying ages, from juvenile to senescent, were available. We selected a male, drove it close to the female, and took notes on the fine components of their behavioral interactions. Results showed that estrous females were sexually receptive to any male, irrespective of age, size, or social status if access to them was provided (Gentry 1998).[4] Females usually mate with prime adult males, not because they prefer them but because prime males actively exclude non-prime males from the land sites that females predictably use for pupping. Similar manipulative experiments could be conducted on other otariid species.

Philopatry begins within the first 30 days of life, and may be based on where the pup suckled, not on where it was born. Northern fur seal pups that had been born on

[4]Estrus females that are physically restrained will sexually present lordosis to humans if they are touched near the pelvic area.

breeding areas were taken with their mothers to our holding facility and held for 30 days while the mothers were tested for the duration of estrus. Thereafter, the pups and their mothers were returned to their original birth site and released. The mothers immediately left on foraging trips but the pups did not remain there. They walked overland (pups of that age do not swim) about 2 km to the holding facility where they had last suckled. They called for their mothers near the compound and, not finding her, walked through the village still calling. We returned these pups to their birth sites as many as nine times but they persisted in returning to the holding facility. This was not a planned experiment but an accidental finding. It could be systematically tested with other otariid species if the basis of philopatry is of interest.

1.8 Research on Diving Behavior

The development and first use of dive recorders is fully reviewed in the book *Diverse Divers* (Kooyman 1989). Kooyman's goal was to study the relationship between diving behavior and diving physiology. He was not getting that information from his first Time-Depth Recorder (TDR) because it ran for only 1 h and, when deployed on Weddell seals, recorded only a single dive during that interval. After seeing my 1974 data on the predictable land-sea movements of female northern fur seals, he decided that an instrument that ran for a week would record all dives the females made on a trip to sea. He obtained a grant, found an instrument maker, built units having that capability, and we deployed them in 1975 (Kooyman et al. 1976) using handling methods developed for that purpose (Gentry and Holt 1982).

By 1983, Kooyman had used the instrument to measure diving in northern, Antarctic, South African, South American, and Galapagos fur seals, and the Galapagos sea lion. The colleagues with whom he worked on these projects gathered to present and discuss their collective data, and to decide on the concepts to be published. Chapter writing followed and the book appeared in 1986 (Gentry and Kooyman 1986).

The first low voltage electronic components needed to measure diving became available in 1984. Kooyman used them to build a completely different kind of dive instrument from the more physical one used before. It weighed a few ounces instead of 1 kg, and it stored the data in digital format. These features made it possible to record many aspects of diving other than time and depth. The first new parameter he measured was swim speed; a small paddle-wheel on the nose of the instrument provided that information, and the data were stored in memory chips.

The change from analog to digital format for recording diving behavior revolutionized the study of marine mammals. Dive instruments could now be built by anyone familiar with electronic components, and researcher demand for them created a new instrument industry. Instruments could now be made to measure many new aspects of foraging such as water temperature, prey capture attempts (via jaw movements), location at sea via satellite, prey finding using echolocation (in cetaceans) and, most recently, electrocardiograms on free-swimming whales

(Goldbogen et al. 2019). A scan of the recent literature suggests that these instruments have caused a major change in the questions that field workers are asking about marine mammals. The earlier research emphasis on social structure and behavior of otariids on land has been largely replaced by measuring behavior, ecology, and physiology of marine animals at sea.

1.9 Research Using Isotopes

Injecting animals with isotopes, especially animals that are carrying dive recorders, greatly increased the amount of physiological and environmental information obtainable for otariids. The first species studied using this combination was the northern fur seal in the early 1980s. The isotopes used were Oxygen 18, tritium, and deuterium-labeled water (Costa and Gentry 1986). Since that initial study, most of the world's otariid species have been studied using this combination of methods. The dive recorder gives information on at-sea behavior, and the isotopes reveal the physiological changes that accompany that behavior.

The details of how this combination of isotopes and dive recorders produces new physiological and environmental information is presented in Costa (1987). The biological factors that have been measured this way include milk production by mothers, milk intake by offspring, growth efficiency, body composition, energy expenditure, maternal investment in the young, metabolic fuel use, foraging energetics, foraging ecology, mass changes and metabolism during the perinatal fast, ontogeny of oxygen stores, physiological dive capabilities, activity-specific metabolic rates for diving, transit time and rest at sea, field metabolic rates, temporal patterns of milk production, parental attendance related to environmental fluctuations, and foraging energy expenditure in relation to environmental changes (specifically El Niño events). Some of these studies involved data from accelerometers contained in the dive instruments (see du Dot and Guinet, Chap. 4).

1.10 Concluding Remarks

Otariids are interesting from the standpoint of terrestrial social behavior as well as foraging behavior at sea. The way students can join in this area of research has changed greatly over the years. Decades ago, just learning about some topic and then finding a project where that information was needed was a viable way to find a research position. No counselor or teacher suggested that I read up on echolocation or pointed me toward Poulter's laboratory. Most of what I learned during those earliest research years came from working with other established researchers. A chance visit to an otariid breeding group steered me toward field research because it fit my interests better than did laboratory research. I spent three breeding seasons studying Steller sea lions, inventing data collection methods, and learning the vast difference between laboratory and field research. That study raised questions that could only be answered by comparing other otariid species that did not live in the climate of central California. I started by studying the New Zealand fur seal on a post-doctoral fellowship at the University of Adelaide, Australia. Afterward, because of my interest in otariids, a U.S. government agency hired me to study the northern fur seal in the Bering Sea, a project that was well-funded and lasted for 20 years. The position let me explore all topics that interested me, provided funds to help Gerry Kooyman develop his time depth recorder, and gave Dan Costa a chance to use dive recorders in combination with his isotope studies (Costa and Gentry 1986). This combination of dive recorders and isotopes seems to have completely changed the questions that modern researchers are asking.

My advice to students is as follows: do not try to enter science the way I did. Read broadly, then work with a counselor or a researcher who can steer you toward appropriate job opportunities. Simply blind-shopping for such an opportunity, as I did, will not work in today's world because of increased competition for jobs. No graduate students in the early 1960s were interested in marine mammals; today there are thousands. A good way to find openings in science is to attend professional conferences in your field of interest. Meet other students and make them your community; they could last throughout your entire career. Learn from working professionals how they fund their research, which funding agencies they use, and how to write proposals that will succeed. Publish only meaningful professional papers, and give conference talks only when you have solid results to report. Befriend older, more experienced researchers and learn from them how your career can mature. Vic Scheffer, my personal hero, worked on fur seals in midlife, retired, led seniors on an expedition to the Galapagos Islands at age 90, and at 103 gave philosophical lectures on the human condition in a series entitled, "Random Thoughts While Shaving." Never stop learning, teaching, or laughing. Discover whether you prefer pure science, i.e. deep, fundamental questions about evolution, ecology behavioral development etc., or applied science- applying scientific principles to management and/or conservation problems, but know that both require the same creative mental skills.

References

Berta A, Churchill M (2012) Pinniped taxonomy: a review of current recognized species and sub-species, and evidence for their description. Mamm Rev 42(3):207–234

Boness DJ, Francis JM (1991) The effect of thermoregulatory behavior on the mating system of the Juan Fernandez fur seal, *Arctocephalus phillipi*. Behaviour 119(1/2):104–126

Costa DP (1987) Isotopic methods for quantifying material and energy intake of free-ranging marine mammals. In: Huntley AC, Costa DP, Worthy GAJ et al (eds) Approaches to marine mammal energetics. Allen Press, Lawrence, pp 43–66

Costa DP, Gentry RL (1986) Free-ranging energetics of northern fur seals. In: Gentry RL, Kooyman GL (eds) Fur seals: maternal strategies on land and at sea. Princeton University Press, Princeton, pp 79–101

Crawley MC, Cameron DB (1972) New Zealand Sea lions, *Phocarctos hookeri*, on the Snares Islands. NZ J Mar Freshw Res 6:127–132

Croxall JP, Gentry RL (eds) (1984) The status, biology, and ecology of fur seals. In: Proceedings of the first international symposium on fur seal biology, Cambridge, UK. NOAA Technical Report No. 51, 212 pp

Elliott W (1882) A monograph of the seal islands of Alaska. U.S. Commercial Fisheries Special Bulletin, U.S. Government Printing Office

Gentry RL (1966) Underwater auditory localization in the California Sea lion. J Audit Res 7:187–193

Gentry RL (1970) Social behavior of the Steller Sea lion. PhD Dissertation, University of California, Santa Cruz. 113 pp

Gentry RL (1980) Seawater drinking in eared seals. Comp Biochem Physiol 68:81–86

Gentry RL (1998) Behavior and ecology of the northern Fur seal. Princeton University Press, Princeton

Gentry RL, Holt JR (1982) Equipment and techniques for handling Northern fur seals. NOAA Technical Report NMFS SSRF-758. Available from NOAA

Gentry RL, Kooyman GL (eds) (1986) Fur seals: maternal strategies on land and at sea. Princeton University Press, Princeton

Goldbogen JS, Cade DE, Calambokidis J et al (2019) Extreme bradycardia and tachycardia in the world's largest animal. PNAS. https://doi.org/10.1073/pnas.1914273116

Griffin DR (1959) Listening in the dark; the acoustic orientation of bats and man. Yale University Press, London

Hubbs CL (1956) Back from oblivion. Guadalupe fur seal still a living species. Pac Discov 9:14–21

Illiger JKW (1811) Prodromus Systematis Mammalium et Avium. Sumptibus C Salfeld 70:138–139

Kooyman GL (1989) Diverse divers; physiology and behavior. Springer, Berlin

Kooyman GL, Gentry RL, Urquhart DL (1976) Northern fur seal diving behavior; a new approach to its study. Science 193:411–412

LeBoeuf BJ, Würsig B (1985) Beyond bean counting and whale tales. Mar Mamm Sci 1 (2):128–148

McCann TS (1980) Territoriality and breeding behaviour of adult male Antarctic fur seals. J Zool 192(3):295–310

Peterson RS (1965) Behavior of the northern fur seal. M Sc Thesis, Johns Hopkins University, Baltimore

Peterson RS, Bartholomew GA (1967) The natural history and behavior of the California Sea lion. Am Soc Mammal Spec Publ 1:1–79. https://doi.org/10.5962/bhl.title.39535

Poulter TC (1963) Sonar signals of the sea lion. Science 139:753–754

Scammon CM (1884) The marine mammals of the northwestern coast of North America together with an account of the American whale fishery. Dover Publications 1968, New York

Scheffer VB (1958) Seals, sea lions, and walruses; a review of the pinnipedia. Stanford University Press, Palo Alto

Schein MW (1975) Social hierarchy and dominance in Benchmark Papers in Animal Behavior, vol. 3, 401pp. Dowden, Hutchinson & Ross, Stroudsburg

Schusterman RJ (1981) Behavioral capabilities of seals and sea lions. A review of their hearing, visual, learning, and diving skills. Psych Rec 31:125–143

Schusterman RJ, Gentry RL (1971) Development of a fatted male phenomenon in California Sea lions. Dev Psychobiol 4(4):333–338

Steller GW (1749) Mémoires de l'Académie Impériale des Sciences de St. Pétersbourg (in Latin). Translation by W. Miller, 1899. In: Jordan DS (ed) The fur seals and fur seal islands of the North Pacific Ocean, part 3, 179–218. U.S. Government Printing Office, Washington, DC

Wolf JBW, Trillmich F (2007) Beyond habitat requirements; individual fine-scale site fidelity in a colony of Galapagos sea lions (*Zalophus wollebaeki*) creates concentrations for social structuring. Oecologia Beh Oecol 152:553
Wolf JBW, Trillmich F (2008) Kin in space; social viscosity in a spatially and genetically sub-structured network. Proc R Soc B Biol Sci 275(1647):2063–2069
Wolf JBW, Mawdsley D, Trillmich F et al (2007) Social structure in a colonial mammal: unravelling hidden structural layers and their foundations by network analysis. Anim Behav 74:1293–1302

Chapter 2
When Physiology and Ecology Meet: The Interdependency Between Foraging Ecology and Reproduction in Otariids

Daniel P. Costa and Ana M. Valenzuela-Toro

Abstract Otariids exhibit a semiaquatic lifestyle, feeding in the water and breeding and resting on land. Fur seals and sea lions, the two groups of otariids, exhibit an overall income breeding system where the females alternate between trips to the foraging grounds and periods at the breeding colony to feed the pup. How far and how long lactating females can be away from their foraging trips is ultimately dictated by ecological and physiological tradeoffs. In this chapter, we examine the interrelations between behavior and physiology and how they enable or constrain the reproductive, foraging biology, and life history of otariids. For example, income breeding limits otariid females to forage near their reproductive colonies, which constrains them to inhabit highly productive oceanographic regions. Further, as sea lions are larger than fur seals they are capable of deeper and longer dives. As a result, fur seals tend to feed closer to the surface in the epipelagic regions, while sea lions tend to forage in deeper environments on benthic, epipelagic and mesopelagic prey. We review the patterns of resource acquisition and allocation, including an examination of the energetics of reproduction, milk composition, foraging behavior, and differences between fur seals and sea lions. Further, we assess how these factors may have led to their current distribution and demographic trends during the last decades. We finish by discussing how these physiological and ecological tradeoffs would have influenced the evolutionary history of otariids in the deep time.

Keywords Foraging ecology · Otariids · Energetics · Diving behavior · Reproduction · Life histories

D. P. Costa (✉) · A. M. Valenzuela-Toro
Department of Ecology and Evolutionary Biology, Institute of Marine Sciences, University of California, Santa Cruz, CA, USA
e-mail: costa@ucsc.edu; anmavale@ucsc.edu

C. Campagna, R. Harcourt (eds.), *Ethology and Behavioral Ecology of Otariids and the Odobenid*, Ethology and Behavioral Ecology of Marine Mammals,
https://doi.org/10.1007/978-3-030-59184-7_2

2.1 Introduction

The origin of the amphibious lifestyle of pinnipeds occurred at a time when coastal upwelling intensified (White et al. 1992), resulting in an abundant, diverse marine food resource for which there was likely decreased competition from other endothermic tetrapods (Vermeij and Dudley 2000; Lipps and Mitchell 1976). For pinnipeds, the necessity to return to land to mate and provision their young required a separation of feeding from breeding in both time and space (Bartholomew 1970; Cassini 1999). In response, otariids evolved a modified "income breeding" system (Costa 1991b; Costa 1991a; Schulz and Bowen 2005; Stephens et al. 2009). Under this strategy, production of milk is supported by many short foraging bouts across a prolonged lactation period (Fig. 2.1). The need to return to the colony to suckle the pup limits the amount of time a lactating otariid mother can forage at sea and thus constrains her to foraging on resources that are relatively close to the breeding colony (Costa 1993; Boyd 1998; Trillmich and Weissing 2006; Houston et al. 2007; Stephens et al. 2014). As a result, reproductive success in otariids is closely connected to the abundance and availability of local prey, creating a strong linkage between local environmental fluctuations (e.g. El Niño events) and population status (Trillmich et al. 1991; Costa 2008).

In the sections that follow, we examine how the interdependency between the requirement to provision the pup on shore with the need to forage at sea fundamentally constrains the reproductive pattern of otariids. This fundamental life-history constraint results in some common patterns that vary across the ecological conditions of the different habitats. We examine (1) the energetic challenges of income breeding, (2) the connection between habitat diversity and diversity of foraging patterns and breeding systems; (3) the origins and evolutionary implications of this reproductive strategy; and (4) future directions for research into otariid reproductive and foraging energetics.

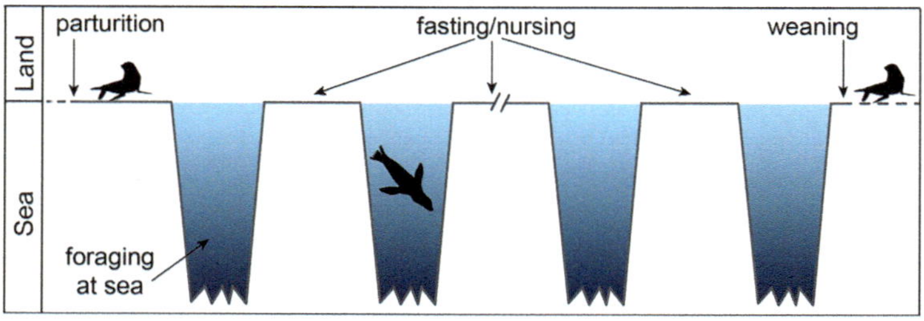

Fig. 2.1 Otariids are income breeders that make intermitted trips foraging at sea and then returning to their pups onshore

2.2 Energetics of Growth, Reproduction and an Amphibious Lifestyle

Growth and reproduction require an animal to acquire more energy and nutrients than are necessary to support vital maintenance. In most mammals, males do not invest energy in parental care of the offspring; instead the cost of reproduction is associated with finding and maintaining access to estrous females and with inseminating as many females as possible. Specifically, in male otariids, reproductive success is associated with large body size. Thus, there is a substantial investment in growth, as large body size confers an advantage in fighting, dominance and territory maintenance as large animals can fast longer, allowing them to maintain their harem longer (Bartholomew 1970; Cassini 1999). The larger size of males is apparent at birth and is maintained throughout life (Payne 1979; Mattlin 1981; Costa and Gentry 1986; Trillmich 1986; Oftedal et al. 1987; Costa et al. 1988; Higgins et al. 1988; Ono and Boness 1996; Georges and Guinet 2001; Weise and Costa 2007). However, in some species (e.g. Guadalupe, *Arctocephalus townsendi*; Juan Fernandez, *A. philippii* and Galapagos, *A. galapagoensis* fur seals and California sea lions, *Zalophus californianus*), the physiological need to access cool water in warmer climates reduces the advantage of large body size (relative heat loss decrease along with increasing body size), thus sexual size dimorphism is reduced in those species that maintain harems in water, or that need access to water (Gentry 1973; Boness and Francis 1991).

For female mammals, reproductive costs are associated with investment in offspring growth through gestation and lactation. Gestation costs come in the form of the energy contained in fetal tissues, as well as the energy to fuel the metabolic processes associated with gestation—also known as the "heat of gestation" (Brody 1945). After a relatively long gestation period, otariid pups are born larger (representing 3.5%–16.5% of the maternal body size) than terrestrial carnivores (0.5%–3%), which is consistent with giving birth to a highly precocial pup. Nevertheless, their lactation duration is comparable to other carnivores (Costa and Maresh 2017). Lactation costs come in the form of the energy and nutrients contained in milk, as well as the energy to fuel milk synthesis. Among pinnipeds, otariids exhibit an income breeding system. Under this strategy, the energy and nutrients required for milk production are obtained during intermittent bouts of foraging, interspersed between shore visits to suckle the pup. For example, female California sea lions can make 50 trips to sea during an 11-month lactation period, typically spending ~3 days foraging within 85 km of their breeding colonies (Kuhn and Costa 2014) (Fig. 2.2).

The shortest lactation durations are 4 months and are found in the Antarctic (*Arctocephalus gazella*) and northern (*Callorhinus ursinus*) fur seals, which inhabit high latitudes. Most otariids have a lactation period of about 11 months. The longest lactation periods are found in Australian sea lions (*Neophoca cinerea*) and Galapagos fur seals. Those species have lactation durations of 17.5 months (Higgins and Gass 1993) and 2–3 years (Trillmich 1990), respectively. These prolonged lactation durations are thought to be associated with a less productive benthic ecosystem in Australia (Costa and Gales 2003) and the highly seasonal and unpredictable nature of prey resources in the Galapagos (Trillmich 1990). Thus,

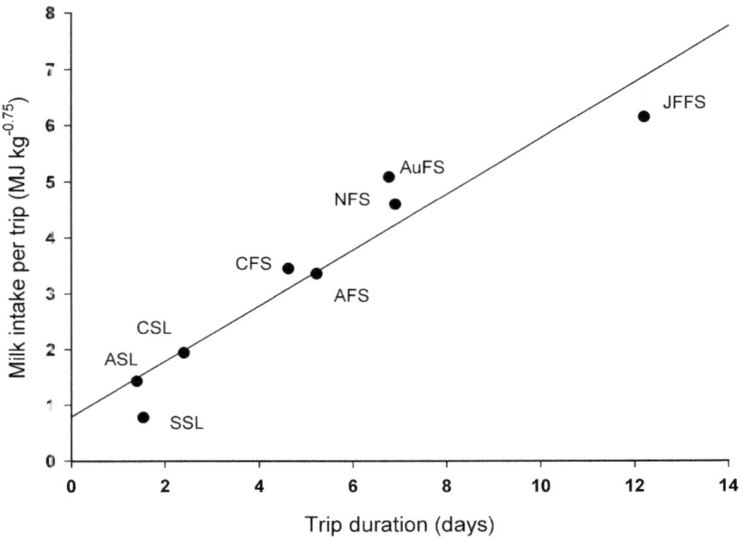

Fig. 2.2 The amount of milk energy delivered to the pup per shore visit relative to the time spent at sea foraging is presented for 8 otariids. Milk energy delivery rates were normalized for differences in female body mass$^{-0.75}$. Data are for Australian sea lion (ASL) Costa unpublished, California sea lion (CSL) (Oftedal et al. 1987), Steller sea lion (SSL) (Higgins et al. 1988), Antarctic fur seal (AFS) (Arnould and Boyd 1995), Cape fur seal (CFS) (Gamel et al. 2005), Australian fur seal (AuFS) (Arnould and Hindell 2002), and Juan Fernandez Fur Seal (JFFS) (Ochoa-Acuña 1995) and northern fur seal (NFS) (Costa and Gentry 1986)

the income breeding strategy couples reproductive success with local prey abundance (Costa 1993; Boyd 1998), creating a strong link between local environmental fluctuations and population status (e.g. El Niño events) (Trillmich et al. 1991; Costa 2008). As a result, income breeders tend to be limited to breeding in highly productive regions (Costa 1993), and the few species that breed in less productive regions compensate with longer lactation periods. Abbreviations: AFS, Antarctic fur seals; AuFS, Australian fur seals; CFS, Cape fur seals; GFS, JFFS, Juan Fernandez fur seals; NFS, northern fur seals; ASL, Australian sea lions; CSL, California sea lions; SSL, Steller sea lions.

2.2.1 *Tradeoffs in Time on Land and at Sea*

In otariids, the timing and pattern of maternal investment are optimized in response to their different habitats. Sea lions tend to forage nearshore and make shorter trips (e.g. Steller sea lions, *Eumetopias jubatus:* ~ 2 days, Australian sea lions: ~2 days, southern sea lions, *Otaria flavescens:* 1.6 days, on average) than fur seals that forage further offshore and make longer trips (e.g. northern fur seals: 7.4 days, Juan Fernandez fur seals: 12.3 days, subantarctic fur seals, *A. tropicalis:* 16 days,

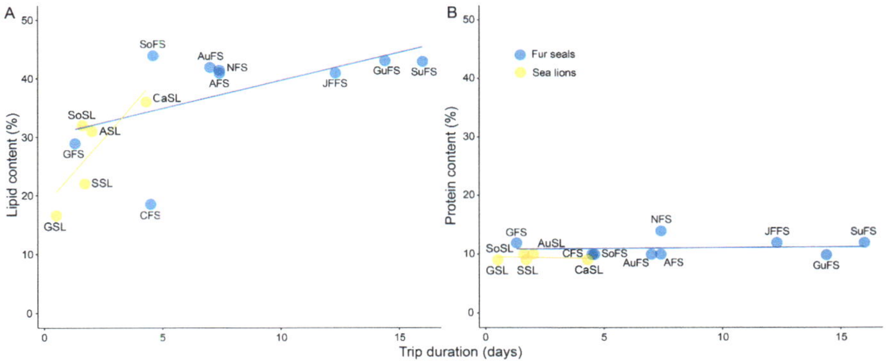

Fig. 2.3 Lipid (**a**) and protein (**b**) content of maternal milk in species of fur seals (blue) and sea lions (yellow). Abbreviations: AFS Antarctic fur seals, AuFS Australian fur seals, CFS Cape fur seals, GFS Galapagos fur seals, GuFS Guadalupe fur seals, JFFS Juan Fernandez fur seals, NFS northern fur seals, SoFS South American fur seals, SuFS SubAntarctic fur seals, ASL Australian sea lions, CaSL California sea lions, GSL Galapagos sea lions, SoSL South American sea lions, SSL Steller sea lions

Antarctic fur seals: 7.4 days, on average) (Gentry and Holt 1986; Ono et al. 1987; Boyd et al. 1991; Higgins and Gass 1993; Georges and Guinet 2000; Milette and Trites 2003; Soto et al. 2006). This is even observed within a species as northern fur seals at San Miguel Island where the shelf break is only 2 km away make 3.8 day foraging trips, compared to 7.1–9.8 day foraging trips on the Pribilof Islands where the shelf break is 115 km (Gentry 1998). Such a pattern follows central place foraging theory, which predicts that the optimal strategy is to make longer trips when prey is distant from the central place (reproductive colony) and shorter trips when prey is nearby (Orians and Pearson 1977). Further, as females have longer foraging trips, it is optimal to provide more milk energy to the offspring per shore visit. This is not only more efficient; it also allows otariid mothers to provide similar amounts of energy while making fewer visits. Indeed, empirical data show that those species that make longer trips to sea provide proportionally more milk energy to their pups during their shore visit (Fig. 2.2). The ability to provide more milk energy over a short shore visit is made possible by the high fat content of their milk, which ranges between ~20% and 45% (Fig. 2.3a). In contrast, terrestrial mammals produce milk relatively low in fat. For example, cows and humans produce milk containing 4% and 8% fat, respectively. Nevertheless, there is a disadvantage to an elevated milk fat content, as it is associated with a reciprocal decrease in water content, with no corresponding change in protein content (Fig. 2.3b). Therefore, the protein-to-energy ratio of pinniped milk is lowest in the most energy-dense milk. While pups provisioned with high fat milk receive sufficient energy to fuel their metabolism, they receive relatively less protein that is essential for lean body growth. The issue of how a young pup who is still developing its physiological competence can maintain water balance when they receive relatively less water (as the milk they ingest is so lipid rich) remains unknown.

2.2.2 Diving Physiology and Capacity

Despite their external similarities, fur seals and sea lions differ in their morphology, physiology, foraging ecology, and life history (Arnould and Costa 2006). For instance, insulation in fur seals is mainly provided by a dense and impermeable fur coat (with an underfur layer), whereas in sea lions, insulation relies on a thick blubber layer (Liwanag et al. 2012a; Liwanag et al. 2012b). While blubber maintains its insulating quality with depth, the insulation of fur seals declines as the seal dives as the air contained in fur compresses (Repenning 1976). Furthermore, just as in sea otters, an air layer would add buoyancy that may require more effort during the descent phase of the dive.

Body size also differs between these groups, as sea lions are larger than fur seals, which results in differences in their physiology and foraging ecology. The larger body size of sea lions facilitates longer and deeper dives (Ponganis 2016). Thus, while sea lions tend to dive deeper and longer often foraging at or near the benthos on the continental shelf, fur seals tend to exploit the upper reaches (epipelagic) of the water column often offshore (Table 2.1 and Fig. 2.4) (Villegas-Amtmann et al. 2013). Diving capacity is, in part, controlled by body size, as the metabolic rate (oxygen demand) scales to body mass$^{0.75}$ while oxygen stores (supply) scale with body mass$^{1.0}$. Therefore, larger animals have a lower mass-specific metabolism for a relatively constant proportion of oxygen storage capacity (Ponganis 2016). All things being equal, large animals should be able to dive longer and deeper than small ones based on body size alone (dive ability scales with body mass $^{0.25}$). This pattern holds when we compare sea lions to fur seals, as sea lions are, in general, larger and are more capable divers than fur seals (Fig. 2.4). However, if we just compare sea lions, we find the largest sea lion species, the southern and Steller sea lions, make the shortest-shallowest dives (Hückstädt et al. 2016) while the smallest, the Galapagos sea lion makes the longest dives (Figs. 2.5 and 2.6) (Villegas-Amtmann and Costa 2010). The exceptional diving ability of Galapagos sea lions is made possible by having the highest oxygen storage capacity (greater blood volume and muscle myoglobin) of any otariid, equal to that of many deep diving phocids (Villegas-Amtmann and Costa 2010).

Different foraging patterns require different physiological capabilities as deeper dives are also longer. In air breathing vertebrates, the primary determinant of dive duration is the aerobic dive limit, which is determined by the animals' oxygen stores coupled with the rate at which oxygen is utilized (Ponganis 2016). The oxygen storage capacity of mesopelagic and benthic divers (e.g. sea lions) is higher, enabling them to make longer dives compared to epipelagic foragers (e.g. fur seals) who have relatively lower oxygen stores and consequently shorter dives (Fig. 2.5). Greater oxygen stores are achieved by a synergistic effect between increases in hematocrit (proportion of red blood cells in whole blood), blood volume, and muscle myoglobin concentration (Ponganis 2016). Nevertheless, there is a tradeoff as an elevated hematocrit also increases blood viscosity, reducing the ability of blood to transport oxygen optimally (Hedrick and Duffield 1991). As otariids exhibit highly aerobic

Table 2.1 Summary of life-history attributes of otariid species. Average values are presented

Common name	Scientific name	Population trend	Mean dive depth (m)	Mean dive duration (min)	Trip duration (days)
Antarctic fur seal	*Arctocephalus gazella*	Stable/ Increasing	53.0	1.6	7.4
Australian fur seal	*Arctocephalus pusillus doriferus*	Stable/ Increasing	58.0	2.9	7.0
Cape/Brown fur seal	*Arctocephalus pusillus pusillus*	Increasing	74.0	2.5	4.5
Galapagos fur seal	*Arctocephalus galapagoensis*	Declining	26	N/I	1.3
Guadalupe fur seal	*Arctocephalus townsendi*	Increasing	17	N/I	14.4
Juan Fernandez fur seal	*Arctocephalus philippii*	Increasing	26.4	1.8	12.3
New Zealand fur seal	*Arctocephalus forsteri*	Increasing	33.3	1.5	~7
Northern fur seal	*Callorhinus ursinus*	Declining	18.3	1.0	7.4
South American fur seal	*Arctocephalus australis*	Declining	61	2.1	4.6
Subantarctic fur seal	*Arctocephalus tropicalis*	Stable/ Increasing	39.7	1.6	16
Australian sea lion	*Neophoca cinerea*	Stable	61.0	3.3	< 2
California sea lion	*Zalophus californianus*	Increasing	58.2	2.2	4.3
Galapagos sea lion	*Zalophus wollebaeki*	Declining	97.4	3.7	0.5
New Zealand sea lion	*Phocarctos hookeri*	Declining	117.1	3.4	1.7
South American sea lion	*Otaria flavescens*	Declining	21.7	2.0	1.6
Steller sea lion	*Eumetopias jubatus*	Declining	29.6	1.8	1.7

Common name	Dominant Foraging behavior	Body length males (m)	Body length females (m)	Body mass males (kg)	Body mass females (kg)	Reference
Antarctic fur seal	Epipelagic	1.8	1.3	165	39.3	1, 8, 20, 28
Australian fur seal	Benthic	2.2	1.6	289	77	1, 7, 27
Cape/Brown fur seal	Epipelagic	2.2	1.4	229	54	1, 14, 25

(continued)

Table 2.1 (continued)

Common name	Dominant Foraging behavior	Body length males (m)	Body length females (m)	Body mass males (kg)	Body mass females (kg)	Reference
Galapagos fur seal	Epipelagic	1.6	1.2	64	28.9	1, 5, 11, 20
Guadalupe fur seal	Epipelagic	2.2	1.5	165	49.1	1, 2, 9
Juan Fernandez fur seal	Epipelagic	2.0	1.4	142.1	48.1	1, 4, 10
New Zealand fur seal	Epipelagic	2.0	1.3	150	38.1	1, 3, 7, 20
Northern fur seal	Epipelagic	2.1	1.5	270	41.4	1, 19, 20, 29
South American fur seal	Epipelagic	2	<1.5	125	41.7	1, 6, 19, 20
Subantarctic fur seal	Epipelagic	1.8	1.4	117.5	49.6	1, 13, 20, 30, 31
Australian sea lion	Benthic	2.2	1.6	215	80	1, 7, 20, 26
California sea lion	Epipelagic/ Mesopelagic/ Benthic	2.5	1.6	523	84	1, 16, 23
Galapagos sea lion	Epipelagic/ Mesopelagic/ Benthic	N/I	1.6	200	77	1, 17, 22
New Zealand sea lion	Mesopelagic/ Benthic	2.4	1.9	375	112	1, 12, 15, 24, 26
South American sea lion	Benthic	2.4	1.8	325	131	1, 15, 24
Steller sea lion	Benthic	3.3	2.5	1000	275	1, 18, 20, 21

N/I no information

References: [(1)] Kovacs et al. (2012), [(2)] Gallo-Reynoso et al. (2008), [(3)] Harcourt et al. (2001), [(4)] Francis et al. (1998), [(5)] Kooyman and Trillmich (1986), [(6)] Baylis et al. (2018), [(7)] Kirkwood and Goldsworthy (2013), [(8)] Hofmeyr (2016), [(9)] Aurioles-Gamboa (2015a), [(10)] Aurioles-Gamboa (2015b), [(11)] Trillmich (2015), [(12)] Chilvers (2015), [(13)] Hofmeyr (2015a), [(14)] Hofmeyr (2015b), [(15)] Cárdenas-Alayza et al. (2016a), [(16)] Aurioles-Gamboa and Hernández-Camacho (2015), [(17)] Gelatt and Sweeney (2016), [(18)] Gelatt et al. (2015), [(19)] Cárdenas-Alayza et al. (2016b), [(20)] Schulz and Bowen (2004), [(21)] Rehberg et al. (2009), [(22)] *Villegas*-Amtmann and Costa (unpublished data), [(23)] *McHuron* and Costa (unpublished data), [(24)] Hückstädt and Costa (unpublished data), [(25)] Kirkman et al. (2019), [(26)] Costa (unpublished data), [(27)] Arnould and Hindell (2001), [(28)] Lea et al. (2002), [(29)] Kuhn et al. (2014), [(30)] Georges and Guinet (2000), [(31)] Luque et al. (2007)

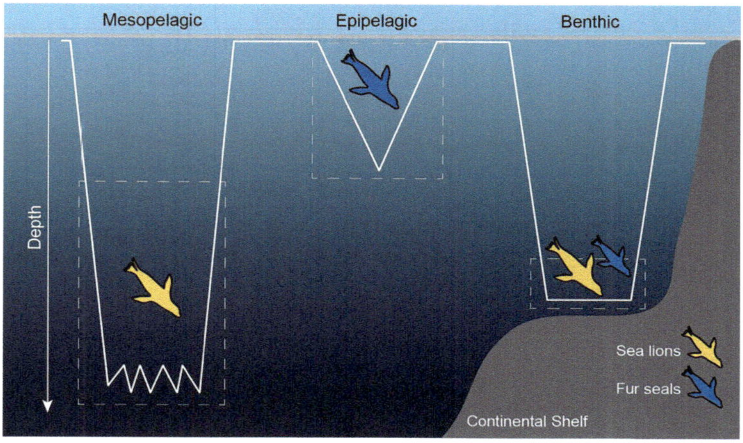

Fig. 2.4 General marine foraging patterns exhibited by otariids. Note that there are some exceptions to this classification, including California sea lions which also exhibit epipelagic foraging. The dashed boxes represent the potential area where they are likely able to search and capture prey. The relative size of fur seals and sea lions is not in scale. Modified from Costa et al. (2006) and Gallagher et al. (2015)

energetic foraging behavior, their hematocrit is lower than phocids but is optimal for maximizing oxygen delivery while maintaining elevated blood oxygen stores. Diving behavior and physiological capacity also vary within a species; for instance, individuals of Galapagos sea lions that make longer-deeper dives also have greater oxygen stores than those that make shorter-shallower dives (Villegas-Amtmann and Costa 2010). Further, the oxygen stores of California sea lions increase as they undergo seasonal increases in dive depth and duration (Villegas-Amtmann et al. 2012). Variation of oxygen stores within a species suggests that oxygen stores are plastic and may increase in response to periodic or prolonged bouts of hypoxia as a "training effect".

In most cases, for most species, individual sea lions that dive longer also exhibit greater post-dive surface intervals (Fig. 2.7). Due to the need to maximize bottom time searching and/or pursuing prey, mesopelagic and benthic foraging otariids appear to be working closer to their physiological limits, often exceeding their aerobic dive limit (Costa and Gales 2000, 2003; Costa et al. 2000a, 2004). Because these deep diving species are already working at or near their maximum physiological capacity, they may have a limited ability to further increase their dive duration and/or foraging effort in order to respond to changes in their environment and resource limitations. Therefore, benthic-foraging species might be particularly sensitive to changes in their habitat resulting from climate change or interactions with fisheries, which remove the larger size classes of fish upon which they depend (Chilvers and Wilkinson 2009). In fact, species such as Australian and New Zealand sea lions, which specialize on benthic or demersal prey, have endangered populations, while the California sea lion, a generalist that feeds on prey

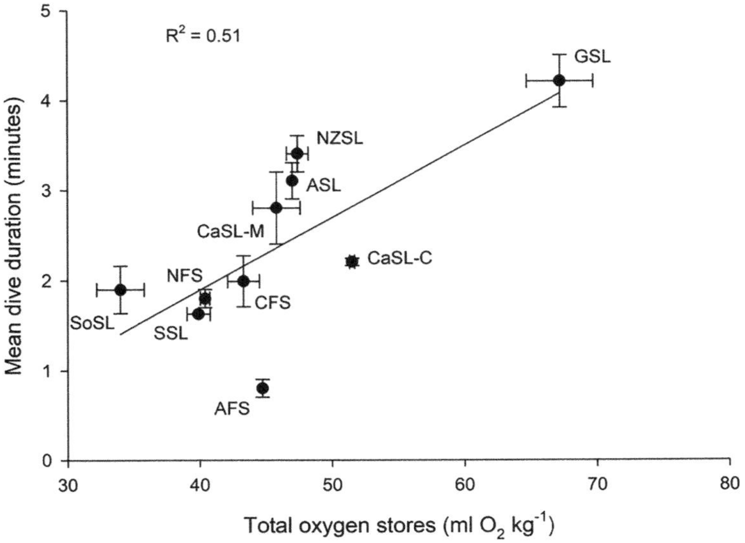

Fig. 2.5 The mean dive duration (mean ± se) is plotted as a function of the total oxygen stores (mean ± se) for South American sea lions (SoSL), Australian sea lions (ASL), California sea lions from Mexico (CaSL-M), California sea lions from California (CaSL-C), Antarctic fur seals (AFS), Cape fur seals (CFS), New Zealand sea lions (NZSL), Galapagos sea lions (GSL), Steller sea lions (SSL), and northern fur seals (NFS). O_2 stores and dive duration data were measured in the same individuals for ASL, NZSL, CaSL-M, SoSL, AFS, GSL and CFS. Data are from Riet-Sapriza et al. (2013), Costa et al. (1998, 2001), Costa and Gales (2000), Kuhn and Costa (2014), Villegas-Amtmann and Costa (2010), Weise and Costa (2007), Richmond et al. (2006), Rehberg et al. (2009), Shero et al. (2012), Hückstädt et al. (2016), and Kirkman et al. (2019)

throughout the water column, is now thriving and has recovered from previous exploitation (Chilvers and Meyer 2017; Hamer et al. 2013; Laake et al. 2018; Costa et al. 2006).

Further, the pressure imposed by fisheries would present an even more significant challenge for juvenile sea lions and fur seals as they are smaller, have less experience and are still acquiring their oxygen stores (Fowler et al. 2007b; Weise and Costa 2007; Shero et al. 2012). Not surprisingly, juvenile sea lions and fur seals make significantly shorter and typically shallower dives compared to adults (Loughlin et al. 2003; Pitcher et al. 2005; Fowler et al. 2006; Jeglinski et al. 2012; Leung et al. 2014) and they forage in different areas or within a restricted region of the same habitat as adults (Fowler et al. 2007a; Jeglinski et al. 2013; Hückstädt et al. 2014; McHuron et al. 2018a; Salton et al. 2019; Zeppelin et al. 2019). These factors make juveniles more susceptible to resource limitations in general and particularly so for benthic-foraging species, and are likely associated with the low recruitment in these declining populations (Costa et al. 2004; Arnould and Costa 2006).

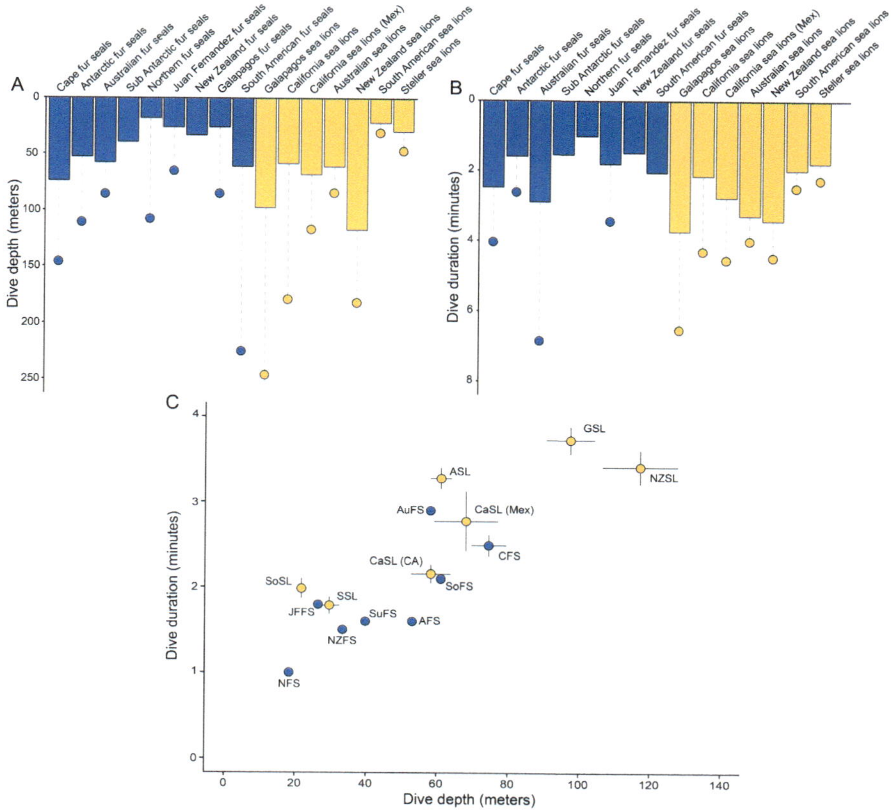

Fig. 2.6 Mean dive depth (**a**) and mean dive duration (**b**) for species of fur seals (blue) and sea lions (yellow). Dots represent the maximum values for dive depth and dive duration, respectively. (**c**) Plot showing the maximum dive duration as a function of diving depth for selected species of otariids. Abbreviations: AFS Antarctic fur seals, AuFS Australian fur seals, CFS Cape fur seals, GFS Galapagos fur seals, GuFS Guadalupe fur seals, JFFS Juan Fernandez fur seals, NFS northern fur seals, SoFS South American fur seals, SuFS SubAntarctic fur seals, ASL Australian sea lions, CaSL California sea lions, GSL Galapagos sea lions, SoSL South American sea lions, SSL Steller sea lions. Data are from Kirkman et al. (2019), Rehberg et al. (2009), Riet-Sapriza et al. (2013), Villegas-Amtmann and Costa (2010), Villegas-Amtmann et al. (2012), Costa and Gales (2000), Hückstädt et al. (2016), Costa et al. (2001), Kuhn and Costa (2014), and Baylis et al. (2016, 2018)

2.2.3 *Field Metabolic Rates*

Field metabolic rate (FMR) measures an animal's total energy expenditure after all constituent costs are supported, and provide insight into the energetic strategies used by marine mammals (Costa 2008). Several approaches have been used to study the FMR of animals at sea and onshore. Time budget analysis sums the daily metabolic

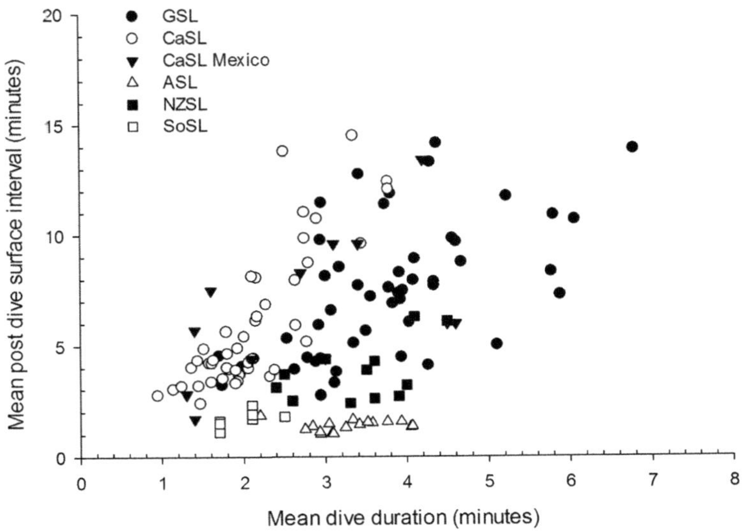

Fig. 2.7 Each point of the figure is the mean post dive surface as a function of the mean dive duration for that individual. Dive data are for individual Southern sea lions (SSL), Australian sea lions (AuSL), California sea lions in Mexico (CaSL-M), California sea lions in California (CaSL), New Zealand sea lion (NZSL) and for Galapagos sea lions (GSL). Data are from Riet-Sapriza et al. (2013), Villegas-Amtmann and Costa (2010), Villegas-Amtmann et al. (2012), Costa and Gales (2000), Hückstädt et al. (2016), Kirkman et al. (2019), Costa et al. (2001), and Kuhn and Costa (2014)

costs associated with various activities (Williams et al. 2004, 2007). Other methods rely on predictive relationships between FMR and physiological variables such as heart rate or ventilation rate, between FMR and changes in body mass and composition, or between FMR and bio-mechanical power (Young et al. 2011; Fahlman et al. 2013; Maresh et al. 2015). The breeding system of otariids where mothers periodically forage at sea and return to the pup on shore facilitates measurements of female foraging behavior and energy expenditure. Measurements of FMR using the doubly labelled water method have taken advantage of the need for the mother to return to her pup onshore, as this facilitates the likelihood of recovering instruments and getting final blood samples required for FMR measurements. Thus, FMR measurements using doubly-labelled water have been carried out on a number of species, including northern, Antarctic, and Galapagos fur seals and California, Galapagos, Australian, and New Zealand sea lions (Costa and Gentry 1986; Costa et al. 1989; Costa and Gales 2000, 2003; Jeanniard-du-Dot et al. 2017a, b, c; Fowler et al. 2007b; McDonald et al. 2012; McHuron et al. 2017b, 2019; Arnould et al. 1996). These studies indicate that otariids have an expensive lifestyle, expending energy at ~6 times the predicted basal metabolic rate. Such a high field metabolic rate is consistent with a fast pace of life, where a high rate of energy expenditure is associated with a high rate of prey energy acquisition (Schmitz and Lavigne 1984;

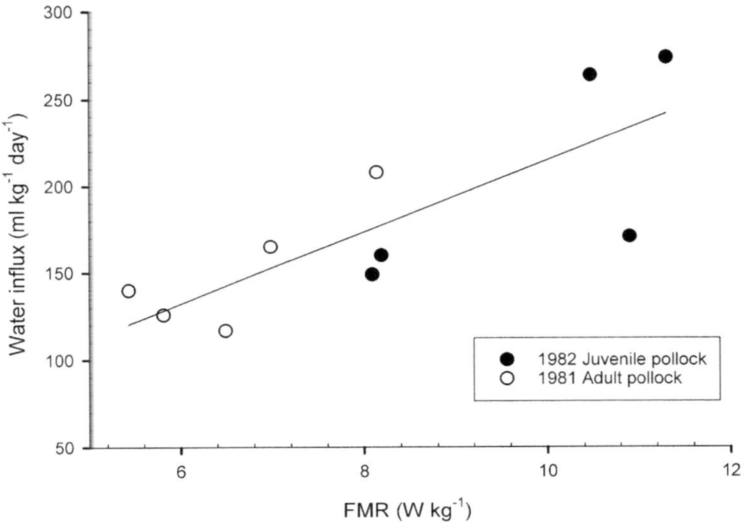

Fig. 2.8 Plot of FMR as a function of water influx of northern fur seal females foraging between two seasons (Costa and Gentry 1986). Fur seals in 1981 (open circles) were feeding primarily on adult pollock and during 1982 (solid circles) they were feeding on 0 age pollock (Sinclair et al. 1996)

Wright et al. 2018), which in turns enables a greater investment of energy into reproduction. This is in marked contrast to the lower field metabolic rate (1.5-3 times) predicted for elephant (*Mirounga* spp.) and Weddell seals (*Leptonychotes weddellii*), phocids which also have a slower pace of life (Maresh et al. 2015).

FMRs are quite variable between and within species and are associated with year-to-year changes in both the abundance and availability of prey (Costa 2008). For example, California sea lions significantly increased their foraging effort in response to reductions in prey availability during the 1983 El Niño event (see below for more insights on the El Niño events). Foraging effort also varies with the type of prey consumed. In fact, northern fur seals, foraging on mature pollock (*Theragra chalcogramma*) during 1981, had lower metabolic rates than fur seals feeding predominately on juvenile pollock during 1982 (Fig. 2.8). Further, the importance of the thermal environment on field metabolic rate can also be seen in Galapagos fur seals and sea lions, which due to the warm equatorial climate, have substantially reduced field metabolic rates compared to other otariids (Trillmich and Kooyman 2001; Villegas-Amtmann et al. 2017).

2.3 Habitat Diversity, Foraging Patterns and Reproductive Strategies

Otariids occupy a wide range of habitats in the Pacific, South Atlantic and Southern Oceans, but are completely absent from the North Atlantic Ocean, which is consistent with both the historical and paleontological records of this family (Berta et al. 2018) (Fig. 2.9). They breed on predator-free islands and, in some cases, mainland colonies from the equatorial regions of the Galapagos to the sub-polar regions of the Antarctic and the Bering Sea. In contrast to true seals and walruses (see Miller and Kochnev, Chap. 22), otariids do not breed on ice. However, some species like Antarctic fur seals often haul out on ice floes to rest (Fig. 2.10). Within these regions, their distribution is limited to highly productive habitats, mostly associated with upwelling systems (Fig. 2.9). This distributional pattern results from the physiological constraints associated with income breeding, an energetically expensive strategy that requires abundant and predictable prey resources relatively near the colony (Costa 1993; Boyd 1998; Trillmich and Weissing 2006; Stephens et al. 2014).

2.3.1 *Foraging Behavior Relative to Habitats and Energetics*

Different habitats are associated with three distinct marine foraging patterns: epipelagic, where animals forage in the upper water column between 0 and 200 m;

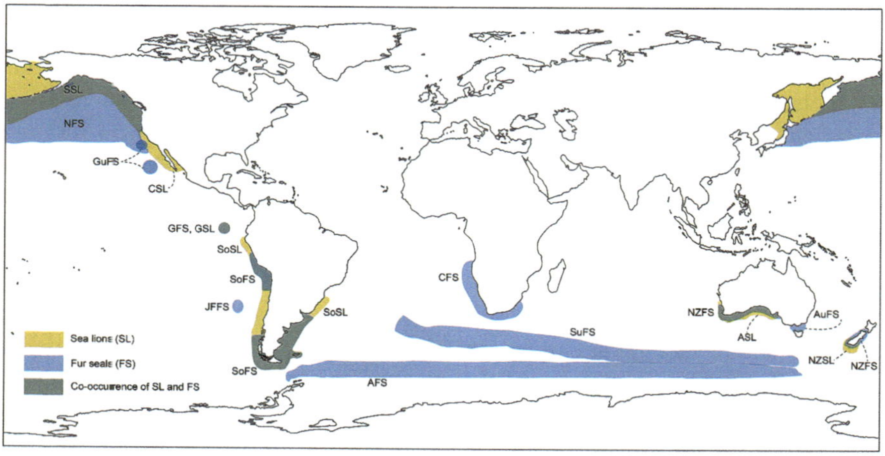

Fig. 2.9 Primary geographic distribution of Otariids around the globe. Abbreviations: AFS Antarctic fur seals, AuFS Australian fur seals, ASL Australian sea lions, CFS Cape fur seals, CSL California sea lions, GFS Galapagos fur seals, GuFS Guadalupe fur seals, GSL Galapagos sea lions, JFFS Juan Fernandez fur seals, NFS Northern fur seals, NZFS New Zealand fur seals, NZSL New Zealand sea lions, SoFS, South American fur seals, SoSL, South American sea lions, SuFS Sub Antarctic fur seals, SSL Steller sea lions

Fig. 2.10 Male Antarctic fur seal resting on an ice flow in the Antarctic Peninsula. Photo by Dan Costa

mesopelagic, where animals forage deeper between 200 and 1000 m, and benthic, where animals forage at or near the bottom of the seafloor (Costa et al. 2001, 2004, 2006; Arnould and Costa 2006) (Fig. 2.4). Benthic and mesopelagic dives are generally longer as animals must transit to depth before they can begin searching and pursuing prey and, once the prey has been captured or pursuit discontinued, they must then transit back to the surface (Fig. 2.4). As epipelagic dives are shallow, animals may initiate searching and pursuit at or near the surface or soon after the dive commences within little or no transit phase. Thus, epipelagic dives can be shorter than either mesopelagic or benthic dives as a greater proportion of the dive is spent searching or pursuing prey. Furthermore, these foraging strategies also affect the diet as epipelagic foragers can capture numerous small prey per dive, whereas benthic foragers tend to pursue single large prey per dive, and mesopelagic foragers pursue many small or a few large prey per dive (Costa 1988, 1991a, 1993; Volpov et al. 2016).

The tradeoffs associated with these different diving patterns have been examined using optimality models (Carbone and Houston 1996; Houston et al. 2003; Foo et al. 2016) and show that large prey and/or energy dense prey are preferred unless small or energy poor prey is considerably more abundant/available. However, there is an upper threshold that is determined by the animals' digestive physiology when they can no longer forage as they are satiated and require time to process the ingested prey (Rosen and Trites 2004). Marine mammals have a long gut compared to terrestrial carnivores, and this may increase their capacity to process and absorb nutrients, increasing foraging efficiency (Williams and Yeates 2004; Williams et al. 2001). The occurrence of these foraging and diving patterns varies widely across the otariids (Table 2.1 and Fig. 2.4). Nevertheless, behavioral plasticity has been recognized in some species including Galapagos sea lions which exhibit all three foraging patterns (Villegas-Amtmann et al. 2008). Other species like Australian sea lions forage primarily at or near the benthos (Costa and Gales 2003), while New Zealand (*Phocarctos hookeri*) sea lions, the deepest diving sea lion, forage in both the mesopelagic and benthic regions (Chilvers et al. 2006; Chilvers 2017).

Foraging patterns also vary with habitat and colony location. For example, Antarctic fur seal females feed primarily on krill around the Antarctic Peninsula (Osman et al. 2004) and South Georgia. However, at Kerguelen and Heard Islands, they forage deeper primarily on fish (Lea et al. 2008; Staniland et al. 2010). Dives are shallower when feeding on small prey, such as krill, compared to larger prey such as fish, which are larger and energetically more beneficial (Staniland et al. 2010). Similarly, female California sea lions foraging in the Southern California Bight are generalists foraging epipelagically on a variety of fish and squid (McHuron et al. 2016, 2018b). Still, in the Sea of Cortez (= Gulf of California), they forage both in the epipelagic and mesopelagic zones (Villegas-Amtmann et al. 2011).

Furthermore, benthic foraging species regularly undertake dive durations exceeding their calculated aerobic dive limit (cADL), whereas the epipelagic foraging species rarely dive longer than their cADL (Costa et al. 1999, 2004). Thus, benthic species might have a lower capacity to increase foraging effort in times of nutritional stress than epipelagic species. This reduced capacity or resilience to change could result in reductions in reproductive output, offspring growth, and survival (Arnould and Costa 2006). This pattern is supported by the lower reproductive rate observed in the Australian fur seals (*A. pusillus doriferus*) (and benthic feeding sea lions) compared to the epipelagic conspecific South African fur seal (*A. pusillus pusillus*) and California sea lions. Indeed, the difference between the mean birth rate of all benthic (61.7 ± 4.0%) and epipelagic (79.1 ± 1.1%) foragers approached significance ($t^2 = 4.15$, $P = 0.053$). The low reproductive rate of Australian fur seals may explain their very slow recovery from commercial sealing in comparison to the rapid recovery of the conspecific South African fur seals, which is now the most numerous otariid. Similarly, California sea lions were hunted prior to their protection under the Marine Mammal Protection Act of 1972 and have since recovered, unlike anything observed in *benthic* foraging sea lions (Laake et al. 2018).

2.3.2 Consequences for Population Dynamics

Associated with these divergent trends of *epipelagic* and *benthic* foraging behavior are differences in the population dynamics of sea lions and fur seals. All species of otariid seals throughout the world were subject to extensive and, in most cases, excessive hunting pressure during the eighteenth and nineteenth centuries (Wickens and York 1997) (Crespo, Chap. 24). By the late 1800s, however, most species were protected or were subject to sustainable harvests. Despite this protection, populations of the various sea lion species have experienced minimal recovery and, in some cases, are declining, whereas many fur seal species have recovered or are recovering (Wickens and York 1997; Costa et al. 2006; Kovacs et al. 2012). This suggests that there may be life-history consequences associated with the different foraging modes, which influence population dynamics (i.e., specific foraging mode may be more efficient?).

2.3.3 Role of Food Availability

The impact of environmental variability on marine mammal populations is best documented by their response to variations to the El Niño Southern Oscillation (ENSO). ENSO events are characterized by a reduction in upwelling favorable winds along the West Coasts of North and South America and teleconnections to other regions, including the Southern Ocean (Murphy et al. 2007; Clarke 2014). Among the changes that occur are the deepening of the thermocline depth and associated reductions in primary production due to the loss of upwelling. Most notable are the 1983 and 1998 ENSO events, where pinnipeds along the west coasts of North and South America faced significant reductions in prey availability, leading to increased foraging times, reduced prey captures rates, reduced pup growth rates, lower weaning weights, increased pup mortality, and in some cases reduced adult survival (Trillmich and Limberger 1985; Trillmich and Ono 1991; Boyd and Roberts 1993; Melin et al. 2012; Sielfeld et al. 2018). The impact of the 1983 and 1998 ENSO events on pup production in California sea lions is a remarkable example (Laake et al. 2018). Pup production significantly declined during the 1983 ENSO event and took almost a decade to recover to pre-ENSO values. This contrasts with the 1998 ENSO event, where although pup production declined during that event, it returned to near normal levels the following year. The difference is likely due to the mortality of both pups and breeding females during the 1983 ENSO event compared to just a loss of pups during the 1998 ENSO event. While the two events were of similar intensity, the 1983 event was of longer duration. There are also multidecadal trends such as the Pacific Decadal Oscillation, which is likened to a longer-term ENSO event on the order of 10–20 years of warm *vs.* cold water regimes (Mestas-Nunez and Miller 2006; Pennington et al. 2006) that are also associated with changes in community composition (Chavez et al. 2003). These longer-term variations in climate have been suggested as a factor affecting the decline of Steller sea lions (Rodionov et al. 2005).

Successful reproduction by otariids requires a foraging pattern that optimizes the amount of time spent feeding at-sea with the amount of milk energy delivered to the pup waiting at the rookery. Studies of females with dependent young show that as food resources decrease, mothers first respond by increasing the intensity of their foraging effort, and if this is insufficient, by increasing the time spent at-sea. Females might be able to capture more prey by increasing the diving frequency and/or the time spent in foraging bouts (also resulting in decreased resting time) (Boyd 1996). In addition, females could increase the time at sea, allowing more foraging time and

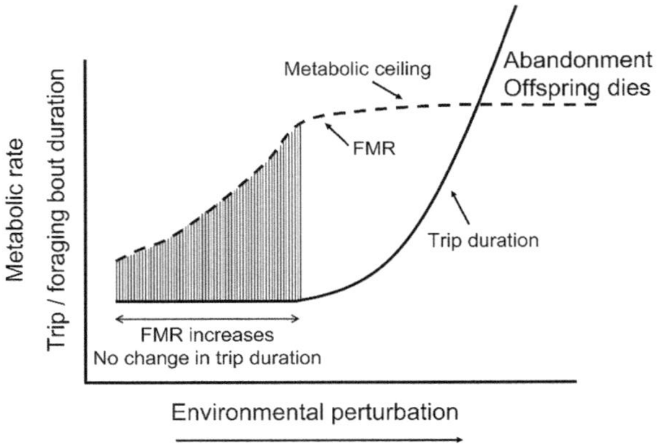

Fig. 2.11 A conceptual model of how a female otariid should respond to environmental perturbation that causes a reduction in prey availability. The dashed line represents FMR and the solid line is trip duration. From Costa (2008)

greater travel distances. For example, during a warm water event, lactating California sea lions traveled a mean of 117 ± 27 se km away from the colony compared to a mean of 68 ± 6.9 se km during normal years (Kuhn and Costa 2014). Attempts to increase foraging effort or intensity would be preferred, as longer foraging trips increase the time between visits to the pup. As trip durations increase, a greater proportion of the pup's energy budget is directed to maintenance rather than growth resulting in slower pup growth (Costa 2008, 2012) (Fig. 2.11). Some species are already working at their maximum capability and have little if any ability to increase their effort without negatively impacting pup growth (Costa et al. 2000b; McHuron et al. 2019). However, success averaged over a series of foraging events is more important than the success over a single foraging trip. This is supported by the observation that weaning mass of northern fur seals was not linked to success over an individual foraging trip (Goebel 2002).

2.4 Evolutionary Implications of Otariid Foraging Patterns

Pinnipeds have a globally distributed and well represented fossil record (Valenzuela-Toro and Pyenson 2019; Berta et al. 2018); however, their macroevolutionary history and ecological transitions, including the evolution of reproductive and foraging strategies, have been comparatively understudied. To date, our knowledge of pinniped reproductive strategies comes primarily from comparative morphology of fossils, including aspects of skull morphology and size (Cullen et al. 2014). Thus, it has been inferred that basal pinnipeds, such as *Enaliarctos*, known from the late Oligocene to early Miocene (~25 million years ago) of the eastern North Pacific (Berta et al. 1989), were sexually dimorphic. This attribute suggests that a polygynous breeding system was the ancestral state for the group (Wyss 1994; Cullen et al. 2014). These small fur seal-like sized early carnivores (150–170 cm) (Churchill et al. 2015) most likely made short trips to feed in the coastal ocean during a period when there was increased upwelling of cold nutrient rich water (White et al. 1992). Enhanced upwelling increased all levels of biological productivity (Lipps and Mitchell 1976), making prey species more available. In such systems, endothermic tetrapod predators would have faced reduced competition for the now abundant prey (Vermeij and Dudley 2000). Like all mammals, *Enaliarctos* and other basal pinnipeds would have high constant body temperature, which given an inferred relatively dense coat of fur, could be maintained even in cold water. As endotherms, these early pinnipeds would have maintained an elevated body temperature with a high aerobic metabolism that, in turn, allowed their muscles to perform optimally even when operating in cold water, similar to extant otariids. Considering that upwelled water is cold, the muscles of ectothermic prey are not as efficient and would thus be susceptible to predation from the faster, more efficient endothermic predators (Cairns et al. 2008; Grady et al. 2019). Therefore, the congruence of abundant prey and the predatory advantage of air-breathing endothermy would make it possible for these early pinnipeds to successfully inhabit coastal marine environments. As these taxa evolved, they diverged into different lineages, including the modern Otariidae as well as the morphologically and ecologically derived Phocidae and Odobenidae and the extinct seal-like taxa, Desmatophocidae. Subsequent transformations of their foraging and diving skills would allow them to forage further and deeper offshore, reducing competition with coexisting coastal nearshore marine mammals. Foraging further away from the colony would have resulted in an increase in the duration of their foraging trips along with a consequent decrease in their number and frequency. In this regard, increased milk lipid content in some groups (e.g. phocids and fur seals) would have facilitated rapid milk energy transfer, decreasing the time females needed to spend with their pups onshore.

Otariids were constrained to remain in the North Pacific Ocean until the Central American Seaway closed as it presented an ecological barrier to an income breeding otariid given their foraging and reproductive pattern are tightly linked in space and time to highly productive upwelling regions (Briscoe et al. 2017, 2018). Once the seaway closed during the late Pliocene (between 3 and 4 million years ago; Duque-Caro 1990; Hoorn and Flantua 2015; Montes et al. 2015; Kirby et al. 2008) oceanographic conditions, including the expansion of coastal upwelling conditions in the Southeast Pacific (Ibaraki 1997) developed. This would have allowed otariids to cross the equator into the Southern Hemisphere (Churchill et al. 2014). Once they invaded the South Eastern Pacific, otariids would have once again encountered an eastern boundary current with upwelling conditions that favored their energy expensive life history. It is interesting that before the closure of the Central American Seaway, there were several fossil phocids in the Southern Hemisphere during the Miocene and Pliocene (Valenzuela-Toro et al. 2013). This shows that phocid seals were able to disperse into the Southern Hemisphere before the Central American Seaway closed, probably because they had evolved a capital breeding system that allowed a spatial and temporal separation of feeding from breeding, increasing their dispersal capability (Costa 1993).

Once otariids dispersed into the Southern Hemisphere their more energy intensive lifestyle could take advantage of the highly productive upwelling regions as the reproductive output (total energy invested in offspring) is greater for species with higher metabolic rates and or that have a faster pace of life (Hennemann 1983; Schmitz and Lavigne 1984; Dmitriew 2011; Wright et al. 2018). It then follows that otariids with a high metabolic rate and thus a faster pace of life could obtain and invest more resources into reproduction and therefore were able to outcompete phocids who, as capital breeders, have a more economical lifestyle and slower pace of life. Other factors such as local and global changes in sea level and subsequent suitable habitat transformations might also play a role in this faunal turnover; however, new fossil findings are required to resolve this issue. Finally, the lack of otariids in the North Atlantic is consistent with the presence of an ecological barrier in the form of the equatorial region between the Gulf of Mexico and the Mediterranean Sea, preventing their dispersal in this region along with their dispersal into the Southern Hemisphere. Further, the orientation of land masses also prevents favorable cold-water oceanographic conditions that would facilitate the expansion of otariids from the South Atlantic across the equator into the North Atlantic.

2.5 Future Directions

While we have acquired a significant amount of information on the reproductive and foraging biology and energetics of otariids, considerable gaps remain. Most of the research has been carried out on tractable species such as Antarctic and Galapagos fur seals and Australian and New Zealand sea lions. The difficulty of accessing mothers and pups on rocky shorelines and their skittish nature has limited measurements on some of the most common species of otariids, such as Steller and California sea lions.

For most species, we lack fundamental bioenergetics measurements, such as prey assimilation efficiencies and metabolic rates. We also lack data on the birth mass, weaning mass, growth rate, and lactation duration. Milk intake and milk composition data have only been completed on a few species, and milk intake over the entire lactation interval has only been completed on Antarctic, Australian and northern fur seals. We have only scratched the surface of the physiological adaptations that allow otariid pups to rapidly process lipid-rich milk. Such data are critical to parametrize models that predict how environmental disturbance will affect behavior and ultimately, population dynamics (Costa 2012; McHuron et al. 2017a; Pirotta et al. 2018). The broad distribution of otariids presents a compelling argument that income breeding, as typified by otariids, is a more successful life history pattern than capital breeding, as exemplified in phocids. This appears to be the case, at least when prey is abundant. Further comparisons between different species in different habitats would help elucidate the eco-evolutionary dynamics that drove the emergence of the costs and benefits of these very different reproductive strategies. Integration of paleontological and neontological knowledge would help elucidate the ecological transitions, including the origin of the reproductive strategies and changes in the foraging ecology along the evolutionary history of otariids, and pinnipeds in general. A better understanding of the energetic constraints of different foraging patterns would allow better prediction on how changing ocean conditions and anthropogenic disturbance will affect otariid populations (McHuron et al. 2017a; Pirotta et al. 2018).

Acknowledgements We thank the many collaborators, students and colleagues who over the years have shaped our thinking on reproductive patterns, energetics and life history patterns. These individuals have freely shared their ideas, thoughts and data. DPC thanks Mary I Zavanelli whose support over his career made his work possible, he only wishes she was here to read this. We thank Arina Favilla for comments on drafts of this paper. The writing of this chapter was supported by the Office of Naval Research N00014-18-1-2822, the E&P Sound and Marine Life Joint Industry Project of the International Association of Oil and Gas Producers (JIP2207–23); and the NSF Polar Programs and Environments award 1644256 to D. Costa. A. Valenzuela-Toro was funded by ANID PCHA/Becas Chile, Doctoral Fellowship; Grant No. 2016-72170286.

References

Arnould JPY, Boyd IL (1995) Temporal patterns of milk production in Antarctic fur seals (*Arctocephalus gazella*). J Zool 237(1):1–12. https://doi.org/10.1111/j.1469-7998.1995.tb02741.x

Arnould JPY, Boyd IL, Speakman JR (1996) The relationship between foraging behaviour and energy expenditure in Antarctic fur seals. J Zool 239:769–782. https://doi.org/10.1111/j.1469-7998.1996.tb05477.x

Arnould JPY, Costa DP (2006) Sea lions in drag, fur seals incognito: insights from the otariid deviants. Sea Lions of the World

Arnould JP, Hindell MA (2001) Dive behaviour, foraging locations, and maternal-attendance patterns of Australian fur seals (*Arctocephalus pusillus doriferus*). Can J Zool 79(1):35–48

Arnould JPY, Hindell MA (2002) Milk consumption, body composition and pre-weaning growth rates of Australian fur seal (*Arctocephalus pusillus doriferus*) pups. J Zool 256:351–359. https://doi.org/10.1017/s0952836902000389

Aurioles-Gamboa D (2015a) *Arctocephalus townsendi*. The IUCN red list of threatened species 2015: e.T2061A45224420. https://doi.org/10.2305/IUCN.UK.2015-2.RLTS.T2061A45224420.en. Accessed 04 Jan 2020

Aurioles-Gamboa D (2015b) *Arctocephalus philippii*. The IUCN red list of threatened species 2015: e.T2059A61953525. https://doi.org/10.2305/IUCN.UK.2015-2.RLTS.T2059A61953525.en. Accessed 05 Jan 2020

Aurioles-Gamboa D, Hernández-Camacho J (2015) *Zalophus californianus*. The IUCN red list of threatened species 2015: e.T41666A45230310. https://doi.org/10.2305/IUCN.UK.2015-4.RLTS.T41666A45230310.en. Accessed 05 Jan 2020

Bartholomew GA (1970) A model for the evolution of pinniped polygyny. Evolution 24 (3):546–559. https://doi.org/10.1111/j.1558-5646.1970.tb01790.x

Baylis AMM, Orben RA, Costa DP, Arnould JPY, Staniland IJ (2016) Sexual segregation in habitat use is smaller than expected in a highly dimorphic marine predator, the southern sea lion. Mar Ecol Prog Ser 554:201–211. https://doi.org/10.3354/meps11759

Baylis AMM, Tierney M, Orben RA, Staniland IJ, Brickle P (2018) Geographic variation in the foraging behaviour of south American fur seals. Mar Ecol Prog Ser 596:233–245. https://doi.org/10.3354/meps12557

Berta A, Ray CE, Wyss AR (1989) Skeleton of the oldest known Pinniped, *Enaliarctos mealsi*. Science 244(4900):60–62. https://doi.org/10.1126/science.244.4900.60

Berta A, Churchill M, Boessenecker RW (2018) The origin and evolutionary biology of pinnipeds: seals, sea lions, and walruses. Ann Rev Earth Planet Sci 46:203–228. https://doi.org/10.1146/annurev-earth-082517-010009

Boness DJ, Francis JM (1991) The effect of thermoregulatory behaviour on the mating system of the Juan Fernández fur seal, *Arctocephalus philippii*. Behaviour 119(1-2):104–126. https://doi.org/10.1163/156853991x00391

Boyd IL (1996) Temporal scales of foraging in a marine predator. Ecology 77(2):426–434. https://doi.org/10.2307/2265619

Boyd IL (1998) Time and energy constraints in pinniped lactation. Am Nat 152(5):717–728. https://doi.org/10.1086/286202

Boyd IL, Roberts JP (1993) Tooth growth in male Antarctic fur seals (*Arctocephalus gazella*) from South Georgia: an indicator of long-term growth history. J Zool 229(2):177–190. https://doi.org/10.1111/j.1469-7998.1993.tb02630.x

Boyd IL, Lunn NJ, Barton T (1991) Time budgets and foraging characteristics of lactating Antarctic fur seals. J Anim Ecol 60(2):577–592. https://doi.org/10.2307/5299

Briscoe DK, Hobday AJ, Carlisle A, Scales K, Eveson JP, Arrizabalaga H, Druon JN, Fromentin JM (2017) Ecological bridges and barriers in pelagic ecosystems. Deep Sea Res Part II 140:182–192. https://doi.org/10.1016/j.dsr2.2016.11.004

Briscoe DK, Fossette S, Scales KL, Hazen EL, Bograd SJ, Maxwell SM, McHuron EA, Robinson PW, Kuhn C, Costa DP, Crowder LB, Lewison RL (2018) Characterizing habitat suitability for a central-place forager in a dynamic marine environment. Ecol Evol 8(5):2788–2801. https://doi.org/10.1002/ece3.3827

Brody S (1945) Bio-energetics and growth with special reference to the efficiency complex in domestic animals. Bio-energetics and growth with special reference to the efficiency complex in domestic animals. Reinhold Publishing Corporation, New York

Cairns DK, Gaston AJ, Huettmann F (2008) Endothermy, ectothermy and the global structure of marine vertebrate communities. Mar Ecol Prog Ser 356:239–250. https://doi.org/10.3354/meps07286

Carbone C, Houston AI (1996) The optimal allocation of time over the dive cycle: an approach based on aerobic and anaerobic respiration. Anim Behav 51:1247–1255. https://doi.org/10.1006/anbe.1996.0129

Cárdenas-Alayza S, Crespo E, Oliveira L (2016a) *Otaria byronia*. The IUCN red list of threatened species 2016: e.T41665A61948292. https://doi.org/10.2305/IUCN.UK.2016-1.RLTS.T41665A61948292.en. Accessed 05 Jan 2020

Cárdenas-Alayza S Oliveira L, Crespo E (2016b) *Arctocephalus australis* ssp. *australis*. The IUCN red list of threatened species 2016: e.T2064A72050370. https://doi.org/10.2305/IUCN.UK.2016-1.RLTS.T2064A72050370.en. Accessed 05 Jan 2020

Cassini MH (1999) The evolution of reproductive systems in pinnipeds. Behav Ecol 10(5):612–616. https://doi.org/10.1093/beheco/10.5.612

Chavez FP, Ryan J, Lluch-Cota SE, Niquen M (2003) From anchovies to sardines and back: multidecadal change in the Pacific Ocean. Science 299(5604):217–221. https://doi.org/10.1126/science.1075880

Chilvers BL (2015) *Phocarctos hookeri*. The IUCN Red List of Threatened Species 2015: e.T17026A1306343. https://doi.org/10.2305/IUCN.UK.2015-2.RLTS.T17026A1306343.en. Accessed 05 Jan 2020

Chilvers BL (2017) Stable isotope signatures of whisker and blood serum confirm foraging strategies for female New Zealand Sea lions (*Phocarctos hookeri*) derived from telemetry. Can J Zool 95(12):955–963. https://doi.org/10.1139/cjz-2016-0299

Chilvers BL, Meyer S (2017) Conservation needs for the endangered New Zealand Sea lion, *Phocarctos hookeri*. Aquat Conserv 27(4):846–855. https://doi.org/10.1002/aqc.2742

Chilvers BL, Wilkinson IS (2009) Diverse foraging strategies in lactating New Zealand Sea lions. Mar Ecol Prog Ser 378:299–308. https://doi.org/10.3354/meps07846

Chilvers BL, Wilkinson IS, Duignan PJ, Gemmell NJ (2006) Diving to extremes: are New Zealand Sea lions (*Phocarctos hookeri*) pushing their limits in a marginal habitat? J Zool 269 (2):233–240. https://doi.org/10.1111/j.1469-7998.2006.00059.x

Churchill M, Boessenecker RW, Clementz MT (2014) Colonization of the southern hemisphere by fur seals and sea lions (Carnivora: Otariidae) revealed by combined evidence phylogenetic and Bayesian biogeographical analysis. Zool J Linnean Soc 172(1):200–225. https://doi.org/10.1111/zoj.12163

Churchill M, Clementz MT, Kohno N (2015) Cope's rule and the evolution of body size in Pinnipedimorpha (Mammalia: Carnivora). Evolution 69(1):201–215. https://doi.org/10.1111/evo.12560

Clarke AJ (2014) El Nino physics and El Nino predictability. Annu Rev Mar Sci 6(1):79–99. https://doi.org/10.1146/annurev-marine-010213-135026

Costa DP (1988) Methods for studying the energetics of freely diving animals. Can J Zool 66:45–52

Costa DP (1991a) Reproductive and foraging energetics of high-latitude penguins, albatrosses and Pinnipeds - implications for life-history patterns. Am Zool 31(1):111–130. https://doi.org/10.1093/icb/31.1.111

Costa DP (1991b) Reproductive and foraging energetics of pinnipeds: implications for life history patterns. In: The behaviour of pinnipeds. Springer, Dordrecht, pp 300–344

Costa DP (1993) The relationship between reproductive and foraging energetics and the evolution of the Pinnipedia. In: Boyd IL (ed) Marine mammals: advances in behavioural and population biology, vol 66. Symposia. Zoological Society of London. Oxford University Press, Symposium Zoological Society of London, Oxford, pp 293–314

Costa DP (2008) A conceptual model of the variation in parental attendance in response to environmental fluctuation: foraging energetics of lactating sea lions and fur seals. Aquat Conserv Mar Freshwat Ecosyst 17(S1):S44–S52

Costa DP (2012) A bioenergetics approach to developing the PCAD model. In: Popper AN, Hawkins A (eds) The effects of noise on aquatic life. Advances in experimental medicine and biology. Springer, New York, pp 423–426. https://doi.org/10.1007/978-1-4419-7311-5_96

Costa DP, Gales NJ (2000) Foraging energetics and diving behavior of lactating New Zealand Sea lions, *Phocarctos hookeri*. J Exp Biol 203. (Pt 23:3655–3665

Costa DP, Gales NJ (2003) Energetics of a benthic diver: seasonal foraging ecology of the Australian sea lion, *Neophoca cinerea*. Ecol Monogr 73(1):27–43

Costa DP, Gentry RL (1986) Free-ranging energetics of northern fur seals. In: Gentry RL, Kooyman Gerald L (eds) Fur Seals: maternal strategies on land and at sea. Princeton University Press, Princeton, NJ, pp 79–101

Costa DP, Maresh JL (2017) Energetics. Encyclopedia of marine mammals, 3rd edn. Academic, New York

Costa DP, Trillmich F, Croxall JP (1988) Intraspecific allometry of neonatal size in the antarctic fur seal (*Arctocephalus galapagoensis*). Behav Ecol Sociobiol 22(5):361–364. https://doi.org/10.1007/bf00295105

Costa DP, Croxall JP, Duck CD (1989) Foraging energetics of antartic fur seals in relation to changes in prey availability. Ecology 70(3):596–606

Costa DP, Crocker DE, Gales NJ (1998) Blood volume and diving ability of the New Zealand sea lion, *Phocarctos hookeri*. Physiol Zool 71(2):208–213

Costa DP, Webb PM, Crocker DE, Goebel ME (1999) The role of physiology in the behavior of diving mammals: insights from animals in nature. Comp Biochem Physiol A Mol Integr Physiol 124(SUPPL):S21

Costa DP, Gales NJ, Goebel MA (2000a) The role of physiology in the behavior of diving mammals. Am Zool 40(6):980–981

Costa DP, Goebel ME, Sterling JE (2000b) Foraging energetics and diving behavior of the Antarctic fur seal. *Arctocephalus gazella* at Cape Shirreff, Livingston Island. In: Davisons W, Howard-Williams C, Broady P (eds) Antarctic ecosystems: models for wider ecological understanding. New Zealand Natural Sciences Press, Christchurch, New Zealand, pp 77–84

Costa DP, Gales NJ, Goebel ME (2001) Aerobic dive limit: how often does it occur in nature? Comp Biochem Physiol A Mol Integr Physiol 129(4):771–783
Costa DP, Kuhn CE, Weise MJ, Shaffer SA, Arnould JPY (2004) When does physiology limit the foraging behaviour of freely diving mammals? Int Cong Ser 1275:359–366. https://doi.org/10.1016/j.ics.2004.08.058
Costa DP, Weise MJ, Arnould JPY (2006) Potential influences of whaling on the status and trends of pinniped populations. Whales Whaling Ocean Ecosyst 2007:344–359
Cullen TM, Fraser D, Rybczynski N, Schroder-Adams C (2014) Early evolution of sexual dimorphism and polygyny in Pinnipedia. Evolution 68(5):1469–1484. https://doi.org/10.1111/evo.12360
Dmitriew CM (2011) The evolution of growth trajectories: what limits growth rate? Biol Rev 86 (1):97–116. https://doi.org/10.1111/j.1469-185X.2010.00136.x
Duque-Caro H (1990) Neogene stratigraphy, paleoceanography and paleobiogeography in Northwest South America and the evolution of the Panama seaway. Paleogeogr Palaeclimatol Palaeoecol 77:203–234
Fahlman A, Svard C, Rosen DAS, Wilson RP, Trites AW (2013) Activity as a proxy to estimate metabolic rate and to partition the metabolic cost of diving vs. breathing in pre- and post-fasted Steller Sea lions. Aquat Biol 18(2):175–184. https://doi.org/10.3354/ab00500
Foo D, Semmens JM, Arnould JPY, Dorville N, Hoskins AJ, Abernathy K, Marshall GJ, Hindell MA (2016) Testing optimal foraging theory models on benthic divers. Anim Behav 112:127–138. https://doi.org/10.1016/j.anbehay.2015.11.028
Fowler SL, Costa DP, Arnould JPY, Gales NJ, Kuhn CE (2006) Ontogeny of diving behaviour in the Australian sea lion: trials of adolescence in a late bloomer. J Anim Ecol 75(2):358–367. https://doi.org/10.1111/j.1365-2656.2006.01055.x
Fowler SL, Costa DP, Arnould JPY (2007a) Ontogeny of movements and foraging ranges in the Australian Sea Lion. Mar Mamm Sci 23(3):598–614. https://doi.org/10.1111/j.1748-7692.2007.00134.x
Fowler SL, Costa DP, Arnould JPY, Gales NJ, Burns JM (2007b) Ontogeny of oxygen stores and physiological diving capability in Australian sea lions. Funct Ecol 21(5):922–935. https://doi.org/10.1111/j.1365-2435.2007.01295.x
Francis J, Boness D, Ochoa-Acuña H (1998) A protracted foraging and attendance cycle in female Juan Fernandez fur seals. Mar Mamm Sci 14(3):552–574
Gallagher AJ, Hammerschlag N, Cooke SJ, Costa DP, Irschick DJ (2015) Evolutionary theory as a tool for predicting extinction risk. Trends Ecol Evol 30(2):61–65. https://doi.org/10.1016/j.tree.2014.12.001
Gallo-Reynoso JP, Figueroa-Carranza AL, Le Boeuf BJ (2008) Foraging behavior of lactating Guadalupe fur seal females. In: Lorenzo C, Espinoza E, Ortega J (eds) Avances en el Estudio de los Mamíferos de México II. Publicaciones Especiales, vol II. Asociación Mexicana de Mastozoología, A. C., Centro de Investigaciones Biológicas del Noroeste, El Colegio de la frontera Sur, Instituto Politécnico Nacional, Universidad Autónoma de Morelos, Universidad Autónoma Metropolitana, Universidad Autónoma de Chiapas y Universidad Veracruzana, San Cristóbal de Las Casas, Chiapas, pp 595–614
Gamel CM, Davis RW, David JH, Meÿer MA, Brandon E (2005) Reproductive energetics and female attendance patterns of cape fur seals (*Arctocephalus pusillus pusillus*) during early lactation. Am Midl Nat 153(1):152–170. https://doi.org/10.1674/0003-0031(2005)153[0152:Reafap]2.0.Co;2
Gelatt T, Sweeney K (2016) *Eumetopias jubatus*. The IUCN red list of threatened species 2016: e.T8239A45225749. https://doi.org/10.2305/IUCN.UK.2016-1.RLTS.T8239A45225749.en. Accessed 05 Jan 2020
Gelatt T, Ream R, Johnson D (2015) *Callorhinus ursinus*. The IUCN red list of threatened species 2015: e.T3590A45224953. https://doi.org/10.2305/IUCN.UK.2015-4.RLTS.T3590A45224953.en. Accessed 05 Jan 2020

Gentry RL (1973) Thermoregulatory behavior of eared seals. Behaviour 46(1):73–93. https://doi.org/10.1163/156853973x00175

Gentry RL (1998) Female attendance pattern. In: Gentry RL (ed) Behavior and ecology of the northern Fur seal. Princeton University Press, Princeton, NJ, p 221

Gentry RL, Holt JR (1986) Attendence behavior in northern fur seals. In: Gentry RL, Kooyman GL (eds) Fur seals: maternal strategies on land and at sea. Princeton University Press, Princeton, NJ, pp 41–60

Georges J-Y, Guinet C (2000) Maternal care in the subantarctic fur seals on Amsterdam Island. Ecology 81(2):295–308. https://doi.org/10.1890/0012-9658(2000)081[0295:Mcitsf]2.0.Co;2

Georges J-Y, Guinet C (2001) Prenatal investment in the subantarctic fur seal, *Arctocephalus tropicalis*. Can J Zool 79(4):601–609

Goebel ME (2002) Northern fur seal lactation, attendance and reproductive success in two years of contrasting oceanography. Dissertation, University of California, Santa Cruz

Grady JM, Maitner BS, Winter AS, Kaschner K, Tittensor DP, Record S, Smith FA, Wilson AM, Dell AI, Zarnetske PL, Wearing HJ, Alfaro B, Brown JH (2019) Metabolic asymmetry and the global diversity of marine predators. Science 363(6425):366. https://doi.org/10.1126/science.aat4220

Hamer DJ, Goldsworthy SD, Costa DP, Fowler SL, Page B, Sumner MD (2013) The endangered Australian sea lion extensively overlaps with and regularly becomes by-catch in demersal shark gill-nets in south Australian shelf waters. Biol Conserv 157(0):386–400. https://doi.org/10.1016/j.biocon.2012.07.010

Harcourt RG, Bradshaw CJ, Davis LS (2001) Summer foraging behaviour of a generalist predator, the New Zealand fur seal (Arctocephalus forsteri). Wildl Res 28(6):599–606

Hedrick MS, Duffield DA (1991) Hematological and rheological characteristics of blood in 7 marine mammal species - physiological implications for diving behavior. J Zool 225:273–283. https://doi.org/10.1111/j.1469-7998.1991.tb03816.x

Hennemann WW III (1983) Relationship among body mass, metabolic rate and the intrinsic rate of natural increase in mammals. Oecologia 56:104–108

Higgins LV, Gass L (1993) Birth to weaning - parturition, duration of lactation, and attendance cycles of Australian sea lions (*Neophoca cinerea*). Can J Zool 71(10):2047–2055. https://doi.org/10.1139/z93-290

Higgins LV, Costa DP, Huntley AC, Boeuf BJ (1988) Behavioral and physiological measurements of maternal Investment in the Steller Sea Lion, *Eumetopias jubatus*. Mar Mamm Sci 4(1):44–58. https://doi.org/10.1111/j.1748-7692.1988.tb00181.x

Hofmeyr GJG (2015a) *Arctocephalus tropicalis*. The IUCN red list of threatened species 2015: e.T2062A45224547. https://doi.org/10.2305/IUCN.UK.2015-4.RLTS.T2062A45224547.en. Accessed 05 Jan 2020

Hofmeyr GJG (2015b) *Arctocephalus pusillus*. The IUCN red list of threatened species 2015: e.T2060A45224212. https://doi.org/10.2305/IUCN.UK.2015-4.RLTS.T2060A45224212.en. Accesses 05 Jan 2020

Hofmeyr GJG (2016) *Arctocephalus gazella*. The IUCN red list of threatened species 2016: e.T2058A66993062. https://doi.org/10.2305/IUCN.UK.2016-1.RLTS.T2058A66993062.en. Accessed 05 Jan 2020

Hoorn C, Flantua S (2015) An early start for the Panama land bridge. Science 348(6231):186–187. https://doi.org/10.1126/science.aab0099

Houston AI, McNamara JM, Heron JE, Barta Z (2003) The effect of foraging parameters on the probability that a dive is successful. Proc Biol Sci 270(1532):2451–2455. https://doi.org/10.1098/rspb.2003.2540

Houston AI, Stephens PA, Boyd IL, Harding KC, McNamara JM (2007) Capital or income breeding? A theoretical model of female reproductive strategies. Behav Ecol 18(1):241–250. https://doi.org/10.1093/beheco/arl080

Hückstädt LA, Quiñones RA, Sepúlveda M, Costa DP (2014) Movement and diving patterns of juvenile male south American sea lions off the coast of Central Chile. Mar Mamm Sci 30(3):1175–1183. https://doi.org/10.1111/mms.12085

Hückstädt LA, Tift MS, Riet-Sapriza F, Franco-Trecu V, Baylis AM, Orben RA, Arnould JP, Sepulveda M, Santos-Carvallo M, Burns JM, Costa DP (2016) Regional variability in diving physiology and behavior in a widely distributed air-breathing marine predator, the south American sea lion (*Otaria byronia*). J Exp Biol 219 (Pt 15):2320–2330. https://doi.org/10.1242/jeb.138677

Ibaraki M (1997) Closing of the central American seaway and neogene coastal upwelling along the Pacific coast of South America. Tectonophysics 281(1–2):99–104. https://doi.org/10.1016/S0040-1951(97)00161-3

Jeanniard-du-Dot T, Guinet C, Arnould JPY, Speakman JR, Trites AW (2017a) Accelerometers can measure total and activity-specific energy expenditures in free-ranging marine mammals only if linked to time-activity budgets. Funct Ecol 31(2):377–386. https://doi.org/10.1111/1365-2435.12729

Jeanniard-du-Dot T, Trites AW, Arnould JPY, Guinet C (2017b) Reproductive success is energetically linked to foraging efficiency in Antarctic fur seals. PLoS One 12(4):e0174001. https://doi.org/10.1371/journal.pone.0174001

Jeanniard-du-Dot T, Trites AW, Arnould JPY, Speakman JR, Guinet C (2017c) Activity-specific metabolic rates for diving, transiting, and resting at sea can be estimated from time-activity budgets in free-ranging marine mammals. Ecol Evol 7(9):2969–2976. https://doi.org/10.1002/ece3.2546

Jeglinski JWE, Werner C, Robinson PW, Costa DP, Trillmich F (2012) Age, body mass and environmental variation shape the foraging ontogeny of Galapagos Sea lions. Mar Ecol Prog Ser 453:279–296. https://doi.org/10.3354/meps09649

Jeglinski JW, Goetz KT, Werner C, Costa DP, Trillmich F (2013) Same size–same niche? Foraging niche separation between sympatric juvenile Galapagos Sea lions and adult Galapagos fur seals. J Anim Ecol 82(3):694–706. https://doi.org/10.1111/1365-2656.12019

Kirby MX, Jones DS, MacFadden BJ (2008) Lower Miocene stratigraphy along the Panama Canal and its bearing on the central American peninsula. PLoS One 3(7):e2791. https://doi.org/10.1371/journal.pone.0002791

Kirkman SP, Costa DP, Harrison AL, Kotze PGH, Oosthuizen WH, Weise M, Botha JA, Arnould JPY (2019) Dive behaviour and foraging effort of female cape fur seals *Arctocephalus pusillus pusillus*. R Soc Open Sci 6(10):191369. https://doi.org/10.1098/rsos.191369

Kirkwood R, Goldsworthy S (2013) Fur seals and sea lions. Csiro Publishing, Melbourne

Kooyman GL, Trillmich F (1986) Diving behavior of Galapagos fur seals. In: Fur seals

Kovacs KM, Aguilar A, Aurioles D, Burkanov V, Campagna C, Gales N, Gelatt T, Goldsworthy SD, Goodman SJ, Hofmeyr GJG, Harkonen T, Lowry L, Lydersen C, Schipper J, Sipila T, Southwell C, Stuart S, Thompson D, Trillmich F (2012) Global threats to pinnipeds. Mar Mamm Sci 28(2):414–436. https://doi.org/10.1111/j.1748-7692.2011.00479.x

Kuhn CE, Costa DP (2014) Interannual variation in the at-sea behavior of California Sea lions (*Zalophus californianus*). Mar Mamm Sci 30(4):1297–1319. https://doi.org/10.1111/mms.12110

Kuhn CE, Ream RR, Sterling JT, Thomason JR, Towell RG (2014) Spatial segregation and the influence of habitat on the foraging behavior of northern fur seals (Callorhinus ursinus). Can J Zool 92(10):861–873

Laake JL, Lowry MS, DeLong RL, Melin SR, Carretta JV (2018) Population growth and status of California Sea lions. J Wildl Manag 82(3):583–595. https://doi.org/10.1002/jwmg.21405

Lea MA, Guinet C, Cherel Y, Hindell M, Dubroca L, Thalmann S (2008) Colony-based foraging segregation by Antarctic fur seals at the Kerguelen archipelago. Mar Ecol Prog Ser 358:273–287. https://doi.org/10.3354/meps07305

Lea MA, Hindell M, Guinet C, Goldsworthy S (2002) Variability in the diving activity of Antarctic fur seals, *Arctocephalus gazella*, at Iles Kerguelen. Polar Biol 25(4):269–279

Leung ES, Chilvers BL, Nakagawa S, Robertson BC (2014) Size and experience matter: diving behaviour of juvenile New Zealand Sea lions (*Phocarctos hookeri*). Polar Biol 37(1):15–26. https://doi.org/10.1007/s00300-013-1405-6

Lipps JH, Mitchell E (1976) Trophic model for the adaptive radiations and extinctions of pelagic marine mammals. Paleobiology 2(2):147–155. https://doi.org/10.1017/s0094837300003420

Liwanag HEM, Berta A, Costa DP, Abney M, Williams TM (2012a) Morphological and thermal properties of mammalian insulation: the evolution of fur for aquatic living. Biol J Linn Soc 106 (4):926–939. https://doi.org/10.1111/j.1095-8312.2012.01900.x

Liwanag HEM, Berta A, Costa DP, Budge SM, Williams TM (2012b) Morphological and thermal properties of mammalian insulation: the evolutionary transition to blubber in pinnipeds. Biol J Linn Soc 107(4):774–787. https://doi.org/10.1111/j.1095-8312.2012.01992.x

Luque SP, Arnould JP, Miller EH, Cherel Y, Guinet C (2007) Foraging behaviour of sympatric Antarctic and subantarctic fur seals: does their contrasting duration of lactation make a difference?. Mar Biol 152(1):213–224

Loughlin TR, Sterling JT, Merrick RL, Sease JL, York AE (2003) Diving behavior of immature Steller Sea lions (*Eumetopias jubatus*). Fish Bull 101(3):566–582

Maresh JL, Adachi T, Takahashi A, Naito Y, Crocker DE, Horning M, Williams TM, Costa DP (2015) Summing the strokes: energy economy in northern elephant seals during large-scale foraging migrations. Mov Ecol 3(1):22. https://doi.org/10.1186/s40462-015-0049-2

Mattlin RH (1981) Pup growth of the New Zealand fur seal *Arctocephalus forsteri* on the Open Bay islands. New Zealand J Zool 193(Mar):305–314

McDonald BI, Goebel ME, Crocker DE, Costa DP (2012) Biological and environmental drivers of energy allocation in a dependent mammal, the Antarctic fur seal pup. Physiol Biochem Zool 85 (2):134–147. https://doi.org/10.1086/664948

McHuron EA, Robinson PW, Simmons SE, Kuhn CE, Fowler M, Costa DP (2016) Foraging strategies of a generalist marine predator inhabiting a dynamic environment. Oecologia 182 (4):995–1005. https://doi.org/10.1007/s00442-016-3732-0

McHuron EA, Costa DP, Schwarz L, Mangel M (2017a) State-dependent behavioural theory for assessing the fitness consequences of anthropogenic disturbance on capital and income breeders. Methods Ecol Evol 8(5):552–560. https://doi.org/10.1111/2041-210x.12701

McHuron EA, Mangel M, Schwarz LK, Costa DP (2017b) Energy and prey requirements of California Sea lions under variable environmental conditions. Mar Ecol Prog Ser 567:235–247. https://doi.org/10.3354/meps12041

McHuron EA, Block BA, Costa DP (2018a) Movements and dive behavior of juvenile California Sea lions from Ano Nuevo Island. Mar Mamm Sci 34(1):238–249. https://doi.org/10.1111/mms.12449

McHuron EA, Hazen E, Costa DP (2018b) Constrained by consistency? Repeatability of foraging behavior at multiple timescales for a generalist marine predator. Mar Biol 165(8). https://doi.org/10.1007/s00227-018-3382-3

McHuron EA, Sterling JT, Costa DP, Goebel ME (2019) Factors affecting energy expenditure in a declining fur seal population. Conserv Physiol 7(1). https://doi.org/10.1093/conphys/coz103

Melin SR, Orr AJ, Harris JD, Laake JL, DeLong RL (2012) California Sea lions: an indicator for integrated ecosystem assessment of the California current system. California Cooperative Oceanic Fisheries Investigations Reports 53:140–152

Mestas-Nunez AM, Miller AJ (2006) Interdecadal variability and climate change in the eastern tropical Pacific: a review. Prog Oceanogr 69(2–4):267–284. https://doi.org/10.1016/j.pocean.2006.03.011

Milette LL, Trites AW (2003) Maternal attendance patterns of Steller Sea lions (*Eumetopias jubatus*) from stable and declining populations in Alaska. Can J Zool 81(2):340–348. https://doi.org/10.1139/z03-008

Montes C, Cardona A, Jaramillo C, Pardo A, Silva JC, Valencia V, Ayala C, Perez-Angel LC, Rodriguez-Parra LA, Ramirez V, Nino H (2015) Middle Miocene closure of the central American seaway. Science 348(6231):226–229. https://doi.org/10.1126/science.aaa2815

Murphy EJ, Trathan PN, Watkins JL, Reid K, Meredith MP, Forcada J, Thorpe SE, Johnston NM, Rothery P (2007) Climatically driven fluctuations in Southern Ocean ecosystems. Proc R Soc B Biol Sci 274(1629):3057–3067. https://doi.org/10.1098/rspb.2007.1180

Ochoa-Acuña H (1995) Ecological and physiological factors that influence pup birth weight and postnatal growth of Juan Fernandez fur seals, *Arctocephalus philippii*. University of Florida, Gainesville, FL

Oftedal OT, Iverson SJ, Boness DJ (1987) Milk and energy intakes of suckling California Sea lion *Zalophus californianus* pups in relation to sex, growth, and predicted maintenance requirements. Physiol Zool 60(5):560–575. https://doi.org/10.1086/physzool.60.5.30156130

Ono AK, Boness JD (1996) Sexual dimorphism in sea lion pups: differential maternal investment, or sex-specific differences in energy allocation? Behav Ecol Sociobiol 38(1):31–41. https://doi.org/10.1007/s002650050214

Ono KA, Boness DJ, Oftedal OT (1987) The effect of a natural environmental disturbance on maternal investment and pup behavior in the California Sea lion. Behav Ecol Sociobiol 21 (2):109–118. https://doi.org/10.1007/Bf02395438

Orians GH, Pearson NE (1977) On the theory of central place foraging. In: Horn DJ, Stairs GR, Mitchel RD (eds) Analysis of ecological systems, 3rd edn. Biosciences Colloquuim, Ohio State University Press, Columbus, OH, pp 155–177

Osman LP, Hucke-Gaete R, Moreno CA, Torres D (2004) Feeding ecology of Antarctic fur seals at Cape Shirreff, South Shetlands, Antarctica. Polar Biol 27(2):92–98. https://doi.org/10.1007/s00300-003-0555-3

Payne MR (1979) Growth in the Antarctic fur seal *Arctocephalus gazella*. J Zool 187(Jan):1–20

Pennington JT, Mahoney KL, Kuwahara VS, Kolber DD, Calienes R, Chavez FP (2006) Primary production in the eastern tropical Pacific: a review. Prog Oceanogr 69(2–4):285–317. https://doi.org/10.1016/j.pocean.2006.03.012

Pirotta E, Booth CG, Costa DP, Fleishman E, Kraus SD, Lusseau D, Moretti D, New LF, Schick RS, Schwarz LK, Simmons SE, Thomas L, Tyack PL, Weise MJ, Wells RS, Harwood J (2018) Understanding the population consequences of disturbance. Ecol Evol 8(19):9934–9946. https://doi.org/10.1002/ece3.4458

Pitcher KW, Rehberg MJ, Pendleton GW, Raum-Suryan KL, Gelatt TS, Swain UG, Sigler MF (2005) Ontogeny of dive performance in pup and juvenile Steller Sea lions in Alaska. Can J Zool 83(9):1214–1231. https://doi.org/10.1139/Z05-098

Ponganis PJ (2016) Diving physiology of marine mammals and seabirds. Cambridge University Press, Cambridge, UK

Rehberg MJ, Andrews RD, Swain UG, Calkins DG (2009) Foraging behavior of adult female Steller Sea lions during the breeding season in Southeast Alaska. Mar Mamm Sci 25 (3):588–604. https://doi.org/10.1111/j.1748-7692.2008.00278.x

Repenning CA (1976) Adaptive evolution of sea lions and walruses. Syst Zool 25(4):375–390. https://doi.org/10.2307/2412512

Richmond JP, Burns JM, Rea LD (2006) Ontogeny of total body oxygen stores and aerobic dive potential in Steller sea lions (*Eumetopias jubatus*). J Comp Physiol B:176(6):535–545

Riet-Sapriza FG, Costa DP, Franco-Trecu V, Marín Y, Chocca J, González B, Beathyate G, Louise Chilvers B, Hückstadt LA (2013) Foraging behavior of lactating south American sea lions (*Otaria flavescens*) and spatial–temporal resource overlap with the Uruguayan fisheries. Deep Sea Res II Top Stud Oceanogr 88–89:106–119. https://doi.org/10.1016/j.dsr2.2012.09.005

Rodionov SN, Overland JE, Bond NA (2005) Spatial and temporal variability of the Aleutian climate. Fish Oceanogr 14(s1):3–21. https://doi.org/10.1111/j.1365-2419.2005.00363.x

Rosen DAS, Trites AW (2004) Satiation and compensation for short-term changes in food quality and availability in young Steller Sea lions (*Eumetopias jubatus*). Can J Zool 82(7):1061–1069. https://doi.org/10.1139/z04-082

Salton M, Kirkwood R, Slip D, Harcourt R (2019) Mechanisms for sex-based segregation in foraging behaviour by a polygynous marine carnivore. Mar Ecol Prog Ser 624:213–226. https://doi.org/10.3354/meps13036

Schmitz OJ, Lavigne DM (1984) Intrinsic rate of increase, body size, and specific metabolic rate in marine mammals. Oecologia 62(3):305–309. https://doi.org/10.1007/BF00384261

Schulz TM, Bowen WD (2004) Pinniped lactation strategies: evaluation of data on maternal and offspring life history traits. Mar Mamm Sci 20(1):86–114. https://doi.org/10.1111/j.1748-7692.2004.tb01142.x

Schulz TM, Bowen WD (2005) The evolution of lactation strategies in pinnipeds: a phylogenetic analysis. Ecol Monogr 75(2):159–177. https://doi.org/10.1890/04-0319

Shero MR, Andrews RD, Lestyk KC, Burns JM (2012) Development of the aerobic dive limit and muscular efficiency in northern fur seals (*Callorhinus ursinus*). J Comp Physiol B 182 (3):425–436. https://doi.org/10.1007/s00360-011-0619-6

Sielfeld W, Barraza J, Amado N (2018) Local feeding patterns of the south American sea lion *Otaria byronia*: case of Punta Patache, northern Chile. Rev Biol Mar Oceanogr 53(3):307–319. https://doi.org/10.22370/rbmo.2018.53.3.1356

Sinclair EH, Antonelis GA, Robson BW, Ream RR, Loughlin TR (1996) Northern fur seal, *Callorhinus ursinus,* predation on juvenile walleye ollock, *Theragra chalcogramma.* Fish Bull Tech Rep 126:167–178

Soto KH, Trites AW, Arias-Schreiber M (2006) Changes in diet and maternal attendance of south American sea lions indicate changes in the marine environment and prey abundance. Mar Ecol Prog Ser 312:277–290. https://doi.org/10.3354/meps312277

Staniland IJ, Gales N, Warren NL, Robinson SL, Goldsworthy SD, Casper RM (2010) Geographical variation in the behaviour of a central place forager: Antarctic fur seals foraging in contrasting environments. Mar Biol 157(11):2383–2396. https://doi.org/10.1007/s00227-010-1503-8

Stephens PA, Boyd IL, McNamara JM, Houston AI (2009) Capital breeding and income breeding: their meaning, measurement, and worth. Ecology 90(8):2057–2067. https://doi.org/10.1890/08-1369.1

Stephens PA, Houston AI, Harding KC, Boyd IL, McNamara JM (2014) Capital and income breeding: the role of food supply. Ecology 95(4):882–896. https://doi.org/10.1890/13-1434.1

Trillmich F (1986) Maternal investment and sex allocation in the Galapagos fur seal, *Arctocephalus galapagoensis*. Behav Ecol Sociobiol 19(3):157–164. https://doi.org/10.1007/bf00300855

Trillmich F (1990) The behavioral ecology of maternal effort in Fur seals and sea lions. Behaviour 114(1–4):3–20. https://doi.org/10.1163/156853990x00022

Trillmich F (2015) *Zalophus wollebaeki*. The IUCN red list of threatened species 2015: e. T41668A45230540. https://doi.org/10.2305/IUCN.UK.2015-2.RLTS.T41668A45230540.en. Accessed 05 Jan 2020

Trillmich F, Kooyman GL (2001) Field metabolic rate of lactating female Galapagos fur seals (*Arctocephalus galapagoenis*): the influence of offspring age and environment. Comp Biochem Physiol Part A 129(4):741–749. https://doi.org/10.1016/S1095-6433(01)00343-9

Trillmich F, Limberger D (1985) Drastic effects of El Niño on Galapagos pinnipeds. Oecologia 67 (1):19–22. https://doi.org/10.1007/BF00378445

Trillmich F, Ono KA (1991) Pinnipeds and El Nino: responses to enviornmental stress, vol 88. Ecological studies, 1st edn. Springer, New York

Trillmich F, Weissing FJ (2006) Lactation patterns of pinnipeds are not explained by optimization of maternal energy delivery rates. Behav Ecol Sociobiol 60(2):137–149. https://doi.org/10.1007/s00265-005-0150-4

Trillmich F, Ono KA, Costa DP, DeLong RL, Feldkamp SD, Francis JM, Gentry RL, Heath CB, LeBoeuf BJ, Majluf P (1991) The effects of El Niño on Pinniped populations in the eastern Pacific. In: Trillmich F, Ono K (eds) Effects of El Niño on Pinnipeds. Springer, Berlin, pp 247–270

Valenzuela-Toro A, Pyenson ND (2019) What do we know about the fossil record of pinnipeds? A historiographical investigation. R Soc Open Sci 6(11). https://doi.org/10.1098/rsos.191394

Valenzuela-Toro AM, Gutstein CS, Varas-Malca RM, Suarez ME, Pyenson ND (2013) Pinniped turnover in the South Pacific Ocean: new evidence from the Plio-Pleistocene of the Atacama Desert, Chile. J Vertebr Paleontol 33(1):216–223. https://doi.org/10.1080/02724634.2012.710282

Vermeij GJ, Dudley R (2000) Why are there so few evolutionary transitions between aquatic and terrestrial ecosystems? Biol J Linn Soc 70(4):541–554. https://doi.org/10.1111/j.1095-8312.2000.tb00216.x

Villegas-Amtmann S, Costa DP (2010) Oxygen stores plasticity linked to foraging behaviour and pregnancy in a diving predator, the Galapagos Sea lion. Funct Ecol 24(4):785–795. https://doi.org/10.1111/j.1365-2435.2009.01685.x

Villegas-Amtmann S, Costa DP, Tremblay Y, Salazar S, Aurioles-Gamboa D (2008) Multiple foraging strategies in a marine apex predator, the Galapagos Sea lion *Zalophus wollebaeki*. Mar Ecol Prog Ser 363:299–309. https://doi.org/10.3354/meps07457

Villegas-Amtmann S, Simmons SE, Kuhn CE, Hückstädt LA, Costa DP (2011) Latitudinal range influences the seasonal variation in the foraging behavior of marine top predators. PLoS One 6 (8):e23166. https://doi.org/10.1371/journal.pone.0023166

Villegas-Amtmann S, Atkinson S, Paras-Garcia A, Costa DP (2012) Seasonal variation in blood and muscle oxygen stores attributed to diving behavior, environmental temperature and pregnancy in a marine predator, the California Sea lion. Comp Biochem Physiol A Mol Integr Physiol 152(4):413–420. https://doi.org/10.1016/j.cbpa.2012.04.019

Villegas-Amtmann S, Jeglinski JW, Costa DP, Robinson PW, Trillmich F (2013) Individual foraging strategies reveal niche overlap between endangered galapagos pinnipeds. PLoS One 8(8):e70748. https://doi.org/10.1371/journal.pone.0070748

Villegas-Amtmann S, McDonald BI, Paez-Rosas D, Aurioles-Gamboa D, Costa DP (2017) Adapted to change: low energy requirements in a low and unpredictable productivity environment, the case of the Galapagos Sea lion. Deep Sea Res Part II 140:94–104. https://doi.org/10.1016/j.dsr2.2016.05.015

Volpov BL, Rosen DAS, Hoskins AJ, Lourie HJ, Dorville N, Baylis AMM, Wheatley KE, Marshall G, Abernathy K, Semmens J, Hindell MA, Arnould JPY (2016) Dive characteristics can predict foraging success in Australian fur seals (*Arctocephalus pusillus doriferus*) as validated by animalborne video. Biol Open 5(3):262–271. https://doi.org/10.1242/bio.016659

Weise MJ, Costa DP (2007) Total body oxygen stores and physiological diving capacity of California Sea lions as a function of sex and age. J Exp Biol 210(2):278–289. https://doi.org/10.1242/jeb.02643

White LD, Garrison RE, Barron JA (1992) Miocene intensification of upwelling along the California margin as recorded in siliceous facies of the Monterey formation and offshore DSDP sites. Geol Soc Lond Spec Publ 64(1):429–442. https://doi.org/10.1144/gsl.Sp.1992.064.01.28

Wickens P, York AE (1997) Comparative population dynamics of fur seals. Mar Mamm Sci 13 (2):241–292. https://doi.org/10.1111/j.1748-7692.1997.tb00631.x

Williams TM, Yeates L (2004) The energetics of foraging in large mammals: a comparison of marine and terrestrial predators. Int Congr Ser 1275:351–358. https://doi.org/10.1016/j.ics.2004.08.069

Williams TM, Haun J, Davis RW, Fuiman LA, Kohin S (2001) A killer appetite: metabolic consequences of carnivory in marine mammals. Comp Biochem Physiol A Mol Integr Physiol 129(4):785–796. https://doi.org/10.1016/s1095-6433(01)00347-6

Williams TM, Fuiman LA, Horning M, Davis RW (2004) The cost of foraging by a marine predator, the Weddell seal *Leptonychotes weddellii*: pricing by the stroke. J Exp Biol 207 (6):973–982. https://doi.org/10.1242/jeb.00822

Williams TM, Rutishauser M, Long B, Fink T, Gafney J, Mostman-Liwanag H, Casper D (2007) Seasonal variability in otariid energetics: implications for the effects of predators on localized prey resources. Physiol Biochem Zool 80(4):433–443. https://doi.org/10.1086/518346

Wright J, Bolstad GH, Araya-Ajoy YG, Dingemanse NJ (2018) Life-history evolution under fluctuating density-dependent selection and the adaptive alignment of pace-of-life syndromes. Biol Rev Camb Philos Soc 94(1):230–247. https://doi.org/10.1111/brv.12451

Wyss AR (1994) The evolution of body size in phocids: some ontogenetic and phylogenetic observations. In: Contributions in marine mammal paleontology honoring Frank C. Whitmore, vol 29. University of California, Santa Barbara, pp 69–75

Young BL, Rosen DA, Hindle AG, Haulena M, Trites AW (2011) Dive behaviour impacts the ability of heart rate to predict oxygen consumption in Steller Sea lions (*Eumetopias jubatus*) foraging at depth. J Exp Biol 214(Pt 13):2267–2275. https://doi.org/10.1242/jeb.047340

Zeppelin T, Pelland N, Sterling J, Brost B, Melin S, Johnson D, Lea MA, Ream R (2019) Migratory strategies of juvenile northern fur seals (*Callorhinus ursinus*): bridging the gap between pups and adults. Sci Rep 9. https://doi.org/10.1038/s41598-019-50230-z

Chapter 3
Maternal Behavior in Otariids and the Walrus

Maritza Sepúlveda and Robert G. Harcourt

Abstract The most energetically costly phases of female reproduction in placental mammals are gestation and lactation. For the young, the success of this relationship is based on obtaining as much maternal care as they can take in, even if this may result in a cost to future reproductive success of the mother. On the other hand, the mother should retain sufficient resources to be able to invest in future offspring, thus ensuring her own future reproductive success. These competing demands lead to a conflict between mothers and their young that eventually results in the weaning of the young, often initiated by the mother. In this review, we describe the main strategies of lactation and maternal care in fur seals, sea lions and walrus. The maternal care strategy in income-breeding pinnipeds with prolonged lactation periods, such as otariids and walrus, is influenced by group-living, and these two factors have been subject to strong selection. Maternal care in pinnipeds is influenced by both social and environmental constraints. This information plays a major role in our understanding of the social and exploratory behavior of the young, as well as into the interaction between maternal care and the development of offspring.

Keywords Maternal care · Income breeding · Lactation · Mother-offspring conflict · Otariid · Fur seal · Sea lion · Walrus

---

M. Sepúlveda (✉)
Centro de Investigación y Gestión de Recursos Naturales (CIGREN), Facultad de Ciencias, Universidad de Valparaíso, Valparaíso, Chile
e-mail: maritza.sepulveda@uv.cl

R. G. Harcourt
Marine Predator Research Group, School of Biological Sciences, Macquarie University, Sydney, NSW, Australia
e-mail: robert.harcourt@mq.edu.au

C. Campagna, R. Harcourt (eds.), *Ethology and Behavioral Ecology of Otariids and the Odobenid*, Ethology and Behavioral Ecology of Marine Mammals,
https://doi.org/10.1007/978-3-030-59184-7_3

3.1 Introduction

In placental mammals, gestation and lactation, periods during which young are energetically dependent on their mothers, represent the two most energetically costly phases of female reproduction. For most mammals, caring for the young is principally carried out by the mother, not only through the provision of milk, but also by other behaviors associated with care such as social interactions, recognition and nurturing, protection aggressive conspecifics, vigilance, anti-predator defense, and in some, teaching of tools required for their survival once independent. For the young, the success of this relationship is based on obtaining as much maternal care as they can take in, increasing its chance of survival at a time when mortality is typically high (Caughley 1966), even if this may result in a cost to the future reproductive success of the mother. On the other hand, the mother is selected to retain sufficient resources to be able to invest in future offspring, thus ensuring her own future reproductive success. These competing demands lead to a conflict between mothers and their young that eventually results in the weaning of the young, often initiated by the mother.

In this chapter, we describe the main strategies of lactation and maternal care in income breeding pinnipeds, the fur seals, sea lions and the walrus (*Odobenus rosmarus*). Much of our knowledge on this topic comes from their habit of giving birth and nursing on land or ice, as this has allowed researchers to examine strategies used by mothers to ensure survival and welfare of their young.

3.2 Gestation Length and Maternal Attendance Patterns

A limited amount of data currently exists regarding physiological changes that occur during gestation of marine mammals, particularly those in their natural habitat. Due to the high-energy cost of maintaining a developing fetus while engaging in thermoregulation and foraging activities, most pinniped females carry a single fetus. Few exceptions have been reported, but occasionally twin pups do survive to weaning and this has been reported in subantarctic (*Arctocephalus tropicalis*, Bester and Kerley 1983; de Bruyn et al. 2010), Antarctic (*A. gazella,* Doidge 1987; Bonin et al. 2012) and New Zealand fur seals (*A. forsteri*, Dowell et al. 2008) and Steller sea lions (*Eumetopias jubatus*, Maniscalco and Parker 2009). Bonin et al. (2012) demonstrated genetically that not all putative twins were in fact siblings; rather, there was a mix of full siblings, half siblings and adoptions. Although rare, reports of twin fetuses typically result in spontaneous abortions by the female at some point during gestation (Spotte 1982).

Development rates *in utero* seem relatively fixed, and to coordinate timing of breeding, otariids and walrus are not able to simply adjust developmental rates. In fur seals and sea lions, gestation lasts 8–10.6 months, with a delayed implantation / embryonic diapause of 1.4–4 months post conception, and this allows them to

Table 3.1 Lactation in fur seals, sea lions and walrus

Species	Lactation duration
Otariids	
Guadalupe fur seal	8–11 months
Juan Fernandez fur seal	7–10 months
Galapagos fur seal	18–36 months
South American fur seal	8–12 months. . .12–24 months
New Zealand fur seal	10–12 months
Antarctic fur seal	3.5–4 months
Subantarctic fur seal	9–11 months
South African fur seal	9–11 months
Australian fur seal	11–12 months
Northern fur seal	4 months
New Zealand Sea lion	6–12 months
Australian sea lion	17 months
South American sea lion	6–24 months
California Sea lion	6–12 months
Galapagos Sea lion	10–12 months
Steller Sea lion	12–36 months
Odobenids	
Atlantic walrus	24 months
Pacific walrus	24 months

Adapted from Riedman (1990) and Berta et al. (2006)

complete their cycle annually, or in the case of the Australian sea lion (*Neophoca cinerea*) 17-18 months and walrus 2.5 years (Boyd 1992; Boness et al. 2002; McIntosh and Pitcher, Chap. 26). Walrus have a 15–16 month gestation including a 3–4 month delayed implantation, resulting in a 2 year breeding cycle, the longest of all the pinnipeds (Boness et al. 2002; Miller and Kochnev, Chap. 22).

Like gestation length, maternal care pattern and lactation duration vary greatly among fur seals, sea lions and walrus (Table 3.1) (Trillmich 1990; Boness et al. 2002). Lactation strategies comprise either mothers leaving their offspring periodically and regularly, alternating foraging cycles of female foraging at sea with fasting nursing periods ashore (Costa and Valenzuela, Chap. 2), or females and their offspring staying together on land and at sea, with their offspring nursing in the water and staying with the mother while she forages.

Otariids exhibit an alternating foraging/fasting cycle, but the duration of lactation is highly variable. At its one extreme, the near-polar living Antarctic fur seal (Doidge et al. 1984) and northern fur seal (*Callorhinus ursinus,* Loughlin et al. 1987), both of which migrate away from the breeding area during the harsh polar winters, have relatively short lactation periods of 4 months. By contrast, Galapagos fur seals, *A. galapagoensis* (Trillmich 1986) and Steller sea lions (Gentry 1970) regularly suckle their pups for up to 3 years. Many of the temperate fur seals display a lactation that lasts 300–360 days (Table 3.1), but Australian sea lions suckle their pup until the birth of their subsequent offspring approximately 18 months later (Higgins and Gass

1993; see McIntosh and Pitcher, Chap. 26, for explanatory hypotheses). Otariid females give birth 1–2 days after arrival in the colony. They remain with their newborn pup for a period of 7–14 days before departing for their first foraging trip (Bonner 1984). Over the course of the entire duration of the maternal attendance period, females alternate suckling while ashore with these foraging trips to sea. Milk ingestion by the pup is intermittent, with intensive suckling periods interspersed with fasting periods while the mother is away foraging (Gentry and Kooyman 1986).

The duration of foraging trips, and thus the duration of the pup's fasting, varies within and across species, and is influenced by intrinsic factors such as age of the pup (Oftedal et al. 1987), and external factors such as productivity of the area (Bailleul et al. 2005), which may vary seasonally (Harcourt et al. 2002) or with unpredictable environmental oscillations such as El Niño (Soto et al. 2006). For instance, the duration of foraging trips in subantarctic fur seal females increases during the lactation period, starting from 15 days on average during the summer to 30 days on average during the winter (Beauplet et al. 2004). These seasonal differences are likely explained by a seasonal variability in food availability, combined with the growing pup's enhanced ability to fast.

Walrus alone exhibit the strategy of aquatic nursing, characterized by a long lactation phase during which the females feed. Both North Atlantic and Pacific walrus subspecies exhibit this strategy. Mothers nurse their calf for 1–2 years but this can be extended to a third year or even longer if another pregnancy does not occur (Fay 1982). Females are diestrous, with a behavioral estrus in late summer and a second heat through February.

Calves are relatively precocious, able to swim competently within a few hours of birth, and stay in permanent contact with their mothers (Stewart and Fay 2001). While moving to foraging grounds or to find floating ice, females often carry their young calf on their back.

3.3 Parental Care

In pinnipeds, care of the young (i.e. protection and nurturance) is provided only by the mother, as males are not involved in the pup's life. For terrestrially breeding colonial species, mothers must protect their pups from three main threats, neighboring females, bulls, and in some colonies, predators (Fig. 3.1). Aggressive females may bite neighboring pups, sometimes severely, especially in high density colonies (Marlow 1972; Harcourt 1992a, b; Cassini 2001). Danger to the pup is compounded by movements of males through the colony, whether to secure a breeding position or for thermoregulation (Marlow 1972; Carey 1992; Harcourt 1992a; Hofmeyr et al. 2007). Male-male competition often leads to pups being trampled and suffering crush injuries as the bull pursues a rival (Marlow 1975). Mothers attempt to protect their pups from invading predators; for example, South American sea lion (*Otaria byronia*) males raid South American fur seal (*A. australis*) colonies in Peru and Uruguay to steal pups, and mothers either flee (Peru) or defend (Uruguay) their

Fig. 3.1 Subadult male of the South American sea lion (*Otaria byronia*) kidnapping and biting a newborn pup. Photo by Guido Pavez

offspring (Harcourt 1992a; Cassini 2001). For young pups, these raids may result in mother-pup separation and this can be fatal if the mother-pup bond has not yet formed. All together, aggressive/defensive behaviors from adult females and disruptive male behavior contribute to a high proportion of pup mortality. For these reasons, mothers spend much of their time protecting an area around their pup, especially during the perinatal period (Harcourt 1992a, b; Phillips 2003). Such maternal protection seems to be correlated with colony density (Le Boeuf and Briggs 1977; Doidge et al. 1984; Harcourt 1992b; Cassini 2001; Hofmeyr et al. 2007), and also with the quality of the substrate, including proximity to water, tide pools, presence of shade, or a flat area (Doidge et al. 1984; Fernández-Juricic and Cassini 2007).

In walrus, mothers are strongly bonded with their calves and are highly protective against any predator, including humans (Fay 1982; Nowak 2003). If females perceive danger, they push their calf into the water and join it. Interestingly, other conspecifics also become actively protective of the calves. It is common to see both female and male walruses forming tight formations i.e. *herding* when a calf produces distress calls (Fay 1982; Nowak 2003).

Fostering, allosuckling or allonursing are the different terms used in the literature for nursing of a non-filial pup by a female, once or for a more prolonged time of several days. In some species of pinnipeds, allosuckling behavior has been described. But, overall, allomaternal care in otariids is extremely rare, except for New Zealand sea lions (*Phocarctos hookeri*) (Fig. 3.2) (6%: Childerhouse and Gales 2001) and Antarctic fur seals (7%–11%: Lunn 1992; Gemmell 2003).

In otariids, most species have developed a highly reliable mother-pup recognition system (higher vocal stereotypy and occurrence of mutual recognition—see

Fig. 3.2 Allomaternal carers: Australian sea lion (*Neophoca cinerea*) female simultaneously nursing two young pups. No aggressive behavior was observed for any of the pups (from Pitcher et al. 2011). Photo by Clarence Kennedy

Charrier, Chap. 14) and so non-specificity of recognition of young by mothers is unlikely to explain the higher rates observed in New Zealand sea lions or Antarctic fur seals. Instead, age of the female may be an important contributing factor in the frequency of allonursing. Young and/or primiparous females reportedly accept non-filial pups more easily than older and/or multiparous females (Georges et al. 1999), with multiparous females highly aggressive towards non-filial pups (Trillmich 1981). Fostering unrelated pups may provide young females opportunities to gain maternal experience, especially those having lost their pup (Riedman 1982).

However much more commonly, pups may attempt to steal milk (sneak-suckling) from non-mother females (e.g. Roux 1986; Lunn 1992; Porter and Trites 2004; Pitcher et al. 2011). This behavior is likely intentional and pups do not suckle the wrong female by mistake. In most reports, the thieving pup quietly approaches a female that is asleep, suckling until the female becomes aware that the suckling pup is not her pup and rejects it (Porter and Trites 2004). Sometimes the female's own pup alerts the mother that an unrelated pup is present (Sepúlveda, pers. obs.). This behavior is frequently observed in pups that are nutritionally stressed when they are separated from their mother for a long period of time, a common phenomenon given the alternating feeding cycle in otariids. Such attempts can be quite risky since females are highly aggressive towards non-filial pups (Harcourt 1992a), suggesting that the benefit of gaining even a small supply of milk may outweigh the risk of immediate injury from bites, since starvation is the likely alternative (Roux 1986). Early research considered milk stealing by pups to be uncommon. However a growing number of observations suggest that this behavior is widespread (de Bruyn et al. 2010), and requires detailed further investigation.

3.4 Mother-Offspring Conflicts and Weaning

That conflicts are likely to occur between mother and her offspring is a tenet of evolutionary theory. Trivers (1974) proposed a conflict for the amount of energy transfer between mother and her offspring will occur intensively at the end of lactation, i.e. near weaning. Thus, weaning bestows an advantage for mother but not for the suckling offspring. For offspring, acquiring energetic resources for a longer period may enhance their chance of survival and thus increase their own reproductive success. From the mother's perspective, providing nutrition for a longer period of time will further decrease her own body resources and so may impact survival of her next offspring and therefore her own future reproductive success (Desprez et al. 2014). Such theory predicts an increase in mother's aggressive behavior towards offspring near weaning and a decrease in successful suckling bouts.

In many otariids, parent-offspring conflict occurs when mothers nurse their young for an extended period, and older pups are still present in the colony when mothers have their next offspring (Fig. 3.3). This parent-offspring conflict can be exacerbated when food resources are highly seasonal and unpredictable. There are frequent observations of mothers having to defend themselves against their offsprings' aggressive begging; for example, in older pups of Galapagos fur seals and sea

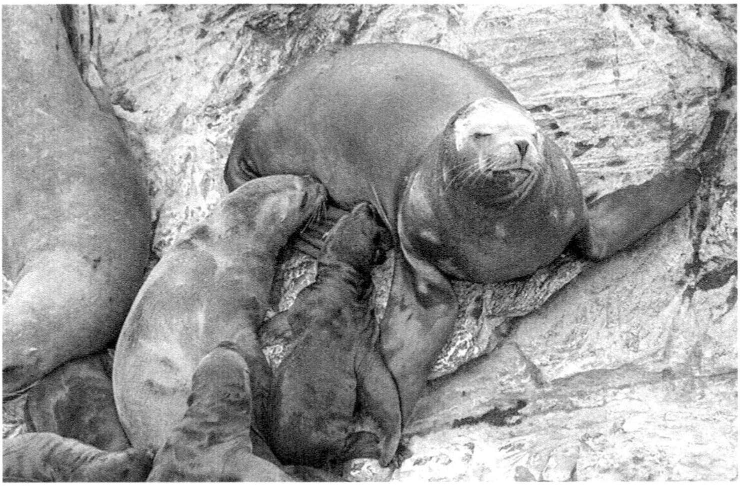

Fig. 3.3 South American sea lion (*Otaria byronia*) female simultaneously nursing a juvenile born in the previous year and a newborn pup. Photo by Guido Pavez

lions (*Zalophus wollebaeki*, Trillmich and Wolf 2007), South American fur seals (Majluf 1987), Australian sea lions (Marlow 1975) and South American sea lions (Sepúlveda, pers.obs). Moreover, sibling conflict occurs in some otariids when females with yearlings give birth to lighter newborns that are at greater risk of mortality due to the presence of older siblings (Trillmich and Wolf 2007). In contrast to these tropical otariids, the Galapagos fur seals and sea lions, no evidence for aggressive behavior or decrease in suckling has been observed in other species showing prolonged lactation, including New Zealand fur seals (Hasse 2004) and Steller sea lions (Marcotte 2006), with young weaning themselves. Self-weaning has been also observed in northern fur seals (Macy 1982) and Antarctic fur seals (Doidge et al. 1984), two species with a short lactation period, where when ready, pups leave the colony to forage before their mothers return from a trip. In most otariids, the absence of parent-offspring conflict may be explained by the fact that young complement their milk diet with solid food obtained during short foraging trips, and thus learn the location of foraging grounds and prey capture techniques. While it has been suggested that pups might learn from their mothers, recent findings on Australian sea lions have shown that pups do not have similar foraging patterns to their mothers. Given that in all otariids, pups complement their milk diet with solid food obtained in short feeding trips, this may presumably be where they first learn to forage - whether independently or by observation of unrelated conspecifics (Lowther et al. 2012).

3.5 Conclusions

Habitat, establishment of social units, and mating system have clear impacts on pinniped maternal care and the relationship between mothers and their young. Lactation strategy and duration are a good illustration of the complexity of maternal care patterns which consider effects of social and environmental constraints, as well as the development of a reliable mother-young recognition system. In otariids, lactation is prolonged, forcing mothers to undertake feeding trips, which along with the usual high density of animals on land leads to the development of a complex and precise recognition system. Maternal care strategies appear complex in group-living species. In pinnipeds, it is unknown whether contact between the mother and her offspring continues after weaning. Long-term recognition is present in some otariid species which, combined with natal site fidelity (i.e. philopatry), provides a mechanism for the development of some degree of association with relatives. However, in other species, kin associations are only weak compared to philopatric drivers of social associations. Further information on post-weaning interactions between mother and offspring will play a major role in our understanding of the social and exploratory behavior of the young, as well as into the interaction between maternal care and the development of offspring.

References

Bailleul F, Luque S, Dubroca L et al (2005) Differences in foraging strategy and maternal behaviour between two sympatric fur seal species at the Crozet Islands. Mar Ecol Prog Ser 293:273–282
Beauplet G, Dubroca L, Guinet C et al (2004) Foraging ecology of subantarctic fur seals *Arctocephalus tropicalis* breeding on Amsterdam Island: seasonal changes in relation to maternal characteristics and pup growth. Mar Ecol Prog Ser 273:211–225
Berta A, Sumich JL, Kovacs KM (2006) Marine mammals: evolutionary biology, 2nd edn. Elsevier Academic Press, San Diego, CA
Bester MN, Kerley GI (1983) Rearing of twin pups to weaning by subantarctic fur seal *Arctocephalus tropicalis* female. S Afr J Wildl Res 13:86–87
Boness D, Clapham PJ, Mesnick SL (2002) Life history and reproductive strategies. In: Hoelzel R (ed) Marine mammal biology: an evolutionary approach. Wiley, Hoboken, NJ, pp 278–324
Bonin CA, Goebel ME, O'Corry-Crowe GM (2012) Twins or not? Genetic analysis of putative twins in Antarctic fur seals, *Arctocephalus gazella*, on the South Shetland Islands. J Exp Mar Biol Ecol 412:13–19
Bonner WN (1984) Lactation strategies in pinnipeds: problems for a marine mammalian group. Symp Zool Soc Lond 51:253–272
Boyd IL (1992) Environmental and physiological factors controlling the reproductive cycles of pinnipeds. Can J Zool 69:1135–1148
Carey PW (1992) Agonistic behaviour in female New Zealand fur seals, *Arctocephalus forsteri*. Ethology 92:70–80
Cassini MH (2001) Aggression between females of the southern fur seal (*Arctocephalus australis*) in Uruguay. Mammal Rev 31:169–172
Caughley G (1966) Mortality patterns in mammals. Ecology 47:906–918
Childerhouse S, Gales N (2001) Fostering behaviour in New Zealand Sea lions *Phocarctos hookeri*. New Zeal J Zool 28:189–195
de Bruyn PJN, Cameron EZ, Tosh CA et al (2010) Prevalence of allosuckling behaviour in Subantarctic fur seal pups. Mamm Biol 75:555–560
Desprez M, Harcourt R, Hindell M et al (2014) Age-specific cost of first reproduction in female southern elephant seals. Biol Lett 10:5
Doidge DW (1987) Rearing of twin offspring to weaning in Antarctic fur seals, *Arctocephalus gazella*. US Department of Commerce, NOAA Technical Report, NMFS
Doidge DW, Croxall JP, Ricketts C (1984) Growth rates of Antarctic fur seal *Arctocephalus gazella* pups at South Georgia. J Zool 203:87–93
Dowell SA, Boren LJ, Negro SS et al (2008) Rearing two New Zealand fur seal (*Arctocephalus forsteri*) pups to weaning. Aust J Zool 56:33–39
Fay F (1982) Ecology and biology of the Pacific walrus, *Odobenus rosmarus divergens* Illiger. North Am Fauna 74:1–279
Fernández-Juricic E, Cassini MH (2007) Intra-sexual female agonistic behaviour of the south American sea lion (*Otaria flavescens*) in two colonies with different breeding substrates. Acta Ethol 10:23–28
Gemmell NJ (2003) Kin selection may influence fostering behaviour in Antarctic fur seals (*Arctocephalus gazella*). Proc R Soc B 270:2033–2037
Gentry RL (1970) Social behaviour of the Steller Sea lion. Dissertation, University of California
Gentry RL, Kooyman GL (1986) Fur seals: maternal strategies on land and at sea. Princeton University Press, Princeton, NJ
Georges JY, Sevot X, Guinet C (1999) Fostering in a subantartic fur seal. Mammalia 63:384–388
Harcourt R (1992a) Factors affecting mortality in the south American fur seal (*Arctocephalus australis*) in Peru: density-related effects and predation. J Zool 226:259–270
Harcourt R (1992b) Maternal aggression in the South American fur seal in Peru. Can J Zool 70:320–325

Harcourt RG, Bradshaw CJ, Dickson K, Davis LS (2002) Foraging ecology of a generalist predator, the New Zealand fur seal. Mar Ecol Prog Ser 227:11–24
Hasse TJ (2004) The determinants of weaning in the New Zealand fur seal. Dissertation, La Trobe University
Higgins LV, Gass L (1993) Birth to weaning: parturition, duration and lactation and attendance cycles of the Australian sea lion (*Neophoca cinerea*). Can J Zool 71:2047–2055
Hofmeyr GG, Bester MN, Pistorius PA et al (2007) Median pupping date, pup mortality and sex ratio of fur seals at Marion Island. Afr J Wildl Res 37:1–9
Le Boeuf BJ, Briggs KT (1977) The cost of living in a seal harem. Mammalia 41:167–196
Loughlin TR, Bengtson JL, Merrick RL (1987) Characteristics of feeding trips of female northern fur seals. Can J Zool 65:2079–2084
Lowther AD, Harcourt RG, Goldsworthy SD, Stow A (2012) Population structure of adult female Australian sea lions is driven by fine-scale foraging site fidelity. Anim Behav 83:691–701
Lunn N (1992) Fostering behaviour and milk stealing in Antarctic fur seals. Can J Zool 70:837–839
Macy SK (1982) Mother-pup interactions in the northern fur seal. Dissertation, University of Washington
Majluf P (1987) Reproductive ecology of female South American fur seals at Punta San Juan, Peru. Dissertation, University of Cambridge
Maniscalco JM, Parker P (2009) A case of twinning and the care of two offspring of different age in Steller Sea lions. Mar Mamm Sci 25:206–213
Marcotte M (2006) Steller watch: timing of weaning and seasonal patterns in numbers and activities of Steller Sea lions (*Eumetopias jubatus*) at a year-round haulout site in Southeast Alaska. Dissertation, The University of British Columbia
Marlow BJ (1972) Pup abduction in the Australian sea lion, *Neophoca cinerea*. Mammalia 36:161–165
Marlow BJ (1975) The comparative behaviour of the Australasian Sea lions *Neophoca cinerea* and *Phocarctos hookeri* (Pinnipedia: Otariidae). Mammalia 39:159–230
Nowak R (2003) Walker's marine mammals of the world. JHU Press, Baltimore, MA
Oftedal OT, Iverson SJ, Boness DJ (1987) Milk and energy intakes of suckling California Sea lion *Zalophus californianus* pups in relation to sex, growth, and predicted maintenance requirements. Physiol Zool 1987:560–575
Phillips AV (2003) Behavioral cues used in reunions between mother and pup South American fur seals. J Mammal 84:524–535
Pitcher B, Ahonen H, Charrier I, Harcourt RG (2011) Allosuckling behavior in the Australian sea lion (*Neophoca cinerea*): an updated understanding. Mar Mamm Sci 27:881–888
Porter BT, Trites AW (2004) Suckling attempts during winter by two non-filial Steller Sea lion pups (*Eumetopias jubatus*). Mammalia 68:23–26
Riedman ML (1982) The evolution for alloparental care and adoption in mammals and birds. Q Rev Biol 57:405–435
Riedman ML (1990) The pinnipeds. Seals, sea lions and walruses. University of California Press, Berkeley, CA
Roux J-P (1986) A successful subantarctic fur seal milk-thief. Mammalia 50:403–405
Soto KH, Trites AW, Arias-Schreiber M (2006) Changes in diet and maternal attendance of South American sea lions indicate changes in the marine environment and prey abundance. Mar Ecol Prog Ser 312:277–290
Spotte S (1982) The incidence of twins in pinnipeds. Can J Zool 60:2226–2233
Stewart REA, Fay FH (2001) Walrus. In: Macdonald D (ed) The new encyclopedia of mammals. Oxford University Press, Oxford, pp 174–179
Trillmich F (1981) Mutual mother–pup recognition in Galapagos fur seals and sea lions: cues used and functional significance. Behaviour 78:21–42
Trillmich F (1986) Maternal investment and sex-allocation in the Galapagos fur seal, *Arctocephalus galapagoensis*. Behav Ecol Sociobiol 19:157–164

Trillmich F (1990) The behavioural ecology of maternal effort in fur seals and sea lions. Behaviour 114:3–20
Trillmich F, Wolf JBW (2007) Parent–offspring and sibling conflict in Galápagos fur seals and sea lions. Behav Ecol Sociobiol 62:363–375
Trivers RL (1974) Parent–offspring conflict. Am Zool 14:249–264

Part II
Foraging Behavior

Chapter 4
Foraging Capacities, Behaviors and Strategies of Otariids and Odobenids

Tiphaine Jeanniard-du-Dot and Christophe Guinet

Abstract Fur seals, sea lions and the walrus (*Odobenus rosmarus*) are breath-hold divers that rely on swimming at depth to feed at sea. As their diving capacities are more limited than phocids, otariids and odobenids are geographically constrained to highly productive environments and relatively shallow dive depths. They are also mostly coastal species, central place foragers with relatively limited foraging ranges. Diving patterns and strategies are diverse among the otariid group—although fur seals tend to be more pelagic and sea lions more benthic divers—, and driven by extrinsic factors such as the type of habitat they occupy, environmental factors, intra- or inter-specific density-dependent competition, predation risk and the behavior of the prey they feed on; as well as intrinsic factors such as age, sex, reproduction status, size and experience. There are usually several foraging strategies present within a species, and individuals tend to specialize to one of these strategies, with a degree of adaptability to changing conditions possible. Diving behaviors and strategies define the feeding success and foraging efficiency of individuals, and as such their capacities to successfully survive and reproduce in their environment. The diversity of these behaviors within otariid and odobenid populations are likely evolutionary stable strategies that provide a buffer under changing environmental conditions.

Keywords Diving physiology · Diving behavior · Energetics · Foraging strategies · Otariids · Odobenids · Fur seals · Sea lions · Walrus

---

T. Jeanniard-du-Dot (✉)
Department of Fisheries and Oceans Canada, Institut Maurice Lamontagne, Mont-Joli, QC, Canada

Centre d'Etudes Biologiques de Chizé - La Rochelle Université, UMR 7372, CNRS, Villiers-en-Bois, France
e-mail: tiphaine.jeanniard-du-dot@cebc.cnrs.fr

C. Guinet
Centre d'Etudes Biologiques de Chizé - La Rochelle Université, UMR 7372, CNRS, Villiers-en-Bois, France
e-mail: christophe.guinet@cebc.cnrs.fr

C. Campagna, R. Harcourt (eds.), *Ethology and Behavioral Ecology of Otariids and the Odobenid*, Ethology and Behavioral Ecology of Marine Mammals,
https://doi.org/10.1007/978-3-030-59184-7_4

4.1 Introduction

Accurately determining dive behavior and underwater foraging of marine predators, including fur seals, sea lions and walruses (Fig. 4.1) is inherently difficult as they cannot be directly observed in the open oceans, and yet it is essential to determine, explain, and predict their foraging success, the use of prey resources in their environment, as well as their consequences on energetics and fitness (Costa et al.

Fig. 4.1 Example of sea lions and fur seals of the family otariidae (**a** Steller sea lions, *Eumetopias jubatus,* **b** Antarctic fur seals, *Arctocephalus gazella*), and the family odobenidae (**c** walruses, *Odobenus rosmarus*). (Picture credits: **a**, **b** Tiphaine Jeanniard du Dot, **c** Brian Battaile)

1989; Boyd 2002; Austin et al. 2006; Bost et al. 2007; New et al. 2014). However, technological advancements in the field of bio-logging and bio-telemetry, i.e. archival and satellite-transmitting devices attached to animals while they go at sea, has made it possible to collect information on these diving behaviors at an increasingly fine scale.

The first bio-loggers used on marine mammals were time-depth recorders (TDR) that recorded low resolution 2D dive patterns over a relatively short period of time (Kooyman 1965). Nowadays, new and miniaturized sensors, as well as improved processing capacities, large memory capabilities and battery life allows marine mammal scientists to record data about dive behavior and foraging success at a much finer scale and for up to several months. For example, tri-axial accelerometers, magnetometers or gyroscopes, recording data at 20 Hz, allow reconstruction of 3D underwater tracks of the animals. They can also provide indirect indices of swimming effort, feeding attempts and foraging success (Viviant et al. 2014; Jeanniard du Dot et al. 2016b; Ladds et al. 2017b). The field of bio-logging is in rapid development (the International Bio-logging Society was born in 2015, https://www.bio-logging.net). Currently, most loggers deployed on otariids incorporate pressure, temperature and light sensors. More recently other types of sensors (passive or active acoustics, salinity etc.) have allowed scientists to study animal behavior in relation to their immediate environment. This has shed light on the links between foraging behavior, feeding success and the oceanographic conditions.

Biologging-based studies of otariid diving behavior have exponentially increased over the last 30 years (McIntyre 2014). Some species have been intensely studied: Antarctic fur seals (*Arctocephalus gazella*), northern fur seals (*Callorhinus ursinus*), followed by New Zealand sea lions (*Phocarctos hookeri*), Steller sea lions (*Eumatopias jubatus*), and California sea lions (*Zalophus californianus*). On the other end of the spectrum, there is little information on dive behavior of Juan Fernandez fur seals (*Arctocephalus phillippii phillippii*) (Francis et al. 1998), Guadalupe fur seals (*Arctocephalus phillippii townsendi*) (Gallo-Reynoso et al. 2008) or southern sea lions (*Otaria flavescens*) (Werner and Campagna 1995; Mueller 2004). In addition, a majority of studies have focused on diving behavior of adult lactating females during the breeding season when they become central place foragers. Significantly less information is available for males or juveniles, even though efforts have been made to start bridging these gaps (Boyd et al. 1998; Baylis et al. 2017; Knox et al. 2018; Salton et al. 2019). Consequently, general knowledge of diving behavior of otariids and odobenids is inherently biased towards the taxa, season, and sex/age groups from which we have the most information.

Fur seals and sea lions are mostly coastal species, especially during breeding season, and their yearly movements can cover little variation in habitat use (Australian sea lion, *Neophoca cinerea*) to thousands of km for those with a pelagic phase (northern and Antarctic fur seals during their 8-month migration). Their distribution is concentrated in areas of high productivity, mostly in temperate to sub-polar latitudes, or near areas with cold upwellings for more tropical species (i.e. Humboldt or Cromwell current, Fig. 4.2). This pattern is a consequence of their relatively expensive lifestyle and their subsequent needs for high feeding rates to

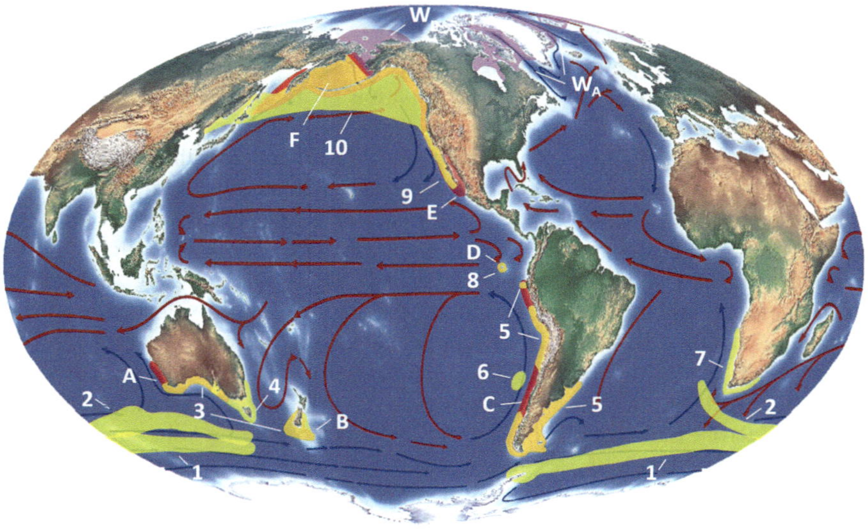

Fig. 4.2 Distribution of otariid and odobenid species worldwide. Fur seals species are shown in yellow and with numbers; sea lion species in red (with overlap between sympatric species resulting in orange) and in single letters, and walruses in purple and double letters. Otariid species are distributed worldwide in areas of cold ocean currents (blue arrows) or upwellings rather than warm ocean currents (red arrows). Legends corresponding to numbers and letters are as follow: 1. Antarctic fur seals—Arctocephalus gazella, 2. Sub-Antarctic fur seals—*Arctocephalus tropicalis*, 3. New Zealand fur seals—*Arctocephalus forsteri*, 4. Australian fur seals—*Arctocephalus pusillus doriferus*, 5. South American fur seals—*Arctocephalus australis*, 6. Juan Fernandez fur seals—*Arctocephalus philippii*, 7. Cape fur seals—*Arctocephalus pusillus pusillus*, 8. Galapagos fur seals—*Arctocephalus galapagoensis*, 9. Guadalupe fur seals—*Arctocephalus townsendi*, 10. Northern fur seals—*Callorhinus ursinus*, A. Australian sea lions—*Neophoca cinerea*, B. New Zealand (Hooker's) sea lions—*Phocarctos hookeri*, C. South American sea lions—*Otaria flavescens*, D. Galapagos sea lions—*Zalophus californianus wollebaeki*, E. California sea lions—*Zalophus californianus californianus*, F. Steller sea lions—*Eumetopias jubatus*, W_P. Pacific walruses—*Odobenus rosmarus divergens*. W_A. Atlantic walruses—*Odobenus rosmarus rosmarus*

balance their energy budget (Boyd 2002; Costa and Toro, Chap. 2). As fur seals and sea lions travel and rest at the surface of the ocean, the main purpose of their diving is to feed at depth. Intrinsic factors such as phylogeny, sex, size, age, experience and reproductive status, as well as extrinsic factors including habitat characteristics, environmental factors, intra- or inter-specific density-dependent competition, predation pressure, food web structure and the distribution and behavior of the prey they feed upon, all influence diving behavior in otariids (Schoener 1974; Gentry and Kooyman 1986; Boyd 1996; Harcourt et al. 2002; Jeglinski et al. 2012; Benoit-Bird et al. 2013; Leung et al. 2014). Ultimately, these factors will shape the foraging efficiency of animals, i.e. the cost/benefit ratio of their foraging at sea and hence their body condition, survival and reproductive success and the resulting trends in population.

This chapter focuses on diving and foraging activities of fur seals and sea lions as a group—and to a lesser extent the walrus (*Odobenus rosmarus*)—, emphasizing phylogenetic similarities or differences between them or with the other major pinniped family, the phocidae. We look at how otariid diving is limited by physiological capacities, what diving strategies they display in the wild and what biological or environmental factors affect them. Finally, we discuss how diving patterns and strategies can be used to assess and understand foraging energetics and efficiency and ultimately how it shapes the capacity of fur seals and sea lions to survive and reproduce successfully in their environment.

4.2 Morphological and Physiological Diving Capacities of Fur Seals and Sea Lions

Diving behavior of fur seals and sea lions is constrained by their overall capacity to balance the energetic costs of moving effectively through water, a medium ~800 times denser than air and 25 times more heat conductive, while on a limited oxygen supply during apnea. The quantity of oxygen available in the body and the rate at which it is consumed (and carbon dioxide produced) by metabolic processes during a dive controls how long animals can remain underwater, given their total oxygen stores. Fur seals and sea lions have developed morphological and physiological adaptations to simultaneously decrease the costs of exercising underwater or diving metabolic rate and increase the magnitude of body oxygen stores and transport. Combined, these adaptations allow fur seals and sea lions to prolong the time spent at depth to forage, and so is essential to understanding diving behavior and foraging efficiencies of eared seals.

Morphologically, otariids have streamlined body shapes that reduce the drag created by, and thus the energetic cost of, moving through water (Fig. 4.3). Their body shape presents an average fineness ratio (~5.5, compared to the walrus with a fineness ratio ~1 for example) and a position of the maximum diameter of their body (shoulders area at ~40% of the total body length) close to the optimum for minimal drag given the body volume (Feldkamp 1987). Unlike phocids and odobenids who use caudal propulsion and lateral or vertical oscillations of their rear appendages to move through water, fur seals and sea lions use their pectoral flippers for propulsion in a 4-phase stroke pattern creating a horizontal thrust and vertical lift with little to no resulting distortion of the body (Feldkamp 1987). Otariid fore-flippers are hydrofoil-shaped and dorso-ventrally compressed which reduces pressure drag and improves lift during propulsion underwater (Fig. 4.3). Fore-flipper propulsion also offers high maneuverability at depth, as well as great speed and turning angle. These adaptations ensure efficient mechanical performance for swimming underwater at a much reduced energetic cost of swimming at depth (at optimum speed) compared to terrestrial mammals of similar sizes (Costa and Williams 1999). Other particularities of otariids amongst marine mammals are their limited fat content (especially in fur

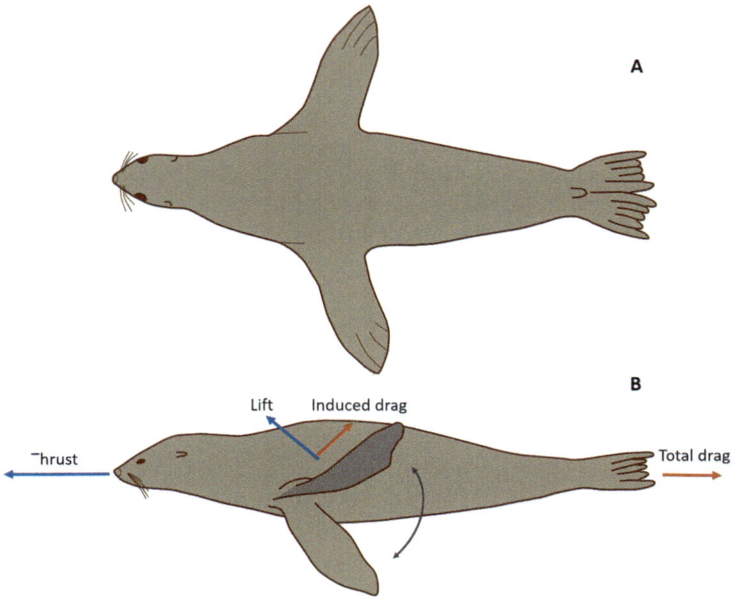

Fig. 4.3 Morphological features of otariids with a view from above (**a**) or from the side (**b**). Notice the streamline body shape closely resembling a technical body of revolution designed to minimize drag underwater. The hydrofoil-shape fore-flippers are used for propulsion in a 4-phase stroke pattern and delivers a forward thrust and lift movement with little to no resulting distortion of this streamlined body shape (**b**)

seals) and their diving with inflated lungs, which both affect buoyancy and thus energetic costs of diving.

In addition to their morphology, fur seal and sea lion physiology has evolved to adapt to breath-hold diving. They possess a series of physiological adaptations which decrease the rate at which body oxygen stores are depleted underwater to prolong dive duration. These physiological adjustments are ubiquitous amongst marine mammals—although not to the same extent—and are together called the 'dive response' (Davis 2014). It mostly involves peripheral vasoconstriction and a decrease in heart rate (bradycardia) compared to 'resting' heart rates in response to asphyxia (diving heart rates decreasing to ~25% to 50% compared to eupneic levels, Ponganis 2015). Together these adjustments reduce the perfusion of blood to 'non-essential' organs—while maintaining adequate blood pressure and cardiac output to the heart and the brain—thereby limiting the overall rate of oxygen consumption. Free-ranging mammals (i.e. non-controlled and unrestrained conditions) show a wide range of physiological plasticity while diving and the extent of the 'dive response' cardiovascular adjustments is usually greater in longer dives (i.e. requiring greater conservation of O_2) (Andrews et al. 1997).

Diving physiology has mostly been studied in deep diving phocid species, and few studies have measured these responses in the shallow and shorter-diver

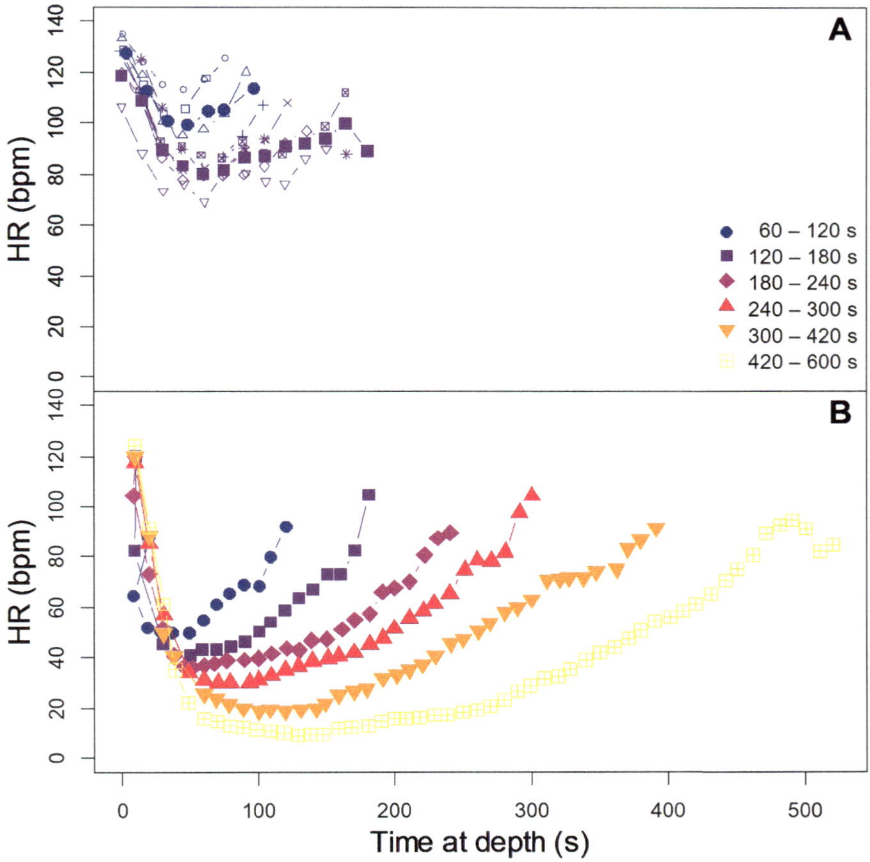

Fig. 4.4 Changes in heart rate (bpm) measured in free-ranging epipelagic Antarctic fur seals (upper graph, data from Boyd et al. 1999) and mesopelagic California sea lions (lower graph, data from McDonald and Ponganis 2014) during dives of various durations (see color-coded legend). For easier comparison, the dives of Antarctic fur seals lasting between 60 and 120 s (in blue open symbols) and between 120 and 180 s (in purple open symbols) were averaged (closed symbols) to match the time bins for California sea lions

otariids—mostly California sea lions, Steller sea lions and Antarctic fur seals (Andrews et al. 1997; Boyd et al. 1999; Hindle et al. 2010). Nevertheless, similar responses have been found in California sea lions which show extreme bradycardia to the same extent as the deepest diving phocid seals during very long dives (McDonald and Ponganis 2014). In pelagic shallow diving fur seals, however, the extent of these adjustments is not always as pronounced (Fig. 4.4). Even during long dives, heart rate is reduced by only ~33% of surface resting heart rate in Antarctic fur seals compared to up to 75% in other pinnipeds (Boyd et al. 1999). Interestingly, while California sea lions show intense bradycardia in deep dives, the profile of heart

Table 4.1 Average oxygen carrying capacities of otariids compared to phocid seals

	Otariids	Phocids
Mass-specific total oxygen stores (ml O_2/kg)	49.3 ± 11.0[a]	69.6 ± 16.4[c]
Hb concentration (g Hb/dl blood)	17.8 ± 2.3[a]	22.9 ± 3.3[c]
Mb concentration (g Mb/100g muscle)	3.7 ± 1.2[a]	6.5 ± 1.7[c]
Oxygen affinity, P_{50} (mmHg)	28 ± 2[b]	26.9 ± 1.2[d]/30.5 ± 1.2[e]

Species used in calculating these averages are indicated in footnote below. Values used to calculate these means can be found in Lenfant et al. (1970), Costa et al. (1998, 2001), Richmond et al. (2006), Fowler et al. (2007), Weise and Costa (2007), Villegas-Amtmann and Costa (2010), Hückstädt et al. (2016), Kirkman et al. (2019), Qvist et al. (1981), Kooyman (1989), Meir et al. (2009) and McDonald and Ponganis (2013). Values for phocids can be found in Table 4.2, 4.7 and 4.10 in Ponganis (2015). Values for the walrus *O. rosmarus* are 38 ml O_2/kg, 16 g Hb/dl blood, and 3 g Mb/100 g muscle (Lenfant et al. 1970)

[a]*Z. californianus, Z. wollebaeki, E. jubatus, N. cinerea, C. ursinus, A. gazella, A. pusillus pusillus, O. byronia*

[b]*Z. californianus*

[c]*P. vitulina, P. groenlandica, P. hispida, P. sibirica, L. carcinophagus, L. weddellii, H. leptonyx, H. grypus, H. fasciata, C. cristata, M. angustirostris*

[d]*L. weddelli*

[e]*M. leonina*

rate decrease is slower than typical of phocid seals. This might be due to otariids inhaling before diving and so gas exchange continues underwater, while phocid seals exhale before diving (Hooker et al. 2005; McDonald and Ponganis 2014).

In addition to oxygen conservation mechanisms, oxygen stores in otariids are primarily located in the lungs, blood and muscles, and depends on their volume/mass as well as on hemoglobin (Hb) and myoglobin (Mb) concentrations in the body (Ponganis 2015). However, amongst pinnipeds, otariids have more limited diving capacities (shallower dive depths, shorter dive durations on average) than comparably sized phocids, which as a group possess larger oxygen stores relative to their mass than similar size otariids (Table 4.1, t-test $p = 0.01$). This results from a lower Hb concentration despite relatively similar blood volumes ($p = 0.008$) and arguably similar O_2 affinity (Qvist et al. 1981; Kooyman 1989; Meir et al. 2009; McDonald and Ponganis 2013) as well as lower Mb concentration ($p = 0.001$). Together, blood and muscles represent ~75% to 90% of oxygen storage in fur seal and sea lions' bodies (Ponganis 2015), and the remaining ~10% to 25% reside in the lungs compared to only ~3% to 10% in phocids. The sixfold higher proportion of oxygen in otariids' lungs compared to blood—1:2.5 in otariids vs 1:12 in phocids—indicates substantial differences in diving strategy (Hooker et al. 2005).

Interestingly, total O_2 stores have a scaling factor of 1 with body mass while metabolic rate a scaling factor of 0.75 (Costa 1993). This means that breath-hold diving capacities increase with body size for a given mass-specific O_2 store (Fig. 4.5). Consequently, larger sea lions and fur seals should have greater diving capacities than smaller ones (Baylis et al. 2015b). Alternatively, smaller animals should increase their mass-specific O_2 stores to maintain similar diving capacities as larger ones. This was observed intra-specifically among females in southern sea lions

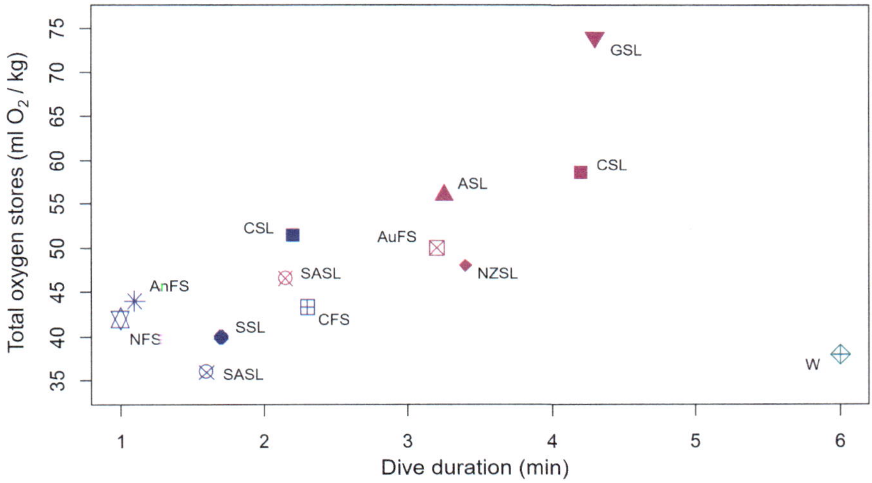

Fig. 4.5 Mass-specific total oxygen stores (ml O_2/kg) related to dive duration (min) in 10 otariid species as well as the walrus for comparison. Different symbols indicate the species (NFS northern fur seal, AnFS Antarctic fur seal, SSL Steller sea lion, SALS South American sea lions, CFS Cape fur seal, CSL California sea lion, AuFS Australian fur seal, NZSL New Zealand sea lion, ASL, Australian sea lion, GSL Galapagos sea lion, and W Walrus). The symbol color indicates epi/mesopelagic otariid species in blue and benthic species in pink, with the walrus shown in green. California sea lions and South American sea lions are shown twice as shallow epipelagic and mesopelagic/benthic strategies are displayed in the population. Total oxygen store data from Lenfant et al. (1970), Costa et al. (1998, 2001), Richmond et al. (2006), Weise and Costa (2007), Fowler et al. (2007), Villegas-Amtmann and Costa (2010), Ponganis (2011) and dive duration data from Kuhn and Costa (2014), Costa and Gales (2003, 2000), Villegas-Amtmann et al. (2017), Jeanniard du Dot et al. (2018), Georges et al. (2000b), Gjertz et al. (2001). Australian fur seal and meso/benthic California sea lion data are taken from Costa et al. (2004), South American sea lion data from Hückstädt et al. (2016), and Cape fur seal data from Kirkman et al. (2019)

(Hückstädt et al. 2016), as well as in California sea lions where females compensated for their smaller body sizes and thus lower absolute oxygen stores, by having higher mass-specific O_2 stores compared to the larger males (Weise and Costa 2007). Similarly, the Galapagos sea lion is the smallest of all sea lion species (60–95 kg for females, Villegas-Amtmann et al. 2017), yet has the highest mass-specific oxygen stores (74 ml O_2/kg, Villegas-Amtmann and Costa 2010)—and thus greatest diving capacities—, while the lowest mass-specific O_2 store is found in Steller sea lion (40 ml O_2/kg, Richmond et al. 2006), the largest of otariids (150–300 kg for females, Calkins et al. 1998) (Fig. 4.5).

The time required to consume O_2 stores via aerobic metabolism while diving is thought to be the major determinant of diving performance (Boyd and Croxall 1996; Costa et al. 2004). The aerobic dive limit (ADL) is defined as the maximum dive duration before blood lactic acid levels rise, because of an increase in anaerobic metabolism. When seals exceed this aerobic threshold, the relative post-dive surface interval increases greatly as extra time is necessary to clear lactic acid accumulated

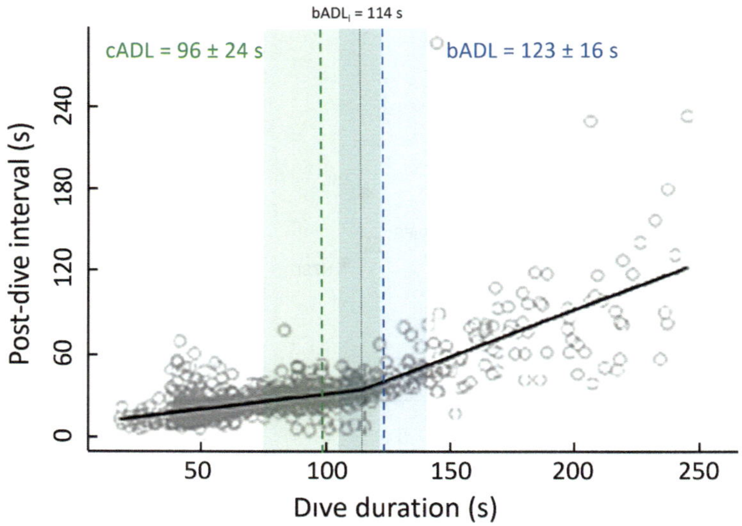

Fig. 4.6 Example of post-dive interval durations (s) as a function of dive duration (s) of a female Antarctic fur seal during a foraging trip at sea used to determine, via broken-line models (solid black lines), the individual behavioral aerobic dive limit (bADL, dotted grey line, data from Viviant et al. 2016). For comparison, the average (±SD) bADL for 11 Antarctic fur seal females in Viviant et al. (2016) is shown in blue dashed line and shaded area, and the average (±SD) calculated aerobic dive limit (cADL) for 15 Antarctic fur seal females in Costa et al. (2004) is shown in green dashed line and shaded area

during the previous dive (Burns 1999; Horning 2012) resulting in a disproportionate increase in surface time and a decrease in foraging efficiency.

To overcome the technical difficulties of measuring blood lactate in free-ranging seals, two indirect estimates have been widely used in eared seal studies. Based on observations that most dives in wild seals are shorter than the measured ADL (Kooyman et al. 1980, 1983), the behavioral aerobic dive limit (ADL_B), has been estimated as the dive duration below which 95%–97% of dives occur and after which surface intervals begin to disproportionately increase (Burns and Castellini 1996; Hindle et al. 2011). Alternatively, the calculated aerobic dive limit (ADL_C) represents the total O_2 stores divided by the diving metabolic rate (Costa et al. 2004) (Fig. 4.6).

Since its first mention in the 1980s (Kooyman et al. 1983), ADL has become a fundamental concept in the interpretation of diving physiology, diving behavior, and more widely foraging ecology of diving animals. Irrespective of the differences due to the ADL estimation method (2.0 min ADL_B versus 1.6 min ADL_C in Antarctic fur seals (Costa et al. 2004; Viviant et al. 2016); 2.3 min ADL in captive California sea lions versus 2.7–3.8 min ADL_C in wild counterparts (Ponganis et al. 1997; Costa et al. 2004)), there is a wide variation in ADL between species of eared seals from 1.3 to 1.7 min in southern sea lions and Australian fur seals (Costa et al. 2004;

Hückstädt et al. 2016) to 3.2 min in Galapagos fur seals (Horning 2012)—and 10.5–14.8 min for walrus (Wiig et al. 1993; Noren et al. 2015) (Table 4.2).

More interestingly, the difference in tendencies to exceed ADL between otariid species appears to depend on their foraging ecology (pelagic versus benthic divers) rather than their phylogeny (fur seals versus sea lions). Benthic fur seal and sea lion divers tend to exceed their ADL more often than pelagic divers (Costa and Gales 2000; Costa et al. 2004; Hückstädt et al. 2016). For example, benthic Australian sea lions exceed their ADL_C in almost 80% of their dives (Costa and Gales 2003),—likely to maximize time on the sea floor—, while the pelagic Antarctic fur seal only surpass it in ~16% of their dives (Viviant et al. 2016).

ADL should also not be considered a fixed unalterable threshold (see Costa and Valenzuela Toro, Chap. 2). In California and South American sea lions, exposure to different habitats and geographical locations dramatically altered individual ADL and overall diving physiology (Hückstädt et al. 2016). Shallow epipelagic California sea lions from San Nicholas island had an ADL of 2.7 min compared to 3.8 min for deeper mesopelagic counterparts off Los Islotes (Costa et al. 2004), while benthic southern sea lions in Southern Chile had greater mass-specific O_2 stores than shallow diver counterparts in Uruguay (Hückstädt et al. 2016). This 'training effect' or 'local adaptation' from greater hypoxic exposure indicates that the physiology of fur seals and sea lions can be adjusted and improved to a certain extent depending on the ecological conditions to which they are routinely exposed. Consequently, diving behavior and overall foraging efficiency result not only from physiological constraints and capacities outlined above but from a complex interaction between genetics, physiology and the environment.

4.3 Diving Strategies of Fur Seals and Sea Lions

Diving strategies are diverse among the otariid group and can be classified as epi-/meso-pelagic (i.e. within the water column) or benthic (on the seafloor) (see also Costa and Toro, Chap. 2). Pelagic divers typically perform bouts of short shallow dives (10–60 m depth for 0.5–3.0 min) with a diurnal variation in dive depths (greater at dawn and dusk and shallow at night) that reflect the circadian migration of their prey to the surface. They spent 20%–35% of their time at sea diving. On the other hand, benthic divers forage mostly on continental shelf areas perform relatively deep and long dives with no obvious diurnal pattern (25–400 m depth for 1.5–7.0 min on average) with long periods at the bottom of their dive. They invest 50%–60% of their time at sea diving, performing 10–20 dives/h with bottom time representing 45%–55% of each dive (Werner and Campagna 1995; Thompson et al. 1998; Costa and Gales 2000, 2003) (Fig. 4.7 and Table 4.2). The maximization of bottom time for deeper dives can be achieved by faster descent rates through 'burst and glide' transit without an increase in swimming costs (Crocker et al. 2001). The deepest diving otariid is thought to be the benthic-feeding New Zealand sea lion, which can dive as deep as 600 m and as long as 20 min (Chilvers 2008b; Chilvers,

Table 4.2 Summary of average size and foraging metrics for otariid and odobenid species

Species	Sex	Strategy	Location	Body mass (kg)	Body length (cm)	Mean dive depth (m)	Max dive depth (m)	Mean dive duration (min)	Max dive duration (min)	Trip duration (d)	Time spent diving (%)	Distance traveled (km)	ADL (min)	FMR (W/kg)	Note	References
Antarctic fur seal	F	Epi	Whole range	27.3–32.7	117–121	12.6–53.0	208	0.7–1.9	10.0	3.1–7.6	16.9–34.5	294–665	1.6–3.5	6.3–9.8		[1–9]
			Kerguelen/Crozet								*22.5–34.5*	*665*				
			South Georgia/ Bouvetøya								*16.9–20.4*	*~300*				
	M	Epi/ Meso	South Georgia	133	180	100.0	354	2.3	9.2	5.4	22.2	~140				[6, 10]
Subantarctic fur seal	F	Epi	Whole range	30.5–52.0	117–135	10.9–39.7	208	0.7–1.5	6.5	4.9–32.2	14.6–33.4	190–2248	2.3		*i*	[3, 11–18]
New Zealand Fur seal	F	Epi	New Zealand/Australia	36.4–42.4	135–138	37.0–53.0	318	1.9–2.7	11.2	2.9–16.6	25.5–37.6	171–727	3.5			[19–22]
	M	Epi/ Meso	Australia	106.4	170	106.4	>380	3.6	14.8	9.1	31.7–33.2					[20]
Australian fur seal	F	Benthic	Australia	64.0–77.7	140–152	58.0–75.2	164	2.9–3.5	10.1	2.7–7.3	40.1–40.7	250–360	4.1		*ii*	[23–27]
	M	Benthic	Australia	138–238	175–201	79.8	283	3.8	9.8	3.4–6.9	39.3	530				[27–29]
Galapagos fur seal	F	Epi	Galapagos	27.8–28.8	105.2	22.1–32.0	169	1.3–1.7	6,5	0.8	23.7–24	44	3.2	5.0		[30–33]
South American fur seal	F	Epi	Falklands/Uruguay	35–51	128–137	43.0–61.0	170	1.8–2.8	7.1	2.1–15.2						[34–36]
Juan Fernandez fur seal	F	Epi	Juan Fernandez	45.0–49.8	142–148	4–12.3	169	0.4–0.8	5.1	12.3–28.0		1394–1778				[37, 38]
Cape fur seal	F	Epi/ Meso	South Africa	53.8	136	59.6	454	2.3	9.8	5.4			3.8			[39, 40]
Guadalupe fur seal	F	Epi	Mexico	47.9–52.5	146	17.0	82	2.6	18.0	14.4	36.2	2375	3.8			[41]

Northern fur seal	F	All	Whole range	35.7–39.4	126–129	11.6–37.2	170	0.5–2.2	12.8	2.8–9.7	15.1–33.3	112–892	2.6	6.2–8.1	*iii*	[42–51]
		Epi/ Benthic	*Pribilof*			*26.8–37.5*		*1.3–2.2*		*7.1–7.9*	*28.4–33.3*	*585–655*				
		Epi	*Pribilof*			*16.9–22.0*		*0.6–1.6*		*7.4–9.7*	*24.7–28.4*	*690–892*				
		Epi	*Bogoslof/Lovushki*			*11.6–18.4*		*0.5–0.7*		*1.8–3.3*	*15.1–29.2*	*112–208*				
South American Sea lion	F	Benthic	Whole range	92.7–126.2	162–191	18.0–75.0	325	1.3–3.1	12.1	0.5–4.4	52.7	103–752	1.3–4.7		*iv*	[52–59]
			Northern Chile/ Uruguay			*18.0–18.7*		*1.4–1.9*		*1.5*		*103*				
			Patagonia/Falklands			*37.2–75.0*		*1.3–3.1*		*0.5–4.4*		*206–752*				
	M	Benthic	Patagonia/Falklands	250	221					2.9–5.7		591				[57]
Galapagos sea lion	F	All	Galapagos	58.4–85.9	156	19.7–153.0	593	1.8–4.9	11.5	0.3–2.5	43.7–65.7	14–76		4.1		[33, 60–66]
		Epi				*19.7–45.5*		*1.8–2.9*			*62.0–65.7*					
		Meso				*53.9–100.9*		*2.5–4.5*			*43.7–56.2*					
		Benthic				*103.0–153.0*		*3.6–4.9*			*45.0–59.0*					
California sea lion	F	All	California/Mexico	80.0–86.7	161–165	32.4–194.2	500+	1.7–5.0	9.9	2.2–5.4	23.3–50.2	167–348	1.2–5.8	4.9–5.9	*v*	[65, 67–73]
		Epi				*32.4–55.1*				*5.4*	*36.3*	*348*				
		Epi/ Benthic				*35.9–72.5*				*5.1*	*38.5*	*304*				
		Meso/ Benthic				*133.6–152.2*				*3.5*	*23.3*	*16*				
	M	All	California	175–187	206	27.7–35.8		1.6–2.2		0.4–2.4	16.7–21.2	47–124	3.8–7.9			
Australian sea lion	F	Benthic	Australia	77.8–89.2	160	32.9–71.0	105	2.7–3.5	8.3	1.2–3.6	52.7–60.7	68–88	2.3	6.8–7.1		[5, 74–78]

(continued)

Table 4.2 (continued)

Species	Sex	Strategy	Location	Body mass (kg)	Body length (cm)	Mean dive depth (m)	Max dive depth (m)	Mean dive duration (min)	Max dive duration (min)	Trip duration (d)	Time spent diving (%)	Distance traveled (km)	ADL (min)	FMR (W/kg)	Note	References
New Zealand Sea lion	F	Benthic	New Zealand	112–125	178	124.0–175.0	612	2.5–4.7	11.5	2.1–4.5	32.5–61.6	45–423	2.3–3.2	6.6–6.9	*vi*	[79–85]
			Otago			*20.2*		*1.8*		*0.5*		*26*				
			Stewart			*59.6*		*2.5*		*0.5*		*45*				
Steller Sea lions	F	Epi	Alaska/Russia	200–300	220–247	24.0–56.0	328	1.9–3.1	21.3	0.3–1.0	22.0–45.0	29	3.0–3.3			[45, 86–91]
Walrus	M	Benthic	Whole range	1350	320	~20	462	~6	49	2.3–7.9	44.0–56.3		10.5–15.0	3.4		[92–97]

Values are means or ranges of means (except for max dive duration and depth) available in the literature for adult free-ranging wild animals. Aerobic dive limit (ADL) values include cADL and bADL so ranges reflect differences due to methodology. At-sea field metabolic rate (FMR) only refers to measurements done on wild animals (ranges also reflecting differences of measurements and calculation methods). Strategies are either epipelagic (Epi), mesopelagic (Meso), or benthic and 'All' indicates every strategy displayed within the population. Values in italic indicate notable differences for specific parameters depending on location, foraging strategies within the population or environmental/seasonal conditions

i. Seasonal variations in foraging trip duration and distance traveled

ii. Seasonal variations

iii. Differences in foraging behaviors depending on bathymetry around breeding island—Pribilof Islands: shallow shelf, epi/benthic foraging; Bogoslof/Lavushki islands: oceanic deep basin, epipelagic foraging

iv. Variations in strategies throughout the range: more pelagic in northern Chile, mixed benthic/pelagic in southern Chile and the Falkland's and more benthic in Uruguay. Seasonal variations observed as well

v. Variations within the population (3 different strategies) and depending on the environmental conditions (cold versus warm years) seen in males and females

vi. Overall averages of dive metrics consistent between studies to the exception of 2 colonies, Otago and Stewart, mentioned for comparison

1. Jeanniard du Dot et al. (2017a), 2. Lea et al. (2002), 3. Bailleul et al. (2005), 4. Costa et al. (2000), 5. Costa et al. (2001), 6. Staniland and Robinson (2008), 7. Costa et al. (1989), 8. Boyd and Croxall (1992), 9. Arnould et al. (1996), 10. Würsig et al. (2017), 11. Georges et al. (2000a), 12. Georges et al. (2000b), 13. Goldsworthy et al. (1997), 14. Beauplet et al. (2004), 15. Luque et al. (2007), 16. Kirkman et al. (2016), 17. de Bruyn et al. (2009), 18. Robinson et al. (2003), 19. Mattlin et al. (1998), 20. Page et al. (2005), 21. Baylis (2008), 22. Baylis et al. (2012), 23. Hoskins and Arnould (2014), 24. Hoskins et al. (2015), 25. Hoskins et al. (2017), 26. Arnould and Hindell (2001), 27. Kernaléguen et al. (2015b), 28. Knox et al. (2018), 29. Kirkwood et al. (2006), 30. Horning and Trillmich (1997), 31. Trillmich and Kooyman (2001), 32. Gentry and Kooyman (1986), 33. Villegas-Amtmann et al. (2013), 34. Thompson et al. (2003), 35. Baylis et al. (2018), 36. Trillmich et al. (1986), 37. Francis et al. (1998), 38. Osman (2007), 39. Kirkman et al. (2019), 40. David and Rand (1986), 41. Gallo-Reynoso et al. (2008), 42. Nordstrom et al. (2013), 43. Kuhn et al. (2010a), 44. Kuhn (2011), 45. Waite et al. (2012), 46. Jeanniard du Dot et al. (2018), 47. Skinner et al. (2012), 48. Battaile et al. (2015b), 49. Battaile et al. (2015a), 50. Shero et al. (2012), 51. Costa and Gentry (1986), 52. Baylis et al. (2015b), 53. Campagna et al. (2001), 54. Werner and Campagna (1995), 55. Hückstädt et al. (2016), 56. Sepúlveda et al. (2015), 57. Baylis et al. (2016), 58. Thompson et al. (1998), 59. Riet-Sapriza et al. (2013), 60. Jeglinski et al. (2012), 61. Paez-Rosas et al. (2017), 62. Villegas-Amtmann et al. (2017), 63. Villegas-Amtmann and Costa (2010), 64. Villegas-Amtmann et al. (2008), 65. Villegas-Amtmann et al. (2011), 66. Trillmich et al. (2014), 67. McHuron et al. (2016), 68. McHuron et al. (2018), 69. Kuhn and Costa (2014), 70. Melin et al. (2008), 71. Weise and Costa (2007), 72. Feldkamp et al. (1989), 73. Ponganis et al. (1997), 74. Hindell et al. (2016), 75. Costa and Gales (2003), 76. Lowther and Goldsworthy (2011), 77. Lowther et al. (2011), 78. Fowler et al. (2007), 79. Chilvers (2017), 80. Chilvers (2009), 81. Chilvers and Wilkinson (2009), 82. Chilvers (2018), 83. Chilvers et al. (2005), 84. Augé et al. (2011), 85. Crocker et al. (2001), 86. Gerlinsky et al. (2014), 87. Winship et al. (2001), 88. Merrick and Loughlin (1997), 89. Merrick et al. (1994), 90. Olivier (2015), 91. Lander et al. (2020), 92. Gjertz et al. (2001), 93. Jay et al. (2001), 94. Acquarone et al. (2006), 95. Garde et al. (2018), 96. Lowther et al. (2015), 97. Noren et al. (2015)

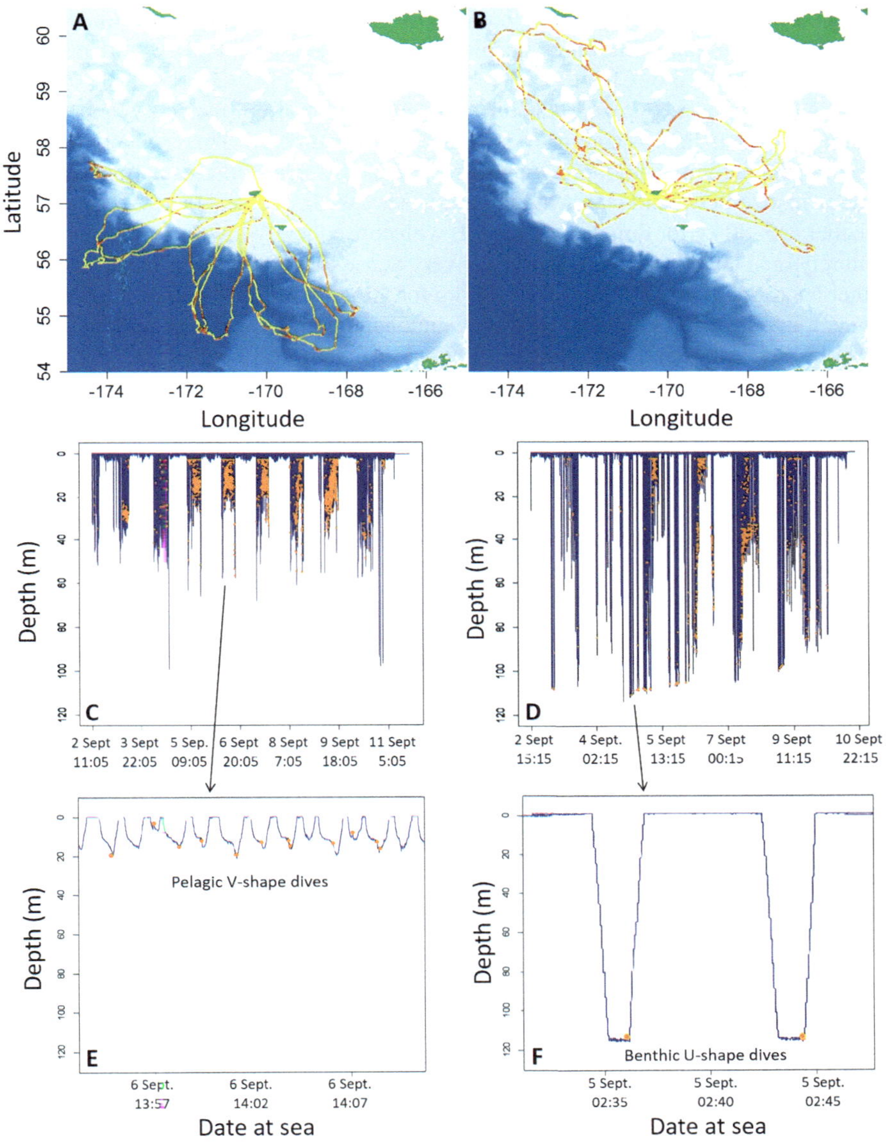

Fig. 4.7 Foraging tracks (**a**, **b**) and dive profiles over a full foraging trip (**c**, **d**), and a 5-min period of diving (**e**, **f**) for lactating northern fur seals from Reef Rookery on St. Paul Island in the Bering Sea in 2011. Fur seals foraged either pelagically off the shelf in oceanic waters where bathymetry reaches 3000 m, (left panels, n = 8) or benthically on the shallow (~100 m deep) shelf in neritic waters (right panels, n = 12). Notice that pelagic divers display a typical circadian diving cycle (only at night, shallow at night and deeper at dawn and dusk, **c**), while benthic divers dive night and day at similar depths (**d**). Orange dots show spatial distribution of prey capture attempts in panels **a**, **b** (with yellow representing no prey capture attempt) and distribution of prey capture attempts over time and depth profiles in panels (**c–f**). Figure from Jeanniard du Dot et al. (2018)

Chap. 25), although recent studies have shown that the Galapagos sea lion can also perform dives within this range despite its smaller size (Villegas-Amtmann et al. 2017; Costa and Toro, Chap. 2).

It is generally recognized that fur seals are nocturnal epi-/meso-pelagic divers, while most sea lions—and walruses—are benthic divers. This general pelagic/benthic distinction is thought to result from a difference in insulation capacity and/or in size between taxa (Arnould and Costa 2006). Fur seals are indeed overall smaller (25–80 kg for females, Table 4.2) with consequently lower diving capacities, and rely on a layer of air trapped in their very dense fur for thermoregulation. On the other hand, sea lions are larger (70–300 kg for females, Table 4.2) with concomitant greater diving capacities and rely on a blubber layer for insulation. This taxon-derived dichotomy in diving strategy is however not entirely so clear cut. Australian fur seals are exclusive benthic divers (Kirkwood and McIntosh, Chap. 27) while California sea lions are epi/meso-pelagic divers (Weise et al. 2010; Kuhn and Costa 2014, Costa and Toro, Chap. 2).

In addition, as the number of bio-logging studies increases and more individuals of different sex/age are tracked repeatedly over time, our understanding of intra-individuals/inter-populations variability gets refined. Several species of fur seals and sea lions seem to employ more than one of these diving strategies (Arnould and Hindell 2001; Chilvers and Wilkinson 2009; Kuhn et al. 2010b; Villegas-Amtmann et al. 2013; Baylis et al. 2015b). For example, northern fur seal females display epipelagic, benthic or mixed strategies when foraging in the Bering sea (Gentry et al. 1986; Kuhn et al. 2010a; Jeanniard du Dot et al. 2018) (Fig. 4.7 and Table 4.2). On the other hand, adult female southern sea lions breeding in northern Patagonia or the Falklands (Werner and Campagna 1995; Baylis et al. 2015b), New Zealand sea lions (Chilvers and Wilkinson 2009) and Galapagos sea lions (Villegas-Amtmann et al. 2008) have all been reported to show a pelagic strategy in tracked animals (Mattlin et al. 1998). Rarely have three foraging patterns (epipelagic, mesopelagic and benthic) been found in one species, with the exception of sea lions from the *Zalophus* genus (California and Galapagos sea lions) that seem to have high behavioral flexibility (Villegas-Amtmann et al. 2008, 2011; McHuron et al. 2016).

Individuals from the same species have a wide range of diving behaviors at their disposal likely driven by both intrinsic and extrinsic circumstances. Age, sex, and size can affect diving behavior. Younger animals are ontogenetically limited in their diving capacities and dive shallower than adults (Horning and Trillmich 1997; Baylis et al. 2005; Fowler et al. 2006; Verrier et al. 2011), while larger adult individuals have greater diving capacities than smaller ones as mentioned earlier (Boyd et al. 1998; Lea et al. 2006; Baylis et al. 2015b). There is also evidence that males have a very different foraging ecology compared to females in Antarctic and New Zealand fur seal males for example, but not necessarily in others species (Fea et al. 1999; Staniland and Robinson 2008; Kernaléguen et al. 2015b; Drago et al. 2016; de Albernaz et al. 2017). Alternatively, variation in diving strategies can be attributed to habitat characteristics, targeted prey type, behavior or prey availability in these habitats, and competition or predation pressure (Chilvers and Wilkinson 2009; Baylis et al. 2015b; Arthur et al. 2016; Paez-Rosas et al. 2017; Jeanniard du

Dot et al. 2018). For example, Galapagos sea lions foraging west of the archipelago displayed pelagic diving and fed on mesopelagic fish such as the Galapagos sardine (*Opisthonema berlangai*), while females foraging benthically south west of the Galapagos archipelago preyed upon larger fish from the Serranidae and Scorpenidae families in deep areas with rocky substrates (Villegas-Amtmann et al. 2017).

While several diving strategies exist amongst fur seal and sea lion species, it appears that adult individuals tend to remain faithful to specific diving strategies. A high degree of specialization was for example observed in female New Zealand sea lions (both within and between years, Chilvers 2008a, b, 2017), Australian sea lions (Lowther et al. 2013), Australian fur seals (Kernaléguen et al. 2015a) and Antarctic fur seals (Arthur et al. 2015), although it is less clear for New Zealand fur seals and California sea lions (Melin et al. 2008; Kuhn and Costa 2014; McHuron et al. 2016). Individual specialization is expected to occur more frequently in generalist top marine predators with a wide range of potential prey as it provides the ecological opportunity for different strategies to occur (Araújo et al. 2011), and when competition for resources is strong. This can be the case for breeding females nursing their offspring. As central place foragers, they are constrained to forage within limited areas around the breeding grounds and experience stronger intraspecific competition for resources, resulting in foraging niche segregation. Foraging niche segregation also occurs inter-specifically for sympatric otariid species such as Galapagos sea lions and Galapagos fur seals, New Zealand fur seals and Australian fur seals, and southern sea lions and South American fur seals with one of the two species being benthic and the other pelagic foragers.

It is difficult to know whether different diving strategies within a population have recently emerged as a result of changes in habitat characteristics or have been retained over time as Evolutionary Stable Strategies (Hines 1987). Regardless, they entail different trade-offs in terms of risks and benefits. Benthic foragers dive deeper for longer periods of time which results in animals surpassing their ADL more frequently with likely greater overall energy expenditure (Costa et al. 2004). However, one of the benefits of preying upon benthic prey is that they tend to be evenly distributed, predictable, relatively large and overall less susceptible to acute environmental changes or events such as El Niño (Miller and Sydeman 2004). By contrast, pelagic divers perform short and shallow dives rarely exceeding ADL, and prey upon small size fish that occur in high densities when they find them. However, mesopelagic prey tend to be patchily distributed, less predictable and more sensitive to oceanographic perturbations (Boyd 1996; Harcourt et al. 2002). Consequently, the different strategies might be more or less beneficial depending upon annual environmental conditions, and the existence of diversity in diving behaviors and strategies within population provides a buffer for species occupying changeable habitats (Villegas-Amtmann et al. 2008; Lowther et al. 2012; McHuron et al. 2016).

An essential component of understanding these diving strategies and the trade-offs they entail is to estimate the energetic costs associated with them. Studies on northern and Antarctic fur seals, and on Galapagos, Australian and New Zealand sea lions that simultaneously measured field metabolic rates (FMR) and foraging strategies in otariids show that energy expenditures of foraging otariids are influenced by the variability of their behaviors at sea, i.e. their time-activity budget comprised of diving, traveling, and resting (Arnould et al. 1996; Costa and Gales 2000; Jeanniard du Dot et al. 2017b). Diving was found to be the most energetically expensive activity at sea in free-ranging northern and Antarctic fur seals (however, see Arnould et al. (1996)), in semi-captive Steller sea lions and others (Butler et al. 1995; Goundie et al. 2015; Jeanniard du Dot et al. 2017b), and deeper dives more costly than shallower dives of similar duration (Halsey et al. 2006). In addition, the most energetically efficient pattern of diving resides in sequences or bouts of multiple dives close to the ADL with short surface intervals (Goundie et al. 2015).

Consequently, benthic divers that dive deeper, for longer, and that overall exceed their ADL more often than pelagic divers, should have greater rates of energy expenditure at sea. It is, however, difficult to make definite conclusions about the relative energetic costs of each strategy in free ranging studies. First, fur seals and sea lions all seem to operate within a narrow range of FMR (6.25–7.05 W/kg, Table 4.2) with the exception of the deep diving Galapagos sea lions (4.08 W/kg, closer to the 3.36 W/kg of walruses) (Arnould et al. 1996; Costa and Gales 2000, 2003; Acquarone et al. 2006; Jeanniard du Dot et al. 2017a, 2018; Villegas-Amtmann et al. 2017). Second, finer patterns of field energy expenditures are not always consistent between and within otariids. Galapagos and California sea lions show a slight decrease in FMR when their time spent diving increases (Hurley and Costa 2001; Villegas-Amtmann et al. 2017). Deep diving New Zealand Sea lions experience lower FMR than shallow diving conspecifics (Costa and Gales 2000), whereas deep diving and shallow diving northern fur seals and California sea lions have similar FMR (Jeanniard du Dot et al. 2018; McHuron et al. 2018).

Otariids can adapt to different foraging habitat and/or environmental changes either behaviorally or physiologically. They tend to increase their time at sea rather than their FMR in poor foraging conditions (Trillmich 1990; Boyd 1999) (noting that foraging trip duration does not appear to impact FMR in either northern fur seals or New Zealand sea lions, Costa and Gales 2000; Jeanniard du Dot et al. 2018). Pelagic divers that forage at night on prey with a circadian migration pattern are limited in the opportunities to increase their diving effort in years of poor foraging conditions, as prey are not accessible to them during daylight. They have to respond by increasing their total time at sea. Deeper divers, or divers with more flexible foraging strategies such as California sea lions, can access prey during the day and increase their diving effort locally without the need to extend their total time at sea. Nevertheless, even for the few species that can increase their FMR, total metabolic changes remain limited. The overall low metabolic plasticity indicates that fur seals and sea lions most probably operate close to their metabolic ceiling with important implications for the scope of their adaptive capacities to environmental changes (Costa and Gales 2000, 2003; Arnould and Costa 2006; Ladds et al. 2017c; Villegas-Amtmann et al. 2017).

4.4 Dive Metrics to Infer Behaviors and Energetics

Unlike phocids who travel and sleep underwater, fur seals and sea lions transit and rest at the surface of the ocean. The function of their diving is thus mainly to find and access patchily distributed prey resources in a three dimensional and dynamic environment. Ecologists have long attempted to transpose dive profiles into foraging behaviors, as well as use dive metrics—such as dive duration, maximum dive depth, surface interval duration, bottom duration, 2D dive shape, time-at-depth, ascent and descent rates, residuals of dive depth-duration models indicative of longer than average dives for a given depth, etc.—to quantify and predict energetic costs of diving and feeding success. The relationship between dive shape and behavior may however not be consistent between species, age classes or life history stages (Kuhn et al. 2009a; Carter et al. 2016).

More recently with high resolution datasets, different methods have emerged to partition and quantify time-activity budgets, foraging efforts and diving activities: the vertical area-restricted search, the vertical sinuosity at the bottom of dives, the vertical velocity used in hidden Markov models or behavioral state-space models for example (Joy et al. 2015; Arthur et al. 2016; Heerah et al. 2017) (Fig. 4.8). It is important to keep in mind however, that the application and the accuracy of each of

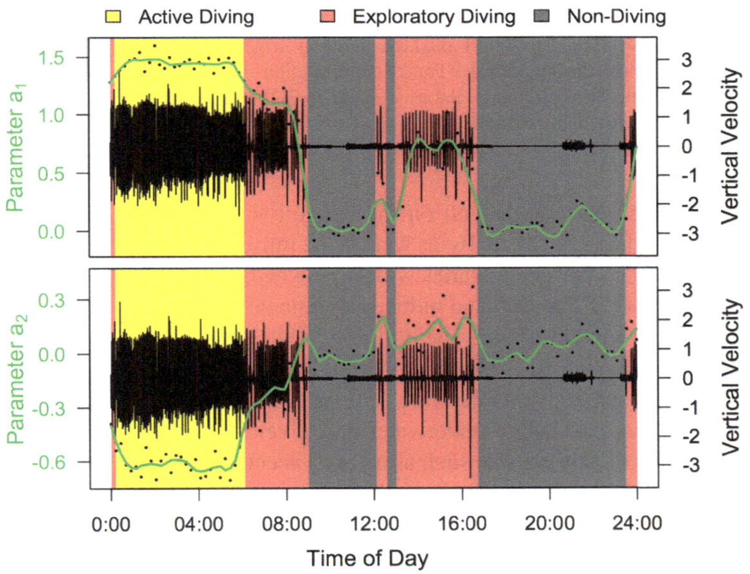

Fig. 4.8 Foraging behaviors of a northern fur seal female breeding on St Paul Island, in the Bering sea during the breeding season 2006 inferred from a behavioral state-space model (results from model parameters a_1 and a_2 are shown with the green lines and original estimates with black dots), based on time series of vertical velocity (black lines). The yellow, pink and grey blocks correspond to active diving, exploratory diving, and non-diving, respectively (as diagnosed from values of a_1 and a_2 from the model). Figure from Joy et al. (2015).

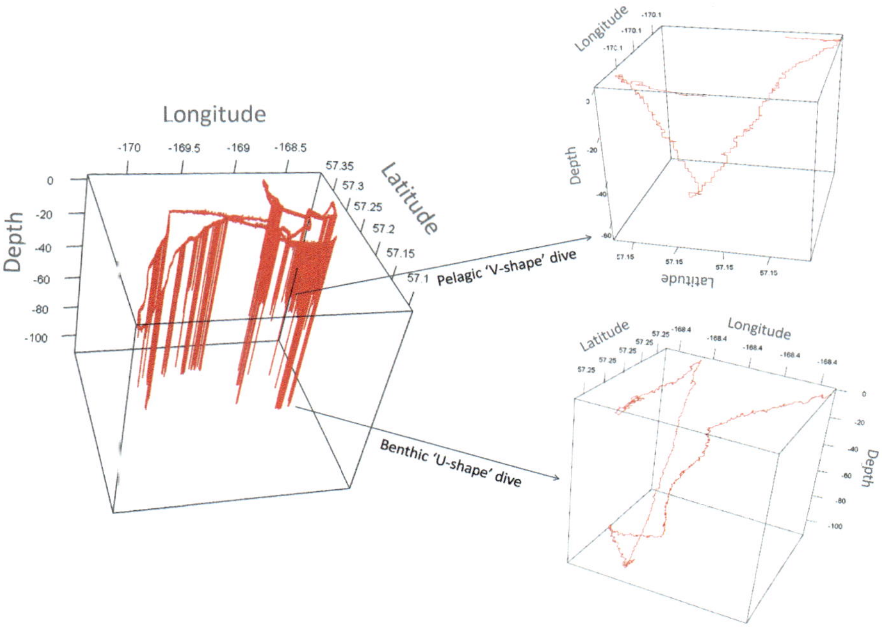

Fig. 4.9 Example of the 3D track of a foraging trip at sea from a female northern fur seal from St. Paul Island reconstructed using tri-axial acceleration, dive depths and GPS data from high-resolution biologger (and using package TrackReconstruction in R colored by depth). Detailed 3D paths during a benthic (typically classified as U-shape in 2D) and a pelagic (typically classified as V-shape in 2D) are shown as examples

these methods and metrics depend on the resolution of the data and the temporal scale of analysis (Carter et al. 2016) as well as the minimum depth taken into account to define a dive. This parameter varies widely between studies (2–6 m minimum dive to define a dive, see Table 4.2) and induces significant variations in the mean diving depth and the percentage of time spent diving.

In addition, inferring tri-dimensional behaviors from two-dimensional information is inherently subjective as 2D dive profiles do not account for lateral displacement—or lack thereof—while underwater at a given depth (Harcourt et al. 2000). The development of new sensors such as tri-axial accelerometers and magnetometers have provided means to derive dive profile of animals in 3D (Figs. 4.9 and 4.10), as well as their body position in the water column, pitch, roll and yaw angles, as well as relative swimming effort (Fig. 4.10), and detailed behaviors using varied methods for example spectral analyses of tri-axial acceleration with unsupervised signal categorization (using available ethographer software, Battaile et al. 2015a), classification tree algorithms (Jeanniard du Dot et al. 2017b), or random forest models (Ladds et al. 2016). This new but complex information has shown that pinniped foraging behavior can be varied and complex and that 2D profiles used on their own

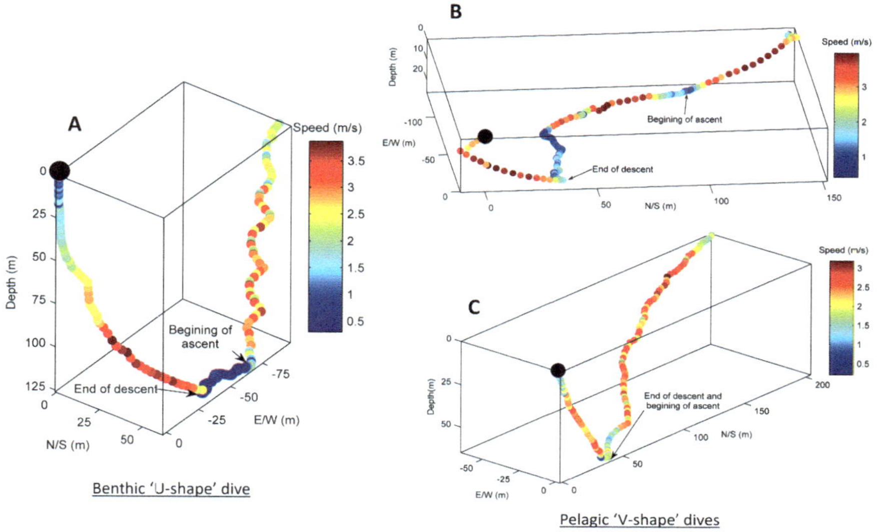

Fig. 4.10 Example of dives performed by a female Steller sea lion from Lovushki Island, Russia, displayed in 3D, color-coded by swim speed and with the black dots representing the beginning of the dives. Graph (**a**) shows a typical benthic 'U-shape' dive with steep descent and ascent angles. The animal speeds up from surface to bottom of the dive, likely due to buoyancy-driven gliding strategy, displays a slow swimming speed at the benthos before barrel-rolling back to the surface (as observed in several species of otariids), slowing down towards the end of the ascent. Graphs (**b**, **c**) illustrate pelagic dives with more moderate descent and ascent angles. Graph (**b**) shows a dive where changes in swim speed are typical of a prey chase behavior. Graph (**c**) illustrates a typical pelagic 'V-shape' dive with a smaller range of swim speeds displayed. Figures from Olivier (2015)

may be overly simplistic (Davis et al. 1999; Harcourt et al. 2000; Simpkins et al. 2001; Hindell et al. 2002).

In addition, without direct validation, it remains unclear whether any of these dive and behavioral characteristics can be used to differentiate successful from unsuccessful foraging in diving pinnipeds. Stomach temperature pills (Kuhn and Costa 2006; Kuhn et al. 2009b) and head- or back-mounted accelerometers have been combined with TDR and video data to identify prey capture attempts in several fur seal and sea lion species (Fig. 4.11, and see Fig. 4.7 for example at sea) (Viviant et al. 2010; Volpov et al. 2016). The combination of dive profiles with these new indices of feeding attempts or events have provided opportunities to ground-truth time-at-depth dive metrics as indicators of foraging behaviors and success (Sala et al. 2011). For example Viviant et al. (2016), deployed accelerometers (measuring jaw-openings as a proxy for prey capture attempts) in conjunction with TDRs in Antarctic fur seals to test dive metrics including bottom duration, ascent and descent rates and maximum dive depth as predictors of foraging success. They concluded that the metrics to best predict foraging success depends on the timescale considered (from single dives, to bout or to nights).

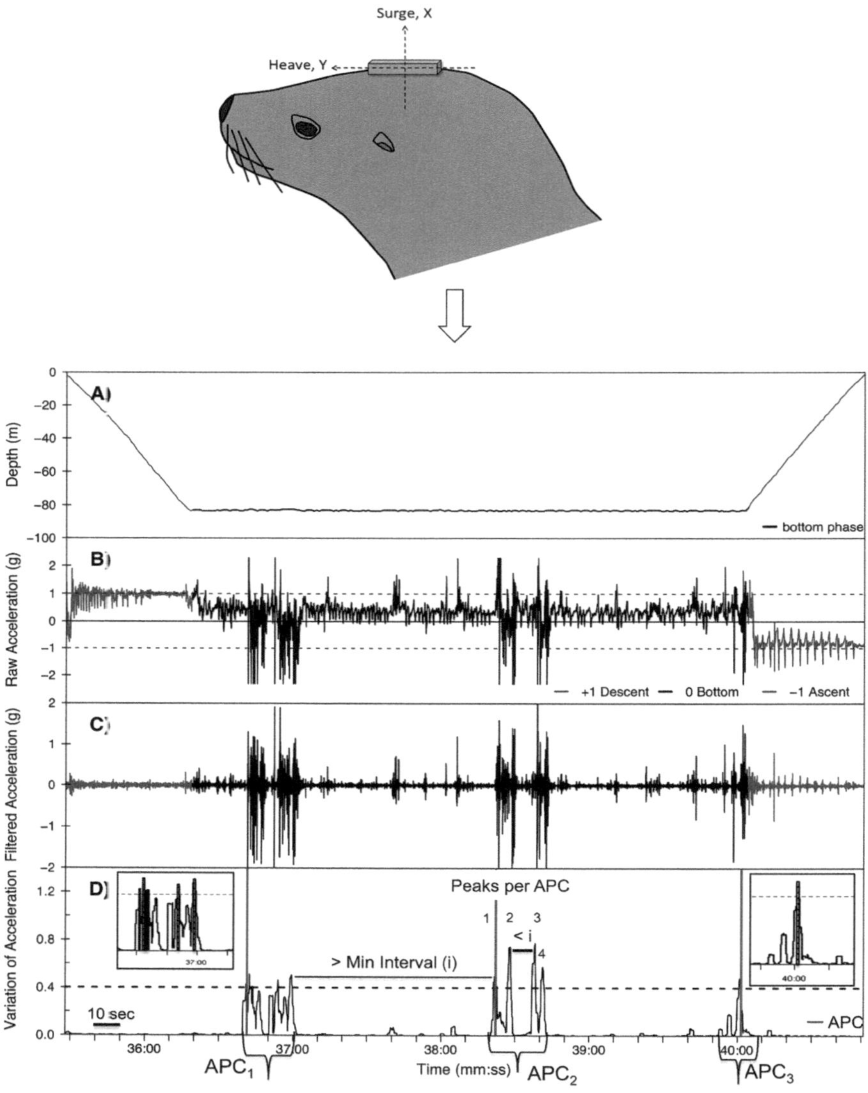

Fig. 4.11 Example of an otariid equipped with a head accelerometer showing the surge (X) and heave (Y) axes, and the process of identification of attempted prey captures (APC) from the recorded signal of depth and acceleration. Dive depth for a single dive is shown in panel (**a**), and raw surge acceleration (in g) in panel (**b**). Head movements were isolated from body movements with a 3 Hz high-pass filter (panel **c**) and variance of acceleration (g^2) was calculated for each individual dive (panel **d**). Peaks in variance of acceleration above a threshold and within a minimum time interval (i) were used to estimate attempted prey capture (APC, panel **d**). Figure from Volpov et al. (2015)

Dynamic tri-axial acceleration metrics such as ODBA or VeDBA (Overall—or Vectorial—dynamic body acceleration) were hypothesized to be directly related to metabolic rate in free ranging animals (Gleiss et al. 2011). While studies have shown promising relationships between metabolic rate and ODBA on terrestrial animals (Halsey et al. 2009), this relationship is more controversial in captive or wild fur seals and sea lions (Fahlman et al. 2008; Dalton et al. 2014; Ladds et al. 2017a). In Antarctic and northern fur seals, ODBA did not correlate with doubly-labeled water measures of field metabolic rates over a full foraging trip at sea (Jeanniard du Dot et al. 2016a). This was argued to be due, amongst other reasons, to the difference in time-activity budgets between individuals by the authors, but it was later shown that relationships between ODBA/VeDBA and energy expenditure metrics was likely found significant due to a 'time trap', i.e. time being inherently conflated into the 2 tested variables (Halsey 2017; Ladds et al. 2017a). Whether or not ODBA/VeDBA can accurately measure field metabolic rate in free-ranging fur seals and sea lions, dynamic acceleration can certainly provide a relative index of swimming effort underwater between phases of dives, between dives, and over total foraging trips for a given individual (Miller et al. 2012). This refines the scale at which changes in swimming effort during dives can be estimated to an unprecedented level.

In addition to using dive profiles, ecologists have built theoretical frameworks to infer foraging efficiency in diving animals. These models were specifically adapted to diving animals from the Optimal Foraging Theory (OFT, MacArthur and Pianka 1966) taking into account the physiological needs to regularly return to breath at the surface of the water, i.e. leave the prey patch. OFT stipulates that natural selection favors animals that forage more efficiently, with foraging efficiency defined as the ratio of energy gained to energy expended to acquire food per unit of time. Consequently, individuals should maximize rate of energy intake while minimizing the rate of energy expenditure associated with prey searching, capture and handling, and thus concentrate their time in areas of successful feeding. Given this and the previously mentioned assumption that most foraging occurs during the bottom phase of dives, Optimal Diving models hypothesized that divers should maximize their time at the bottom phase, i.e. at the feeding depth, effectively increasing the probability of capturing prey while minimizing the time and energy spent traveling from and to the surface (descent and ascent phase), as well as the time recovering at the surface after the dive (post-dive surface interval) (Kramer 1988; Houston and Carbone 1992; Thompson et al. 1993). Within this framework, foraging success should thus increase with dive duration, bottom phase duration and dive frequency.

This was however not verified for wild benthic diving Australian fur seals regularly diving at the fairly constant benthos of the oceans. Neither bottom duration nor post-dive surface intervals were indicative of whether prey capture attempts occurred during dives, while descent rate was the best predictor of successful feeding dives (Volpov et al. 2016). In Australian and Antarctic fur seals, bottom time duration was not the best predictor of successful foraging which rather depended on the temporal scale at which the analyses occurred from single dives, to bout, nights etc. (Iwata et al. 2015; Foo et al. 2016; Viviant et al. 2016). This highlights the importance of the time scale and initial research question since predictors of foraging success for a particular dive bout may not perform as well when applied across an entire foraging trip.

The distribution, accessibility and quality/density of prey patches, the level of competition for a given prey patch, as well as the physiological needs, capacities or degree of plasticity of diving animals are likely to influence their foraging strategies. Adding inferences about prey patch density and quality in the decision making process of diving seals (Marginal Value Theorem, Charnov 1976; Thompson and Fedak 2001) hypothesized that individuals should terminate a dive earlier (i.e. shorter dive and bottom durations) when there are no prey present or prey density is low, but that the benefit of early dive termination is reduced for deeper dives. These assumptions are however not clearly validated in wild otariids either. Antarctic fur seal females increased their foraging effort by diving more and spending more time searching for prey at the bottom of dives in poor-quality patches (Mori and Boyd 2004; Viviant et al. 2014). Similarly, Australian fur seals decreased bottom duration with increasing prey encounter rate at the scale of a dive, possibly to come back to the surface to consume larger prey items, but not at the scale of a bout (Foo et al. 2016). It is however interesting to note that Antarctic fur seals did tend to dive and forage at a depth shallower than the depth with the highest rate of prey capture attempts (Viviant et al. 2016). This is in accordance with Mori's model (1998) that postulates that if species are physically capable of reaching depths of highest prey densities, they will tend to dive at depths slightly shallower than the maximum prey density as a trade-off with physiological constraints of diving. This indicates that species favor foraging efficiency (i.e. the net energy gain taking into account diving costs) rather than just maximizing prey intake.

The wide breadth of these results highlights the complexity of the decision making process of fur seals and sea lions in the wild. They modify their dive behavior based on real-time evaluation of prey encounter rates during a given dive (especially if the first prey encounter occurs early in the dive, Foo et al. 2016), or during several preceding dives (Iwata et al. 2015; Viviant et al. 2016; Volpov et al. 2016) all while within their physiological constraints. Information gathered at any point in the foraging experience is thus essential in the multifactorial decision making process, and current theoretical foraging models using dive metrics only may be too simplistic to accurately represent such a complex ecological system.

To conclude, there is no doubt that dive-related data are a powerful resource to assess underwater behavior and quantify foraging efficiency in animals which cannot be observed directly. This is particularly relevant given the new types of data accessible from a wide range of additional sensors. However, these data should be used with a clear understanding of their limitations given the wide diversity of behaviors and the complexity of factors affecting the decision making process of individual animals even within the single otariid taxon itself.

4.5 Conclusion: Impacts of Diving Strategies on Fitness and Population Trends

The identification and behavioral characterization of foraging strategies is an important first step towards understanding the energetic consequences of the variability in foraging behaviors of fur seals and sea lions. In a population of Antarctic fur seal with a single foraging strategy, the time lactating females spend ashore was positively correlated with pup weaning mass (Doidge and Croxall 1989; Lunn et al. 1994); and mothers with diving patterns yielding a greater foraging efficiency during their trips at sea produced heavier pups more likely to survive during their first year at sea (Jeanniard du Dot et al. 2017a). Consequently, the resulting foraging efficiency ultimately impacts survival and reproductive capacities of individuals and the trend of their population both intra- and inter-specifically.

It has been suggested that the benthic feeding strategy inherently results in a higher foraging cost for likely similar energy gain (thus lower foraging efficiency) (Costa et al. 2004). In addition, benthic foragers usually exhibit smaller population sizes and lower population growth rates compared with those that feed in the water column (Arnould and Costa 2006). In northern fur seals and California sea lions, the pelagic strategy offers greater foraging efficiency than the benthic strategy, but females often have to travel further and spend more time at sea away from their pup to reach their foraging ground (McHuron et al. 2016; Jeanniard du Dot et al. 2018). Pup growth rates of benthic and pelagic females could not be measured in these two studies so it is difficult to tease apart what strategy would be more beneficial, if any. However, the population of northern fur seals from Bogoslof island, a breeding ground where only the pelagic diving strategy exists, is increasing while the Pribilof Island breeding population where both the pelagic and benthic strategies can be seen has been decreasing by 3.5% per year for the last 20 years (Muto et al. 2018). Similarly, adult female southern sea lions in northern Patagonia feeding on 'offshore' pelagic prey had higher pup growth rates than females with an 'inshore' benthic prey diet (in this particular study year, Dragon et al. 2010). Southern sea lions which have only a benthic feeding strategy in Uruguay are decreasing, while the population in the Falklands where some individuals display limited pelagic feeding is showing signs of recovery (Baylis et al. 2015a). Beyond the benthic/pelagic dichotomy, the deeper dive depths of Antarctic fur seals breeding on Heard Island compared to fur seals from Bird Island is hypothesized to be a reason for their low population growth rates and limited population recovery (Staniland et al. 2010). All these studies indicate that behavioral decisions related to diving and foraging affect pup growth and survival for income-breeding species, and shape population trends and structure.

Despite evidence that pelagic strategies seem to lead to higher population growth rates than benthic ones, the fact that diverging strategies are retained and co-exist within and between populations indicates that they have valuable benefits on the evolution time-scale. While the benthic diving strategy may inhibit rapid population growth compared to pelagic feeding, benthic prey are more predictable and stable and so benthic strategies may buffer populations during times of drastic environmental change (see also Costa and Toro, Chap. 2). Given that our oceans are now facing major biochemical, biological and trophic alteration of marine ecosystems, there is a growing need to understand which factors influence long-term persistence of different strategies within and between species and hence what the broader implications of foraging decisions are over evolutionary time.

References

Acquarone M, Born EW, Speakman JR (2006) Field metabolic rates of Walrus (*Odobenus rosmarus*) measured by the doubly labeled water method. Aquat Mamm 32:363–369

Andrews RD, Jones DR, Williams JD, Thorson PH, Oliver GW, Costa DP, LeBoeuf BJ (1997) Heart rates of northern elephant seals diving at sea and resting on the beach. J Exp Biol 200:2083–2095

Araújo MS, Bolnick DI, Layman CA (2011) The ecological causes of individual specialisation. Ecol Lett 14:948–958

Arnould J, Costa D (2006) Sea lions in drag, fur seals incognito: insights from the otariid deviants. In: Sea lions of the world: proceedings of the symposium sea lions of the world: conservation and research in the 21st century, Citeseer

Arnould JPY, Hindell MA (2001) Dive behaviour, foraging locations, and maternal-attendance patterns of Australian fur seals (*Arctocephalus pusillus doriferus*). Can J Zool 79:35–48

Arnould JPY, Boyd IL, Speakman JR (1996) The relationship between foraging behaviour and energy expenditure in Antarctic fur seals. J Zool (Lond) 239:769–782

Arthur B, Hindell M, Bester M, Trathan P, Jonsen I, Staniland I, Oosthuizen WC, Wege M, Lea M-A (2015) Return customers: foraging site fidelity and the effect of environmental variability in wide-ranging Antarctic fur seals. PLoS One 10:e0120888

Arthur B, Hindell M, Bester MN, Oosthuizen WC, Wege M, Lea M-A (2016) South for the winter? Within-dive foraging effort reveals the trade-offs between divergent foraging strategies in a free-ranging predator. Funct Ecol 30:1623–1637

Augé AA et al (2011) In the shallow end: diving behaviour of recolonising female New Zealand sea lions (Phocarctos hookeri) around the Otago Peninsula. Can J Zool 89(12):1195–1205

Austin D, Bowen WD, McMillan JI, Iverson SJ (2006) Linking movement, diving, and habitat to foraging success in a large marine predator. Ecology 87:3095–3108

Bailleul F et al (2005) Differences in foraging strategy and maternal behaviour between two sympatric fur seal species at the Crozet Islands. Mar Ecol Prog Ser 293:273–282

Battaile BC, Sakamoto KQ, Nordstrom CA, Rosen DAS, Trites AW (2015a) Accelerometers identify new behaviors and show little difference in the activity budgets of lactating northern fur seals (*Callorhinus ursinus*) between breeding islands and foraging habitats in the eastern Bering Sea. PLoS One 10:e0118761

Battaile BC et al (2015b) Foraging a new trail with northern fur seals (Callorhinus ursinus): Lactating seals from islands with contrasting population dynamics have different foraging strategies, and forage at scales previously unrecognized by GPS interpolated dive data. Mar Mamm Sci 31(4):1494–1520

Baylis AMM (2008) Seasonal and colony differences in the foraging ecology of New Zealand fur seals (Arctocephalus forsteri)

Baylis AMM, Page B, Peters K, McIntosh R, Mckenzie J, Goldsworthy S (2005) The ontogeny of diving behaviour in New Zealand fur seal pups (*Arctocephalus forsteri*). Can J Zool 83:1149–1161

Baylis AMM et al (2012) Individual foraging site fidelity in lactating New Zealand fur seals: continental shelf vs. oceanic habitats. Mar Mamm Sci 28(2):276–294
Baylis AMM, Orben RA, Arnould JPY, Christiansen F, Hays GC, Staniland IJ (2015a) Disentangling the cause of a catastrophic population decline in a large marine mammal. Ecology 96:2834–2847
Baylis AMM, Orben RA, Arnould JPY, Peters K, Knox T, Costa DP, Staniland IJ (2015b) Diving deeper into individual foraging specializations of a large marine predator, the southern sea lion. Oecologia 179:1053–1065
Baylis AMM et al (2016) Sexual segregation in habitat use is smaller than expected in a highly dimorphic marine predator, the southern sea lion. Mar Ecol Prog Ser 554:201–211
Baylis AMM, Orben RA, Costa DP, Tierney M, Brickle P, Staniland IJ (2017) Habitat use and spatial fidelity of male south American sea lions during the non-breeding period. Ecol Evol 7:3992–4002
Baylis AMM et al (2018) Geographic variation in the foraging behaviour of South American fur seals. Mar Ecol Prog Ser 596:233–245
Beauplet G et al (2004) Foraging ecology of subantarctic fur seals Arctocephalus tropicalis breeding on Amsterdam Island: seasonal changes in relation to maternal characteristics and pup growth. Mar Ecol Prog Ser 273:211–225
Benoit-Bird KJ, Battaile BC, Heppell SA, Hoover B, Irons D, Jones N, Kuletz KJ, Nordstrom CA, Paredes R, Suryan RM, Waluk CM, Trites AW (2013) Prey patch patterns predict habitat use by top marine predators with diverse foraging strategies. PLoS One 8:e53348
Bost CA, Handrich Y, Butler PJ, Fahlman A, Halsey LG, Woakes AJ, Ropert-Coudert Y (2007) Changes in dive profiles as an indicator of feeding success in king and Adélie penguins. Deep Sea Res II 54:248–255
Boyd IL (1996) Temporal scales of foraging in a marine predator. Ecology 77:426–434
Boyd IL (1999) Foraging and provisioning in Antarctic fur seals: interannual variability in time-energy budgets. Behav Ecol 10:198–208
Boyd IL (2002) Energetics: consequences for fitness. In: Hoelzel AR (ed) Marine mammal biology: an evolutionary approach. Blackwell Sciences, Oxford, Malden
Boyd IL, Croxall JP (1992) Diving behaviour of lactating Antarctic fur seals. Can J Zool 70 (5):919–928
Boyd IL, Croxall JP (1996) Dive durations in pinnipeds and seabirds. Can J Zool 74:1696–1705
Boyd IL, McCafferty DJ, Reid K, Taylor R, Walker TR (1998) Dispersal of male and female Antarctic fur seals (*Arctocephalus gazella*). Can J Fish Aquat Sci 55:845–852
Boyd IL, Bevan RM, Woakes AJ, Butler PJ (1999) Heart rate and behavior of fur seals: implications for measurements of field energetics. Am J Phys 276:H844–H857
Burns JM (1999) The development of diving behavior in juvenile Weddell seals: pushing physiological limits in order to survive. Can J Zool 77:737–747
Burns JM, Castellini MA (1996) Physiological and behavioral determinants of the aerobic dive limit in Weddell seal (*Leptonychotes weddellii*) pups. J Comp Physiol B 166:473–483
Butler PJ, Bevan RM, Woakes AJ, Croxall JP, Boyd IL (1995) The use of data loggers to determine the energetics and physiology of aquatic birds and mammals. Braz J Med Biol Res 28:1307–1317
Calkins DG, Becker EF, Pitcher KW (1998) Reduced body size of female Steller Sea lions from a declining population in the Gulf of Alaska. Mar Mamm Sci 14:232–244
Campagna C et al (2001) Movements and location at sea of South American sea lions (Otaria flavescens). J Zool (Lond) 257:205–220
Carter MID, Bennett KA, Embling CB, Hosegood PJ, Russell DJ (2016) Navigating uncertain waters: a critical review of inferring foraging behaviour from location and dive data in pinnipeds. Movement Ecol 4:25
Charnov EL (1976) Optimal foraging, the marginal value theorem. Theor Popul Biol 9:129–136
Chilvers B (2008a) Foraging site fidelity of lactating New Zealand Sea lions. J Zool 276:28–36

Chilvers BL (2008b) New Zealand Sea lions *Phocarctos hookeri* and squid trawl fisheries: bycatch problems and management options. Endang Species Res 5:193–204
Chilvers BL (2009) Foraging locations of female New Zealand sea lions (Phocarctos hookeri) from a declining colony. N Z J Ecol 33(2):106–113
Chilvers B (2017) Stable isotope signatures of whisker and blood serum confirm foraging strategies for female New Zealand Sea lions (*Phocarctos hookeri*) derived from telemetry. Can J Zool 95:955–963
Chilvers BL (2018) Preliminary assessment of the foraging behaviour and population dynamics of a cryptic population of the endangered New Zealand sea lion. N Z J Ecol 42(1):48–57
Chilvers BL, Wilkinson IS (2009) Diverse foraging strategies in lactating New Zealand Sea lions. Mar Ecol Prog Ser 378:299–308
Chilvers B et al (2005) Summer foraging areas for lactating New Zealand sea lions Phocarctos hookeri. Mar Ecol Prog Ser 304:235–247
Costa C (1993) The relationship between reproductive and foraging energetics and the evolution of the pinnipedia. Symp Zool Soc Lond 66:293–313
Costa D, Gales N (2000) Foraging energetics and diving behavior of lactating New Zealand Sea lions, *Phocarctos hookeri*. J Exp Biol 203:3655–3665
Costa DP, Gales NJ (2003) Energetics of a benthic diver: seasonal foraging ecology of the Australian sea lion, *Neophoca cinerea*. Ecol Monogr 73:27–43
Costa DP, Gentry RL (1986) Free-ranging energetics of northern fur seals. In: Gentry RL, Kooyman GL (eds) Fur seals: maternal strategies on land and at sea. Princeton University Press, Princeton, NJ, pp 79–101
Costa DP, Williams TM (1999) Marine mammals energetics. In: Reynolds JE, Rommel SA (eds) Biology of marine mammals. Smithsonian Institution Press, Washington, DC
Costa DP, Croxall JP, Duck CD (1989) Foraging energetics of Antarctic fur seals in relation to changes in prey availability. Ecology 70:596–606
Costa DP, Gales NJ, Crocker DE (1998) Blood volume and diving ability of the New Zealand Sea lion, *Phocarctos hookeri*. Physiol Zool 71:208–213
Costa DP, Goebel ME, Sterling JT (2000) Foraging energetics and diving behavior of the Antarctic fur seal, Arctocephalus gazella at Cape Shirreff, Livingston Island. In: Davisons W, Howard-Williams C, Broady P (eds) Antarctic ecosystems: models for a wider ecological understanding. New Zealand Natural Sciences Press, Christchurch, New-Zealand, pp 77–84
Costa DP, Gales NJ, Goebel ME (2001) Aerobic dive limit: how often does it occur in nature? Comp Biochem Physiol A 129:771–783
Costa DP, Kuhn CE, Weise MJ, Shaffer SA, Arnould JPY (2004) When does physiology limit the foraging behaviour of freely diving mammals? Int Congr Ser 1275:359–366
Crocker D, Gales NJ, Costa DP (2001) Swimming speed and foraging strategies of New Zealand Sea lions (*Phocarctos hookeri*). J Zool (Lond) 254:267–277
Dalton AJM, Rosen DAS, Trites AW (2014) Season and time of day affect the ability of accelerometry and the doubly labeled water methods to measure energy expenditure in northern fur seals (*Callorhinus ursinus*). J Exp Mar Biol Ecol 452:125–136
David JHM, Rand RW (1986) Attendance behaviour of South African fur seals. In: Gentry RL, Kooyman GL (eds) Fur seals: maternal strategies on land and at sea. Princeton University Press, Princeton, NJ, pp 126–141
Davis RW (2014) A review of the multi-level adaptations for maximizing aerobic dive duration in marine mammals: from biochemistry to behavior. J Comp Physiol B 184:23–53
Davis RW, Fuiman LA, Williams TM, Collier SO, Hagey WP, Kanatous SB, Kohin S, Horning M (1999) Hunting behavior of a marine mammal beneath the Antarctic fast ice. Science 283:993–996
de Albernaz TL, Secchi ER, de Oliveira LR, Botta S (2017) Ontogenetic and gender-related variation in the isotopic niche within and between three species of fur seals (genus *Arctocephalus*). Hydrobiologia 787:123–139

de Bruyn PJN et al (2009) Bathymetry and frontal system interactions influence seasonal foraging movements of lactating subantarctic fur seals from Marion Island. Mar Ecol Prog Ser 394:263–276

Doidge DW, Croxall JP (1989) Factors affecting weaning weight in Antarctic fur seals, *Arctocephalus gazella* at South Georgia. Polar Biol 9:155–160

Drago M, Franco-Trecu V, Cardona L, Inchausti P, Tapia W, Páez-Rosas D (2016) Stable isotopes reveal long-term fidelity to foraging grounds in the Galapagos Sea lion (*Zalophus wollebaeki*). PLoS One 11. https://doi.org/10.1371/journal.pone.0147857

Dragon AC, Monestiez P, Bar-Hen A, Guinet C (2010) Linking foraging behaviour to physical oceanographic structures: southern elephant seals and mesoscale eddies east of Kerguelen Islands. Prog Oceanogr 87:61

Fahlman A, Wilson R, Svard C, Rosen DAS, Trites AW (2008) Activity and diving metabolism correlate in Steller Sea lion *Eumetopias jubatus*. Aquat Biol 2:75–84

Fea NI, Harcourt R, Lalas C (1999) Seasonal variation in the diet ofNew Zealand fur seals (*Arctocephalus forsteri*) at Otago peninsula, New Zealand. Wildl Res 26:147–160

Feldkamp SD (1987) Swimming in the California Sea lion: morphometrics, drag and energetics. J Exp Biol 131:117–135

Feldkamp SD, DeLong RL, Antonelis GA (1989) Diving patterns of California sea lions, Zalophus californianus. Can J Zool 67:872–883

Foo D, Semmens JM, Arnould JPY, Dorville N, Hoskins AJ, Abernathy K (2016) Testing optimal foraging theory models on benthic divers. Anim Behav 112:127

Fowler SL, Costa DP, Arnould JPY, Gales NJ, Kuhn CE (2006) Ontogeny of diving behaviour in the Australian sea lion: trials of adolescence in a late bloomer. J Anim Ecol 75:358–367

Fowler SL, Costa DP, Arnould JPY, Gales NJ, Burns JM (2007) Ontogeny of oxygen stores and physiological diving capability in Australian sea lions. Funct Ecol 21:922–935

Francis J, Boness D, Ochoa-Acuna H (1998) A protected foraging and attendance cycle in female Juan Fernandez fur seals. Mar Mamm Sci 14:552–574

Gallo-Reynoso JP, Figueroa-Carranza A-L, Le Boeuf B (2008) Foraging behavior of lactating Guadalupe fur seal females. In: Lorenzo C, Espinoza E, Ortega J (eds) Avances en el Estudio de los Mamíferos de México, vol 2. Publicaciones Especiales, Mexico, pp 595–614

Garde E et al (2018) Diving behavior of the Atlantic walrus in high Arctic Greenland and Canada. J Exp Mar Biol Ecol 500:89–99

Gentry RL, Kooyman GL (eds) (1986) Fur seals - maternal strategies on land and at sea. Princeton University Press, Princeton, NJ

Gentry RL, Kooyman GL, Goebel ME (1986) Feeding and diving behavior of nothern fur seals. In: Gentry RL, Kooyman GL (eds) Fur seals: maternal strategies on land and at sea. Princeton University Press, Princeton, pp 61–78

Georges JY, Bonadonna F, Guinet C (2000a) Foraging habitat and diving activity of lactating Subantarctic fur seals in relation to sea-surface temperatures at Amsterdam Island. Mar Ecol Prog Ser 196:291–304

Georges J-Y, Tremblay Y, Guinet C (2000b) Seasonal diving behaviour in lactating subantarctic fur seals on Amsterdam Island. Polar Biol 23:59–69

Gerlinsky CD, Trites AW, Rosen DAS (2014) Steller sea lions (Eumetopias jubatus) have greater blood volumes, higher diving metabolic rates and a longer aerobic dive limit when nutritionally stressed. J Exp Biol 217(5):769–778

Gjertz I, Griffiths D, Krafft BA, Lydersen C, Wiig O (2001) Diving and haul-out patterns of walruses *Odobenus rosmarus* on Svalbard. Polar Biol 24(5):314–319

Gleiss AC, Wilson RP, Shepard ELC (2011) Making overall dynamic body acceleration work: on the theory of acceleration as a proxy for energy expenditure. Methods Ecol Evol 2:23–33

Goldsworthy S, Hindell M, Crowley H (1997) Diet and diving behaviour of sympatric fur seals Arctocephalus gazella and Arctocephalus tropicalis at Macquarie Island. In: Marine mammal research in the Southern Hemisphere

Goundie ET, Rosen DA, Trites AW (2015) Dive behaviour can predict metabolic expenditure in Steller Sea lions. Conserv Physiol 3:1

Halsey LG (2017) Relationships grow with time: a note of caution about energy expenditure-proxy correlations, focussing on accelerometry as an example. Funct Ecol 31:1176–1183

Halsey L, Blackburn T, Butler P (2006) A comparative analysis of the diving behaviour of birds and mammals. Funct Ecol 20:889–899

Halsey LG, Green JA, Wilson RP, Frappell PB (2009) Accelerometry to estimate energy expenditure during activity: best practice with data loggers. Physiol Biochem Zool 82:396–404

Harcourt R, Hindell M, Bell DG, Waas JR (2000) Three-dimensional dive profiles of free-ranging Weddell seals. Polar Biol 23:479–487

Harcourt RG, Bradshaw CJA, Dickson K, Davis LS (2002) Foraging ecology of a generalist predator, the female New Zealand fur seal. Mar Ecol Prog Ser 227:11–24

Heerah K, Woillez M, Fablet R, Garren F, Martin S, De Pontual H (2017) Coupling spectral analysis and hidden Markov models for the segmentation of behavioural patterns. Movement Ecol 5:20

Hindell MA, Harcourt R, Waas JR, Thompson D (2002) Fine-scale three-dimensional spatial use by diving, lactating female Weddell seals *Leptonychotes weddellii*. Mar Ecol Prog Ser 242:275–284

Hindell MA, McMahon CR, Bester MN, Boehme L, Costa D, Fedak MA et al (2016) Circumpolar habitat use in the southern elephant seal: implications for foraging success and population trajectories. Ecosphere 7(5):1

Hindle AG, Young BL, Rosen DAS, Haulena M, Trites AW (2010) Dive response differs between shallow- and deep-diving Steller Sea lions (*Eumetopias jubatus*). J Exp Mar Biol Ecol 394:141–148

Hindle AG, Mellish JAE, Horning M (2011) Aerobic dive limit does not decline in an aging pinniped. J Exp Zool A Ecol Genet Physiol 315:544–552

Hines WGS (1987) Evolutionary stable strategies: a review of basic theory. Theor Popul Biol 31:195–272

Hooker SK, Miller PJO, Johnson MP, Cox OP, Boyd IL (2005) Ascent exhalations of Antarctic fur seals: a behavioural adaptation for breath–hold diving? Proc R Soc Lond B 272:355–363

Horning M (2012) Constraint lines and performance envelopes in behavioral physiology: the case of the aerobic dive limit. Front Physiol 3. https://doi.org/10.3389/fphys.2012.00381

Horning M, Trillmich F (1997) Ontogenty of diving behaviour in the Galapagos fur seal. Behaviour 134:1211–1257

Hoskins AJ, Arnould JPY (2014) Relationship between long-term environmental fluctuations and diving effort of female Australian fur seals. Mar Ecol Prog Ser 511:285–295

Hoskins AJ, Costa DP, Arnould JPY (2015) Utilisation of intensive foraging zones by female Australian fur seals. PLoS One 10(2):e0117997

Hoskins AJ et al (2017) Foraging niche separation in sympatric temperate-latitude fur seal species. Mar Ecol Prog Ser 566:229–241

Houston AI, Carbone C (1992) The optimal allocation of time during the diving cycle. Behav Ecol 3:255–265

Hückstädt LA, Tift MS, Riet-Sapriza F, Franco-Trecu V, Baylis AM, Orben RA, Arnould JP, Sepulveda M, Santos-Carvallo M, Burns JM (2016) Regional variability in diving physiology and behavior in a widely distributed air-breathing marine predator, the south American sea lion (*Otaria byronia*). J Exp Biol 219:2320–2330

Hurley JA, Costa DP (2001) Standard metabolic rate at the surface and during trained submersions in adult California Sea lions (*Zalophus californianus*). J Exp Biol 204:3273–3281

Iwata T, Sakamoto KQ, Edwards EWJ, Staniland IJ, Trathan PN, Goto Y, Sato K, Naito Y, Takahashi A (2015) The influence of preceding dive cycles on the foraging decisions of Antarctic fur seals. Biol Lett 11:20150227

Jay CV, Farley SD, Garner GW (2001) Summer diving behavior of male walruses in Bristol Bay, Alaska. Mar Mamm Sci 17(3):617–631

Jeanniard du Dot T, Guinet C, Arnould JPY, Speakman JR, Trites AW (2016a) Accelerometers can measure total and activity-specific energy expenditures in free-ranging marine mammals only if linked to time-activity budgets. Funct Ecol 31:377–386

Jeanniard du Dot T, Trites AW, Arnould JPY, Speakman JB, Guinet C (2016b) Flipper strokes can predict energy expenditure and locomotion costs in free-ranging northern and Antarctic fur seals. Sci Rep 6:33912

Jeanniard du Dot T, Trites AW, Arnould JPY, Guinet C (2017a) Reproductive success is energetically linked to foraging efficiency in Antarctic fur seals. PLoS One 12:e0174001

Jeanniard du Dot T, Trites AW, Arnould JPY, Speakman JR, Guinet C (2017b) Activity-specific metabolic rates for diving, transiting, and resting at sea can be estimated from time–activity budgets in free-ranging marine mammals. Ecol Evol 7:2969–2976

Jeanniard du Dot T, Trites AW, Arnould JPY, Speakman JR, Guinet C (2018) Trade-offs between foraging efficiency and pup feeding rate of lactating northern fur seals in a declining population. Mar Ecol Prog Ser 600:207–222

Jeglinski JW, Werner C, Robinson PW, Costa DP, Trillmich F (2012) Age, body mass and environmental variation shape the foraging ontogeny of Galapagos Sea lions. Mar Ecol Prog Ser 453:279–296

Joy R, Dowd MG, Battaile BC, Lestenkof PM, Sterling JT, Trites AW, Routledge RD (2015) Linking northern fur seal dive behavior to environmental variables in the eastern Bering Sea. Ecosphere 6:art75

Kernaléguen L, Arnould JPY, Guinet C, Cherel Y (2015a) Determinants of individual foraging specialization in large marine vertebrates, the Antarctic and subantarctic fur seals. J Anim Ecol 84:1081–1091

Kernaléguen L, Cherel Y, Knox TC, Baylis AM, Arnould JP (2015b) Sexual niche segregation and gender-specific individual specialisation in a highly dimorphic marine mammal. PLoS One 10 (8). https://doi.org/10.1371/journal.pone.0133018

Kirkman SP et al (2016) Foraging behavior of subantarctic fur seals supports efficiency of a marine reserve's design. PLoS One 11(5):e0152370

Kirkman SP, Costa DP, Harrison A-L, Kotze PGH, Oosthuizen WH, Weise M, Botha JA, Arnould JPY (2019) Dive behaviour and foraging effort of female Cape fur seals *Arctocephalus pusillus pusillus*. R Soc Open Sci 6(10):191369

Kirkwood R et al (2006) At-sea movements and habitat use of adult male Australian fur seals (Arctocephalus pusillus doriferus). Can J Zool 84(12):1781–1788

Knox TC, Baylis AM, Arnould JP (2018) Foraging site fidelity in male Australian fur seals. Mar Biol 165:108

Kooyman GL (1965) Techniques used in measuring diving capacities of Weddell seals. Polar Rec 12:391–394

Kooyman GL (1989) Diverse divers: physiology and behavior. Springer, Berlin; New York

Kooyman GL, Wahrenbrock EA, Castellini MA, Davis RW, Sinnett EE (1980) Aerobic and anaerobic metabolism during voluntary diving in Weddell seals: evidence of preferred pathways from blood chemistry and behavior. J Comp Physiol 138:335–346

Kooyman GL, Castellini MA, Davis RW, Maue RA (1983) Aerobic diving limits of immature Weddell seals. J Comp Physiol 151:171–174

Kramer DL (1988) The behavioral ecology of air breathing by aquatic animals. Can J Zool 66:89–94

Kuhn CE (2011) The influence of subsurface thermal structure on the diving behavior of northern fur seals (Callorhinus ursinus) during the breeding season. Mar Biol 158(3):649–663

Kuhn CE, Costa DP (2006) Identifying and quantifying prey consumption using stomach temperature change in pinnipeds. J Exp Biol 209:4524–4532

Kuhn C, Costa D (2014) Interannual variation in the at-sea behavior of California Sea lions (*Zalophus californianus*). Mar Mamm Sci 30. https://doi.org/10.1111/mms.12110

Kuhn CE, Crocker DE, Tremblay Y, Costa DP (2009a) Time to eat: measurements of feeding behaviour in a large marine predator, the northern elephant seal *Mirounga angustirostris*. J Anim Ecol 78:513–523
Kuhn CE, Johnson DS, Ream RR, Gelatt TS (2009b) Advances in the tracking of marine species: using GPS locations to evaluate satellite track data and a continuous-time movement model. Mar Ecol Prog Ser 393:97–109
Kuhn CE, Tremblay Y, Ream RR, Gelatt TS (2010a) Coupling GPS tracking with dive behavior to examine the relationship between foraging strategy and fine-scale movements. Integr Comp Biol 50:E94
Kuhn CE, Tremblay Y, Ream RR, Gelatt TS (2010b) Coupling GPS tracking with dive behavior to examine the relationship between foraging strategy and fine-scale movements of northern fur seals. Endang Species Res 12:125–139
Ladds MA, Thompson AP, Slip DJ, Hocking DP, Harcourt RG (2016) Seeing it all: evaluating supervised machine learning methods for the classification of diverse otariid behaviours. PLoS One 11:e0166898
Ladds MA, Rosen DA, Slip DJ, Harcourt RG (2017a) Proxies of energy expenditure for marine mammals: an experimental test of "the time trap". Sci Rep 7:1–10
Ladds MA, Rosen DA, Slip DJ, Harcourt RG (2017b) The utility of accelerometers to predict stroke rate in captive fur seals and sea lions. Biol Open 6:1396–1400
Ladds MA, Slip DJ, Harcourt RG (2017c) Intrinsic and extrinsic influences on standard metabolic rates of three species of Australian otariid. Conserv Physiol 5:cow074
Lander ME et al (2020) Mixing it up in Alaska: habitat use of adult female Steller sea lions reveals a variety of foraging strategies. Ecosphere 11(2):e03021
Lea MA et al (2002) Variability in the diving activity of Antarctic fur seals, Arctocephalus gazella, at Iles Kerguelen. Polar Biol 25:269–279
Lea MA, Guinet C, Cherel Y, Duhamel G, Dubroca L, Pruvost P, Hindell M (2006) Impacts of climatic anomalies on provisioning strategies of a Southern Ocean predator. Mar Ecol Prog Ser 310:77–94
Lenfant C, Johansen K, Torrance JD (1970) Gas transport and oxygen storage capacity in some pinnipeds and the sea otter. Resp Physiol 9:277–286
Leung ES, Chilvers BL, Nakagawa S, Robertson BC (2014) Size and experience matter: diving behaviour of juvenile New Zealand Sea lions (*Phocarctos hookeri*). Polar Biol 37:15–26
Lowther AD, Goldsworthy SD (2011) Detecting alternate foraging ecotypes in Australian sea lion (Neophoca cinerea) colonies using stable isotope analysis. Mar Mamm Sci 27(3):567–586
Lowther AD et al (2011) Creatures of habit: foraging habitat fidelity of adult female Australian sea lions. Mar Ecol Prog Ser 443:249–263
Lowther AD, Harcourt RG, Goldsworthy SD, Stow A (2012) Population structure of adult female Australian sea lions is driven by fine-scale foraging site fidelity. Anim Behav 83:691–701
Lowther AD, Harcourt RG, Page B, Goldsworthy SD (2013) Steady as he goes: at-sea movement of adult male Australian sea lions in a dynamic marine environment. PLoS One 8:e74348
Lowther AD et al (2015) Identification of motivational state in adult male Atlantic walruses inferred from changes in movement and diving behavior. Mar Mamm Sci 31(4):1291–1313
Lunn NJ, Boyd IL, Croxall JP (1994) Reproductive performance of female Antarctic fur seals: the influence of age, breeding experience, environmental variation and individual quality. J Anim Ecol 63:827–840
Luque SP et al (2007) Foraging behaviour of sympatric Antarctic and subantarctic fur seals: does their contrasting duration of lactation make a difference? Mar Biol 152(1):213–224
MacArthur RH, Pianka ER (1966) On optimal use of a patchy environment. Am Nat 100:603–609
Mattlin RH, Gales NJ, Costa DP (1998) Seasonal dive behaviour of lactating New Zealand fur seals (*Arctocephalus forsteri*). Can J Zool 76:350–360
McDonald BI, Ponganis PJ (2013) Insights from venous oxygen profiles: oxygen utilization and management in diving California Sea lions. J Exp Biol 216:3332–3341

McDonald BI, Ponganis PJ (2014) Deep-diving sea lions exhibit extreme bradycardia in long-duration dives. J Exp Biol 217:1525–1534
McHuron EA, Robinson PW, Simmons SE, Kuhn CE, Fowler M, Costa DP (2016) Foraging strategies of a generalist marine predator inhabiting a dynamic environment. Oecol 182 (4):995–1005
McHuron EA, Peterson SH, Hückstädt LA, Melin SR, Harris JD, Costa DP (2018) The energetic consequences of behavioral variation in a marine carnivore. Ecol Evol 8:4340–4351
McIntyre T (2014) Trends in tagging of marine mammals: a review of marine mammal biologging studies. Afr J Mar Sci 36:409–422
Meir JU, Champagne CD, Costa DP, Williams CL, Ponganis PJ (2009) Extreme hypoxemic tolerance and blood oxygen depletion in diving elephant seals. Am J Phys Regul Integr Comp Phys 297:R927–R939
Melin S, DeLong R, Siniff D (2008) The effects of El Niño on the foraging behavior of lactating California Sea lions (*Zalophus californianus californianus*) during the nonbreeding season. Can J Zool 86:192–206
Merrick RL, Loughlin TR (1997) Foraging behavior of adult female and young-of-year Steller sea lions in Alaskan waters. Can J Zool 75(5):776–786
Merrick RL et al (1994) Use of satellite-linked telemetry to study Steller sea lion and northern fur seal foraging. Polar Res 13(1):105–114
Miller AK, Sydeman WJ (2004) Rockfish response to low-frequency ocean climate change as revealed by the diet of a marine bird over multiple time scales. Mar Ecol Prog Ser 281:207–216
Miller PJO, Biuw M, Watanabe YY, Thompson D, Fedak MA (2012) Sink fast and swim harder! Round-trip cost-of-transport for buoyant divers. J Exp Biol 215:3622–3630
Mori Y (1998) The optimal patch use in divers: optimal time budget and the number of dive cycles during bout. J Theor Biol 190:187–199
Mori Y, Boyd IL (2004) The behavioral basis for nonlinear functional responses and optimal foraging in Antarctic fur seals. Ecology 85:398–410
Mueller G (2004) The foraging ecology of south American Sea lions (*Otaria flavescens*) on the Patagonian shelf. PhD thesis, Christian-Albrechts-Universitaet zu Kiel, Germany
Muto MM, Helker VT, Angliss RP, Allen BA, Boveng PL, Breiwick JM, Cameron MF, Clapham PJ, Dahle SP, Dahlheim ME (2018) Alaska marine mammal stock assessments, 2018. In: Commerce USDo (ed) NOAA Technical Memorandum Book NMFS-AFSC-393
New LF, Clark JS, Costa DP, Fleishman E, Hindell MA, Klanjšček T, Lusseau D, Kraus S, McMahon CR, Robinson PW, Schick RS, Schwarz LK, Simmons SE, Thomas L, Tyack P, Harwood J (2014) Using short-term measures of behaviour to estimate long-term fitness of southern elephant seals. Mar Ecol Prog Ser 496:99–108
Nordstrom CA et al (2013) Foraging habitats of lactating northern fur seals are structured by thermocline depths and submesoscale fronts in the eastern Bering Sea. Deep Sea Res II Top Stud Oceanogr 88–89:78–96
Noren SR, Jay CV, Burns JM, Fischbach AS (2015) Rapid maturation of the muscle biochemistry that supports diving in Pacific walruses (*Odobenus rosmarus divergens*). J Exp Biol 218 (20):3319–3329
Olivier P (2015) Foraging ecology of lactating Steller Sea lions (*Eumetopias jubatus*) at Lovushki Island, Russia. PhD, Texas A & M University, Texas, 95pp
Osman LP (2007) Population status, distribution and foraging ecology of Arctocephalus philipii (Peters 1866) at Juan Fernandez archipelago. In: Faculdad de Ciencias. Universidad Austral de Chile, Valdivia, p 107
Paez-Rosas D, Villegas-Amtmann S, Costa D (2017) Intraspecific variation in feeding strategies of Galapagos Sea lions: a case of trophic specialization. PLoS One 12:e0185165
Page B, McKenzie J, Goldsworthy SD (2005) Inter-sexual differences in New Zealand fur seal diving behaviour. Mar Ecol Prog Ser 304:249–264
Ponganis PJ (2011) Diving mammals. Wiley, Hoboken, NJ

Ponganis PJ (2015) Diving physiology of marine mammals and seabirds. Cambridge University Press, Cambridge

Ponganis PJ, Kooyman GL, Winter lM, Starke LN (1997) Heart rate and plasma lactate responses during submerged swimming and trained diving in California Sea lions, *Zalophus californianus*. J Comp Physiol B 167:9–16

Qvist J, Weber RE, Zapol WM (1981) Oxygen equilibrium properties of blood and hemoglobin of fetal and adult Weddell seals. J Appl Physiol 50:999–1005

Richmond JP, Burns JM, Rea LD (2006) Ontogeny of total body oxygen stores and aerobic dive potential in Steller Sea lions (*Eumetopias jubatus*). J Comp Physiol B 176:535–545

Riet-Sapriza FG et al (2013) Foraging behavior of lactating South American sea lions (Otaria flavescens) and spatial–temporal resource overlap with the Uruguayan fisheries. Deep Sea Res II Top Stud Oceanogr 88-89:106–119

Robinson SA et al (2003) The foraging ecology of two sympatric fur seal species, Arctocephalus gazella and Arctocephalus tropicalis, at Macquarie Island during the austral summer. Mar Freshw Res 53(7):1071–1082

Sala JE, Quintana F, Wilson RP, Dignani J, Lewis MN, Campagna C (2011) Pitching a new angle on elephant seal dive patterns. Polar Biol 34:1197

Salton M, Kirkwood R, Slip D, Harcourt R (2019) Mechanisms for sex-based segregation in foraging behaviour by a polygynous marine carnivore. Mar Ecol Prog Ser 624:213–226

Schoener TW (1974) Resource partitioning in ecological communities. Science 185:27–39

Sepúlveda M et al (2015) Using satellite tracking and isotopic information to characterize the impact of South American Sea Lions on Salmonid aquaculture in Southern Chile. PLoS One 10 (8):e0134926

Shero MR et al (2012) Development of the aerobic dive limit and muscular efficiency in northern fur seals (Callorhinus ursinus). J Comp Physiol B 182(3):425–436

Simpkins MA, Kelly BP, Wartzok D (2001) Three-dimensional analysis of search behaviour by ringed seals. Anim Behav 62:67

Skinner JP, Burkanov VN, Andrews RD (2012) Influence of environment, morphology, and instrument size on lactating northern fur seal Callorhinus ursinus foraging behavior on the Lovushki Islands, Russia. Mar Ecol Prog Ser 471:293–308

Staniland IJ, Robinson SL (2008) Segregation between the sexes: Antarctic fur seals, *Arctocephalus gazella*, foraging at South Georgia. Anim Behav 75:1581–1590

Staniland I, Gales N, Warren N, Robinson S, Goldsworthy S, Casper R (2010) Geographical variation in the behaviour of a central place forager: Antarctic fur seals foraging in contrasting environments. Mar Biol 157:2383–2396

Thompson D, Fedak MA (2001) How long should a dive last? A simple model of foraging decisions by breath-hold divers in a patchy environment. Anim Behav 61:287–296

Thompson D, Hiby AR, Fedak MA (1993) How fast should I swim? Behavioural implications of diving physiology. Symp Zool Soc Lond 66:349–368

Thompson D, Duck CD, McConnell BJ, Garrett J (1998) Foraging behaviour and diet of lactating female southen sea lions (*Otaria flavescens*) in the Falkland Islands. J Zool Lond 246:135–146

Thompson D, Moss SEW, Lovell P (2003) Foraging behaviour of South American fur seals Arctocephalus australis: extracting fine scale foraging behaviour from satellite tracks. Mar Ecol Prog Ser 260:285–296

Trillmich F (1990) The behavioral ecology of maternal effort in fur seals and sea lions. Behaviour 114:3–20

Trillmich F, Kooyman GL (2001) Field metabolic rate of lactating female Galápagos fur seals (Arctocephalus galapagoensis): the influence of offspring age and environment. Comp Biochem Physiol A Mol Integr Physiol 129(4):741–749

Trillmich F et al (1986) Attendance and diving behavior of South American fur seals during El Nino in 1983. In: Gentry RL, Kooyman GL (eds) Fur seals: maternal strategies on land and at sea. Princeton University Press, Princeton, NJ, pp 153–167

Trillmich F et al (2014) The Galapagos sea lion: adaptation to spatial and temporal diversity of marine resources within the archipelago. In: The Galapagos marine reserve. Springer, New York, pp 61–70
Verrier D, Guinet C, Authier M, Tremblay Y, Schaffer S, Costa DP, Groscolas R, Arnould JPY (2011) The ontogeny of diving abilities in subantarctic fur seal pups: developmental trade-off in response to extreme fasting? Funct Ecol 25:818
Villegas-Amtmann S, Costa DP (2010) Oxygen stores plasticity linked to foraging behaviour and pregnancy in a diving predator, the Galapagos Sea lion. Funct Ecol 24:785–795
Villegas-Amtmann S, Costa DP, Tremblay Y, Salazar S, Aurioles-Gamboa D (2008) Multiple foraging strategies in a marine apex predator, the Galapagos Sea lion *Zalophus wollebaeki*. Mar Ecol Prog Ser 363:299–309
Villegas-Amtmann S, Simmons SE, Kuhn CE, Huckstadt LA, Costa DP (2011) Latitudinal range influences the seasonal variation in the foraging behavior of marine top predators. PLoS One 6 (8):e23166
Villegas-Amtmann S, Jeglinski JW, Costa DP, Robinson PW, Trillmich F (2013) Individual foraging strategies reveal niche overlap between endangered Galapagos pinnipeds. PLoS One 8:e70748
Villegas-Amtmann S, McDonald BI, Páez-Rosas D, Aurioles-Gamboa D, Costa DP (2017) Adapted to change: low energy requirements in a low and unpredictable productivity environment, the case of the Galapagos Sea lion. Deep Sea Res II Top Stud Oceanogr 140:94–104
Viviant M, Trites AW, Rosen DAS, Monestiez P, Guinet C (2010) Prey capture attempts can be detected in Steller Sea lions and other marine predators using accelerometers. Polar Biol 33:713–719
Viviant M, Monestiez P, Guinet C (2014) Can we predict foraging success in a marine predator from dive patterns only? Validation with prey capture attempt data. PLoS One 9:e88503
Viviant M, Jeanniard-du-Dot T, Monestiez P, Authier M, Guinet C (2016) Bottom time does not always predict prey encounter rate in Antarctic fur seals. Funct Ecol 30:1843–1844
Volpov BL, Hoskins AJ, Battaile B, Viviant M, Wheatley KE, Marshall GJ, Abernathy K, Arnould JPY (2015) Identification of prey captures in Australian fur seals (*Arctocephalus pusillus doriferus*) using head-mounted accelerometers: field validation with animal-borne video cameras. PLoS One 10:e0128789
Volpov BL, Rosen DA, Hoskins AJ, Lourie HJ, Dorville N, Baylis AM, Wheatley KE, Marshall G, Abernathy K, Semmens J (2016) Dive characteristics can predict foraging success in Australian fur seals (*Arctocephalus pusillus doriferus*) as validated by animal-borne video. Biol Open 5:262–271
Waite JN et al (2012) Resource partitioning by sympatric Steller sea lions and northern fur seals as revealed by biochemical dietary analyses and satellite telemetry. J Exp Mar Biol Ecol 416–417:41–54
Weise MJ, Costa DP (2007) Total body oxygen stores and physiological diving capacity of California Sea lions as a function of sex and age. J Exp Biol 210:278–289
Weise MJ, Harvey JT, Costa DP (2010) The role of body size in individual-based foraging strategies of a top marine predator. Ecology 91:1004–1015
Werner R, Campagna C (1995) Diving behaviour of lactating southern sea lions (*Otaria flavescens*) in Patagonia. Can J Zool 73:1975–1982
Wiig O, Gjertz I, Griffiths D, Lydersen C (1993) Diving patterns of an Atlantic walrus *Odobenus Rosmarus Rosmarus* near Svalbard. Polar Biol 13:71–72
Winship AJ, Trites AW, Calkins DG (2001) Growth in body size of the Steller sea lion (Eumetopias jubatus). J Mammal 82:500–519
Würsig B, Thewissen J, Kovacs KM (2017) Encyclopedia of marine mammals. Academic Press, New York

Chapter 5
Prey Capture and Processing in Fur Seals, Sea Lions and the Walrus

David P. Hocking, Travis Park, James P. Rule, and Felix G. Marx

Abstract Otarioids (fur seals, sea lions and the walrus, *Odobenus rosmarus*) are an ancient group of marine mammals that have adapted to feeding in water in a variety of ways. Fur seals and sea lions (otariids) primarily feed on fish and cephalopods, but opportunistically target a wide range of prey types and sizes, including sharks, penguins, and even their own kind. Like their terrestrial carnivoran relatives, otariids primarily rely on their teeth to catch and process their food. Suction—the ability to lower the pressure inside the oral cavity to draw in water and prey—also plays an important role, however, especially when ingested items are small. Osteological adaptations for suction are seemingly absent, but the behavior is nonetheless facilitated by the shape of the soft tissues surrounding the mouth. Walruses are suction specialists, as reflected in their robust skull, muscular lips and strong throat muscles. They primarily feed on benthic bivalves, gastropods and annelids, but sometimes also target larger prey, including birds and other pinnipeds. Their foraging activities affect vast areas of the (sub)Arctic seafloor, affecting the structure of benthic communities and leading to major increases in nutrient flux. These large-scale effects, plus a voracious appetite, make walruses a major ecosystem engineer.

Keywords Feeding · Foraging behavior · Marine mammal · Otariidae · Odobenidae · Pinnipedia

---

D. P. Hocking (✉)
School of Biological Sciences, Monash University, Clayton, VIC, Australia
e-mail: david@dphocking.com

T. Park
Department of Earth Sciences, University of Oxford, Oxford, UK

J. P. Rule
School of Biological Sciences, Monash University, Clayton, VIC, Australia

Department of Anatomy and Developmental Biology, Monash University, Clayton, VIC, Australia

F. G. Marx
Museum of New Zealand Te Papa Tongarewa, Wellington, New Zealand

C. Campagna, R. Harcourt (eds.), *Ethology and Behavioral Ecology of Otariids and the Odobenid*, Ethology and Behavioral Ecology of Marine Mammals,
https://doi.org/10.1007/978-3-030-59184-7_5

5.1 Introduction

5.1.1 *Challenges of Feeding in Water*

All animals must consume enough food to support both their growth and reproduction. For predators, this means having the anatomical tools and behavioral repertoire required to find, capture and consume other animals. Often, such adaptations are variations on a theme, suited to broad environmental contexts that remain stable for millions of years. Occasionally, however, profound transitions between different environmental realms (e.g. from land to water) require a major evolutionary 'rethink'. The emergence of pinnipeds is one such example.

Like all marine mammals, pinnipeds had to master the challenge of adapting to life in water after more than 250 million years of evolution on land. Pinnipeds are carnivorans, and thus share a common origin with cats, dogs, wolves, and bears (Arnason et al. 2006). Most of these predators live in a largely two-dimensional world, with the ground beneath their feet guiding both their movements and that of their quarry. In this setting, sharp claws and teeth (including a set of specialised slicing teeth known as 'carnassials') have emerged as the primary means to catch, kill and dismember prey.

Pinnipeds have inherited much of the basic anatomy of their terrestrial cousins, yet face a set of fundamentally different challenges. Their world is three-dimensional, without the limitations and support of *terra firma*. With prey potentially escaping in any direction, considerable speed and agility are essential for a successful hunt. Even once dead, prey suspended in water can still float away, complicating its manipulation and processing (Taylor 1987).

Perhaps the most fundamental challenges, however, are posed by the water itself. Like all mammals, pinnipeds must hold their breath when diving, thus limiting the amount of time available for hunting. As they swim, the body creates a bow wave that inadvertently pushes prey away from the mouth, making it yet more difficult to catch (Motta and Wilga 2001; Werth 2000). When the prey *is* finally caught, water enters the mouth along with it, and must be expelled prior to swallowing. Finally, prey processing is also restricted by the need to breathe, and furthermore affected by the greater resistance of water relative to air.

5.1.2 *Origin and Structure of Otarioid Feeding Anatomy*

The oldest uncontroversial pinniped, occurring as early as 30 million years ago, is *Enaliarctos* (Berta 1991; Berta et al. 2018; Paterson et al. 2020). Despite its considerable age, *Enaliarctos* already showed clear adaptations for swimming, including broad, paddle-like hind limbs likely used to generate underwater propulsion (Bebej 2009; Berta et al. 1989, 2018). Its feeding apparatus, on the other hand, was rather archaic. Like terrestrial carnivorans, its jaws bore carnassial teeth capable

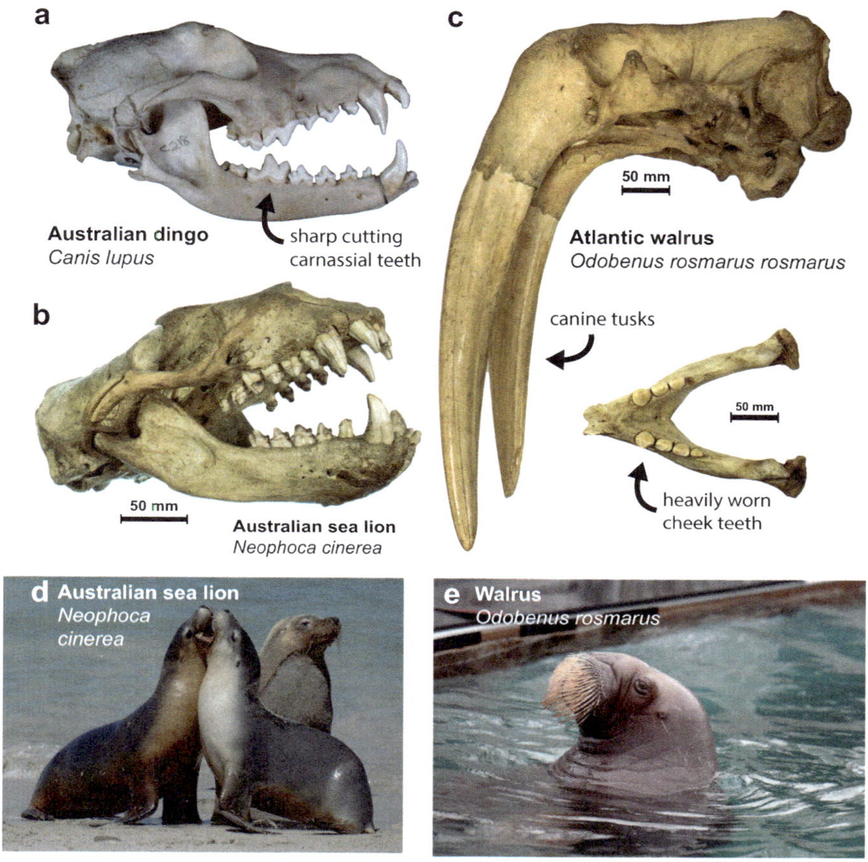

Fig. 5.1 Overview of otarioid feeding anatomy. (**a**) Skull of a typical terrestrial carnivoran, the Australian dingo (*Canis lupus*, NMV C218); note the well-developed carnassial teeth. (**b**) Skull of an Australian sea lion (*Neophoca cinerea*, NMV C1996), characterised by equally-sized postcanine teeth and no carnassials. (**c**) Skull of the walrus (*Odobenus rosmarus rosmarus*, NHMUK 1926.12.2.6) with simplified, peg-like teeth and enlarged vertical tusks. (**d**, **e**) Sea lions and the walrus in the flesh. Photos of live animals by Robert Harcourt

of cutting and slicing (Mitchell and Tedford 1973). Likewise, the forelimb retained strong claws and flexible digits, which would have enabled it to grasp prey during processing even while floating in water. A similar behavior occurs in living phocine seals, which can hold large food items between their clawed paws while ripping off smaller pieces for swallowing (Hocking et al. 2018).

From these early beginnings arose true seals (Phocidae) and otarioids: fur seals and sea lions (Otariidae) and walruses (Odobenidae) (Fig. 5.1). Unlike *Enaliarctos*, otariids propel themselves with their forelimbs, which makes them exceptionally agile and fast. Their postcanine teeth are simplified, though sharp, with no carnassials and no differences between the molars and premolars (Hocking et al. 2017b;

Repenning 1976). Their paws have turned into streamlined flippers, with virtually immobile fingers and—crucially—no claws (English 1976; Kuhn and Frey 2012).

Otariids first appear around 17–15 Ma, in the form of small-bodied species (*Eotaria* spp.) with an archaic but already simplified dentition incapable of effectively slicing flesh (Boessenecker and Churchill 2015; Vélez-Juarbe 2017). Nothing is known about the postcranial skeleton of these early forms, but the rocks in which they are found suggest that they mostly lived offshore, perhaps feeding on evasive shoaling prey, as do living fur seals (Arnould and Costa 2006; Kohno 2004). Forelimb propulsion and homodonty, both implying essentially modern foraging behavior, became established by the late Miocene (Deméré and Berta 2005; Repenning and Tedford 1977).

The story is more complex for odobenids, whose sole living survivor, the present-day walrus (*Odobenus rosmarus*), neither reflects the former diversity or ecology of the clade. Modern walruses are large-bodied, restricted to the Arctic region, and stand out for their distinctive anatomy, including elongated tusks, well-developed lips and vibrissae, peg-like teeth, and a strongly vaulted palate (Kastelein and Gerrits 1990; Kastelein et al. 1991). They are specialist benthic feeders, and use suction to uncover and extract molluscs buried in the sea floor (Levermann et al. 2003).

By contrast, extinct odobenids (first appearing around 17–16 Ma) were far more diverse, more widely distributed, variably sized, and often lacked tusks (Kohno et al. 1995a; Magallanes et al. 2018). Well-developed cuspate teeth may suggest a fish-based diet for several of these early species (Boessenecker and Churchill 2013). Nevertheless, like otariids, odobenids lost their carnassials early during their evolution, and thus also the ability to cut large prey into swallowable pieces (Deméré and Berta 2001). Long-tusked walruses with vaulted palates and simplified or reduced teeth first appear around the Miocene-Pliocene boundary, and likely mark the emergence of suction feeding and a mollusc-based diet (Berta et al. 2018; Kohno et al. 1995b; Biewer et al. 2020).

5.2 The Feeding Cycle

Feeding in mammals, as in all tetrapods, involves prey capture, processing, and swallowing: the main stages of the feeding cycle (Hiiemae and Crompton 1985; Schwenk 2000; Schwenk and Rubega 2005). Aquatic species like otarioids, however, face an additional challenge: as they catch prey in their jaws, they are also forced to ingest the (usually salty) water surrounding it. Simply swallowing the latter would put a burden on their excretory system, and likely affect both speed and feeding performance. Instead, water is thus normally expelled from the mouth prior to deglutition.

Overall, the feeding cycle of aquatic mammals thus comprises four stages (Fig. 5.2). (I), prey is captured via movements of the jaws or the entire body. This stage is further subdivided into an approach phase, during which the prey is chased or otherwise brought close to the mouth; and a prehension phase, during the which

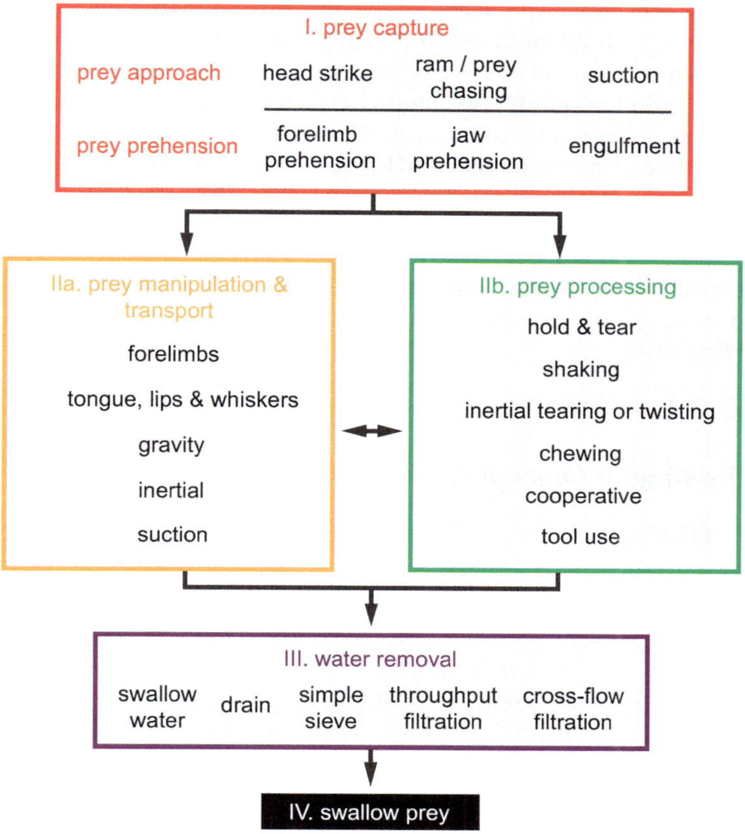

Fig. 5.2 Overview of the aquatic mammal feeding cycle. Each stage (I–IV) may involve a range of different foraging behaviors, which together combine into a feeding strategy. Adapted from Hocking et al. (2017c, d)

the prey is seized. (II), prey is manipulated, transported, and—if necessary—processed into swallowable pieces. Prey transport and processing often occur in an alternating fashion, and mostly inside the oral cavity. (III), water ingested with the food is expelled from the oral cavity. Finally, (IV), the prey is swallowed (Hocking et al. 2017c, d).

Except for swallowing, all of these stages may involve a variety of individual foraging behaviors. For example, prey can be captured with the teeth, or else be sucked directly into the oral cavity. Food can be processed via chewing, but also be violently thrown about to cause it to tear. Further, several behaviors can be combined as part of a single feeding event. For example, small fish may first be drawn close to the mouth via suction, but then be captured with a bite.

Together, all of the behaviors involved in the capture and ultimate consumption of a prey item cluster into a *feeding strategy* (Hocking et al. 2017c). Otarioids primarily employ just two strategies, namely, raptorial feeding, which involves capturing prey with the teeth and consuming it underwater; and suction feeding, during which prey is sucked directly into the mouth and then swallowed whole. Sometimes, however, prey is also captured on land (terrestrial feeding) or consumed while floating at the surface (semi-aquatic feeding).

Definitions of what constitutes a strategy are somewhat arbitrary, and not always consistent (e.g. Adam and Berta 2002; Hocking et al. 2017c, d; Kienle et al. 2017). Nevertheless, they provide a useful shorthand to describe a broad continuum of feeding modes suited to different environments and prey types. Importantly, a species need not be restricted to a single strategy only. Instead, many otarioids can switch between two or more strategies, depending on the foraging scenario (Hocking et al. 2014, 2016).

5.3 Feeding in Otariids

5.3.1 Diet

Otariids target a variety of prey types and sizes, depending on the habitat and species. Most important are bony fishes, cephalopods and, in some areas, crustaceans (Arnould and Costa 2006; Bonner 1968; Deagle et al. 2009; Emami-Khoyi et al. 2016; Peters et al. 2015). In addition, several species have been observed to attack birds, rays, sharks, and even other pinnipeds (Bradshaw et al. 1998; Fallows et al. 2015; Harcourt 1993; Kernaléguen et al. 2016; Mathews and Adkison 2010; Visser et al. 2008; Womble and Conlon 2010).

The full range of prey species taken by otariids has long been difficult to establish, as studies had to rely on scats, regurgitates and gut contents (Aurioles-Gamboa and Camacho-Ríos 2007; Dellinger and Trillmich 1999; George-Nascimento et al. 1985; Harcourt et al. 2002; McIntosh et al. 2006; Reisinger et al. 2018). The latter provide a reasonable overview of prey with hard parts (e.g. fish otoliths and squid beaks), but often fail to record cartilaginous (e.g. rays and sharks) and soft-bodied species. More recent techniques have employed analyses of faecal DNA to gain a more complete picture of otariid feeding habits (Deagle et al. 2009; Emami-Khoyi et al. 2016; Peters et al. 2015).

Their generalist diet affords otariids a fair amount of ecological flexibility, but is not without risk. Some of their prey are well-defended, and can seriously wound or even kill their would-be predators. One striking example is provided by a subadult New Zealand fur seal (*Arctocephalus forsteri*), which washed up in south-eastern Australia with numerous chimaera fin spines and stingaree barbs embedded in its face (Fig. 5.3) (Hocking et al. 2020).

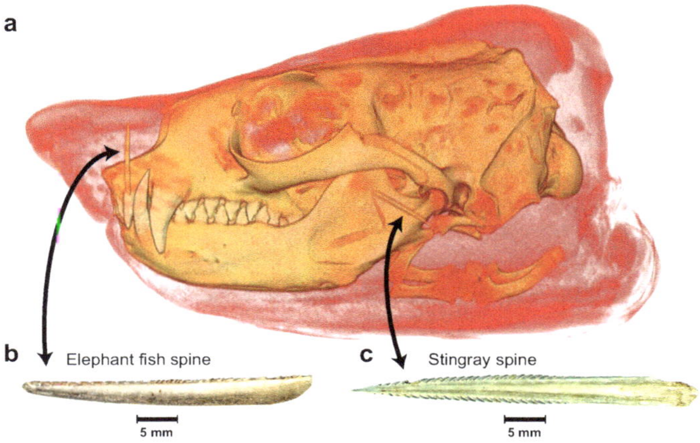

Fig. 5.3 Dangers of targeting well-defended prey. (**a**) Head of a wild New Zealand fur seal (*Arctocephalus forsteri*, NMV C38315) with numerous injuries caused by elephant fish (**b**) and stingray (**c**) spines. Adapted from Hocking et al. (2020)

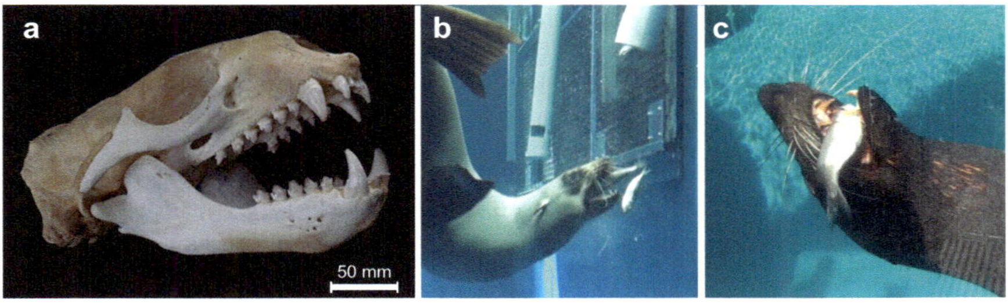

Fig. 5.4 Example of raptorial biting in otariids. (**a**) Skull and teeth of an Australian fur seal (*Arctocephalus pusillus doriferus*, NMV C5717). (**b**) Captive Australian fur seal approaching and biting a fish. (**c**) Captive New Zealand fur seal (*Arctocephalus forsteri*) capturing a fish by biting

5.3.2 *Semi-Aquatic and Raptorial Feeding*

Raptorial and semi-aquatic feeding both involve the capture of prey underwater using teeth and/or claws (Hocking et al. 2017c). With claws absent in living otariids, they must rely on their teeth alone to pierce and trap prey (Fig. 5.4). This style of feeding is what most people associate with fur seals and sea lions, and indeed it seems to be their preferred strategy. When captive Australian (*Arctocephalus pusillus doriferus*) and subantarctic (*A. tropicalis*) fur seals were fed 144 whole prey items as part of a behavioral experiment, they caught 134 by biting them with their teeth (Hocking et al. 2016).

For raptorial feeding to work, the jaws of the predator must first overtake the prey. Otariids achieve this either by actively pursuing the prey with their entire body, or by striking at it with the head only. The latter is made possible by the highly flexible

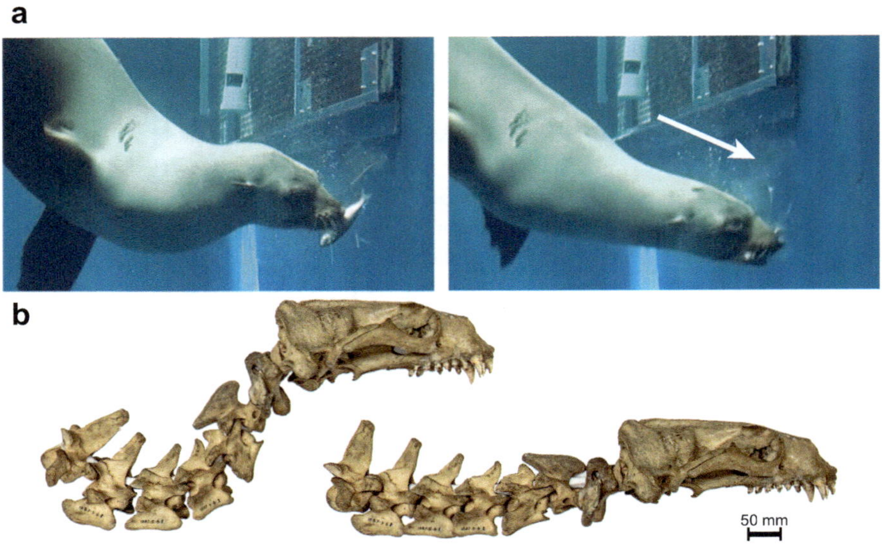

Fig. 5.5 Australian fur seal (*Arctocephalus pusillus doriferus*) performing a head strike. (**a**) Captive individual striking at fish floating in the water. (**b**) Same movement replicated using the skull and neck vertebrae of a museum specimen (NHMUK 1887.5.6.1)

neck, which is held curved during swimming. As the seal approaches its prey, the neck is rapidly straightened, thereby accelerating the head forward (Fig. 5.5). Head strikes are common, and detection of this behavior via animal-mounted accelerometers may be a useful indicator of prey capture in wild individuals (Skinner et al. 2009; Viviant et al. 2010; Volpov et al. 2015). Beyond increasing the forward reach of the feeding apparatus, a flexible neck also increases the ability of the head to move from side to side to capture highly evasive prey.

Movements of the head and body create a compressive bow wave that may alert prey and/or push it away from the mouth. To compensate, many otariids use suction alongside biting to draw prey within range of the teeth (Hocking et al. 2014, 2016). Suction is generated via retraction of the tongue, which expands the oral cavity and draws in water through the mouth just like a piston would in a syringe. Pursing of the lips further enhances this process by creating a small, circular mouth opening which helps to direct and strengthen the suction forces (Werth 2006).

Compared to most other carnivorans, otariids target relatively small prey—around 16 cm long on average (Etnier and Fowler 2010)—that can be swallowed whole. Long but narrow prey can also be swallowed this way, albeit in more than one go. When the body depth of the prey exceeds a certain limit, however, it must be processed into swallowable chunks (Hocking et al. 2016). Terrestrial carnivorans achieve this by slicing or tearing with their sharp teeth, claws, and mobile paws. Similar behaviors are also employed by certain true seals and the Australian sea lion *Neophoca cinerea*, which can hold prey between their paws while tearing at it with

Fig. 5.6 Otariids processing large prey items. (**a–c**) Captive Australian sea lion (*Neophoca cinerea*) tearing a fish between its jaws and flippers. (**d–g**) Wild Australian fur seal (*Arctocephalus pusillus doriferus*) thrashing about and tearing an angelshark. Photos of wild fur seal by Robert Harcourt

their teeth (Fig. 5.6) (Hocking et al. 2017a, 2018). Overall, however, the stiff, clawless flippers of otariids make this approach rather ineffective. Instead, large prey is often grasped between the teeth and then violently thrashed about via movements of the head and neck, until it eventually rips apart (Hocking et al. 2016, 2017a; Visser et al. 2008; Womble and Conlon 2010). Prior to shaking, the prey is commonly chewed to create weak points that facilitate tearing and/or make it easier to swallow (Hocking et al. 2017a).

Prey processing in general can occur either underwater or in air at the surface. Processing at the surface is the primary behaviour distinguishing raptorial from semi-aquatic feeding, and brings with it several advantages: it allows the animal to keep breathing during lengthy processing bouts, facilitates the shaking and tearing of prey thanks to the lower resistance of air, and allows excess water to drain away from the prey (Hocking et al. 2017c). There is also a danger, however, that the associated splashing will attract predators or competitors, potentially facilitating kleptoparasitism (Hocking et al. 2016).

Once prey or suitable sized prey chunks are finally captured in the jaws, they may need to be reoriented or transported backwards to aid swallowing. Such transport can occur via direct manipulation with the tongue and jaws, and/or be aided again by suction (Werth 2000). As the prey moves towards the throat, the tongue is pushed against the palate to force excess water out of the mouth. This water expulsion is often visible as a cloud of small bubbles squirting from both sides of the mouth (Hocking et al. 2014; Marshall et al. 2015), while the prey itself remains trapped by the tongue and the 'cage' created by the jaws and teeth (Norris and Møhl 1983).

5.3.3 Suction Feeding

In addition to biting, otariids can capture and consume prey purely via suction. The process starts with the pursing of the lips to create a round, pipette-like mouth opening framed by the canines and incisors (Fig. 5.7). Piston-like movements of the tongue and hyoid then draw water into the mouth, carrying along any prey swimming or floating within it. Once inside the oral cavity, the prey is held with the tongue or trapped behind the teeth and jaws, while the surrounding water is expelled. Finally, the prey is swallowed whole, without any processing.

Unlike biting or grasping, suction relies on the inherent properties of water (i.e. its greater density relative to air), making it a truly 'aquatic' feeding strategy (Hocking et al. 2017d). Instead of being actively overtaken, prey is transported towards the predator, which can be located some distance away (e.g. in the water column, or above the ocean floor). Nevertheless, suction is normally preceded by chasing and/or other prey approach behaviors, such as head strikes (Hocking et al. 2016).

Suction feeding independently evolved in a variety of marine mammals, including whales, dolphins and phocid seals (Hocking et al. 2017d; Johnston and Berta 2011). In the most specialized species, it correlates with a variety of osteological adaptations, such as a blunt snout to improve water flow, a vaulted roof of the mouth to increase the size of the oral cavity, robust pterygoid hamuli to reinforce the soft palate, or an enlarged hyoid apparatus for the attachment of well-developed gular muscles (Adam and Berta 2002; Bloodworth and Marshall 2007; Kastelein and Gerrits 1990; Reidenberg and Laitman 1994; Werth 2006; Werth 2007).

Except in the southern sea lion *Otaria byronia*, suction in otariids has not led to obvious adaptations of the skeleton (Adam and Berta 2002). Instead, it is mostly facilitated by soft tissue structures, such as an expandable oral cavity and flesh lobes beside the tooth rows helping to close off the lateral gape (Fig. 5.7). Nevertheless, suction is widespread, and appears to be used by all species tested so far except the northern fur seal *Callorhinus ursinus* (Hocking et al. 2014, 2016; Marshall et al. 2015).

When suction feeding is used depends on the foraging scenario. Australian and subantarctic fur seals primarily employ it to capture small prey that can be swallowed

Fig. 5.7 Anatomical adaptations for suction feeding in otariids. (**a**) The small rounded mouth opening is framed by the incisors and canines in the skull of an Australian fur seal (*Arctocephalus pusillus doriferus*, NMVC5717). (**b**, **c**) During feeding, the sides of the mouth are occluded by the well-developed lips and fleshy lobes running along the outside of the toothrows, as shown here in (**d**, **e**) an Australian sea lion (*Neophoca cinerea*), (**f**, **g**), a New Zealand fur seal (*Arctocephalus forsteri*) and (**h**, **i**) a California sea lion (*Zalophus californianus*)

whole (Hocking et al. 2016). This pattern makes sense, as such prey may be too small to be bitten effectively. It also reflects a constraint, however, in that only prey that fits into the oral cavity can realistically be captured through suction alone. Larger items need to be held between the jaws and teeth, and often require processing before they can be swallowed.

Suction also comes into play when prey is hidden or difficult to reach. In a series of behavioral experiments, Australian fur seals and Steller sea lions (*Eumetopias jubatus*) were able to suck small fish out of elongate tubes, giving them access to food sources that could not have been captured by biting (Fig. 5.8) (Hocking et al. 2014; Marshall et al. 2015). In the wild, this ability may allow otariids to extract prey from small spaces created by rocks or man-made structures, such as aquaculture facilities and various types of artificial reef (Arnould et al. 2015; Kemper et al.

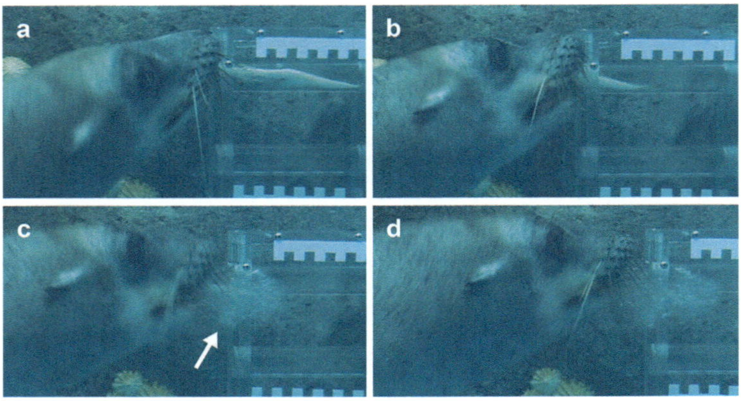

Fig. 5.8 Otariid suction feeding. Captive California sea lion (*Zalophus californianus*) sucking small fish out of a tube. Arrow points to the cloud of water being expelled from the mouth following suction

2003). In theory, suction might also allow them to capture prey buried in the seafloor. This scenario is unlikely to occur in practice, however, given that otariids generally lack the extreme abrasive tooth wear typical of benthic suction feeders.

Finally, suction allows the capture of multiple, closely-spaced food items in quick succession—up to three per second in Australian and subantarctic fur seals (Hocking et al. 2016). Wild Antarctic fur seals (*Arctocephalus gazella*) also repeatedly open their mouth during the bottom phase of a dive, suggesting the capture of multiple prey items (Iwata et al. 2012). This behavior may correlate with their dietary preference for krill (Bonner 1968), and has sometimes been considered a type of filter feeding (Riedman 1990). However, Antarctic fur seals only feed on krill throughout the South Atlantic part of their range, with fish dominating their diet in the Indian Ocean (Green et al. 1989; Reisinger et al. 2018; Robinson et al. 2003). They also lack the specialized tooth morphology shown by the filter-feeding leopard (*Hydrurga leptonyx*) and crabeater (*Lobodon carcinophaga*) seals (Hocking et al. 2013; King 1961), and instead have postcanines that are even more simplified than in other otariids. In California sea lions (*Zalophus californianus*) unaccustomed to extremely small prey, similarly unspecialized postcanines perform poorly in retaining small food items inside the oral cavity (Hocking et al. 2013). This may suggest that otariids primarily rely on behavioral (e.g. movements of the tongue), rather than anatomical adaptations, when retaining small prey inside the mouth during feeding.

5.4 Feeding in Walruses

5.4.1 Diet

The walrus has traditionally been thought of as a bivalve specialist, but its' diet is in fact surprisingly varied and, besides clams, also heavily draws on snails, polychaetes and, to a lesser extent, crustaceans, sea cucumbers, squid and jellyfish (Fay 1982; Sheffield and Grebmeier 2009). Additionally, some individuals have been observed to feed on larger vertebrate prey, including birds and other marine mammals (Fay 1982; Gjertz 2009; Lowry and Fay 1984; Mallory et al. 2004). Opportunistic consumption of large prey is also supported by stable isotope data (Seymour et al. 2014a, b), but it is possible that at least some of the reported attacks instead reflect social play behavior (Giljov et al. 2017).

5.4.2 Suction Feeding

Unlike most otariids, the walrus is a specialized suction feeder with a short rostrum, notably arched palate, and robust pterygoid hamuli (Adam and Berta 2002; Kastelein and Gerrits 1990). Except for the elongate tusks, its teeth are reduced to simple pegs, and grind, rather than shear, against each other when the jaw is closed (Cobb 1933). The snout is muscular and covered in a dense mat of highly mobile, tactile vibrissae, which appear to function during food identification, excavation, manipulation, and processing (Fay 1982; Kastelein et al. 1991).

Walruses primarily use suction to capture bivalves buried in the seafloor (Vibe 1950). When feeding, they position themselves face down at an angle of 10–90 degrees, while bracing their flippers against the seafloor (Fay 1982; Kastelein and Mosterd 1989; Levermann et al. 2003). Once in position, they slowly swim forwards, dragging their elongate tusks along the bottom like a sledge (Fig. 5.9) (Levermann et al. 2003). As they move, they scan for prey using their eyes and tactile vibrissae. This can create problems, as sediment stirred up by their movements tends to reduce visibility. In response, wild walruses have been observed to create water currents using one of their flippers, which helps to clear the area in front of the head (Levermann et al. 2003).

To excavate the bivalves, walruses either root around the bottom with their muzzle, or use their flipper to fan away the surrounding sediment. Like humans, and indeed some other marine mammals (Karenina et al. 2016; MacNeilage 2014), they appear to be right-biased, with the right flipper being both somewhat larger and used preferentially during feeding (Levermann et al. 2003). In addition, walruses can use their strong gular musculature to create a targeted jet of water, which may be strong enough to expose buried prey. Indeed, hydraulic jetting was the main excavation technique observed in an early captive study (Kastelein and Mosterd

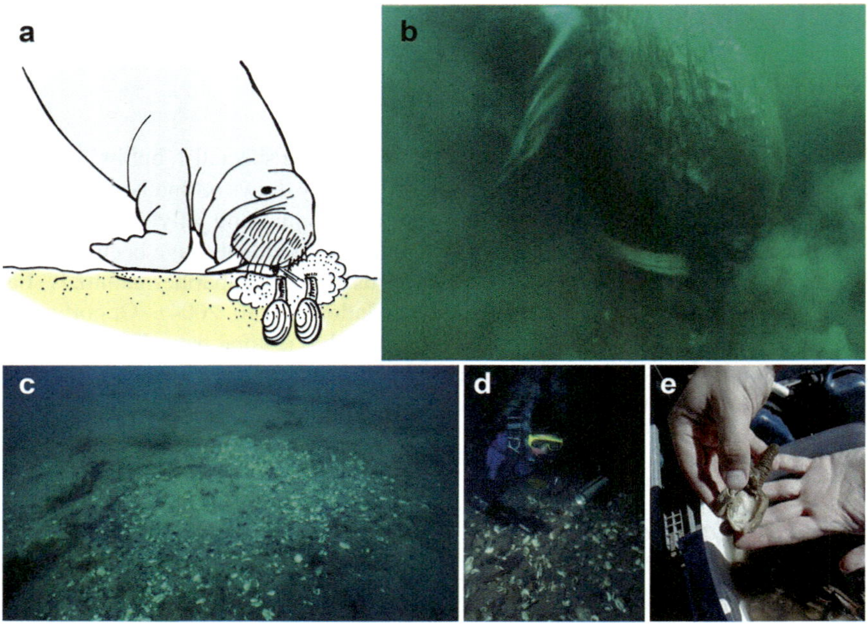

Fig. 5.9 Suction feeding in walruses (*Odobenus rosmarus*). (**a**) Schematic drawing of a walrus excavating bivalves buried in the seafloor via a jet of water from its mouth. (**b**) Photo a wild walrus feeding on the seafloor. (**c–e**) Walrus feeding patch littered with empty shells. (**a**) modified from Kastelein and Mosterd (1989); (**b–e**) copied from Levermann et al. (2003) under a Creative Commons Attribution Licence 4.0

1989), but appears to be less common in the wild—possibly because of differences in sediment coarseness and/or depth (Levermann et al. 2003).

Once excavated, the bivalves are manipulated with the stiff vibrissae and positioned between the lips. The latter are pursed into a small circular opening and tightly grip the shell, while piston-like movements of the tongue decrease the pressure inside the oral cavity by as much as -1.188 Bar (108 kPa). This pressure difference results in a strong suction force that draws out the soft tissue of the prey, leaving behind the shells (Fay 1982; Kastelein et al. 1991, 1994; Oliver et al. 1983).

Walrus suction feeding is highly efficient, with single individuals ingesting up 6–9 clams per minute, and up to 36 kg and thousands of individual prey items per day (Kastelein et al. 1994; Oliver et al. 1983; Ray et al. 2006). This voracious appetite makes walruses a major consumer, accounting, for example, for roughly 3% of the total yearly biomass turnover in the Bering Sea (Ray et al. 2006). The resulting disturbance of vast areas of seafloor (estimated at 3000–5000 km^2 in Beringia alone) furthermore affects benthic community structure, and increases nutrient flux from the sediment to water column by as much as two order of magnitude (Oliver et al. 1983, 1985, Ray et al. 2006). Together these large-scale effects make walruses a major ecosystem engineer.

5.4.3 Feeding on Large Prey

Walruses occasionally target larger prey in a variety of ways and settings. Birds often appear to be approached underwater, following which they are rapidly grasped with the teeth, stabbed with the tusks, or perhaps even sucked below the surface (Fox et al. 2010; Giljov et al. 2017; Gjertz 2009; Mallory et al. 2004). The carcass is then dismembered either underwater, or with the help of the flippers while the walrus floats at the surface (Fox et al. 2010; Mallory et al. 2004).

Reports of direct attacks on other seals are rarer, although Fay (1960) relayed traditional Inuit lore describing a rare case of 'rogue' bulls feeding primarily on vertebrates. They were described as:

> relatively lean and slender, with shoulders and forelimbs appearing unusually large and powerfully developed; the chin, neck, and breast are impregnated with oil from frequent contact with seal blubber, and the oxidized oil imparts an amber colour to these regions and to the tusks (cf. Brooks 1954, p. 57). The tusks are exceptionally long, slender, and sharp-pointed, and their labial surfaces are covered with scratches. (Fay 1960: p. 112)

Despite relatively frequent finds of seal remains in walrus stomachs, active predation remained unconfirmed until the publication of accounts describing fresh prey remains indicative of recent kills (Fay et al. 1990; Lowry and Fay 1984). These reports were coincident with an apparent rise in the incidence of 'carnivorous' behavior, perhaps in response to an increasing walrus population or a trend towards a more generalist diet (Lowry and Fay 1984; Seymour et al. 2014a). Both direct observations and stable isotope analyses have since shown that seal-eating is not confined to bulls, and appears to be equally common across both sexes (Lowry and Fay 1984; Seymour et al. 2014b).

From what little is known, at least some attacks by walruses seem to target seals resting on ice floes. Rising from the water beside the floe, the walrus seems to use its tusks to stab or strike its victim. Once dead, the prey is held down by the flipper, opened with the tusks, eviscerated via suction, and finally torn into swallowable chunks (Fig. 5.10) (Lowry and Fay 1984).

5.5 Conclusions

Over the course of their evolutionary history, otarioids have adapted to feeding in water in a variety of ways. Despite resembling their terrestrial ancestors far more than other marine mammals like cetaceans, fur seals and sea lions are agile marine predators targeting everything from fishes, to molluscs, crustaceans, birds and even other mammals. They are adept at both biting and suction, employing either or both behaviors depending on the foraging scenario. Walruses are more specialized in their anatomy, yet also capable of feeding on a surprisingly broad range of prey. Their large size, voracious appetite and invasive benthic foraging strategy all conspire to make them an engineer and keystone species of the Arctic marine ecosystem.

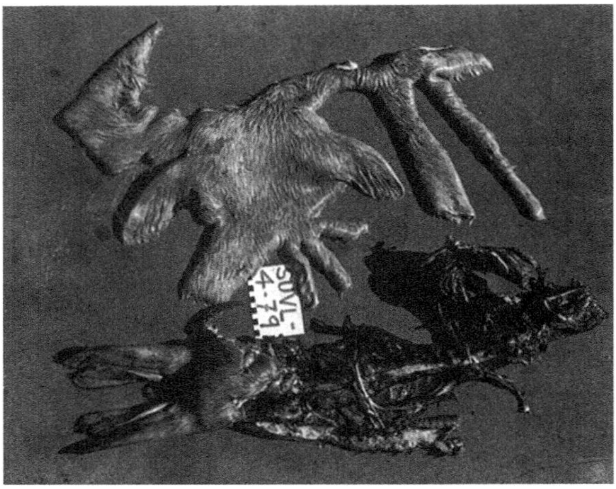

Fig. 5.10 Remains of a spotted seal pup consumed by a walrus (*Odobenus rosmarus*). Copied with permission from Lowry and Fay (1984)

Overall, otarioids provide a remarkable example of adaptation, having successfully met the challenges posed by the marine realm. From humble terrestrial beginnings, they evolved into some of the most versatile predators of the modern ocean, with the tools and skills to exploit its remarkable range of ecological opportunities.

Acknowledgements Thanks to Roberto Portela Miguez and Richard Sabin for providing access to specimens at the Natural History Museum UK (NHMUK); Kevin Rowe, Karen Roberts and Ricky-Lee Erickson for providing access to specimens at Museums Victoria (NMVC); Robert Harcourt for providing the live animal photographs in Fig. 5.1; Tony Mitchell and the Victorian Department of Environment, Land, Water and Planning for collection of the fur seal specimen in Fig. 5.3; Alexander McDonald, Justin Adams and Michael de Veer for CT Scanning the specimen, Michael de Veer is a Fellow of the National Imaging Facility, a National Collaborative Research Infrastructure Strategy (NCRIS) capability at Monash Biomedical Imaging (MBI), which is a Technology Research Platform at Monash University; Justin Bell and Martin Gomon for identifying the fish spines embedded within it.

References

Adam PJ, Berta A (2002) Evolution of prey capture strategies and diet in the Pinnipedimorpha (Mammalia, Carnivora). Oryctos 4:83–107

Arnason U, Gullberg A, Janke A, Kullberg M, Lehman N, Petrov EA, Väinölä R (2006) Pinniped phylogeny and a new hypothesis for their origin and dispersal. Mol Phylogenet Evol 41:345–354

Arnould JPY, Costa DP (2006) Sea lions in drag, fur seals incognito: insights from the otariid deviants. In: Sea lions of the world: conservation and research in the 21st century. University of Alaska Fairbanks, Fairbanks, pp 309–323

Arnould JPY, Monk J, Ierodiaconou D, Hindell MA, Semmens J, Hoskins AJ, Costa DP, Abernathy K, Marshall GJ (2015) Use of anthropogenic sea floor structures by Australian fur seals: potential positive ecological impacts of marine industrial development? PLoS One 10: e0130581

Aurioles-Gamboa D, Camacho-Ríos FJ (2007) Diet and feeding overlap of two otariids, *Zalophus californianus* and *Arctocephalus townsendi*: implications to survive environmental uncertainty. Aquat Mamm 33:315–326

Bebej RM (2009) Swimming mode inferred from skeletal proportions in the fossil pinnipeds *Enaliarctos* and *Allodesmus* (Mammalia, Carnivora). J Mamm Evol 16:77–97

Berta A (1991) New *Enaliarctos* (Pinnipedimorpha) from the Oligocene and Miocene of Oregon and the role of "enaliarctids" in pinniped phylogeny. Smithson Contrib Paleobiol (69):1–33

Berta A, Ray CE, Wyss AR (1989) Skeleton of the oldest known pinniped, *Enaliarctos mealsi*. Science 244:60–62

Berta A, Churchill M, Boessenecker RW (2018) The origin and evolutionary biology of pinnipeds: seals, sea ions, and walruses. Annu Rev Earth Planet Sci 46:203–228

Biewer JN, Velez-Juarbe J, Parham JF (2020) Insights on the dental evolution of walruses based on new fossil specimens from California. J Vertebr Paleontol 40:e1833896. https://doi.org/10.1080/02724634.2020.1833896

Bloodworth BE, Marshall CD (2007) A functional comparison of the hyolingual complex in pygmy and dwarf sperm whales (*Kogia breviceps* and *K. sima*), and bottlenose dolphins (*Tursiops truncatus*). J Anat 211:78–91

Boessenecker RW, Churchill M (2013) A reevaluation of the morphology, paleoecology, and phylogenetic relationships of the enigmatic walrus *Pelagiarctos*. PLoS One 8:e54311

Boessenecker RW, Churchill M (2015) The oldest known fur seal. Biol Lett 11:20140835

Bonner WN (1968) The fur seal of South Georgia. Br Antarct Surv Sci Rep 16:56–81

Bradshaw CJA, Lalas C, McConkey S (1998) New Zealand Sea lion predation on New Zealand fur seals. N Z J Mar Freshwater Res 32:101–104

Cobb WM (1933) The dentition of the walrus, *Odobenus obesus*. Proc Zool Soc Lond 103:645–668

Deagle BE, Kirkwood R, Jarman SN (2009) Analysis of Australian fur seal diet by pyrosequencing prey DNA in faeces. Mol Ecol 18:2022–2038

Dellinger T, Trillmich F (1999) Fish prey of the sympatric Galápagos fur seals and sea lions: seasonal variation and niche separation. Can J Zool 77:1204–1216

Deméré TA, Berta A (2001) A reevaluation of *Proneotherium repenningi* from the Miocene Astoria formation of Oregon and its position as a basal odobenid (Pinnipedia: Mammalia). J Vertebr Paleontol 21:279–310

Deméré TA, Berta A (2005) New skeletal material of *Thalassoleon* (Otariidae: Pinnipedia) from the late Miocene–early Pliocene (Hemphillian) of California. Bull Fla Mus Nat Hist 45:379–411

Emami-Khoyi A, Hartley DA, Paterson AM, Boren LJ, Cruickshank RH, Ross JG, Murphy EC, Else T-A (2016) Identifying prey items from New Zealand fur seal (*Arctocephalus forsteri*) faeces using massive parallel sequencing. Conserv Genet Resour 8:343–352

English AW (1976) Functional anatomy of the hands of fur seals and sea lions. Am J Anat 147:1–17

Etnier MA, Fowler CW (2010) Size selectivity in marine mammal diets as a guide to evolutionarily enlightened fisheries management. N Am J Fish Manage 30:588–603

Fallows C, Benoît HP, Hammerschlag N (2015) Intraguild predation and partial consumption of blue sharks *Prionace glauca* by Cape fur seals *Arctocephalus pusillus pusillus*. Afr J Mar Sci 37:125–128

Fay FH (1960) Carnivorous walrus and some Arctic zoonoses. Arctic 13:111–122

Fay FH (1982) Ecology and biology of the Pacific walrus, *Odobenus rosmarus divergens* Illiger. N Am Fauna 74:1–279

Fay FH, Sease JL, Merrick RL (1990) Predation on a ringed seal, *Phoca hispida*, and a black guillemot, *Cepphus grylle*, by a Pacific walrus, *Odobenus rosmarus divergens*. Mar Mamm Sci 6:348–350

Fox AD, Fox GF, Liaklev A, Gerhardsson N (2010) Predation of flightless pink-footed geese (*Anser brachyrhynchus*) by Atlantic walruses (*Odobenus rosmarus rosmarus*) in southern Edgeøya, Svalbard. Polar Res 29:455–457

George-Nascimento M, Bustamante R, Oyarzun C (1985) Feeding ecology of the South American sea lion *Otaria flavescens*: food contents and food selectivity. Mar Ecol Prog Ser 21:135–143

Giljov A, Karenina K, Kochnev A (2017) Prey or play: interactions between walruses and seabirds. Acta Ethol 20:47–57

Gjertz I (2009) Walrus predation of seabirds. Polar Rec 26:317–317

Green K, Burton HR, Williams R (1989) The diet of Antarctic fur seals *Arctocephalus gazella* (Peters) during the breeding season at Heard Island. Antarct Sci 1:317–324

Harcourt R (1993) Individual variation in predation on fur seals by southern sea lions (*Otaria byronia*) in Peru. Can J Zool 71:1908–1911

Harcourt RG, Bradshaw CJA, Dickson K, Davis LS (2002) Foraging ecology of a generalist predator, the female New Zealand fur seal. Mar Ecol Prog Ser 227:11–24

Hiiemae KM, Crompton AW (1985) Mastication, food transport, and swallowing. In: Hildebrand M, Bramble D, Liem K, Wake D (eds) Functional vertebrate morphology. The Belknap Press of Harvard University Press, Cambridge, MA, pp 262–290

Hocking DP, Evans AR, Fitzgerald EMG (2013) Leopard seals (*Hydrurga leptonyx*) use suction and filter feeding when hunting small prey underwater. Polar Biol 36:211–222

Hocking DP, Salverson M, Fitzgerald EMG, Evans AR (2014) Australian fur seals (*Arctocephalus pusillus doriferus*) use raptorial biting and suction feeding when targeting prey in different foraging scenarios. PLoS One 9:e112521

Hocking DP, Fitzgerald EMG, Salverson M, Evans AR (2016) Prey capture and processing behaviors vary with prey size and shape in Australian and subantarctic fur seals. Mar Mamm Sci 32:568–587

Hocking DP, Ladds M, Slip D, Fitzgerald EMG, Evans AR (2017a) Chew, shake, and tear: prey processing in Australian sea lions (*Neophoca cinerea*). Mar Mamm Sci 33:541–557

Hocking DP, Marx FG, Fitzgerald EMG, Evans AR (2017b) Ancient whales did not filter feed with their teeth. Biol Lett 13:20170348

Hocking DP, Marx FG, Park T, Fitzgerald EMG, Evans AR (2017c) A behavioural framework for the evolution of feeding in predatory aquatic mammals. Proc R Soc B 284:20162750

Hocking DP, Marx FG, Park T, Fitzgerald EMG, Evans AR (2017d) Reply to comment by Kienle et al. 2017. Proc R Soc B 284:20171836

Hocking DP, Marx FG, Sattler R, Harris RN, Pollock TI, Sorrell KJ, Fitzgerald EMG, McCurry MR, Evans AR (2018) Clawed forelimbs allow northern seals to eat like their ancient ancestors. R Soc Open Sci 5:172393

Hocking DP, Marx FG, Parker WMG, Rule JP, Cleuren SGC, Mitchell AD, Hunter M, Bell JD, Fitzgerald EMG, Evans AR (2020) Inferring diet, feeding behaviour and causes of mortality from prey-induced injuries in a New Zealand fur seal. Dis Aquat Org 139:81–86. https://doi.org/10.3354/dao03473

Iwata T, Sakamoto KQ, Takahashi A, Edwards EWJ, Staniland IJ, Trathan PN, Naito Y (2012) Using a mandible accelerometer to study fine-scale foraging behavior of free-ranging Antarctic fur seals. Mar Mamm Sci 28:345–357

Johnston C, Berta A (2011) Comparative anatomy and evolutionary history of suction feeding in cetaceans. Mar Mamm Sci 27:493–513

Karenina K, Giljov A, Ivkovich T, Malashichev Y (2016) Evidence for the perceptual origin of right-sided feeding biases in cetaceans. Anim Cogn 19:239–243

Kastelein RA, Gerrits NM (1990) The anatomy of the walrus head (*Odobenus rosmarus*). Part 1: the skull. Aquat Mamm 16:101–119

Kastelein RA, Mosterd P (1989) The excavation technique for molluscs of Pacific walrusses (*Odobenus rosmarus divergens*) under controlled conditions. Aquat Mamm 15:3–5

Kastelein RA, Gerrits NM, Dubbeldam JL (1991) The anatomy of the walrus head (*Odobenus rosmarus*). Part 2: description of the muscles and of their role in feeding and haul-out behaviour. Aquat Mamm 17(3):156–180

Kastelein RA, Muller M, Terlouw A (1994) Oral suction of a Pacific walrus (*Odobenus rosmarus divergens*) in air and under water. Z Säugetierkunde 59:105–115

Kemper CM, Pemberton D, Cawthorn M, Heinrich S, Mann J, Würsig B, Shaughnessy P, Gales R (2003) Aquaculture and marine mammals: co-existence or conflict? In: Gales N, Hindell M,

Kirkwood R (eds) Marine mammals: fisheries, tourism, and management issues. CSIRO Publishing, Clayton, pp 208–225

Kernaléguen L, Dorville N, Ierodiaconou D, Hoskins AJ, Baylis AMM, Hindell MA, Semmens J, Abernathy K, Marshall GJ, Cherel Y, Arnould JPY (2016) From video recordings to whisker stable isotopes: a critical evaluation of timescale in assessing individual foraging specialisation in Australian fur seals. Oecologia 180:657–670

Kienle SS, Law CJ, Costa DP, Berta A, Mehta RS (2017) Revisiting the behavioural framework of feeding in predatory aquatic mammals. Proc R Soc B 284:20171035

King JE (1961) The feeding mechanism and jaws of the crabeater seal (*Lobodon carcinophagus*). Mammalia 25:462–466

Kohno N (2004) Ecological shift in the otariid pinnipeds from pelagic to inshore: evidence from the Middle Miocene record of the North Pacific. J Vertebr Paleontol 24:79

Kohno N, Barnes LG, Hirota K (1995a) Miocene fossil pinnipeds of the genera *Prototaria* and *Neotherium* (Carnivora; Otariidae; Imagotariinae) in the North Pacific Ocean: evolution, relationships and distribution. Island Arc 3:285–308

Kohno N, Tomida Y, Hasegawa Y, Furusawa H (1995b) Pliocene tusked odobenids (Mammalia: Carnivora) in the Western North Pacific, and their paleobiogeography. Bull Natl Sci Mus Ser C (Geol Paleontol) 21:111–131

Kuhn C, Frey E (2012) Walking like caterpillars, flying like bats—pinniped locomotion. Palaeobiodivers Palaeoenviron 92:197–210

Levermann N, Galatius A, Ehlme G, Rysgaard S, Born EW (2003) Feeding behaviour of free-ranging walruses with notes on apparent dextrality of flipper use. BMC Ecol 3:9

Lowry LF, Fay FH (1984) Seal eating by walruses in the Bering and Chukchi Seas. Polar Biol 3:11–18

MacNeilage PF (2014) Evolution of the strongest vertebrate rightward action asymmetries: marine mammal sidedness and human handedness. Psychol Bull 140:587–609

Magallanes I, Parham JF, Santos G-P, Velez-Juarbe J (2018) A new tuskless walrus from the Miocene of Orange County, California, with comments on the diversity and taxonomy of odobenids. PeerJ 6:e5708

Mallory ML, Woo K, Gaston AJ, Davies WE, Mineau P (2004) Walrus (*Odobenus rosmarus*) predation on adult thick-billed murres (*Uria lomvia*) at Coats Island, Nunavut, Canada. Polar Res 23:111–114

Marshall CD, Rosen DAS, Trites AW (2015) Feeding kinematics and performance of basal otariid pinnipeds, Steller Sea lions and northern fur seals: implications for the evolution of mammalian feeding. J Exp Biol 218:3229–3240

Mathews EA, Adkison MD (2010) The role of Steller Sea lions in a large population decline of harbor seals. Mar Mamm Sci 26:803–836

McIntosh RR, Page B, Goldsworthy SD (2006) Dietary analysis of regurgitates and stomach samples from free-living Australian sea lions. Wildl Res 33:661–669

Mitchell ED, Tedford RH (1973) The Enaliarctinae: a new group of extinct aquatic Carnivora and a consideration of the origin of the Otariidae. Bull Am Mus Nat Hist 151:201–284

Motta PJ, Wilga CD (2001) Advances in the study of feeding behaviors, mechanisms, and mechanics of sharks. Environ Biol Fish 60:131–156

Norris KS, Møhl B (1983) Can odontocetes debilitate prey with sound? Am Nat 122:85–104

Oliver JS, Slattery PN, O'Connor EF, Lowry LF (1983) Walrus, *Odobenus rosmarus*, feeding in the Bering Sea: a benthic perspective. Fish Bull 81:501–512

Oliver JS, Kvitek RG, Slattery PN (1985) Walrus feeding disturbance: scavenging habits and recolonization of the Bering Sea benthos. J Exp Mar Biol Ecol 91:233–246

Paterson RS, Rybczynski N, Kohno N, Maddin HC (2020) A total evidence phylogenetic analysis of pinniped phylogeny and the possibility of parallel evolution within a monophyletic framework. Front Ecol Evol 7:457. https://doi.org/10.3389/fevo.2019.00457

Peters KJ, Ophelkeller K, Bott NJ, Deagle BE, Jarman SN, Goldsworthy SD (2015) Fine-scale diet of the Australian sea lion (*Neophoca cinerea*) using DNA-based analysis of faeces. Mar Ecol 36:347–367

Ray GC, McCormick-Ray J, Berg P, Epstein HE (2006) Pacific walrus: benthic bioturbator of Beringia. J Exp Mar Biol Ecol 330:403–419

Reidenberg JS, Laitman JT (1994) Anatomy of the hyoid apparatus in Odontoceti (toothed whales): specializations of their skeleton and musculature compared with those of terrestrial mammals. Anat Rec 240:598–624

Reisinger RR, Landman M, Mgibantaka N, Smale MJ, Bester MN, De Bruyn PJN, Pistorius PA (2018) Overlap and temporal variation in the diets of sympatric Antarctic and Subantarctic fur seals (*Arctocephalus* spp.) at Marion Island, Prince Edward Islands. Polar Res 37:1451142

Repenning CA (1976) Adaptive evolution of sea lions and walruses. Syst Biol 25:375–390

Repenning CA, Tedford RH (1977) Otarioid seals of the neogene. US Geol Surv Prof Pap 992:1–93

Riedman M (1990) The pinnipeds: seals, sea lions and walruses. University of California Press, Berkeley

Robinson SA, Goldsworthy SG, Jvd H, Hindell MA (2003) The foraging ecology of two sympatric fur seal species, *Arctocephalus gazella* and *Arctocephalus tropicalis*, at Macquarie Island during the austral summer. Mar Freshw Res 53:1071–1082

Schwenk K (2000) An introduction to tetrapod feeding. In: Schwenk K (ed) Feeding: form, function and evolution in tetrapod vertebrates. Academic, San Diego, pp 21–61

Schwenk K, Rubega M (2005) Diversity of vertebrate feeding systems. In: Starck JM, Wang T (eds) Physiological and ecological adaptations to feeding in vertebrates. Science Publishers, Enfield, pp 1–41

Seymour J, Horstmann-Dehn L, Wooller MJ (2014a) Inter-annual variability in the proportional contribution of higher trophic levels to the diet of Pacific walruses. Polar Biol 37:597–609

Seymour J, Horstmann-Dehn L, Wooller MJ (2014b) Proportion of higher trophic-level prey in the diet of Pacific walruses (*Odobenus rosmarus divergens*). Polar Biol 37:941–952

Sheffield G, Grebmeier JM (2009) Pacific walrus (*Odobenus rosmarus divergens*): differential prey digestion and diet. Mar Mamm Sci 25:761–777

Skinner JP, Norberg SE, Andrews RD (2009) Head striking during fish capture attempts by Steller Sea lions and the potential for using head surge acceleration to predict feeding behavior. Endanger Spec Res 10:61–69

Taylor MA (1987) How tetrapods feed in water: a functional analysis by paradigm. Zool J Linn Soc Lond 91:171–195

Vélez-Juarbe J (2017) *Eotaria citrica*, sp. nov., a new stem otariid from the "Topanga" formation of Southern California. PeerJ 5:e3022

Vibe C (1950) The marine mammals and the marine fauna in the Thule district (Northwest Greenland) with observations on ice conditions in 1939-41. Medd Grøn 150:1–115

Visser IN, Drennan MP, White RW, MacLean SF, Lagerstrom LC, Francis JM (2008) Antarctic fur seals (*Arctocephalus gazella*) observed predating Adélie (*Pygoscelis adeliae*) and chinstrap penguins (*P. antarctica*), Antarctic Peninsula. Aquat Mamm 34:193–199

Viviant M, Trites AW, Rosen DAS, Monestiez P, Guinet C (2010) Prey capture attempts can be detected in Steller Sea lions and other marine predators using accelerometers. Polar Biol 33:713–719

Volpov BL, Hoskins AJ, Battaile BC, Viviant M, Wheatley KE, Marshall G, Abernathy K, Arnould JPY (2015) Identification of prey captures in Australian fur seals (*Arctocephalus pusillus doriferus*) using head-mounted accelerometers: field validation with animal-borne video cameras. PLoS One 10:e0128789

Werth AJ (2000) Feeding in marine mammals. In: Schwenk K (ed) Feeding: form, function and evolution in tetrapods. Academic, San Diego, pp 487–526

Werth AJ (2006) Mandibular and dental variation and the evolution of suction feeding in Odontoceti. J Mammal 87:579–588

Werth AJ (2007) Adaptations of the cetacean hyolingual apparatus for aquatic feeding and thermoregulation. Anat Rec 290:546–568

Womble JN, Conlon S (2010) Observation of Steller Sea lion (*Eumetopias jubatus*) predation on a harbor seal (*Phoca vitulina richardii*) in the Glacier Bay Region of southeastern Alaska. Aquat Mamm 36:129–137

Part III
Reproductive Behavior

Chapter 6
Polygyny in the Era of Molecular Biology: Revisiting Bartholomew's Model

Marcelo H. Cassini

Abstract The theory of sexual selection hypothesizes that, when females are selected to perform most parental investment, males compete for access to mates and polygyny evolves. Polygyny should result in high variance in male reproductive success, with a few males fathering a disproportionate number of pups. I tested this prediction by reviewing the literature on paternity, available for seven otariids. I identified a total of eight publications with data on paternity in seven species at 12 colonies. Overall, harem size produced an estimated mean of 20.7 pups per dominant male (assuming that the male fathered all pups born within its harem), but genetic analysis revealed that territorial males fathered 32.5% of pups born in colonies, resulting in an estimated two pups per territorial male, i.e. behavioural estimations of reproductive success was tenfold (20/2) higher than genetic ones; many males were unsuccessful in obtaining mates, even when following the dominant strategy; size of harem did correlate with paternity by territorial males but the slope was significantly lower than expected from behavioural estimations; differences in paternity between territorial and satellites males were not statistically significant in most studies. Generalizations of classical theories on the evolution of polygyny may need qualification in the light of evidence provided by molecular ecology.

Keywords Fur seals · Sea lions · Sexual behaviour · Sexual dimorphism

6.1 Introduction

Darwin's (1871) theory of sexual selection hypothesised that physical and behavioral traits that provide advantages in male–male competition are, over time, selected for. Otariids (fur seals and sea lions) and elephant seals have been considered the

M. H. Cassini (✉)
Laboratorio de Biología del Comportamiento, Instituto de Biología y Medicina Experimental, CONICET, Vuelta de Obligado, Buenos Aires, Argentina

C. Campagna, R. Harcourt (eds.), *Ethology and Behavioral Ecology of Otariids and the Odobenid*, Ethology and Behavioral Ecology of Marine Mammals,
https://doi.org/10.1007/978-3-030-59184-7_6

most illustrative examples of this theory (Bonnes 1991; Cassini 1999, 2017). The theoretical approach is expressed in this sentence: 'intense and bloody male aggression, extremely skewed male mating success, and a resulting extraordinary size dimorphism provide convincing evidence of the powers of sexual selection' (Lindenfors et al. 2002, p. 188).

Trivers (1972) realized that differences in parental effort between the sexes are correlated with differences in sexual behavior and the reproductive system of a species. Polygyny is prevalent in species where males are freed from parental care duties. In other words, there is a phylogenetic potential for polygyny. In mammals, females are constrained by gestation to invest more parental effort, thus polygyny is expected to be the most common mating system in this taxon.

In another seminal paper written almost half a century ago, it was proposed that, in addition to phylogenetic factors, there are ecological features that determine the intensity of sexual selection in different species and populations (Emlen and Oring 1977). The environmental potential for polygyny depends on the degree to which multiple mates, or resources critical to gaining multiple mates, are economically defendable by males. Authors argued that the operational sex ratio predicts the intensity of sexual selection, defined as the average ratio of fertilizable females to sexually active males at any given time.

The development of molecular biology techniques during the last five decades has revolutionized most fields of biology, including the study of behaviour, ecology and evolution. Individual DNA profiling or fingerprinting became a standard tool after the development of the polymerase chain reaction (PCR) technique (Bartlett and Stirling 2003). Until the 1980s, it was assumed that genetic and behavioural levels of polygyny were correlated. At one extreme, in a monogamous system, a male should be the only father of the offspring born to his partner, while at the other, in a polygynous mating system, male mating success should be directly proportional to the number of females with which they are able to associate following successful competition with other males (Gonzalez-Suarez and Cassini 2014). The introduction of molecular markers provided opportunities to test these assumptions. Before analyzing the case of the Otariids, it is instructive to describe results from paternity estimates using genetic methods in birds.

Monogamy is the most common mating system in birds; traditionally over 90% of avian species were defined as monogamous (Lack 1968; Emlen and Oring 1977). Theory predicted that both parents are required for successful rearing of young, and that the costs of courting and mating with additional mates overcome benefits (Emlen and Oring 1977). Yet, molecular markers indicated that the incidence of extra-pair paternity was unexpectedly common. For example, in a review by Griffith (2007) of approximately 160 studies of avian parentage, extra-pair copulations occurred in over 130 bird species. Accordingly, in birds: (1) extra-pair paternity is recurrent, (2) variance in reproductive success (RS) among males is much higher than expected, and (3) less than 25% of avian species are actually genetically monogamous (Sheldon and Ellegren 1999; Whittingham and Dunn 2005).

This discoveries produced an impact that was partly responsible for what was defined as the "the first major triumph of modern sexual selection research" (Jones

Fig. 6.1 Image showing sexual dimorphism in southern sea lions

and Ratterman 2009, p. 10002): extra-pair paternity exerts great sexual selection pressure on males even in assumed monogamous species; for females, it is an opportunity for active mate choice. The aim of sexual selection theory became to discover the mechanisms that could explain the evolution of these female preferences (Jones and Ratterman 2009). There was a proliferation of models of mate choice and quantitative approaches to measure selection differentials (e.g., Mead and Arnold 2004; Kotiaho and Puurtinen 2007; Kokko et al. 2007).

In birds, the use of molecular markers for determining paternity become a cornerstone of sexual selection research, changing our understanding of mating patterns in bird natural populations (Jones and Ratterman 2009). This then raises the question of how this may apply to mammals. Here I will review the impact of the use of molecular markers on our understanding of the role of sexual selection in the evolution of mating systems in the Otariids (sea lions and fur seals), one of the most polygynous taxa among mammals.

Size sexual dimorphism and high levels of behavioural polygyny are traits shared by all 14 species of Otariids (Berta 2018). Males are on average three times heavier, but can become up to five or more times heavier, than females (Fig. 6.1). An example of a study in northern fur seal (*Callorhinus ursinus*) illustrates the extreme polygyny of this taxon: a single male was observed to mate with 161 females while hundreds of other males were excluded from mating (Mesnik and Ralls 2018). Most male Otariids typically defend territories containing resources needed by females, while lekking has been proposed in at least four species: *Zalophus californianus*, *Arctocephalus australis*, *Phocarctos hookeri* (Mesnik and Ralls 2018 and references therein), and *Otaria flavescens* (Soto and Trites 2011). The mating system of *O. flavescens* has been also been defined as direct defence of females and of territories, often present in the same colony (Campagna and Le Boeuf 1988; Franco-Trecu et al. 2014).

In his pioneering work, Bartholomew (1970) provided an explanation of the evolution of mating systems in Otariids. Even today, his model remains the core paradigm to explain Otariid polygyny (Miller 2018). Otariids are the only mammalian family that lives part of the time at sea but copulate and give birth on land.

Otariid females aggregate and synchronize reproduction, providing to males high environmental potential for polygyny. Dominant males, capable of occupying a territory or defending a position among female aggregates, have a great chance of multiple copulations; subordinate males, conversely, are excluded from mating. Males can gain reproductive advantages during pre-copulatory contest competitions if large and belligerent, which may have selected for sexual size dimorphism (Bartholomew 1970). Subordinate males may resort to alternative strategies, which would yield lower reproductive success than dominant congeners.

In recent years, several studies have been published that estimated paternity based on molecular markers. Using their findings, I test the main assumptions of sexual selection theory applied to the evolution of polygyny in Otariids: that males show a large variance in reproductive success, and that the most successful mating strategy is performed by dominant males that aggressively defend a space or position within an aggregation of females.

6.2 Methods

A literature search in the Web of Knowledge database was conducted in August 2018 to find behavioural and genetic studies of RS and paternity in Otariids.

Behavioural Measures Two reviews summarized results on behavioural measures of reproductive success in 14 species (Bonnes 1991; Lindenfors et al. 2002; Table 6.1). Bonnes (1991) discussed as indicators the operational sex ratio, the intensity of sexual selection (I_s), and the maximum number of observed copulations by a single male (Table 6.1). I_s is the variance in fitness scaled in relation to the square of the average fitness, which requires a measure of fitness for each adult male (Wade and Arnold 1980). Wade and Arnold (1980) applied this concept to data from a population of birds, using offspring/adult male and mates/adult male to calculate fitness. Bonnes (1991) reviews its use in Otariids and concludes that I_s is the least reliable measure, because its failure to obtain a representative sample can produce extremely misleading values. I_s is also very sensitive to the count of males that fail to reproduce. Lindenfors et al. (2002) provided data on harem size for most species of Otariids (Table 6.1).

Genetic Measures Eight studies provided data on paternity in seven species of Otariids (*Arctocephalus australis*, Franco-Trecu et al. 2014; *A. forsteri*, Caudron et al. 2010; *A. gasella*, Gemmell et al. 2001; Hoffman et al. 2003; *Callorhinus ursinus*, Kiyota et al. 2008; *Neophoca cinerea,* Ahonen 2013; *Zalophus californianus*, Flatz et al. 2012; *Z. wollebaeki*, Porschmann et al. 2010; Table 6.2). Two studies were conducted at the same study site (Gemmell et al. 2001; Hoffman et al. 2003) and shared some results. Three studies (Caudron et al. 2010; Ahonen 2013; Flatz et al. 2012) provided information on more than one study site; these were analysed separately, bringing the total sample size to 12. When data covered several years, the information was pooled. All studies used microsatellites as molecular

Table 6.1 Behavioral estimates of RS

Species	Harem size[a]	SSD (kg)[a]	OSD[b]	I_s[b]	Maximum mates[b]
A .australis	6	3.3	–	–	–
A philippii	140	2.8	–	–	–
A. forsteri	6.2	3.0	7.9	6.5	30
A. galapagoensis	5.9	2.4	6	3.3	14
A. gazella	10.3	4.1	10	–	14
A. pusillus	28.9	3.9	7.5	–	–
A. pusillus	10.5	3.7	53	–	–
A. townsendi	6.2	2.9	10	–	–
A. tropicalis	6.5	3.1	6.6	–	–
C. ursinus	36.4	5.1	39	45.9	161
E. jubatus	11.8	3.5	10	9.7	32
N. cinerea	3.8	3.8	–	–	–
O. flavescens	6	2.1	4.3	2.3	19
P. hookeri	12.9	2.0	–	3.3	19
Z. californianus	16.2	3.4	25	5.9	27
Z. wollebaeki	13.5	2.6	19	–	–
Mean for Otariidae	20.1	3.2	16.5	11.0	39.5

Variables described in methods
[a]From Lindenfors et al. (2002)
[b]From Bonness (1991)

markers. Table 6.2A provides information on the number of highly polymorphic microsatellite loci, and the minor value of confidence level in CERVUS (some studies used two levels), which reflect the accuracy of the methods used. Total pups/year, total males/year and total territories/year provided an estimation of the population size at the study site (Table 6.2B). Sampled effort per study site was estimated from the percentage of total males sampled, study years, and number and percentage of genotyped pups (Table 6.2C). Total pup paternity was the total number of offspring identified with fathers in the colony. Few studies discriminated between territorial and satellite fathers, others sampled only territorial males (Table 6.2D). Mean pups/male was calculated as the ratio of total pups with paternity per total candidate males. There are several measures that provide an estimation of variation in paternity within the samples (Table 6.2F): mean number of pups/male, number of fathers with more than two pups, and the maximum number of pups assigned to one father (together with the number of fathers with this level of paternity). The final estimation of RS of males was obtained by extrapolating paternity of sample males to all males of the colony. This was obtained as: total pups in the colony per year * proportion of pups sired by males/total males in the colony per year. Some studies measured paternity only for territorial males. Variance in paternity was obtained from Gonzalez-Suarez and Cassini (2014), with the exception of *A. australis*, which was provided by Franco-Trecu et al. (2014).

Table 6.2 Results of genetic analysis of male paternity for seven species and 11 colonies from eight studies

			A. forsteri		*A. gasella*	*A. gasella*		*N. cinerea*			*Z. californianus*			
	Species	*A. australis*	Study site	South	(Gemmel)	(Hoffman)	*C. ursinus*	Olive	Lilliput	Blefuscu	San Jorge	Los islotes	*Z. wollebaeki*	Mean
A	Loci	10	10		8	8	5	16			14		19	11.3
	Minor confidence	80	80		–	80	99	80			95		95	87.0
B	Total pups/year	260	44	119	700	635	100	196	66	104	968	143	143.5	289.9
	Total males/year	139	50	–	–	–	–	112	40	59	–	–	173	70.6
	Territories/year	36	7	–	147	101	9	–	–	–	250	50		85.6
C	Years	1	1	1	2	7	5	2	2	2	3	2	2	2.5
	Genotyped pups	220	39	25	184	660	85	202	102	119	82	40	287	170.4
	% genotyped pups	84.6	88.0	21.0	13.1	14.8	17.0	51.5	77.3	57.2	2.8	14.0	100.0	45.1
	Genotyped males	36	15	8	243	415	37	58	28	37	65	34	374	112.5
	Dominant males	26	3	5	243	415	37	–	–	–	–	–	–	121.5
	Satellite males	10	12	3	–	–	–	–	–	–	–	–	–	8.3
D	Pups with paternity (N)	103	12	6	65	372	83	79	32	23	15	3	154	78.9
	Dominant males	86	9	1	65	372	45	–	–	–	15	3	–	74.4

	Satellite males	17	3	6	–	–	11	–	–	–	0	0	0	5.3
	Pups with paternity (%)	46.8	30.8	24.0	35.3	56.4	97.6	39.1	31.4	19.3	18.3	7.5	53.7	38.3
	Dominant males	39.1	23.1	4.0	35.3	56.4	94.8	–	–	–	18.3	7.5	–	29.1
	Satellite males	7.7	7.7	24.0	–	–	5.2	–	–	–	–	–	–	13.1
E	Pups/candidate male	2.9	0.8	0.8	0.3	0.9	2.2	1.4	1.1	0.6	0.2	0.1	0.4	1.0
	Dominant males	3.4	3.0	0.0	0.3	0.9	1.2	–	–	–	–	–	0.4	1.3
	Satellite males	1.9	0.3	2.0	–	–	–	–	–	–	–	–	–	1.4
	Fathers w. >2 pups (N)	23	1	0	20	78	–	6	2	0	1	0	9	12.7
	(%)	63.9	6.7	0.0	8.2	18.8	–	–	–	–	1.5	0.0	2.4	12.7
	Maximal pups/male	13	4	1	3	8	–	4	5	4	5	1	4	4.7
	RS	0.9	0.3					2.6	1.5	1.1			0.4	1.1
	Dominant males	2.8	1.5	–	1.7	3.6	10.5	–	–	–	0.7	0.2	–	2.0
	Variance	1.9	1.9	0.2	0.2	2.9	–	1.0	1.4	0.9	0.3	0.2	0.2	1.0

References and variables described in methods

6.3 Summary of Results from Paternity Studies

In *A. australis*, Franco-Trecu et al. (2014) found that the overall average number of pups per putative father was 3.0 (SD = 2.9, range = 0–13). They also showed that the mean number did not significantly differ (Welch $t = -1.9$, $p = 0.06$) between territorial (mean ± SD = 3.4 ± 3.1) and satellite (1.9 ± 1.5) males.

Three breeding areas were studied in *A. forsteri* (Caudron et al. 2010): 13, 4, and 7 pups were assigned a known father in the Ohau Point, North and South study areas, respectively. Sampled males included territorial ($n = 3$) and non-territorial ($n = 12$) individuals at Ohau, and 5 territorial males at the South area. The 8 territorial males fathered a total of 9 (4-2-2-1-0-0-0-0) pups born in their territories, while 8 non-territorial males of Ohau fathered a total of 4 pups (1-1-1-1-0-0-0-0). Five additional pups sired by 3 non-territorial males were assigned to South or North study sites (2-2-1), while 2 of the 3 territorial males of Ohau fathered 4 pups from these two colonies (2-2). For the purpose of the present analysis, the first 5 pups were considered as extra-harem copulations, while the last 4 pups were excluded from the analysis (although they could be considered extra-harem paternities because they are not related to stable females of the harem; they were probably related to females visiting male territories or to ex-territorial males acting as satellites in other colonies). In summary, intra-harem and extra-harem paternities coincided in 1.12 (9/8) pups per territorial/satellite male. Total territorial males in the main study site was seven and all were sampled. None of territorial males genotyped in the South colony fathered any of the offspring sampled; the estimate of reproductive success of the territorial males of that colony was zero.

Gemmell et al. (2001) studied paternity in *A. gazella* territorial males in a colony with an average of 264 males and 700 pups per season. They sampled 243 territorial males and 184 pups, and they assigned a total of 65 paternities (35%) to the males (190 males with no paternities, 45 with 1, 4 with 2, and 4 with 3 paternities/male), given a RS of 0.34 pups/territorial male. Hoffman et al. (2003) studied the same species in the same colony and reported: (i) a mean number of pups assigned per male of 0.93, (ii) 58.8% of pups with assigned paternity (with the least conservative method), and (iii) 2.49 pups/sampled father.

Callorhinus ursinus was studied by Kiyota et al. (2008). They collected genetic samples from 37 males and 85 pups in 5 years in a colony with an average of 9 territories and 100 pups per year. Territorial males were classified into the following categories, according to their timing and duration of territory tenure: primary males (males that held territories before July 1 before the arrival of most females); secondary males (males that held territories after June 30); and temporary males (males that stayed in the breeding area for <1 day). They differentiated between associate and non-associate (primary or secondary) males, based on whether pups were identified to born or not in their territories. 98.2% of paternity corresponded to territorial males associated (45 pups) or not associated (10 pups) to females, and average reproductive success of territorial males was 10.9 pups/male (100 pups * 0.98/9 territorial males). However there were 29 pups without assigned males, 27 of which had a candidate father but the authors did not state whether they were territorial or temporary males. Had the 29 pups been sired by satellite males, paternity of territorial males would drop to 64.7%, with a reproductive success of 7.2 pups/territorial male. This result is significantly higher than those in other species, probably due to two characteristics of the study: (i) the study area was an isolated cove where satellite males would have low chances to interact with females, and (ii) the study focused more on older females who came ashore in the beginning of the reproductive season, and were more likely to copulate with territorial males than with satellite males (M. Kiyota, pers. commun.).

Ahonen (2013) studied three colonies of *Neophoca cinerea* during two seasons. Assigned paternities were 39.1%, 31.4%, and 19.3%, while total pups-total males were 196-112, 66-40, and 104-59 for Olive, Lilliput and Blefuscu sites, respectively. Estimated reproductive success of territorial males was: 0.59 (77/112), 0.52 (21/40), and 0.34 (20/59) pups/territorial male, respectively.

Flatz et al. (2012) studied two colonies of *Zalophus californianus*. Overall, they found considerably lower variation in male RS than expected in a species that exhibits behavior associated with strongly polygynous mating. With one exception, no male was identified as the father of more than two pups, and mean RS for San Jorge and Los Islotes colonies was 0.71 and 0.21 respectively (Table 6.2).

Zalophus wollebaeki was studied by Porschmann et al. (2010) in the Galapagos islands. They obtained 53.7% of pup paternity assignments. They sampled all males and pups in the colony, thus RS estimation was straightforward: 0.41 (154/374) pups/male.

Overall, harem size produced an estimation of 20.1 pups per dominant male (Table 6.1), while territorial males fathered only 32.5% of pups born in colonies, giving an estimation of 2.0 pups per territorial male (Table 6.2). These results indicated that behavioural estimations of reproductive success in Otariids are more than tenfold higher than genetic ones.

6.4 Does Theory Meet Facts?

The most obvious and characteristic feature of Otariids is the remarkable sexual dimorphism in size. The scene of two territorial males fighting in a colony saturated with females is spectacular and fascinates anyone interested in animal behaviour. It seems evident that the size and aggressiveness of these males serve for intra-sexual competition for access to reproductive partners. Added to these conspicuous features, Otariids fit in the theory, which explains the evolution of mating systems by sexual selection: environmental potential for polygyny is high because most species display female spatial clustering during a synchronous birthing season, giving some males the opportunity to gather numerous matings in a short time.

However, this review indicates that, in the face of paternity data, the main assumptions of sexual selection theory applied to explain the evolution of mating systems of Otariids are not met for several studies and species. Variance in the reproductive success of males is substantially lower than what was believed when the data to measure it was behavioural (operational sex ratio, copulations, harem size). A territorial male can be surrounded by females and even copulate with a large number of them, but that does not mean that he will sire many offspring. Also, there are numerous offspring whose paternities were not identified, which are probably the result of alternative strategies used by satellite males.

In summary, although only seven species have been surveyed, a preliminary conclusion is that research on genetic paternity challenges traditional theory on the evolution of mating systems in Otariids, as it did in studies of birds. Unexpected high levels of extra-harem paternities were found, similarly to the extra-pair paternities discovered in birds. However the effect in terms of male reproductive variance of these extra-paternities is exactly the opposite: while in birds this variance is higher than expected by behaviour, in Otariids it was significantly lower. If variance is low, and there are no large differences in the breeding success of dominant and satellite males, then the intensity of sexual selection would be considerably less than it was assumed in traditional models.

Larger males appear to be the most successful in maintaining a position within the colony (Lidgard et al. 2005; Lindenfors et al. 2002; Meise et al. 2014). This pattern has been interpreted as the result of sexual selection operating on large males that provides success in agonistic interactions, dominance and efficient territorial defence (Lindenfors et al. 2002). Porschmann et al. (2010) postulated an alternative hypothesis. They proposed that fasting ability selects for male size rather than just fighting ability and dominant status. Larger males have greater fat reserves, allowing them to extend tenure without the need to go to the sea for food. Large attendance alone would be enough to access to more estrus females.

Cassini (2017, Chap. 6) proposed that the size of Otariid males is determined primarily by natural selection, due to the advantages it could bring to foraging in cold seas. This hypothesis complements that proposed by Porschmann et al. (2010) because the largest males can access the best foraging areas (or any other advantage that body size could provide during foraging), they may build more fat reserves, so they would be better prepared to follow a strategy based on a long attendance at a site where females aggregate. On the other hand, small males cannot stay long in the colony, as they are forced to go foraging.

This natural selection hypothesis produces the same prediction that sexual selection hypothesis: large males will use the territorial strategy, and the small males will use alternative strategies. The most important difference between the two hypotheses is that natural selection does not need to assume that 'dominant' males will have greater fitness per reproductive season than satellite males. This hypothesis interprets territorial defence as an opportunistic strategy of males with large reserves of fat, and also as a secondary evolutionary product that originates from the advantages that large body size provides for survival at sea.

Krüger et al. (2014) found support to the hypothesis that male body size may have evolved by natural selection pressures. They analyzed 11 life-history traits in 35 pinniped species to determine their co-evolutionary dynamics and inferred their most likely evolutionary trajectories in the evolution of this taxon. They found that sexual size dimorphism had evolved prior to changes in the mating system. Only subsequently did polygyny evolve, leading to further coevolution as the strength of sexual selection intensified. They called for the inclusion of ecological variables when studying sexual selection and argue for caution when assuming causality between coevolving traits. They proposed that differences in body sizes between sexes could be the consequence of niche partitioning during aquatic foraging, or in combination with sexual selection on males to enforce copulations on females. Assuming that natural selection pressures are a determinant of body size, a question remains as to why females are smaller than males. There are at least four hypotheses that explain the origin of sexual size dimorphism by processes different to sexual selection (for an explanation, see Chap. 7): (1) non-adaptive processes (Leutenegger and Cheverud, 1982, 1985), (2) inter-sexual competition, (3) bimodal niches, and (4) dimorphic niches (Slatkin 1984; Cassini 2017).

Several alternative strategies have been described in Otariids. Some of them were not observed but inferred after discovering the significant occurrence of extra-harem paternities. In *A. australis*, satellite males stayed intermittently in a fraction of the study area farthest from the coastline that was partially used by females and pups whenever the tide level rose (Franco-Trecu et al. 2014). In *O. flavescens*, non-territorial males group raid territories and breeding aggregations of females, or they try stop females entering the colony from the sea (Campagna et al. 1988, Chap. 9). In *A. gazella*, Gemmell et al. (2001) suggested the existence of aquatic or nocturnal mating by satellite males. In *Z. wollebaeki*, a fraction of males seems to be successful by paying only short visits to the colony (Porschmann et al. 2010). In *Z. californianus*, Flatz et al. (2012) suggested that copulations may be occurring outside the boundaries of the colonies, such that some males intercept females traveling to and from the rookery on foraging trips. In *Z. wallobaeki*, Meise et al. (2014) propose that intermediate-sized males can hide among females when the territorial male is close by, thus avoiding being spotted and chased away.

Research has focused on territorial males, while the study of satellite males has been notably less intense. There are very few studies measuring the reproductive success of alternative strategies, probably due to two reasons: more theoretical interest in the dominant strategy, and more difficulty in studying satellite males. It has been proposed that satellite males would incur lower energetic costs from male–male interactions, thermoregulation, and site tenure than territorial males, thus being potentially able to invest in post-copulatory traits by augmenting both the quantity and quality of their sperm and seminal fluids (Simmons and Fitzpatrick 2012; Franco-Trecu et al. 2014). A possible aspect of low variance in male reproductive success that could be studied in the future relates to the effectiveness of fertilization. Although the frequency of copulations in dominant males seems to be high, fertilizations are lower than expected. Could the cause be in the low production of sperm by the large males? It maybe that smaller males ultimately may be more successful in postcopulatory competition.

6.5 A Mixed Model of the Evolution of Sexual Size Dimorphism and Mating Systems in Mammals

Recent studies that estimates intensity of sexual selection using molecular measures of male reproductive success have challenged the hypothesis that sexual selection is the only selective pressure involved in the evolution of mammalian mating system. Taxa that show higher levels of polygyny and dimorphism (and that have been used as biological models of the theory of sexual selection) have shown lower variances in male fitness than expected by behavior or dimorphism (in pinnipeds this chapter and Gonzalez-Suarez and Cassini (2014), in primates Cassini (2020a), in artiodactyls, Cassini (2020b)). Cassini (2020c) developed a theoretical model to explain these differences between what was obtained by ethological studies and what was obtained with molecular techniques (Fig. 6.2). The mixed model of the evolution of sexual size dimorphism and mating systems in mammals proposes an initial phase where natural selection operates on body size, followed by increases in sexual dimorphism (Fig. 6.3). Ecological factors and male disturbance promote female grouping. Males compete for access to these groups according to two possible evolutionary scenarios comprising directional or equilibrium sexual selection, where both produce a similar type of behavioral polygyny, but they differ in terms of the intensity of intra-male pre-copulatory sexual selection.

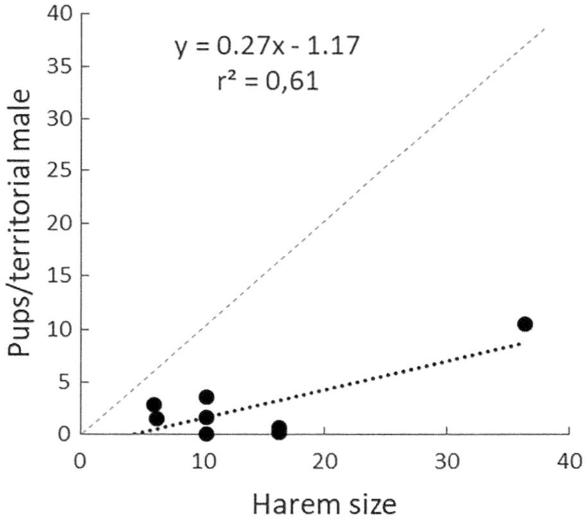

Fig. 6.2 Relationship between genetic estimation (average paternity of territorial males) and behavioural estimation (harem size) of RS

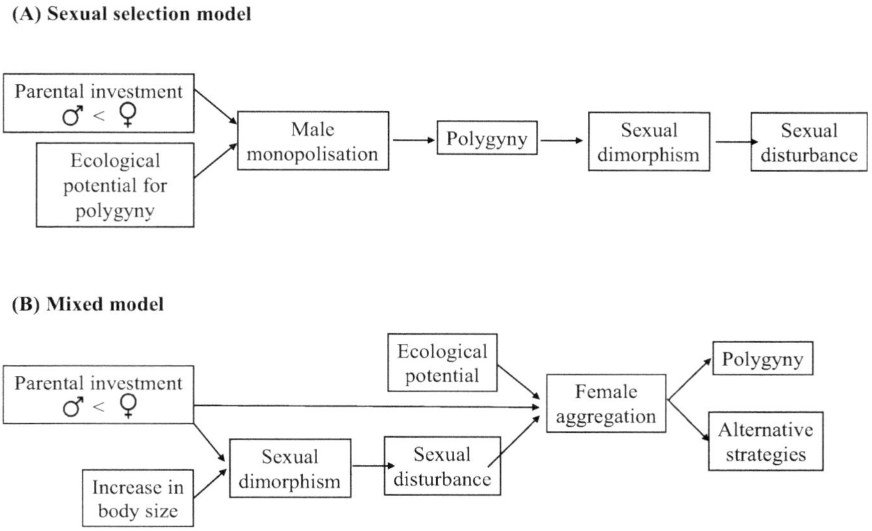

Fig. 6.3 (**a**) Sequence of evolutionary events postulated by the classical sexual selection model of the evolution of polygyny and sexual size dimorphism in mammals. (**b**) Mixed model of the evolution of sexual size dimorphism and mating systems in mammals (described in text)

6.6 Final Remarks

The theory of sexual selection applied to mating systems has been previously challenged by new discoveries originating in molecular techniques. The first major challenge occurred with monogamous species, when it was discovered that many animals that form long-term pair bonds produce extra-pair offspring as a norm. Studies employing molecular markers have demonstrated that in many cases levels of extra-pair parentage cannot accurately be estimated from behavioural observations alone (Hughes 1998). These results stimulated research in the field of mate selection by females, which became one of the most fruitful fields within behavioural ecology research during the late twentieth century.

At present, we are facing a similar phenomenon regarding sexual selection in polygynous mating systems. In this study, I reviewed published studies on genetic paternity in Otariids and concluded that classical theories on the evolution of polygyny in mammals should be revised in light of the new results provided by molecular ecology, specifically the high level of extra-harem paternity. I proposed a new approach that relativizes the role of sexual selection in favour of a more holistic approach, in which the influence of different forms of selection is evaluated. Efforts should be made to increase the number of studies on genetic paternity, and on the behaviour of adults of both sexes who move between areas of a colony and between colonies.

Acknowledgements I thank Claudio Campagna for the invitation that this work originated from. MHC is a researcher of the Argentinean Research Council CONICET.

References

Ahonen H (2013) Population structure and mating system of the Australian sea lion (*Neophoca cinerea*). Neurons and Cognition [q-bio.NC]. Université Paris Sud - Paris XI

Bartholomew GA (1970) A model for the evolution of pinniped polygyny. Evolution 24:546–559

Bartlett JMS, Stirling D (2003) A short history of the polymerase chain reaction. Methods Mol Biol 226:3–6

Berta A (2018) Pinnipeds. In: Würsig B, Thewissen JGM, Kovacs K (eds) Encyclopedia of marine mammals. Academic, New York, pp 734–740

Bonnes DJ (1991) Determinants of mating systems in the Otariidae (Pinnipedia). In: Renouf D (ed) The behaviour of Pinnipeds. Chapman & Hall, New York, pp 1–44

Campagna C, Le Boeuf BJ (1988) Reproductive behaviour of southern sea lions. Behaviour 104:233–260

Campagna C, Le Boeuf BJ, Cappozzo HL (1988) Group raids: a mating strategy of male southern sea lions. Behaviour 105:224–249

Cassini MH (1999) The evolution of reproductive systems in pinnipeds. Behav Ecol 10(5):612–616

Cassini MH (2017) Role of fecundity selection on the evolution of sexual size dimorphism in mammals. Anim Behav 128:1–4

Cassini MH (2020a) A mixed model of the evolution of polygyny and sexual size dimorphism in mammals. Mammal Rev. https://doi.org/10.1111/mam.12171

Cassini MH (2020b) Sexual size dimorphism and sexual selection in primates. Mammal Rev. https://doi.org/10.1111/mam.12191

Cassini MH (2020c) Sexual size dimorphism and sexual selection in Artiodactyles. Behav Ecol. https://doi.org/10.1093/beheco/araa017

Caudron AK, Negro SS, Fowler M, Boren L, Poncin P, Robertson BC, Gemmell NJ (2010) Alternative mating tactics in the New Zealand fur seal (*Arctocephalus forsteri*): when non-territorial males are successful too. Aust J Zool 57(6):409–421

Darwin C (1871) The descent of man, and selection in relation to sex. Wiley, London

Emlen ST, Oring LW (1977) Ecology, sexual selection and the evolution of mating systems. Science 197:215–223

Flatz R, González-Suárez M, Young JK, Hernández-Camacho CJ, Immel AJ, Gerber LR (2012) Weak polygyny in California Sea lions and the potential for alternative mating tactics. PLoS One 7:e33654

Franco-Trecu V, Costa P, Schramm Y, Tassino B, Inchausti P (2014) Sex on the rocks: reproductive tactics and breeding success of South American fur seal males. Behav Ecol 25(6):1513–1523

Gemmell NJ, Burg TM, Boyd IL, Amos W (2001) Low RS in territorial male Antarctic fur seals (*Arctocephalus gazella*) suggests the existence of alternative mating strategies. Mol Ecol 10:451–460

Gonzalez-Suarez M, Cassini MH (2014) Variance in male RS and sexual size dimorphism in pinnipeds: testing an assumption of sexual selection theory. Mammal Rev 44:88–93

Griffith SC (2007) The evolution of infidelity in socially monogamous passerines: neglected components of direct and indirect selection. Am Nat 169(2):274–281

Hoffman JI, Boyd IL, Amos W (2003) Male reproductive strategy and the importance of maternal status in the Antarctic fur seal *Arctocephalus gazella*. Evolution 57:1917–1930

Hughes C (1998) Integrating molecular techniques with field methods in studies of social behavior: a revolution results. Ecology 79:383–399

Jones AG, Ratterman NL (2009) Mate choice and sexual selection: what have we learned since Darwin? Proc Natl Acad Sci 106(Suppl 1):10001–10008. https://doi.org/10.1073/pnas.0901129106

Kiyota M, Insley SJ, Lance S (2008) Effectiveness of territorial polygyny and alternative mating strategies in northern fur seals, *Callorhinus ursinus*. Behav Ecol Sociobiol 62:739–746

Kokko H, Jennions MD, Houde A (2007) Evolution of frequency-dependent mate choice: keeping up with fashion trends. Proc R Soc B Biol Sci 274(1615):1317–1324

Kotiaho JS, Puurtinen M (2007) Mate choice for indirect genetic benefits: scrutiny of the current paradigm. Funct Ecol 21(4):638–644

Krüger O, Wolf JBW, Jonker RM, Hoffmann JI, Trillmich F (2014) Disentangling the contribution of sexual selection and ecology to the evolution of size dimorphism in pinnipeds. Evolution. https://doi.org/10.1111/evo.12370

Lack D (1968) Ecological adaptations for breeding in Birds. Methuen, London

Leutenegger W, Cheverud J (1982) Correlates of sexual dimorphism in primates: ecological and size variables. Int J Primatol 3:387–402

Leutenegger W, Cheverud J (1985) Sexual dimorphism in Primates. The effects of size. In: Jungers WL (ed) Size and scaling in primate biology. Plenum, New York, pp 33–60

Lidgard DC, Boness OJ, Bowen WD, McMillan JI (2005) Statedependent male mating tactics in the grey seal: the importance of body size. Behav Ecol 16:541–549

Lindenfors P, Tullberg BS, Biuw M (2002) Phylogenetic analyses of sexual selection and sexual size dimorphism in pinnipeds. Behav Ecol Sociobiol 52:188–193

Mead LS, Arnold SJ (2004) Quantitative genetic models of sexual selection. Trends Ecol Evol 19 (5):264–271

Meise K, Piedrahita P, Kruger O, Trillmich F (2014) Being on time: size-dependent attendance patterns affect male RS. Anim Behav 93:77–86. https://doi.org/10.1016/j.anbehav.2014.04.019

Mesnik SL, Ralls K (2018) Mating systems. In: Würsig B, Thewissen JGM, Kovacs K (eds) Encyclopedia of marine mammals. Academic, New York, pp 712–719

Miller EH (2018) Territorial behavior. In: Würsig B, Thewissen JGM, Kovacs K (eds) Encyclopedia of marine mammals. Academic, New York, pp 1156–1166

Porschmann U, Trillmich F, Müller B, Wolf JBW (2010) Male RS and its behavioural correlates in a polygynous mammal, the Galapagos Sea lion (*Zalophus wollebaeki*). Mol Ecol 19:2574–2586

Sheldon BC, Ellegren H (1999) Sexual selection resulting from extrapair paternity in collared flycatchers. Anim Behav 57(2):285–298

Simmons LW, Fitzpatrick JL (2012) Sperm wars and the evolution of male fertility. Reproduction 144(5):519

Slatkin M (1984) Ecological causes of sexual dimorphism. Evolution 38:622–630

Soto KH, Trites AW (2011) South American sea lions in Peru have a lek-like mating system. Mar Mamm Sci 27:306–333

Trivers RL (1972) Parental investment and sexual selection. In: Campbell B (ed) Sexual selection and the descent of man, 1871–1971. Aldine-Atherton, Chicago, pp 136–179

Wade MJ, Arnold SJ (1980) The intensity of sexual selection in relation to male sexual behavior, female choice, and sperm precedence. Anim Behav 28:446–461

Whittingham LA, Dunn PO (2005) Effects of extra-pair and within-pair reproductive success on the opportunity for selection in birds. Behav Ecol 16(1):138–144

Chapter 7
The Role of Females in the Evolution of Sexual Dimorphism and Mating Systems

Marcelo H. Cassini

Abstract A hypothetical evolutionary process that may have given rise to sexual dimorphism and the mating system of pinnipeds was proposed half a century ago by Bartholomew (Evolution 24:546–559, 1970), yet it still remains the most accepted theory that links limited resources, body shape and behavior in the group (see Cassini, Chap. 6). Under this sexual selection approach, females are viewed as passive actors in the evolution of sexual characters and mating systems. The objective of this chapter is to propose a more active role of females in the evolutionary ecology of pinnipeds than that traditionally described. In this attempt to prioritize the role of females, the role of natural selection in the evolution of the sexual characteristics of these marine mammals becomes more relevant. The first section analyses the evolution of body size in pinnipeds, based on paleontological records, and describes the numerous advantages of large body size for a marine life. Each of the remaining four sections describe a different aspect of females that would have had an influence on the evolution of some unique pinniped trait. I argue that: (i) sexual size dimorphism skewed towards males can be explained by at least five different evolutionary mechanisms not related to sexual selection, (ii) that energy constraints of lactation has been an influential selective force that explains why phocids are on average significantly larger than otariids, (iii) that female avoidance of sexual disturbance is one of the most important determinant of the evolution of mating systems in pinnipeds, and (v) that mate choice by females has a more significant role in sexual selection than previously thought.

Keywords Body size · Pinnipeds · Sexual disturbance · Energy constraints · Natural selection

M. H. Cassini (✉)
Laboratorio de Biología del Comportamiento, IBYME, CONICET, Buenos Aires, Argentina

C. Campagna, R. Harcourt (eds.), *Ethology and Behavioral Ecology of Otariids and the Odobenid*, Ethology and Behavioral Ecology of Marine Mammals,
https://doi.org/10.1007/978-3-030-59184-7_7

7.1 Introduction

Sexual selection is proposed to explain the evolution of traits that provide advantages in reproductive fitness (Darwin 1874). While females are limited by gamete production and parental investment, leading to their reproductive success (RS) being directly linked to survival and health (Bateman 1948; Trivers 1972), male RS is determined by the number of mating partners that can be acquired per unit of time (Bateman 1948; Trivers 1972; Clutton-Brock 1988; Dubuc et al. 2014). Thus, it is expected to find stronger sexual selection operating in males than in females. This difference should be particularly pronounced in large mammals: females show lower reproductive rate and higher costs in gestation and lactation, while males do not often invest in offspring, concentrating their reproductive investment to mating effort, which in turn creates a strong opportunity for sexual selection (Trivers 1972).

Sexual selection theory predicts that intra-male competition for mates should be strong in mammals (Darwin 1874). Dominant males are expected to monopolize fertile females and exclude other males from reproduction, leading to a high variance in male RS within populations (Wade 1979; Andersson 1994; Arnold and Duvall 1994). This, in turn, selects for traits that provide advantages in male-male contests, such as large male body size and various forms of weaponry (e.g., antlers, long sharp canines; Alexander et al. 1979; Loison et al. 1999; Dubuc et al. 2014). Sexual selection theory also predicts that the most common mating system in mammals is polygyny. Females are distributed according to the resources that they need for breeding, they group and make possible for males to monopolize access to them or the resources that females need (Emlen and Oring 1977).

Among the three pinniped families, phocids, otariids and odobenids, the Otariids are characterized as highly polygynous and sexually dimorphic, biased towards males (Table 7.1). Three species of phocids share these features with the otariids, i.e., extreme polygyny and males much larger than females: southern elephant seal, *Mirounga leonina*, northern elephant seal, *M. angustirostris* and gray seal, *Halichoerus grypus*. The Odobenidae family, represented only by the walrus, *Odobenus rosmarus*, is less well known; copulations are difficult to observe, and the mating system seems to be slightly polygynous according to Fay (1982), and

Table 7.1 Characteristics of the three families of pinniped

Characteristic	Otariids	Most phocids	*Mirounga* spp. and *H. grypus*
Body size	Smaller	Larger	
No. of species	15	15	3
Gregariousness	High	Low or solitary	High
Mating system	Polygyny	Slight polygyny	Polygyny
Dimorphism	High	Low	High
Breeding	Land	Ice or land	Land
Mating	Land	Water	Land
Lactation	Long	Short	Short

may involve female defense (Sjare and Stirling 2011). Due to paucity of data, this analysis excludes the walrus.

A hypothetical evolutionary process that may have given rise to sexual dimorphism and the mating system of pinnipeds was proposed half a century ago by Bartholomew (1970), yet it still remains as the most accepted theory that links limited resources, body shape and behavior in the group (see Cassini, Chap. 6). This theory places fundamental importance on the availability of breeding space (Stirling 1983): land-breeding females aggregate due to limited space or anti-predator behaviour while ice-breeding females, with access to potentially larger breeding space, are more dispersed and so less susceptible to being monopolized (Lindenfors 2002). This would explain polygyny in the otariids, while most phocids are closer to monogamy. This interpretation also conforms to Emlen and Oring (1977) one-way model, where the first link is the environmental potential for polygyny (see Cassini, Chap. 6). Under this sexual selection approach, the place for females in pre-natal reproduction is to increase the potential for polygyny when they aggregate in sites with resources they need for breeding. Thus, females are viewed as more passive actors in the evolution of sexual characters and mating systems, at least for these mammals.

The objective of this chapter is to propose a more active role of females in the evolutionary ecology of pinnipeds than that traditionally described. In this attempt to prioritize the role of females, the role of natural selection in the evolution of the sexual characteristics of these marine mammals becomes more relevant. The first section analyses the evolution of body size in pinnipeds, based on paleontological records, and describes the numerous advantages of large body size for a marine life. Each of the remaining four sections describes a different aspect of females that would have had an influence on the evolution of sexual characters and mating systems in pinnipeds. I argue that: (i) sexual size dimorphism skewed towards males can be explained by at least five different evolutionary mechanisms not related to sexual selection, (ii) energy constraint of lactation has been an influential selective force that explains why phocids are on average significantly larger than otariids, (iii) female avoidance of sexual disturbance is one of the most important determinants of the evolution of mating systems in pinnipeds, and (iv) mate choice by females has a more significant role in sexual selection than previously thought. This last point is further developed in Cassini, Chap. 6 and illustrated empirically in Kiyota, Chap. 8 and Sinclair et al., Chap. 13 of this book.

7.2 Evolution of Body Size in Pinnipeds

Sexual dimorphism in body size was an early acquisition in the evolution of pinnipeds (Krüger et al. 2014). Contrary to predictions grounded in sexual selection, it has been demonstrated that sexual dimorphism did not evolve as a consequence of male-male competition for mates but appeared prior to the evolution of polygyny (Krüger et al. 2014). Early evolution of sexual dimorphism was probably a

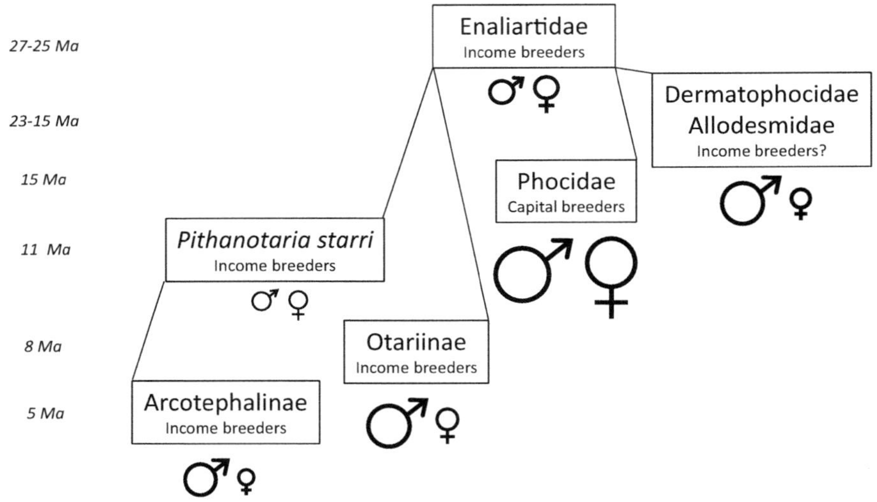

Fig. 7.1 Size, sexual size dimorphism, and maternal strategy of extinct and extant pinnipeds (size and SSD are represented by the symbol size). Numbers represent date of origin in millions of years

consequence of natural selection mechanisms operating during periods when environmental conditions favored the evolution of a large body size in pinnipeds, mainly during the Miocene.

Palaeontologists have described that both large body size and sexual size dimorphism appeared early in pinniped evolution (Fig. 7.1). There is consensus that it occurred due to climate cooling during much of the Miocene (Repenning 1976; Cullen et al. 2014; Schulz and Bowen 2005; Berta et al. 2015). *Enaliarctos mealsi* is considered the ancestral pinniped (Berta et al. 1989). Enaliarctids were small animals that show no evidence of sexual size dimorphism and would have had dense underfur, retained by living fur seals. They possessed heterodont dentition, suggesting coastal feeders because they had to haul their prey ashore to dismember and consume it, as was done by their fissiped ancestors (Schulz and Bowen 2005). Cullen et al. (2014) suggested that the Miocene-aged basal pinniped taxon *E. emlongi* showed male-biased size dimorphism, yet scant evidence supports this view (Berta et al. 2015). It turns out that the only specimen used to estimate female size may have been an immature individual (Berta (1991, p. 13). Other authors (e.g.: Repenning 1976; Schulz and Bowen 2005) did not come to the conclusion that these early ancestors were sexually dimorphic.

The earliest known otariid, *Pithanotaria starri*, occurred about 11 million years ago and was very small. All known specimens are about the same size, suggesting lack of size dimorphism. They had homodont dentition, presumably to more effectively capture marine prey, which could be swallowed whole at sea (Repenning 1976). The first unquestionable evidence of an increase in body size and the occurrence of dimorphism to a degree similar to living species in otariids comes from the North Pacific, about 8 million years ago (Repenning 1976). Similarly, the first lineage of ancestral phocids of the early to mid-Miocene period, *Desmatophoca* and *Allodesmus*, evolved into large species exhibiting size dimorphism (Berta 2002). I will return to body size of phocids in a later section.

Several benefits have been described related to having a large body size. Air-breathing homeotherms that dive for aquatic prey are presented with behavioral and physiological problems that are largely absent for animals foraging in a terrestrial environment, or on the surface. Marine mammals have developed several adaptations to diving, one of which is increased body size. Size is the main predictor of maximum diving duration in cetaceans (Marino et al. 2006). Large size allows for more oxygen storage relative to a lean body mass, hence greater capacity for deep diving (Weise et al. 2010). Another advantage of having larger body size with a marine existence is that it allows for more insulation in cold waters (Mayr 1963; Watts et al. 1993). A large body size and a thick blubber layer also provides fat reserves that permit long trips to distant, rich foraging patches, such as polar regions (Le Boeuf et al. 2000). Finally, large body size may represent an advantage when it comes to protection against predators (Churchill et al. 2015).

7.3 Role of Females in the Evolution of Sexual Size Dimorphism: Fecundity Selection

The previous section suggested that sexual dimorphism in body size is an early acquisition in the evolution of pinnipeds closely associated to the increase of body size during the Miocene. I postulate that: (i) sexual dimorphism may have evolved as a consequence of the increase in body size of males due to intra-sexual competition, but it was prior to the evolution of polygyny (Cassini 2020), and (ii) the evolution of male-biased size dimorphism may be explained by natural selection operating on both sexes in favor of large body size in pinnipeds, during the Miocene, and fecundity selection operated only in females against large body size (Cassini 2017). Since body size is positively related to maturation age in many species, selection for early maturation in females may lead to a decrease in body size in females in relation to males (Clutton-Brock and Harvey 1978).

What about the evolutionary mechanism that was responsible for early evolution of large body size? The most accepted hypothesis for larger males is that it is the result of intra-sexual selection due to competition for mates: larger males win fights and obtain more females. Fitzpatrick et al. (2012) supported this hypothesis in a study that used comparative phylogenetic methods to analyze the correlates of phenotypic diversification in body mass in pinnipeds. However, they adopted harem size as a measure of intensity of sexual selection, while genetic studies indicate that there is little correlation between harem size and paternity in Otariids (see Bonin, Chap. 11). Further, Gonzalez-Suarez and Cassini (2014) showed a lack of correlation between paternity and male dimorphism in pinnipeds.

Cullen et al. (2014) were the first to incorporate data from the fossil record into the analyses of evolution of mating systems in extant pinnipeds. They suggested that polygyny evolved early in pinnipeds because paleontological records appeared to suggest that the taxon showed sexual dimorphism, i.e., they inferred polygyny from dimorphism. However, the assertion that *Enialarctos* was polygynous "following a near-exclusive association between a polygynous mating system and dimorphism in extant relatives" (Cullen et al. 2014, p. 1478) is problematic as correlation may be interpreted as causality. They argued that polygyny was the mating system of the ancestor based on a correlation between large males and polygyny in extant relatives; thus, the conclusion that polygyny in extant relatives originated in this ancestor is not justified.

Important evidence against the hypothesis of sexual selection as the evolutionary mechanisms that resulted in larger males than females is provided by Krüger et al. (2014). There are no records of the mating system, thus it is not possible to know, using paleontological methods, if size dimorphism evolved prior to or following polygyny. However, by using phylogenetic comparative methods, Krüger et al. (2014) concluded that sexual size dimorphism does appear to have evolved before polygyny in pinnipeds. In this seminal work, the authors proposed that the differences in body size between the sexes could have evolved as a consequence of niche partitioning during aquatic foraging, and not due to intra-male sexual competition. Their analysis using approaches other than paleontological evidence suggested that ancestral pinnipeds were sexually monomorphic, formed small female groups and showed a promiscuous mating system with equal variance in RS in both sexes. Their evolutionary sequence analysis recognized three changes before the evolution of size dimorphism and mating system: the evolution of deep diving, the colonization of breeding habitats without terrestrial predators, and a shift from aquatic copulation to copulation on land/ice. They proposed that deeper dives would select for larger size, as previously described by traditional phylogenetic studies (Repenning 1976; Berta et al. 2018).

In summary, male size dimorphism could have evolved prior to polygyny, and coevolved with body size. The next question is why females did not evolve to similar sizes as males. In the literature, many mechanisms have been proposed for the evolution of sexual size dimorphism based on natural selection. Slatkin (1984) proposed dimorphic niches, bimodal niches and competitive displacement as ecological causes of sexual dimorphism. Dimorphic niches arise "if there are intrinsic differences between males and females because of their different energetic needs to ensure successful reproduction or because of their different social roles, then it is possible that there is a different optimum value of the trait in each sex" (Slatkin 1984, p. 623). The bimodal niche model is distinguished from the previous model by the fact that sexes do not show differences in energetic needs but the environment provides with two optima to which they could evolve separately. For example, niche partitioning may result when the sexes compete for resources.

Another approach to the evolution of dimorphism investigates the causes and effects of variance dimorphism within the sexes. Leutenegger and Cheverud (1982) proposed that size dimorphism may evolve solely as a consequence of differences between the sexes in the phenotypic variance in body size. They postulated that, if there were variance in dimorphism in the ancestral species and selection occurred for larger body size in both sexes due to new ecological pressures (e.g. decrease in ambient temperature or increase in predation pressure), sexual dimorphism would evolve together with mean body size. Thus selection for larger body size would also increase dimorphism, as an "unintended" or indirect result. They applied their approach to the evolution of dimorphic males in primates, where a transition between arboreal to terrestrial habits was followed by the increases both in body size and by larger males.

In summary, there are at least five competing hypotheses on the evolutionary origin of male size dimorphism in mammals. Yet, another question remains: why is it that pinniped females are always the smallest sex when there is strong skew in size? Only one of the five mechanisms can explain why the bias is always towards males and not to females: the dimorphic niche hypothesis requires that males and females show fundamental differences that cause the divergence. In mammals, these dissimilarities are in energetic costs to ensure successful reproduction, i.e. gestation and lactation. It has been demonstrated that (i) energetic costs increase, and (ii) fecundity (measured as litter size x litter per year, biomass produced per unit of adult body size, etc) decreases with body size (Allainé et al. 1987; Boyce 1988; Lee et al. 1991; Purvis and Harvey 1995). These two patterns are explained by trade-offs between reproduction and maintenance/growth. On the one hand, gestation and lactation are particularly expensive considering that mammals are homeotherms and females must not only produce food for their offspring, but perform self-maintenance: large bodies (both of the mother and of the offspring) require more energy than small ones (Rogowitz 1996). On the other hand, there is a trade-off between growth and reproduction: growth to a large size takes time and energy, so life-history relationships with body mass are often thought of as physiologically-constrained allometries, with resulting trade-offs between body size and reproduction (Charnov 1993).

Cassini (2017) developed a model that explains the simultaneous evolution of body size and male dimorphism, without resorting to sexual selection and a polygynous system, where intra-sexual competition operates (Fig. 7.2). When natural selection promotes large body size in mammals, females do not reach the same size as males due to fecundity constraints. Larger size can mean benefits in terms of a greater dive time or greater insulation from the cold, but it also implies a higher cost in terms of total higher energy requirements that would be especially costly for lactating females.

7.4 Role of Females in the Evolution of Body Size in Pinnipeds: Lactation Strategies

Theory described in previous sections can explain the origin of male dimorphism in size in otariids as in most large mammalian taxa. I will here address why most phocid females are of similar size to males.

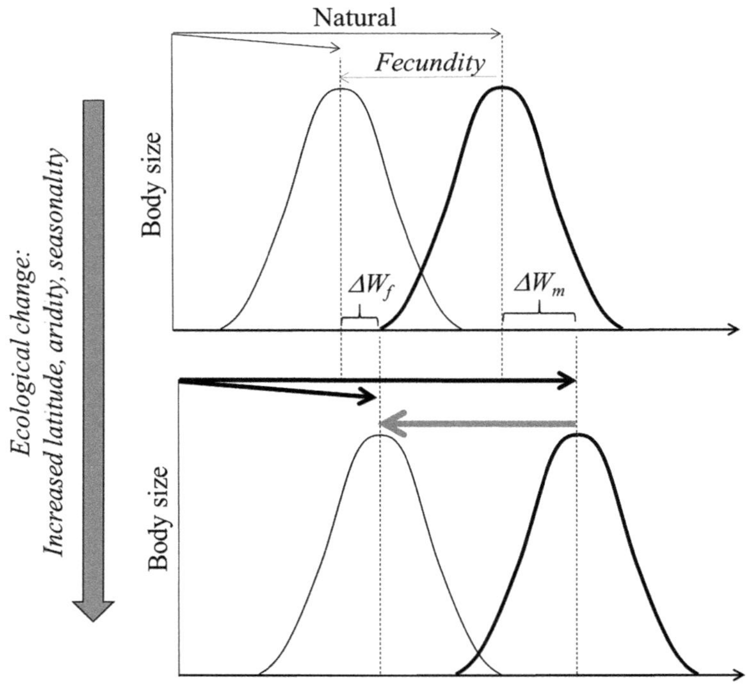

Fig. 7.2 Natural selection model of sexual size dimorphism (from Cassini 2017). Gaussian distribution of body size for females (fine line) and males (gross line) with (**a**) low and (**b**) high natural selection shown. The thickness of the arrow represents the strength of natural and fecundity selection. When an ecological change favors a large body size, such as increased latitude, aridity or seasonality, the strength of natural selection increases

An important overall characteristic of phocids is their larger body size, compared to otariids (Wyss 1994). Differences between families in body size may be explained by an early evolution of different lactation strategies. Maternal strategies in these families have been classified into two categories, according to the way in which the costs of reproduction are allocated (or 'financed'): 'capital breeding' refers to a situation in which reproduction is based upon stored-energy reserves. 'Income breeding' refers to the use of current energetic income to sustain reproduction and maternal attendance (Houston et al. 2007). Endotherms tend to be categorized as income breeders, as opposed to ectotherms that are considered to be mainly capital breeders, because their low metabolism favors the accumulation of body reserves to fuel reproductive effort (Bonnet et al. 1998).

Each pinniped family has developed a different maternal strategy, where both phylogeny and ecology appear to be important in their evolution (Berta et al. 2015). Female phocids can be considered capital breeders with a large body mass of phocid mothers that allows the storage of sufficient fat reserves to invest heavily in offspring during an exceptionally short lactation period and without supplemental feeding

(Boyd 2000). In contrast, otariid females follow an income strategy: they lack the body mass necessary to bring ashore fat reserves that are sufficient to support their reproductive investment during lactation; they follow a foraging-cycle strategy with a few days at sea foraging, interspersed with shore-based nursing (reviewed by Berta et al. 2015; Sepulveda and Harcourt, Chap. 3).

But how did these remarkable differences in the breeding patterns among families of pinnipeds arise? To answer this question, we must look at evolutionary history. Enaliarctids may have exhibited a form of the otariid-foraging-cycle lactation strategy and retained a lactation length of several months, both of which are typical of many terrestrial carnivores (Costa 1993). Thus, the breeding pattern of numerous short-duration feeding trips that maximize energy and nutrient delivery to the pup, appears to be ancestral to both otariids and phocids, being most closely linked to their terrestrial common ancestors (Berta et al. 2015).

So, how did the maternal strategies of phocids and otariids diverge from this ancestral strategy? Schulz and Bowen (2005) used phylogenetic comparative statistical methods to account for phylogenetic effects on correlations between traits related to lactation strategies. They combined the result of these analyses with information on pinniped paleobiology and paleoclimatology to produce a conceptual model on the evolution of maternal strategies in pinnipeds. They concluded that ancestral phocids and otariids responded to selective pressures in different ways, with the result of two different maternal strategies and body sizes.

Females of the earliest known otariid, *Pithanotaria starri*, may have likely bred in isolated rookeries that probably reduced exposure to terrestrial predators and allowed proximity to highly productive local prey resources (Berta et al. 2015). They would have retained the ancestral maternal strategy (Schulz and Bowen 2005). Climate cooling during much of the Miocene would also have selected for increased body size, which allows more efficient thermoregulation and foraging. Otariid females retained the foraging-cycle lactation strategy, increasing the duration but reducing the frequency of foraging trips during lactation (Costa 1993). Reduced frequency of milk provisioning to offspring selected for lower offspring growth rate and extended duration of lactation to ensure sufficient transfer of energy (Schulz and Bowen 2005). The southern fur seal genus *Arctocephalus* represent the otariid prototype, and spread to the southern hemisphere at least 5 million years ago. Sea lions, the most recent development of otariids, appear to have spread to the southern hemisphere less than 3 million years ago, with an increase in size, and a loss of insulating fur. Although sea lion species have evolved larger body size, they still retain the foraging strategy, high metabolic rates, and hence the prolonged lactation that likely characterized early otariids (Schulz and Bowen 2005).

Ancestral phocids of the early Miocene period, Desmatophocids, had acquired homodont dentition and evolved into large body size, suggesting that they could efficiently exploit distant and less predictable prey, and store energy in the form of blubber (Deméré and Berta 2002). Large body size also would allow females to change their foraging strategy: an increasing reliance on stored maternal energy to support milk production meant that nutrient transfer to young would be optimized by concentrating milk energy and reducing maternal overhead by minimizing the duration of lactation (Costa 1993). Thus, the fundamental characteristics of the phocid lactation strategy appeared early in pinniped evolution (Schulz and Bowen 2005), in close association with an increase in body size. While otariids increased lactation duration, female phocids took advantage of sea-ice for parturition and nursing, which in turn selected for an abbreviated duration of lactation, as well as high rates of milk energy output and pup mass gain.

The model described in the previous section (see also Cassini 2017) may explain differences in size dimorphism between phocids and otariids. The difference in maternal strategies suggests that the influence of fecundity selection on female body size should have followed different paths. In otariids, the role of fecundity selection is probably the same as in the rest of mammals, i.e. females are constrained when it comes to an increase in body size due to the trade-off between survival and reproduction. On the other hand, the maternal strategy of phocid females resembles that of oviparous species: both make a simultaneous maternal investment in the production of the fetus and its nutrients, and then make a relatively short and less expensive post-birth investment. Even when a negative correlation between body size and pup production exists, its effect would be counterbalanced by the advantages that a larger body size has when it comes to the accumulation of nutrients that are invested, both in survival at sea and in capital breeding strategy. Trillmich (1996) described how fertility (and perhaps mortality) effects are more pronounced in otariids than in phocids, and he related this fact to the difference in maternal strategies between taxa.

7.5 Role of Females in the Evolution of Mating Systems: Avoidance of Male Harassment

As discussed in the previous section, Krüger et al. (2014) suggested that male dimorphism evolved prior to polygyny in pinnipeds, and that size dimorphism would have evolved related to dominance exercised on females. Trillmich and Trillmich (1984) argued for the importance that male harassment may have on the mating system of pinnipeds. They proposed that females may choose to reproduce in an area defended by a strong territorial male that may protect them from copulation attempts by marginal males. In this section, I postulate how the avoidance of sexual harassment may have been the main force responsible for the evolution of polygyny in otariids but not in phocids.

Male harassment is widespread in pinnipeds, takes several forms, and is mainly performed by satellite males. 'Herding' is probably the most common type of harassment in the otariids. It occurs when males position themselves between the female and her goal if she starts moving away (Marlow 1975). There is a difference in the intensity of this behavior, depending on whether performed by a satellite male, a male with one female, or a dominant male that displays the harassment. Herding is rare in males with large harems, except when females are in estrus. In contrast, herding may be intense, even violent, when males control a single female. This active herding frequently results in separation of a female and her pup.

'Interception' is another type of harassment (Cassini and Vilá 1990; Chilvers et al. 2005). It occurs when females arrive at the colony from the sea. Arriving females move slowly until one or more nearby satellite males approach her (the first males attract distant others). Females may respond by hurrying to reach the colony or escaping back to sea; other times they cannot avoid interception. Interception involves males smelling the female, preventing other males from approaching, and herding her towards a place where she can be controlled from escaping. Interception also occurs during 'group raids', in which ousted males attack the breeding area, attempting to seize females from resident, territorial males, as has been first described in southern sea lions (*O. flavescens*) (Campagna et al. 1988a, b, Chap. 9). Individual males with positions on the periphery of existing territories can also attempt to seize females, but they tend to be less successful. Male harassment involves time and energy investment in avoidance behavior, and it may cause mother-pup separation, pup injury and death, or injuries and death of the female (Le Boeuf and Briggs 1977; Vilá and Cassini 1990; Campagna et al. 1988a, b; Francis 1987; Chilvers et al. 2004).

There are features in the reproduction of social pinnipeds that predispose females to be sexually harassed. Sexual dimorphism is one. Significantly larger and stronger males allow them to take and carry a female in their mouth, or flip them in the air, from a place to a better one for further controlling her movement (Marlow 1975; Francis 1987; Campagna and Le Boeuf 1988; Ryazanov, Chap. 10). Low mobility on land makes it difficult for the females to escape male coercion. Females so subdued may give birth and a few days later become sexually receptive, so male–female interactions occur when the pups still require attendance from the mother. The temporal relationship between these two life history components predisposes female RS to be affected by male coercion (Cassini 1999).

For females staying in a breeding group under the influence of a resident male may reduce the reproductive cost of harassment by satellite males, since resident males defend their harems against intruders (Trillmich and Trillmich 1984; Campagna et al. 1992; Cassini 1999, 2000; Kiyota, Chap. 8). However, males holding female groups can also harass the females within their groups. They show diverse forms of interaction with females: olfactory inspections, mount attempts, aggressive displays, fights and the physical blocking of female movements when females search for their pups or run away (Vaz-Ferreira 1975; Cassini 1985; Campagna and Le Boeuf 1988). These male behaviors can disturb breeding females and are the main cause of mother–pup separations (Le Boeuf 1974; Vilá and Cassini 1990; Harcourt 1992). Therefore, even when a female in a group can reduce harassment by satellite males, they can still suffer from harassment by the dominant male.

Female gregariousness may work to reduce harassment by resident males with two mechanisms (Fig. 7.3): by (1) a 'dilution effect' (Hamilton 1971), that reduces the probability of female interactions with the resident male as the breeding group size increases, or (2) pairing with successful resident males that indirectly protect

Fig. 7.3 Females aggregated around a male in *O. byronia*

females from coercion by satellite males (Cassini 1999, 2000). Males in large groups show fewer interactions with females as they devote time to other activities (e.g. rejecting intruders), or because males that harass less attract more females.

Campagna et al. (1992) directly tested the effect of harassment of satellite males on female RS. They compared mortality between pups born in the colony and those born in isolated pairs formed by a female and a marginal male. Mortality was 0.7% in pups born within the colony at the peak of the reproductive season, whereas it was 60% in pups born to females separated from the colony. These authors also demonstrated that the main cause of pup mortality in solitary pairs was male harassment by satellite males. Another test of the effect of male harassment was conducted by Cappozzo et al. (2008). To investigate the relationship between male harassment and female gregariousness in relation to the size of breeding groups, they analyzed the behavior of dominant males and their females in a breeding colony of southern sea lions. Females in large breeding groups received less harassment by resident males due to the dilution effect, i.e. the probability of a female receiving sexual coercion in a group decreases proportionally to the number of females in the group (Fig. 7.4).

In summary, evidence indicates that male harassment is widespread in otariids, that it can affect female RS, and that females do not respond passively but they join groups with other females to reduce the probability of being the target of male coercion. Traditional theory proposed that females aggregate in colonies due to the scarcity of breeding sites (Bartholomew 1970). Cassini (1999, 2000, 2011) proposed that females join harems to avoid sexual harassment, and built an ideal free distribution model with Allée effect (Fretwell and Lucas 1970) that predicts female distribution when there are social benefits of aggregation.

Ideal free distribution theory is a game theory set of models about population distribution, and the Allée effect refers to an increasing per capita growth rate with density at low rates (Cassini 2011). In behavioral terms, what these models predict is that, for example, a female who has to decide to give birth in one of two colonies

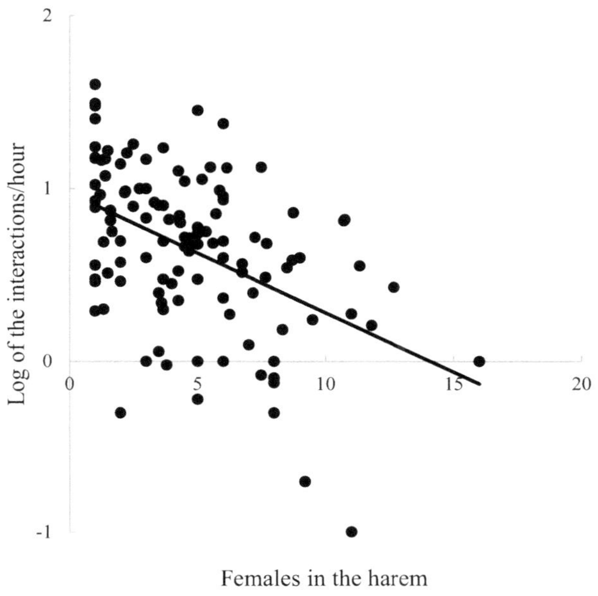

Fig. 7.4 Logarithms of female–male interaction rates in relation to the number of females in the breeding group. Log-transformation reduced the positive skew of the residuals and improved the value of r. From Cappozzo et al. (2008)

with different animal densities, will go to the most populated colony when the colony density is low, but she will go to the less dense colony when the mean density is high. In other words, at low densities being part of a group is beneficial (due to avoidance of sexual coercion), but at high densities the cost of intra-female competition overcomes that benefit. Fretwell and Lucas's (1970) model on population distribution can be applied to female habitat selection during the breeding season, with (1) the probability of offspring survival as currency, (2) the increase in female–female competition for resources required for breeding as the cost of increasing density, and (3) a reduction in male coercion as the benefit. The model predicts that male coercion forces females to form denser aggregations than expected based on the distribution of resources alone, because a reduction in male harassment compensates for the increase in female competition.

The most important prediction of this model is that social behavior (in otariids: male harassment avoidance), rather than resource distribution, is the main factor responsible for female otariid aggregation. This prediction is also important because it shows how female behavior could be the most important factor in the evolution of male mating strategies: without size dimorphism in males that sustains sexual coercion, polygyny will not necessarily evolve. This 'social potential for polygyny' also explains why only the dimorphic phocids, elephant seals and gray seals, show high levels of polygyny (Table 7.1). In summary, male size dimorphism evolved prior to polygyny and it also created the condition for its evolution: larger males may harass females more effectively for their purposes, and females defended themselves by forming groups.

7.6 Role of Females in Sexual Selection: Extra-Harem Paternity, Lekking and Mate Choice

Studies using molecular techniques indicate that a large proportion of pups that are born in a reproductive colony of otariids are not fathered by the territorial male associated with the mother (see Bonin, Chap. 11). The occurrence of extra-harem copulations opens up the possibility of female choice of mates that are not necessarily the harem holder.

Mate choice is a most important research topic in behavioral ecology (Jones and Ratterman 2009), yet the genetic mechanisms underlying its evolution remain the subject of debate (Andersson and Simmons 2006). There are two broad types of benefit: direct benefits or good resources (nest sites or food) and indirect benefits or good genes (Davies et al. 2012). Mate choice can occur both before and after copulation, and sperm competition has been extensively studied since Parker (1970) postulated it as a mechanism.

However, mate choice has been neglected in studies of otariid mating behavior (Gemmell et al. 2001). It was assumed that a female would most probably mate with the male occupying the territory that she is in when she becomes receptive, and that she will mate once. Even though "extra-harem" copulations are observed, it may be questioned if these are related to female choice. For example, Caudron et al. (2009) suggested that extra-harem copulations occur only because nulliparous females come into estrus after the peak of the breeding season, when the territorial males have left, or it may be a response to females being young and inexperienced hauling out to breed in areas away from the main territorial areas. Kiyota et al. (2008) proposed that extra-harem copulations were mainly the consequence of male alternative mating strategies, such as satellite males that force females to mate. This type of study suggests that females that have high RS copulate with the territorial male with which they are associated, while unsuccessful females are those that are engaged in extra-harem copulations. The single instance of mate choice that is traditionally accepted is in which the female chooses her pupping site. In this case, the female may be choosing a male that provides resources for thermoregulation (e.g. pools) or protection against the harassment of other males. She could be choosing competitive skills that could be genetically transferred to her pup (e.g. body size) (Renouf 1991). In other words, only territorial males will be selected.

However it is now recognized that mate choice is an important and complex behavior in Otariids. Goldsworthy et al. (1999) studied mate choice in Antarctic fur seals *A. gazella* and Subantarctic fur seals *A. tropicalis*. They found that females show a preference for males that shared the same species-typical characteristics. Non-territorial males fathered 34% of the pups. Following on, Sinclair et al. (Chap. 13) experimentally demonstrated that females can recognize the species of the territory holder, they can decide to look for extra-harem copulations with co-phenotypic males, and they can enhance their mate choice by leaving the territory (where they gave birth) prior to estrus. Goldsworthy et al. (1999) also observed that copulations do not ensure paternity and that females may copulate with more than one male. Multi mating strategies have also been described in New Zealand fur seals, also in the context of hybridization between *Arctocephalus* species (Lancaster et al. 2006, 2007a, b).

Hoffman et al. (2007) found that, in Antarctic fur seals females show active choice for males in a balance between high heterozygosity and low relatedness, both of which tend to increase offspring heterozygosity. They showed that female movements were related to mate choice. There was a negative relationship between the distance that females moved and the father's relatedness. They concluded that females move more in order to maximize the heterogeneity of their partners. Investigating the same species, Gemmell et al. (2001) went further and suggested that female choice is an integral component of the Antarctic fur seal mating system. Their data suggested that most males that sire pups on their study beach reside elsewhere, and consequently that females move to search for mates. They also found that multiple copulations are the norm among females. It is possible that most successful copulations occur after the females have left the beach on their first foraging trip. They deduced that mate choice is carried out during aquatic copulation. In the water, the females probably have more control over encounters and can choose which males they want to mate with. It is possible that aquatic mating may only be suitable for the very best (or at least preferred) males.

In a long-term study with Steller sea lions (*Eumetopia jubatus*), Parker and Manicalco (2014) found that approximately one-half the postpartum females copulated with a male in a different territory from where they gave birth. Their analysis indicated that females preferred to copulate with males holding centrally located coastal territories with pathways to the sea. A similar figure of extra-harem copulations was found in southern sea lions (Soto and Trites 2011). Most of these females moved freely from where they remained with their pup after giving birth (which was close to their birth males) and walked toward males located as far as ten neighboring males away. Females were observed to solicit mounting by exposing their genitalia to the chosen male. Males responded by smelling or licking the vulvae before mounting a female.

There is an indirect line of evidence for the existence of mate choice in Otariids: a growing number of authors have suggested a lek-like mating system for their species (California sea lions, *Zalophus californianus*, Flatz et al. 2012; Antarctic fur seals, Hoffman et al. 2007; Southern sea lions, Soto and Trites 2011; South American fur seals, *A. australis,* Franco-Trecu et al. 2014). In lekking systems, males cluster to defend very small territories that rarely contain significant resources but are located in areas commonly used by females (Emlen and Oring 1977). Females benefit by male clumping since they can better compare a large number of potential mates in a minimum period of time, and so mate choice is an integral part of lekking.

These preliminary investigations show that mate choice is more complex than expected by the traditional theory, when the behavior of females around the estrus period is tracked. It remains to investigate the reproductive behavior of females in their first trip to the sea after parturition. The possibility that females seek extra-harem copulations remains open, considering that females can have multiple copulations and smaller, satellite males may have the greater abilities when it comes to post-copulatory competition.

7.7 Conclusions

Intra-sexual selection among males was the monolithic dominant explanation of the evolution of the sexual characteristics and behavior of mammals in general and of pinnipeds in particular, which has male competition for mates as the main selective force. In this chapter, I have tried to question this somewhat "male chauvinist" vision and explore the role played by females in the evolution of sexual characters and mating systems of pinnipeds.

Three recent main discoveries are fundamental to reconstruct this alternative approach: (i) difference in body size between phocids and otariids appeared early in paleontological records, (ii) sexual size dimorphism evolved prior to polygyny, and (iii) behavioral mating system is poorly related to the genetic measures of paternity, with extra-harem paternity as a common phenomenon. Throughout the chapter, I propose a series of hypotheses that assign to maternal strategies, fecundity constraints, male harassment avoidance and mate choice a more significant role in the evolution of pinnipeds than it was previously proposed.

References

Alexander RMN, Jayes AS, Maloiy GMO, Wathuta EM (1979) Allometry of limb bones of mammals from shrews (Sorex) to elephant (Loxodonta). J Zool Lond 189:305–314

Allainé D, Pontier D, Gaillard JM, Lebreton JD, Trouvilliez J, Clobert J (1987) The relationship between fecundity and adult body weight in homeotherms. Oecologia 73:478–480

Andersson M (1994) Sexual selection, vol 72. Princeton University Press

Andersson M, Simmons LW (2006) Sexual selection and mate choice. Trends Ecol Evol 21 (6):296–302

Arnold SJ, Duvall D (1994) Animal mating systems: a synthesis based on selection theory. Am Nat 143(2):317–348

Bartholomew GA (1970) A model for the evolution of pinniped polygyny. Evolution 24:546–559

Bateman AJ (1948) Intra-sexual selection in Drosophila. Heredity 2(3):349–368

Berta A (1991) New Enaliarctos∗ (Pinnipedimorpha) from the Miocene of Oregon and the role of "Enaliarctids" in pinniped phylogeny. Smithson Contrib Paleobiol (69):1–32

Berta A (2002) Pinniped evolution. In: Perrin WF, Wursig B, Thewissen JGM (eds) Encyclopedia of marine mammals. Academic Press, SanDiego, CA

Berta A, Ray CE, Wyss AR (1989) Skeleton of the oldest known pinniped, *Enaliarctos mealsi*. Science 244:60–62

Berta A, Sumich JL, Kovacs KM (2015) Marine mammals: Evolutionary biology, 3rd edn. Academic. San Diego, pp 35–58

Berta R, Churchill M, Boessenecker RW (2018) The origin and evolutionary biology of pinnipeds: seals, sea lions, and walruses. Annu Rev Earth Planet Sci 46:203–228

Bonnet X, Bradshaw SD, Shine R (1998) Capital versus income breeding: an ectothermic perspective. Oikos 83:333–341
Boyce MS (ed) (1988) Evolution of life histories of mammals: theory and pattern. Yale University Press, London
Boyd I (2000) State-dependent fertility in pinnipeds: contrasting capital and income breeders. Funct Ecol 14:623–630
Campagna C, Le Boeuf BJ (1988) Reproductive behaviour of southern sea lions. Behaviour 104(3–4):233–261
Campagna C, Le Boeuf BJ, Cappozzo HL (1988a) Group raids: a mating strategy of male southern sea lions. Behaviour 105:224–249
Campagna C, Le Boeuf BJ, Cappozzo HL (1988b) Pup abduction and infanticide in southern sea lions. Behaviour 107:44–60
Campagna C, Bisioli C, Quintana F, Perez F, Vila A (1992) Group breeding in sea lions: pups survive better in colonies. Anim Behav 43(4):541–548
Cappozzo HL, Tunez JI, Cassini MH (2008) Sexual harassment and female gregariousness in the Southern Sea lion *Otaria flavescens*. Naturwissenschaften 95:625–630
Cassini MH (1985) Comportamiento social de las hembras adultas del Lobo Marino del Sur, *Otaria byronia* (Blainville, 1820), durante la temporada de cría. Rev Latinoam Psicol 17:339–350
Cassini MH (1999) The evolution of reproductive systems of pinnipeds. Behav Ecol 10:612–616
Cassini MH (2000) A model on female breeding dispersion and the reproductive systems of pinnipeds. Behav Process 51:93–99
Cassini MH (2011) Consequences of local Allee effects in spatially structured populations. Oecologia 165:547–552
Cassini MH (2017) Role of fecundity selection on the evolution of size sexual dimorphism in mammals. Anim Behav 128:1–4
Cassini MH (2020) A mixed model of the evolution of polygyny and sexual size dimorphism in mammals. Mammal Rev 50:112–120
Cassini MH, Vilá BL (1990) Male mating behaviour of the southern sea lions. Bull Mar Sci 46 (2):555–559
Caudron AK, Negro SS, Fowler M, Boren L, Poncin P, Robertson BC, Gemmell NJ (2009) Alternative mating tactics in the New Zealand fur seal (*Arctocephalus forsteri*): when non-territorial males are successful too. Aust J Zool 57:409–421
Charnov EL (1993) Life history invariants: some explorations of symmetry in evolutionary ecology, vol 6. Oxford University Press, USA
Chilvers BL, Delean S, Gales NJ, Holley DK, Lawler IR, Marsh H, Preen AR (2004) Diving behaviour of dugongs, Dugong dugon. J Exp Mar Biol Ecol 304(2):203–224
Chilvers BL, Wilkinson IS, Duignan PJ, Gemmell NJ (2005) Summer foraging areas for lactating New Zealand sea lions Phocarctos hookeri. Mar Ecol Prog Ser 304:235–247
Churchill M, Clementz MT, Kohno N (2015) Cope's rule and the evolution of body size in Pinnipedimorpha (Mammalia: Carnivora). Evolution 69(1):201–215
Clutton-Brock TH (1988) Reproductive success. University Of Chicago Press, Chicago
Clutton-Brock TH, Harvey PH (1978) Mammals, resources and reproductive strategies. Nature 273:191–195
Costa DP (1993, October) The relationship between reproductive and foraging energetics and the evolution of the Pinnipedia. Symp Zool Soc Lond 66:293–314
Cullen TM, Fraser D, Rybczynski N, Schröder-Adams C (2014) Early evolution of sexual dimorphism and polygyny in Pinnipedia. Evolution 68(5):1469–1484
Darwin C (1874) The descent of man, and selection in relation to sex. John Murray, London
Davies NB, Krebs JR, West SA (2012) An introduction to behavioural ecology. Wiley-Blackwell, Oxford
Deméré TA, Berta A (2002) The pinniped Miocene *Desmatophoca oregonensis* Condon, 1906 (Mammalia: Carnivora) from the Astoria formation, Oregon. Smithson Contrib Paleobiol 93:113–147

Dubuc C, Ruiz-Lambides A, Widding A (2014) Variance in male lifetime reproductive success and estimation of the degree of polygyny in a primate. Behav Ecol 25(4):878–889
Emlen ST, Oring LW (1977) Ecology, sexual selection, and the evolution of mating systems. Science 197(4300):215–223
Fay FH (1982) Ecology and biology of the Pacific walrus, *Odobenus rosmarus divergens* Illiger. N Am Fauna 74:1
Francis JM (1987) Inter-female aggression and spacing in the northern fur seal, *Callorhinus ursinus*, and the California Sea lion, *Zalophus californianus* (PhD dissertation). University of California, Santa Cruz
Fretwell SD, Lucas HL (1970) On territorial behaviour and other factors influencing habitat distribution in birds. Acta Biotheor 19:16–36
Fitzpatrick JL, Almbro M, Gonzalez-Voyer A, Kolm N, Simmons LW (2012) Male contest competition and the coevolution of weaponry and testes in pinnipeds. Evolution 66:3595–3604
Flatz R, González-Suárez M, Young JK, Hernández-Camacho CJ, Immel AJ, Gerber LR (2012) Weak polygyny in California sea lions and the potential for alternative mating tactics. PLoS One 7:e33654
Franco-Trecu V, Costa P, Schramm Y, Tassino B, Inchausti P (2014) Sex on the rocks: reproductive tactics and breeding success of South American fur seal males. Behav Ecol 25:1513–1523
Gemmell NJ, Burg TM, Boyd IL, Amos W (2001) Low RS in territorial male Antarctic fur seals (Arctocephalus gazella) suggests the existence of alternative mating strategies. Mol Ecol 10:451–460
Goldsworthy SD, Boness DJ, Fleischer RC (1999) Mate choice among sympatric fur seals: female preference for conphenotypic males. Behav Ecol Sociobiol 45:253–267
Gonzalez-Suarez M, Cassini MH (2014) Variance in male RS and sexual size dimorphism in pinnipeds: testing an assumption of sexual selection theory. Mammal Rev 44:88–93
Hamilton WD (1971) Geometry for the selfish herd. J Theor Biol 31(2):295–311
Harcourt R (1992) Factors affecting mortality in the south American fur seal (*Arctocephalus australis*) in Peru: density-related effects and predation. J Zool (Lond) 226:259–270
Hoffman JI, Forcada J, Trathan PN, Amos W (2007) Female fur seals show active choice for males that are heterozygous and unrelated. Nature 445:912–914
Houston AI, Stephens PA, Boyd IL, Harding KC, Mcnamara JM (2007) Capital or income breeding? A theoretical model of female reproductive strategies. Behav Ecol 18:241–250
Jones AG, Ratterman NL (2009) Mate choice and sexual selection: what have we learned since Darwin? PNAS 106:10001–10008
Kiyota M, Insley SJ, Lance S (2008) Effectiveness of territorial polygyny and alternative mating strategies in northern fur seals, Callorhinus ursinus. Behav Ecol Sociobiol 62:739–746
Krüger O, Wolf JBH, Jonker RM, Hoffman JI, Trillmich R (2014) Disentangling the contribution of sexual selection and ecology to the evolution of size dimorphism in pinnipeds. Evolution 68 (5):1485–1496
Lancaster ML, Gemmell NJ, Negro S, Goldsworthy S, Sunnucks P (2006) Ménage à trois on Macquarie Island: hybridization among three species of fur seal (*Arctocephalus* spp.) following historical population extinction. Mol Ecol 15(12):3681–3692
Lancaster ML, Goldsworthy SD, Sunnucks P (2007a) Multiple mating strategies explain unexpected genetic mixing of New Zealand fur seals with two congenerics in a recently recolonized population. Mol Ecol 16(24):5267–5276
Lancaster ML, Bradshaw CJ, Goldsworthy SD, Sunnucks P (2007b) Lower reproductive success in hybrid fur seal males indicates fitness costs to hybridization. Mol Ecol 16(15):3187–3197
Le Boeuf BJ (1974) Male-male competition and reproductive success in elephant seals. Am Zool 14 (1):163–176
Le Boeuf BJ, Briggs KT (1977) The cost of living in a seal harem. Mammalia 41:167–195
Lee PC, Majluf P, Gordon IJ (1991) Growth, weaning and maternal investment from a comparative perspective. J Zool (Lond) 225:99–114

Le Boeuf BJ, Crocker DE, Costa DP, Blackwell SB, Webb PM, Houser DS (2000) Foraging ecology of northern elephant seals. Ecol Monogr 70(3):353–382
Leutenegger W, Cheverud J (1982) Correlates of sexual dimorphism in primates: ecological and size variables. Int J Primatol 3:387–402
Lindenfors P (2002) Sexually antagonistic selection on primate size. J Evol Biol 15:595–607
Loison A, Gaillard JM, Pélabon C, Yoccoz NG (1999) What factors shape sexual size dimorphism in ungulates? Evol Ecol Res 1(5):611–633
Marino L, Sol D, Thoren K, Lefebvre L (2006) Does diving limit brain size in cetaceans? Mar Mamm Sci 22:413–425
Marlow BJ (1975) The comparative behaviour of the Australasian sea lions Neophoca cinerea and Phocarctos hookeri (Pinnipedia: Otariidae). Mammalia 39(2):159–230
Mayr E (1963) Populations, species, and evolution. Belknap Press, Cambridge, MA
Parker GA (1970) Sperm competition and its evolutionary consequences in the insects. Biol Rev 45:525–567
Parker P, Maniscalco JM (2014) A long-term study reveals multiple reproductive behavior strategies among territorial adult male Steller sea lions (Eumetopias jubatus). Can J Zool 92 (5):405–415
Purvis A, Harvey PH (1995) Mammal life-history evolution-a comparative test of Charnov's model. J Zool (Lond) 237:259–283
Renouf D (1991) The behavior of pinnipeds. Chapman & Hall, New York, p 410
Repenning CA (1976) Adaptive evolution of sea lions and walruses. Syst Zool 25:375–390
Rogowitz GL (1996) Trade-offs in energy allocation during lactation. Am Zool 36:197–204
Schulz T, Bowen W (2005) The evolution of lactation strategies in pinnipeds: a phylogenetic analysis. Ecol Monogr 75:159–177. https://doi.org/10.1890/04-0319
Stirling I (1983) The evolution of mating systems in pinnipeds. Adv Study Mamm Behav 7:489–527
Sjare B, Stirling I (2011) The breeding behavior of Atlantic walruses, *Odobenus rosmarus rosmarus*, in the Canadian high Arctic. Can J Zool 74(5):897–911
Slatkin M (1984) Ecological causes of sexual dimorphism. Evolution 38:622–630
Soto KH, Trites AW (2011) South American sea lions in Peru have a lek-like mating system. Mar Mamm Sci 27:306–333
Trillmich F (1996) Parental investment in pinnipeds. In: Advances in the Study of Behavior, vol 25. Academic, pp 533–577
Trillmich F, Trillmich KGK (1984) The mating systems of pinnipeds and marine iguanas: convergent evolution of polygyny. Biol J Linn Soc 21:209–216
Trivers RL (1972) Parental investment and sexual selection. In: Campbell B (ed) Sexual selection and the descent of man, 1871–1971. Aldine-Atherton, Chicago, USA, pp 136–179
Vaz-Ferreira R (1975) Behavior of the southern sea lion, *Otaria flavescens* (Shaw) in the Uruguayan islands. Rapp P-V Reun Cons Int Explor Mer 169:219–227
Vilá BL, Cassini MH (1990) Aggressiveness between females and mother-pup separation in the southern sea lion, in Chubut, Argentina. Rev Chil Hist Nat 63:169–176
Wade MJ (1979) Sexual selection and variance in reproductive success. Am Nat 114(5):742–747
Watts P, Hansen S, Lavigne DM (1993) Models of heat loss by marine mammals: thermoregulation below the zone of irrelevance. J Theor Biol 163(4):505–525
Weise MJ, Harvey JT, Costa DP (2010) The role of body size in individual-based foraging strategies of a top marine predator. Ecology 91:1004–1015
Wyss AR (1994) The evolution of body size in phocids: some ontogenetic and phylogenetic observations. Proc San Diego Soc Nat Hist 29:69–75

Chapter 8
Site Fidelity, Male Harassment and Female Gregariousness: Factors Shaping the Highly Polygynous Mating System of the Northern Fur Seal (*Callorhinus ursinus*)

Masashi Kiyota

Abstract The northern fur seal, *Callorhinus ursinus*, is a highly dimorphic, polygynous otariid. Early studies based on observations of reproductive colonies indicated that males compete for places where females aggregate for pupping. Longitudinal observations of individual northern fur seals provide a somewhat more complex view of how the mating system is organized. Both males and females show natal philopatry and breeding site fidelity. Territory acquisition by males is a site-specific multi-year process with males returning to the same place for several years and progressively rising in reproductive status; from non-territorial invaders through peripheral or late-coming territorial males, to finally earning the status of early-coming territory holders. Genetic paternity analysis supports the effectiveness of the main mating strategy, but also suggests that alternative mating strategies such as female stealing (kleptogyny) and extra-territory copulations occur occasionally and provide some mating success for less successful males. Non-territorial males repeatedly sneak into the central breeding areas, interact with females and pups, disturb breeding colonies, and occasionally cause injury and mortality. Adult females can avoid male harassment of their pups and themselves indirectly by aggregating in areas defended by dominant males. Females show an aggregative tendency when selecting breeding microhabitats and prefer to enter a territory where four or more females already exist. Such aggregative behavior of females favors territorial males that establish their territories early in the season and maintain their territories undisturbed through vigilance against adjacent territorial males and invasive non-territorial males. Thus, natal philopatry and breeding-site fidelity, aggregative behavior of females, monopolization of receptive females by early-coming territory holders, and male harassment of females and pups are likely to form a feedback loop that enhances the development of the highly polygynous mating system in the northern fur seal.

M. Kiyota (✉)
Graduate School of Environmental and Fishery Sciences, Nagasaki University, Nagasaki, Japan
e-mail: m.kiyota@nagasaki-u.ac.jp

C. Campagna, R. Harcourt (eds.), *Ethology and Behavioral Ecology of Otariids and the Odobenid*, Ethology and Behavioral Ecology of Marine Mammals,
https://doi.org/10.1007/978-3-030-59184-7_8

Keywords Alternative mating strategy · Territoriality · Female aggregation · Infanticide · Kleptogyny · Male harassment · Philopatry · Polygyny · Reproductive success

8.1 Introduction

The northern fur seal (*Callorhinus ursinus*) is the most abundant otariid in the subarctic zone of the North Pacific Ocean. These fur seals disperse across oceanic areas to forage in winter and spring and in summer breeding animals gather ashore in large aggregations on a number of remote islands (Gentry 1998, 2009).

Northern fur seals have a long history of human exploitation, with a large trade ramping up in the eighteenth century for their luxuriant fur. Scientific knowledge of their biology has accumulated through the activities of national and multilateral management organizations such as the North Pacific Fur Seal Commission (Bonner 1982; Scheffer et al. 1984). Northern fur seals exhibit the most extreme level of sexual dimorphism in body size and polygyny in the Otariidae (Boness 1991; Lindenfors et al. 2002; Cullen et al. 2014). Adult breeding males guard territories in Central Breeding Areas (CBAs) of the rookeries to mate with receptive females (Fig. 8.1). Dominant males in high-density colonies gather up to 100 females in their territories (Bartholomew and Hoel 1953).

Fig. 8.1 Photo of breeding territories in the CBA of a large rookery (the Northeast Point breeding area, St. Paul Island, Alaska). Territorial males (dark colored large animals in the middle of the photo), adult females (light-colored medium-sized animals), and pups (dark colored small animals) attended by their mothers (lower middle) or aggregated in a pup pod (lower right) are discernible. Three temporary males in the intertidal zone and a few waterfront males are also seen in the upper right corner

Historically, the mating system was defined as one of males guarding "harems," by which was meant that a group of females was associated with a single male for a prolonged period. But Peterson (1968) pointed out that female northern fur seals move through rookeries rather freely. Since then, the northern fur seal mating system has been regarded more as resource-defense polygyny mediated by male territoriality and female gregariousness in breeding locations (Boness 1991).

On the basis of a longitudinal study of northern fur seal mating behavior, I came to recognize that male competition for breeding sites is only one of many elements that contribute to their complex mating system. In this chapter, I demonstrate that repeated return to the same breeding sites, male harassment on females and pups, and aggregative behavior of females, all play important roles in the formation of highly concentrated, polygynous, breeding aggregations. Studies like the one summarized here illustrate the classical ethological work that established the foundations of behavioral ecology with a detailed, single species focus.

Following pioneering animal behavior research from the 1950s to the 1970s (e.g., Bartholomew and Hoel 1953; Peterson and Bartholomew 1967; Peterson 1968; Gentry 1974), intensive field studies of the terrestrial behavior of otariids allowed comparison among species (see Gentry, Chap. 1). This also laid the foundation for the first genetic approaches in the 1980s and 1990s that deepened our understanding of behavioral descriptions (see Cassini, Chap. 6). Simultaneously, the first comprehensive records of otariid behavior at sea were being gathered. It is only following the door being opened by these studies that we can today integrate the biology on land and at sea for these semi-aquatic mammals (see Costa and Toro, Chap. 2).

8.2 Longitudinal Behavioral Study

The research outlined here was conducted at the Polovina Cliffs and Polovina rookeries of St. Paul Island, Pribilof Islands, Alaska (57°10'N, 170°10'W) from 1993 to 1999 (Kiyota 2005). More details of the entire research program are reported in Kiyota (2005), Kiyota and Okamura (2005), and Kiyota et al. (2008). The Polovina Cliffs rookery has high cliffs that extend along the beach for 1.4 km, and CBAs are formed on boulder beach and lava shelves between the cliff and the shoreline. The two continuous rookeries are subdivided into nine sections, each spanning about 200 m, that facilitate area-specific records for population census (Antonelis 1992). Regular behavioral observations were conducted in the study cove located in the southernmost section of the Polovina Cliffs rookery. The study cove was approximately 20 m wide and 10 m deep, surrounded by 2–5 m high cliffs on three sides, and open to the sea to the east, and regularly occupied by 6–12 territorial males during the breeding season (Fig. 8.2).

Individual fur seals were identified by using tags, artificial bleach marks, and/or natural marks. Identifiable animals were located in the CBAs, peripheral areas and haul-out resting areas around the study cove. The place, age class and social status of each identified animal were systematically and repeatedly recorded. Some seals had

Fig. 8.2 View from the water's edge of the study cove in the Polovina Cliffs rookery

metal flipper tags which were attached at their natal rookeries from 1987 to 1990 when they were pups (Fowler et al. 1994). Resighting records of these seals were used to determine natal site fidelity. Tissue samples were collected from identifiable territorial males, adult females and their pups, and used for paternity analysis.

Males were classified into three age classes according to body size: adult (males whose body sizes are equivalent to territorial males, estimated age 7 years or older); subadult (males smaller than territorial males but larger than adult females, estimated age 4–6 years); juvenile (immature males as small as or smaller than adult females, estimated age 2–3 years). Male social status was classified as follows: territorial with females (males holding territories and females in the CBAs); territorial without females (males holding territories in the CBAs but without adult females); peripheral (males settled in the peripheral area and defending a specific spot; Fig. 8.3); invasive (males attempting to intrude into the CBAs); haul-out (males landing on the resting area without showing any territoriality; Fig. 8.3). Territorial males were classified into three categories according to their timing of arrival and duration of territory tenure: early-comers (males that established territories prior to July 1 before the arrival of most females); late-comers (males that established territories after June 30); temporary males (males that stayed in the CBA for less than 1 day).

Female breeding site/mate choice behavior was analyzed using the multi-year occurrence records of identifiable adult females and males in the study cove. The study cove was subdivided into seven sub-sectors (north waterfront; north cliffside; middle waterfront; middle intermediate; middle innermost; south waterfront; south cliffside). On the basis of the attendance records of identifiable females, location of territorial males, and daily maximum number of adult females in each territory, five hypotheses of female choice behavior were considered: (i) random, a female chooses the breeding site randomly; (ii) female aggregative response, a female chooses a territory already occupied by other females; (iii) mate fidelity, a female chooses a

Fig. 8.3 An adult male showing moderate territoriality in the peripheral area behind the study cove. Behind him are adult and subadult males resting in the grassy haul-out area

territory held by the same male with whom she spent the peri-estrus period in the previous year; (iv) site fidelity, a female chooses the same sub-sector where she spent the peri-estrus period in the previous year; (v) site preference, a female prefers waterfront sub-sectors to inland sub-sectors. Discrete choice models were used for the analysis which allows variable resource availability (Arthur et al. 1996; Manly et al. 2002). The effects of the hypotheses were compared by model selection using Akaike's information criterion (Burnham and Anderson 2002).

8.3 Phenology of Reproductive Behavior

Adult males began to come ashore at the CBA in late May. The number of territorial males increased further in July and in early August because late-coming males entered the CBA and occupied their own territories. Adult females typically landed on the CBAs in early July, gave birth (c.a. 2 days after landing) and nursed pups, came into estrus and copulated with an adult male (c.a. 6 days post-partum), and left the breeding area to go to sea for their first feeding trip (about 1 day after copulation). The number of adult females increased steadily from late June to middle July, reached a peak (50–60 females in the study cove) around July 16–21, and fluctuated thereafter because females repeated the cycle of foraging at sea and nursing pups on land. Females are mono-estrus and usually come into estrus a few days after first landing on the rookery. Pupping started from the end of June, and the number of pups increased up to early August. After that, daily pup counts showed large fluctuations because pups form aggregations called pup pods, walk around the whole area, enter the surf zone, and travel to adjacent CBAs, in contrast to neonate pups that remain with their mothers.

8.3.1 Natal Philopatry of Males

Adult males showed strong homing behavior to the natal colony (natal philopatry) and moved within the natal colony in search of appropriate breeding sites. Among the 36 metal-tagged males resighted in the study area, 34 (94%) were born in the Polovina Cliffs and Polovina rookeries. The 34 tagged males spread over all sections of the Polovina and Polovina Cliffs rookeries. Judging from resights of metal tags, males first settled in the peripheral area at age 7 and held territories in the CBAs in the late breeding season. Males age 7–9 can apparently hold territories with adult females in CBAs from the beginning of the breeding season.

Earlier studies also reported natal philopatry in this species, as well as occasional but relatively infrequent emigration to non-natal breeding sites (Kenyon and Wilke 1953, Nagasaki and Matsumoto 1957, Baker et al. 1982). Repenning (1976) hypothesized that homing instincts to natal rookeries evolved in offshore feeding pinnipeds to ensure the reunion of breeding individuals. At the same time, juvenile males may possibly play the role of predominant dispersers to avoid inbreeding as known in many other polygynous mammals (Dobson 1982; Perrin and Mazalov 2000).

8.4 Territory Acquisition

Repeated sightings of bleach-marked territorial and peripheral males indicate that territory acquisition is a multi-year process. Males return repeatedly to the same breeding sites i.e. they show breeding-site fidelity, but they only acquire territories in later breeding seasons and/or later years. Among the 68 peripheral males successfully bleach marked within the study cove, 21 (31%) were resighted as territorial males later in the same breeding season, and 22 (32%) and 18 (27%) were resighted as territorial and non-territorial males in the next breeding season, respectively. In the third year, 10 (48%) territorial and four (19%) non-territorial males were resights of the 21 traceable individuals that had been remarked. Location of the territories of the resighted males were within the same section where they were marked as peripheral males except for two individuals that held territories 500 m north of the original locations (Kiyota 2005).

Physical fights were infrequent through the process of territory acquisition and abandonment. Many males initiated their territory tenure by occupying empty places in May and June, before the onset of female return ($n = 32$, 46% of observed tenure). Most of the late-comers, who entered the CBA in July and August, acquired territories without serious fights ($n = 28$, 40%). They first occupied the waterfront zone, and slowly infiltrated the gaps between existing territories (see also Campagna et al. Chap. 9) or occupied abandoned territories (Fig. 8.4). Many early-comers abandoned their territories spontaneously, without fights ($n = 26$, 59%) either by entering the water or climbing up the slope out of the CBA and into the haul-out area. Late-coming males generally moved into the abandoned territories or competed with

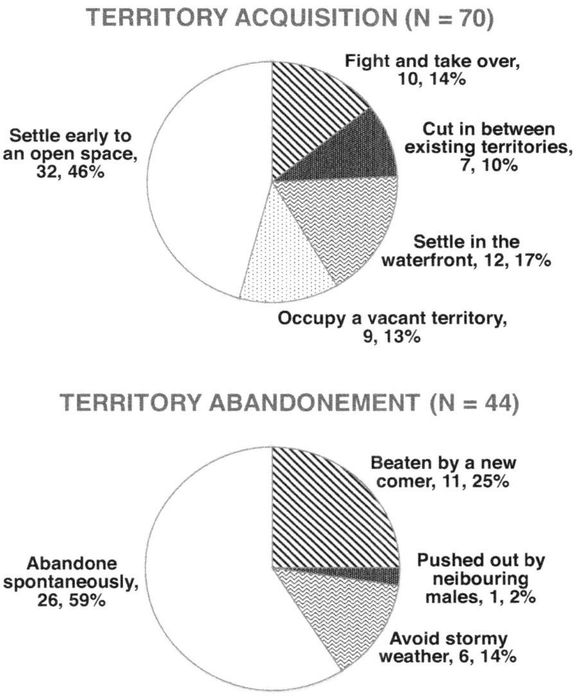

Fig. 8.4 Composition of observed incidences for territory acquisition (upper panel) and abandonment (lower panel). Data from Kiyota (2005)

neighboring males, and shifted or expanded their territories in the late breeding season. However, replacement of territorial males by physical fights was not common at the initiation and termination of the territory tenure (14% and 25% of initiation and termination, respectively). Replacement of territorial males by physical fights occurred more frequently in late July than in other periods (70% of tenure initiation and 55% of tenure termination) because early-coming males were vigorous and vigilant in early July, and because many early-comers became exhausted and ready to leave their territories spontaneously in August (Kiyota 2005).

Many territorial males held territories in the same places for two or more successive breeding seasons. In the study cove, five males held territories for a single breeding season, 31 males held territories for multiple breeding seasons of up to 5 years (14 individuals for two seasons, 11 for three seasons, five for four seasons, and one for five seasons). The average territory tenure for the territorial males was 2.2 breeding seasons. These males held territories earlier in the second year of their territory tenure than in the first year. Only three males shifted their territories outside the study cove in their second or third tenure years. Their original territory locations in the cove were occupied by those males that held territories there in the late breeding season of previous years (Kiyota 2005).

Earlier studies based on cross-sectional rather than longitudinal surveys suggested breeding-site fidelity of male northern fur seals. Bryant and Allen

(1870) reported long ago that old males appeared to return to the same rock each year and established territories by the middle of June. Kenyon (1960) conducted a single-year observation of male reproductive behavior and found that older bulls first came ashore to the breeding area and established territories. He also speculated that selection of territory location was the result of homing to natal rookery and re-occupation of previous territory location. Gentry (1998, Chap. 1) conducted multi-year observations and reported that experienced males started to hold territories before June 30 and that peripheral males moved into CBAs in the late breeding season. Other polygynous pinnipeds show similar patterns of territory acquisition and tenure (McCann 1987).

8.5 How Males Find Mating Partners

Adult females are distributed unevenly among male territories so that a small number of early-coming males are able to monopolize receptive females in the CBA. Figure 8.5 shows an example of the male territory tenure and the number of females in their territories in the study cove over a breeding season. Early-coming males appeared in the CBA before the arrival of the first female. Among the early-coming males, FX (3rd year of his territory tenure), IV (3rd year) and TA (1st year) occupied most females in early July, as many as 30 females in a territory. But on July 14, FX and TA suddenly lost females because they were challenged by waterfront males (DI, ED, and LA) and this allowed the intrusion of DI in-between FX and HI. To avoid the violent male-male fights and subsequent intrusion of non-territorial males triggered by the loss of vigilance of the attacked territory holders, most females moved to the adjacent territories (HI and KI) away from the water's edge. In addition, stormy weather and large surf the next day accelerated the escapement of females from these waterfront territories. After successful early-comers (FX, TA and IV) left the CBA in late July, the remaining early-comers (HI, KI and JA) and some late-comers (e.g. DI, ED, LA) kept adult females.

A male's index of mating success (IMS), was calculated from the observed number of females in his territory and the daily probability of an adult female becoming receptive in a reproductive season (Kiyota 2005). Comparison of the index demonstrated that apparent mating success was not shared evenly among the territorial males but was monopolized by a few early-comers. Figure 8.6 shows the annual IMS of early-coming males who held territories in the study cove from the beginning of breeding seasons. Five to eight early-comers established their territories in the study cove in each breeding season. Most of the successful males were experienced males who held territories in the same cove in two or more breeding seasons (such as HI, FX, and IV in 1997). Other males (YE in 1994, RB in 1995, BI in 1996, KI and TA in 1997) held territories from the beginning of the breeding season in their first tenure year but could access fewer females than those experienced males over the peak of female receptiveness in middle July. Some of them returned to the same territory in the following year(s) (e.g., YE in 1995–1996 and KI

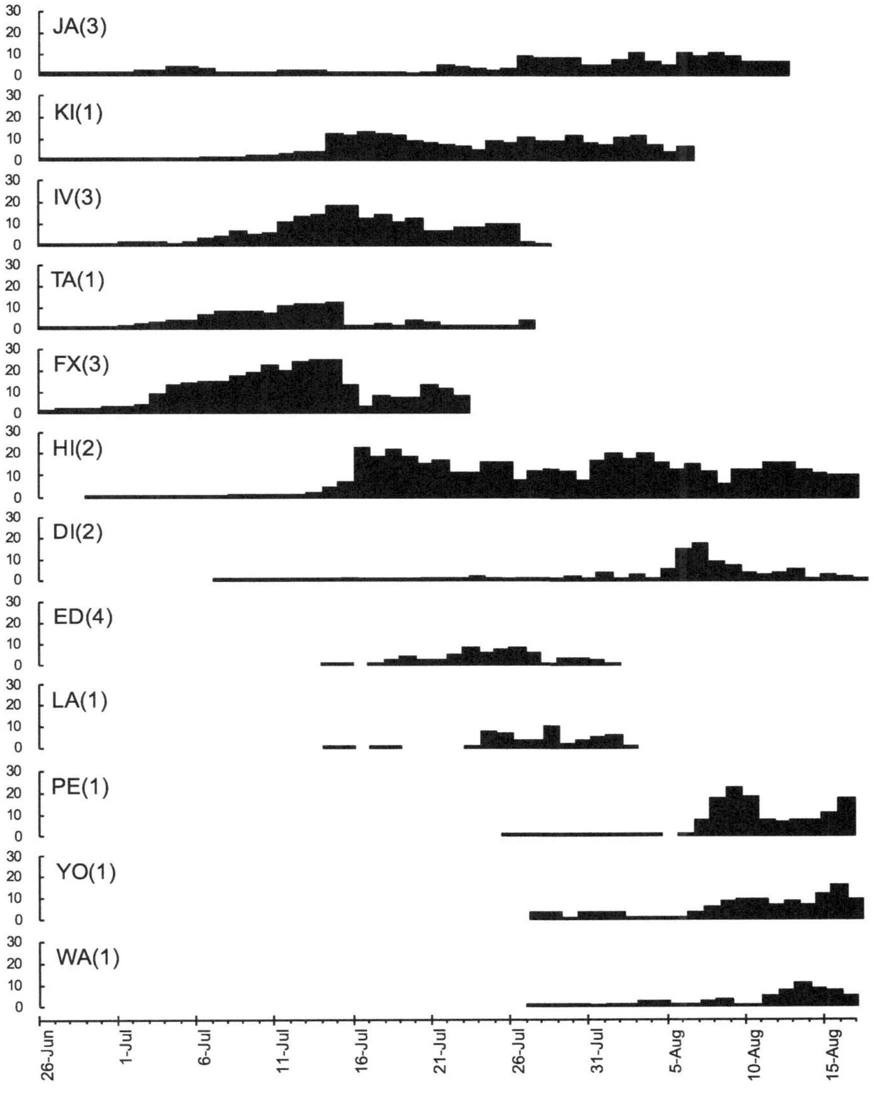

Fig. 8.5 An example of the tenure of identifiable territorial males and number of females in their territories in the study cove for the 1997 breeding season. Two capital letters and numbers in parentheses represent IDs of territorial males and their tenure years, respectively

in 1998) and became the most successful male in the study cove in their second or third years of tenure.

As shown above, males gradually moved their location from the periphery to the center of the breeding area, and shifted territory tenure from late breeding season to the peak of female receptivity. This demonstrates that acquisition of territories and

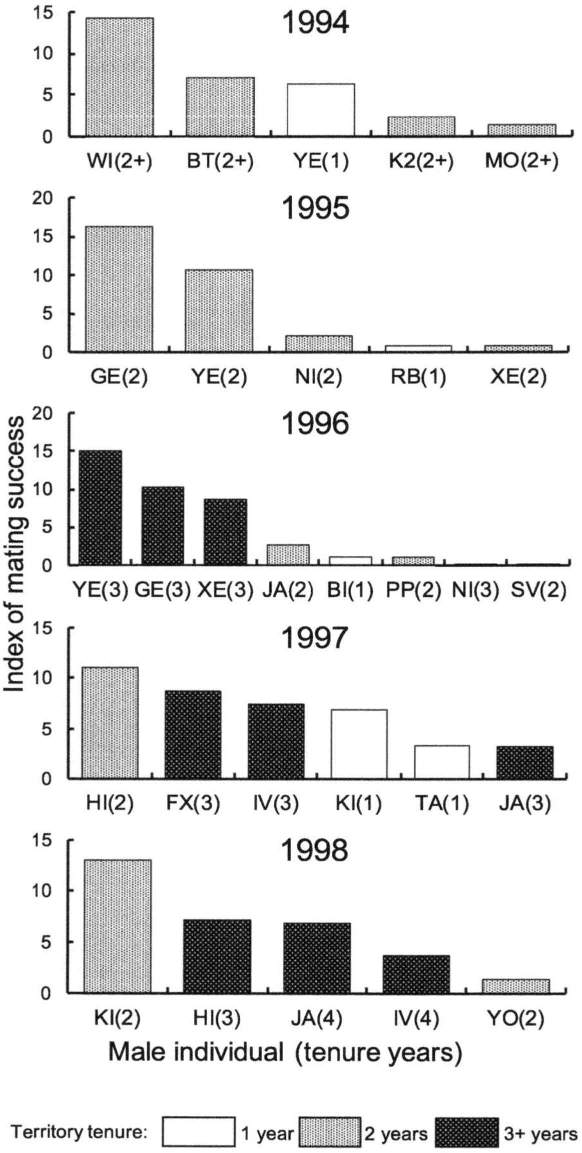

Fig. 8.6 IMS of early-coming territorial males in the study cove for the 1994–1998 breeding seasons. Two capital letters beneath columns and numbers in parentheses represent IDs of territorial males and their tenure years, respectively. Modified from Kiyota (2005)

mates is a multi-year site-specific process. Males can maximize their mating success if they hold territories at the peak of female estrus, because post-partum estrus of female northern fur seals is highly synchronized in the early reproductive season (Trites 1992; Kiyota 2005). A male could acquire mating success if he invaded the CBAs and take over a territory through fighting at the peak season of female estrous. However and contrary to this expectation, replacement of territorial males by physical fights was infrequent in the early reproductive season. Late-coming males generally occupied the periphery of established territories first (e.g. waterfront males in Fig. 8.1), and gradually moved to the central area and expanded their territories. Peterson (1968) suggested that experience was more important for males to acquire reproductive success than random competition. Greenwood (1980) further explained that familiarity with local conditions developed through prior residence could provide an advantage during the territory acquisition process in mammals. In male northern fur seals, prior experience from perennial returns to the same breeding sites, in combination with female site fidelity and aggregative behavior (described in detail below), appears to be an important element for territory acquisition and mating success of males. The topography of the study site might enhance the effect of site fidelity and experience in this study. The CBA in the study area was located in a cove surrounded by cliffs (Fig. 8.2) and more protected against invasion of non-territorial males than in other CBAs that are in wide open areas (as shown in Fig. 8.1). This unique situation favored males with established territories, and probably led to the longer average territory tenure in this study (2.2 year) than in other studies (1.5 year in Peterson 1968; 1.68 year in Vladimirov 1987).

8.6 Male-Female Association and Paternity

As shown so far, the primary mating strategy of a male northern fur seal is to hold a territory in the CBAs from the early breeding season, and hold as many adult females as possible in his territory during the peri-estrus periods. The term peri-estrus period is defined herein as the on-land period of a pregnant female from parturition to the first feeding trip or that of a non-pregnant female from the first landing to the first feeding trip. The term associate male(s) is defined as the male(s) in whose territory a female spent her peri-estrus period. Results of the genetic analysis generally support the results of behavioral observation, but also indicate a certain degree of non-associate paternity as an alternative strategy for less successful males see below.

On the basis of behavioral observations, a female's pup conceived in the breeding season and born in the following year is expected to be sired by the associated male (s). Nuclear DNA samples from identifiable females ($n = 50$), their pups ($n = 85$), and adult males ($n = 37$) that appeared in the study cove during the breeding season when the pups were conceived were analyzed for five micro-satellite loci (Kiyota et al. 2008). The genetic analysis demonstrated that 83 pups (98%) were sired by territorial males who appeared in the study cove during the year of conception. Among the 56 genotyped pups for which attendance records of mothers were available, 45 (80%) pups were sired by the associate males. Among the 11 non-associate fathers, five were territorial males who held territories adjacent to those of the associate males, five were temporary males located on the intertidal waterfront of the study cove and established territories later in the same breeding season, and one was a late-coming territorial male (territory tenure = 5 days, maximum number of females = 2) who left the territory the day before the female's arrival. In all the cases of non-associate paternity (n = 11), the genetic fathers held significantly fewer females than the associate territorial males during the mother's peri-estrus period.

The high percentage of associate paternity found here supports the idea that territory defense in the female-aggregation area is the most effective mating strategy in male northern fur seals (see also Cassini, Chap. 6). A significant amount (20%) of non-associate paternity suggests there is also effective mating in territorial males adjacent to dominant males or even in non-territorial males temporarily resident in the intertidal area. There are three possible explanations for the occurrence of non-associate paternity: (i) mate choice, where females choose more attractive and/or successful males as mates (see Sinclair et al. Chap. 13); (ii) alternate mating strategies where non-associate males obtain access to receptive females through stealing, interception, and sneak copulation without direct confrontation with the associate males; (iii) inbreeding avoidance, where females tend to mate outside the philopatric breeding area to avoid inbreeding. Although it is difficult to conclusively reject any of these alternative explanations, the alternative mating strategy explanation is more consistent with the results of this study because all the non-associate paternity were assigned to neighboring or temporary territorial males that held fewer females than the associate males.

8.7 Kleptogyny and Other Alternative Strategies of Less Successful Males

Past field observations suggested that male northern fur seals might adopt alternative mating strategies in the form of: (i) female stealing (kleptogyny), where territorial males steal females from the adjacent territories to obtain access to receptive females (Gentry 1998; see also Campagna et al. Chap. 9); (ii) sneak copulation or interception, where peripheral and waterfront males surreptitiously mate with females while females move out of the associate males' territory after copulation (Bartholomew and Hoel 1953); and (iii) aquatic copulation, where males copulate with females in water (Baker 1989). Interception/sneak copulations and aquatic copulations are also observed in other rookery-breeding pinnipeds. The frequent occurrence of kleptogyny is rather specific to northern fur seals. The unique combination of extreme sexual dimorphism in body size, synchrony of estrus, limited female mobility on land, and high adult density in the CBAs based on female gregariousness and territory proximity possibly generates a condition liable to cause kleptogyny (Fig. 8.7).

In this study, a total of 95 incidences of female stealing attempts were observed: 28 approaches; 16 physical contacts; and 51 actual steals (Kiyota et al. 2008). The stolen females remained in the stealer's territories at least for a short period time on 44 of the 51 occasions. Six females returned to their original territories immediately after being released by the stealers, and one female moved to a third male's territory

Fig. 8.7 A territorial male (DI) stealing a female from the adjacent territory

after release. A minimum of eight females were confirmed to be injured by the bite and seizure and were seen to be bleeding. The stealers held fewer females than the targeted males in their territories. For example, in the 1997 breeding season, territorial males JA and KI often stole females from FX, the most successful male during the early phase of the breeding season. But kleptogyny does not always lead to genetic success. In two cases in which identifiable females were stolen by adjacent territorial males, the neonates delivered in the following year were not sired by the stealers but by the associate males.

8.8 Harassment of Females and Pups by Non-territorial Males

Although battles against established territorial males are infrequent events for non-territorial males, adult, subadult and juvenile males sneak into CBAs repeatedly and interact with females and pups. Harassment and abduction of pups by invasive males occurs regularly in the late breeding season, and sometimes lead to external trauma, mother-pup separation and emaciation, although cases of infanticides are rare. Harassment and abduction of pups appears to be an extension of play in juvenile males, and the misplacement or reorientation of sexual motivation in subadult and adult males (Kiyota and Okamura 2005; see also Ryazanov Chap. 10).

Interactions between invasive males and pups were typically observed in the waterfront zone of a CBA, starting with sniffing and nasal contacts. Their interactions sometimes proceeded to agonistic and violent behavior (Fig. 8.8) or to sexual behavior with pups. Invading subadult and juvenile males often mounted pups with pelvic thrusts, but intromission was not confirmed. Abduction of pups occurred

Fig. 8.8 An invasive subadult male approaching the aggregation of pups and sniffing a pup. The territorial male lying in the upper left did not notice the invasive male

Fig. 8.9 A subadult male grabbing and carrying a pup in the waterfront zone of the study cove. Please notice that the territorial male lying behind him is unresponsive to the pup abductor

either at the escalated stage of agonistic and violent interactions, or immediately after encounter. Pups were carried outside the breeding territories on land, to shallow intertidal water, or offshore (Fig. 8.9). Among the 13 cases of observed offshore abduction, five pups returned to the CBA by themselves, six pups landed on a hauling beach within 200 m of their original locations (four of them were confirmed to return to the CBA within 2 days), and two were taken away from the beach and their fate could not be confirmed (Kiyota and Okamura 2005). Invasive males were not choosy of the sex of the pups they attacked. The sex ratio of 11 abducted pups (male:female = 5:6) was not significantly different from the sex ratio of pups on Pribilof Islands (0.55:0.45) estimated by Antonelis et al. (1994).

Killing of pups by non-territorial males occurred accidentally as byproducts of male aggression and abduction. Three cases of pup mortality caused by male harassment were observed. The first pup was captured in the intertidal area by a subadult male, grabbed and swung around, and slammed headfirst into a rock, and suffered a skull fracture. The second pup was suddenly abducted by an adult male and taken offshore, landed on a beach 100 m away from the CBA, and died of emaciation because it could not find its mother. The third pup was abducted by a subadult male, taken offshore, and drowned after being submerged underwater repeatedly. The abductors did not eat the dead pups in any of these cases. The observed pup mortality caused by male harassment accounted for 0.6% of total pup production in the study cove (Kiyota and Okamura 2005).

The behavior of pups helped limit mortality caused by invasive males. Newborn pups stay close to their mothers for the first few days after birth (Bartholomew 1953). Since territorial males are quite active and vigilant during the peak of pupping season. The behavior of pups staying close to their mothers minimized the risk of being harassed by invasive males during the early breeding season. As pups grew, they aggregated together and wandered around the breeding colony and this may have increased the risk of encountering invasive males. However, pups changed their behavior in response to aggressive males. They learned to escape from approaching males, hid in rock shelters to deflect the perpetrator's attention, and eventually became fast enough to elude the attacks of invasive males.

Although many territorial males paid little attention to the intrusion of subadult and juvenile males (see Figs. 8.8 and 8.9), territory defense behavior of some territorial males may have provided protection for pups indirectly. Vigilant territorial males keep non-territorial males away from their territories and suppress aggressive interactions of invasive males toward females and pups. Later in the breeding season when early-coming territorial males were replaced by late-comers, the vigilance of territory holders against young males became lax, allowing young males to enter the CBAs. Thus, the risk to pups from the harassment of invasive males depended to a great degree on the ontogeny of pup behavior and the timing of replacement of territorial males in addition to the abundance and composition of non-territorial males.

Adult females are not effective in protecting pups against invasive males. When adult females attend their pups within the breeding territories, females defend pups from subadult and juvenile males. But once pups are taken out of the territories, they no longer pursue the abducted pups (see also Campagna et al. 1988). Adult females escape from an invasive adult male without protecting their pups because the primary targets of invasive adult males are females rather than pups. However, adult females may protect pups indirectly by aggregating in rocky areas defended by dominant males. Campagna et al. (1992) observed that survival of South American sea lion pups was better in group breeding females than in females in solitary mating pairs, and suggested that pups were more protected in colonies than in isolation against harassment and infanticide caused by subordinate males. Bradshaw et al. (1999) demonstrated that high-density colonies of New Zealand fur seal (*Arctocephalus forsteri*) were formed on such terrain that contained more rocks, crevices and ledges and provided shelters to pups. Most of the northern fur seal rookeries on St. Paul Island are located on beaches that contain lava or rocks. Female aggregation in such places may help to avoid male harassment and abduction of pups.

8.9 Female Choice of Breeding Sites

Multi-year occurrence records of identifiable adult females in the study cove demonstrated the breeding site fidelity of females. Among the 50 identifiable females, 40 returned to the study cove in the next year, and 34 returned in the following year. Annual rates of return were 0.80 and 0.85, respectively. The actual rate of return could be even higher because two females were resighted in the CBA adjacent to the study cove, and because two other females lost tags and only had tag holes.

Fine-scale breeding site selection of females analyzed by the discrete choice model revealed that adult females have a strong tendency to enter a territory where many females were already present (female aggregative response), and where the females reproduced in the previous year (site fidelity), but that females had a slight tendency to avoid waterfront territories (site preference). The effect of female aggregative response was also confirmed through the observation of the female landing process. Females were more likely to enter a territory if there were already females there; the likelihood that a female would enter a territory where four or more females had already entered was close to 1, whereas the rate was as low as 0.3 in a territory without females (Fig. 8.10).

One reason for the aggregative behavior of females may be to minimize the risk for a female of receiving aggressive interference by territorial males. Territorial males, especially younger inexperienced males, check the estrus condition of females repeatedly and prevent females from leaving their territory with aggressive threatening and physical blocking (Fig. 8.11). Therefore, there is a high frequency of harassment by the territory holder when there are only one or two females in a territory. Another possible reason is the stability of the territory. A territorial male can hold a large number of females by avoiding disturbance triggered by invasion of non-territorial males and kleptogyny of adjacent territorial males, which means that a territory with many females is stable and so safer for females. Females can therefore avoid harassment caused by territorial males and non-territorial males by choosing a territory with more females (see also Campagna et al. 1992).

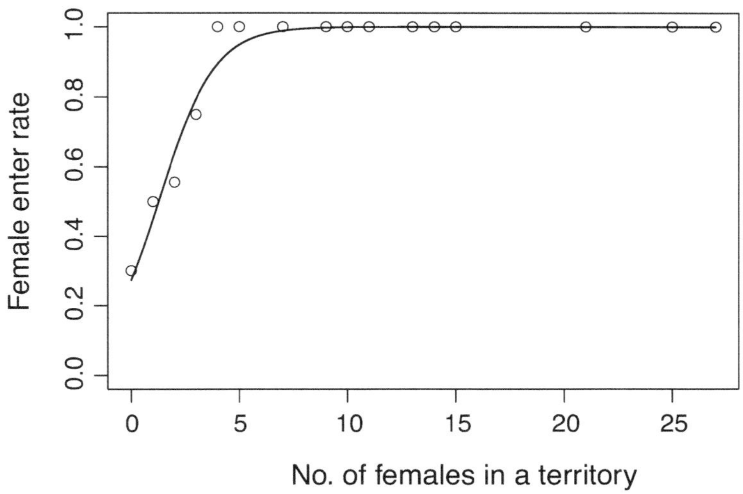

Fig. 8.10 A logistic curve fitted to the relationship between the probability of a female to enter a territory in her first landing in a breeding season and the number of females in the territory

Fig. 8.11 A territorial male aggressively blocking the movement of the one and only female in his territory. Two invasive subadult males are also seen in the waterfront zone behind him

8.10 Extreme Polygyny

This longitudinal analysis of individual northern fur seals demonstrates that the dominant mating strategy in male northern fur seals is to hold a territory in the CBA from early in the reproductive season. Males return to the same section of a rookery for several years and raise their reproductive status starting as non-territorial invaders through peripheral or late-coming territorial males, to finally earn the status of early-coming territory holders. Genetic paternity analysis supports the effectiveness of the major mating strategy, but alternative mating strategies such as kleptogyny and extra-territory copulation occur occasionally and provide mating opportunities to less successful males. Thus, natal philopatry, breeding-site fidelity, and concentration of receptive females in narrow spatio-temporal zones are key elements that shape the polygynous mating system in northern fur seals. Bartholomew (1970) hypothesized that terrestrial parturition and offshore marine feeding necessitated a mechanism to ensure the reunion for reproduction of adults on land and thus enhanced the evolution of gregariousness and polygyny in some pinniped species. However, the contrast of terrestrial breeding and oceanic feeding does not fully explain the evolution of this extreme polygyny in northern fur seals, and it may have appeared in recent times. The northern fur seal used to be the most common pinniped widely distributed along the coast of the North Pacific Ocean. Their limited present-day distribution appears to be due to a combination of historical hunting and other human disturbances, including climate change (Burton et al. 2001; Newsome et al. 2007). This may have driven the remnant fur seals to extreme high densities, and their present breeding behavior, see below. Supporting this hypothesis is that establishment of a new breeding colony (e.g. Bogoslof Island; Loughlin and Miller 1989) and recolonization to once extirpated breeding islands (e.g., Kuril and Farallon Islands) have occurred in recent decades.

From the sociobiological perspective, several hypotheses have been proposed for the evolution of female gregariousness and polygyny in otariids: (i) marginal male effect hypothesis, females prefer to mating with dominant males in the central part of the breeding area (Cassini 1999, 2000); (ii) ecological marginal male effect hypothesis, females avoid injury and mortality of pups and themselves caused by peripheral subordinate males (Trillmich and Trillmich 1984, Boness 1991); (iii) harassment avoidance hypothesis, females avoid excessive interference from the territory-holding male (Cassini 1999, 2000); (iv) selfish herd hypothesis, females avoid attacks by predators in the marginal breeding area (Boness 1991). Although these hypotheses are not mutually exclusive, results of this study support (ii) and (iii); namely, avoidance of harassment caused by territorial and non-territorial males appears to be an important factor which enhances the development of female gregariousness (see also Cassini, Chap. 7). Female aggregative response is hypothesized to have the largest effects on breeding site choice of females, and females prefer to enter a territory with at least four females.

The term mate choice copying refers to non-independent mate choice behavior affected by the choice of other individuals, and pseudo-copying refers to the situation where females copy the choice of others as a result of adaptive behavior other than mate choice such as herding (Dugatkin 1996; Brooks 1998; Vakirtzis 2011). In the northern fur seal, aggregative behavior of females appears to help avoid male harassment and promotes pseudo-copying of breeding site/mate choice. High breeding success of early-coming territorial males can be explained by this pseudo-copying behavior of females. It is risky for a female to walk among territories to choose appropriate mates and sites because territorial males will block female movement and sometimes injure them. By pseudo-copying, a female can lower the risk of injury and select a safe place for pup delivery and nursing. In the later breeding season when young primiparous females come ashore for their first reproduction, pseudo-copying also helps inexperienced females to detect a stable territory. Pseudo-copying behavior of primiparous females may have further implications in choosing their future mates on the basis of the behavior of experienced multiparous females. Since northern fur seals are known to have the ability to recognize individuals for a long-term period (Insley 2000, 2001; Insley et al. 2003), breeding-site fidelity and pseudo-copying may allow females to transfer mate choice to their daughters.

Males have very different options. A male will be more successful if he can hold a territory and collect at least four females in the early breeding season. Aggregative behavior of females favors early-coming territorial males that keep their territories undisturbed through vigilance against adjacent territorial males and invasive non-territorial males. A rapid increase in female numbers for a few early-coming territorial males (e.g., FX and KI in Fig. 8.4) in early July may be explained by the above context. Experienced early-comers hold territories in advance of other males without physical fights and play a waiting game to attract pregnant females early in a breeding season. Aggregative behavior of adult females favors early initiation of male territory tenure, and might enhance large body size with rich energy storage that enables early and long territory tenure and constant vigilance behavior. In this way, male harassment and female gregariousness form a positive feedback loop that enhances the development of the polygynous mating system in this species. Cassini (2000) and Cassini and Fernández-Juricic (2003) postulated a model in which female gregariousness is accelerated as a counter strategy against harassment of territorial males and accelerated polygyny in otariid pinnipeds (see also Cassini, Chap. 7). This study demonstrates that harassment by non-territorial males is another important element in the evolution of polygyny in otariids.

The frequency and intensity of male harassment vary depending on sex ratio, and the ratios of adult, subadult and juvenile males. Commercial harvests of this species are likely to have made strong impacts on the social structure, survival, and productivity of the breeding population. In particular, a female-based harvest, that was conducted in the 1990s to attain maximum sustainable production based on a simple population dynamics model (York and Hartley 1981), caused large disturbances at breeding aggregations, intensified male aggressiveness, and increased injury and mortality of pups and adult females. Since this species has a complex social structure formed by breeding site fidelity, male harassment and female gregariousness, management and conservation of the population should be made with careful monitoring of social structure and function of the breeding colonies, in addition to monitoring of population sizes.

Acknowledgements I thank S. J. Insley, S. L. Lane, H. Okamura, T. R. Loughlin, C. W. Fowler, G. A. Antonelis, J. D. Baker, B. Robson, C. Gburski, N. Baba, S. Yonezaki, S. Kitani, Y. Hamano, E. O. Tamura, A. Kobayashi, C. G. Kawai, R. Wada, and many other colleagues who supported and participated in the prolonged field survey and sample analysis. I also thank C. Campagna and

R. Harcourt for their constructive comments and advice on the earlier versions of the manuscript. This study was conducted under U.S. MMPA Scientific Research Permits No.868 and No.782-1455. Partial funding for the genetic analysis came from the Colby College Department of Biology and INBRE grant #P20 RR-016463.

References

Antonelis GA (1992) Northern fur seal research techniques manual. U.S. Department of Commerce, NOAA Technical Memorandum, NMFS-F/NWC 214

Antonelis GA, Ragen TJ, Rooks NJ (1994) Male-biased secondary sex ratios of northern fur seals on the Pribilof Islands, Alaska, 1989 and 1992. U.S. Department of Commerce, NOAA Technical Memorandum NMFS-AFSC 45:84–49

Arthur SM, Manly BFJ, McDonald LL et al (1996) Assessing habitat selection when availability changes. Ecology 77:215–227. https://doi.org/10.2307/2265671

Baker JD, Antonelis GA, Fowler CW et al (1982) Natal site fidelity in northern fur seals, *Callorhinus ursinus*. Anim Behav 50:237–247. https://doi.org/10.1006/anbe.1995.0236

Baker JD (1989) Aquatic copulation in the northern fur seal, *Callorhinus ursinus*. Northwest Nat 70:33–36

Bartholomew GA (1953) Behavioral factors affecting social structure in the Alaska fur seal. Trans 18th North Am Wildl Conf 481–502

Bartholomew GA (1970) A model for the evolution of pinniped polygyny. Evolution 24:546–559. https://doi.org/10.1111/j.1558-5646.1970.tb01790.x

Bartholomew GA, Hoel PG (1953) Reproductive behavior of the Alaska Fur seal, *Callorhinus ursinus*. J Mammal 34:417–436. https://doi.org/10.2307/1375858

Boness DJ (1991) Determinants of mating systems in the Otariidae (Pinnipedia). In: Renouf D (ed) Behaviour of pinnipeds. Chapman & Hall, London, pp 1–44

Bonner WN (1982) Seals and man: a study of interactions. University of Washington Press, Seattle

Bradshaw CJA, Thompson CM, Davis LS et al (1999) Pup density related to terrestrial habitat use by New Zealand fur seals. Can J Zool 77:1579–1586. https://doi.org/10.1139/z99-125

Brooks R (1998) The importance of mate copying and cultural inheritance of mating preferences. Trends Ecol Evol 13:45–46. https://doi.org/10.1016/S0169-5347(97)01253-6

Bryant C, Allen JA (1870) On the habits of the northern fur seal (*Callorhinus ursinus*, Gray), with a description of the Pribyloff group of islands. Bull Museum Comp Zool 2:89–108

Burnham K, Anderson D (2002) Model selection and multimodel inference: a practical information theoretic approach, 2nd edn. Springer, New York

Burton RK, Snodgrass JJ, Gifford-Gonzalez D et al (2001) Holocene changes in the ecology of northern fur seals: insights from stable isotopes and archaeofauna. Oecologia 128:107–115. https://doi.org/10.1007/s004420100631

Campagna CB, Le Boeuf J, Cappozzo HL (1988) Pup abduction and infanticide in southern sea lions. Behaviour 135:1005–1012. https://doi.org/10.1163/156853988X00188

Campagna C, Bisioli C, Quintana F et al (1992) Group breeding in sea lions: pups survive better in colonies. Anim Behav 43:541–548. https://doi.org/10.1016/S0003-3472(05)81014-0

Cassini MH (1999) The evolution of reproductive systems in pinnipeds. Behav Ecol 10:612–616. https://doi.org/10.1093/beheco/10.5.612

Cassini MH (2000) A model on female breeding dispersion and the reproductive systems of pinnipeds. Behav Process 51:93–99. https://doi.org/10.1016/S0376-6357(00)00121-2

Cassini MH, Fernández-Juricic E (2003) Costs and benefits of joining south American sea lion breeding groups: testing the assumptions of a model of female breeding dispersion. Can J Zool 1160:1154–1160. https://doi.org/10.1139/Z03-098

Cullen TM, Fraser D, Rybczynski N et al (2014) Early evolution of sexual dimorphism and polygyny in Pinnipedia. Evolution 68:1469–1484. https://doi.org/10.1111/evo.12360
Dobson FS (1982) Competition for mates and predominant juvenile male dispersal in mammals. Anim Behav 30:1183–1192. https://doi.org/10.1016/S0003-3472(82)80209-1
Dugatkin LA (1996) Copying and mate choice. In: Heyes CM, Galef BG (eds) Social learning in animals: the roots of culture. Academic, London, pp 85–105
Fowler CW, Antonelis GA, Baker JD (1994) Studies of juvenile males tagged as pups and resighted during roundups in 1992. U.S. Department of Commerce, NOAA Technical Memorandum NMFS-AFSC 45:48–70
Gentry RL (1974) The development of social behavior through play in the Steller sea lion. Am Zool 14:391–403
Gentry RL (1998) Behavior and ecology of the northern Fur seals. Princeton University Press, Princeton
Gentry RL (2009) Northern fur seal: *Callorhinus ursinus*. In: Perrin WF, Würsig B, Thewissen JGM (eds) Encyclopedia of marine mammals, 2nd edn. Academic, London, pp 788–791. https://doi.org/10.1016/B978-0-12-373553-9.00183-8
Greenwood PJ (1980) Mating systems, philopatry and dispersal in birds and mammals. Anim Behav 28:1140–1162. https://doi.org/10.1016/S0003-3472(80)80103-5
Insley SJ (2000) Long-term vocal recognition in the northern fur seal. Nature 406:404. https://doi.org/10.1038/35019064
Insley SJ (2001) Mother-offspring vocal recognition in northern fur seals is mutual but asymmetrical. Anim Behav 61:129–137. https://doi.org/10.1006/anbe.2000.1569
Insley S, Phillips AV, Charrier I (2003) A review of social recognition in pinnipeds. Aquat Mamm 29:181–201
Kenyon KW (1960) Territorial behavior and homing in the Alaska fur seal. Mammalia 24:431–444
Kenyon KW, Wilke F (1953) Migration of the northern fur seal, *Callorhinus ursinus*. J Mamm 34:86–98
Kiyota M (2005) Site fidelity, territory acquisition and mating success in male northern fur seals (*Callorhinus ursinus*). Mamm Study 30:19–27. https://doi.org/10.3106/1348-6160(2005)30[19:sftaam]2.0.co;2
Kiyota M, Okamura H (2005) Harassment, abduction, and mortality of pups by nonterritorial male northern fur seals. J Mammal 86:1227–1236. https://doi.org/10.1644/04-MAMM-A-031R2.1
Kiyota M, Insley SJ, Lance SL (2008) Effectiveness of territorial polygyny and alternative mating strategies in northern fur seals, *Callorhinus ursinus*. Behav Ecol Sociobiol 62:739–746. https://doi.org/10.1007/s00265-007-0499-7
Lindenfors P, Tullberg BS, Biuw M (2002) Phylogenetic analyses of sexual selection and sexual size dimorphism in pinnipeds. Behav Ecol Sociobiol 52:188–193. https://doi.org/10.1007/s00265-002-0507-x
Loughlin TR, Miller RV (1989) Growth of the northern fur seal colony on Bogoslof island, Alaska. Arctic 42:368–372
Manly BFJ, McDonald LL, Thomas DL et al (2002) Resource selection by animals: statistical design and analysis for field studies, 2nd edn. Kluwer Academic Publishers, Dordrecht
McCann TS (1987) Male fur seal tenure. U.S. Department of Commerce, NOAA Technical Report NMFS 51:197–198
Nagasaki F, Matsumoto K (1957) Homing tendency of the bachelor seals to their native rookeries. Bull Tokai Reg Fish Res Lab 18:15–19
Newsome SD, Etnier MA, Gifford-Gonzalez D et al (2007) The shifting baseline of northern fur seal ecology in the Northeast Pacific Ocean. Proc Natl Acad Sci 104:9709–9714. https://doi.org/10.1073/pnas.0610986104
Perrin N, Mazalov V (2000) Local competition, inbreeding, and the evolution of sex-biased dispersal. Am Nat 155:116–127. https://doi.org/10.1086/303296

Peterson RS (1968) Social behavior of pinniped with particular reference to the northern fur seal. In: Harrison RJ, Hubbard RC, Peterson RC, Rice CE, Shusterman RJ (eds) The behavior and physiology of pinnipeds. Appleton-Century-Crofts, New York, pp 3–67

Peterson RS, Bartholomew GA (1967) The natural history and behavior of the California Sea lion. American Society of Mammalogists, Special Publications 1, 79pp

Repenning CA (1976) Adaptive evolution of sea lions and walruses. Syst Zool 25:375–390. https://doi.org/10.2307/2412512

Scheffer VB, Fiscus CH, Todd EI (1984) History of scientific study and management of the Alaskan fur seal, *Callorhinus ursinus*, 1786-1964. U.S. Department of Commerce, NOAA Technical Memorandum, NMFS-SSRF-780

Trillmich F, Trillmich KGK (1984) The mating systems of pinnipeds and marine iguanas: convergent evolution of polygyny. Biol J Linn Soc 21:209–216. https://doi.org/10.1111/j.1095-8312.1984.tb02062.x

Trites AW (1992) Reproductive synchrony and the estimation of mean date of birth from daily counts of northern fur seal pups. Mar Mamm Sci 8:44–56. https://doi.org/10.1111/j.1748-7692.1992.tb00124.x

Vakirtzis A (2011) Mate choice copying and nonindependent mate choice: a critical review. Ann Zool Fenn 48:91–108. https://doi.org/10.5735/086.048.0202

Vladimirov VA (1987) Age specific reproductive behavior in northern fur seals on Commander Islands. US Department of Commerce, NOAA Technical Reports NMFS 51:467–476

York AE, Hartley JR (1981) Pup production following harvest of female northern fur seals. Can J Fish Aquat Sci 38:84–90. https://doi.org/10.1139/f81-011

Chapter 9
Mating Games and Raiding Parties in Southern Sea Lions

Claudio Campagna, Burney Le Boeuf, and Claudio Bisioli

Abstract We review mating systems of the South American or southern sea lion, *Otaria byronia*, in Argentina and Uruguay, over the last five decades. Social and environmental variables determine whether males defend territories containing females or defend females directly. Variables that influence the mating strategy include: the geophysical substrate, ambient temperature, access to water and cooling, sex composition and population status. Males defend territories if the substrate is suitable, such as containing ledges and outcroppings that serve as boundaries, flat surfaces as parturition sites for females, and tidepools for females to cool off when the temperature rises. When these conditions do not exist, and the breeding area is uniform, males defend up to 4–5 females directly. Peripheral, bachelor males attempt to acquire females by raiding the breeding units, individually or in groups. Raids are disruptive and injurious to females and pups, and few raiders acquire females. Raids occur less frequently, involve fewer males, and are less disruptive today than the group raids of the 1980s. The difference is due primarily to the changes in physical configuration of breeding areas, which forces females to spread out horizontally, decreasing opportunities for raiding. We conclude that male southern sea lions defend territories when possible but can switch to defending females directly. The physical conditions and environmental context determines the optimal social strategy.

Keywords Southern sea lion · *Otaria byronia* · *Otaria flavescens* · Group raids · Mating systems · Female defense polygyny

C. Campagna (✉)
Wildlife Conservation Society, Marine and Argentina Programs, Buenos Aires, Argentina
e-mail: ccampagna@wcs.org

B. Le Boeuf
Ecology & Evolutionary Biology, University of California, Santa Cruz, CA, USA
e-mail: leboeuf@ucsc.edu

C. Bisioli
Germinal Biotech, Buenos Aires, Argentina

C. Campagna, R. Harcourt (eds.), *Ethology and Behavioral Ecology of Otariids and the Odobenid*, Ethology and Behavioral Ecology of Marine Mammals,
https://doi.org/10.1007/978-3-030-59184-7_9

9.1 Genesis

She had eaten her fill of hake and was sleeping at the high-tide mark at Punta Norte on the Península Valdés, in Patagonia, Argentina. He was cruising in the shallow waters near the surf line looking for love and not having much luck until he saw her. He surfed in on a wave and galumphed straight for her. She woke, startled, and jerked away but he prevented her escape. She tried again but he grabbed her in his jaws by the scruff of the neck and hurled her into the air, landing with a thump beside him. This went on for an hour or so until she relented, settled down, and went back to sleep. She gave birth a few days later and nursed her pup several times a day until she had to eat. She made several half-hearted attempts to enter the water but the male blocked her exit. Then, suddenly, her mood changed and they copulated briefly. The male let his guard down and she slithered into the water at high tide. After foraging for 3 days, she returned, found her pup and nursed it. The male was gone.

Next December she returned with her juvenile in tow, followed by two other females. He was already there and promptly corralled all three of them close to him. Another male watched, tentatively approached but was repulsed by his vocal threats. The juvenile moved about freely but remained within a few meters of its mother. In a few days, all three females gave birth. While the mothers nursed or slept next to their pups, the resident male roared in an aggressive challenge when his neighbor showed interest in approaching. This was a harbinger of things to come.

A year later, there were 10 females on the beach, but males now outnumbered them. The founder male could not hold them all while fending off two or three neighbors and multiple intruders. The largest of the challenging males herded some of the females and took them as his own. The fierce competition with his neighboring males and with peripheral males continued; when a female escaped her keeper, she was held by another resident male.

Years later, this site had turned into a flourishing colony, with a breeding area that contained hundreds of females and resident males, and with many males on the periphery eager to enjoy their share of the mating.

* * *

We do not know if this is how the breeding colony of southern sea lions, *Otaria byronia*, got started at Punta Norte, Península Valdés, Argentina (Fig. 9.1).[1] Perhaps males arrived first, as it occurs today in December of each year, signaling the onset of the breeding season. Perhaps it was the young males that established the sites that females later populated slowly. Origins are always murky and uncertain, even for humans who keep and are able to provide written records. But the mating system observed and documented by us at Punta Norte in the mid-1980s, is considerably different from the mating systems described for other otariids such as the Steller sea

---

[1]The species is also referred as Otaria flavescens. We followed the taxonomic list of The Society for Marine Mammalogy (https://marinemammalscience.org/species-information/list-marine-mammal-species-subspecies/) but see Rodríguez and Bastida (1993).

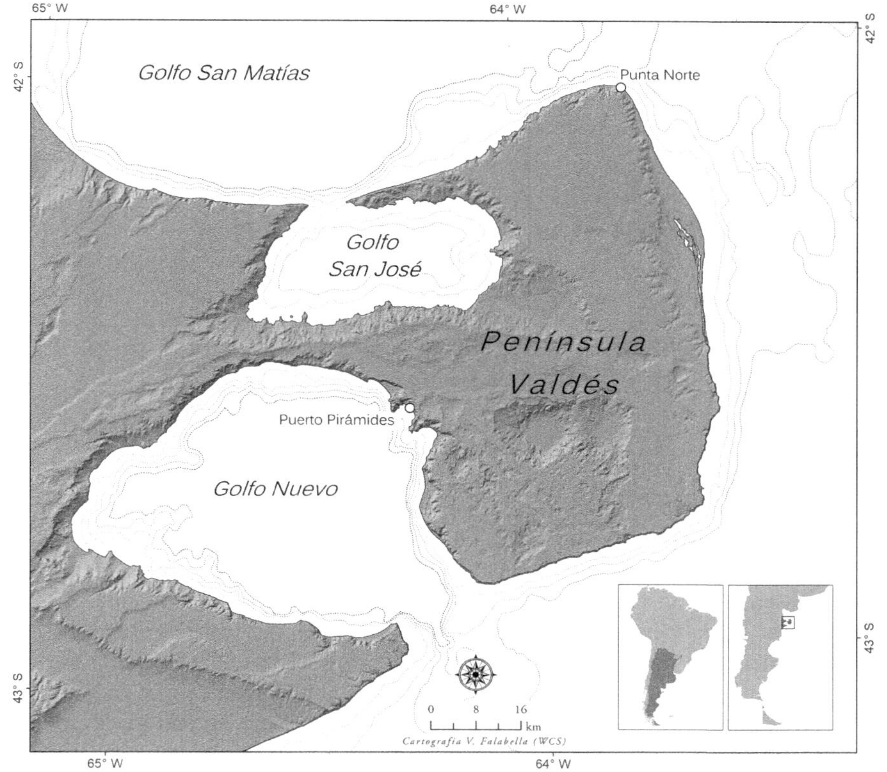

Fig. 9.1 Map of the area with the location of the two main study sites

lions, *Eumetopias jubatus,* (see Trites, Chap. 24) and northern fur seals, *Callorhinus ursinus* (Gentry 1975; see Kiyota, Chap. 8). So much so that one wonders how it began and what keeps it going (Campagna 1985; Campagna and Le Boeuf 1988a, b; Campagna et al. 1988a, b, 1992). The social behavior of breeding males and females is costly for both males and females, and seemingly not adaptive from a social evolution perspective. One aspect of this mating strategy is that males guard females and keep them close (Fig. 9.2). This contrasts with the other principal strategy, typical of most otariids, in which males defend territories and, through these territories, gain access to females. When males defend females directly, harassment of females and their pups is more common than when they defend territories. The sexual conflict between the sexes stands out in sharp relief and the behavior of males is costly to females (see Cassini, Chap. 7).

Fig. 9.2 Three adult males guarding one female each. Photo: Catalina Bisioli

9.2 Objectives

We provide a brief summary of the elements that differentiate the mating system of southern sea lions observed at Punta Norte during the 1980s from mating systems observed at other locations in the area, at other times, and in other otariids. By mating system, we mean the manner in which polygynous males compete to mate with as many females as possible. We discuss the essential elements of male mating behavior at this study site in the 1980s such as mate guarding and group raiding parties, the physical and social variables associated with these strategies, and the consequences for both sexes. We then compare with observations conducted in 2020 and review the behaviors observed at other sites. We make the case that the mating system in southern sea lions is highly flexible and depends in large part on the social and physical context in which mating occurs.

9.3 The Mating System of the 1980s

At Punta Norte, and at most sites in Península Valdés, sea lions group along a stretch of shoreline at the high tide mark (Fig. 9.3). In the 1980s, the central breeding area (CBA) of Punta Norte extended for about 200 m, forming the core of the colony, and was where most reproductive animals gathered to mate. The beach was composed of small pebbles that became extremely hot (45 °C) during the Austral summer months when sea lions reproduce, from late December to early February. The CBA was a narrow zone where the animals could cope with the high temperatures because they were splashed with surf at high tide or burrowed down into the wet pebbles at low tide. The beach was uniform, with no tide pools and no shade. The tidal range in the area was 5 m. In the space between the high tide and inland boundaries of the CBA

Fig. 9.3 Typical coastal landscape for Southern sea lion colonies at Península Valdés: pebble beaches limited inland by cliffs. Photo: Claudio Campagna

about three to four peripheral males guarded four to five females, generally about one each. Males did not tolerate the close vicinity of neighbors, but high tides and the hot substrate limited their vertical movements within the CBA. Consequently, the animals were located in the splash zone at high tide and as far as 70 m from the water at low tide. The angle from the water to the top of the berm was steep, an estimated 20–25 degrees for less than 100 m. The CBA, and the males and females in it, did not move with the ebb and flow of the tides throughout the breeding season; they remained on the edge of the berm.

Adult males arrived first. They defended space from each other but as females came ashore each male competed to herd, hold, and defend as many females as possible until they were ready to mate. On average, males could only partially control the movements of up to five females. They forced females to stay near them by biting, pushing and even lifting them in their jaws and hurling them violently into place; this caused numerous injuries to females who were either pregnant or nursing. Adult males did this easily, being about three times larger than females, with an immense head and neck making up about one third of the male's entire body (Fig. 9.2).

At peak season, around the third week of January, there were 140 males and 350 females on the beach. All females were in or near the CBA and among them were the resident adult males, each one guarding those females closest to him. Itinerant or peripheral males, most of them subadults, were scattered around the CBA looking on. Most of the time, resident males prevented them from approaching closer.

The critical element in this mating system was that males hoarded and held as many females as possible, with the ultimate aim of mating with them. Accessing a female reproductively could take up to a week, from the time the female arrived, gave birth and became ready to mate. The struggle to control females put them in a cruel bind; they were torn, on the one hand, between preventing neighboring resident males, or peripheral males, from capturing and absconding with one or more of their females, and, on the other, preventing their females from escaping. Within the colony, it was as if the resident male was watching a tennis match, his head and eyes directed to the intruder and then to the females near him. Back and forth. Action to ward off challenging males was always brief, lasting only a few seconds. This was a compromise for resident males but it benefited the challenging males who were not severely punished for their transgressions, had little to lose, and so were encouraged to try again and again. Over time, the situation drained the energy of the resident males, who frequently had to chase the peripheral males, each time moving a few meters away from their central position, risking that females would flee to other places and fall under the control of neighbors.

9.4 Group Raids and Its Consequences

To make matters worse for resident males, the peripheral males distributed around the CBA were strongly motivated to acquire females of their own. These had two options if they were to secure matings: (1) acquire a place in the CBA with one or more females, or (2) steal a female from the CBA and herd her to the periphery, where, if lucky, he might hold and defend her from other males and eventually copulate with her. Peripheral males often tried to herd and hold females arriving from the ocean, even though often some were pregnant and some coming back from a foraging trip to attend their pups, and thus already mated. Most of these females escaped the males and entered the CBA.

The most effective opportunity for acquiring females by ousted males was during group raids into the CBA (Fig. 9.4; Campagna et al. 1988a). Most raids began when an adult peripheral male rushed into the CBA and attempted to appropriate a female. This incited all nearby peripheral males nearby to rush in after the instigator and do what they could to acquire a female. Raids were conducted with great speed, noise, and a cloud of dust; they were disruptive, upset the status quo, and resulted in a spatial rearrangement of males and females in the CBA. In these raids, one got the impression that the surge of peripheral males into the CBA was synchronous, following instigation by the lead male, and suggested some coordination among those in the group. That is, individuals were aware and anticipated what others were doing or were going to do.

When raids occurred, females fled from the onrushing raiders, moving several meters laterally (along the high tide line) away from their resident holders. The progress of raiders was checked somewhat by the aggressive counter-attacks of resident males. In the mêlée, one or more of the largest raiders typically singled

Fig. 9.4 The central breeding area congregates most of the reproductive males and females. Photo: Ángel García

Fig. 9.5 Female abducted by an adult male during a raid. Photo: William Conway

out a female and held her while fighting off the resident males, or he attempted to carry her in his jaws to the periphery (Fig. 9.5). Resident males at the point of attack alternated between trying to halt the escape of females and defending their females with threats or bites to the encroaching raiders. They mounted females, seized females from each other, engaged in a tug of war with the female in the middle, and bit at the intruders. As females stampeded away from the raiders, their pups were left behind. In the tumult, one or two wandering pups were usually seized by subadult male raiders and carried away from the breeding group, where they were guarded, held and treated like adult females or were taken to the ocean where some drowned. After a few minutes of rushing about, fighting and milling around, most young raiders were rebuffed and driven out of the CBA. Usually, most raiders failed to acquire a female. On occasion, one or two adult raiders remained in the CBA for a few minutes, fighting off resident males for a place or a female. Sometimes, one or two succeeded in establishing residency. In about 30 min, things calmed down, there was a gradual return to pre-raiding conditions and females called to and, if fortunate, reunited with their pups. Some females, however, were now aligned with a different male, and some had lost their pup.

Raids were frequent, 120–160 during the 6-week breeding season with as many as a dozen raids per day. They occurred during all hours of the day and night and were most frequent at high tide. On average, 10–15 males were involved but raids composed of up to 40 males were observed at times. The same individuals participated in many raids. Indeed, if a resident male lost all females during a raid, he was most likely to join a subsequent raid to attempt to recapture what he had lost. Raids were incited by many different events: an adult rushing into the CBA to grab a female, females moving about, males fighting, at times even a thunderclap. If a raiding male succeeded in acquiring a female, it was usually the largest, oldest male. If he abducted a female and took her to the inland side of the CBA, after a few hours in the sun, the heat became too intense and, if mating was not possible, he was then forced to abandon her to go to the water to cool off.

There was great inequality in mating among male southern sea lions. Group raids were one way that males with poor prospects for mating had some access to females; it was like a revolt of the masses. The odds of winning a female were low, especially for subadults, and they often compromised by stealing pups. Perhaps the practice paid off in the long run, but this was often injurious to pups and a costly to the mother if she lost her pup of the year (see Ryazanov, Chap. 10).

9.5 Group Raids from the Male and Female Perspective

Males When a group raid occurred, the resident male, like Odysseus facing Scylla and Charybdis, had to make a difficult choice, neither optimal. This choice was either to fight each male attempting to steal females or to focus on actively preventing females from leaving. If he chose to fight and fought too long, some or all of the females would leave and then be captured by other males. If he primarily focused on preventing females from leaving, the intruding male(s) or a neighboring male then had a better chance of capturing one or more of them. The stress level for these rutting sea lions must be enormous. Indeed, if any male mammal gets stress-related ulcers, we would wager that it would be southern sea lion.

Clearly, there are limits to reproductive success for males in this situation; a male can hold only a few females and he may lose some or all of them at any time. Even if he copulates with all of them, this is low level of reproductive success compared to males of some other polygynous pinnipeds, e.g., northern fur seals (Gelatt and Gentry 2018; Mesnick and Ralls 2018) or northern elephant seals, *Mirounga angustirostris* (Le Boeuf 1974). Sexual dimorphism is however marked in southern sea lions, which suggests that there has been strong selection pressure for great size in males in the past (see Cassini, Chap. 6). Another possibility is that successful males have inseminated many females but under different circumstances. We will return to this possibility and this raises other important questions. For instance, how long can males maintain high activity levels during the breeding season while fasting and enduring high solar radiation and temperatures up to 40 °C during the day. What are the energetic costs? How long can males continue to compete over the years?

Females Females are forced to make the best of a bad situation. They seek a place to give birth where their pups are safe and where, a few days later, they can safely get re-inseminated during the post-partum estrus. They can best do this in the company of other females. If they give birth alone on the beach they are constantly harassed by several males, which disturbs nursing and their pups are less likely to survive (Campagna et al. 1992). Thus, females are effectively forced into the CBA. Females sustain injuries in trying to escape from a male or during group raids. By the end of the breeding season, one in four females shows injuries from male bites. Moreover, resident males often keep mothers away from their pups. After giving birth and nursing, females—although heat-stressed and hungry—cannot escape from the male to reach the water to cool off and to feed. This is when females copulate. This raises the question of whether they copulate at this time because estrus is associated with a spontaneous ovulation, or whether females copulate to buy safe passage to sea, with copulation inducing ovulation? Female elephant seals, who have already copulated with the alpha or a dominant bull, copulate again with peripheral males to buy safe passage to the water (Le Boeuf and Mesnick 1991; Mesnick and Le Boeuf 1991). We wonder whether female sea lions exploit the chaos of raids to align with another (perhaps better) male? If so, what is the basis for choice and is it effective?

9.6 Mating Systems in Other Places

The behavior observed at Punta Norte in the 1980s was also seen at other southern sea lion colonies along the beaches of Península Valdés and in nearby places where the topography precluded demarcation of defensible territories. The system was similar to the first description of group raids at Isla de Lobos in Uruguay, during the 1960s and 1970s (Vaz-Ferreira 1965, 1975). At all of these sites, uniform beaches precluded males from defending defined territories and this put a priority on males guarding females and not on defending territories. Casual observations report group raids, pup abductions and infanticide occur in South American sea lion colonies in Chile as well (H. Paves and M. Sepúlveda, pers. comm.). However, group raids have not been observed in a well studied South American sea lion colony in Peru (P. Majluf, pers. comm.). The physical environment there differs from that of Punta Norte, with large boulders and other complex topography.

Fig. 9.6 A view of Puerto Pirámides breeding colony, showing no pebble beaches but a rocky sandstone shelf. Photo: Catalina Bisioli

Similarly, the mating system observed at Puerto Pirámides, about 70 km from Punta Norte, was quite different (Campagna and Le Boeuf 1988b; Cassini and Vila 1990) and more like that found in most otariids. At Puerto Pirámides, males defended a site, not females directly. Territorial defense has also been observed as one strategy in recent studies at Isla de Lobos in Uruguay. There, males that mated near tide pools would defend territories, while males that mated along the tide line defended females (Franco-Trecu et al. 2015a). The topography at Puerto Pirámides consists of an extended rocky shelf backed by nearby cliffs, with several tide pools, big boulders, and a few places where females can access the rookery at low tide (Fig. 9.6). Males would fight to control the narrow passages where females came in from the sea and also established territories near boulders that offered shade, or pools filled with water. These pools afforded females and their pups a respite from high temperatures. Females could move freely in and through male territories in contrast to the restriction of their movements at Punta Norte. The topography is such that peripheral males could congregate and approach females from inland or from the sea at very high tide. Consequently, few males were found in the vicinity of the breeding animals most of the day. Among the 16 territorial males and 100 females observed in 1986–1987, only nine group raids were recorded during the last 3 weeks of January (Campagna and Le Boeuf 1988b). This figure is low compared with the Punta Norte rookery, where nine raids occurred in any single day near or at peak season. At Pirámides, all intrusions were repulsed by the territorial males. Consequently, raids there were not only infrequent but had little influence on the behavior of females or pups (but see Cassini and Vila 1990). No raiders gained a territory or displaced a territorial male.

9.7 Mating System at Punta Norte in 2020

We returned to Península Valdés in January 2020 to compare the mating behavior with that of the 1980s. While changes were not apparent at Puerto Pirámides, where males on the rocky shelf continue to defend territories and raids were rare, the situation was vastly different at Punta Norte in 2020 compared to the 1980s. A 3-day study at peak season during high tides, showed that raids still occurred but were weaker and less disruptive than in the past. The main difference today is to do with: (i) configuration of the beach, (ii) demographics of the population, (iii) number of breeding colonies, and (iv) distribution of mating pairs and peripheral males. We discuss each of these in turn.

Physical Variables The physical difference is most apparent (Fig. 9.7). In the present day, there is a 1–2 m ledge that prevents the sea lions from moving away from high surf and further inland. At high tide, females in the CBA are restricted to a narrow line along the berm, incapable of moving vertically by this ledge on the high side and the water on the low side. Instead, the females are distributed horizontally along the beach up against the ledge. That is, to accommodate the number of animals requiring space, the CBA has spread out over 300 m along the beach; 50% more than the 200 m it covered during the 1980s. Animals also moved a few hundred meters to the south. The slope of the beach from the water line at low tide to the top is now much steeper than in the 1980s. However, the breeding system continues to be female defense. The uniform pebbly substrate, with no tide pools and no large rocks or ridges has not changed. This homogeneous substrate is still unsuitable for the development of territorial behavior (Campagna et al. 1988a, b; Miller 2018). Most of the 200 m of beach occupied in the 1980s did not have animals in 2020.

Population Expansion Group raids in the 1980s were first observed during a period when the population was in the early stages of recovering from a severe population reduction (see Crespo, Chap. 23). Approximately 500,000 southern sea lions were killed in Argentine Patagonia alone, for commercial purposes, from the 1920s to the early 1960s (Grandi et al. 2012; Crespo, Chap. 23). The culling was especially severe at Península Valdés, where more than 30,000 adults of both sexes were killed in some years and almost a quarter of a million in a few decades. By the time the killing stopped, less than 10% of the original population size before the harvest remained in northern Patagonia. A reduction in population size of this magnitude is expected to disrupt the social structure and affect the form of male-male competition (see Franco-Trecu et al. 2015b), as well as the incidence of group raids. Today, 300,000 to 400,000 southern sea lions inhabit northern and central Patagonia (Crespo, Chap. 23).

The increase in the population may be associated with changes in the sex ratio and age structure of breeding animals. In female defense polygyny systems, a relative lack of females is expected to escalate competition among adult males. Many males become peripheral males, not resident males, and these peripheral adults are the main instigators of group raids. Subadult males escalate raids by adding chaos in numbers,

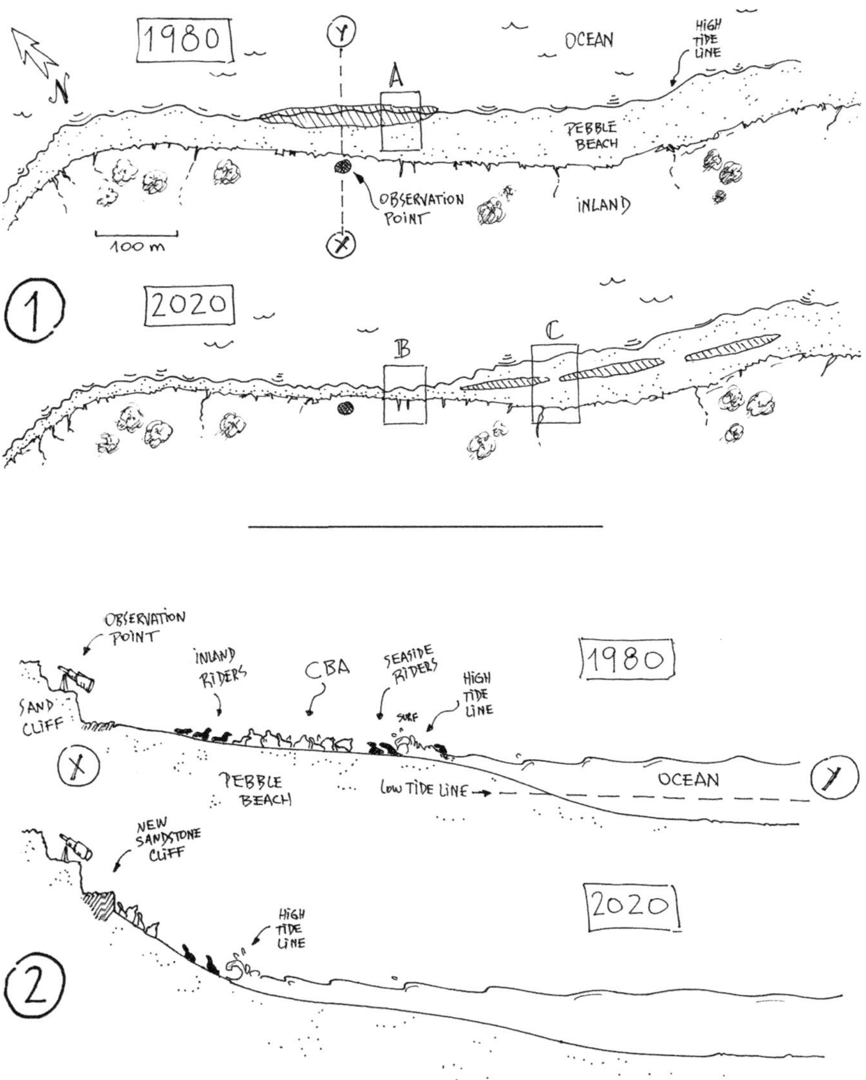

Fig. 9.7 Sketch of Punta Norte in the 1980s and in 2020. (1) View of a small proportion of the beach at the level of the observation point. The amplitude of the pebble beach at high tide was highly reduced in 2020, compared to decades before (2) Profile of the beach at the level of x – y in (1). An increase in the slope is evident in 2020, as well as a barrier of sandstone, that makes it difficult the presence of animals inland of the CBA (indicated with diagonal lines). Rectangles A, B and C are areas enlarged in Fig. 9.8. Sketches by Claudio Bisioli

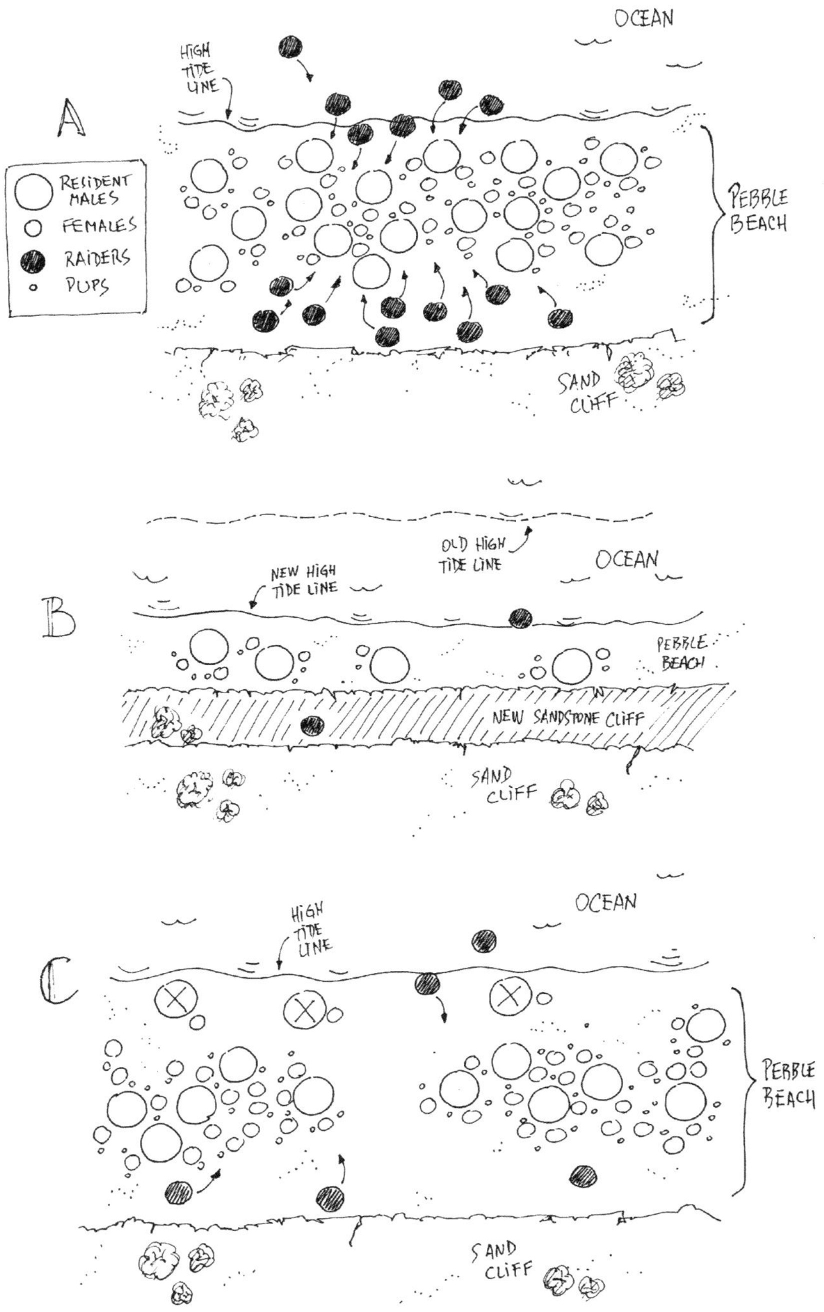

Fig. 9.8 (**a**) The CBA during a group raid in the 1980s at the level of Fig. 9.7/1 rectangle A. There is more space for the CBA at high tide than in 2020 and more peripheral males gathered in the ocean and inland peripheries. (**b**) The breeding colony in 2020 (rectangle B of Fig. 9.7/1. The beach is

but the main benefit, achieving a place in the CBA or abducting a female to the periphery, is only for the most tenacious males. In the 1980s, the number of reproductive females per male was approximately 2.8:1 (Campagna 1985). Data collected by Crespo and collaborators (Chap. 23) showed a ratio of 3.5:1 females per male in the 1990s, and this increased to 4.9:1 in 2000–2010. This suggests that the number of females available per male during the 1980s was considerably lower than in 2015. Group raids have decreased commensurately with the sex ratio as the population has grown.

New Breeding Colonies As the populations have recovered, new colonies have been founded or resettled. In the 1980s, there were three breeding aggregations along a 10-km coastline, today there are at least five. The CBA at Punta Norte is fragmented into subgroups separated by 10–50 m gaps. A similar number of males and females that occupied 200 m in the 1980s occupies 300 or more meters today. In addition, there are two additional breeding aggregations near Punta Norte. The physical change makes for a different social context and this influences behavior. Raids still occur but at lower frequency and involve fewer males and females. Violence and disturbance is more moderate than in the 1980s. These changes have been confirmed by rangers as having been the *status quo* at least during the last decade (I. Peinecura and P. Ciraolo, pers. comm.).

Distribution of Peripheral Males and Occurrence of Mating Pairs Changes in the configuration of the beach significantly restrict the space inland in some areas where the CBA used to be. This is most evident at high tide and with high surf (Fig. 9.7). The probability of intrusions by raiding males from the land also decreases as peripheral males cannot find space either to start a raid or to hold a female nearby. There are fewer potential raiders, no more than ten where there used to be up to 40. In contrast, while many adult males once were able to secure a female on the ocean side, a few meters of from the CBA, this is no longer possible. These males in the ocean front struggle with tides but not with high temperatures. Young males congregate in all-male groups, away from CBAs, as in the 1980s, but today, due to the expanded population, they represent, in the long run, the founders of new colonies. Males now seem to be more attached to place as is characteristic of the early days of the breeding cycle.

In summary, an increased number of animals, particularly of breeding females relative to males (Crespo, Chap. 23), has facilitated the establishment of new CBAs. Males engaged in these incipient colonies, with fewer peripheral males, instigate fewer raids. Attacks into the CBA still occur but with negligible consequences. A smaller number of males are involved in raids and intrusions from inland are more limited (Fig. 9.8). Males of the age group that used to be most involved in pup abductions now seek to start new colonies. If raiding involves males learning to raid, training from a young age, then the context for continued raids has decreased. In fact, group raids have not been observed at Isla de Lobos, Uruguay, during the period 2004 and 2013, although individual attacks by juvenile males and pup abductions continue to occur (Franco-Trecu, pers. comm.), as they do at Punta Norte.

Fig. 9.8 (continued) narrow and there are fewer animals. (**c**) The CBA at the level of rectangle C in Fig. 9.7/1. Many males (open circles with an X) hold one female in the ocean side of the CBA, compared to the 1980s. Sketches by Claudio Bisioli

9.8 Female Safety in 2020

How does the 2020 mating system compare to the 1980s from the female perspective? Females are mobile and can get away from males to attend their pups, to cool off, to go to sea to feed, or possibly to join a neighboring male, so they cannot be easily guarded. Nevertheless, females group together for protection from males and to improve their pup's chances of surviving (Campagna et al. 1992; Drago et al. 2011; Franco-Trecu et al. 2015a; see also Cassini, Chap. 7).

In the 1980s, females and pups were highly vulnerable to raids. Today, the combination of new CBAs nearby, lower density of the study site due to the horizontal dispersion of females, fewer ousted males in the vicinity of the CBA, lack of space to hold females inland, and more males consorting with single females at the ocean front but not in the periphery of the CBA, result in CBA females being less harassed by ousted males. In contrast, residents today focus on the females, fighting neighbors but less so peripherals. As the number of subadults near the CBAs is low, pup abduction is less frequent. It seems that, in 2020, a female-defense mating system is possible with greater overall social harmony.

9.9 Other Species

Female defense and group raids made the mating strategy of southern sea lions unusual compared to fur seals and other otariids, which are the paradigmatic examples of males holding territories that affords access to females. Another otariid in which males sequester females and participate in behaviors similar to raids is the Australian sea lion, *Neophoca cinerea* (Marlow 1975). At Dangerous Reef, where, decades ago about 90 females and 30 males reproduced, the number of females in territories changed constantly because groups of subadult males attempted to isolate females, causing mother-pup separation and pup mortality (see Ryazanov, Chap. 10). Nevertheless, the raids were of lesser magnitude and frequency than those described at Punta Norte in the 1980s. In the New Zealand sea lion, *Phocarctos hookeri*, territorial males may simultaneously attack and displace a neighboring male, which may have to move to another breeding colony for an opportunity to breed (in Robertson et al. 2006). Likewise, early in the season, fights that involve two or three males may spread across the colony. Soon, all the colony is in a turmoil, not dissimilar in effect to a group attack in southern sea lions (Chilvers, per. comm.).

Most otariids, such as northern fur seals (Bartholomew 1953; Bartholomew and Hoel 1953), Steller sea lions and the several *Arctocephalus* species defend territories against other males (Miller 2018). The topography of the breeding sites allows males to delineate territorial boundaries. In territorial species, fights are almost exclusively between neighboring males and they last minutes. Copulations are of long duration and are never interrupted by neighboring males. Nevertheless, Bartholomew (1953) reported the invasion of territories by groups of two to four "idle bull" northern fur seals; the males seized, herded or tossed females from neighboring territories. Gentry (1998) reported that northern fur seal adult males do not act in concert when establishing a territory, but subadults engage in opportunistic "forays" that are brief, and involve multiple males rushing into a breeding area. These appear to be simultaneous intrusions of several individuals:

> Forays involving multiple males are not a result of group action but are nearly simultaneous movements of unaffiliated individuals. They appear simultaneous because the factor that usually prevents them from occurring—the attention to boundaries of territorial males—can be suspended in all parts of the breeding area at once by a sudden disturbance. The simultaneous movement of many males onto the breeding site can give the false impression of concerted group action. (Gentry 1998, page 212).

Forays, or simultaneous intrusions by subadults northern fur seals, were observed on St. George Island during the 1970s but not on nearby St. Paul Island (Gentry, pers. comm.). This difference may be explained by the lack of subadult males on St. George owing to extensive culling of males for commercial purposes, suggesting here again the importance of the sex ratio in influencing the social context. Populations changes may also have a disruptive effect in socially learned behaviors (Van Schaik 2002).

The aggressive herding and group raids we observed in southern sea lions during the 1980s resembles the behavior of Hamadryas baboons, *Papio hamadryas hamadryas*, and *P. hamadryas anubis* (Kummer 1968; Swedell et al. 2014) and langurs, *Presbytis entellus* (Hrdy 1974, 1977). Both are collective revolts by the outsiders in male-male competition against the winning males who are associated with females and thus do most of the mating. In the baboons, a male coerces females to follow him wherever he goes and punishes her if she does not comply. A subadult or peripheral male "plans for the future" and may adopt juvenile females into his female group so that he can mate with them when they mature months or years later. Males impose significant costs on females (Swedell et al. 2014). Female coercion, for example, is the "*use by a male of force, that functions to increase the chances that a female will mate with him at a time when she is likely to be fertile, and to decrease the chances that she will mate with other males, at some costs to the female*" (Smuts and Smuts 1993). Males overcome female resistance and force mating, they may kill infants sired by competing males and constrain female movements, herding them or preventing them from joining her kin (Smuts and Smuts 1993). We humans too, have a long record of coercing women for sex and participating in group raids where women are abducted (Chagnon 1988; Stark 2007; Muller and Wrangham 2009; Alcorn 2014).

9.9.1 Human Raids

Competition for mates, or sex, in humans takes a variety of forms. The mass and violent rape of women occurs in wars, ethnic cleansing genocides, nationalistic conflicts, and foreign invasions (e.g., Alcorn 2014). The symbolic representation of these behaviors is depicted in myth and art, an example being the The Rape of the Sabines.

Rape finds roots in the Latin word "*raptio*," that may be understood as an abduction: the action of forcibly taking someone away (without sexual consent), or as the action of forcing someone into sexual intercourse. The *raptio* of the Sabine women by the Romans was represented by many artists in large scenographic paintings (e.g., Fig. 9.9). They imagined women struggling against men warriors, their feminine nude torsos and twisted bodies subdued by them, in an attempt to either prevent their abductees from going back to their men, or to take them to places where they did not want to go.

It is easy to see similarities between the *raptio* of the Sabines and the abduction of female sea lions during group raids. In both, females are victimized against their will and disputed by coercion. Courtship is not involved. Sabines were abducted *en masse* by a concerted action of a group of men. As in animals, the leader—the strongest most powerful male—had priority of choice (or so it was in the Iliad). In sea lions, raids rarely involve more than one abduction and are not a concerted effort with division of roles for individuals. In antique human history, sexual abuses were common in tyrannical societies. In a way, raids represent a momentary interruption of the absolute dominance of few adult males, those residing in the CBA, over the female population. For a short time, the structure of the breeding colony is disrupted and dominant males cannot cope with revolting competitors.

More common that abductions *en masse* was female kidnapping, the most eminent example being the kidnapping of Helen of Troy by Paris, which triggered the Trojan War. In *The Abduction of Proserpine* (1570), Florentine painter

Fig. 9.9 The Rape of the Sabine Women by Nicolas Poussin (Public domain photography)

Fig. 9.10 The Abduction of Proserpine by Alessandro Allori (Public domain photography)

Alessandro Allori portrayed Pluto, the god of the Underworld, seizing the daughter of Ceres, the Roman goddess of agriculture and fertility (Fig. 9.10).

In more recent times, and closer to reality, Argentine painter Ángel Della Valle (1882) recreates the scenario of an abduction of a woman during a raid by Pampas aborigines (https://www.bellasartes.gob.ar/coleccion/obra/6658/). A very similar representation of a kidnap is depicted in *A Raid by Kurds,* by Swiss illustrator Frank Feller.

9.10 General Conclusions

Southern sea lions adapt to circumstances at hand. It is misleading to speak of a standard mating system that applies to all of these sea lions. The manner in which the mating game is played out is context dependent. It is more prudent to describe the mating system observed and tie it to time, place and context. The language must fit reality. It is more appropriate to speak of several forms of male-male competition that we describe as a mating system. Which one prevails depends on variables such as the physical configuration of the beach in terms of substrate, slopes, or range of tides, environmental variables such as temperature and solar radiation, the composition of the colony, particularly the sex ratio of reproductive animals, and the colony history. Conclusions depend to a large extent on where and when one observes, i.e., the "window in time." Under certain circumstances, the mating game can take strange forms, like the one observed at Punta Norte in the 1980s. Mating systems are plastic, and males and females adjust to breed under the prevailing conditions.

Sexual conflict is inherent in heterosexual behavior because males pursue mate quantity while females benefit from mate selectivity. Conflict escalates and may become lethal when males coerce mating at a cost to females. When males are larger and stronger than females, they can treat females as a resource. This is most dangerous to females and their offspring when a male raiding party forcibly abducts females from a group kept by other males.

In southern sea lions, male behavior that increases mating success (forceful herding of females to mate with them and raiding parties) is selected even though it imposes significant reproductive costs to females (bodily injury, pup mortality after the investment in gestation and nursing). It is important to determine the conditions under which males keep females for mating by vigilance, threats, and physical aggression, as well as by participation in spontaneous or coordinated group raids.

Sexual selection has not spared the female sex. To reproduce females must suffer the harsh treatment of the males that would father their pups. Some females are thrown about like a dog flinging a rag doll (Fig. 9.5). Moreover, this is when the female is either pregnant or has just given birth and is nursing. Nevertheless, a female is better off being kept with other females by a male in the CBA. She is pestered less in this situation, and her pup is more likely to survive than if she bred in isolation (Campagna et al. 1992). So, she must tolerate the behavior of her keeper to bear his pup as well as to survive the periodic chaos, disruption and rearrangement of the mating system caused by raids. In the conflict between the sexes in this species, males "win" by imposing costs on females, and females "win" by enduring those costs and producing viable pups, but also by creating fertile conditions for male competition, which she might exploit. Swedell et al. (2014) concluded that in hamadryas baboons, (*Papio hamadryas hamadryas*): "the reproductive strategies benefitting males can evolve despite substantial costs to females" (in terms of "increasing the chance of infant mortality while delaying subsequent conception"). "These costs may be mitigated over the long term, however, by female counterstrategies and protective behaviour by males".

Long-term studies spanning many years and covering many locations at different historical times are required to document the adaptability of mating strategies in animals. This analysis of the mating systems of South American sea lions shows what is possible, and what variables are important. Similar plasticity in mating games is likely to be found in many other mammals.

A note about humans: in some respects, we are not much different from the sea lions. When sultans control a harem, like that of Süleyman the Magnificent of the Ottoman Empire, there is no need to raid or kidnap, like when a bull sea lion defends a territory with resources that several females like or need (Peirce 1993). Yet, nonconsensual bride kidnappings, forced marriage, marriage by abduction, or marriage by capture has been reported in dozen of cultures, and may today affect, still today, thousands of women (Farr 2009). The human female, however, is not as helpless as sea lions. In Roman times at least, she charmed her abductor for a while, then poisoned his soup.

Acknowledgements We are thankful to Enrique Crespo for providing demographic data from which the sex ratios of adult animals were estimated for the Punta Norte breeding area. Rangers Inés Peinecura and Paula Ciraolo provided information regarding the occurrence of raids at Punta Norte in recent years, and Valentina Franco-Trecu did it for Isla de Lobos, Uruguay. Julieta Campagna, Catalina Bisioli and Valentina Bisioli provided field assistance during the 2020 observations.

References

Alcorn T (2014) Responding to sexual violence in armed conflict. Lancet 383(9934):2034–2037

Bartholomew GA Jr (1953) Behavioral factors affecting social structure in the Alaska fur seal. Trans 18th N Am Wildl Conf, pp 481–502

Bartholomew GA Jr, Hoel PG (1953) Reproductive behavior of the Alaska Fur seal, *Callorhinus ursinus*. J Mammal 34(4):417–436

Campagna C (1985) The breeding cycle of the Southern Sea lion, *Otaria byronia*. Mar Mamm Sci 1(3):210–218

Campagna C, Le Boeuf BJ (1988a) Reproductive behavior of Southern Sea lions. Behaviour 104:233–261

Campagna C, Le Boeuf BJ (1988b) Thermoregulatory behavior of Southern Sea lions and its effect on mating strategies. Behaviour 107:72–90

Campagna C, Le Boeuf BJ, Cappozzo HL (1988a) Group raids: a mating strategy of male Southern Sea lions. Behaviour 105:224–249

Campagna C, Le Boeuf BJ, Cappozzo HL (1988b) Pup abduction and infanticide in Southern Sea lions. Behaviour 107:44–60

Campagna C, Bisioli C, Quintana F et al (1992) Group breeding in sea lions: pups survive better in colonies. Anim Behav 43:541–548

Cassini MH, Vila BL (1990) Male mating behavior of the Southern Sea lion. Bull Mar Sci 46(2):555–559

Chagnon N (1988) Life histories, blood revenge, and warfare in a tribal population. Science 239:985–992

Drago M, Cardona L, García N et al (2011) Influence of colony size on pup fitness and survival in Southern Sea lions. Mar Mamm Sci 27:167–181. https://doi.org/10.1111/j.1748-7692.2010.00402.x

Farr K (2009) Extreme war rape in today's civil-war-torn states: a contextual and comparative analysis. Gend Issues 26:1–41

Franco-Trecu V, Costa-Urrutia P, Schramm Y et al (2015a) Tide line versus internal pools: mating system and breeding success of Southern Sea lion males. Behav Ecol Sociobiol 69:1985–1996. https://doi.org/10.1007/s00265-015-2010-1

Franco-Trecu V, Drago M, Baladán C et al (2015b) Postharvesting population dynamics of the Southern Sea lion (*Otaria byronia*) in the southwestern Atlantic. Mar Mamm Sci 31(3):963–978. https://doi.org/10.1111/mms.12197

Gelatt TS, Gentry R (2018) Northern fur seal *Callorhinus ursinus*. In: Würsig B, JGM T, Kovacs KM (eds) Encyclopedia of marine mammals, 3rd edn. Academic Press, New York, pp 645–648

Gentry RL (1975) Comparative social behavior of eared seals. Rapp P-V Réun Cons Int Explor Mer 169:189–194

Gentry RL (1998) Behavior and ecology of the northern Fur seal. Princeton University Press, Princeton

Grandi MF, de Oliveira LR, Dans SL et al (2012) A hunted population in recovery: effective population size for Southern Sea lions from Patagonia. Anim Biol 62:433–450. https://doi.org/10.1163/157075612X642941

Hrdy SB (1974) Male-male competition and infanticide among the langurs (*Presbytis entellus*) of Abu, Rajasthan. Folia Primatol 22:19–58

Hrdy SB (1977) The langurs of Abu: female and male strategies of reproduction. Harvard University Press, Cambridge

Kummer H (1968) Social organization of hamadryas baboons: a field study. University of Chicago Press, Chicago

Le Boeuf BJ (1974) Male-male competition and reproductive success in elephant seals. Am Zool 14 (1):163–176

Le Boeuf BJ, Mesnick SL (1991) Sexual behavior of male northern elephant seals: I. Lethal injuries to adult females. Behaviour 116:143–162

Marlow BJ (1975) The comparative behavior of the Australasian Sea lions, *Neophoca cinerea* and *Phocarctos hookeri* (Pinnipedia: Otariidae). Mammalia 39:159–230

Mesnick SL, Le Boeuf BJ (1991) Sexual behavior of male northern elephant seals: II female response to potentially injurious encounters. Behaviour 117:262–280

Mesnick SL, Ralls K (2018) Mating systems. In: Würsig B, JGM T, Kovacs KM (eds) Encyclopedia of marine mammals, 3rd edn. Academic Press, New York, pp 586–592

Miller T (2018) Territorial behavior. In: Würsig B, JGM T, Kovacs KM (eds) Encyclopedia of marine mammals, 3rd edn. Academic Press, New York, pp 983–990

Muller MN, Wrangham RW (2009) Sexual coercion in primates and humans: an evolutionary perspective on male aggression against females. Harvard University Press, Cambridge

Peirce LP (1993) The imperial harem: women and sovereignty in the ottoman empire. Oxford University Press, New York

Robertson BC, Chilvers LB, Duignan PJ et al (2006) Dispersal of breeding, adult male *Phocarctos hookeri*: implications for disease transmission, population management and species recovery. Biol Conserv 127(2):227–236. https://doi.org/10.1016/j.biocon.2005.08.011

Rodriguez DH, Bastida RO (1993) The southern sea lion, *Otaria byronia* or *Otaria flavescens*? Mar Mamm Sci 9:372–381

Smuts BB, Smuts RW (1993) Male aggression and sexual coercion of females in non-human primates and other mammals: evidence and theoretical implications. Adv Study Behav 22:1–63

Stark E (2007) Coercive control: how men entrap women in personal life. Oxford University Press, New York

Swedell L, Leedom L, Saunders J et al (2014) Sexual conflict in a polygynous primate: costs and benefits of a male-imposed mating system. Behav Ecol Sociobiol 68:263–273. https://doi.org/10.1007/s00265-013-1641-3

Van Schaik CP (2002) Fragility of traditions: the disturbance hypothesis for the loss of local traditions in orangutans. Int J Primatol 23:527–538

Vaz-Ferreira R (1965) Comportamiento antisocial en machos subadultos de *Otaria byronia* (lobo marino de un pelo). Revista de la Facultad de Humanidades y Ciencias, Montevideo 22:203–207

Vaz-Ferreira R (1975) Behaviour of the Southern Sea lion, *Otaria flavescens* (Shaw) in the Uruguayan islands. Rapp P-V Reun Cons Int Explor Mer 169:219–227

Chapter 10
Infanticide in Eared Seals

Sergey D. Ryazanov

Abstract Infanticide is a complex behavior found in many mammalian species. It may be intentional, when the infant killer obtains a benefit or when it kills due to pathological behavior; or unintentional—as a by-product of behaviors not directed to the killing of infants. Intentionality is a key element in understanding the nature and adaptive meaning of infanticide. I here categorize intra- and inter-specific infanticide by the criterion of intentionality. Intentional intra-specific infanticide occurs in the otariids as food resource utilization and social pathology or, perhaps, in combination. This behavior is very rare. Intentional inter-specific infanticide is low-frequency, predatory behavior that sometimes may have a demographic effect for the species that is preyed upon. Unintentional infanticide is the by-product of many behaviors, often related to intra-sexual competition, but also as a consequence of misdirected aggression, adoption avoidance, competition for misrecognized pups, rough handling by mothers, and incidental trampling. Unintentional infanticide may cause an impact in a population, but usually yield no gains for the infant killer.

Keywords Infanticide · Pup abuse · Killing of young · Otariidae

The killing of young occurs in many species for a number of reasons and within a diverse range of social interactions (e.g. Hausfater and Hrdy 1984; Lukas and Huchard 2014). Infanticide may be intentional, when directed killing of infants confers adaptive benefits, or may be a consequence of pathological behavior. It may also be unintentional, when the death of the infant is a by-product of abuse or negligent behavior by the killer.

It is difficult to determine when infanticide in the wild is intentional or not; the purpose of the infanticidal animal is unknown to the observer. Intentionality may be inferred by context and the aftermath of the occurrence, and by the behavior of the killer and other members of the social group. Despite risks of misinterpretations,

S. D. Ryazanov (✉)
V.I. Il'ichev Pacific Oceanological Institute, FEB RAS, Vladivostok, Russia

C. Campagna, R. Harcourt (eds.), *Ethology and Behavioral Ecology of Otariids and the Odobenid*, Ethology and Behavioral Ecology of Marine Mammals,
https://doi.org/10.1007/978-3-030-59184-7_10

hypothesizing the intentionality of infanticide is critical because it helps evaluate potential reasons and perhaps adaptive meanings of the phenomenon.

Work by Hrdy (1979), proposed several explanations for infanticide: (1) exploitation of young as a resource, i.e. when the killer consumes the infant to obtain energetic benefits, uses it as a protective buffer against aggression by a third party, or as a source of maternal-like experience; (2) elimination of competitors, i.e. when the killer, or the killer's offspring, benefit by having access to more resources; (3) sexual selection, i.e. when killing of the infant decreases reproductive success of competitors and increases reproductive opportunities for the killer, usually by shortening intervals to next ovulation of the infant's mother; (4) parental manipulation, i.e. when a parent kills the infant to terminate parental investment and thereby improve chances for survival of the mother or other existing offspring; and (5) social pathology—maladaptive behavior, which conveys no apparent benefits.

Among the afore-mentioned explanations, the first one (exploitation) should be separated into two categories by the criterion of intentionality: (a) exploitation as a food resource, and (b) other forms of exploitation. Killing for food is intentional infanticide. Infanticide that results from using the young as protective buffer for aggression, or as a training substitute for maternal experiences, would be the by-product of abusive behaviour. The death of the infant may not happen, and thus this form of infanticide is unintentional. Infanticide as by-product of exploitation of young as a female sex substitute to obtain reproductive-like behavioral experience by the killer also belongs to the unintentional category (e.g. Campagna et al. 1988a, b).

In species that reproduce seasonally, infanticide might not reduce the inter-birth interval (van Schaik 2000). The typical mating system of otariids is polygyny. Males copulate with multiple females during the breeding season (e.g. Bartholomew 1970; Cassini, Chap. 6), and some females may copulate several times with a few males, i.e., a system tending towards polygynandry (e.g. Marlow 1975; Gisiner 1985; Altukhov 2012). Thus, it is unlikely that males are able to identify the paternal origin of an infant with sufficient confidence to commit infanticide for the purpose of reducing reproductive success of a competitor. Under these circumstances, sexual selection might not be considered as the explanation of infanticide.

Otariid males do not display parental care (Bartholomew 1970); therefore, they cannot terminate their investment through infanticide. Females, on the other hand, would gain benefits through termination of their investments under some circumstances, such as the risk to their own survival. Thus, the parental manipulation explanation in eared seals can, theoretically, be attributed only to infanticide by females.

Pierotti (1991) and Ebensperger (1998) discussed the hypothesis that infanticide may be perpetrated to avoid adoption (mis-recognition of alien pups as their own, or milk stealing by alien pups). This form of infanticide may occur in otariid females (e.g. Marlow 1975; Trillmich 1981), although female aggression toward alien infants rarely leads to death. Thus, infanticide in the cases of adoption-avoidance in otariids seems to be unintentional.

In summary, intentional intra-specific infanticide in eared seals may theoretically result from: (1) exploitation as a food resource, (2) elimination of competitors, (3) parental manipulation by females, and (4) social pathology.

The same explanations may apply to inter-specific pup or young killing, with the exception of parental manipulation, and with one addition. Janssen et al. (2002) argued that inter-specific infanticide might be performed by representatives of a prey species during vulnerable development stages of the predator, and this might be one way to reduce risk of future predation. Some otariids do prey on other otariids (e.g. Gentry and Johnson 1981), and so they may theoretically engage in inter-specific infanticide to avoid future predation.

Behaviors that are deemed unintentional are intra- and inter-specific infanticide due to exploitation as protective buffer, use of a substitute of a mate or to obtain maternal-like experience, or as a consequence of misdirected aggression, misdirected sexual behavior, aggression to avoid adoption, or negligent behavior.

10.1 Intentional Infanticide

10.1.1 Intra-Specific

In two otariids—the New Zealand sea lion (*Phocarctos hookeri*) and Steller sea lion (*Eumetopias jubatus*), infanticide seems to be intentional (Table 10.1). Pup abductions by New Zealand sea lion males were recorded twice at Dundas Island, and males were seen consuming conspecific pups nine times (Wilkinson et al. 2000). Males abducted pups, violently thrashed them from side to side, killed and then ate them. Once, the mother of the abducted pup unsuccessfully confronted the abductor. Infanticidal males were not individually recognized, but judging by their age, the authors concluded that they were different individuals. Wilkinson et al. (2000) calculated that one pup could provide an excess of the adult male's daily energy requirements. Thus, they attributed the events to exploitation as a food resource, although social pathology was not ruled out as an explanation or co-explanation.

Intentional infanticide in Steller sea lions was described for two territorial males at Kuril (Permyakov 2006 and pers. comm.) and Commander (Ryazanov et al. 2018) Islands. The male at the Kuril Islands grabbed pups and smashed them against the rocks (Fig. 10.1a). After that, he crushed the carcasses with his teeth for some time, but did not consume them. This male killed three pups during the same reproductive season. At the Commander Islands, another adult male seized one pup, threw it to his territory, and tore the body by violently shaking it from side to side. The male ate the stomach, intestine and perhaps other organs. Three more dead pups were recorded torn in two in the territory of the male during two reproductive seasons. During killings, mothers were not in the vicinity, while other sea lions and fur seals at the rookeries appeared indifferent to the episodes.

Of the potential explanations for these events, only that for pathological behavior applies in this case. The Commander Islands male consumed only a small part of the

Table 10.1 Intentional infanticide in Otariidae

Species	Sex-age group of killers	Sites	Victims	Explanation	Source of information
Intra-specific					
New Zealand sea lion	Adult male, Subadult	Dundas Islands	Conspecific pups	Food resource utilization (social pathology not dismissed)	Wilkinson et al. (2000)
Steller sea lion	Adult males	Commander and Kuril Islands	Conspecific pups	Social pathology	Permyakov (2006) and Ryazanov et al. (2018)
Inter-specific					
Steller sea lion	Adult male	Año Nuevo Island	California sea lion yearling	Food resource utilization	Byrnes and Hood (1994)
	Juvenile males	St. George Island	Northern fur seal pups	Food resource utilization	Gentry and Johnson (1981)
South American sea lion	Unknown, Adult males Subadult male	Falkland (Malvinas) Island, Punta San Juan (Peru), Guafo Island (Chile)	South American fur seal pups	Food resource utilization	Gentry and Johnson (1981) (by personal communication of S. Sommerhays), Harcourt (1992, 1993), Pavés (2008)
New Zealand sea lion	Subadult	Macquarie Island	Antarctic and subantarctic fur seal pups	Food resource utilization	Robinson et al. (1999)
	Unknown[a]	Snares Island, South Island of New Zealand	New Zealand fur seal pup	Food resource utilization	Mattlin (1978) (by personal communication of M.W. Cawthorne and D.S. Horning), Bradshaw et al. (1998)

[a]Multiple findings in regurgitations of sea lions. Active predation events were not observed

Fig. 10.1 Eared seal behaviour: (**a**) Steller sea lion adult male killing conspecific pup. Photo by P.A. Permyakov; (**b**) Northern fur seal bachelor carrying abducted conspecific pup. Photo by A.V. Chetvergov; (**c**) South American sea lion subadult male with abducted South American fur seal pup. Photo by V. Franco-Trecu; (**d**) Northern fur seal bachelor mounting and performing pelvic thrusts at conspecific pup. Photo by author; (**e**) South American sea lion male carrying South American fur seal pup out to the sea. Photo by V. Franco-Trecu; (**f**) Steller sea lion female throwing off a strange pup. Photo by P.A. Permyakov; (**g**) Emaciated northern fur seal pup stealing milk from

carcass, insufficient for any valuable energetic benefit. Observers also did not find evidence of misdirected territorial aggression. Therefore, intentional intra-specific infanticide in otariids occurred only in males as a form of exploitation of food resource and social pathology, or perhaps a combination of both.

Among pinnipeds, infanticide followed by cannibalism has also been reported in gray seals (*Halichoerus grypus*) (Bedard et al. 1993; Kovacs et al. 1996; Bishop et al. 2016; Brownlow et al. 2016,; van Neer et al. 2019) and southern elephant seals (*Mirounga leonina*) (Wilkinson et al. 2000; Campagna pers. comm.). In other marine mammals, cannibalistic infanticide is well known in polar bears (*Ursus maritimus*) (e.g. Derocher and Wiig 1999; Stirling and Ross 2011). In addition, there are some reported cases of combination of food resource utilization and parental manipulation—polar bear females have killed and eaten their own cubs (see Taylor et al. 1985 for review). Other reasons for intentional intra-specific infanticide in marine mammals may be as part of sexual selection (killing young to make the female sexually receptive) proposed for toothed whales (e.g. Diaz Lopez et al. 2017; Towers et al. 2018); and its combination with food resource utilization proposed for polar bears, when a male kills and consumes a cub and then copulates with the cub's mother which has become sexually receptive due to the infanticide (Taylor et al. 1985).

10.1.2 Inter-Specific

Inter-specific intentional infanticide has only been attributed to food resource utilization (Table 10.1). This form of inter-specific interaction was observed in three species: Steller, southern (*Otaria byronia*) and New Zealand sea lions.

Steller sea lion juvenile males at St. George Island caught northern fur seal (*Callorhinus ursinus*) pups in the water, within 50 m offshore, approaching them from beneath the surface or, in large surf conditions, riding waves into the midst of pup groups (Gentry and Johnson 1981). After seizing a pup, sea lions sharply shook their victims from side to side. They often dove with the prey and resurfaced further offshore. The sea lions tore the carcasses into pieces and ate parts. A Steller sea lion territorial male at Año Nuevo Island caught a California sea lion (*Zalophus californianus*) yearling in the water, close to his territory (Byrnes and Hood 1994). The male shook the young violently from side to side and drowned it. Then the male moved back to his territory, with the California sea lion in his mouth, slapped it against the rocks several times and ate the flesh. Among other pinniped infants,

Fig. 10.1 (continued) a female. Photo by author; (**h**) Steller sea lion female nursing adopted northern fur seal pup with her own pup. Photo by author; (**i**) Steller sea lion female misrecognises an alien pup as her own and competes for the pup with its mother. Photo by P.A. Permyakov; (**j**) Steller sea lion adult male and trampled northern fur seal pup carcass. Photo by author

Steller sea lion males and females are also reported to prey on harbor seals (*Phoca vitulina*) (Pitcher and Fay 1982; Mathews and Adkison 2010; Womble and Conlon 2010). Males also prey on ringed (*Pusa hispida*) (Tikhomirov 1959) and bearded seal (*Erignathus barbatus*) pups, and spotted seal (*Phoca largha*) yearlings (Calkins 1998).

Early observations of southern sea lions report males feeding on South American fur seal (*Arctocephalus australis*) neonates and females in the Falkland (Malvinas) Islands (Gentry and Johnson 1981 reported an observation by S. Sommerhays). Adult males at Punta San Juan (Peru) surfed into the rookery and attacked South American fur seal pups, or in some cases trapped them in caves (Harcourt 1992, 1993). Pups were killed and consumed. Such predation was observed for one subadult southern sea lion male in Guafo Island (Chile) (Pavés 2008). Southern sea lion adult males are also known to kill southern elephant seal pups (Harcourt 1993 on a personal communication by C. Duck).

A New Zealand sea lion subadult male fed on Antarctic and subantarctic fur seal pups (*Arctocephalus gazella*, *Arctocephalus tropicalis*) at Macquarie Island (Robinson et al. 1999). The male violently thrashed the pup's carcass to tear off strips of flesh, and then swallowed it. Mattlin (1978), on personal communication by M.W. Cawthorne and D.S. Horning, reported findings of New Zealand fur seals (*Arctocephalus forsteri*) including one small pup in regurgitations of New Zealand sea lions from the Snares Islands. Bradshaw et al. (1998) reported multiple findings of New Zealand fur seal pups in regurgitations and their hair in scats of New Zealand sea lions at the haul-out beaches in Otago (South Island, New Zealand). Based on G. Clark's personal communication, they also reported the case of an abduction of a New Zealand fur seal pup by a New Zealand sea lion, suggesting active predation.

Overall, inter-specific intentional infanticide in eared seals seems to be predation that occurs rarely and is carried out only by some individuals. Gentry and Johnson (1981) found that the energy that is accessible for Steller sea lions through predation on northern fur seal pups would be insufficient to sustain the entire population of predators for a long period. Harcourt (1993) estimated that only 7%–8% of about 2000 southern sea lion males could prey on fur seals at Punta San Juan, making it a successful hunting strategy only for a small group of individuals. Conversely, this type of predation sometimes may have a significant impact on the prey. Gentry and Johnson (1981) estimated that in 104 days at St. George Island, Steller sea lions killed from 3.4% to 6.8% of all 79,708 fur seal pups estimated to have been alive at the beginning of the observed hunting period. Robinson et al. (1999) supposed that only one subadult New Zealand sea lion killed up to 43% of all 130 Antarctic and Subantarctic fur seals pups born during one reproductive period.

Beyond otariids, predation on young pinnipeds is a common feeding strategy for leopard seals (*Hydrurga leptonyx*) (e.g. Hall-Aspland and Rogers 2004), and rarely by walruses (*Odobenus rosmarus*) (e.g. Mansfield 1958). Intentional inter-specific infanticide for reasons other than predation has not been described in pinnipeds. However, it is difficult to evaluate elimination of competitors and social pathology, particularly taking into account that both reasons could appear together with food resource utilization. There are no observations of intentional inter-specific infanticide by pinnipeds as a strategy to prevent future predation. However, Ovsyanikov and Rheborg (2015) observed an aggressive approach by an adult walrus to a swimming female polar bear with a cub of the year near Wrangel Island (Russia), which is compatible with the anti-future predation notion.

10.2 Unintentional Infanticide

10.2.1 By-Product of Sexual Behavior and Intra-Sexual Competition

Abuse of infants that ends in their death in a context of sexual behavior and intra-sexual competition is relatively common (e.g. Campagna 2002). Juvenile and subadult males of some otariids may abduct pups during "group raids," or singly during the reproductive season (e.g. Chelnokov 1974; Campagna et al. 1988a) (Table 10.2). Pups are frequently taken away from the breeding grounds (Fig. 10.1b) and sometimes carried to sea (Fig. 10.1e). Males may hold pups close to them, biting, grabbing, and pressing them to the ground, or shaking and tossing them into the air (e.g. Chelnokov 1974; Cassini 1998). At times, males mount pups and engage in pelvic thrusting movements (Fig. 10.1d). Pups are occasionally taken to the water and drowned. Young bachelor males sometimes try to steal the abducted pups from each other. Overall, this behavior simulates that of female defense by adults (Campagna et al. 1988b). The pups occasionally become wounded or killed during this abusive behaviour.

Young males may also abduct infants of other otariid species. Southern sea lions abduct, abuse and mount South American fur seal pups (Fig. 10.1c), which may lead to death of the pups (Harcourt 1993; Cassini 1998; Pavés 2008, V. Franco-Trecu pers. comm.). Northern fur seal subadult males abuse and mount Steller sea lion pups at sympatric rookeries (personal observations), despite the fact that the pups may be larger than their abductors. But, this abuse apparently does not end in the death of the young.

Although pup abduction is mostly carried out by subadults, adult males may sometimes be involved in abusive behaviours and killing (Cassini 1998; Kiyota and Okamura 2005). Australian and New Zealand sea lion adult males were reported mounting pups, and performing pelvic thrusts resulting in pups' death by crushing (Marlow 1975). A Guadalupe fur seal (*Arctocephalus townsendi*) adult male carried a dead California sea lion pup around and mounted it (Stewart et al. 1987). They suggested that the pup was killed by the copulatory efforts of the male.

Females are rarely able to retrieve their pups from male-abductors (Cassini 1998; Kiyota and Okamura 2005), but their efforts to do so increase the probabilities of abducted pups' survival (Cassini 1998). In addition, by breeding in large aggregations, females may reduce the risk of abduction and increase the probability of

Table 10.2 Unintentional infanticide in otariids

Species	Sex-age group of killers	Sites	Victims	Explanation	Source of information
Southern sea lion	Young males, subadult and adult males	Argentina, Peru, Uruguay, Chile, Falkland (Malvinas) Islands	Conspecific pups, South American fur seal pups	By-product of sexual behavior and intrasexual competition	Campagna et al. (1988a, b), Harcourt (1993), Cassini (1998), Pavés (2008), Campagna (2009), V. Franco-Trecu pers. com.
	Females	North Ballestas Islands	Conspecific pups	By-product of female competition for misrecognised pups	Soto et al. (2004)
Australian sea lion	Subadult and adult males	Dangerous Reef	Conspecific pups	By-product of sexual behavior and intrasexual competition	Marlow (1975)
	Adult males	Dangerous Reef, Kangaroo Island	Conspecific pups	By-product of misdirected aggression	Marlow (1975), Higgins and Tedman (1990)
	Females	Dangerous Reef	Conspecific pups	By-product of adoption avoidance behavior	Marlow (1972, 1975)
	Females	Dengerouth Reef	Conspecific pups	By-product of rough handling by mothers	Marlow (1975)
New Zealand sea lion	Adult males	Enderby Island	Conspecific pups	By-product of sexual behavior	Marlow (1975)
Northern fur seal	Subadult males	St. Paul Island, Commander Islands, Kuril Islands, Tuleny Island	Conspecific pups	By-product of sexual behavior and intrasexual competition	Chelnokov (1974), Vladimirov (1998), Kiyota and Okamura (2005), P.A. Permyakov pers. com., unpublished data
Steller sea lion	Females	Marmot Island, Unknown	Conspecific pups	By-product of adoption avoidance behavior	Gisiner (1985), Pierotti (1991)

(continued)

Table 10.2 (continued)

Species	Sex-age group of killers	Sites	Victims	Explanation	Source of information
	Female	Kuril Island.	Conspecific pups	By-product of rough handling by mothers	P.A. Permyakov pers. com.
	Adult male	Commander Islands.	Conspecific pups	By-product of misdirected aggression	Ryazanov et al. (2018)
Guadalupe fur seal	Adult male	San Miguel Island	California sea lion pup	By-product of sexual behavior	Stewart et al. (1987)
Antarctic fur seal	Subadult males	Macquarie Island	Conspecific pups and subantarctic fur seal pups	[a]By-product of sexual behavior and intrasexual competition	Robinson et al. (1999)
Galapagos fur seal	Females	Galapagos Islands	Conspecific pups	By-product of adoption avoidance behavior	Trillmich (1981)
	Female	Galapagos Islands	Conspecific pups	By-product of rough handling by mothers	Trillmich and Wolf (2008)
South American fur seal	Subadult males	Isla de Lobos (Uruguay)	Conspecific pups	By-product of sexual behavior	V. Franco-Trecu pers. com.
	Females	Punta San Juan (Peru)	Conspecific pups	By-product of adoption avoidance behavior	Harcourt (1992)
	Females	Punta San Juan (Peru)	Conspecific pups	By-product of female competition for misrecognised pups	Harcourt (1992)
Steller sea lion, New Zealand fur seal, New Zealand sea lion, Australian sea lion, Antarctic fur seal, South American sea lion, South American fur	Adults and subadults	Multiple sites	Infants	Incidental trampling and crashing	e.g. Orr and Poulter (1967), Stirling (1971), Marlow (1975), Mattlin (1978), Doidge et al. (1984), Campagna et al. (1988b), Harcourt (1992), De Villiers and

(continued)

Table 10.2 (continued)

Species	Sex-age group of killers	Sites	Victims	Explanation	Source of information
seal, Cape fur seal, Northern fur seal, Sub-antarctic fur seal					Roux (1992), Vladimirov (1998), Georges and Guinet (2000), Castinel et al. (2007)

[a]Presumably by-product of sexual behaviour, the reason was not described by the authors

survival of their pups, compared to females that breed solitarily or in harems away from the rookeries (Campagna et al. 1992; Drago et al. 2011).

Misdirected aggressive sexual behavior appears to be an epiphenomenon of high sexual motivation in males during the reproductive season (Campagna et al. 1988a, b). This kind of aggression is exhibited not only towards pups, but also towards adults and subadults, both intra- and inter-specific, and can result in their death (e.g. Miller et al. 1996; Altukhov et al. 2012). At the same time, in young bachelors such behavior may play an adaptive role. Pups begin to mount each other and perform pelvic thrusts as a play element from five days old (Gentry 2002). Young males may mount each other or even adults, and perform pelvic thrusts (Marlow 1975). A young Antarctic fur seal male mounted and tried to copulate with an adult king penguin (*Aptenodytes patagonicus*) (de Bruyn et al. 2008). Subadult northern fur seal males were occasionally observed trying to mount adult Steller sea lions (personal observations). Young bachelors of this species were seen mounting and performing pelvic thrusts movements towards stones and logs (S.M. Artemyeva pers. comm.). Marlow (1975) described New Zealand sea lion male performing pelvic thrusts on sandy substrate. In light of the foregoing, sexual abuse of pups by young males seems to be a part of the ontogenesis of reproductive behavior. Bachelors may obtain reproductive-like experience by practicing with pups as female substitutes, and competing for them with each other – and thereby increasing, perhaps, their own chances for reproduction in the future (Campagna et al. 1988a, b; Kiyota and Okamura 2005).

Among other pinnipeds, sexual abuse with killing pups has been described in northern (*Mirounga angustirostris*) and southern elephant seals (Rose et al. 1991; Le Boeuf and Campagna 1994; Hayward 2003), gray seals (reported in Campagna 2002 as pers. comm. by D. Boness and P. Pomeroy) and Hawaiian monk seals (*Monachus schauinslandi*) (Hiruki et al. 1993; Campagna 2002 reported as pers. comm. by M. Craig). Notable cases of inter-specific sexual harassment of young were also found in southern sea otters (*Enhydra lutris nereis*). Males of this species aggressively forced copulations with juvenile harbor seals (*Phoca vitulina richardsi*); leading to their death (Harris et al. 2010).

10.2.2 By-Product of Misdirected Aggression

Misdirected aggression appears primarily during the reproductive period, when animals are most motivated to protect their territories. Territorial males may mis-recognise infants as threats and attack them. Infants are bitten and thrown several meters distance. The consequences may be severe injuries and death of the infant. This form of infanticide was described for Australian sea lions (Marlow 1975; Higgins and Tedman 1990). Eight attacks were performed by six different males. Two attacks occurred immediately after threatening other males. After the attacks, the males returned to their usual positions on their territories. A Steller sea lion territorial male at the Commander Islands was seen attacking a gull at the bordering zone of his territory. When the gull flew away, the attacker turned to a pup. The pup was caught and smashed against the rocks. The pup died and the killer did not show further interest in the carcass (Ryazanov et al. 2018).

10.2.3 By-Product of Adoption Avoidance Behavior

Otariid females recognize their pups by olfactory, vocal and visual cues (see Insley et al. 2003 for review, Charrier Chaps. 14 and 15). During the first days of a pup's life, mother and pup stay close and engage in intensive contact to learn the characteristics of each other (e.g. Trillmich 1981). To avoid misrecognition of alien pups as their own, with the following incorrect "parental" investments and to prevent milk stealing (Fig. 10.1g) females become aggressive (e.g. Bartholomew and Hoel 1953; Marlow 1972). Sometimes females attack pups that smell or are visually different from their own (e.g. Trillmich 1981; Wierucka et al. 2017). They violently bite, shake and toss strange pups (Fig. 10.1f). The pups occasionally may die owing to these attacks. This form of intra-specific unintentional infanticide has been reported in several otariids (Table 10.2). Inter-specific adoption occasionally occurs in rookeries where more than one species congregates (Fig. 10.1h) (Artemyeva et al. 2013), and females also exhibit aggression toward hetero-specific pups. This has been observed in mixed rookeries of Steller sea lions and northern fur seals (unpublished observations). Thus, inter-specific infanticide as a by-product of adoption avoidance may be a plausible explanation of pup killing.

10.2.4 By-Product of Competition for Mis-recognised Pups

Despite the fact that females exhibit adoption avoidance behavior, the mis-recognition of alien pups sometimes occurs in otariids (e.g. Insley et al. 2003; Pitcher et al. 2011). In some cases, alien females try to abduct pups from their mothers (Fig. 10.1i). This creates conflict between females. Harcourt (1992) described tags-of-war that result in the actual dismembering of the pup's body. Abduction attempts of alien pups may more frequently involve females that recently lost their own offspring (Marlow 1972).

Soto et al. (2004) reported that a year after the 1997–1998 El Niño, southern sea lion pup production at Ballestas Islands (Peru), was about one third of the pre El Niño period. In this context, females engaged in the unusual behavior of raiding the rookery in groups from four to six animals and abducting pups. In most cases, females injured pups while trying to keep them away from other females. It was estimated that 2.5% to 8.4% of pup mortality in the season was caused by this behavior. Such behaviors were likely related to the low number of females that successfully gave birth. A year later, when the birth rate was more typical, the abduction of pups by females phased out almost completely. Thus, both pup mis-recognition and mis-recognition avoidance can result in infanticide.

10.2.5 By-Product of Rough Handling by Mothers

Rough and negligent handling of neonates by their mothers may cause pup death, especially when young, inexperienced females are involved. Young Australian sea lion mothers at Dangerous Reef accidentally killed their pups when charging at males, or dragging pups violently when attempting to change places (Marlow 1975). Galapagos fur seal females were seen dragging their pups vigorously to move them out of reach of older siblings, and by doing so they wounded and killed the pups (Trillmich and Wolf 2008). A young Steller sea lion female at the Kuril Islands frequently moved her neonate from one place to another, perhaps looking for a better site. The pup was fatally injured in the process (P.A. Permyakov pers. comm.).

10.2.6 Incidental Trampling and Crushing

Otariid males usually ignore pups. Therefore, pups can occasionally be trampled or crushed when males move over them or sit on pups as if there were nobody underneath (Fig. 10.1j) (e.g. Orr and Poulter 1967; Stirling 1971; Marlow 1975; Campagna pers. comm.). Vigorously moving males can also trample pups during fights. Trampling may also occur during stampedes of many animals during a mass-perceived danger (Sandegren 1970). Mothers attempt to defend pups by attacking approaching males, and moving their pups away from them.

Infanticide by neglect may be a significant cause of pup mortality. For example, Orr and Poulter (1967) found that crushing was the second most important cause of death, after drowning, of Steller sea lion pups at Año Nuevo Island, California. Incidental trampling and crushing have been described for a majority of the otariids (Table 10.2), and presumably can occur in all of them. Le Boeuf and Campagna (1994) reported that trampling occurs in species where males are significantly larger than pups and their movements are frequently close to infants. Le Boeuf and Briggs (1977) suggested that trampling infants by northern elephant seal males seems more effective for the males than avoiding young when males chase competitors or evade dominant males. This explanation seems also to apply to otariid males.

Besides infanticide, harassing and infant abuse may result in mother-pup separation, with the subsequent "trauma-starvation syndrome" and subsequent death of the pups (e.g. Marlow 1975; Le Boeuf and Campagna 1994; Campagna 2002). These cases can also play an important role in pup mortality.

10.3 Conclusions

There are presently few anecdotal and at times incomplete descriptions of infanticide in Otariids. The accessible observations are not always supplemented by unequivocal context or can be misinterpreted. Thus, some cases listed in this review may be misclassified, or there may be several reasons to explain a single case of infanticide. Nevertheless, it is possible to make some generalizations. The intentional infanticide among eared seals occurs as a form of food resource utilization, or as a social pathology. These reasons may co-occur. Unintentional infanticide happens as a by-product of varieties of behavioral stimuli and acts. Despite the apparent rarity—or at least infrequency—of infanticide, in some cases infanticide can be a significant factor in the demography of prey populations.

Infanticide is a subject of lengthy disputes on the issue of its potential adaptive role or lack (see Sommer 2000 for review). The available information on eared seals shows that it can be both adaptive and non-adaptive at least under the criterion of intentionality. Unintentional infanticide seemingly does not play an adaptive role, because animals benefit by the behavior that can entail an infant's death, but they do not gain by killing the young itself. In eared seals, intentional inter-specific infanticide seems to form part of a typical predator-prey relationship, as well as preying on other warm-blooded animals. Predation may play an adaptive role of acquisition of energetic benefits. Intentional intra-specific infanticide is found only in males during the reproductive period. Infanticide followed by cannibalism can bring energetic benefits and so also can be adaptive.

Acknowledgements The author is grateful to P.A. Permyakov, S.M. Artemyeva, C. Campagna, V. Franco-Trecu, H.J. Pavés, A.V. Chetvergov, B. Würsig and R. Harcourt for their useful comments to the manuscript and providing photos and observations of eared seals behavior.

References

Altukhov AV (2012) Reproductivnoe povedenie sivucha (*Eumetopias jubatus* Schreber 1776) (Reproductive behaviour of the Steller sea lion (*Eumetopias jubatus* Schreber 1776)). Dissertation. Moscow State University

Altukhov AV, Permyakov PA, Andrews RD et al (2012) Adult Steller sea lion mortality on breeding rookeries in the Russian Far East in the years 2002–2010. Russ J Mar Biol 38:442–447

Artemyeva S, Permyakov P, Ryazanov S et al (2013) Inter-specific interactions: northern fur seal (*Callorhinus ursinus*) pups nursed by Steller sea lions (*Eumetopias jubatus*) and Steller sea lion pups nursed by northern fur seal females. In: de Mar E (ed) Abstract book of the 27th Annual Conference of the European Cetacean Society, Setubal, Portugal, 8–10 April 2013, p 253

Bartholomew G (1970) The model of the evolution of pinniped polygyny. Evolution 24(3):546–559

Bartholomew GA, Hoel PG (1953) Reproductive behavior of the Alaska fur seal, *Callorhinus ursinus*. J Mammal 34(4):417–436

Bedard C, Kovacs K, Hammill M (1993) Cannibalism by grey seal, *Halichoerus grypus*, on Amet Island, Nova Scotia. Mar Mamm Sci 9:421–424

Bishop AM, Onoufriou J, Moss S et al (2016) Cannibalism by a male gray seal (*Halichoerus grypus*) in the North Sea. Aquat Mamm 42:137–143

Bradshaw CJA, Lalas C, Mcconcey S (1998) New Zealand sea lion predation on New Zealand fur seals. N Z J Mar Fresh Res 32(1):101–104

Brownlow A, Onoufriou J, Bishop A et al (2016) Corkscrew seals: Grey seal (*Halichoerus grypus*) infanticide and cannibalism may indicate the cause of spiral lacerations in seals. PLoS One. https://doi.org/10.1371/journal.pone.0156464

Byrnes PE, Hood WR (1994) First account of Steller sea lion (*Eumetopias jubatus*) predation on a California sea lion (*Zalophus californianus*). Mar Mamm Sci 10:381–383

Calkins DG (1998) Prey of Steller sea lions in the Bering Sea. Biosphere Cons 1:33–44

Campagna C (2002) Infanticide and abuse of young. In: Perrin WF, Wursing B, Thewissen JGM (eds) Encyclopedia of marine mammals. Academic Press, San Diego, pp 625–628

Campagna C (2009) Aggressive behaviour, intra-specific. In: Perrin WF, Wursing B, Thewissen JGM (eds) Encyclopedia of marine mammals, 2nd edn. Academic Press, San Diego, pp 18–24

Campagna C, Le Boeuf BJ, Cappozzo HL (1988a) Group raids: a mating strategy of male southern sea lions. Behaviour 105:224–248

Campagna C, Le Boeuf BJ, Cappozzo HL (1988b) Pup abduction and infanticide in southern sea lions. Behaviour 107:44–60

Campagna C, Bisioli C, Quintana F et al (1992) Group breeding in sea lions: pups survive better in colonies. Anim Behav 43(4):541–548

Cassini MH (1998) Inter-specific infanticide in South American otariids. Behaviour 135:1005–1012

Castinel A, Duignan PJ, Pomroy WE et al (2007) Neonatal mortality in New Zealand sea lions (*Phocarctos hookeri*) at Sandy Bay, Enderby Island, Auckland Islands from 1998 to 2005. J Wildl Dis 43(3):461–474

Chelnokov FG (1974) O vzaimootnosheniyah holostyakov i detenishey morskih kotikov na garemnih territoriyah (About relationship of bachelors and pups of northern fur seal at harem territory). In: Issledovaniya po biologii rib i promislovoy okeanografii, vol 5. TINRO, Vladivostok, pp 137–140

de Bruyn PJN, Tosh CA, Bester MN (2008) Sexual harassment of a king penguin by an Antarctic fur seal. J Ethol 26(2):295–297

De Villiers DJ, Roux J-P (1992) Mortality of newborn pups of the South African fur seal *Arctocephalus pusillus pusillus* in Namibia. S Afr J Mar Sci 12:881–889

Derocher AE, Wiig Ø (1999) Infanticide and cannibalism of juvenile polar bears (*Ursus Maritimus*) in Svalbard. Arctic 52(3):307–310

Diaz Lopez B, Lopez A, Methion S et al (2017) Infanticide attacks and associated epimeletic behaviour in free-ranging common bottlenose dolphins (Tursiops truncatus). J Mar Biol Assoc UK. https://doi.org/10.1017/S0025315417001266
Doidge DW, Croxall JP, Baker JR (1984) Density-dependent pup mortality in the Antarctic fur seal *Arctocephalus gazella* at South Georgia. J Zool Lond 202(3):449–460
Drago M, Cardona L, Garcia N et al (2011) Influence of colony size on pup fitness and survival in South American sea lions. Mar Mamm Sci 27(1):167–181
Ebensperger LA (1998) Strategies and counterstrategies to infanticide in mammals. Biol Rev 73:321–346
Gentry RL (2002) Northern fur seal *Callorhinus ursinus*. In: Perrin WF, Wursing B, Thewissen JGM (eds) Encyclopedia of marine mammals. Academic Press, San Diego, pp 813–817
Gentry RL, Johnson JH (1981) Predation by sea lions on northern fur seal neonates. Mammalia 45:423–430
Georges J-Y, Guinet C (2000) Early mortality and perinatal growth in the subantarctic fur seal (*Arctocephalus tropicalis*) on Amsterdam Island. J Zool Lond 251(3):277–287
Gisiner RC (1985) Male territorial and reproductive behaviour of the Steller sea lion, *Eumetopias jubatus*. Dissertation. University of California
Hall-Aspland SA, Rogers TL (2004) Summer diet of leopard seals (*Hydrurga leptonyx*) in Prydz Bay, Eastern Antarctica. Polar Biol 27:729–734
Harcourt R (1992) Factors affecting early mortality in the South American fur seal (*Arctocephalus australis*) in Peru: density-related effects and predation. J Zool Lond 226(2):259–270
Harcourt R (1993) Individual variation in predation on fur seals by southern sea lions (*Otaria byronia*) in Peru. Can J Zool 71(9):1908–1911
Harris HS, Oates SC, Staedler MM et al (2010) Lesions and behaviour associated with forced copulation of juvenile Pacific harbor seals (*Phoca vitulina richardsi*) by Southern Sea Otters (*Enhydra lutris nereis*). Aquat Mamm 36(4):331–341
Hausfater G, Hrdy SB (eds) (1984) Infanticide: comparative and evolutionary perspectives. Aldine Press, New York
Hayward JL (2003) Sexual aggression by a male northern elephant seal on harbor seal pups in Washington. Northwestern Nat 84(3):148–150
Higgins LV, Tedman RA (1990) Effect of attacks by male Australian sea lions, *Neophoca cinerea*, on mortality of pups. J Mammal 71(4):617–619
Hiruki LM, William GG, Becker BL et al (1993) Wounding in Hawaiian monk seals (*Monachus schauinslandi*). Can J Zool 71(3):458–468
Hrdy SB (1979) Infanticide among animals: a review, classification, and examination of the implications of the reproductive strategies of females. Ethol Sociobiol 1:13–40
Insley SJ, Phillips AV, Charrier I (2003) A review of social recognition in pinnipeds. Aquat Mamm 29(2):181–201
Janssen A, Faraji F, van der Hammen T et al (2002) Inter-specific infanticide deters predators. Ecol Lett 5:490–494
Kiyota M, Okamura H (2005) Harassment, abduction, and mortality of pups by nonterritorial male northern fur seals. J Mammal 86:1227–1236
Kovacs K, Lydersen C, Hammill M (1996) Grey seal cannibalism. Mar Mamm Sci 12:161
Le Boeuf BJ, Briggs KT (1977) The cost of living in a seal harem. Mammalia 41(2):167–195
Le Boeuf BJ, Campagna C (1994) Protection and abuse of young in pinnipeds. In: Parmigiani S, vom Saal FS (eds) Infanticide and parental care. Harwood Academic Publications, London, pp 257–276
Lukas D, Huchard E (2014) The evolution of infanticide by males in mammalian societies. Science 346:841–844
Mansfield AW (1958) The biology of the Atlantic walrus, *Odobenus rosmarus rosmarus* (Linnaeus) in the eastern Canadian Arctic. Fisheries Research Board of Canada. Manuscript report series (Biological) 653, 146 pp

Marlow BJ (1972) Pup abduction in the Australian sea lion, *Neophoca cinerea*. Mammalia 36 (1):161–165
Marlow BJ (1975) The comparative behaviour of the Australian sea lions *Neophoca cinerea* and *Phocarctos hookeri* (*Pinnipedia: Otariidae*). Mammalia 39(2):159–230
Mathews EA, Adkison MD (2010) The role of Steller sea lions in a large population decline of harbor seals. Mar Mamm Sci 26:803–836
Mattlin RH (1978) Pup mortality of the New Zealand fur seal (*Arctocephalus forsteri* Lesson). N Z J Ecol 1:138–144
Miller EH, de Leon AP, DeLong RL (1996) Violent inter-specific sexual behaviour by male sea lions (*Otariidae*): evolutionary and phylogenetic implications. Mar Mamm Sci 12(3):468–476
Orr RT, Poulter TC (1967) Some observations on reproduction, growth, and social behaviour in the Steller sea lion. Proc Cal Acad Sci 35:193–226
Ovsyanikov N, Rheborg A (2015) Walrus in water attacked and killed swimming rein deer. Why? In: Marine mammals of the holarctic. Collection of scientific papers, after the 9th International conference, 22–27 September 2014, Saint Petersburg, pp 339–341
Pavés H (2008) Conducta Reproductiva, Mortalidad Neonatal y Tendencia Poblacional De Una Agrupación Reproductiva Del Lobo Fino Austral (*Arctocephalus australis*, Zimmermann, 1783) En La Isla Guafo, Chiloé, Chile (Reproductive Behaviour, Neonatal Mortality and Population Tendency of a Reproductive Group of the Southern Fur Seal (*Arctocephalus australis*, Zimmermann, 1783) in Isla Guafo, Chiloé, Chile). Dissertation. Universidad Austral de Chile
Permyakov PA (2006) Cases of direct infanticide of Steller sea lion's (*Eumetopias jubatus*) pups by territorial male on rookery of Brat Chirpoev Island (Kuril Islands). In: Marine mammals of the Holarctic, conference proceedings of 4th International Conference, 10–14 September 2006, Saint Petersburg, pp 417–419
Pierotti R (1991) Infanticide versus adoption: an intergenerational conflict. Am Nat 138 (5):1140–1158
Pitcher KV, Fay FH (1982) Feeding by Steller sea lion on harbor seals. Murrelet 63:70–71
Pitcher BJ, Ahonen H, Charrier I, Harcourt R (2011) Allosuckling behaviour in the Australian sea lion (*Neophoca cinerea*): an updated understanding. Mar Mamm Sci 27:881–888
Robinson S, Wynen L, Goldsworthy S (1999) Predation by a Hooker's sea lion (*Phocarctos hookeri*) on a small population of fur seals (*Arctocephalus* sp.) at Maquarie Island. Mar Mamm Sci 15(3):888–893
Rose NA, Deutsch CJ, Le Boeuf BJ (1991) Sexual behaviour of male northern elephant seals: III. The mounting of weaned pups. Behaviour 119(3-4):171–192
Ryazanov SD, Kirillova AD, Laskina NB et al (2018) Infanticide and cannibalism in Steller sea lions (*Eumetopias jubatus*). Mar Mammal Sci 34(1):200–207
Sandegren FE (1970) Breeding and maternal behaviour of the Steller sea lion (*Eumetopias jubatus*) in Alaska. Dissertation. University of Alaska
Sommer V (2000) The holy wars about infanticide. Which side are you on? And why? In: van Schaik CP, Janson CH (eds) Infanticide by males and its implications. Cambridge University Press, Cambridge, pp 9–26
Soto KH, Trites AW, Arias-Schreiber M (2004) The effects of prey availability on pup mortality and the timing of birth of South American sea lions (*Otaria flavescens*) in Peru. J Zool Lond 264 (4):419–428
Stewart BS, Yochem PK, DeLong RL et al (1987) Interactions between Guadalupe furs seal and California sea lions at San Nicolas and San Miguel Islands, California. In: Croxall JP, Gentry RL (eds) Status, biology, and ecology of fur seals. Proceedings of an International Symposium and Workshop. Cambridge University Press, Cambridge, pp 103–109
Stirling I (1971) Studies on the behaviour of the South Australian fur seal, *Arctocephalus forsteri* (Lesson). I. Annual cycle, postures and calls, and adult males during the breeding season. Austral J Zool 19(3):243–266

Stirling I, Ross JE (2011) Observations of cannibalism by polar bears (*Ursus maritimus*) on summer and autumn sea ice at Svalbard, Norway. Arctic 64(4):478–482
Taylor M, Larsen T, Schweinsburg RE (1985) Observations of Intra-specific aggression and cannibalism in polar bears (*Ursus maritimus*). Arctic 38(4):303–309
Tikhomirov EA (1959) K voprosu o pitanii sivucha teplokrovnimi zhivotnimi (To the issue of feeding on warm blooded animals by Steller sea lion). Izvestiya TINRO 47:185–186
Towers JR, Halle MJ, Symonds HK et al (2018) Infanticide in mammal-eating killer whale population. Sci Rep 8:4366. https://doi.org/10.1038/s41598-018-22714-x
Trillmich F (1981) Mutual mother-pup recognition in Galapagos fur seals and sea lions: cues used and functional significance. Behaviour 78(1/2):21–42
Trillmich F, Wolf JBW (2008) Parent-offspring and sibling conflict in Galapagos fur seals and sea lions. Behav Ecol Sociobiol 62(3):363–375
van Neer A, Gross S, Kesselring T et al (2019) Behavioural and pathological insights into a case of active cannibalism by a grey seal (*Halichoerus grypus*) on Helgoland, Germany. J Sea Res 148–149:12–16
van Schaik CP (2000) Infanticide by male primates: the sexual selection hypothesis revised. In: van Schaik CP, Janson CH (eds) Infanticide by males and its implications. Cambridge University Press, Cambridge, pp 27–60
Vladimirov VA (1998) Struktura populyaciy I povedenie severnih morskih kotikov (Structure of populations and behaviour of the northern fur seals). In: Sokolov VE, Aristov AA, Lisitzina TYu (eds) Severniy morskoy kotik: sistematika, morfologiya, ekologiya, part 2, pp 555–722
Wierucka K, Pitcher BJ, Harcourt R et al (2017) The role of visual cues in mother-pup reunions in a colonially breeding mammal. Biol Lett 13:20170444. https://doi.org/10.1098/rsbl.2017.0444
Wilkinson IS, Duignan PJ, Gulland FMD (2000) Infanticide and cannibalism in the New Zealand sea lion, *Phocarctos hookeri*. Mar Mamm Sci 16:494–500
Womble JN, Conlon S (2010) Observation of Steller sea lion (*Eumetopias jubatus*) predation on a harbor seal (*Phoca vitulina richardii*) in the Glacier Bay region of Southeastern Alaska. Aquat Mamm 36:129–137

Chapter 11
Genetic Consequences of Dispersal, Philopatry and Reproductive Behaviors

Carolina A. Bonin

Abstract Otariid behaviors, such as decisions to disperse or return to breed at the natal site and certain reproductive behaviors, when repeated over generations, profoundly shape population ecological and evolutionary processes. In this chapter, genetic consequences of these behaviors are reviewed, highlighting their implications for conservation. A few patterns emerge: (i) there are varying degrees of population genetic structure among otariids, reflecting variability in rate of female philopatry and levels of male-mediated gene flow; (ii) male reproductive skew is generally high in otariids but is largely balanced by territorial male turnover and nuances in female behavior that preclude them from remating often; (iii) although recolonization is largely a positive density-dependent process, alternative scenarios are emerging, where pioneer individuals may come from less densely populated areas; and (iv) the intense era of exploitation in the nineteenth century was survived by most species with relatively low impact to their genetic make-up. These findings exemplify the strong connections among behavior, ecology and evolution.

Keywords Gene flow · Breeding site fidelity · Genetic diversity · Recolonization · Polygyny

11.1 Introduction

Recurrent behaviors over generations have population-level consequences, and therefore dictate the rate of evolutionary change (Kappeler et al. 2013). The decisions to either disperse or return to breed at the natal site, give birth at a new suitable spot on the beach and other reproductive behaviors, such as female mate choice, when repeated over time, can affect: (i) the rate at which genetic differences may become permanent, or "fixed", and eventually lead to new evolutionary lineages;

C. A. Bonin (✉)
Department of Marine and Environmental Sciences, Hampton University, Hampton, VA, USA
e-mail: carolina.bonin@hamptonu.edu

C. Campagna, R. Harcourt (eds.), *Ethology and Behavioral Ecology of Otariids and the Odobenid*, Ethology and Behavioral Ecology of Marine Mammals,
https://doi.org/10.1007/978-3-030-59184-7_11

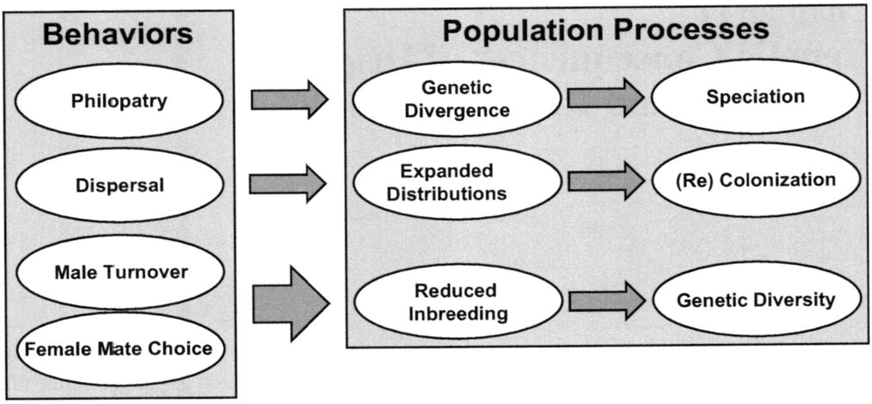

Fig. 11.1 Recurrent individual behaviors over generations have population-level consequences. On the left side are examples of otariid behaviors known to impact evolutionary processes. These behaviors determine a population's evolutionary potential such as their ability to recolonize, maintain genetic diversity over time or branch into distinct evolutionary units

(ii) species-level genetic diversity; and (iii) the chances for population distribution expansion and recolonization (Fig. 11.1). The social structure impact on population processes has been studied extensively since the late 1980s, through application of population genetic theory to behavioral ecology data. When these efforts are employed in otariid species, they lead to a surprisingly clear picture regarding the processes that shape the fate of fur seal and sea lion populations.

Although much conservation focus is placed on species, all relevant ecological and evolutionary processes occur at the level of populations (Moritz 1994; Ryder 1986). Therefore, populations and not species are typically the target of conservation plans and actions. This is due to the fact that by the time concerns are raised regarding species conservation status, their fate has already been sealed and trend reversal is often no longer possible. In other words, large-scale issues such as environmental shifts may lead to the extirpation of populations without triggering a conservation action. An emphasis on population-level biological processes is paramount at a time when environmental change may lead populations to their permanent demise. Otariid behaviors' impact on population-level evolutionary processes are analyzed in this chapter with the goal of highlighting their importance for conservation practices.

11.2 Male Harassment as a Central Driver of Otariid Gregariousness

The image of a crowded beach or rocky outcrop is an icon of otariid ecology and behavior. We refer to these aggregations as "colonies" or "rookeries". During the breeding season, female otariids are gregarious and their breeding cycle is typically highly synchronized. Females haul out on beaches or rocky foreshores where predictable resources are available for nearby foraging. However, otariid female gregariousness is often facilitated by male harassment (Cappozzo et al. 2008; Cassini 2000) and the avoidance behaviors affiliated with escaping male aggression (see also Kiyota, Chap. 8). Males can trample newborn pups (Doidge et al. 1984) when aggressively pursuing mothers or disputing with other males (see Ryazanov, Chap. 10). Direct violence toward females is also observed in some species. In New Zealand sea lions, *Phocarctos hookeri*, permanent scars from male bites are found in as many as 84% of the adult female population (Chilvers et al. 2005). In high numbers, females can weaken/dilute the negative effects of male coercion and aggressive behavior, which also help provide protection for their pups from this aggression (Campagna et al. 1992). Therefore, female reproductive synchronicity is also perceived as a product of male harassment; strong evidence for this argument can be found in phocids. In grey seals, *Halichoerus grypus*, the relative number of aggressive male disturbances toward mothers is significantly higher in females that give birth later in the season (three times more often), relative to mothers that give birth in the beginning of the pupping season, when the ratio of females to males can be nearly double (Boness et al. 1995).

The synchronicity of otariid breeding behavior typically leads to a sequence of events that happens across a colony within a breeding season. This well-timed annual cycle repeats itself during the reproductive lives of each individual. For example, in Antarctic fur seals, *Arctocephalus gazella*, breeding occurs within 2 months: females arrive at breeding beaches in late November and December to birth a pup and become receptive 6–7 days after giving birth (Fig. 11.2; Doidge et al. 1986; Lunn and Boyd 1991). Similarly, 90% of New Zealand fur seal, *Arctocephalus forsteri*, pups are born within a period of 34 days every year between Dec 3rd and Jan 6th (Goldsworthy and Shaughnessy 1994). Tight synchronicity in breeding is more typical of polar and sub-polar otariid species, where the breeding season is constrained by climate, so females can ensure survivable environmental conditions for newborn pups. In tropical otariids, breeding has lower seasonality and females are not particularly limited by extreme seasonal shifts in climate or foraging opportunities (Villegas-Amtmann et al. 2009). Synchronicity is absent in the peculiar case of the Australian sea lion (see McIntosh and Pitcher, Chap. 26).

Fig. 11.2 The synchronicity of breeding events in otariids is illustrated in this picture of Antarctic fur seal mothers and their newborn pups. Photo credit: Carolina Bonin. Marine Mammal Protection Permit No. 774-1847-03 granted by the National Marine Fisheries Service of the United States

11.3 Gregariousness and the Metapopulation Model

The gregarious nature of otariids leads them to live colonially, at least for part of the year. Gregariousness makes them suited to the testing of hypotheses rooted in the *metapopulation* model, where metapopulation is an assembly of discrete groupings of individuals (often referred as "local populations") that interact on occasion via migration (Fig. 11.3). This idea was first established by Sewall Wright, Ronald Fisher, and John Haldane, who fathered the field of population genetics in the early-mid 1900s (Hanski and Gilpin 1991).

The "island model of population structure" proposes that local populations are connected by migration (individuals moving among colonies) and gene flow (exchange of genetic material through breeding). Wright's model originally assumed that local populations were permanent over time and their sizes were qualitatively similar; assumptions easily violated. Building upon this model, Levins later proposed that metapopulations are highly dynamic, continuously affected by extinctions and recolonization events (Hanski and Gilpin 1991). These events shape the partitioning of genetic variation within and between local populations and are therefore fundamental determinants of genetic diversity within species (Slatkin 1987). The decades 1970–1990 were exciting for evolutionary biologists, with scientists investigating the genetic effects of local extinctions and the source of migrants in the genetic pool of new local populations (e.g., Slatkin 1977, 1987). These ideas were effectively expressed using mathematical models and the resulting body of theoretical work comprises the basis for most current population genetic analyses.

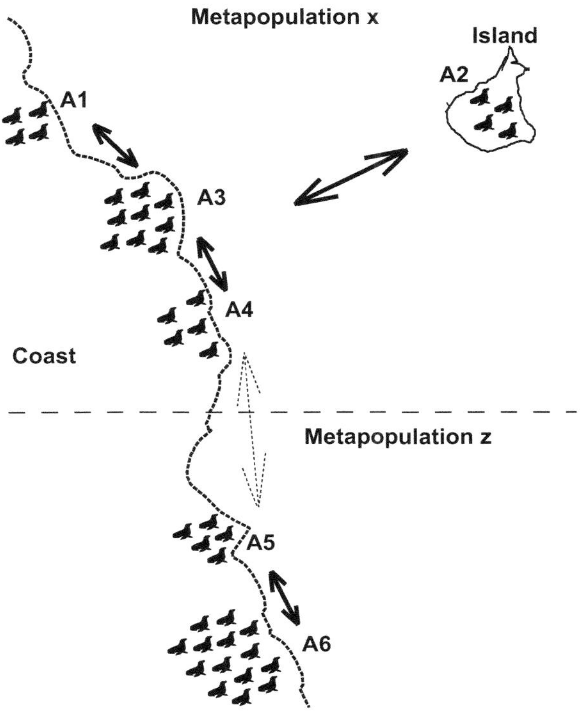

Fig. 11.3 The metapopulation concept applied to otariid populations. Individuals rarely move between metapopulations (Metapopulations x and z), indicated by the dashed arrow; but migrate more frequently between local populations (A1–A4 and A5–A6), indicated by the solid bold arrow. The dashed line indicates a genetic break in this hypothetical otariid species

Otariids provide biologists with an opportunity to study evolutionary processes comparatively due to certain key characteristics, particularly their gregarious breeding habits and relative catchability when on land. This means that females of many species can be captured, tagged, identified, and tracked so researchers can document an individual's life history (e.g., fecundity, long-term health, survivorship). Pups can easily be marked and sampled while mothers leave them unattended during foraging events (Fig. 11.4). In addition, for most species, pups do not typically move among colonies until weaned; thereby providing a true sample of the local population. This has been of great relevance for the advanced understanding of pinniped population dynamics. Importantly, otariid generation time is well-known for most species, and this allows for the calibration of molecular clocks and reliable genetic inferences of past demographics. In the scientific context, otariid populations distributed in semi-discrete colonies provide a great model with which to empirically test and apply metapopulation concepts.

Fig. 11.4 Antarctic fur seal pups rest while their mothers are at sea. Marked pups have been sampled for genetic study. Photo credit: Carolina Bonin. Marine Mammal Protection Permit No. 774-1847-03 granted by the National Marine Fisheries Service of the United States

11.4 Natal Philopatry and Breeding Site Fidelity

Breeding site fidelity and more extensively, natal philopatry, have long-lasting implications for population trends (Greenwood 1980). Furthermore, natal philopatry can be detrimental to a species' ability to expand in distribution. For example, extreme natal side fidelity of loggerhead sea turtles, *Caretta caretta*, limits this species' ability to colonize new areas, a concern for declining insular populations (Carreras et al. 2018). Individuals must break the pattern of site fidelity to establish new colonies (Matthiopoulos et al. 2005). At times, the interplay between natal philopatry and site fidelity rates largely determines the fate of local populations. In southeastern Alaska rookeries of Steller sea lions, *Eumetopias jubatus*, natal philopatry rates are moderately high (0.776–0.859), but breeding site fidelity rate is nearly 1.0, suggesting flexibility in female reproductive strategies with regard to dispersal from natal sites; however, once females find a suitable site, they tend to keep returning to it, presumably due to familiarity with topography and predictability of foraging sites (Hastings et al. 2017). Similarly, in Australian sea lions, *Neophoca cinerea*, female breeding site fidelity is closely related to fine-scale detection of suitable sites for pupping and is associated with foraging specializations in predictable areas or habitats where individuals traditionally find prey (e.g., Lowther et al. 2012).

In otariids, 60%–86% of females return to their natal site to breed (Chilvers and Wilkinson 2008; Gentry 1998; Hoffman and Forcada 2012; Lunn and Boyd 1991). Therefore, it is widely accepted that genetic mixing among populations is accomplished primarily by male dispersal (e.g., Robertson et al. 2006), as observed in most polygynous mammals (Dobson 1982; Greenwood 1980). Formal studies of male natal philopatry are relatively scarce in otariids, but limited genetic evidence

Fig. 11.5 Male Antarctic fur seals at a beach in Livingston Island, Antarctica. Males arrive before females and fight to establish breeding territories before the female's arrival. Photo credit: Carolina Bonin. Marine Mammal Protection Permit No. 774-1847-03 granted by the National Marine Fisheries Service of the United States

supports it being lower than for females (González-Suárez et al. 2009). Even with lower natal philopatry relative to females, otariid males often have high rates of breeding site fidelity during their reproductive years once they have established a territory. In Antarctic fur seals, males and females tend to return to the same breeding site year after year (Hoffman et al. 2006b; Hoffman and Forcada 2012; Lunn and Boyd 1991), and males are remarkably precise: most successful males return within a body length of the exact spot they occupied the previous breeding season (Hoffman et al. 2006b; Fig. 11.5). Similarly, in Australian sea lions, males and females are highly philopatric with limited dispersal, and even if animals disperse between close colonies, it is a maximum of 110 km, with a tendency for males to move further than females (Ahonen et al. 2016).

11.4.1 The Consequences of Dispersal on Genetic Variation

Varying levels of genetic structuring are observed in otariids and largely reflect sex differences in dispersal rates. Accordingly they tend to display strong, and often significant, genetic differences among populations in maternally inherited mitochondrial DNA (mtDNA) (*genetic differentiation*), while bi-parentally inherited, microsatellite markers (DNA fragments in which nucleotide DNA motifs are consecutively repeated;) show weak levels of differentiation. The most extreme example effect of female natal philopatry in otariids is found in Australian sea lions, where it has led to large and significant differences in mtDNA haplotype frequencies among colonies only ~60 km apart, which suggests very few occasions when females disperse to breed away from the site where they were born (Campbell

2003; Campbell et al. 2007). Galapagos fur seals, *Arctocephalus galapagoensis*, also present high matrilineal population structure (Lopes et al. 2015) driven by female philopatry, not matched by microsatellite data.

Despite a lack of congruency between mtDNA and microsatellite/other nuclear markers, most studies propose the delineation of stocks when strong mtDNA breaks are detected and most individuals can be confidently assigned to a putative population of origin using nuclear markers (typically >80%). This is the case in Steller sea lion eastern and western stocks (Bickham et al. 1996; Hoffman et al. 2006a), southern sea lion (*Otaria flavescens*) Pacific and Atlantic stocks (de Oliveira et al. 2008, 2017), California sea lions in Mexico: Upper Gulf of California, Southern Baja Peninsula, and Upper Pacific Coast of Baja (González-Suárez et al. 2009); and to lesser extent, Antarctic and Subantarctic fur (*A. tropicalis*) seals (Bonin et al. 2013; Wynen et al. 2000). Significant differences in the pool of mtDNA sequences have been detected in Galapagos fur seals and Australian sea lions, but the lack of support from nuclear DNA data limits clear delineation of evolutionary lineages (Campbell 2003; Lopes et al. 2015). Employment of next generation sequencing technologies should increase the resolution for detecting fine-scale genetic differences (Cammen et al. 2016; Rosel et al. 2017) but may over-split taxa and so require cautious interpretation (Coates et al. 2018).

11.4.2 Fixing Genetic Differences: Evolution at Work

Patterns of genetic structuring may intensify over generations, leading to speciation. Detection of this process is at the core of conservation management practices because it allows managers to put policies and conservation practices into place, and prioritize conservation efforts (Moritz 1994). Defining evolutionary lineages below the taxonomic level of species is problematic for marine mammals. Obvious geographic breaks do not always correspond to genetic breaks and there is evidence that ecological niche partitioning may drive speciation, as exemplified by fur seal species in Australia (Hoskins et al. 2017).

For extant otariid species, the most recent taxonomic split proposed is for the genus *Zalophus*. The former subspecies *Z. californianus* and *Z. wollebaeki* are now recognized as two species (Berta et al. 2018). These considerations were based on a combination of characters and coincide with *Z. californianus* and *Z. wollebaeki* geographic distributions; they are also supported by findings of significant breaks in the genetic make-up of *Zalophus* (Wolf et al. 2007). Importantly, this splitting of *Zalophus* has implications for conservation management because Z. *wollebaeki* and *Z. californianus* show contrasting population growth trends; *Z. wollebaeki* is currently listed as "Endangered" by the International Union for Conservation of Nature Red List while *Z. californianus* is classified under "Least Concern" (Aurioles-Gamboa and Hernández-Camacho 2015). The only extant species with currently recognized subspecies is the Afro-Australian fur seal (*Arctocephalus pusillus*), which also has clear geographic variants: the Cape fur seal (*A. pusillus pusillus*) and Australian fur seal (*A. pusillus doriferus*; Berta et al. 2018). In both *Zalophus* and *A. pusillus* cases, clear geographic breaks coincide with other supporting criteria for the designation of subspecies.

11.4.3 Overcoming Natal Philopatry: Otariid Recolonization and Its Nuances

Pinniped colonization is generally regarded as a positive density-dependent process (Fig. 11.6a; Gaggiotti et al. 2002). For example, research on expanding populations of southern sea lions suggests that high population density leads to limited physical space and intraspecific harassment forces young breeders to disperse to suitable breeding sites nearby (Grandi et al. 2008). With less competition, young animals may have enhanced reproductive success by venturing away from large colonies. The establishment of smaller colonies in the vicinity of large ones suggests a 'spill-over' effect, as observed in New Zealand fur seals along the Otago Peninsula (Bradshaw et al. 2000). Breeding site fidelity of young pioneering breeders along with natal philopatry seemingly promotes colony establishment in new areas.

Although density dependence is traditionally viewed as the major driving force for colonization and recolonization in mammals, some otariid studies suggest a more complex process. This is exemplified by Steller sea lions in the North Pacific. Two

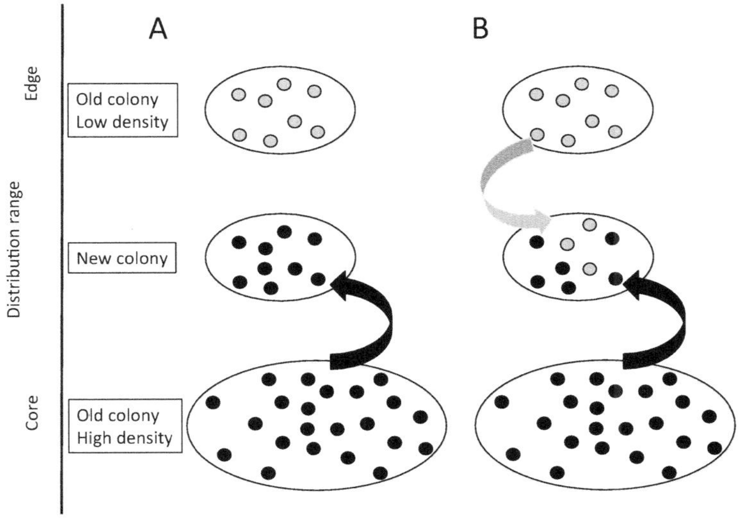

Fig. 11.6 Otariid recolonization hypotheses. (**a**) Positive density-dependent: an older colony with a high density of individuals is the main origin of pioneer individuals; (**b**) Alternative scenarios: pioneer individuals come from an older colony with a high density of individuals but also from colonies with low density of breeders

genetic stocks of Steller sea lions are currently recognized, to the East and West of the 144°W meridian, and they show contrasting population growth trends: the Western stock has been listed as Endangered under the United States Endangered Species Act (ESA) since 1997 while the Eastern stock is not. Surprisingly, newly established rookeries at the border of the genetic break have been colonized with individuals from both the endangered declining Western stock of Steller sea lions, and the Eastern stock, where population growth is positive (O'Corry-Crowe et al. 2014). This contradicts the hypothesis that pioneering individuals mainly originate from areas of population expansion (Fig. 11.6b). Genetic studies of Antarctic fur seal recolonization also do not accord with density-dependent theorising. It appears that the main source of pioneer breeders that recolonized the South Shetland Islands (SSI) following extinction in the mid nineteenth century, was not its nearest, largest neighboring colony on South Georgia. Rather it appears that the fur seals that recolonized SSI probably arrived from small, remnant colonies from remote and unexplored areas of Antarctica that survived the period of intense hunting (Bonin et al. 2013; Wynen et al. 2000). In a similar fashion, recolonization of New Zealand fur seals along the New Zealand coastline has been fueled primarily by immigration from individuals in remote locations (Dussex et al. 2016).

11.4.3.1 Implications for Management

Wildlife managers tend to focus conservation efforts on the "core" of a species geographic distribution, due to the assumption that geographical range should contract towards its center, where population density should remain higher over time. This is based on population densities being typically higher where the environment is more suitable for a given species. However, empirical and theoretical studies suggest a plausible alternative scenario. Meta-analyses revealed that 98% of 245 endangered species distributed across several taxonomic groups maintained a portion of their range at the periphery (contracted less than expected), which is defined by half the distance between the central historical range to the distribution's edge (Channell and Lomolino 2000a). Further, modelling suggests that extinction factors are not predictably influenced just by population size (its reduction), but rather, that extinction factors spread along the landscape like a *contagion* (Lomolino and Channell 1995). In this scenario, the last areas impacted within the species' distributions are those most isolated from the areas where the contagion started (Channell and Lomolino 2000b). Due to climate change, "edge" populations are likely to play an important role in maintaining and expanding species distributions (Rehm et al. 2015), as well as their evolutionary potential and otariid recolonization provides clear examples of this.

The relative high importance of low density and small colonies in the demographic history of otariids has tremendous implications for our understanding of how genetic diversity is maintained in wild populations. Together, these findings indicate that positive density-dependent population expansion is not an exclusive process in otariid recolonization. From a practical standpoint, this information can guide managers to consider smaller demographic components of the population at the edge of species distributions, even where colony inumbers are small.

11.5 Polygyny

In polygyny, a low number of males are hypothesized to have the majority of paternities. In theory, this has lasting negative population consequences: less genetic variability is passed on to following generations, potentially decreasing population-level fitness over time. Such predictions have been verified experimentally in *Drosophila* where harem mating resulted in a significant loss of genetic diversity in a controlled environment (Briton et al. 1994).

However where measured in otariids, genetic variability in wild polygynous populations is not lost at this predictable rate. There is evidence that: (i) sexual selection of heterozygote individuals and/or female choice for outbred mates may compensate for a reduction in effective population size (Amos et al. 2001; Hoffman et al. 2007; Pérez-González et al. 2009), and (ii) variability in the degree in which males effectively monopolize females and/or nuances in female individual behavior can counteract supposed negative effects of polygyny (Bonin et al. 2016). This is supported by the prediction that mating systems tend to evolve towards maximum efficiency, minimizing inbreeding, when males are the dispersing sex (Sugg et al. 1996). This is precisely the case in otariids.

11.5.1 The Many Shades of Polygyny

Antarctic fur seals are the best studied otariid regarding mating system, and these studies provide interesting clues to the intricacies of polygynous systems. In Antarctic fur seals, one male may defend up to 10 females (Arnould and Duck 1997) and in low density colonies, 2–3 males can father ~80% of the pups (Bonin et al. 2014). In this species, both sexes tend to return to the same breeding site annually (Hoffman et al. 2006b; Lunn and Boyd 1991) leading to the prediction that not only is there high male reproductive skew, but individuals also tend to remate often. Over generations, one would hypothesize that high levels of polygyny and remating would lead to the erosion of genetic diversity.

However, Antarctic fur seal populations retain high levels of heterozygosity (Bonin et al. 2013; Hoffman et al. 2011; Wynen et al. 2000). Admittedly, the historically large sizes of Antarctic fur seal populations contribute to high heterozygosity. Yet, it is plausible that nuances in breeding site fidelity and other reproductive behavior contribute to the maintenance of genetic variation in Antarctic fur seals: (i) Females rarely mate with the same male (Bonin et al. 2016). Although mostly faithful to their pupping sites, females change sites within the breeding beach more often than males (Hoffman et al. 2006b; Lunn and Boyd 1991), and these small scale movements alone could interfere with the probability of re-mating. (ii) Males father the majority of pups during the first or second year of territorial tenure (Hoffman et al. 2003). Therefore, although there is high male reproductive skew, older males are continuously replaced over generations. (iii) Female choice for outbred males may occur in some otariid populations. Antarctic fur seals on Bird Island, South Georgia, may cross a crowded colony in search of highly heterozygous mates and most unrelated to them (Hoffman et al. 2007), suggesting that mate choice for outbred males may occur in this species (see also Kiyota, Chap. 8 and Sinclair et al., Chap. 13). (iv) Polygyny has tremendous variability when considered across local populations and environments. Controlling females on a crowded beach is challenging (Bonin et al. 2014; Kiyota et al. 2008); thus the form of polygyny varies with population density and local conditions. For instance, southern sea lion males breeding on the tide line of a colony tend to defend females (female defense polygyny), while males breeding on internal pools tend to defend territories (resource defense polygyny; Franco-Trecu et al. 2015).

In summary, despite some species' historically large population sizes conferring a high level of genetic variation, the complexity of otariid reproductive behavior counteracts theoretical predictions of genetic diversity.

11.6 Otariid Behavior and Sealing

The gregarious behavior of otariids during the breeding season made them particularly susceptible to hunting in the eighteenth and nineteenth centuries. Sealing was driven by the high market value of fur seal pelts as they were used to make hats and coats. Once sealers settled in an area, they tended to harvest all animals in a given location. For example, in the South Shetland Islands, approximately 250,000 Antarctic fur seals were harvested within 3 years of the Islands' discovery (McCann and Doidge 1987). The intense harvesting targeted most fur seal and sea lion species and by the end of the eighteenth century, nearly all species were locally extinct in parts of their ranges (Gerber and Hilborn 2001). One otariid species, *Zalophus japonicus*, a species most closely related to *Z. californianus*, is extinct (Sakahira and Niimi 2007).

Sealing activities only ceased when conservation laws and practices came into effect in the mid to late twentieth century (see Crespo, Chap. 24). Most species have recolonized parts of their original distribution range; an ongoing process for many species (Hucke-Gaete et al. 2004). Post-sealing recoveries provide study systems for investigating the consequences of extirpation and recolonization on current genetic diversity, which can rarely be explored in the natural world.

The most prominent conclusion from several genetic studies is that recolonization promoted genetic mixing, and thus had an important "homogenizing" effect in the genetic make-up of some species such as New Zealand fur seals (Dussex et al. 2016) and northern fur seals (Dickerson et al. 2010). This occurred because the recovery of local populations was largely fueled by immigration rather than intrinsic growth (Hucke-Gaete et al. 2004). In addition, pioneer individuals of colonies appear to have immigrated from multiple remnant colonies (Bonin et al. 2013; Wynen et al. 2000) and so recolonization promoted gene-flow, which diluted pre-existing genetic structure.

11.6.1 The Long-Term Consequences of Sealing: What Have We Learned?

The amount of genetic variation within a species is related to the species' capacity to adapt to a changing environment and persist through catastrophic events (Van Dyke 2008). Genetic diversity provides the framework for adaptive evolution to occur and determines the species' potential to generate other species as a basis for biological diversity. For these reasons, conservation biologists are particularly concerned with modes of measuring and tracking within-species genetic variation.

Post sealing genetic diversity in otariids has been discussed in several studies, where comparative tables for mtDNA and microsatellite diversity parameters are available (see Lopes et al. 2015; Robertson and Chilvers 2011). These comparisons depict a complex scenario. Despite past severe demographic declines, the mtDNA control region of the Antarctic fur seal is considered one of the highest among mammal species with nucleotide diversity of 4.8% (Wynen et al. 2000). These extremely high levels of mtDNA diversity are not matched by other otariid species, which hold much lower levels of nucleotide diversity, for example only 1% in Galapagos fur seals (Lopes et al. 2015) and 0.1% in Cape fur seals (Matthee et al. 2006). Nucleotide diversity is heavily affected by historical population sizes and tends to reflect genetic bottlenecks (Tajima 1989).

A systematic evaluation of 30 pinniped species using nearly all microsatellite data available did not find support for a genetic bottleneck within some of the most severely exploited pinniped species, such as southern sea lions and northern elephant seals, which recovered well from near extinction (Stoffel et al. 2018). For context, nucleotide diversity in northern elephant seals is approximately 0.6% (Hoelzel et al. 1993; Weber et al. 2000). Stoffel et al. (2018) also found that genetic diversity varies nearly five-fold across pinniped species, with otariids at an intermediate position between ice breeding seals (higher heterozygosity) and species with near-extinction trajectories such as northern elephant and monk seals (low heterozygosity). Overall, it seems (i) That populations need to be reduced to an extremely small number of breeding individuals to significantly lose genetic diversity (Stoffel et al. 2018); (ii) Demographic declines need to be not only intense, but also sustained for a prolonged period to substantially reduce genetic diversity (Amos 1996); and (iii) The trajectory of a population pre-bottleneck matters: pre-bottleneck populations that were large and in expansion were not impacted as severely (Weber et al. 2004).

As intense as sealing was for many species, it is well established that a large number of individuals survived the sealing era (Bonin et al. 2013): sealing became economically unviable before local populations reached their bottleneck thresholds;

it was intense, but relatively short in duration (Hoffman et al. 2011), or prolonged at lower intensity. It is also probable that pre-sealing censuses underestimated the number of animals or did not encompass remote locations, as otariids can often be difficult to reach in caves or at the base of cliffs (Goldsworthy et al. 2000). Hence, the remoteness of their breeding habitats combined with efficient foraging abilities (Costa and Valenzuela-Toro, Chap. 2) and complex breeding habits may be responsible for their phenomenal success in recovering from intense over-exploitation.

11.7 Living in a Dynamic Environment: The Walrus

Walruses (*Odobenus rosmarus*) breed and primarily haul out on a highly dynamic substrate: the sea ice. They congregate during mid-winter in breeding aggregations near ice free water such as ice cracks and polynyas, which vary annually in location according to sea ice conditions and extent (Sjare and Stirling 1996). In contrast to otariids, parturition does not occur in concert with breeding because females give birth during their northward migration (Fay 1982). Interestingly, walruses also gather in sexually segregated non-breeding groups during summer. In general, this complexity makes the social structure of walruses and its genetic consequences difficult to interpret in comparison to the more predictable terrestrial breeding otariids.

Most studies indicate some level of female philopatry to non-breeding summering grounds (Andersen and Born 2000; Born et al. 2001). These studies exclusively refer to the Atlantic walrus subspecies (*O. r. rosmarus*) emphasizing aggregations in the West and Northwest of Greenland; subdivisions supported by mtDNA data, yet there is moderate gene flow (Andersen and Born 2000). In contrast, there is no indication of philopatry in the Pacific walrus (*O. r. divergens*; Sonsthagen et al. 2012; Scribner et al. 1997; Cronin et al. 1994). Accordingly, genetic datasets do not support population differentiation in the Pacific walrus. This has been attributed to its large population (>200,000 Pacific vs. >25,000 Atlantic; Lowry 2016) and the closer proximity of Pacific walrus aggregations (Sonsthagen et al. 2012). The correlation between geographic distance and genetic distance between Atlantic walrus aggregations is significant (*isolation by distance*; Born et al. 2001). From this perspective, it is plausible that long distances across the Arctic ultimately led to the divergence of Atlantic and Pacific walruses into two subspecies.

Walruses have very different mating behaviors from otariids (Miller and Kochnev, Chap. 22). Fay (1982) and Fay et al. (1984) provide detailed descriptions of walrus mating behavior and report that males do not typically harass females on ice during the breeding season. Instead, they practice courtship in the water, while defending an immediate female group from other males, an aquatic version of "female defense" polygyny (Sjare and Stirling 1996). Bulls perform mating rituals and produce stereotypical vocalizations (Stirling et al. 1987). Females from their "defended herd" enter the water and copulations occur underwater; therefore, male harassment is not the main driver for female gregariousness, but it is rather driven by the need of females to maintain unity in their dynamic habitat (Fay 1982). High levels of genetic variation were detected in walrus populations despite a prolonged period of exploitation (200 years) that led to a drastic reduction in numbers (Sonsthagen et al. 2012). This observation is in line with most otariid species, which were similarly heavily exploited. Importantly, the monopolization of female groups by males during breeding is only temporary (Fay 1982), and high male turnover is likely a contributing factor to the high genetic diversity levels in walruses (e.g., 206 mtDNA haplotypes among the 235 Pacific walrus individuals; Sonsthagen et al. 2012).

The dynamic nature of pack sea ice brings an additional level of complexity to the interpretation of walrus behavior. Obvious challenges to the study of pack ice pinniped breeders remain, as they are among the most elusive marine mammals inhabiting expansive areas of sea ice (e.g., leopard seal; *Hydrurga leptonyx*). In particular, little is known about their breeding behavior and movements.

As sea ice diminishes due to global climate change, walrus populations have become extremely vulnerable due to their specific niche as primary benthic feeders; a short-term northward shift in walrus distribution is predicted along with a population reduction driven by alterations in carrying capacity of arctic shallow water environments (MacCracken 2012). These changes are likely to impact population-level evolutionary processes that influence the long-term architecture of genetic diversity in walruses.

Acknowledgements Dr. Eric Lewallen provided editorial comments for this chapter. The National Oceanic and Atmospheric Administration Living Marine Resources Cooperative Science Center supports C. Bonin's pinniped research at Hampton University (NOAA-LMRCSC-FY2016; Award #NA16SEC4810007)."

References

Ahonen H, Lowther AD, Harcourt RG, Goldsworthy SD, Charrier I, Stow AJ (2016) The limits of dispersal: fine scale spatial genetic structure in Australian sea lions. Front Mar Sci 3:65

Amos B (1996) Levels of genetic variability in cetacean populations have probably changed little as a result of human activities. Rep Int Whal Comm 46:657–658

Amos W, Wilmer JW, Kokko H (2001) Do female grey seals select genetically diverse mates? Anim Behav 62(1):157–164

Andersen LW, Born EW (2000) Indications of two genetically different subpopulations of Atlantic walruses (*Odobenus rosmarus rosmarus*) in west and northwest Greenland. Can J Zool 78 (11):1999–2009

Arnould JP, Duck C (1997) The cost and benefits of territorial tenure, and factors affecting mating success in male Antarctic fur seals. J Zool 241(4):649–664

Aurioles-Gamboa D, Hernández-Camacho, J (2015) *Zalophus californianus*. The IUCN Red List of Threatened Species, e.T41666A45230310

Berta A, Churchill M, Boessenecker RW (2018) The origin and evolutionary biology of pinnipeds: seals, sea lions, and walruses. Annu Rev Earth Planet Sci. https://doi.org/10.1146/annurev-earth-082517-010009

Bickham JW, Patton JC, Loughlin TR (1996) High variability for control-region sequences in a marine mammal: implications for conservation and biogeography of Steller Sea lions (*Eumetopias jubatus*). J Mammal 77(1):95–108

Boness DJ, Bowen WD, Iverson SJ (1995) Does male harassment of females contribute to reproductive synchrony in the grey seal by affecting maternal performance? Behav Ecol Sociobiol 36(1):1–10

Bonin CA, Goebel ME, Forcada J, Burton RS, Hoffman JI (2013) Unexpected genetic differentiation between recently recolonized populations of a long-lived and highly vagile marine mammal. Ecol Evol 3(11):3701–3712

Bonin CA, Goebel ME, Hoffman JI, Burton RS (2014) High male reproductive success in a low-density Antarctic fur seal (*Arctocephalus gazella*) breeding colony. Behav Ecol Sociobiol 68(4):597–604

Bonin CA, Goebel ME, O'Corry-Crowe GM, Burton RS (2016) Impacts of ecology and behavior on Antarctic fur seal remating and relatedness. J Exp Mar Biol Ecol 476:72–77

Born E, Andersen L, Gjertz I, Wiig Ø (2001) A review of the genetic relationships of Atlantic walrus (*Odobenus rosmarus rosmarus*) east and west of Greenland. Polar Biol 24(10):713–718

Bradshaw CJ, Lalas C, Thompson CM (2000) Clustering of colonies in an expanding population of New Zealand fur seals (Arctocephalus forsteri). J Zool 250(1):105–112

Briton J, Nurthen R, Briscoe D, Frankham R (1994) Modelling problems in conservation genetics using Drosophila: consequences of harems. Biol Conserv 69(3):267–275

Cammen KM, Andrews KR, Carroll EL et al (2016) Genomic methods take the plunge: recent advances in high-throughput sequencing of marine mammals. J Hered 107(6):481–495

Campagna C, Bisioli C, Quintana F, Perez F, Vila A (1992) Group breeding in sea lions: pups survive better in colonies. Anim Behav 43(4):541–548

Campbell R (2003) Demography and population genetic structure of the Australian sea lion, *Neophoca cinerea*. University of Western Australia

Campbell R, Gales N, Lento G, Baker C (2007) Islands in the sea: extreme female natal site fidelity in the Australian sea lion, *Neophoca cinerea*. Biol Lett 4(1):139–142

Cappozzo HL, Túnez JI, Cassini MH (2008) Sexual harassment and female gregariousness in the south American sea lion, *Otaria flavescens*. Naturwissenschaften 95(7):625–630

Carreras C, Pascual M, Tomás J et al (2018) Sporadic nesting reveals long distance colonisation in the philopatric loggerhead sea turtle (*Caretta caretta*). Sci Rep 8(1):1435

Cassini M (2000) A model on female breeding dispersion and the reproductive systems of pinnipeds. Behav Process 51(1–3):93–99

Channell R, Lomolino MV (2000a) Dynamic biogeography and conservation of endangered species. Nature 403(6765):84

Channell R, Lomolino MV (2000b) Trajectories to extinction: spatial dynamics of the contraction of geographical ranges. J Biogeogr 27(1):169–179

Chilvers BL, Wilkinson IS (2008) Philopatry and site fidelity of New Zealand Sea lions (*Phocarctos hookeri*). Wildl Res 35(5):463–470

Chilvers BL, Robertson BC, Wilkinson IS, Duignan PJ, Gemmell NJ (2005) Male harassment of female New Zealand Sea lions, *Phocarctos hookeri*: mortality, injury, and harassment avoidance. Can J Zool 83(5):642–648

Coates DJ, Byrne M, Moritz C (2018) Genetic diversity and conservation units: dealing with the species-population continuum in the age of genomics. Front Ecol Evol 6:165

Cronin MA, Hills S, Born EW, Patton JC (1994) Mitochondrial DNA variation in Atlantic and Pacific walruses. Can J Zool 72:1035–1043

De Oliveira LR, Hoffman JI, Hingst-Zaher E et al (2008) Morphological and genetic evidence for two evolutionarily significant units (ESUs) in the south American fur seals *Arctocephalus australis*. Conserv Genet 9(6):1451–1466

De Oliveira LR, Gehara MC, Fraga LD et al (2017) Ancient female philopatry, asymmetric male gene flow, and synchronous population expansion support the influence of climatic oscillations on the evolution of south American sea lion (*Otaria flavescens*). PLoS One 12(6):e0179442

Dickerson BR, Ream RR, Vignieri SN, Bentzen P (2010) Population structure as revealed by mtDNA and microsatellites in northern fur seals, *Callorhinus ursinus*, throughout their range. PLoS One 5(5):e10671
Dobson FS (1982) Competition for mates and predominant juvenile male dispersal in mammals. Anim Behav 30(4):1183–1192
Doidge D, Croxall J, Baker J (1984) Density-dependent pup mortality in the Antarctic fur seal *Arctocephalus gazella* at South Georgia. J Zool 202(3):449–460
Doidge D, McCann T, Croxall J (1986) Attendance behavior of Antarctic fur seals. In: Gentry R, Kooyman G (eds) Fur seals: maternal strategies on land and at sea. Princeton University Press, Princeton, NJ, pp 102–114
Dussex N, Robertson BC, Salis AT, Kalinin A, Best H, Gemmell NJ (2016) Low spatial genetic differentiation associated with rapid recolonization in the New Zealand fur seal *Arctocephalus forsteri*. J Hered 107(7):581–592
Fay FH (1982) Ecology and biology of the Pacific walrus, *Odobenus rosmarus divergens* Illiger. North American Fauna 74:1–279
Fay FH, Ray GC, Kibal'chich AA (1984) Time and location of mating and associated behavior of the Pacific walrus, *Odobenus rosmarus divergens* Illiger. Soviet-American Cooperative Research on Marine Mammals 1:89–99
Franco-Trecu V, Costa-Urrutia P, Schramm Y, Tassino B, Inchausti P (2015) Tide line versus internal pools: mating system and breeding success of south American sea lion males. Behav Ecol Sociobiol 69(12):1985–1996
Gaggiotti OE, Jones F, Lee WM, Amos W, Harwood J, Nichols RA (2002) Patterns of colonization in a metapopulation of grey seals. Nature 416(6879):424–427
Gentry RL (1998) Behavior and ecology of the northern fur seal. Princeton University Press, Princeton, NJ
Gerber LR, Hilborn R (2001) Catastrophic events and recovery from low densities in populations of otariids: implications for risk of extinction. Mammal Rev 31(2):131–150
Goldsworthy S, Shaughnessy PD (1994) Breeding biology and haul-out pattern of the New Zealand fur seal, *Arctopehalus forsteri*, at Cape Gantheaume, South Australia. Wildl Res 21(3):365–375
Goldsworthy S, Francis J, Boness D, Fleischer R (2000) Variation in the mitochondrial control region in the Juan Fernandez fur seal (*Arctocephalus philippii*). J Hered 91(5):371–377
González-Suárez M, Flatz R, Aurioles-Gamboa D, Hedrick P, Gerber L (2009) Isolation by distance among California Sea lion populations in Mexico: redefining management stocks. Mol Ecol 18 (6):1088–1099
Grandi MF, Dans SL, Crespo EA (2008) Social composition and spatial distribution of colonies in an expanding population of South American sea lions. J Mammal 89(5):1218–1228
Greenwood PJ (1980) Mating systems, philopatry and dispersal in birds and mammals. Anim Behav 28(4):1140–1162
Hanski I, Gilpin M (1991) Metapopulation dynamics: brief history and conceptual domain. Biol J Linn Soc 42(1–2):3–16
Hastings KK, Jemison LA, Pendleton GW, Raum-Suryan KL, Pitcher KW (2017) Natal and breeding philopatry of female Steller sea lions in southeastern Alaska. PLoS One 12(6): e0176840
Hoelzel A, Halley J, O'Brien SJ et al (1993) Elephant seal genetic variation and the use of simulation models to investigate historical population bottlenecks. J Hered 84(6):443–449
Hoffman JI, Forcada J (2012) Extreme natal philopatry in female Antarctic fur seals (*Arctocephalus gazella*). Mamm Biol 77(1):71–73
Hoffman JI, Boyd IL, Amos W (2003) Male reproductive strategy and the importance of maternal status in the Antarctic fur seal *Arctocephalus gazella*. Evolution 57(8):1917–1930
Hoffman J, Matson C, Amos W, Loughlin T, Bickham J (2006a) Deep genetic subdivision within a continuously distributed and highly vagile marine mammal, the Steller's sea lion (*Eumetopias jubatus*). Mol Ecol 15(10):2821–2832

Hoffman J, Trathan P, Amos W (2006b) Genetic tagging reveals extreme site fidelity in territorial male Antarctic fur seals *Arctocephalus gazella*. Mol Ecol 15(12):3841–3847

Hoffman J, Forcada J, Trathan P, Amos W (2007) Female fur seals show active choice for males that are heterozygous and unrelated. Nature 445(7130):912

Hoffman J, Grant S, Forcada J, Phillips C (2011) Bayesian inference of a historical bottleneck in a heavily exploited marine mammal. Mol Ecol 20(19):3989–4008

Hoskins AJ, Schumann N, Costa DP, Arnould JP (2017) Foraging niche separation in sympatric temperate-latitude fur seal species. Mar Ecol Prog Ser 566:229–241

Hucke-Gaete R, Osman L, Moreno C, Torres D (2004) Examining natural population growth from near extinction: the case of the Antarctic fur seal at the south Shetlands, Antarctica. Polar Biol 27 (5):304–311

Kappeler PM, Barrett L, Blumstein DT, Clutton-Brock TH (2013) Constraints and flexibility in mammalian social behaviour: introduction and synthesis. Philos Trans R Soc Lond B Biol Sci 368(1618):20120337

Kiyota M, Insley SJ, Lance SL (2008) Effectiveness of territorial polygyny and alternative mating strategies in northern fur seals, *Callorhinus ursinus*. Behav Ecol Sociobiol 62(5):739–746

Lomolino MV, Channell R (1995) Splendid isolation: patterns of geographic range collapse in endangered mammals. J Mammal 76(2):335–347

Lopes F, Hoffman JI, Valiati VH et al (2015) Fine-scale matrilineal population structure in the Galapagos fur seal and its implications for conservation management. Conserv Genet 16 (5):1099–1113

Lowry L (2016) *Odobenus rosmarus*. The IUCN red list of threatened species 2016: e. T15106A45228501.https://doi.org/10.2305/IUCN.UK.20161.RLTS.T15106A45228501.en

Lowther A, Harcourt R, Goldsworthy S, Stow A (2012) Population structure of adult female Australian sea lions is driven by fine-scale foraging site fidelity. Anim Behav 83(3):691–701

Lunn N, Boyd I (1991) Pupping-site fidelity of Antarctic fur seals at Bird Island, South Georgia. J Mammal 72(1):202–206

MacCracken JG (2012) Pacific walrus and climate change: observations and predictions. Ecol Evol 2(8):2072–2090

Matthee C, Fourie F, Oosthuizen W, Meyer M, Tolley K (2006) Mitochondrial DNA sequence data of the cape fur seal (*Arctocephalus pusillus pusillus*) suggest that population numbers may be affected by climatic shifts. Mar Biol 148(4):899–905

Matthiopoulos J, Harwood J, Thomas L (2005) Metapopulation consequences of site fidelity for colonially breeding mammals and birds. J Anim Ecol 74(4):716–727

McCann TS, Doidge DW (1987) Antarctic fur seal, *Arctocephalus gazella*

Moritz C (1994) Defining 'evolutionarily significant units' for conservation. Trends Ecol Evol 9 (10):373–375

O'Corry-Crowe G, Gelatt T, Rea L, Bonin C, Rehberg M (2014) Crossing to safety: dispersal, colonization and mate choice in evolutionarily distinct populations of Steller sea lions, *Eumetopias jubatus*. Mol Ecol 23(22):5415–5434

Pérez-González J, Mateos C, Carranza J (2009) Polygyny can increase rather than decrease genetic diversity contributed by males relative to females: evidence from red deer. Mol Ecol 18 (8):1591–1600

Rehm EM, Olivas P, Stroud J, Feeley KJ (2015) Losing your edge: climate change and the conservation value of range-edge populations. Ecol Evol 5(19):4315–4326

Robertson BC, Chilvers BL (2011) The population decline of the New Zealand Sea lion *Phocarctos hookeri*: a review of possible causes. Mammal Rev 41(4):253–275

Robertson BC, Chilvers BL, Duignan PJ, Wilkinson IS, Gemmell NJ (2006) Dispersal of breeding, adult male *Phocarctos hookeri*: implications for disease transmission, population management and species recovery. Biol Conserv 127(2):227–236

Rosel PE, Taylor BL, Hancock-Hanser BL et al (2017) A review of molecular genetic markers and analytical approaches that have been used for delimiting marine mammal subspecies and species. Mar Mamm Sci 33(S1):56–75

Ryder OA (1986) Species conservation and systematics: the dilemma of subspecies. Trends Ecol Evol 1:9–10
Sakahira F, Niimi M (2007) Ancient DNA analysis of the Japanese sea lion (*Zalophus californianus japonicus* Peters, 1866): preliminary results using mitochondrial control-region sequences. Zool Sci 24(1):81–86
Scribner KT, Hills S, Fain SR, Cronin MA (1997) Population genetics studies of the walrus (*Odobenus rosmarus*): a summary and interpretation of results and research needs. Mol Genet Mar Mamm Spec Publ 3:173–184
Sjare B, Stirling I (1996) The breeding behavior of Atlantic walruses, *Odobenus rosmarus rosmarus*, in the Canadian high Arctic. Can J Zool 74(5):897–911
Slatkin M (1977) Gene flow and genetic drift in a species subject to frequent local extinctions. Theor Popul Biol 12(3):253–262
Slatkin M (1987) Gene flow and the geographic structure of natural populations. Science 236 (4803):787–792
Sonsthagen SA, Jay CV, Fischbach AS, Sage GK, Talbot SL (2012) Spatial genetic structure and asymmetrical gene flow within the Pacific walrus. J Mammal 93(6):1512–1524
Stirling I, Calvert W, Spencer C (1987) Evidence of stereotyped underwater vocalizations of male Atlantic walruses (*Odobenus rosmarus rosmarus*). Can J Zool 65(9):2311–2321
Stoffel M, Humble E, Paijmans A et al (2018) Demographic histories and genetic diversity across pinnipeds are shaped by human exploitation, ecology and life-history. Nat Commun 9(1):4836
Sugg DW, Chesser RK, Dobson FS, Hoogland JL (1996) Population genetics meets behavioral ecology. Trends Ecol Evol 11(8):338–342
Tajima F (1989) The effect of change in population size on DNA polymorphism. Genetics 123 (3):597–601
Van Dyke F (2008) Genetic diversity–understanding conservation at genetic levels. In: Conservation biology: foundations, concepts, applications. Springer, Dordrecht, pp 153–184
Villegas-Amtmann S, Atkinson S, Costa D (2009) Low synchrony in the breeding cycle of Galapagos Sea lions revealed by seasonal progesterone concentrations. J Mammal 90 (5):1232–1237
Weber D, Stewart BS, Garza JC, Lehman N (2000) An empirical genetic assessment of the severity of the northern elephant seal population bottleneck. Curr Biol 10(20):1287–1290
Weber D, Stewart B, Lehman N (2004) Genetic consequences of a severe population bottleneck in the Guadalupe fur seal (*Arctocephalus townsendi*). J Hered 95(2):144–153
Wolf JB, Tautz D, Trillmich F (2007) Galápagos and Californian sea lions are separate species: genetic analysis of the genus *Zalophus* and its implications for conservation management. Front Zool 4(1):20
Wynen LP, Goldsworthy SD, Guinet C et al (2000) Postsealing genetic variation and population structure of two species of fur seal (*Arctocephalus gazella* and *A. tropicalis*). Mol Ecol 9 (3):299–314

Chapter 12
Intergeneric Fertile Hybridization in Otariids

Valentina Franco-Trecu

Abstract Hybridization that results from breeding with non-conspecifics occurs when both isolating mechanism, pre-breeding recognition and post-breeding gene incompatibilities, fail. Behavior often works to prevent hybridization. Despite these mechanisms, hybridization has been documented in the fur seals, among species within the genus Arctocephalus. We report a case of a hybrid female from South American fur seal (*Arctocephalus australis*) and South American sea lion (*Otaria flavescens*), with her offspring. Mitochondrial DNA analysis indicated that mother lineage was from South American fur seal. Nuclear genetic data supported the admixture ancestry between both genera of both individuals, confirming the first intergeneric pinniped hybridization. In addition, relatedness coefficient for the hybrid female-pup pair confirms that the hybrid female was fertile and had produced a live, hybrid pup. Results trigger a discussion on the circumstances in which isolating mechanisms may be circumvented and insights into the importance of hybridization in speciation processes in mammals.

Keywords *Arctocephalus australis* · Isolating mechanism · *Otaria flavescens* · Pinniped hybridization · Speciation · Uruguay

Although there is no agreement concerning the species concept, the biological concept proposed by Mayr (1949) is, in practice, well accepted for organisms with sexual reproduction. According to this, a species represents a community of individuals separated from other communities by reproductive barriers or isolating mechanisms (Dobzhansky 1935; Mayr 1996). Avoiding breeding among non-conspecific involves a number of behaviors to facilitate pre-breeding recognition and/or is penalized by high mortality rates or infertility, avoiding post-breeding costs

V. Franco-Trecu (✉)
Departamento de Ecología y Evolución, Facultad de Ciencias, Universidad de la República, Montevideo, Uruguay
e-mail: vfranco-trecu@fcien.edu.uy

C. Campagna, R. Harcourt (eds.), *Ethology and Behavioral Ecology of Otariids and the Odobenid*, Ethology and Behavioral Ecology of Marine Mammals,
https://doi.org/10.1007/978-3-030-59184-7_12

Fig. 12.1 Picture of the hybrid female (middle) and its parental species; South American fur seal (left) and South American sea lion (right)

Fig. 12.2 Hybrid pup after 1 month (**a**) and 10 months (**b**) of birth. Hybrid pup is marked in red and fur seal pups of the same age in blue

due to genetic incompatibilities that reduce fitness (Darwin 1861; Mayr 1963). However, the production of viable offspring resulting from the crossing of genetically distinguishable taxa (i.e., hybridization), has been documented for about 10% of wild vertebrate species (e.g. Roper and Ryon 1977; Mallet 2005; Kingston and Gwilliam 2007; Willis et al. 2012). Hybridization events are relevant to the understanding of speciation as well as the operation of evolutionary forces such as reproductive isolation and female choice, in a variety of mating systems. Previous studies reported hybridization cases in fur seals, particularly among the genus Arctocephalus (Goldsworthy et al. 1999; Lancaster et al. 2006; Kingston and Gwilliam 2007; Lancaster et al. 2007a, b; Goldsworthy et al. 2009), but also between *Arctocephalus-Zalophus* and *Otaria-Zalophus* (Brunner 2002). Examples of intergeneric hybridization have not been backed by genetic data or evidence of hybrid fertility.

During the 2010 breeding season, a hybrid adult female of a South American fur seal (*Arctocephalus australis*) and South American sea lion (*Otaria flavescens*) occurred at Isla de Lobos, Uruguay (Franco-Trecu et al. 2016), with her offspring. The hybrid female was clearly distinguishable due to conspicuous size, color and pinna appearing to fall midway between species in the way they were expressed phenotypically (Fig. 12.1). This observation was verified using mitochondrial and nuclear genetic data (Franco-Trecu et al. 2016). The hybrid pup was a male (Fig. 12.2), 76.0 cm long and weighed 9.9 kg; its measurements were between the

Fig. 12.3 Picture of a copulation between a South American sea lion male and a South American fur seal female on Isla de Lobos, Uruguay

averages obtained for male pups of both parental species (fur seal: length = 67.9 ± 5.1 cm, weight = 6.9 ± 1.4 kg, n = 82; sea lion: length = 86.0 ± 4.6 cm, weight = 15.8 ± 2.3 kg, n = 25) for the same month of the year.

From analysis of the mitochondrial DNA, it turned out that his mother was a female South American fur seal. This is consistent with observations of subadult or adult South American sea lion males attempting to copulate with South American fur seal females here and elsewhere (Harcourt 1993; Miller et al. 1996, VFT pers. obs.). Nuclear DNA data supported the admixture ancestry of both individuals, a product of the crossing between different pinniped genera (*Arctocephalus* and *Otaria*). In addition, the relatedness coefficient for the hybrid female-pup pair was $R = 0.5$, confirming that the hybrid female was the mother.

The fact that only two additional individuals (fur seals), out of 254 fur seals and 83 sea lions analyzed, had a genetic component of the other species (sea lions), suggests that while hybridization is occurring, it is at low frequency. Difference in body size between species and sexes probably is the most relevant isolating mechanism between them. That is, the similarity in the weights of an adult fur seal male (200 kg) and an adult sea lion female (140 kg on average) would preclude fur seal males mounting reluctant sea lion females, while the dissimilarity in the weights of an adult sea lion male (300 kg) and an adult fur seal female (40 kg) means that copulation is likely to result in a females' death (Harcourt 1993; Miller et al. 1996). However, in some circumstances mating may successfully occur, as when the female fur seal is large and the male sea lion small, as shown in Fig. 12.3. Another potential isolating mechanism is the difference in breeding dates between both species: while fur seal breeding season peaks in mid-December (Franco-Trecu et al. 2014), sea lions breed during January (Franco-Trecu et al. 2015), with no more than 1 week of overlap in the breeding period. These morphological and phenological restrictions could further limit the frequency of hybrids in this Uruguayan colony.

Furthermore, behavioral constraints, in particular the mating systems of the two species, could bias the direction of hybridization (fur seal females with sea lion males). The South American sea lion is a polygynous species with female defense (Campagna and Le Boeuf 1988; Franco-Trecu et al. 2015), which makes it difficult for other males (either of the same species or another) to gain access to reproductive females. In turn, in Uruguay the South American fur seal mating system is classified as a lek, with territorial and satellite tactics coexisting, and females moving freely in the breeding area (Franco-Trecu et al. 2014). In a lekking system, where female fur seals are not defended by territorial males, sea lion males may behave as a fur seal peripheral one, gaining access to females without being expelled by a territorial male. Thus lekking may be more permeable to hybridization, while territorial or female defense systems would not.

In summary, genetic evidence confirms intergeneric pinniped hybridization with the resultant hybrid also fertile, with a live pup produced (Franco-Trecu et al. 2016). We not know how the mechanisms that prevent mating with non-conspecific were circumvented in this case. It may be possible that disparity in size, although not as extreme in this case as it can be between the involved species, resulted in the thwarting of female choice processes. Besides the proximate considerations, and particularly because the hybrid was fertile, this record calls attention to a possible mechanism of speciation in these and other mammals.

References

Brunner S (2002) A probable hybrid sea lion - *Zalophus californianus* x *Otaria byronia*. J Mammal 83:135–144

Campagna C, Le Boeuf BJ (1988) Reproductive behaviour of Southern Sea lions. Behaviour 104:233–261

Darwin CR (1861) On the origin of species by means of natural selection, or the preservation of favoured races in the struggle for life. John Murray, London

Dobzhansky T (1935) A critique of the species concept in biology. Philos Sci 2:344–355

Franco-Trecu V, Costa-Urrutia P, Schramm Y, Tassino B, Inchausti P (2014) Sex on the rocks: breeding tactics and reproductive success of South American fur seal males. Behav Ecol 25 (6):1513–1523

Franco-Trecu V, Costa-Urrutia P, Schramm Y, Tassino B, Inchausti P (2015) Tide line versuss internal pools: mating system and breeding success of South American sea lion males. Behav Ecol Sociobiol 69:1985–1996

Franco-Trecu V, Abud C, Kloetzer G, Casacuberta M, Feijoo M, Costa-Urrutia P (2016) Sex beyond species: intergeneric fertile hybridization in marine mammals, the first genetically analyzed case in pinnipeds. Evol Dev 18(2):127–136

Goldsworthy SD, Boness DJ, Fleischer RC (1999) Mate choice among sympatric fur seals: female preference for conphenotypic males. Behav Ecol Sociobiol 45:253–267

Goldsworthy SD, McKenzie J, Page B, Lancaster ML, Shaughnessy PD, Wynen LP, Robinson SA, Peters KJ, Baylis AMM, McIntosh RR (2009) Fur seals at Macquarie Island: post-sealing colonisation, trends in abundance and hybridisation of three species. Polar Biol 32:1473–1486
Harcourt R (1993) Individual variation in predation on fur seals by southern sea lions (*Otaria byronia*) in Peru. Can J Zool 71:1908–1911
Kingston JJ, Gwilliam J (2007) Hybridization between two sympatrically breeding species of fur seal at Iles Crozet revealed by genetic analysis. Conserv Genet 8:1133–1145
Lancaster ML, Gemmell NJ, Negro S, Goldsworthy S, Sunnucks P (2006) Menage a trois on Macquarie Island: hybridization among three species of fur seal (*Arctocephalus* spp.) following historical population extinction. Mol Ecol 15:3681–3692
Lancaster ML, Bradshaw CJA, Goldsworthy SD, Sunnucks P (2007a) Lower reproductive success in hybrid fur seal males indicates fitness costs to hybridization. Mol Ecol 16:3187–3197
Lancaster ML, Goldsworthy SD, Sunnucks P (2007b) Multiple mating strategies explain unexpected genetic mixing of New Zealand fur seals with two congenerics in a recently recolonized population. Mol Ecol 16:5267–5276
Mallet J (2005) Hybridization as an invasion of the genome. Trends Ecol Evol 20:229–237
Mayr E (1949) The species concept - semantics versus semantics. Evolution 3:371–373
Mayr E (1963) Animal species, evolution, and geographic isolation - reply. Syst Zool 12:204–206
Mayr E (1996) What is a species, and what is not? Philos Sci 63:262–277
Miller EH, Ponce de Leon A, DeLong RL (1996) Violent interspecific sexual behavior by male sea lions (Otariidae): evolutionary and phylogenetic implications. Mar Mamm Sci 12:468–476
Roper TJ, Ryon J (1977) Mutual synchronization of diurnal activity rhythms in groups of red wolf-coyote hybrids. J Zool 182:177–185
Willis PM, Rosenthal GG, Ryan MJ (2012) An indirect cue of predation risk counteracts female preference for conspecifics in a naturally hybridizing fish *Xiphophorus birchmanni*. PLoS One 7:e34802

Chapter 13
Visual Species Recognition and Mate Choice in Fur Seals: An Experimental Study

Jennifer Kingston Sinclair, Simon Goldsworthy, Christophe Guinet, and Robert Harcourt

Abstract Most extant fur seal species have discrete breeding ranges, and it is assumed that prior to sealing they were even more separate. However, post sealing most populations have made dramatic recoveries and in the process of so doing have recolonized their previous ranges and possibly further afield. At three locations, Crozet Is., Macquarie Is. and Marion Is., subantarctic and Antarctic fur seals now breed sympatrically and produce hybrid offspring. Hybridization is rare in mammals with multiple barriers to gene exchange between species. At Îles Crozet hybridization between Antarctic and subantarctic fur seals is significantly less (~3%) than expected if random mating were occurring suggesting that one isolating mechanism may be traits by which females choose conspecific mates. Male subantarctic fur seals have a conspicuous pelage with creamy ventral surface, dark dorsal fur and a prominent sagittal crest, visually very distinct from other fur seals which are uniformly dark in color with no sagittal crest. We tested whether male pelage is a visual cue for species recognition by manipulating eight male subantarctic fur seals to make them resemble Antarctic fur seals and comparing their behavior and mating success with eight control males. Experimental male territories had significantly fewer females than control male territories, suggesting that females may choose conspecifics based on visual cues.

Keywords Fur seals · Hybridization · Mate choice · Pelage · Visual cues · Species specific

J. K. Sinclair (✉) · R. Harcourt
Marine Predator Research Group, School of Biological Sciences, Macquarie University, Sydney, NSW, Australia

S. Goldsworthy
School of Biological Sciences, South Australian Research & Development Institute (SARDI) – Aquatic Sciences, Henley Beach, SA, Australia

C. Guinet
Centre d'Etudes Biologiques de Chizé - La Rochelle Université, UMR 7372, CNRS, Villiers-en-Bois, France

C. Campagna, R. Harcourt (eds.), *Ethology and Behavioral Ecology of Otariids and the Odobenid*, Ethology and Behavioral Ecology of Marine Mammals,
https://doi.org/10.1007/978-3-030-59184-7_13

13.1 Introduction

In most species that defend resources required by breeding females, males with higher quality or greater quantity of resources within their territory have access to more females, and sexual selection acts on traits that enhance a male's ability to procure and retain these resources (Andersson 1994; Emlen and Oring 1977). In many animals, males compete for access to mates or access to resources that attract mates while females exercise mate choice for desirable partner males (Andersson 1994; Bateson 1983; Bradbury and Andersson 1987). Females most often exhibit mate choice for conspicuous male traits, although the exact mechanisms and the evolution of female preferences is not clear, and is a highly controversial issue in the theory of sexual selection. Mate preference is not necessarily a cognitive process (Kirkpatrick and Ryan 1991) and females can choose mates based on a variety of traits including visual characteristics such as coat color and secondary sex ornaments, vocalizations, resources such as nesting sites and territories, chemical signals or genotype.

The evolution of mate choice can be roughly divided into two schools, the 'nonadaptive' school and the 'good genes' school (Kirkpatrick 1987). Kirkpatrick (1987) states that the former postulates that preferences cause male traits to evolve in ways that are not adaptive with respect to their ecological environment while the latter holds that mate choice evolves under selection for females to mate with ecologically adaptive genotypes. Supporters of the nonadaptive school argue that females prefer males with traits that are not favored by natural selection (Kirkpatrick 1987). For instance, females may prefer a trait that also attracts predators such as in the túngara frog (*Physalaemus pustulosus*) where the component of calls used by males to attract females also attracts predatory bats (Ryan 1985). Advocates of the 'good genes' school contend that females prefer male traits that indicate high male quality (Kirkpatrick 1987). In red-winged blackbirds (*Agelaius phoeniceus*) which practice resource defense polygyny, experienced males have higher quality territories and provide better parental care (Yasukawa 1981). Female preference for male phenotype (i.e. intense courtship, song repertoire and large size) may be favored by direct benefits (male territory quality) since these traits increase with age.

Traits that indicate mate quality may not always advertise direct material benefits but rather indicate mate compatibility. Generally, most compatible mates are conspecific. In sympatrically breeding populations, traits males use to attract females can also act as criteria for recognizing compatible mates (Ptacek 2000). Recognition of conspecific signals is a major selective force influencing the evolution of mate choice (Maynard Smith 1987; Ryan 1994; Ptacek 2000) and divergent preferences or traits can serve as reproductive isolating barriers resulting in hetero-specific avoidance and reduced hybridization (Andersson 1994). A divergent trait acting as a reproductive barrier is defined as ethological isolation (Mayr 1963). Ethological mechanisms generally refer to differences in signaling and behavior, including courtship, auditory, chemical and visual stimuli (Andersson 1994). Hence, in species that practice resource defense polygyny, such as fur seals, it is likely that females recognize conspecifics based on male phenotype.

While there is a lack of information on female mating strategies in pinnipeds, there are several indications that inter-specific reproductive barriers exist between subantarctic and Antarctic fur seals. Foremost, hybridization between subantarctic and Antarctic fur seals at Îles Crozet is less than would be expected in a randomly mating population (Kingston and Gwilliam 2007). Inter-specific matings are a rare occurrence despite significant numbers of individuals residing in mixed species breeding territories (Kingston and Gwilliam 2007). An early study conducted by Goldsworthy et al. (1999) at Macquarie Island, where subantarctic and Antarctic fur seals also breed sympatrically, reported an increase in extra-territorial fertilizations in females where the resident male displayed a hetero-specific phenotype, suggesting female preference for conspecific traits. Habitat segregation and fine scale site fidelity appeared to help reduce rates of hybridization (Lancaster et al. 2010). Crucially hybridization incurred fitness costs, and a greater proportion of females in hybrid male territories conceived extra-territorially than those in territories of pure-species males (Lancaster et al. 2007). These extra-territorial conceptions combined with female movement between territories suggests that males possess physical or behavioral traits that females choose. Heath (1989) observed California sea lion (*Zalophus californianus*) females changing pupping sites from 1 year to the next to remain within the territory of a specific individual male and Kiyota (Chap. 8) observed high site fidelity to male breeding sites in Northern fur seals (also see Chap. 6 by Cassini). Indirect female mate choice through aggression toward unwanted suitors has been demonstrated in two phocid species, the northern elephant seal (*Mirounga angustirostris*) (Cox and Le Boeuf 1977) and gray seal (*Halichoerus grypus*) (Boness et al. 1982), and may occur in fur seals as females often display aggression to their suitors (Harcourt 1992).

Fur seals exhibit a number of divergent traits that could potentially act as mechanisms for species recognition including morphology, pelage and vocalizations. Subantarctic fur seals exhibit a greater pattern of divergence, i.e. they are more different from other closely related species than the other species are from each other, in vocalizations (Page et al. 2002) and pelage compared to other fur seal species (see Fig. 13.1). Fur seals of the genus *Arctocephalus* are essentially monomorphic in appearance with the exception of the subantarctic fur seal (Fig. 13.1), males of which have a unique pelage pattern that consists of a creamy yellow chest and face, dark dorsal fur and crest. No other explanation has been given for the unique pelage pattern of the subantarctic fur seal that distinguishes it from all other species in the genus. We propose that the unique pelage characteristics of the subantarctic fur seal have been under strong sexual selection and mate choice for species recognition.

We tested whether subantarctic fur seal females recognize conspecifics based on pelage characteristics. We designed a mate choice experiment by manipulating the pelage of several subantarctic fur seal males to resemble that of other male fur seals in the genus *Arctocephalus*. The identity of females residing in male territories was monitored and recorded throughout the breeding season. Male reproductive success

Fig. 13.1 Adult male pelage patterns of several southern hemisphere fur seal species of the Genus *Arctocephalus* (clockwise from top left): (**a**) Antarctic fur seal, (**b**) South American fur seal, (**c**) Australian fur seal, (**d**) New Zealand fur seal and (**e**) subantarctic fur seal

was assessed using microsatellite markers. Manipulating the pelage of territorial males should decrease their reproductive success if females use pelage as an obligatory cue for mate recognition but remain unchanged if it does not act as a cue, or if it is not obligate.

13.2 Materials and Methods

This experiment was conducted at a sympatric breeding colony of fur seals on Île de la Possession in the Crozet archipelago during the 2001/2 breeding season. Prior to the start of the subantarctic fur seal breeding season we captured and manipulated the pelage of eight territorial males with black hair dye and by shaving the sagittal crest (Fig. 13.2) so that they resembled males of other fur seal species in the genus *Arctocephalus* (Fig. 13.1). A control group consisting of eight subantarctic territorial males was treated in the same fashion as the experimental group although no visible manipulations to pelage were made.

When designating experimental and control group males we attempted to control for variation in body size to remove any bias that it may have on female choice or reproductive success. This was done by estimating male size *in situ* before capture

Fig. 13.2 An adult male subantarctic fur seal prior to pelage manipulation (left) and 2 months later (right), after the fur had been dyed and the sagittal crest shorn

since body measurements could not be taken until the animal was anesthetized. Once captured, morphometric measurements were taken for body length, girth and testes size. Experimental and control territories were homogeneously distributed throughout the colony to minimize the potential influence of female choice based on territory location. The number and identity of females within experimental and control territories was recorded several times throughout the mating period.

Tissue samples for paternity analysis were collected from adult males and females during the 2001/2 breeding season and from the resulting offspring the following year. Paternity analyses were conducted using the program CERVUS (Marshall et al. 1998) and all individuals were genotyped for species identity.

13.3 Analysis

All statistical analyses were carried out using SYSTAT v10.0 (SPSS Inc., Chicago, USA). Pearson's correlation was used to test for relationships between body size and harem size. Harem size and morphometric data were tested for Lilleifors normal distribution using the Kolmogorov-Smirnov test. T-tests were conducted to test the null hypotheses that experimental and control males were similar in body size, had equal reproductive success and equal numbers of females in their territories.

13.4 Results

Excluding experimental males (to remove experiment bias) there was a significant correlation between girth and harem size ($n = 11$; Pearson's correlation coefficient $= 0.812$; $P = 0.043$). There was no significant correlation b/w length and harem size ($n = 11$; Pearson's correlation coefficient $= 0.211$; $P = 1.00$), body

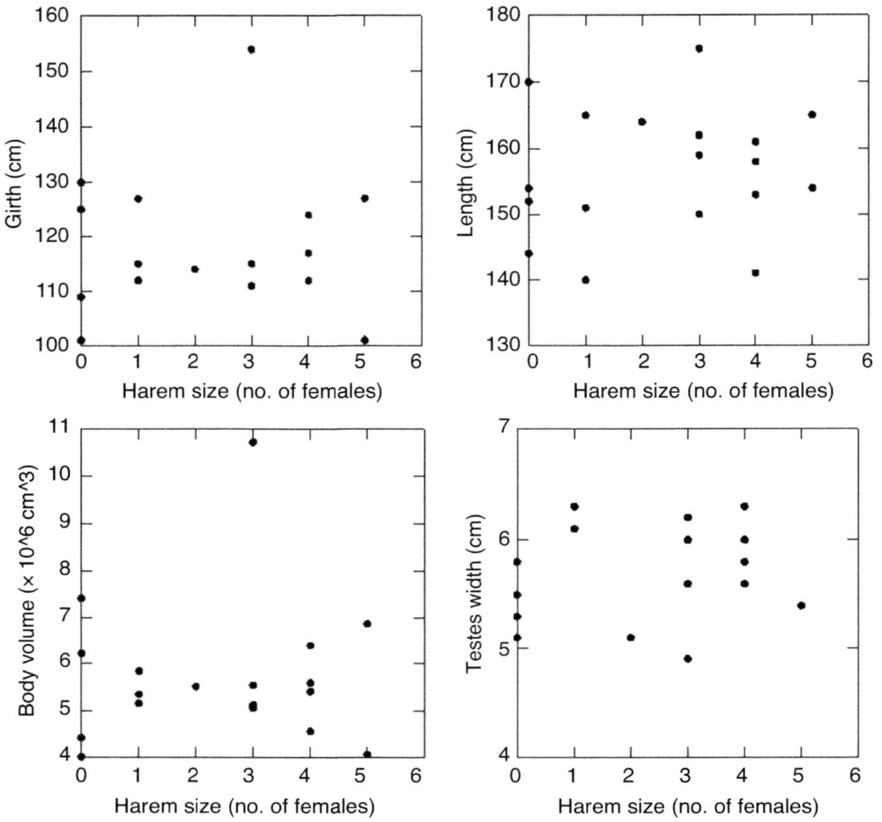

Fig. 13.3 There is a correlation between girth and harem size (top left) but there are no correlations between length (top right), body volume (bottom left) and testes width (bottom right) and harem size in *A. tropicalis*

volume and harem size (n = 11; Pearson's correlation coefficient = 0.711; $P = 0.213$) or testes width and harem size (n = 10; Pearson's correlation coefficient = 0.269; $P = 1.00$) (see Fig. 13.3).

All morphometric data followed a normal distribution (length: n = 16; $P = 1.00$; girth: n = 16; $P = 0.199$; body volume: n = 16; $P = 0.231$; testes: n = 15; $P = 0.471$). There was no significant difference between males of the control and experimental groups for length ($n_{con} = 8$; $n_{exp} = 8$; t = –0.26; $P = 0.80$), body volume ($n_{con} = 8$; $n_{exp} = 8$; t = –1.684; $P = 0.12$) or testes size ($n_{con} = 7$; $n_{exp} = 8$; t = 0.01; $P = 0.99$) (Table 13.1). There was a considerable difference in girth measurements ($n_{con} = 8$; $n_{exp} = 8$; t = –2.09; $P = 0.06$) and power analysis revealed that quadrupling sampling effort (n = 38) would reveal a significant size difference between control and experimental groups (Table 13.1). However girth is dependent on body condition and has been shown to decrease over the period of territorial

Table 13.1 Experimental and control group male morphometric measurements, harem size and reproductive success

Male ID	Species genotype	Experimental/ control	Length (cm)	Girth (cm)	Body Volume (cm^3)	Testes (cm)	No. of females in harem	No. of offspring sired '02	No. of offspring sired '03
Y088	At	E	170	130	7,423,419.4	5.3	0	2	0
Y089	H	E	173	125	6,984,486.8	5.8	2	0	0
Y091	At	E	140	127	5,834,502.8	6.3	1	0	0
Y092	At	E	175	154	10,723,779.2	6.0	3	0	0
Y093	At	E	151	115	5,159,896.6	6.1	1	1	0
Y094	At	E	154	125	6,217,404.4	5.8	0	1	1
Y095	At	E	158	117	5,588,524.8	6.3	4	0	2
Y096	At	E	150	115	5,125,725.1	4.9	3	0	0
Y097	At	C	153	117	5,411,672.7	5.6	4	1	2
Y098	At	C	161	124	6,396,429.5	6.0	4	0	1
Y099	At	C	141	112	4,570,077.1	5.8	4	1	1
Y100	At	C	165	112	5,347,962.6	6.1	1	2	0
Y087	At	C	159	111	5,061,875.4	6.2	3	5	1
Y086	At	C	154	101	4,059,119.5		5	3	1
Y085	At	C	165	127	6,876,378.3	5.4	5	1	1
Y084	At	C	162	115	5,535,783.1	5.6	3	0	0

tenure (Boyd and Duck 1991). Furthermore, although girth is correlated to harem size it has no influence on male reproductive success in subantarctic or Antarctic fur seals. Other measures of size, including length and testes width (unless measured shortly after ejaculation), do not vary over the course of the breeding season and are more reliable indicators of body size. A power analysis of these variables shows that the sample size in this study should be sufficient to detect any differences in length (increase to $n = 1740$) and testes size (increase to $n = 1,242,294$). Body volume, like girth, is also dependent on body condition so is not as reliable as length and testes measurements.

Data on harem size for experimental and control males followed a normal distribution ($P = 0.09$). The mean number of females residing in control male territories (mean = 3.62 ± 1.30) was significantly larger than the number of females residing in experimental male territories (mean = 1.75 ± 1.49; $t = 2.68$; $P = 0.02$) (Table 13.1).

Samples were collected from 42 pups of known mothers during the 2002/3 breeding season. Of these 42 samples 7 were offspring of females residing in an experimental territory and 16 were offspring of females residing in control territories. The remaining 19 were from offspring of females residing in unaltered subantarctic or Antarctic male territories. Paternity was assigned at the 95% confidence level to 5 offspring of females residing in experimental territories, 15 offspring of females residing in control territories and 13 offspring of females residing in territories of unaltered males. One of the experimental males was found to be of hybrid origin and two of the five females whose pups were assigned paternity resided in this territory. Both these females were inseminated by extra-territorial males, although one was inseminated by a con-phenotypic/con-specific (subantarctic origin and appearance) male and the other was inseminated by a hetero-specific male (Antarctic). Of the remaining three females that resided in experimental territories whose pups were assigned paternity, one was of hybrid origin and was fertilized by the territorial male. Another subantarctic female was also fertilized by the territorial (and therefore experimental/hetero-phenotypic) male and another subantarctic female was fertilized by an extra-territorial con-phenotypic/conspecific male (Fig. 13.4).

Only 4 of the 15 females residing in control territories whose pups were assigned paternity were inseminated by their territorial male. Eight of the remaining 11 were fertilized by con-phenotypic/con-specific (subantarctic origin and appearance) extra-territorial males while there was one extra-territory fertilization by a hybrid male and two extra-territory fertilizations were by an experimental subantarctic male.

Excluding the mating between the hybrid female and subantarctic male, 14 of the 19 females whose offspring were assigned paternity were inseminated by con-phenotypic (unaltered in appearance) males. Of these 14 females, 10 were inseminated by an extra-territorial male. There were a total of 14 extra-territorial fertilizations of which 4 females mated with males exhibiting a hetero-specific phenotype.

Although there was a significant difference in harem size, there was no significant difference in the number of offspring sired ($n_{con} = 8$; $n_{exp} = 8$; $t = 1.44$; $P = 0.172$)

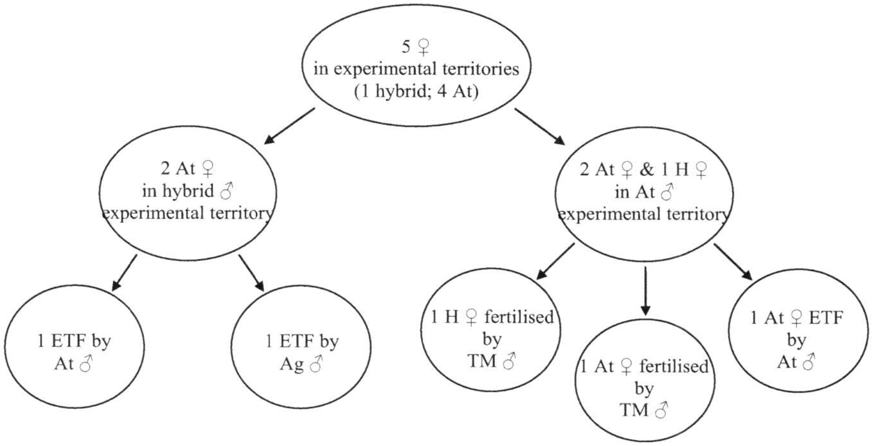

Fig. 13.4 Schematic of the distribution of females whose pups were assigned paternity in experimental territories and the status of the males with whom they mated

by experimental (mean = 0.875 ± 0.641) and control (mean = 0.375 ± 0.744) males (Table 13.1). When the data for the number of offspring sired by con-phenotypic males that were not part of the control group are included in the analysis, there is still no significant difference in the number of offspring sired by experimental males and males exhibiting the subantarctic phenotype (n_{con} = 63; n_{exp} = 8; t = –0.44; *P* = 0.669. Variance in reproductive success was high for both control (mean = 0.875 ± 0.641) and experimental male (mean = 0.375 ± 0.744) groups and a *post hoc* power analysis shows that with increased sampling (to n = 68) a significant difference in reproductive success might become evident. However, an increase in sample size of this magnitude would not have been possible at the Mares aux Elephants colony since there were only 24 subantarctic territorial males present during the 2001/2 breeding season.

13.5 Discussion

Females most often exhibit mate choice for conspicuous male traits, and discriminating characteristics can act as a mechanism for species recognition, preventing mismatings with other species (Andersson 1994). Male subantarctic fur seals, which have pelage characteristics that differ markedly in appearance from other fur seal species, breed sympatrically with congeners at several locations. The low frequency of hybridization in sympatric populations (Goldsworthy et al. 2009; Lancaster et al. 2010) implies the existence of a species recognition mechanism which leads to assortative mating. If female fur seals recognize conspecific mates based on pelage characteristics, then it is expected that subantarctic fur seal females would mate preferentially with males exhibiting the subantarctic phenotype and that hetero-phenotypic males would have low reproductive success. The results of a mate choice experiment conducted in this study show that although con-phenotypic males were not more successful at inseminating female subantarctic fur seals than hetero-phenotypic males, most females mated with conspecific/con-phenotypic males.

Only a small number of females (n = 2), whose pups were assigned paternity, residing in territories of males whose pelage had been altered to resemble hetero-specific adult males were inseminated by the territorial male. One of these females was a hybrid individual so it is difficult to draw conclusions on whether or not she is able to discriminate between species. Most females, including females residing in territories held by control group males, were fertilized by extra-territorial males. This is somewhat surprising considering that subantarctic fur seal males who hold territories have greater reproductive success than males that do not. However, extra-territorial fertilizations and female polyandry are not as uncommon as previously thought (Boness et al. 1993). Several studies have documented both extra-territorial and multiple matings in fur seals (Harcourt 1990; Kuroiwa and Majluf 1989; Miller 1975; Stirling and Warneke 1971; McCann 1980; Francis and Boness 1991; Goldsworthy et al. 1999; Trillmich 1987; Chap. 8 by Kiyoto). In this study, nearly three-quarters of the extra-territorial fertilizations (71.4%; n = 10) that occurred were achieved by subantarctic fur seal males displaying the subantarctic phenotype. Although females appear to have mated preferentially with con-phenotypic males, control group males were not more successful than experimental group males. This finding does not necessarily preclude the role of pelage characteristics as a species recognition mechanism. The mate choice experiment conducted in this study included two-thirds (n = 24) of the territorial subantarctic fur seal males in the colony and there were many non-territorial males present as well. Many of the extra-territorial fertilizations (61.5%; n = 8) were obtained by conspecific males excluded from the experiment. Consequently, the number of offspring sired by con-phenotypic males is underestimated when limited just to offspring of control males. Nonetheless, when con-phenotypic males that were excluded from the experiment were included in the analysis there was still no significant difference in the number of offspring sired by hetero-phenotypic experimental males and males displaying the subantarctic phenotype. However, there were considerably more con-phenotypic males (n = 63) than experimental males (n = 8) and due to the limited number of pups sampled, many were unsuccessful.

In this study, fewer than one-half (44.8%; n = 43) of females tagged in the 2001/2 breeding season were resighted in the 2002/3 season (Luque, pers. comm.). The low number of resightings of females is surprising since there is evidence that female fur seals are highly philopatric (Lunn and Boyd 1991). Consequently, only a small proportion of the pups conceived during the experimental breeding season (2001/2) were sampled. Only half (n = 7) of the offspring of mothers residing in experimental territories were sampled, 5 of which were assigned fathers, and just 55% (n = 16) of the offspring whose mothers resided in control territories were sampled, 15 of which were assigned paternity at the 95% confidence level. Despite the small sample size, two-thirds of extra-territorial fertilizations (n = 2) obtained by females residing in hetero-phenotypic (experimental) territories resulted in fertilization by a con-phenotypic male. This is in accordance with findings of earlier research by Goldsworthy et al. (1999), who found that 68% of all extra-territorial fertilizations of females residing in hetero-phenotypic territories were also inseminated by con-phenotypic males. Of the four subantarctic pure-stock females (the fifth female was hybrid) that resided in experimental territories, two were inseminated by con-phenotypic extra-territorial males (Fig. 13.2).

Further evidence supporting the hypothesis that pelage has a role in female mate choice for conspecifics was found when examining the number of females residing in experimental and control group territories. Due to the potential biases created by the mate choice experiment, we were unable to test the relationship between harem size and reproductive success, but territory ownership provides the greatest opportunity to monopolize matings with females (McCann 1980; Bartholomew 1970). Males with larger territories (number of females) achieve copulations with a greater number of females (Goldsworthy et al. 1999). There are several traits that could potentially influence harem size including body size and territory quality (Carey 1991; Arnould and Duck 1997; McCann 1980, 1987). In this study we attempted to control for the effect of body size and territory quality on mate choice and/or reproductive success. Experimental and control males were of similar size and territories of each group were distributed homogeneously throughout the subantarctic fur seal breeding area. Comparison of the territory sizes of each group shows that control group males had twice as many females residing in their territories as experimental group males. In species which practice resource-defense polygyny, females are predicted to choose mates based on the quality of the resources that they hold (Miller 1975; Bartholomew 1970; Carey 1991). However, in populations that breed in sympatry, male phenotype can serve as a mate recognition mechanism resulting in hetero-specific avoidance and reduced hybridization (Andersson 1994). For instance, in the pygmy swordtail fish, *Xiphophorous pygmaeus*, a species in which males have lost their conspicuous sword (elongation of the caudal fin), females from sympatric populations have lost preference for males with swords, reducing the likelihood of hybridization (Rosenthal and Ryan 2011). Therefore, if female fur seals are likely to be inseminated by the male within whose territory they reside (Clutton-Brock 1989) it makes sense to exhibit choice for conspecifics when choosing where to pup. These results suggest that male phenotype does in fact influence female territory choice and that females prefer to pup in a territory defended by a conspecific male.

Whether pelage characteristics are the sole phenotypic criterion for species recognition is unknown. In some species conspecific mate recognition is based on multiple cues (Candolin 2003). Fur seals exhibit several traits that are divergent between species including vocalizations (Page et al. 2002), scent (Chap. 15 by Charrier) and habitat preference (Hofmeyr et al. 1997). Whether these traits serve as species recognition mechanisms warrants further investigation.

References

Andersson M (1994) Sexual selection. Princeton University Press, Princeton
Arnould JPY, Duck CD (1997) The cost and benefits of territorial tenure, and factors affecting mating success in male Antarctic fur seals. J Zool 241:649–664
Bartholomew GA (1970) A model for the evolution of pinniped polygyny. Evolution 24:546–559
Bateson P (1983) Mate choice. Cambridge University Press, Cambridge
Boness DJ, Anderson SS, Cox CR (1982) Functions of female aggression during the pupping and mating season of grey seals, *Halichoerus grypus* (Fabricus). Can J Zool 60:2270–2278
Boness DJ, Bowen WD, Francis JM (1993) Implications of DNA fingerprinting for mating systems and reproductive strategies of pinnipeds. Zool Symp 66:61–93
Boyd IL, Duck CD (1991) Mass change and metabolism of territorial male Antarctic fur seals (*Arctocephalus gazella*). Physiol Zool 64:375–392
Bradbury JW, Andersson MB (1987) Sexual selection: testing the alternatives. Wiley, New York
Candolin U (2003) The use of multiple cues in mate choice. Biol Rev 78:575–595
Carey PW (1991) Resource-defense polygyny and male territory quality in the New Zealand fur seal. Ethology 88:63–79
Clutton-Brock TH (1989) Mammalian mating systems. Proc Roy Soc Lond B236:339–372
Cox CR, Le Boeuf BJ (1977) Female incitation of male competition: a mechanism in sexual selection. Am Nat 111:317–335
Emlen ST, Oring LW (1977) Ecology, sexual selection and the evolution of mating systems. Science 197:215–223
Francis JM, Boness DJ (1991) The effect of thermoregulatory behaviour on the mating system of the Juan Fernandez fur seal, *Arctocephalus philippii*. Behaviour 119:104–126
Goldsworthy SD, Boness DJ, Fleischer RC (1999) Mate choice among sympatric fur seals: female preference for conphenotypic males. Behav Ecol Sociobiol 45:253–267
Goldsworthy SD, McKenzie J, Page B et al (2009) Fur seals at Macquarie Island: post-sealing colonisation, trends in abundance and hybridisation of three species. Polar Biol 32 (10):1473–1486
Harcourt RG (1990) Maternal influences on pup survival and development in South American fur seal. PhD thesis, Cambridge University
Harcourt R (1992) Maternal aggression in the South-American fur-seal in Peru. Can J Zool Revue Canadienne De Zoologie 70:320–325
Heath CB (1989) The behavioral ecology of the California sea lion. PhD, University of California, Santa Cruz
Hofmeyr GJG, Bester MN, Jonker FC (1997) Changes in population sizes and distribution of fur seals at Marion Island. Polar Biol 17:150–158
Kingston JJ, Gwilliam J (2007) Hybridization between two sympatrically breeding species of fur seal at Iles Crozet revealed by genetic analysis. Conserv Genet 8(5):1133
Kirkpatrick M (1987) Sexual selection by female choice on polygynous animals. Annu Rev Ecol and Syst 18:43–70
Kirkpatrick M, Ryan MJ (1991) The evolution of mating preferences and the paradox of the lek. Nature 350:33–38
Kuroiwa MI, Majluf P (1989) Do South American fur seals in Peru lek? In: Abstracts of biennial conference for marine mammals. Society for Marine Mammalogy, pp 36
Lancaster ML, Bradshaw CJ, Goldsworthy SD et al (2007) Lower reproductive success in hybrid fur seal males indicates fitness costs to hybridization. Mol Ecol 16(15):3187–3197
Lancaster ML, Goldsworthy SD, Sunnucks P (2010) Two behavioural traits promote fine-scale species segregation and moderate hybridisation in a recovering sympatric fur seal population. BMC Evol Biol 10(1):143
Lunn NJ, Boyd IL (1991) Pupping-site fidelity of Antarctic fur seals at Bird Island South Georgia South Atlantic Ocean. J Mammal 72:202–206

Marshall TC, Slate J, Kruuk LEB et al (1998) Statistical confidence for likelihood-based paternity inference in natural populations. Mol Ecol 7:639–655

Maynard Smith J (1987) Sexual selection: a classification of models. In: Bradbury JW, Andersson MB (eds) Dahlem workshop on sexual selection: testing the alternatives. Wiley, New York, pp 9–20

Mayr E (1963) Animal species and evolution. Harvard University Press, Cambridge

McCann TS (1980) Territoriality and breeding behaviour of adult male Antarctic fur seal *Arctocephalus gazella*. J Zool (Lond) 192:295–310

McCann TS (1987) Male fur seal tenure. In: Croxall JP, Gentry RL (eds) Status, biology, and ecology of fur seals. National Oceanic and Atmospheric Administration Technical Report NMFS, vol 51, pp 197–198

Miller EH (1975) Social and evolutionary implications of territoriality in adult male New Zealand fur seals, *Arctocephalus forsteri* (Lesson, 1828), during the breeding season. Rapports et Proces-Verbaux des Reunions, Conseil International pour L'Exploration Scientifique de la Mer 169:170–185

Page B, Goldsworthy SD, Hindell MA et al (2002) Interspecific differences in male vocalizations of three sympatric fur seals (*Arctocephalus* spp.). J Zool 258:49–56

Ptacek MB (2000) The role of mating preferences in shaping interspecific divergence in mating signals in vertebrates. Behav Process 51:111–134

Rosenthal GG, Ryan MJ (2011) Conflicting preferences within females: sexual selection versus species recognition. Biol Lett 7:525–527

Ryan MJ (1985) The Túngara Frog: a study in sexual selection and communication. University of Chicago Press, Chicago

Ryan MJ (1994) Mechanisms underlying sexual selection. In: Real LA (ed) Behavioral mechanisms in evolutionary ecology. University of Chicago Press, Chicago, pp 190–215

Stirling I, Warneke RM (1971) Implications of a comparison of the airborne vocalizations and some aspects of the behaviour of the two Australian fur seals, *Arctocephalus* spp., on the evolution and present taxonomy of the genus. Aust J Zool 19:227–241

Trillmich F (1987) Galapagos fur seal, *Arctocephalus galapagoensis*. In: Croxall JP, Gentry RL (eds) Status, biology, and ecology of fur seals. National Oceanic and Atmospheric Administration Technical Report NMFS 51

Yasukawa K (1981) Male quality and female choice of mate in the red-winged blackbird (*Agelaius phoeniceus*). Ecology 62:922–929

Part IV
Cognitive Behavior, Communication, Personality and Behavioral Development

Chapter 14
Vocal Communication in Otariids and Odobenids

Isabelle Charrier

Abstract Otariids and Odobenids produce copious vocal signals in air and/or under water. The two main behavioral contexts in which they produce vocalizations are during mother-young interactions and male breeding vocal behavior. The social bond between mother and her young is very strong in both otariids and walrus, and their vocal exchanges start at birth. Individual vocal stereotypy has been found in both mothers and pups, and individual vocal recognition seems to be mutual in most otariids species, with the onset of individual identification being established soon after parturition in females, and within 2 weeks of birth in pups. The individual vocal signatures deciphered for mothers and pups involve temporal and spectral features, allowing for a large array of possible vocal 'codes' that can be used to help avoid confusion among individuals. Otariids and walrus males show greater vocal diversity than females, and produce calls both in air and under water, with significant individual vocal stereotypy allowing individual recognition among rivals and mate choice for females. Recognition capabilities appear to depend on their breeding strategies, with males in territorial species that vocally discriminate between neighbors and strangers, whereas mate-guarding males seem to use phenotype-related acoustic cues to assess their rivals.

Keywords Animal communication · Vocalizations · Mother-young recognition · Male competition · Individual signature

I. Charrier (✉)
Institut des Neurosciences Paris-Saclay, Université Paris-Saclay, CNRS (UMR 9197), Université Paris-Sud, Orsay, France
e-mail: isabelle.charrier@u-psud.fr

C. Campagna, R. Harcourt (eds.), *Ethology and Behavioral Ecology of Otariids and the Odobenid*, Ethology and Behavioral Ecology of Marine Mammals,
https://doi.org/10.1007/978-3-030-59184-7_14

14.1 Introduction

14.1.1 Vocal Communication

Animal communication is defined as the transfer of information from an emitter to a receiver (Shannon and Weaver 1949). The receiver will thus adapt its behavior according to the function of the information encoded in the signal. Several sensory modalities can be involved in communication: touch, olfaction, vision and hearing. Sensory cues present different properties for propagation, persistence in the environment, localization and nocturnal use (Alcock 1998). Thus, depending on the biological function and spatial range of the information (i.e., public or private information) encoded in a signal or cue, animals preferentially use a given sensory modality. Vocal signals are efficient at both short and long range, and are easy to detect and localize. In vertebrates, vocal communication serves many vital functions such as territorial defense, mate attraction, parent-offspring interactions, predator alert and group activity coordination (Halliday 1983; Bradbury and Vehrencamp 2011) . To better understand vocal communication systems, it is important to take into account the characteristics of the transmission channel such as environmental constraints: level of biotic and abiotic background noise, obstacles for sound transmission, and risk of confusion with other acoustic signals, including from intra- or inter-specific individuals. Depending on the environmental constraints, animals develop different strategies to optimize their vocal communication.

14.1.2 Vocal Repertoire and Vocalizations

The vocal repertoire represents the different call types produced by a species, with each call type being characterized by its acoustic structure across three dimensions: time, frequency and amplitude. The name of a call type is either related to the acoustic structure or sound of the call (noisy call, whistle, wop, trill, bell sound. . .) or to its biological function (contact call, food call, alarm call, territorial call. . .). When referring to the vocal repertoire, it includes only sound produced by the vocal tract, and thus excludes sound produced by another part of the animal or made with an object, such as wing flapping in birds, tail flapping for birds and marine mammals, knee clicking in antelopes, beak drumming in birds, drumming in primates (Bro-Jørgensen and Dabelsteen 2008; Remedios et al. 2009; Bradbury and Vehrencamp 2011).

Vocalizations may be described by their acoustic properties (Fig. 14.1): their frequency range (infrasound: <20 Hz, audible sound: 20 Hz to 20 kHz or ultrasound: >20 kHz), frequency bandwidth (broadband or narrowband), temporal organization (single or multi-unit calls, Fig. 14.1a), spectral content (noisy or tonal, i.e., showing an harmonic structure, Fig. 14.1b, d), duration (short or long), the presence of temporal modulation (variation of the frequency or the amplitude over time:

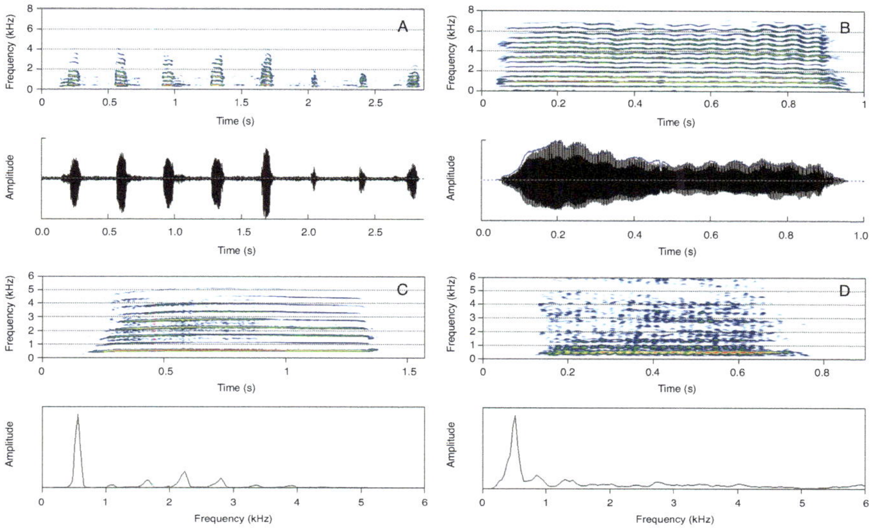

Fig. 14.1 Different visual representations of vocalizations produced by males, pups and females: spectrograms (**a–d**), oscillograms (**e**, **f**) and energy spectra (**g**, **h**). (**a**, **e**) Spectrogram and oscillogram of a series of barks produced by a subantarctic fur seal male; barks are short and tonal calls composed of a fundamental frequency and its series of harmonics. (**b**, **f**) Spectrogram and oscillogram of a mother-attraction call produced by a 1-week old subantarctic fur seal pup, this call is a tonal call with a fundamental frequency and its series of harmonics, and some frequency modulations (i.e., variation of frequency over time) can be seen on the spectrogram (quavering occurring during the first 2 weeks of life), and amplitude modulation (i.e., variation of amplitude over time) can be seen on the oscillogram (the blue line represents the smooth amplitude envelope). (**c**, **d**, **g**, **h**) Spectrograms and average energy spectra of pup-attraction calls produced by Australian sea lion females: the female call shown in **c** is tonal (clear harmonic structure) whereas the female call shown in **d** is noisy (no harmonic structure), and the average spectrum (**g**, **h**) presents the distribution of energy among the frequencies of the call. On the average spectrum of the female tonal call (**g**), the fundamental frequency is represented by the first energy peak and is also the frequency with the highest energy

frequency or amplitude modulation, FM or AM respectively, Fig. 14.1b, f). These characteristics are defined using the three main graphical representations of sound (Fig. 14.1): spectrogram (frequency as function of time, and amplitude coded on a grey or color scale, Fig. 14.1a–d), oscillogram (amplitude as a function of time, Fig. 14.1e, f) and energy spectrum (amplitude as a function of frequency, Fig. 14.1g, h).

14.1.3 Information and Functions of Vocalizations

The description of the vocal repertoire is an important step to understand the role and function of each vocalization, and thus to assess the information embedded in the signal. A variety of information about the sender can be conveyed in vocalizations: identity of the species, group, individual, sex, but also age, body size, motivation and social status (Halliday 1983). Information can be addressed to particular receivers ("private" information) or to a large audience ("public information"), and it needs to be perceived at short or long distance (Halliday 1983). Depending on the address and on the range of detection, information will be encoded in different acoustic features as they present different propagation properties. For instance, in songbirds, the song has the function to both attract a mate and to advertise the territory ownership to rivals for that mate. Primary information is the identity of the species (to attract mates and detect rivals) and of the individual (to discriminate among neighboring and unknown rivals). Species identity is long range information that needs to be encoded using propagation-resistant acoustic features such as slow frequency modulation, whereas individual identity tends to be short-range information (Mathevon et al. 2008). Considering both ecological and environmental constraints, animals encode essential information within different types of acoustic features (i.e., temporal or spectral), and may also use a number of acoustic features to secure the code (redundancy of information, (Shannon and Weaver 1949)).

When studying animal vocal communication, it is essential to understand the function of a vocalization. In an important first step, function is deduced from the context in which the sender is producing it and from the behavioral reaction of the receiver (Shannon and Weaver 1949). Acoustic analyses aim to determine the information encoded in the signal, providing the opportunity to hypothesize about the function of the vocalization. However, only playback experiments can definitively confirm the biological function of the signal, thereby allowing researchers to investigate the perceptual abilities of the receiver (McGregor 1992)

14.2 Vocal Communication in Otariids and Odobenids

Most otariids come on land to breed and all give birth to a single pup, whereas odobenids mate in water during winter and females give birth to a single calf on ice in summer (Riedman 1990; Renouf 1991). Both otariids and odobenids gather in dense colonies or groups, characterized by loud noise levels arising from intensive vocal activity of mothers and their pups, combined with sounds produced by adult males defending their harems or mate-guarding females. Most often, vocalizations are described and named by onomatopoeic words such as grunt, bark, roar, bell-like sound; or by the behavioral context of their production, such as contact, attraction, threat, submissive calls. Depending on their social structure (low to dense colony), mating system (low to high level of polygyny, territorial to mate guarding strategies), and habitat (sandy beaches, boulders, rocky shelves), the vocal repertoire of types and numbers of calls varies among otariids species. The nomenclature of aerial calls is not homogenous in the literature, which can make it difficult to compare across species (see Table 1 in Phillips and Stirling 2001); however, similarities in their call structure along with the behavioral context in which they are produced do facilitate inter-species comparison.

The two main vocal behaviors studied in animals are (1) parent-offspring vocal interactions and (2) breeding vocal behavior. In this chapter, I give an overview of mother-pup vocal communication in otariids and odobenids, as well as male vocal behavior linked to their breeding activities.

14.3 Mother-Pup Vocal Communication

14.3.1 Vocal Repertoire

14.3.1.1 In Otariids

In contrast to most phocids, otariid females are income breeders: they cannot store enough energetic resources to fast during the entire lactation and thus they need to forage at sea throughout the entire pup provisioning period (Chap. 3 by Sepulveda and Harcourt). Females alternate foraging trips with suckling periods on land for a lactation that lasts from 4 to 24 months depending on the species (Riedman 1990; Chap. 3 by Sepulveda and Harcourt). At each of their returns from the ocean, females need to find their pup in the colony and so they call to their young, starting while they are still in the water. Observations of returning females showed that otariid females are highly vocal in spite of being exhausted from their foraging trip which may last more than 3 weeks during winter months in subantarctic (*Arctocephalus tropicalis*) (Guinet and Georges 2000) and New Zealand fur seals (*Arctocephalus forsteri*) (Harcourt et al. 2002). Females produce loud pup attraction calls or pup contact calls (PAC or PCC respectively, the nomenclature varies depending on studies, Fig. 14.2) until they have reunited with their own pup. Often one can see several hungry pups that call back to the returning female and produce mother attraction calls or mother contact calls (MAC or MCC) but usually only one pup keeps calling back and approaches the calling female. When non-filial pups approach a non-mother female, they can be either gently threatened by an open mouth display or a puff sound, or they can be bitten and at times seriously injured (Harcourt 1992). When in close proximity, the female confirms the identity of the pup by performing nose to nose contact and then accepts its vigorous demands to suckle (Trillmich 1981; Renouf 1991).

The vocal exchange between mother and her pup starts as soon as pup birth. Immediately after parturition, otariid mothers and their offspring produce calls at a high rate and so imprint themselves to their respective voices. These mother-pup calls show some grading as they are used in different contexts across mother-young interactions, from reunions to close affirmation of the relationship. They can be produced at higher and lower amplitude, with the distribution of energy among frequencies varying from low frequency for close contacts to high frequency during reunions or when interacting with other conspecifics such as agonistic interactions

Fig. 14.2 Australian sea lion female dragging the skin of her long-dead pup through the colony. Photo Credits: I. Charrier

with juvenile or adult males or with other females. For instance, when in close contact, mothers and pups produce calls similar to those produced during reunions but at a much lower amplitude, with the mouth closed or open. Similar low-pitched and low-amplitude calls have been found in other mammal species such as sheep (Sebe et al. 2010), ungulates (Kiley 1972) and felids (Peters 2002). It has been suggested that these vocalizations are an expression of maternal responsiveness with a calming effect on the young (Vince 1993). The bond between the mother and her pup is very strong, and a female usually still calls her dead pup. Some females stay close to their dead pup's body and defend them for weeks. I have seen an Australian sea lion (*Neophoca cinerea)* mother dragging the skin of her long-dead pup through the colony and down to the water during hot hours, and then carrying it back to the bushes later as the day cools, just as females do with their live pups (Fig. 14.2).

Across all otariid species, female and pup calls show the same general structure (Fig. 14.3). They produce complex calls composed of a fundamental frequency and its series of harmonics. Most of the time calls present a frequency modulation pattern and show an inverted-U or chevron-like shape i.e., ascending FM at the beginning followed by a plateau, and ending with a descending FM (see Fig. 14.3). Calls sometimes present a broadband noisy part and a complex amplitude and frequency modulation (AM/FM pattern) i.e., a fast AM matching the frequency of the FM depending on the species (as shown in Fig. 14.3).

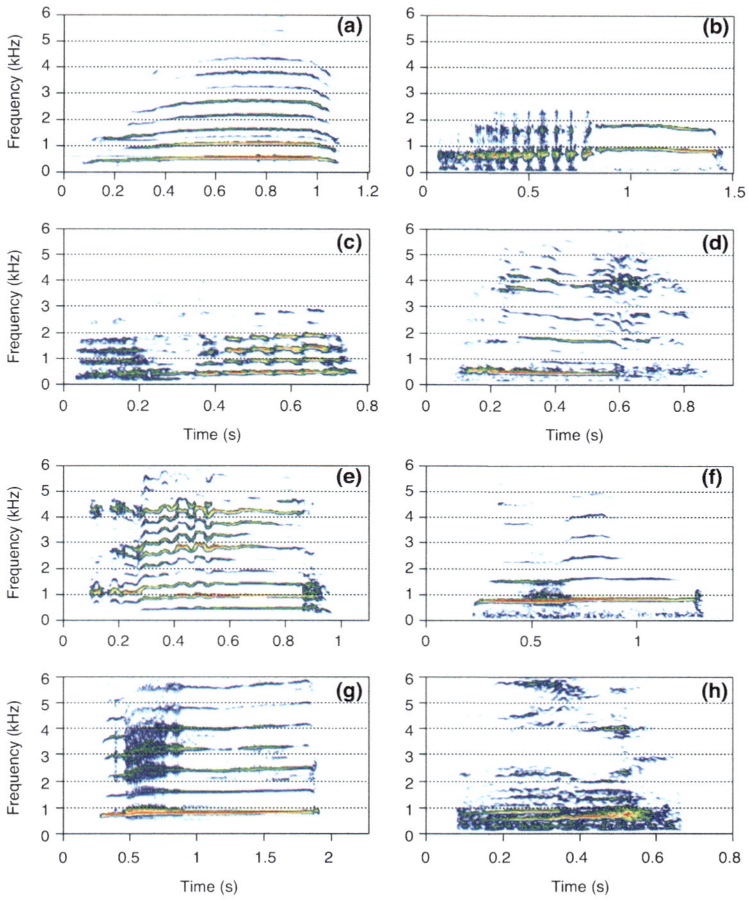

Fig. 14.3 Calls produced by females and pups from four different species of otariids, subantarctic (**a**, **e**) and Antarctic fur seals (**b**, **f**), Australian (**c**, **g**) and Galapagos sea lions (**d**, **h**). Recordings Credits: I. Charrier (Subantarctic fur seal and Australian sea lion), T. Aubin (Antarctic fur seal), Galapagos sea lion project of the University of Bielefeld (Galapagos sea lion)

14.3.1.2 In Odobenid

Walrus (*Odobenus rosmarus*) mothers are among the most protective of mammalian mothers and this strong maternal attachment to the calf extends to females being very attentive and responsive to their calves' calls. Anecdotal stories from Inuit hunters state that mothers have chased hunters' boats for many kilometres to retrieve their dead calf. Walrus females defend their calf even against polar bears. Similarly, calves stay around the mother even if it has been killed by hunters. This strong maternal bond can persist for several years after weaning (Knudtson 1998; Nowak 2003).During lactation mothers and calves almost never separate, even when the

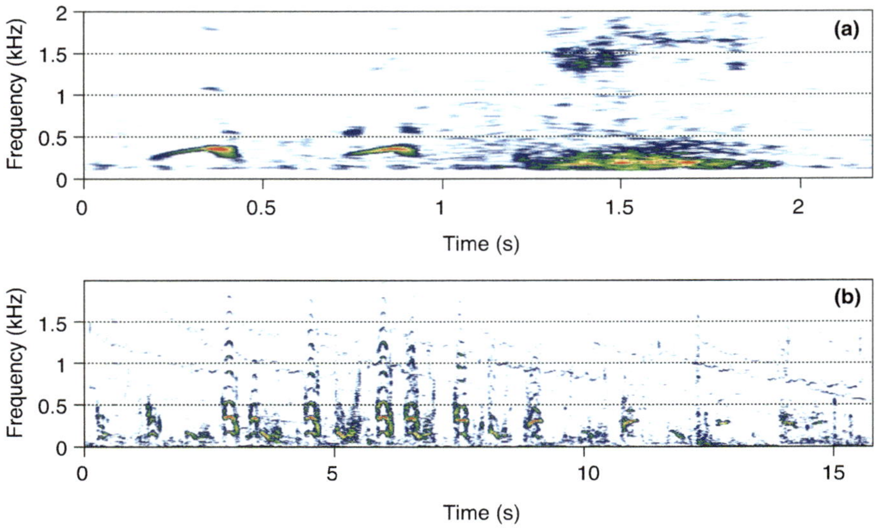

Fig. 14.4 Atlantic walrus mother-calf vocal exchanges (barks) (**a**) in air (2 barks from a calf followed by its mother's bark, and **b**) underwater (higher-pitched barks are produced by the calf, bearded seals trills can be seen in the background). Recordings Credits: I. Charrier

mother forages at sea, so there are no lengthy periods of separations as found in otariids. However, short separations can occur during group movements on unstable floating ice, because of predator attacks or when disturbed by approaching hunters, and occasionally from group movements during foraging. During these separations, mothers and calves produce a series of barks that help them to find each other (Figs. 14.4 and 14.8d). While in close contact females produce single barks at very low amplitude to comfort the calf, these "soft" barks are produced with the mouth closed. While under water, mothers and calves also produce barks to maintain contact (Fig. 14.4).

14.3.2 Vocal Stereotypy and Evidence for Individual Recognition

As soon as they give birth, otariid and walrus mothers are highly vocal with their young, and they exchange vocalizations even when in close proximity. This suggests that vocal imprinting occurs between the mother and her neonate. In otariids, females stay with their pup for about 10–15 days after birth (perinatal period), and then alternate foraging trips at sea with suckling periods ashore. On each return to the colony, females call to relocate and reunite with the pup. Observations of mother-pup reunions in different species suggest that mothers and pups use vocalizations to recognize each other (Bartholomew 1959; Paulian 1964; Rand 1967; Bonner 1968;

Peterson and Bartholomew 1969; Stirling 1972; Sandegren 1970; Stirling 1970; Marlow 1975; McNab and Crawley 1975; Trillmich 1981; Fay 1982). This vocal recognition needs to be efficient shortly after birth and certainly before the first mother-pup separation. A prerequisite for individual vocal recognition is the presence of an individual signature in vocalizations. Many studies have assessed the level of individuality in female and pup calls, and all species studied have exhibited individual stereotypy; however, some species show higher levels of individuality than others (see Table 14.1). While individual-specific vocalizations are mandatory for mother-young recognition, the production of individual calls does not define the presence of individual recognition, rather experimental evidence has to be provided using the playback experiment paradigm. Seven otariid species and one walrus subspecies have been tested so far, but in the majority, only the mother or the pup has been tested. In three species, both members of the dyad have been tested (Table 14.1). Evidence of mutual mother-pup vocalizations comes from northern fur seals (*Callorhinus ursinus*) (Insley 2000, 2001), subantarctic fur seals (Charrier et al. 2002, 2003a), and Australian sea lions (Charrier et al. 2009; Pitcher et al. 2012). In the Atlantic walrus, barks produced by mothers and calves have been shown to be individual-specific and field playback experiments have shown that females can discriminate the calls of their calf from others (Charrier et al. 2010). It is likely that vocal recognition is mutual in this species, but this still remains to be experimentally investigated in walrus calves.

Considering the selective pressures for mother-pup vocal recognition in otariids and odobenids, it is likely that all species show mutual vocal recognition. Differences that may occur across species are likely in the ontogeny of recognition and in characteristics of the individual vocal signature, such as the acoustic features involved in recognition.

14.3.3 Ontogeny of Vocal Recognition

As otariid mothers leave soon after birth to forage at sea, individual recognition has to be established before the first separation. Spatial cues can be used by the mother to find her pup again, for example the birth site or the last suckling spot (Bartholomew 1959; Trillmich 1981; Phillips 2003). However, in dense colonies, spatial cues of this nature may be insufficient to ensure a mother will reliably reunite with her pup, and so vocal recognition becomes important. While vocal recognition has been shown in several otariid species, investigation of the establishment of this discrimination is limited to four species. Using playback experiments, it has been demonstrated that Galapagos sea lion pups (*Zalophus wollebaeki*) recognize their mother's voice 10–30 days after birth (Trillmich 1981), Galapagos fur seal pups (*Arctocephalus galapagoensis*) 10 days after birth (Trillmich 1981), subantarctic fur seal pups 2–5 days after birth (Charrier et al. 2001) and Australian sea lion pups between 10 days and 2 months (Pitcher et al. 2009). Ecological constraints of colony density and duration of maternal absence vary among these species with fur seals

Table 14.1 Vocal Stereotypy and mother-pup recognition of the studied species in regards to their ecological constraints and selective pressures for individual recognition

Species	Pop density	Mating system	Sel. pres. for IR	Cl. rate mothers	Cl. rate pups	Recognition tested and occurring	References for classification rates	References for playback experiments
Subantarctic FS	H	Extreme P	H	84	83	Mutual	Page et al. (2002)	Charrier et al. (2001, 2002)
Northern FS	H	Extreme P	H	83	79	Mutual	Insley (1992)	Insley (2000, 2001)
Antarctic FS	H	Extreme P	H	74	52	Mother by pup	Page et al. (2002)	Aubin et al. (2015)
South Am FS	H	P	H	70	60	–	Phillips and Stirling (2000)	–
Australian FS	H	Extreme P	H	76	75	–	Tripovich et al. (2006)	–
New Zealand FS	H	P	H	88	67	–	Page et al. (2002); Dowell (2005)	–
	H	Extreme P	H	71	–	–	Campbell et al. (2002)	–
Steller SL	H	P	H	95	89	–	Fernandez-juricic et al. (1999)	–
South Am SL				63	53	–		–
	H	P	H	–	–	Mother by pup[a]	Trimble and Charrier (2011)	Hanggi (1992)
California SL	H	P	H	–	–	Mother by pup	–	Trillmich (1981)
Galapagos FS	MM-L	Moderate	M	–	–	Mother by pup	–	Trillmich (1981)
Galapagos SL		PModerate	M	65	77	Mutual	Charrier and Harcourt (2006)	Charrier et al. (2009); Pitcher et al. (2012)
Australian SL		P					–	

Atlanticwalrus	H	P	H	64	67	Pup by mother	Charrier et al. (2010) (DFA)	Charrier et al. (2010)
	H			74	71		Charrier et al. (2010) (ANN)	
Pacific walrus		P	H	–	–	Mutual[a]		Charrier Unpubl. data

Population density: H, high; M, moderate; L, low; mating system: P, polygyny; Selective Pressures for Individual Recognition: H, high; M, moderate; Classification Rate in %

[a]Indicates experiments on captive animals (n = 1)

showing the strongest constraints. Accordingly, this corresponds to the species in which individual vocal recognition of the mother by the pup develops the earliest and most importantly before the first mother-pup separation, at least for the subantarctic fur seal. For the Galapagos fur seal, pups have not been tested at earlier than 10-days old, so we do not yet know if vocal recognition is established before the first separation from the mother. The two sea lion species inhabit low density colonies which lowers the risk of confusion among individuals compared to fur seals, and this may explain why recognition is established at a later stage. Later vocal recognition onset in pups appears to be compensated by early vocal recognition of the pup by the mother, and observations of Galapagos sea lions and fur seals, as well as subantarctic fur seals suggest that females can discriminate their pup's voice a few hours after birth (Trillmich 1981, Charrier unpublished data). Exact onset of pup's voice recognition by its mother has only been systematically investigated in the Australian sea lion, and in an elegant series of experiments, Pitcher et al. (2010b) showed that females can discriminate between calls from their own pup and an unrelated pup within 48 h of parturition. It appears that in species that show mutual vocal recognition, mothers are the first in the dyad to develop recognition. The time difference in onset of recognition between females and pups may vary with selective pressures for individual recognition and a combination of ecological factors (colony density, maternal attendance). In the walrus, such mother-pup vocal recognition is likely to be established very soon after birth as they live in a very unstable environment (pack ice). The likelihood is high of being separated by movements of the herd, of ice or during bad weather conditions, or when a stampede is induced by the presence of predators. However, the ontogeny of recognition in walrus remains to be experimentally investigated.

14.3.4 Vocal Development and Long-Term Recognition

As the duration of lactation ranges from 4 to 24 months in otariids, calls produced by offspring change gradually with age, and significant changes in their spectral characteristics occur (Fig. 14.5). Call features are linked to the anatomy (the source-filter theory, (Titze 1994)) and spectral characteristics ("filter cues") vary with age as do the fundamental frequency and duration features ("source cues"). For instance, calls of subantarctic fur seal pups have been collected from birth to the age of 7 months (Charrier et al. 2003b), and there are significant changes in the distribution of energy among frequencies, with calls of older pups showing lower frequencies than those of young ones. In addition, the frequency modulation pattern changes, with quavering disappearing after 2 weeks (see Fig. 14.5). This rapid change in call structure suggests that mothers have to permanently learn the characteristics of their pup's voice, and update the memorised template to recognize their offspring. This learning and memorization process is highly efficient and appears long lasting in otariid species, with females not only able to remember and recognize different versions of their pups' calls within a breeding season (Charrier et al.

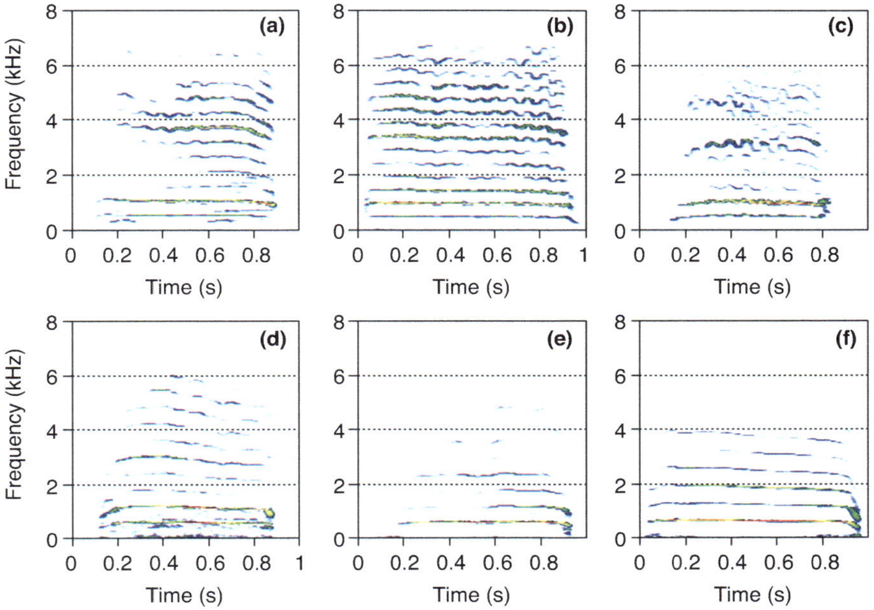

Fig. 14.5 Subantarctic fur seal pup calls of a given individual at different ages (**a**: 1 day, **b**: 1 week, **c**: 2 weeks, **d**: 1 month, **e**: 3 month and **f**: 7 month). Recordings Credits: I. Charrier

2003b), but also to remember the calls of their young several years after weaning (Insley 2000). The young can also remember their mothers' voice several years after the separation, demonstrating that imprinting is long-lasting for both (Trillmich 1981; Insley 2000; Pitcher et al. 2010a). This type of long-term vocal recognition has rarely been demonstrated in wild mammals, the only other example apart from otariids being the highly social African elephant (*Loxodonta africana)* (McComb et al. 2000). This long term memory ability may be a common trait across otariids, and possibly also in the walrus given their impressive social bonds between mothers and calves.

14.3.5 Individual Vocal Signature

Finally we may ask how do mother and pups actually recognize each other? In other words, what are the acoustic features used to make vocal identifications? To answer this question, researchers have performed playback experiments involving modifying calls or broadcasting synthetic calls in which a given acoustic feature has been either modified or removed. In this experimental setup, the behavioral response of the tested animal provides evidence of whether an acoustic feature is essential: if a call broadcast but with this feature removed or modified it still triggers a strong vocal response, then that specific parameter is not used in the identification process. However, if the modification results in diminishment of the behavioral response from the animal, then this parameter is essential for individual recognition. To date, only three otariid species have been tested in this way: the subantarctic fur seal (both mothers and pups), Antarctic fur seal (*Arctocephalus gazella*) (pups only), and Australian sea lion (both mothers and pups). In each of these species, experiments have shown that mothers and pups use a multi-parameter vocal signature to secure the code of the individual identity and so ensure a reliable identification process. Use of multiple parameters mitigates against circumstances where one given parameter might not be entirely reliable, as the second one then secures identification. The most important parameters involved in this recognition process are FM, AM and spectral features (SF; including pitch and timbre) (Charrier et al. 2002, 2003a, 2009; Pitcher et al. 2012, Aubin et al. 2015). However, some differences occur among the studied species, showing high to moderate selective pressures for individual recognition. The two species showing high selective pressures for individual recognition the subantarctic and the Antarctic fur seals use a vocal signature in the time domain based on an amplitude-time (AM) and/or frequency-time (FM) analysis and in the frequency domain based on a timbre analysis (i.e., repartition of energy among the different frequencies of the call), whereas the species with moderate selective pressure (i.e., Australian sea lion) use a vocal signature in the time domain based on AM/FM analysis and in the frequency domain based on a pitch analysis (by analyzing the exact frequency values of the calls, based on the fundamental frequency of the call). In terms of coding possibilities, an individual identity coding based on temporal features such as AM and FM offers a larger set of individual signatures as it combines two dimensions (amplitude and time, frequency and time, respectively) (Greenwalt 1968).A code based on spectral features is constrained by the frequency bandwidth that can be produced by each species. If a timbre coding (i.e., distribution of energy among frequencies) offers a wide range of coding possibilities, a pitch coding shows limited coding possibilities as the fundamental frequency values among individuals are within the same frequency range. It turns out that the species with the highest constraints use an individual vocal signature, offering more combinations and thus limiting the risk of confusion among individuals.

The acoustic parameters within the individual signature differ in their propagation properties—that is, how far they transmit before becoming degraded. With these differences in propagation, the range over which individuals are able to recognize each other is limited, particularly in complex environments. Propagation tests have been performed for calls of different species, and by testing the degradation of the different parameters of the individual signatures in their natural environment, it is possible to determine the theoretical effective range of individual identification. We know that some animals have the ability to extract acoustic signals from noise and these will show detection and recognition range greater than the theoretical propagation range tested. For instance, the King penguin (*Aptenodytes patagonicus*) exemplifies the so-called cocktail-party effect, with mates detecting individual calls from a background of thousands of calling penguins (Aubin and Jouventin

1998). Experimental assessment of call propagation ranges for both otariid and odobenid species show that call propagation encompasses the natural range of mother-pup reunion or is much greater. For instance, in Australian sea lions, mother and pup calls reliably propagate up to 32 and 64 m, respectively, and under natural circumstances, mothers and pups exchange vocalizations over distances of up to 50 m (Charrier et al. 2009; Pitcher et al. 2012). In the Atlantic walrus, individual vocal signatures of barks from mothers and calves efficiently propagate up to 16 and 32 m over ice and up to 128 m above water. These propagation ranges are far above the observed distance at which a mother and her calf can be separated (usually within 5–20 m) (Charrier et al. 2010). Depending on the species, the individual vocal signature shows different propagation efficiency or coding possibilities; however, it appears that such differences are always adapted to the ecological constraints of each species.

14.4 Vocal Communication in Males

14.4.1 The Vocal Repertoire

14.4.1.1 Otariids

Otariid males produce a high diversity of call types, and in some species, males also produce vocalizations underwater. Aerial call types include bark, growl, roar, guttural threat call, high-pitched or full threat call, submissive call, and exhalations. All these calls are mainly produced during the breeding season, while interacting agonistically with other males, holding their harems or mate-guarding a female, before or during a fight, or while trying to escape rivals. Males also produce vocalizations while interacting with females during territorial or mate-guarding activities, and even during the act of mating. Non-vocal productions such as exhalations and puff sounds (air expelled through the mouth and the nostrils, respectively) are often produced agonistically towards a conspecific, or a human. All these vocal productions are composed of frequencies up to 8 kHz, with most of the energy below 4 kHz.

Barks are the most commonly produced calls: they are used by males when holding their harems and to warn rivals off from approaching their females, as well as when males interact with females of their harem or females they mate-guard. A bark is a short vocalization always produced in series (Fig. 14.6), either tonal showing a clear harmonic structure (e.g., in subantarctic fur seal, Fig. 14.6a), or noisy (e.g., Australian sea lion, Fig. 14.6b). The repetition rate of barks varies with the behavioral context and tends to accelerate when the threat level increases (Roux 1986; Tripovich et al. 2008b; Charrier et al. 2011a). Barks can even be produced when males are sleeping (Ahonen et al. 2014).

Territorial calls (named differently according to species: examples include territorial call, trumpeter roar, full threat call, high-pitched calls) are high-pitched

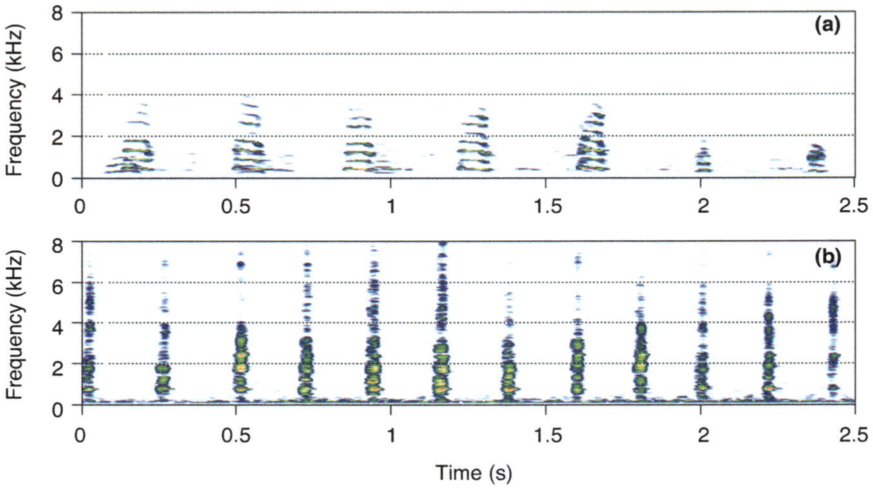

Fig. 14.6 Barks produced by (**a**) subantarctic fur seal and (**b**) Australian sea lion males. Recordings Credits: I. Charrier

vocalizations sounding somewhat like a wolf howl, and are produced by males while holding their harem or territory or when mate-guarding. This vocalization seems to be produced to advertise the presence of the male to other rivals, as well as in response to other males vocalizing in the vicinity.

Submissive calls can be also heard in the colony when a male is defeated in a fight with a rival or when avoiding future conflict. In subantarctic fur seals, juvenile males produce a succession of submissive calls while running through the crowded colony to reach the water. Despite these submissive calls, juveniles still get bitten by territorial males, but bulls usually do not attempt to follow them and instead stay in their territory/harem.

Finally, a female-like call (contact call or pup attraction call) can be produced by males. This is particularly common in Australian sea lions in which males mate-guard females. During the prolonged breeding season (see Chap. 26 by McIntosh and Pitcher) males search for pregnant females or for females that have just given birth. Production of the female-like call seems to be a trick to detect perinatal pups as they respond to any female calls, and this may help males locate the mother of the pup and thus a potential mate.

California (*Zalophus californianus*) and Steller sea lion (*Eumetopias jubatus*) and northern fur seal males produce underwater barks, clicks, whinny sounds, buzzes, bangs and belching (sea lions (Schusterman and Balliet 1969; Schusterman et al. 1970); northern fur seals (Cummings and Fish 1971)). Underwater barks have similar structure and temporal organization as airborne barks. Clicks have been recorded with repetitions rates ranging from 5 to 80 pulses per second, buzzes are fast successions of pulses, and belching sounds are defined as series of low-frequency pulses.

Most of these vocal productions are linked to male territorial defense activity that seems to be expanded in the water, and thus interaction with other males. The production of clicks were suggested to be involved in echolocation and detection of prey while foraging (Poulter 1963) but this hypothesis has long been refuted and clicks are involved in social communication among males (Schusterman et al. 2000; see Chap. 1 by Gentry). Further investigations in other otariid species are needed to assess if underwater production is only limited to these three species, and if there are no other vocalizations produced while foraging, to coordinate hunting or to alert predator, as the latter could be true especially for species under high predation pressure especially by killer whales (*Orcinus orca*) and sharks.

14.4.1.2 In Odobenids

Walruses are among the most vocal of the pinnipeds. They produce sounds when hauling-out on land or ice, when moving at the water surface, and underwater. Walrus males produce an amazing underwater courtship display during the breeding season occurring during winter months (February-March). Males produce knocks, taps, and bell sounds during the breeding season, and in stereotypic order (Fig. 14.7). Knocks and taps are broadband pulses with most energy below 4 kHz (see Fig. 14.7), whereas bell sounds are metallic tonal sounds that might involve laryngeal vocal fold vibrations and resonance of inflated pharyngeal pouches (Fay 1982; Reidenberg and

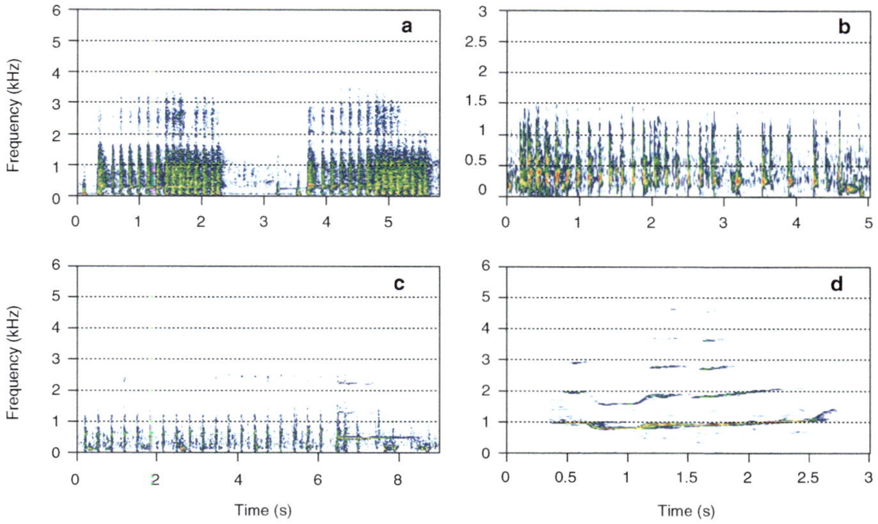

Fig. 14.7 Walrus vocalizations produced by Atlantic walrus males during their underwater courtship display (**a**: coda song, **b**: diving vocalization, **c**: bell-like sound), and aerial whistles by a captive Pacific walrus. Recordings Credits: I. Stirling (Atlantic walrus) & I. Charrier (captive Pacific walrus)

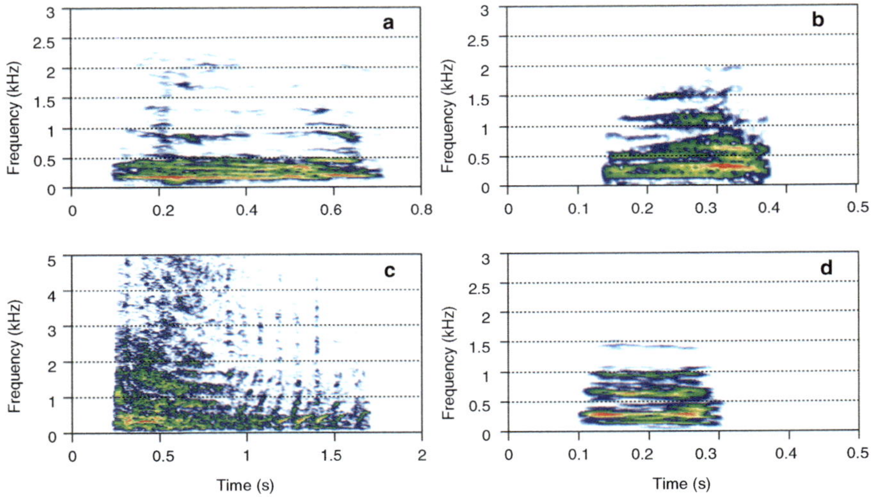

Fig. 14.8 Example of aerial vocalizations produced by adult male and female adult Atlantic walrus (**a**: roar, **b**: grunt, **c**: guttural sound, **d**: bark). Recordings Credits: I. Charrier

Laitman 2010). Atlantic walrus produce two main songs: the coda song and the diving vocalization song, and they differ from each other by their pattern of knocks, taps and bell sounds (Stirling et al. 1987; Sjare et al. 2003). During this courtship display, males also produce a single airborne vocalization, the whistle (Fig. 14.7), using their lips while surfacing. It is a tonal sound with a slow FM-pattern that seems to be individually specific and very stable over time (Charrier et al. 2011b).

Apart from the courtship display of males, walrus of both sexes and of all age classes produce contact calls in air and underwater while swimming or while hauling-out on ice to keep contact among conspecifics. These contact calls are barks, and underwater barks have the same general structure as aerial barks.

The aerial vocal repertoire of walrus is highly diverse, as they produce different types of threat calls, barks (contact calls), and non-vocal sounds (Miller 1985). Adult males and females produce three threat calls (Fig. 14.8): roars, grunts, and guttural sounds in order of increasing threat level. These threat calls are low-frequency calls with most energy below 400 Hz, but they differ in duration. Roars are produced when an unknown individual approaches the group as an alert call, grunts are most often produced in series prior to a more intense threat. Finally, guttural calls are often coupled with expiration and inspiration of breath and produced during high-intensity threats (Miller 1985).

The bark is the most common call type produced by adults and calves, and is a tonal sound composed of a fundamental frequency and its series of harmonics (see Figs. 14.3 and 14.8d). Barks can be produced in an affiliative context (between mother and calf, between social congeners), in submissive contexts between adults, or during stressful contexts such as mother-calf separation or isolation of an individual from the group. Depending on the social context, barks are produced at a more and less fast rate, with different amplitude levels, and showing more and less energy in higher frequencies (i.e., there is more energy in high frequencies when the stress level increases) similar to that found in terrestrial mammals (for review, see Briefer 2012).

Walrus also produce non-vocal sounds such as sneezes, flatulent sounds, coughing sounds and motorboat calls (Miller 1985). The exact social context and meaning of these sounds remain unknown, but they can be heard when walruses haul-out on land or on ice.

14.4.2 Vocal Recognition and/or Rival Assessment

In otariids, males have developed two main breeding strategies: territorial defense in which males hold territories and control a group of females with the size of the group varying across species, from 3 to 5 females on average to up to 60 females in the South African fur seal (*Arctocephalus pusillus pusillus*) (Nowak 2003), and mate-guarding in which males for a few days monopolize 1 or 2 females at a time until they mate. In walrus, Atlantic males breed in fast-ice polynyas and have developed female-defense polygyny, in which the largest and most dominant male has exclusive access to females and mates with many females during his tenure (Nowicki et al. 1997). In contrast, Pacific walrus males breed in pack ice, and show a lek-like system with several males displaying around a herd of females (Ray and Watkins 1975; Fay 1982). Depending on the breeding strategy developed by the males, management of intra-sexual interactions and evaluation of the rival will vary to meet the constraints of these breeding strategies.

14.4.2.1 In Otariids

Territorial males advertise the ownership of territory by producing territorial calls, mostly a series of barks (see Sect. 14.4.1.1 and Fig. 14.6). As in territorial songbirds, there is also evidence of vocal discrimination between neighbors and foreigners in territorial otariid males (Roux 1986; Tripovich et al. 2008a), namely a dear enemy effect (Temeles 1994), where males react more strongly to the barks of foreign males than those of neighboring males. The maintenance of the territory and the network of neighbors are based on an acoustic and spatial memorization of the territorial males. This strategy allows males that fast throughout the entire breeding season to not react to any male in the vicinity of their territories, but only potential rivals, and thus limit excessive energy expenditure. Neighbour/stranger vocal recognition seems to be based only on spectral call characteristics but this is known experimentally only from the Australian fur seal (*Arctocephalus pusillus dorifeus*) (Tripovich et al. 2008a). Observational studies carried out on other otariid species suggest that they too have individual recognition between neighbors/foreigners among territorial males (Gentry 1975; McCann 1980; Gisinier 1985; Fernández-Juricic et al. 2001; Kunc and Wolf

2008) especially as their territorial calls encode individuality (Roux and Jouventin 1987; Fernández-Juricic and Campagna 1999; Tripovich et al. 2011). For non-territorial males, rival assessment seems more complex and one observes different strategies or a mix of different strategies. For instance, in the Australian sea lion, a male monopolizes 1 or 2 females for a few days until mating (Chap. 26 by McIntosh and Pitcher), but as the breeding period lasts several months, and the density of the colonies are rather moderate, interactions between males are not very frequent, and so development for individual vocal recognition might not occur. Experimental work has shown that Australian sea lion males use information related to the male phenotype to gauge their rivals based on the spectral characteristics of the barks. Males reacted more strongly to "higher-pitched" barks mimicking "smaller" males, but do not respond to "lower-pitched" barks simulating larger males (Charrier et al. 2011a). Bark rate also influenced the behavioral response of males as they reacted strongly to a bark series with a fast rate but almost ignored the series with a slow rate. Grading in the bark rate has been reported in many otariid males: males produce their barks with a faster rate when interactions become more aggressive and/or precede a fight (Roux 1986; Fernández-Juricic et al. 2001; Kunc and Wolf 2008). So, the bark rate does not code a phenotypic trait of a given adult male, but rather his motivation and his level of aggressiveness/threat. Australian sea lion male barks also show a strong individuality (Gwilliam et al. 2008; Ahonen et al. 2014) so we cannot totally rule out the hypothesis that males cannot memorize the voice of rivals with whom they have previously interacted, and so use individual vocal recognition to assess competitors. Investigations are needed in more otariid species to improve our understanding of the communication network of otariid males and their recognition mechanisms (individual recognition and/or phenotype-linked vocal assessment).

14.4.2.2 In Odobenids

Our knowledge of rival assessment and vocal recognition in walrus males is quite limited. The courtship songs of Atlantic walrus males have been well described and individual variations have been found to reliably identify mature males by the structure of their surface codas and diving vocalizations (Stirling et al. 1987; Sjare and Stirling 1996; Sjare et al. 2003).However, there are limited behavioral observations and no experimental evidence for individual vocal recognition, nor if some songs' components can be used by males to assess body condition, fighting abilities or breeding success. Sjare et al. (2003) reported that Atlantic males singing the most extensively were those that gained access to females, and they seemed also to be the biggest with the largest blubber reserves, likely with the best body conditions. Taking into account the time spent by males singing during the breeding season, rival males and females have the opportunity to assess mature males with males being able to decide whether to challenge a rival, and with females potentially being able to choose a mate. Further investigations on both subspecies are needed to better understand their communication network. As breeding occurs during winter, and as weather and ice conditions are quite extreme in the Arctic, any behavioral and experimental study is a challenge, and thus the function of walrus singing behavior may remain a mystery to science.

A pilot study performed on two captive Pacific walrus showed that mature males produce individual whistles and were able to discriminate between the whistles (aerial part of the acoustic courtship) of a familiar male and those of a foreign male (Charrier et al. 2011b). Whistles were highly stable within an individual and different between individuals; however, the limited number of males available for the study prevented investigation of whether whistle characteristics could be linked to phenotype cues of males. As Pacific walrus males form a lek-like system, and display in front of a herd of females, aerial whistles could be used by males and females to individually recognize displaying males, allowing females to perform mate choice, and males to better assess rivals.

14.5 Conclusions

Otariids and Odobenids produce vocalizations to regulate their social interactions and thus to manage vital functions such as breeding, mate selection, care of young and predator alerts. Studying their vocal communication system is important to better understand their behavior, and also how communication systems in mammals, as well as in vertebrates in general, are shaped by social and environmental constraints. But this knowledge is also important to better estimate the threats they can be exposed to. Disturbance and acoustic pollution of their natural environment is unfortunately growing, with acoustic marine pollution one of the fastest growing anthropogenic threatening processes (Pirotta et al. 2019; Mikkelsen et al. 2019). If communication is impaired, then their survival may be greatly threatened. For instance, noise pollution may impact hearing abilities or vocal production. If a female can no longer hear, then she will not be able to find her pup in the colony. Since adoption and allo-nursing is extremely rare in otariids, that pup is going to starve to death. With temporary or permanent hearing loss, pinnipeds will not be able to detect many dangers on land or at sea, increasing the risk of injury or death. So, studying and understanding the vocal communication system is important to protect pinnipeds and walrus to help mitigate against threats in future. Knowledge of their social communication should be seen as a tool for conservation and protection.

References

Ahonen H, Stow AJ, Harcourt RG, Charrier I (2014) Adult male Australian sea lion barking calls reveal clear geographical variations. Anim Behav 97. https://doi.org/10.1016/j.anbehav.2014.09.010

Alcock J (1998) Animal behavior: an evolutionary approach. Sinauer, Sunderland, MA

Aubin T, Jouventin P (1998) Cocktail-party effect in king penguin colonies. Proc R Soc Lond Ser B Biol Sci 265:1665–1673
Aubin T, Jouventin P, Charrier I (2015) Mother vocal recognition in Antarctic fur seal Arctocephalus gazella pups: a two-step process. PLoS One 10:1–13. https://doi.org/10.1371/journal.pone.0134513
Bartholomew GA (1959) Mother-young relations and the maturation of pup behaviour in the Alaska fur seal. Anim Behav 7:163–171
Bonner W (1968) The fur seal of South Georgia
Bradbury JW, Vehrencamp SL (2011) Principles of Animal Communication. Sinauer, Sunderland, MA
Briefer EF (2012) Vocal expression of emotions in mammals: mechanisms of production and evidence. J Zool 288:1–20. https://doi.org/10.1111/j.1469-7998.2012.00920.x
Bro-Jørgensen J, Dabelsteen T (2008) Knee-clicks and visual traits indicate fighting ability in eland antelopes: multiple messages and back-up signals. BMC Biol 6:47. https://doi.org/10.1186/1741-7007-6-47
Campbell GSG, Gisiner RRC, Helweg DA, Milette LL (2002) Acoustic identification of female Steller sea lions (*Eumetopias jubatus*). J Acoust Soc Am 111:2920. https://doi.org/10.1121/1.1474443
Charrier I, Harcourt R (2006) Individual vocal identity in mother and pup Australian sea lions (Neophoca cinerea). J Mammal 87:929–938
Charrier I, Mathevon N, Jouventin P (2001) Mother's voice recognition by seal pups. Nature
Charrier I, Mathevon N, Jouventin P (2002) How does a fur seal mother recognize the voice of her pup? An experimental study of *Arctocephalus tropicalis*. J Exp Biol 205:603–612
Charrier I, Mathevon N, Jouventin P (2003a) Vocal signature recognition of mothers by fur seal pups. Anim Behav 19:543–550. https://doi.org/10.1006/anbe.2003.2073
Charrier I, Mathevon N, Jouventin P (2003b) Fur seal mothers memorize subsequent versions of developing pups' calls: Adaptation to long-term recognition or evolutionary by-product? Biol J Linn Soc 80. https://doi.org/10.1046/j.1095-8312.2003.00239.x
Charrier I, Pitcher BJ, Harcourt RG (2009) Vocal recognition of mothers by Australian sea lion pups: individual signature and environmental constraints. Anim Behav 78. https://doi.org/10.1016/j.anbehav.2009.07.032
Charrier I, Aubin T, Mathevon N (2010) Mother-calf vocal communication in Atlantic walrus: a first field experimental study. Anim Cogn 13:471–482. https://doi.org/10.1007/s10071-009-0298-9
Charrier I, Ahonen H, Harcourt RG (2011a) What makes an Australian sea lion (*Neophoca cinerea*) male's bark threatening? J Comp Psychol 125. https://doi.org/10.1037/a0024513
Charrier I, Burlet A, Aubin T (2011b) Social vocal communication in captive Pacific walruses *Odobenus rosmarus divergens*. Mamm Biol 76:622–627. https://doi.org/10.1016/j.mambio.2010.10.006
Cummings W, Fish J (1971) A synopsis of marine animal underwater sounds in eight geographic areas. US Naval Undersea Res. and Dev. Cent. NTIS AD-A068875
Dowell SA (2005) Mother-pup recognition behaviour, pup vocal signatures and allosuckling in the New Zealand fur seal, *Arctocephalus forsteri*. MS Thesis, University of Canterbury, p 156
Fay FH (1982) Ecology and biology of the Pacific walrus, *Odobenus rosmarus divergens* Illiger. North Am Fauna 1–279. https://doi.org/10.3996/nafa.74.0001
Fernández-Juricic E, Campagna C (1999) Vocal communication and individual variation in breeding South American sea lions. Behaviour 136(4):495–517
Fernandez-Juricic E, Campagna C, Ortiz CLEO (1999) Vocal communication and individual variation in breeding South American sea lions. Behaviour 136:495–517
Fernández-Juricic E, Campagna C, Enriquez V, Ortiz CL (2001) Vocal rates and social context in male South American sea lions. Mar Mammal Sci 17:387–396
Gentry RL (1975) Comparative social behavior of eared seals. Rapp Proces-Verbaux des Reun Int pour L'Exploration la Mer 169:188–194

Gisinier R (1985) Male territorial and reproductive behavior in the Steller sea lion, *Eumetopias jubatus*. Ph.D. Diss. Univ. California, St. Cruz, 145p

Greenwalt CH (1968) Bird song: acoustics and physiology. Smithsonian Institution Press, Washington, DC

Guinet C, Georges J-YY (2000) Growth in pups of the subantarctic fur seal (Arctocephalus tropicalis) on Amsterdam Island. J Zool 251:289–296. https://doi.org/10.1111/j.1469-7998.2000.tb01079.x

Gwilliam J, Charrier I, Harcourt R (2008) Vocal identity and species recognition in male Australian sea lions, *Neophoca cinerea*. J Exp Biol 211:2288–2295

Halliday TR (1983) Information and communication. In: Halliday TR, Slater PJB (eds) Animal behaviour: communication. Blackwell, Oxford, pp 43–81

Hanggi E (1992) The importance of vocal cues in mother–pup recognition in a California sea lion. Mar Mammal Sci 8:430–432

Harcourt R (1992) Maternal aggression in the South American fur seal in Peru. Can J Zool 70:320–325. https://doi.org/10.1139/z92-048

Harcourt RG, Bradshaw CJA, Dickson K, Davis LS (2002) Foraging ecology of a generalist predator, the female New Zealand fur seal. Mar Ecol Prog Ser 227:11–24. https://doi.org/10.3354/meps227011

Insley SSJ (1992) Mother-offspring separation and acoustic stereotypy: a comparison of call morphology in two species of pinnipeds. Behaviour 120:103–122. https://doi.org/10.1007/s13398-014-0173-7.2

Insley SJ (2000) Long-term vocal recognition in the northern fur seal. Nature 406:404–405. https://doi.org/10.1038/35019064

Insley S (2001) Mother–offspring vocal recognition in northern fur seals is mutual but asymmetrical. Anim Behav 61(1):129–137

Kiley M (1972) The vocalizations of ungulates, their causation and function. Z Tierpsychol 31:171–222. https://doi.org/10.1111/j.1439-0310.1972.tb01764.x

Knudtson P (1998) The world of the walrus. Sierra Club, San Francisco

Kunc HP, Wolf JBW (2008) Seasonal changes of vocal rates and their relation to territorial status in male Gal??pagos sea lions (*Zalophus wollebaeki*). Ethology 114:381–388. https://doi.org/10.1111/j.1439-0310.2008.01484.x

Marlow B (1975) The comparative behaviour of the Australasian sea lions *Neophoca cinerea* and *Phocarctos hookeri* (Pinnipedia: Otariidae). Mammalia 39:159–230

Mathevon N, Aubin T, Vielliard J et al (2008) Singing in the rain forest: how a tropical bird song transfers information. PLoS One 3:e1580. https://doi.org/10.1371/journal.pone.0001580

McCann TS (1980) Territoriality and breeding behaviour of adult male Antarctic Fur seal, *Arctocephalus gazella*. J Zool 192:295–310. https://doi.org/10.1111/j.1469-7998.1980.tb04235.x

McComb K, Moss C, Sayialel S, Baker L (2000) Unusually extensive networks of vocal recognition in African elephants. Anim Behav 59:1103–1109. https://doi.org/10.1006/anbe.2000.1406

McGregor PK (1992) Quantifying responses to playback: one, many, or composite multivariate measures? In: McGregor PK (ed) Playback and studies of animal communication. Springer, Boston, MA, pp 79–96

McNab A, Crawley M (1975) Mother and pup behaviour of the New Zealand fur seal, *Arctocephalus forsteri* (Lesson). Mauri Ora 3:77–88

Mikkelsen L, Johnson M, Wisniewska DM, et al (2019) Long-term sound and movement recording tags to study natural behavior and reaction to ship noise of seals. Ecol Evol ece3.4923. https://doi.org/10.1002/ece3.4923

Miller EH (1985) Airborne acoustic communication in the walrus *Odobenus rosmarus*. Natl Geogr Res 1:124–145

Nowak R (2003) Walker's marine mammals of the world. JHU Press

Nowicki SN, Stirling I, Sjare B (1997) Duration of stereotyped underwater vocal displays by male Atlantic walruses in relation to aerobic dive limit. Mar Mammal Sci 13:566–575. https://doi.org/10.1111/j.1748-7692.1997.tb00084.x

Page B, Goldsworthy SD, Hindell MA (2002) Individual vocal traits of mother and pup fur seals. Bioacoustics 13:121–143

Paulian P (1964) Contribution à l'étude de l'otarie de l'Ile Amsterdam. Mammalia 28:1–146

Peters G (2002) Purring and similar vocalizations in mammals. Mamm Rev 32:245–271. https://doi.org/10.1046/j.1365-2907.2002.00113.x

Peterson R, Bartholomew G (1969) Airborne vocal communication in the California sea lion, *Zalophus californianus*. Anim Behav 17:17–24

Phillips AV (2003) Behavioral cues used in reunions between mother and pup South American fur seals. J Mammal 84:524–535. https://doi.org/10.1644/1545-1542(2003)084<0524:BCUIRB>2.0.CO;2

Phillips AV, Stirling I (2000) Vocal individuality in mother and pup South American fur seals, *Arctocephalus australis*. Mar Mammal Sci 16:592–616. https://doi.org/10.1111/j.1748-7692.2000.tb00954.x

Phillips A, Stirling I (2001) Vocal repertoire of South American fur seals, *Arctocephalus australis*: structure, function, and context. Can J Zool 79:420–437

Pirotta V, Grech A, Jonsen ID et al (2019) Consequences of global shipping traffic for marine giants. Front Ecol Environ 17:39–47. https://doi.org/10.1002/fee.1987

Pitcher BJ, Ahonen H, Harcourt RG, Charrier I (2009) Delayed onset of vocal recognition in Australian sea lion pups (*Neophoca cinerea*). Naturwissenschaften 96. https://doi.org/10.1007/s00114-009-0546-5

Pitcher B, Harcourt R, Charrier I (2010a) The memory remains: Long-term vocal recognition in Australian sea lions. Anim Cogn 13:771–776. https://doi.org/10.1007/s10071-010-0322-0

Pitcher BJ, Harcourt RG, Charrier I (2010b) Rapid onset of maternal vocal recognition in a colonially breeding mammal, the Australian sea lion. PLoS One 5. https://doi.org/10.1371/journal.pone.0012195

Pitcher BJ, Harcourt RG, Charrier I (2012) Individual identity encoding and environmental constraints in vocal recognition of pups by Australian sea lion mothers. Anim Behav 83. https://doi.org/10.1016/j.anbehav.2011.12.012

Poulter TC (1963) Sonar signals of the sea lion. Science 139:753–755. https://doi.org/10.1126/science.139.3556.753

Rand R (1967) The Cape fur seal (*Arctocephalus pusillus*): general behaviour on land and at sea. Gov Printer, South Africa

Ray GC, Watkins WA (1975) Social function of underwater sounds in the walrus *Odobenus rosmarus* Rapp P-v Réuns Cons int Explor Mer 169:524–526

Reidenberg JS, Laitman JT (2010) Generation of sound in marine mammals. In: Handbook of behavioral neuroscience. pp 451–465

Remedios R, Logothetis NK, Kayser C (2009) Monkey drumming reveals common networks for perceiving vocal and nonvocal communication sounds. Proc Natl Acad Sci U S A 106:18010–18015. https://doi.org/10.1073/pnas.0909756106

Renouf D (1991) The behaviour of pinnipeds. Chapman and Hall, London

Riedman M (1990) The pinnipeds: seals, sea lions, and walruses. University of California Press

Roux J (1986) Sociobiologie de l'Otarie à fourrure d'Amsterdam, Arctocephalus tropicalis. Université de Montpellier

Roux J, Jouventin P (1987) Behavioral cues to individual recognition in the subantarctic fur seal, *Arctocephalus tropicalis*. NOAA Tech Rep NMFS 51:95–102

Sandegren F (1970) Breeding and maternal behavior of the Steller sea lion (*Eumetopias jubata*) in Alaska

Schusterman R, Balliet R (1969) Underwater barking by male sea lions. Nature 222:1179–1181

Schusterman RJ, Balliet RF, St. John S (1970) Vocal displays under water by the gray seal, the harbor seal, and the stellar sea lion. Psychon Sci 18:303–305. doi: https://doi.org/10.3758/BF03331839

Schusterman RJ, Kastak D, Levenson DH et al (2000) Why pinnipeds don't echolocate. J Acoust Soc Am 107:2256–2264

Sebe F, Duboscq J, Aubin T et al (2010) Early vocal recognition of mother by lambs: contribution of low- and high-frequency vocalizations. Anim Behav 79:1055–1066. https://doi.org/10.1016/j.anbehav.2010.01.021

Shannon C, Weaver W (1949) The mathematical theory of communication. University of Illinois Press, Urbana and Chicago

Sjare B, Stirling I (1996) The breeding behavior of Atlantic walruses, *Odobenus rosmarus rosmarus*, in the Canadian High Arctic. Can J Zool 74:897–911. https://doi.org/10.1139/z96-103

Sjare B, Stirling I, Spencer C (2003) Structural variation in the songs of Atlantic walruses breeding in the Canadian High Arctic. Aquat Mamm 29:297–318. https://doi.org/10.1578/016754203101024121

Stirling I (1970) Observations on the behavior of the New Zealand fur seal (*Arctocephalus forsteri*). J Mammal 766–778

Stirling I (1972) Observations on the Australian sea lion, *Neophoca cinerea* (Peron). Aust J Zool 20:271–279

Stirling I, Calvert W, Spencer C (1987) Evidence of stereotyped underwater vocalizations of male Atlantic walruses (*Odobenus rosmarus rosmarus*). Can J Zool 65:2311–2321. https://doi.org/10.1139/z87-348

Temeles EJ (1994) The role of neighbours in territorial systems: When are they "dear enemies"? Anim Behav 47:339–350

Titze IR (1994) Mechanical stress in phonation [Review]. J Voice 8:99–105

Trillmich F (1981) Mutual mother-pup recognition in Galapagos fur seals and sea lions: cues used and functional significance. Behaviour 78(1-2):21–42

Trimble M, Charrier I (2011) Individuality in South American sea lion (*Otaria flavescens*) mother-pup vocalizations: Implications of ecological constraints and geographical variations? Mamm Biol 76:208–216. https://doi.org/10.1016/j.mambio.2010.10.009

Tripovich JS, Rogers TL, Canfield R, Arnould JPY (2006) Individual variation in the pup attraction call produced by female Australian fur seals during early lactation. J Acoust Soc Am 120:502. https://doi.org/10.1121/1.2202864

Tripovich J, Charrier I, Rogers T et al (2008a) Who goes there? Differential responses to neighbor and stranger vocalizations in male Australian fur seals. Mar Mammal Sci 24:941–948. https://doi.org/10.1111/j.1748-7692.2008.00222.x

Tripovich JS, Charrier I, Rogers TL et al (2008b) Acoustic features involved in the neighbour-stranger vocal recognition process in male Australian fur seals. Behav Process 79:74–80. https://doi.org/10.1016/j.beproc.2008.04.007

Tripovich J, Rogers T, Arnould JPY (2011) Species-specific characteristics and individual variation of the bark call produced by male Australian fur seals (*Arctocephalus pusillus doriferus*). Bioacoustics 15:79–96

Vince M (1993) The interaction that leads to sucking. Adv Study Behav 22:239–268

Chapter 15
Non-vocal Communication in Otariids and Odobenids: The Involvement of Visual and Olfactory Cues in Their Social Lives

Isabelle Charrier

Abstract While vocal communication is prominent in pinnipeds, visual and olfactory cues are also important in social interactions, yet they have been poorly investigated. Behavioral observations of mother-pup and male-male interactions suggest the involvement of visual and olfactory cues to facilitate mother-pup reunion and to advertise the presence of competitors or territory ownership. Experimental studies have shown that mothers can recognize their pups on the only basis of olfactory cues, and visual cues such as size and fur color pattern are used by females to sort pups in different age-classes. Otariid males exhibit specific body postures to threaten their competitors and walrus males display elaborate visual courtship display during the breeding season. During the breeding season, mature males develop a specific odor advertising their presence and reproductive status, and experiments show they can discriminate this musky male scent. This chapter relates the importance of olfactory and visual cues in their social interactions, and outlines future investigations needed to better understand the role of these under explored sensory modalities in communication in marine mammals and mammals in general.

Keywords Animal communication · Visual cues · Chemical cues · Mother-offspring · Male assessment

15.1 Introduction

Animals have developed elaborate strategies to communicate with conspecifics. Communication allows animals to transfer information from an emitter to a receiver (Shannon and Weaver 1949), with the receiver adapting its behavior according to the type of information embedded in the signal. Signal or cue? This is an important

I. Charrier (✉)
Institut des Neurosciences Paris-Saclay, Université Paris-Saclay, CNRS (UMR 9197), Université Paris-Sud, Orsay, France
e-mail: isabelle.charrier@u-psud.fr

C. Campagna, R. Harcourt (eds.), *Ethology and Behavioral Ecology of Otariids and the Odobenid*, Ethology and Behavioral Ecology of Marine Mammals,
https://doi.org/10.1007/978-3-030-59184-7_15

question. Cues can provide information about the receiver, but they are passive and non-evolving biological and environmental traits, whereas signals have evolved to actively provide information about the emitter and to impact the behavior of the receiver (Lehmann et al. 2014). Animals use both signals and cues to extract information about emitter and during diverse social interactions. As information about the evolutionary origins of many sensory cues are not always known, and distinguishing between cues and signals can be very difficult especially with chemicals, I will use the term "cues" in this chapter.

Different types of sensory cues can be involved in such information transfer such as acoustic, visual, olfactory and tactile cues, and depending both on the biological function of such cue and its spatial range, animals will use specifically a given sensory modality. As with vocal cues, both visual and olfactory cues are also exposed to attenuation and transmission loss, as well as background noise during their transmission in the environment, and their perception will depend on the perceptual ability of the receiver. However, unlike acoustic cues that can be detected by the receiver when the sender is not visible (in an obstructed environment or in dark conditions), visual cues can only work if the sender is visible to the receiver (Penteriani and Delgado 2017). Also, in contrast to acoustic cues that are produced by animals in a specific behavioral context, most visual cues (e.g., color and size of the feathers in birds, color pattern of the fur in mammals) and some olfactory cues (e.g., body odors) are present continuously and thus totally independent of the behavioral state of the emitter. Therefore, visual and olfactory cues can have other functions than communication (locomotion dynamics, anti-predator strategy).

15.1.1 Visual Cues

Visual cues are represented by three main types: color, shape or size and body posture. Color and shape/size can give information about age, sex, identity, sexual status (i.e., estrus, courtship, sexual maturity) of the animal, whereas body postures will transmit information about its behavioral and emotional state. Both color and size can evolve with age and thus sexual maturity, but they are available to receivers in permanence or for a long period of time (i.e., breeding period, pre-weaning). These visual traits allow receivers to categorize and thus discriminate individuals according to their age-class, sex, sexual status. Sexual dimorphism in mammals is quite common and is based on differences in body size and/or color pattern. For instance, in gorilla (*Gorilla gorilla*), mature adult males are much larger than adult females, and they develop grayish hair on their back and shoulder "silverback" as they get older. In many primate species, females in estrus have a swollen genital area indicating to males that they are ready to mate.

In contrast, body postures are produced in very specific contexts and are highly stereotyped. When threatened, an animal might flee to escape the threatening individual, or it will display a submissive posture, such as a rolling-over in canids, positioning itself as low as possible and exposing a vulnerable part of its body

(Bradbury and Vehrencamp 1998). Such visual displays are often coupled with the production of high-pitched vocalizations. During agonistic interactions, aggressive postures are also displayed to threaten and warn the opponent from an eminent charge or fight. Elephants hold their head above their shoulders ("standing tall"), and spread their ears to put them perpendicular to their body ("ear-spreading") to appear as large as possible (Poole and Granli 2011). Other common visual threat displays include hair erections, teeth or horn displaying.

15.1.2 Olfactory Cues

Saliva, urine, feces and glandular secretions are used as chemical cues, and they can transfer different information about the receiver such as sex, age, identity, emotional and reproductive states (Eisenberg and Kleiman 1972; Wyatt 2003). Chemical cues can be used to locate food sources, detect predators and to socially interact with con- and hetero-specifics, and involve volatile and semi-volatile compounds produced by specialized glands or derive from a mixture of the general body odor.

Chemical cues can be actively placed in the environment by the animal to mark their territory and to advertise their presence and/or their sexual status. Scent marking is performed using urine, feces and glandular secretions, and mammals have a high diversity of glands to scent-mark. For instance, ungulates have pre-orbital, inter-digital, meta-tarsal and tarsal glands to scent-mark their environment with different types of information about the receiver (identity, sex, sexual status) and confer different biological functions (alarm, territorial defense, recognition, sexual selection) (Grau 1976; Wyatt 2003). In some social mammals, chemical cues can also be deposited on conspecifics to label social partners, with members of the group often marked with the scent of the dominant individuals (Wyatt 2014). Chemical cues are also non-intentionally produced by animals, and are the product of different sources (sebaceous glands, microbial activity, diet and environment) generating a general body odor.

15.2 Visual Cues Involved in Social Recognition in Otariids and Odobenids

As mentioned in the previous chapter, both otariids and odobenids use vocalizations in diverse social interactions, however, visual cues such as fur color pattern, shape or size and body posture are also involved in different types social interactions such as mother-pup reunion and recognition, and agonistic interactions between males.

Adult otariid males differ in size dramatically from females and younger animals and this provides a potential source to distinguish them using visual cues (Figs. 15.1, 15.2 and 15.3). Male and female growth curves differ significantly with female

Fig. 15.1 Individual visual differences among adult females in Australian sea lion (top) and Atlantic walrus (bottom) (photographs taken by I. Charrier)

Fig. 15.2 Age-class visual pattern in Australian sea lion pups. Top panel: Pups show a black pelage when < 2-month old, then they turn to a brown/cinnamon fur pattern (2–4 month old), and after molting (> 4-month) they show an adult pelage pattern (silver and beige). Bottom panel: during the pupping season lasting up to 7 months, pups of different age classes can be encountered, and they present different visual aspects (difference in fur color pattern and body size) (photographs taken by I. Charrier)

Fig. 15.3 Difference in visual aspect of adult males in several otariid and odobenid species. From right to left: Subantarctic fur seal adult males with a visible crest and a yellow chest (bib); Australian sea lion males exhibit a blond mane and a dark fur; and Atlantic walrus adult males show thick and long tusks with a chest and a neck with scars and tubercules (photographs taken by I. Charrier)

growth tailing off once they start breeding. By contrast males show a post-pubertal growth spurt during which they develop marked secondary sexual characteristics, a wider chest, strong neck and shoulders, wider muzzle, and in some species, the fur color pattern also changes when they reach sexual maturity. In the Australian sea lion (*Neophoca cinerea*), adult males develop a blond mane and yellow chest, while in the subantarctic fur seal (*Arctocephalus tropicalis*) adult males their chest becomes a white-cream type of bib and they grow a fur crest that resembles a mohawk (Fig. 15.3) (Jefferson et al. 2008; see Chap. 13 by Sinclair et al.). Steller (*Eumetopias jubatus*) and southern sea lion (*Otaria byronia*) males also develop a mane of long and coarse guard hairs. Walrus (*Odobenus rosmarus*) males develop a bulky and scarred chest with tubercles on both chest and neck, and present longer and thicker tusks compared to juveniles and females (Jefferson et al. 2008; Chap. 22 by Miller and Kochnev). Humans can readily differentiate females and young males from these full-sized adult males by their body size and colour fur pattern. It is likely that these visual cues can also be used by males, females and pups of the same species to assess the age of males, and thus the level of threat or breeding status.

15.2.1 Visual Abilities

Individual visual difference among individuals can be encoded in their face traits, color fur pattern, body size, and / or skin scars (Fig. 15.1), and otariids have the visual acuity in air to perform such visual discrimination (Schusterman 1972, 1981; Renouf 1991; Dehnhardt 2002). As vision is adapted to their aquatic lifestyle, pinnipeds develop myopia and astigmatism in air (Wartzok and Ketten 1999), and thus their visual acuity in air is thus mostly efficient at short range (Chap. 16 by Hanke et al.). While pinnipeds are considered color-blind from an anatomic and physiology point of view (Levenson et al. 2006), they can behaviorally show color discrimination (Griebel and Schmid 1992). This inconsistency between physiological and behavioral data have been experimentally investigated, and it has been suggested that the ability to discriminate colors is based on brightness discrimination rather than color vision *per se* (Scholtyssek et al. 2015), as fur seals can functionally perceive fine brightness differences (Scholtyssek and Dehnhardt 2013).

Visual acuity in walrus has been suggested to be lower than in most pinnipeds (Berta et al. 2006), and specialized for short-range use underwater. Indeed, they present a much smaller retina, composed of few rods per unit of surface area, and the lateral position of their eyes implies a limited binocular vision (Fay 1985).

15.2.2 Visual Cues in Mother-Offspring Relationship

Visual or geographical cues can facilitate the reunion between the mother and her young in the colony. Otariid females show a degree of small-scale site fidelity in the colony and tend to go back to the birth site early in their pup's life, or to the last suckling spot when pups get older (Chap. 7 by Cassini). As pups of most species do not move too much from their birth site during their first weeks, this facilitates the reunion process with the mother (Bartholomew 1959; Marlow 1975; Trillmich 1981; Roux 1986; Phillips 2003). While searching the colony, returning females are easily distinguishable from hauling out females or juveniles as they present with a dark, wet and shiny pelage and a distinctive calling posture (i.e., with neck and head outstretched), and these visual cues may help the young to locate its mother among other females of the colony (Stirling 1970; Insley 2001; Phillips 2003).

A recent study in Australian sea lion females has shown that visual cues of pups can facilitate the reunion with the mother. It seems that females are able to discriminate among pups of different age-classes using only the body size and color pattern of the pup (Wierucka et al. 2017). In this species, the pupping period is particularly long, averaging 124 days, which is the longest among pinnipeds (McIntosh et al. 2012); Chap. 26 by McIntosh and Pitcher) thus pups of different ages occur in the colony, and they differ in body size and color fur pattern (from black to brown to a final silver grey/ beige adult pattern; Fig. 15.2). When she returns from a foraging trip, an Australian sea lion female has to find her pup amongst many others, but by focussing on the appropriate age-class (i.e., size and contrast/brightness of color fur pattern) of her pup, she can more easily sort pups and find her own.

No studies have demonstrated individual visual recognition in pinnipeds. Recent studies on other mammal species such as sheep (Kendrick et al. 2001) and chimpanzee (*Pan troglodytes*) (Parr and de Waal 1999) have experimentally shown individual visual discrimination, so this is likely that such individual visual recognition also occurs in pinnipeds.

15.2.3 Visual Cues Used in Male-Male Interactions

During the breeding period, territorial otariid males produce vocalizations (see Chap. 14 by Charrier) but they can also display specific body postures (Fig. 15.4b–d) to advertise to rivals relating to territorial tenure and/or to threaten any intruder (Stirling 1970; Roux 1986). Adjacent territorial males perform boundary display, by facing each other at distance, and usually barking at the same time. In a more aggressive context, such as before engaging in a fight, males face each other, stretch their neck and head to appear as high as possible, erect their whiskers, and produce barks while simultaneously rotating their head (Fig. 15.4b). If neither of the males retreats, a chest to chest contact will occur, followed by a physical fight (Fig. 15.4d) which may result in serious injuries and sometimes in the death of one of the opponents (Miller 1991). Fighting over, mating and defending females or territories can exact a very high toll on otariid males during the breeding season, at its worst 30% of Galapagos fur seal (*Arctocephalus galapagoensis*) males (Trillmich and Trillmich 1984) and 38% of northern fur seal (*Callorhinus ursinus*) (Johnson 1968) may perish over the reproductive season.

During agonistic interactions among males, they sometimes show an open-mouth display, and produce puff sounds by rapid exhalation of air from the mouth. Submissive postures can also be observed in adult or subadult males while retreating

Fig. 15.4 Aggressive displays and postures used by males during agonistic interactions. (**a**) Atlantic walrus threat display with tusks in an upright position, (**b**) fight between two Australian sea lion adult males and chest to chest position, (**c**) Australian sea lion juvenile showing an open-mouth display, (**d**) Australian sea lion adult males showing the typical face to face position, males in an upright position and barking with a stretched neck (photographs taken by I. Charrier)

from an agonistic interaction: the animal lies down with the head in the direction of the dominant/older individual. Male pups and juveniles have been shown to play by displaying male threat behaviors: they stretch their neck and head, barking and pushing their playmates on the chest. This has been observed at least in Steller sea lions (*Eumetopias jubatus*) (Gentry 1974), Australian sea lions (Marlow 1975), and subantarctic fur seals (Charrier, unpublished information). These play displays may be a way to practice fighting (see Llamazares-Martín and Palagi, Chap. 20 on play behavior).

In walrus, males perform an elaborate vocal display during courtship (see Chap. 22 by Miller and Kochnev), but they also exhibit a visual display going under water and coming back to the surface to breath and whistle in front of a herd of females (Fay 1982; Sjare et al. 2003). It is unknown if the visual part of such a complex courtship display plays a role in mate selection. During agonistic interactions among walrus, males or females show an upright position and by stretching their neck they display their tusks in front of their opponent (Fig. 15.4a) (Miller 1985), a threat visual display most of the time coupled with aggressive vocalizations of roars, grunts or guttural sounds.

15.3 Olfactory Cues Involved in Social Recognition in Otariids and Odobenids

A striking fact when approaching a colony of otariids or a group of walrus is the smell, so chemical cues are present and behavioral observations have highlighted the importance of olfaction in their social life (Fig. 15.5). During the breeding season, mothers and pups always perform nose to nose contacts while reuniting (Riedman 1990; Renouf 1991) (Fig. 15.5b), males inspect the uro-genital area of females probably to detect their estrus status (in northern fur seal (Bartholomew et al. 1953); in South African fur seal (*Arctocephalus pusillus pusillus*), (Rand 1955), in California sea lion (*Zalophus californianus*), (Peterson and Bartholomew 1967), in New Zealand fur seal (*Arctocephalus forsteri*), (Miller 1974); (Fig. 15.5c), and juveniles as well as sub-adult males smell the air when approaching conspecifics to detect the presence of mature males in the surroundings (Fig. 15.5).

15.3.1 Olfactory Abilities

Otariids and walrus can detect the smell of humans at tens of meters, and so for skittish species such as California sea lions, South American fur seals (*Arctocephalus australis*) or walrus a careful approach downwind is essential to avoid a stampede into the water. These behavioral observations reflect their ability to detect chemical cues and associated human smells as a threat. The sense of smell in

Fig. 15.5 Olfactory investigations in different social interactions in Australian sea lions. (**a**) Nose to nose contact between two pups during play behavior; (**b**) nose to nose contact between a mother and her pup during reunion; (**c**) mate-guarding male investigating a female to assess estrus during the perinatal period; (**d**) nose to nose contact between juveniles during play behavior (photographs taken by I. Charrier)

pinnipeds has long been underestimated and the role of social olfaction (i.e., olfactory discrimination among conspecifics) unexplored. Histological studies have shown the presence of a typical mammalian olfactory epithelium (Lowell and Flanigan 1980) and all pinnipeds possess a vomeronasal organ (Hoelzel 2002). Neuroanatomic data confirm the presence of functional olfactory receptor genes similarly to terrestrial mammals (Kishida et al. 2007). Experimental studies on non-social chemical cues have shown that South African fur seals can discriminate natural odors and thus were able to distinguish between fish and non-fish odors, but also between two fish odors (Laska et al. 2008). They can also discriminate artificial odors (Laska et al. 2010) distinguishing between members of five chemical classes of aliphatic odorants. This latter experimental work shows the ability of fur seals to discriminate between structurally related aliphatic odorants, and thus highlights their well-developed sense of smell.

15.3.2 Olfactory Cues in Mother-Pup Interactions

During mother-pup reunions, females and pups exchange vocalizations to locate each other, and when in close proximity, mothers perform nose to nose contact (Fig. 15.5b) with their young, apparently as a final check of the pup's identity before suckling (Riedman 1990). This olfactory check has been observed in many otariid species and the walrus (Miller 1985; Insley et al. 2003), but the experimental evidence that mothers perform olfactory recognition of their pup has been only shown only in one species, the Australian sea lion (Pitcher et al. 2011). Australian sea lion females are able to distinguish between the odor of their own pup and that of another without the help of any additional cues. Australian sea lion mothers were used in a two-choice test using two pup models, one impregnated with the smell of the filial pup and another with the smell of a non-filial pup. The females sniffed the pup model with the filial pup's scent significantly more than the model impregnated with the non-filial pup's scent. Moreover, nine out of the ten females tested investigated the model with the filial scent more, two females even gently grabbed the model impregnated with the filial scent, and one female showed clear aggressive behavior (i.e., biting, shaking and throwing the model away) towards the model impregnated with the non-filial scent (Pitcher et al. 2011). All these observed behaviors are consistent with natural behavior of females approaching their own pups or non-filial pups. Knowing that otariid females show the same behavioral pattern while searching their pups in the colony, i.e., exchanging vocalizations and thus smelling it when at close range before initiating suckling, it is likely that such olfactory discrimination also will occur in other otariid and odobenid species. Further experimental investigations are needed to draw firm conclusions, and information about the onset of olfactory recognition will be crucial to better understand the mother-pup binding process in pinnipeds and in mammals in general. In domestic species, such as sheep, experimental studies showed that ewes can discriminate the scent of their own lamb (Alexander 1985) and this occurs as soon as 4 h after birth (Porter et al. 1991). Otariids and odobenids present similar or higher ecological constraints to develop a mother-pup recognition than sheep species, so early onset of olfactory recognition may also occur in these marine mammals.

15.3.3 Olfactory Cues Used in Male-Male Interactions

In otariids, mature males develop a musky smell during the breeding season (Hamilton 1956; Bonner 1968; Marlow 1975; Roux 1986). This smell is easily detectable, and humans as well as other conspecifics can detect the presence of a male at more than 10 m, even before being able to see it. It is common to see juvenile and subadult males screening the air to detect adult males in a given area, and detection of a male smell will scare them away without even seeing the animal (Charrier, unpublished data). A recent study on Australian sea lions (Charrier, unpublished data) showed that adult and subadult males can discriminate between the smell of a mature male and a control smell of seaweed. They investigated the male scent (i.e. sniffs), produced snorts (i.e., aggressive vocalization) and they showed vigilance behavior by changing posture (moving from a laying down to stand up position to visually scanning the surroundings) (Charrier pers. Com.). Chemical analyses reveal that adult males show different chemical profiles during and outside the breeding season, and several chemical compounds contribute to the observed variation. A difference in chemical profile of males belonging to two different breeding colonies has also been found, demonstrating that both genetic and environmental factors can influence male scent (Charrier, pers. Com.). It is still unknown if males can recognize each other using their smell, or if any other phenotype or dominance cues could be encoded in their body odor that could be involved in intra- and inter-sexual assessment. Further investigations are thus needed to better understand the involvement of olfactory cues in mate selection and rival assessment.

15.4 Conclusions

Olfactory and visual cues play an important role in social interaction and social recognition in otariids and odobenids. Along with acoustic cues, these sensory cues deliver similar or different information from the emitter; such redundancy is quite important to secure the information. Understanding how different sensory cues are involved in communication and individual recognition processes has drawn more attention in the last 15 years (Hebets and Papaj 2005), and thus investigating the synergic roles of different sensory cues is crucial for a better knowledge and understanding of communication processes. Recent research on the Australian sea lion has demonstrated the predominance of acoustic cues over visual and/or olfactory cues in the mother-pup recognition context (Wierucka et al. 2018a, b). In a multimodal setting, the active space of each sensory cue is impacted by constraints imposed by the environment of the animal but also by its biology.

As mentioned in the chapter on vocal communication (see Chap. 14 by Charrier), a better knowledge of pinniped communication systems is important to assess which threats may impact these amazing animals. Chemical pollution of our oceans and climate change are major threats for marine mammals. Toxic algal blooms releasing biotoxins such as domoic acid have increased due to pollution of coastal waters and global warming (Mos 2001). In the US, high concentrations of domoic acid are found in molluscs, fishes, crabs and many marine predators such as seabirds; and California sea lions have been poisoned, some fatally with animals showing brain seizure, short-term memory loss, disorientation, ataxia, tetanic convulsion (Mos 2001; Gulland et al. 2002). Recently it has been demonstrated that domoic acid also affects the olfactory abilities of California sea lions by damaging the olfactory bulb and pyriform lobe in their brain (De Maio et al. 2018). Olfactory impairment due to domoic acid impacts the survival of California sea lions by impairing their social interactions, including olfactory recognition of the pup by mother. Rejection of the pup by its own mother is invariably fatal as adoption and allo-suckling is rare in otariids, and females are aggressive towards non-filial pups (Harcourt 1992). The example of the impact of domoic acid on California sea lions illustrates the importance of improving our knowledge of the different sensory modalities involved in animal communication.

References

Alexander G (1985) Odour, and the recognition of lambs by merino ewes. Applied Anim Ethol 153–158. https://doi.org/10.1016/B978-0-444-42444-0.50033-4

Bartholomew GA (1959) Mother-young relations and the maturation of pup behavior in the Alaska fur seal. Anim Behav 7:163–171

Bartholomew GA, Hoel PG, Hoel PG (1953) Reproductive behavior of the Alaska fur seal, *Callorhinus ursinus*. J Mammal 34:417. https://doi.org/10.2307/1375858

Berta A, Sumich JL, Kovacs KM, et al (2006) Marine mammals. Mar Mamm 456–489. https://doi.org/10.1016/B978-012088552-7/50016-3

Bonner W (1968) The fur seal of South Georgia. Cambridge, British Antarctic Survey, 82 p (British Antarctic Survey Scientific Reports, 56)

Bradbury JW, Vehrencamp SL (1998) Principles of animal communication. Sinauer, Sunderland, MA

Dehnhardt G (2002) Sensory systems. In: Hoelzel AR (ed) Marine mammal biology: an evolutionary approach. Blackwell, Oxford, pp 116–141

De Maio LM, Cook PF, Reichmuth C, Gulland FMD (2018) The evaluation of olfaction in stranded California sea lions (*Zalophus californianus*) and its relevance to domoic acid toxicosis. Aquat Mamm 44:231–238. https://doi.org/10.1578/AM.44.3.2018.231

Eisenberg JF, Kleiman DG (1972) Olfactory communication in mammals. Annu Rev Ecol Syst 3:1–32. https://doi.org/10.1146/annurev.es.03.110172.000245

Fay FH (1982) Ecology and biology of the Pacific walrus, *Odobenus rosmarus divergens* Illiger. North Am Fauna:1–279. https://doi.org/10.3996/nafa.74.0001

Fay FH (1985) *Odobenus rosmarus*. Mamm Species:1. https://doi.org/10.2307/3503810

Gentry RL (1974) The development of social behavior through play in the steller sea lion. Am Zool 14:391–403. https://doi.org/10.1093/icb/14.1.391

Grau GA (1976) Olfaction and reproduction in ungulates. In: Doty RL (ed) Mammalian olfaction, reproductive processes, and behavior. Academic, pp 219–242

Griebel U, Schmid A (1992) Color vision in the California sea lion (*Zalophus californianus*). Vision Res 32(3):477–482

Gulland EMD, Haulena M, Fauquier D et al (2002) Domoic acid toxicity in Californian sea lions (*Zalophus californianus*): clinical signs, treatment and survival. Vet Rec 150:475–480. https://doi.org/10.1136/vr.150.15.475

Hamilton JE (1956) Scent of Otariids. Nature 177:900–900. https://doi.org/10.1038/177900c0

Harcourt R (1992) Maternal aggression in the South American fur seal in Peru. Can J Zool 70:320–325. https://doi.org/10.1139/z92-048

Hebets EA, Papaj DR (2005) Complex signal function: Developing a framework of testable hypotheses. Behav Ecol Sociobiol 57:197–214. https://doi.org/10.1007/s00265-004-0865-7

Hoelzel AR (2002) Marine mammal biology: an evolutionary approach. Blackwell

Insley S (2001) Mother–offspring vocal recognition in northern fur seals is mutual but asymmetrical. Anim Behav 61(1):129–137

Insley S, Phillips A, Charrier I (2003) A review of social recognition in pinnipeds. Aquat Mamm 29:181–201

Jefferson TA, Webber MA, Pitman RL (2008) Marine mammals of the world : a comprehensive guide to their identification. Academic

Johnson AM (1968) Annual mortality of territorial male fur seals and its management significance. J Wildl Manage 32:94. https://doi.org/10.2307/3798241

Kendrick KM, da Costa AP, Leigh AE et al (2001) Sheep don't forget a face. Nature 414:165–166. https://doi.org/10.1038/35102669

Kishida T, Kubota S, Shirayama Y, Fukami H (2007) The olfactory receptor gene repertoires in secondary-adapted marine vertebrates: evidence for reduction of the functional proportions in cetaceans. Biol Lett 3:428–430. https://doi.org/10.1098/rsbl.2007.0191

Laska M, Svelander M, Amundin M (2008) Successful acquisition of an olfactory discrimination paradigm by South African fur seals, *Arctocephalus pusillus*. Physiol Behav 93:1033–1038. https://doi.org/10.1016/j.physbeh.2008.01.019

Laska M, Lord E, Selin S, Amundin M (2010) Olfactory discrimination of aliphatic odorants in South African fur seals (*Arctocephalus pusillus*). J Comp Psychol 124:187–193. https://doi.org/10.1037/a0018189

Lehmann KDS, Goldman BW, Dworkin I et al (2014) From cues to signals: evolution of interspecific communication via aposematism and mimicry in a predator-prey system. PLoS One 9: e91783. https://doi.org/10.1371/journal.pone.0091783

Levenson DH, Ponganis PJ, Crognale MA et al (2006) Visual pigments of marine carnivores: pinnipeds, polar bear, and sea otter. J Comp Physiol A 192:833–843. https://doi.org/10.1007/s00359-006-0121-x

Lowell WR, Flanigan WF (1980) Marine mammal chemoreception. Mamm Rev 10:53–59. https://doi.org/10.1111/j.1365-2907.1980.tb00233.x

Marlow B (1975) The comparative behaviour of the Australasian sea lions *Neophoca cinerea* and *Phocarctos hookeri* (Pinnipedia: Otariidae). Mammalia 39:159–230

McIntosh RR, Goldsworthy SD, Shaughnessy PD et al (2012) Estimating pup production in a mammal with an extended and aseasonal breeding season, the Australian sea lion (*Neophoca cinerea*). Wildl Res 39:137–148. https://doi.org/10.1071/WR10222

Miller E (1974) Social behaviour between adult male and female New Zealand fur seals, *Arctocephalus forsteri* (Lesson) during the breeding season. Aust J Zool 22:155–173

Miller EH (1985) Airborne acoustic communication in the walrus *Odobenus rosmarus*. Natl Geogr Res 1:124–145

Miller EH (1991) Communication in pinnipeds, with special reference to non-acoustic signalling. Behav Pinnipeds 128–235

Mos L (2001) Domoic acid: a fascinating marine toxin. Environ Toxicol Pharmacol 9:79–85. https://doi.org/10.1016/s1382-6689(00)00065-x

Parr LA, de Waal FB (1999) Visual kin recognition in chimpanzees. Nature 399:647–648. https://doi.org/10.1038/21345

Penteriani V, Delgado MDM (2017) Living in the dark does not mean a blind life: bird and mammal visual communication in dim light. Philos Trans R Soc B Biol Sci 372:20160064. https://doi.org/10.1098/rstb.2016.0064

Peterson R, Bartholomew G (1967) The natural history and behavior of the California sea lion. Am Soc Mammal 1(Spec Publ No 1):79

Phillips AV (2003) Behavioral cues used in reunions between mother and pup South American fur seals. J Mammal 84:524–535. https://doi.org/10.1644/1545-1542(2003)084<0524:BCUIRB>2.0.CO;2

Pitcher BJ, Harcourt RG, Schaal B, Charrier I (2011) Social olfaction in marine mammals: Wild female Australian sea lions can identify their pup's scent. Biol Lett 7:60–62. https://doi.org/10.1098/rsbl.2010.0569

Poole JH, Granli P (2011) Signals, gestures, and behavior of African elephants. In: Moss CJ, Croze H, Poole JH et al (eds) The Amboseli elephants: a long-term perspective on a long-lived mammal. University of Chicago, pp 109–124

Porter RH, Lévy F, Poindron P et al (1991) Individual olfactory signatures as major determinants of early maternal discrimination in sheep. Dev Psychobiol 24:151–158. https://doi.org/10.1002/dev.420240302

Rand RW (1955) Reproduction in the female Cape Fur Seal, *Arctocephalus pusillus* (Schreber). Proc Zool Soc Lond 124:717–740. https://doi.org/10.1111/j.1469-7998.1955.tb07812.x

Renouf D (1991) The behaviour of pinnipeds. Chapman and Hall, London

Riedman M (1990) The pinnipeds: seals, sea lions, and walruses. University of California Press

Roux J (1986) Sociobiologie de l'Otarie à fourrure d'Amsterdam. Université de Montpellier, Arctocephalus tropicalis

Scholtyssek C, Dehnhardt G (2013) Brightness discrimination in the South African fur seal (*Arctocephalus pusillus*). Vision Res 84:26–32

Scholtyssek C, Kelber A, Dehnhardt G (2015) Why do seals have cones ? Behavioural evidence for colour-blindness in harbour seals. Anim Cogn 551–560. https://doi.org/10.1007/s10071-014-0823-3

Schusterman R (1972) Visual acuity in pinnipeds. In: Behavior of marine animals. Springer, pp 469–492

Schusterman RJR (1981) Behavioral capabilities of seals and sea lions: a review of their hearing, visual, learning and diving skills. Psychol Rec 31:125–143

Shannon C, Weaver W (1949) The mathematical theory of communication. Urbana, IL

Sjare B, Stirling I, Spencer C (2003) Structural variation in the songs of Atlantic walruses breeding in the Canadian High Arctic. Aquat Mamm 29:297–318. https://doi.org/10.1578/016754203101024121

Stirling I (1970) Observations on the behavior of the New Zealand fur seal (*Arctocephalus forsteri*). J Mammal:766–778

Trillmich F (1981) Mutual mother-pup recognition in Galapagos fur seals and sea lions: cues used and functional significance. Behaviour 21–42

Trillmich F, Trillmich KGK (1984) The mating systems of pinnipeds and marine iguanas: convergent evolution of polygyny. Biol J Linn Soc 21:209–216. https://doi.org/10.1111/j.1095-8312.1984.tb02062.x

Wartzok D, Ketten D (1999) Marine mammal sensory systems. In: Reynolds J, Rommel S (eds) Biology of marine mammals, Smithsonian Institution Press, pp 117–175

Wierucka K, Pitcher BJ, Harcourt R et al (2017) The role of visual cues in mother–pup reunions in a colonially breeding mammal. Biol Lett 13:20170444. https://doi.org/10.1098/rsbl.2017.0444

Wierucka K, Charrier I, Harcourt R, Pitcher BJ (2018a) Visual cues do not enhance sea lion pups' response to multimodal maternal cues. Sci Rep 8. https://doi.org/10.1038/s41598-018-28171-w

Wierucka K, Pitcher BJ, Harcourt R et al (2018b) Multimodal mother–offspring recognition: the relative importance of sensory cues in a colonial mammal. Anim Behav 146:135–142. https://doi.org/10.1016/j.anbehav.2018.10.019

Wyatt T (2003) Pheromones and animal behaviour: communication by smell and taste. Cambridge University Press

Wyatt TD (2014) Pheromones and animal behavior: chemical signals and signatures. Cambridge University Press

Chapter 16
The Sensory World of Otariids

Frederike D. Hanke, Colleen Reichmuth, and Peter Cook

Abstract Otariids face many unique challenges with respect to lifestyle and habitat. They need to find suitable foraging areas in the open ocean, detect and capture moving prey in near darkness, identify suitable mating partners in traditional terrestrial breeding areas, and relocate their pups following extended separations. Above all, otariids have to cope with the different physical properties of air and water. This chapter illustrates how the challenges of amphibious living have shaped the sensory systems considered to be the 'antennae' through which otariids gather information about the surrounding world. Our current understanding of the sensory capabilities of otariids comes from studies of both structure (anatomy, neurobiology) and function (sensitivity, performance) of specific sensory modalities. This information helps us to describe what the senses are specialized for and to identify the particular biological tasks they are probably involved in. However, future studies need to explicitly link the senses, behavior, and ecology. Altogether, this knowledge will be informative to behavioral ecologists in their attempts to determine why an otariid behaves the way it does.

Keywords Sea lions · Fur seals · Vision · Audition · Chemoreception · Olfaction · Gustation · Mechanoreception · Sensory adaptation

F. D. Hanke
Neuroethology, Institute for Biosciences, University of Rostock, Rostock, Germany

C. Reichmuth (✉)
Long Marine Laboratory, Institute for Marine Sciences, University of California Santa Cruz, Santa Cruz, CA, USA
e-mail: coll@ucsc.edu

P. Cook
Department of Psychology, New College of Florida, Sarasota, FL, USA

C. Campagna, R. Harcourt (eds.), *Ethology and Behavioral Ecology of Otariids and the Odobenid*, Ethology and Behavioral Ecology of Marine Mammals,
https://doi.org/10.1007/978-3-030-59184-7_16

16.1 Introduction

Behavioral ecology, the topic of this book, is focused on the interactions between animals and their environment (Krebs and Davies 1981). A key role herein is played by the sensory systems—as they are an individual's interface to the physical environment, providing the information that enables and guides specific behaviors. For otariids (as for all animals), sensory biology is thus fundamental to behavior.

Otariids represent a fascinating case of animals evolved to operate with precision within, and to transit effortlessly between, two sensory worlds. They show impressive capability at solving ecological and social problems both in their aquatic and terrestrial habitats—from foraging alone or in multi-species aggregations on an assortment of marine prey to providing extended maternal care in crowded coastal colonies. For the amphibious pinnipeds including the otariids, the competing demands of terrestrial and aquatic environments have strongly shaped their sensory systems and led to adaptations that allow them to efficiently receive relevant airborne and waterborne sensory cues. This is a remarkable feat, as the vastly different physical properties of air and water make it extremely difficult to build a sensor or instrument that is equally effective in both media.

Here, we review much of what is known about the sensory processing abilities of otariids, and contrast their sensory adaptations with those of other marine mammals as well as with terrestrial carnivores. A strong emphasis is placed on the California sea lion (*Zalophus californianus;* Fig. 16.1) as it is the best studied species among otariids with respect to sensory biology. Of course, sensory abilities are closely tied to cognitive operations, which will be described in more detail in the following chapter (Chap. 17). Together, these accounts of information gathering and information processing in otariids should provide behaviorists with an appropriate sensory perspective with which they can better understand the behaviors and strategies of free-ranging animals.

Fig. 16.1 (**a**) The California sea lion *(Zalophus californianus)* is a model species among otariids regarding sensory and cognitive research. The image illustrates the visible sensory structures including the large eyes, two types of vibrissae (mystacial and supraorbital), the external ear, and the nose, which remains tightly closed aside from breathing and sniffing in air. Image: C. Reichmuth/NMFS 14535. (**b**) The right eye of a sea lion showing the central flat region of the cornea. The image also offers a close-up of the mystacial vibrissae which are smooth and stiff. Image: C. Reichmuth/NMFS 14535. (**c**) The left ear of a sea lion showing the outer ear, the pinna. Image: MSC

16.2 The Visual System

The eyes are perhaps the most noticeable sensory structures in sea lions and fur seals (Fig. 16.1a, b). Otariids have large, bulging eyes with axial eye diameters from 30 to 44 mm (Mass and Supin 2005; Miller et al. 2010; Pütter 1903; Turner et al. 2017; West et al. 1991). Pinnipeds have orbits that are proportionally larger than those of related terrestrial species (Debey and Pyenson 2013). Among otariids, there is an interesting correlation between orbit size and body mass, with body mass in turn correlated to maximum diving depth; this correlation points to the fact that large eyes can be considered an adaptation to low light conditions experienced during diving. In California sea lions, several parameters of the visual system generally reflect the rather shallow dive depth of this species (usually less than 80 m; Feldkamp et al. 1989). These include a smaller pupillary range (Levenson and Schusterman 1997) and slower dark adaptation rate (~16 min) than deeper diving seals (Levenson and Schusterman 1999).

Despite not traveling to the depths explored by some other marine mammal species, otariids have eyes that are adapted to function under low ambient illumination. The eye itself is large, which allows the vertical, slit-shaped pupil to dilate widely to capture even small amounts of light. Like many predators that hunt in low-light conditions, otariids have a well-developed reflective layer, the *tapetum lucidum*, underlying the retina in the eye fundus. This tapetum maximizes photon absorbance by reflecting photons that were not absorbed by the photoreceptors at first passage through the retina, thus allowing absorbance at second passage (Miller et al. 2010; Pütter 1903; Turner et al. 2017; West et al. 1991). The tapetum of otariids lines the entire fundus (Miller et al. 2010) to maximize absorbance of photons reaching the eye from all directions, which is crucial for an animal that can rotate its head and body along all axes underwater. Otariids also appear sensitive to small differences in brightness contrasts. A brightness discrimination study of Scholtyssek and Dehnhardt (2013), in contrast to the study of Busch and Dücker (1987), revealed that a South African fur seal *(Arctocephalus pusillus)* was able to perceive brightness differences of ≤ 10 %, which is slightly better than the performance of previously tested phocid pinnipeds (14%; Scholtyssek and Kelber 2008) and even humans (11–14%; Cornsweet and Pinsker 1965; Griebel and Schmid 1997). Finally, the photoreceptors that line the otariid retina are predominantly light-sensitive rods with a wavelength of maximum absorbance at 497–501 nm, within the blue-green portion of the light spectrum that dominates under water (Crescitelli 1958; Lavigne and Ronald 1975a, b; Levenson et al. 2006).

In general, the retina of otariids is densely packed with photoreceptors, about 220,000 cells/mm^2 (Landau and Dawson 1970), a density higher than that of many terrestrial mammals including humans (120,000–160,000 cells/mm^2), although less than that found in fully aquatic mammals such as cetaceans (cell density in the bottlenose dolphin is ~400,000 cells/mm^2; Wartzok and Ketten 1999). Only a tiny fraction of the receptor population are cones (Peichl et al. 2001). Among the cones, only L-cones, with a wavelength of maximum absorbance of 560 nm, are present in

California sea lions (Levenson et al. 2006); no S-cones were found in this species (Peichl et al. 2001). Because of this, cone-based color vision may not be possible. One possibility for otariids to perceive colors would be by comparing the responses of cones and rods under light conditions when both receptor types are active, called mesopic light conditions. Unfortunately, color vision tests to address either cone-based (Busch and Dücker 1987; Griebel and Schmid 1992) or mesopic color vision (Oppermann et al. 2016) have failed thus far to fully control for ambient illumination. Consequently, the conclusion that California sea lions, South African fur seals, and South American fur seals *(Arctocephalus australis)* are able to perceive colors must be considered with caution.

The eyes of otariids show some characteristics reflecting their predatory nature. The viewing axes point to the front (Fig. 16.1a, b), and the visual fields of both eyes most likely overlap to some extent, which is a prerequisite for binocular depth perception. Additionally, otariids have good aerial and underwater visual acuity (Schusterman 1972; Schusterman and Balliet 1970, 1971) that is comparable to terrestrial carnivores (Rahmann 1967). Normally one would expect to find the amphibiously living otariids to be emmetropic (normal-sighted) in one medium and ametropic (short- or far-sighted) in the other, resulting in an inferior visual acuity in one medium. For the pinnipeds, it is usually assumed that they are emmetropic under water, supporting precise visual assessments such as prey localization during hunting. To provide a well-resolved image of the marine environment, all otariids have a spherical lens (Miller et al. 2010; West et al. 1991) which resembles the lenses of fish and differs from the more oblong lens shape of terrestrial mammals. The cornea, which provides focusing power in air, is ineffective at refracting light in water. In terrestrial environments, an eye with a lens adapted to underwater vision should result in myopic (near-sighted) vision due to the combined effects of both the spherical, highly refractive lens and the curved cornea in the aerial medium. A careful examination of the cornea of California sea lions however reveals an ingenious solution to this problem—a completely flattened region of the cornea (~ 6.5 mm diameter; Fig. 16.1b; Dawson et al. 1987; Miller et al. 2010). In contrast to a more typically curved mammalian cornea, this flattened cornea does not contribute much to the overall refractive power of the eye. Instead, this feature renders the cornea optically ineffective in air, allowing light to pass directly into the eye with only the circular lens focusing the light rays as in water. Visual performance reflects this remarkable anatomical adaptation. Behavioral measurements obtained from trained sea lions tested both in air and under water in bright light reveal visual acuity values of 5.5–5.7 cycles/deg in the California sea lion (Schusterman 1972; Schusterman and Balliet 1971) comparable to the visual acuity of felid carnivores (for example, visual acuity in cats: 6–9 cycles/deg; Peichl 1997; Rahmann 1967). Visual acuity was also assessed in behavioral experiments conducted under water in the Steller sea lion *(Eumetopias jubatus)* as 4.2 cycles/deg (Schusterman 1972; Schusterman and Balliet 1970), and in air in the South America fur seal as 4.2 cycles/deg and the South African fur seal as 4.5 cycles/deg (Busch and Dücker 1987). When ambient light levels decline, aerial visual acuity decreases faster than underwater visual acuity (Schusterman and Balliet 1971). Thus, visual adaptations allow sea lions to see clearly in both media with the exception of in dim conditions while on shore.

Behavioral measures of visual acuity obtained from experiments with trained sea lions are in the same range as measures of retinal resolution calculated on the basis of neural anatomy, that is, on the basis of the density of ganglion cells in the best vision zone of the retina, the *area centralis*. Both Steller sea lions and northern fur seals *(Callorhinus ursinus)* have maximum ganglion cell densities corresponding to a resolution of 5.3–5.5 cycles/deg in air and 7.1–7.2 cycles/deg in water (Mass 1992; Mass and Supin 2005). In addition to these species, the density and topography of ganglion cells in the retina was also assessed for the California sea lion (Landau and Dawson 1970). If, in the future, the internal geometry of the sea lions' eyes is characterized, the maximum ganglion cell density of 5,000 cells/mm^2 documented in the latter study can be used to estimate retinal resolution in California sea lions for further comparison with behavioral visual acuity measurements.

In other otariids, at least 260,000 ganglion cells were found in the retinae of northern fur seals (Mass and Supin 1992) and 177,500 ganglion cells in the retinae of Steller sea lions (Mass and Supin 2005). As in all pinnipeds, giant ganglion cells reaching a diameter of up to 50μm in some species (Mass and Supin 1992) accounted for 8–10% of the whole ganglion cell population in otariids (Mass and Supin 2007); these giant cells may serve to mediate motion information as hypothesized by Dawson et al. (1982). The ganglion cells give rise to the optic nerve, which transports the visual information to higher brain regions. The optic nerve of California sea lions comprises 130,000 myelinated axons (Turner et al. 2017). Only a single study has so far described the higher brain areas in sea lions which are involved in the processing of optical stimuli, such as the superior colliculus, the lateral geniculate nucleus, and the visual cortex (Turner et al. 2017). In contrast, many behavioral experiments concerning cognitive processing of visual information have been conducted; the visual stimuli used in these studies have included moving gestural cues given by experimenters, black-and-white two-dimensional shapes presented in air and underwater, and more complex, three-dimensional objects. These experiments are summarized by Cook et al. (Chap. 17).

The visual system of otariids is well-adapted to function in air as well as under water and shows many adaptations for vision under low light conditions. Underwater, predatory otariids may use vision to detect their prey when visibility is high but also under low contrast conditions. Observations of free-ranging individuals have revealed that sea lions mainly approach their swimming prey from below, suggesting that sea lions may silhouette the fish against the bright(er) water surface (Bonnot 1932; Hobson 1966). In this foraging context, future studies on motion vision will be highly informative. In the social domain, visual cues contribute to the recognition of individuals, as summarized by Charrier (Chap. 7) and Sinclair (Chap. 13), and may further function to support recognition of familiar places, an aspect left to be demonstrated in the future.

16.3 Mechanoreception

16.3.1 Haptics and Hydrodynamics

Besides the eyes, the face of otariids is dominated by long vibrissae (colloquially referred to as whiskers; Fig. 16.1a, b). Otariids have 20–40 vibrissae on each side of the snout, called mystacial vibrissae, and 1–2 vibrissa(e) above each eye, called supraorbital vibrissae (Fig. 16.1a, b; King 1983). The vibrissae of otariids are thicker, longer, and stiffer than those of terrestrial carnivores. In California sea lions, the longest of the 38 mystacial vibrissae (Dehnhardt 1994; Sawyer et al. 2016; Sprowls and Marshall 2019) reach a length of at least 12 cm (McHuron et al. 2020; Sawyer et al. 2016), whereas, among the otariids, the longest vibrissae (>48 cm) can be found in South Georgia Antarctic fur seals (*Arctocephalus tropicalis gazelle*; Bonner 1968) and Steller sea lions (C. Reichmuth, unpublished data). In contrast to the phocids, otariids do not have vibrissae close to the nose, called rhinal vibrissae (Ling 1977), and their vibrissae are smooth instead of undulated and slightly compressed in cross-section (oval rather than round; Hanke et al. 2010; Miersch et al. 2011). Comparative morphometric descriptions of otariid vibrissae were published by Ginter et al. (2012).

The vibrissae are composed of non-living keratinized tissue and emerge from follicle sinus complexes (FSCs) in the skin. As in other pinnipeds (Hyvärinen et al. 2009; Marshall et al. 2006; Mattson and Marshall 2016; McGovern et al. 2015), these FSCs have several distinguishing features. In California sea lions, FSCs are long with an upper cavernous sinus in addition to the ring and lower cavernous sinus (Sprowls and Marshall 2019; Stephens et al. 1973). The upper cavernous sinus, usually considered to support the function of the vibrissae in cold temperatures (Mauck et al. 2000), is shorter than in phocids, reflecting that sea lions inhabit temperate waters and are shallow divers. The follicle is innervated by the deep vibrissal nerve, and the innervation per follicle is higher than in terrestrial carnivores, although not as high as in phocid pinnipeds. For comparison, California sea lions have ~86,000 axons per vibrissal pad, while a terrestrial carnivore, the pole cat, has only ~7,500 axons per vibrissal pad (Hyvärinen et al. 2009), and seal species have 60,000–160,000 axons per vibrissal pad (Hyvärinen et al. 2009; Marshall et al. 2006; Mattson and Marshall 2016; McGovern et al. 2015). Sprowls and Marshall (2019) took differences in innervation of different vibrissae in the vibrissal pad into account and thus offer a more conservative estimate in comparison to the estimate provided by Sawyer et al. (2016). The extensive and highly organized neural investment along the pathway that extends from an individual vibrissa to somatosensory regions of the brain suggests significant information transfer. In the California sea lion as well as in the northern fur seal, the brain areas that receive and process this information are enlarged and spatially structured, with the neural tissue somatotopically mapped in the central nervous system (Ladygina et al. 1985, 1992; Sawyer et al. 2016). However, how otariids integrate the incoming information flow from individual vibrissae to develop a sensory representation based on the full sensory array remains poorly understood.

When otariids explore the world, they often protract their vibrissae to bring them into contact with objects in the environment. With their pliable vibrissae, California sea lions can rapidly respond to moving stimuli (Milne and Grant 2014) and discriminate objects on the basis of size and shape by using active touch (Dehnhardt 1990, 1994; Dehnhardt and Dücker 1996). For example, when presented with discs differing in diameter, a trained sea lion wearing a blindfold could reliably detect a size difference of 22% (Dehnhardt 1994); when presented with triangles differing in size, the same sea lion performed similarly well in a discrimination task (Dehnhardt and Dücker 1996). Shape discrimination between triangles and discs with identical surface areas is possible on the basis of the size difference between the longest measurable lines (Dehnhardt and Dücker 1996); after the sea lion had learned to perform this shape choice task visually, she was able to discriminate the objects using only her whiskers even when the size difference was less than 5%, corresponding to an absolute size difference of less than 2 cm. During such difficult size and shape discrimination tests, sea lions use short lateral head movements with protracted vibrissae to investigate and compare objects (Dehnhardt 1994; Dehnhardt and Dücker 1996).

While in water, the vibrissae of otariids seem to function as hydrodynamic sensors, as they do in phocids (Dehnhardt et al. 2014; Hanke et al. 2013), enabling the animals to sense water movements. Preliminary measurements conducted with a trained California sea lion revealed the vibrissae to be very sensitive to water movements generated by a dipole vibrating at 20 and 30 Hz (Dehnhardt and Mauck 2008); the sea lion was even more responsive than a trained harbor seal *(Phoca vitulina)* performing the same task at the same frequencies (Dehnhardt et al. 1998). Like harbor seals, a trained sea lion can find and follow the hydrodynamic trail left behind by a remote-controlled submarine (Gläser et al. 2011). However, the performance of the sea lion was reduced in comparison to harbor seals when tasked with following hydrodynamic trails including directional changes or when following hydrodynamic trails after increasing delay periods (Gläser et al. 2011); this experimental procedure mimics a situation in which a sea lion encounters a hydrodynamic trail of a fish that had swum by some time ago. The vibrissae of otariids are able to transmit external events such as a hydrodynamic trail left behind by a fish or a conspecific, as phocids do; however, structural differences result in noise being substantially reduced in the seal in comparison to the sea lion vibrissa (Hanke et al. 2010; Witte et al. 2012). Nevertheless, both types of vibrissae are able to detect external events; the proposed mechanism is described by Miersch et al. (2011).

From the few experimental studies conducted so far on the ability of California sea lions to perceive haptic and hydrodynamic information, we can confirm a role for active touch, the exploration of objects when in direct contact with surfaces, and for the detection and localization of swimming prey using hydrodynamic cues. Vibrissae may have other important roles as well, for example in nursing behavior or benthic feeding. In social contexts, the vibrissae are part of facial expressions and may signal arousal and provide other important communicative cues (Miller 1975). For example, the whiskers of male Steller sea lions extend forward and touch those of territorial neighbors during boundary displays, and serve as key components of multi-modal communication (Chap. 7).

16.3.2 Audition

Otariids have a visible outer ear, a pinna (Fig. 16.1a, c), which clearly distinguishes them from the phocids (and grants them the moniker of 'eared' seals). The pinna is small and tightly rolled, which reduces both drag and heat loss in water. The bony anatomy of the ear is comparable to that of terrestrial carnivores, with a few exceptions (Nummela 2008; Repenning 1972). The middle ear ossicles (bones) are not enlarged, as they are in other diving mammals (Nummela 1995). Other modifications in comparison to terrestrial animals include the increased size of the round window of the cochlea in comparison to the oval window, a small tympanic membrane and small middle ear cavity (Repenning 1972). As in phocids (and in contrast to terrestrial carnivores), the auditory canal and middle ear cavity are lined with cavernous tissue that apparently engorges with blood to counteract pressure changes during diving (Odend'hal and Poulter 1966; Repenning 1972). Repenning (1972) concluded that the otariid ear is best suited for hearing by bone and tissue conduction under water. In air, hearing seems to be mediated by the normal mammalian auditory pathway via the ear canal and middle ear ossicles.

As with other sensory modalities, most of what is known about the auditory sense in otariids comes from studies of California sea lions. Several studies have described the hearing ability of California sea lions (Fig. 16.2); these include studies using behavioral methods with trained individuals in air and under water, as well as electrophysiological methods with anesthesized individuals in air. Terrestrial hearing (Fig. 16.2a) is most sensitive within the frequency range between 1 and 23 kHz; hearing ability declines below 2 kHz and above 16 kHz, with an upper-frequency hearing limit around 38 kHz (Kastak and Schusterman 1998; Moore and Schusterman 1987; Mulsow et al. 2011; Reichmuth et al. 2013; Reichmuth et al. 2017; Schusterman 1974). While the hearing profile is generally similar to that of other carnivores, the lowest hearing thresholds (~ 0 dB re 20μPa, measured at 12 kHz) are slightly above those of both phocids (as low as –12 dB re 20μPa; Sills et al. 2015) and terrestrial carnivores (as low as –24 dB re 20μPa; Fay 1988); this difference may be due to attenuation by cartilaginous structures of the pinna and auditory canal that keep water from penetrating into the ears. Audiograms that are markedly similar to those measured for California sea lions have been obtained for northern fur seals (Babushina et al. 1991; Moore and Schusterman 1987) and Steller sea lions (Mulsow and Reichmuth 2010). As these three species span both the size range and greatest evolutionary distance among living otariids, it is likely that all otariids share similar hearing abilities (Mulsow and Reichmuth 2010; Southall et al. 2019). The finding that otariids have retained sensitive hearing for airborne sounds

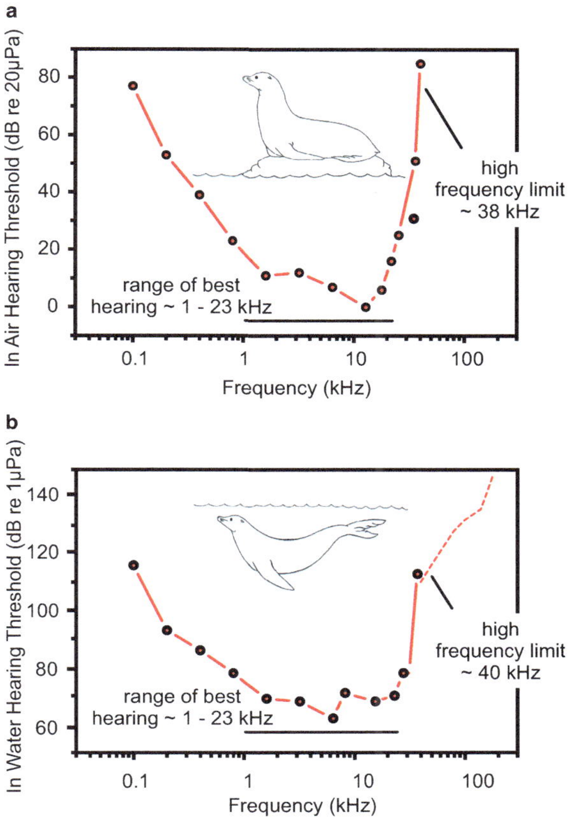

Fig. 16.2 Amphibious hearing profiles for California sea lions tested (**a**) in air (Reichmuth et al. 2017) and (**b**) under water (Reichmuth and Southall 2012), highlighting similar hearing abilities and range of hearing in both media. Panel b further shows sensitivity to sounds above the 'functional' high-frequency hearing limit (dotted line); at high levels, sea lions can detect ultrasound above 40 kHz, and as high as 180 kHz (Cunningham and Reichmuth 2016). Figure drawing: R. Jones

reflects that these species spend considerable amounts of time on land, and most of their social behavior occurs while on shore.

Underwater hearing in California sea lions (Fig. 16.2b) is similar to aerial hearing in terms of frequency sensitivity: audiograms obtained from trained individuals show best hearing between 1 and 23 kHz, with hearing rolling off gradually below 2 kHz and sharply above 30 kHz (Kastak and Schusterman 1998; Mulsow et al. 2012; Reichmuth et al. 2013; Reichmuth and Southall 2012; Schusterman et al. 1972; Southall et al. 2005). The lowest hearing thresholds in water (~ 62 dB re 1µPa, measured at 6 kHz) are less sensitive than those of both phocids (as low as 49 dB re 1µPa; Sills et al. 2015) and fully aquatic cetaceans (as low as 41 dB; Johnson 1967), suggesting that otariids are not quite as adapted for hearing under water. The 'functional' high-frequency limit of hearing in water is similar to that observed in air, around 40 kHz; unlike phocids, otariids do not show an extended range of acute high-frequency underwater hearing (see Reichmuth et al. 2013). Hemilä et al. (2006) proposed that high-frequency hearing in otariids is limited by cochlear constraints, as their hearing range in water is comparable to that of similar-sized carnivores in air.

Interestingly, and despite this constraint, sea lions can detect 'ultrasound' at frequencies extending from 40 to >180 kHz when underwater sounds are of high intensity (Fig. 16.2b; Cunningham and Reichmuth 2016). While otariids cannot discriminate between sound frequencies within this range (Schusterman and Moore 1978), they are still able to detect sounds above their true hearing range in water. Such sounds may include the echolocation clicks of killer whales, *Orcinus orca*, the output of various commercial echosounders, and 'ultrasonic' acoustic tags used to track the movements of fish (Cunningham et al. 2014).

Many other aspects of hearing in otariids have been studied, including temporal processing and sound localization. Perhaps not surprisingly, sea lions and other pinnipeds cannot resolve the temporal separation of sounds at rates as high as echolocating odontocete cetaceans can differentiate, and their temporal processing abilities are instead more similar to those of canids (Mulsow and Reichmuth 2007). The data available thus far suggest that northern fur seals can resolve the spatial origin of underwater sounds in the region of best hearing to within 8° of the source (Babushina and Poliakov 2004), close to the estimate by Gentry (1967) for California sea lions. In air, despite the absence of a true sound-ducting pinna, sound localization performance is even better, with an accuracy of < 5° in the horizontal plane (Holt et al. 2004, 2005; Moore 1975; Moore and Au 1975), comparable to the performance of cats, pigs, and rhesus macaques (5.7°, 4.6°, and 5.0°, respectively; Heffner and Heffner 1992).

A number of studies have examined the effects of noise on otariid hearing in air and under water. The ability of California sea lions and northern fur seals to detect simple and complex signals in the presence of masking noise has been reviewed elsewhere (Erbe et al. 2016; Reichmuth 2012). Studies concerning the temporary and permanent effects of noise exposure on hearing in sea lions have been thoroughly described by Finneran (2015) and Southall et al. (2019). Given high and fluctuating levels of noise in the coastal habitats used by otariids, such research has relevance to assessing risks from human-generated noise exposure, improving knowledge of auditory adaptations, and increasing understanding of how otariids operate in complex soundscapes.

Otariids can rely on their sense of hearing in air to detect threats and support long- and short-range social communication in noisy coastal rookeries (Charrier, Chap. 14). Their acute hearing abilities in water likely aid in alerting them to swimming movements and sounds generated by potential predators and prey, as fish emit sounds (Kasumyan 2008; Wilson et al. 2004) and some predators emit vocalizations and echolocation clicks. Otariids also produce some social sounds under water; for example, territorial male California sea lions bark under water when their territories are tidally submerged, and similar sounds can sometimes be detected at significant depths far offshore (Schusterman and Balliet 1969). In both terrestrial and aquatic environments, orientation and navigation are facilitated by a broad sense of the surrounding acoustic scene (Bregman 1990); this is especially true in the dark, three-dimensional underwater realm where distant acoustic cues may support a sense of space relative to the water surface (Schusterman et al. 2000).

16.4 Chemoreception

In general, the chemoreceptive senses, olfaction and gustation, are the least studied senses in otariids. Regarding olfaction, the external olfactory system and the olfactory bulb are reduced in California and Steller sea lions as well as in northern fur seals in comparison to terrestrial carnivores (Reep et al. 2006). However, Steller sea lions have maintained their olfactory receptor multigene family (Kishida et al. 2007). In line with this finding, Laska et al. (2008) successfully trained South African fur seals to discriminate between olfactory stimuli. The fur seals were able to discriminate between structurally similar chemical compounds, including related aliphatic odorants which differ in the length of their carbon chain (Laska et al. 2010) and between many of the presented enantiomers (Kim et al. 2013). This experimental evidence for sensitivity to airborne olfactory cues is entirely in line with observations of free-ranging otariids, many of which are described by Charrier (Chap. 15). Scent recognition is a critical component of individual recognition, including recognition of dependent pups by their mothers (Pitcher et al. 2011; Wierucka et al. 2018). Among other social contexts, olfactory cues have been shown in field observations and experiments to play a role in reproductive behavior; olfaction is used by adult males to detect the onset of estrous in breeding females, and territorial males emit strong odor cues (Gentry 1998). These data suggest an important role for scent cues in otariid behavior.

Regarding gustation, the apex of the tongue of California sea lions and northern fur seals is divided in two parts, and the two tips of the tongue are rounded (Kubota 1968; Sonntag 1923; Tuckerman 1890), the number of taste buds on the tongue is reduced, and the buds are modified in structure compared to terrestrial mammals. Nevertheless, the California sea lion (Friedl et al. 1990) and Steller sea lion (Kuznetzov 1990) can apparently perceive the primary tastes salty, sour, and bitter, except for sweet, albeit at a substantially higher concentration in comparison to humans. For example, a concentration of 3.6 parts per thousand (ppt) was needed to produce a sensation in one trained California sea lion (Friedl et al. 1990) while humans can perceive the same primary tastes at concentrations as low as 0.18 ppt (Pfaffmann et al. 1971). It remains unknown if the sense of taste is relevant to marine mammals when navigating in the ocean or when consuming fish under water; however, the apparent food preferences of captive individuals and aversion to substances hidden in food (C. Reichmuth, unpublished data) suggest that taste remains an important sense to otariids despite their adaptations for marine living.

16.5 Other Senses

Early research with California sea lions concentrated on putative echolocation abilities. However comprehensive experiments, summarized by Schusterman et al. (2000) and Cook et al. (Chap. 17), failed to demonstrate echolocation in sea lions. So far, there have been no studies on additional possible sensory modalities, such as on magneto- or electroreception, other than those mentioned in the previous paragraphs of this chapter. Although not a sensory dimension *per se*, a recent investigation revealed the ability of a South African fur seal to discriminate time intervals lasting milliseconds to seconds, with precision (Heinrich et al. 2020). A well-developed sense of time might support the classical sensory systems when otariids are making foraging decisions or need to estimate travel duration or distance.

16.6 Conclusions

From this review of available studies concerning the sensory systems of otariids, we find substantial information about the anatomy and function of some of the sensory modalities, especially of vision and audition, whereas others, such as chemoreception, remain poorly studied. The model species among otariids is the California sea lion (*Zalophus californianus*; Fig. 16.1), and there is only limited information available about sensory biology for other sea lions and fur seals. For the future, new insights into the adaptations of the sensory systems could be obtained by intensifying comparative research on otariids with different evolutionary, ecological, and life history characteristics.

Despite the gaps, we have tried to link the current knowledge of sensory processes in otariids to aspects of their ecology and behavior. It is clear that otariid evolution has produced a range of remarkable adaptations. By studying the structure and function of these adaptations we can better understand the environmental cues that support amphibious behavior.

The links between structure, function, and behavior provided in this chapter are mostly preliminary, often speculative hypotheses. We still await the design of new experiments that will relate measures of perception to natural behavior and allow us to consider the most promising paths forward for research. In our view, further developing the sensory ecology approach—which combines sensory physiology and behavioral ecology —should reveal the types of information otariids can obtain from the environment, how they gather and use this information, the role that ecology plays in shaping sensory systems to best acquire and process information, and how sensory abilities influence observed behavior (Dusenbery 1992; Stevens 2013). In addition, continued research to better connect sensory acquisition with sensory processing, perception, and cognition in general will be informative (see Chap. 17 by Cook et al.).

Behavior and cognition are both supported and limited by the sensory world (or 'Umwelt') of an animal (Barett 2011). With this in mind, it may someday be possible to understand how sea lions and other otariids form a multi-modal representation of the external world (whether it be terrestrial or aquatic) through sensory integration. This is one of the key questions of sensory biology (Johnsen 2017) and is of particular interest to neuroscientists studying brain organization, development, and evolution (Ghazanfar et al. 2005). New tools may advance this goal, such as using functional brain imaging to map the connections between different sensory regions and to visualize activated brain areas during multimodal stimulation. Research that enables us to access the perceptual worlds inhabited by amphibious marine mammals is of interest to sensory biologists and behavioral ecologists, but also for neurobiologists, as such studies provide insight into how otariid carnivores have adapted to a lifestyle spanning the shore and the sea.

References

Babushina ES, Poliakov MA (2004) The underwater and airborne horizontal localization of sound by the northern fur seal. Biofizika 49:723–726

Babushina YS, Zaslavskii GL, Yurekevich LI (1991) Air and underwater hearing characteristics of the northern fur seal: audiograms, frequency and differential thresholds. Biophysics 36:909–913

Barett L (2011) Beyond the brain: how body and environment shape animal and human minds. University Press, Princeton

Bonner WN (1968) The fur seal of South Georgia. Br Antarctic Surv Bull 56:1–95

Bonnot P (1932) A note on the fishing of the California sea lion. Calif Fish Game 18:98–99

Bregman AS (1990) Auditory scene analysis: the perceptual organization of sound. Bradford Books, MIT Press, Cambridge

Busch H, Dücker G (1987) Das visuelle Leistungsvermögen der Seebären *(Arctocephalus pusillus und Arctocephalus australis)*. Zool Anz 219:197–224

Cornsweet JC, Pinsker HM (1965) Luminance discrimination of brief flashes under various conditions of adaptation. J Physiol 176:294–310

Crescitelli F (1958) The natural history of visual pigments. Ann N Y Acad Sci 74:230–255

Cunningham KA, Reichmuth C (2016) High-frequency hearing in seals and sea lions. Hear Res 331:83–91

Cunningham KA, Hayes SA, Wargo Rub AM, Reichmuth C (2014) Auditory detection of ultrasonic coded transmitters by seals and sea lions. J Acoust Soc Am 135:1978–1985

Dawson WW, Hawthorne MN, Jenkins RL, Goldston RT (1982) Giant neural systems in the inner retina and optic nerve of small whales. J Comp Neurol 205:1–7

Dawson WW, Schroeder JP, Sharpe SN (1987) Corneal surface properties of two marine mammal species. Mar Mamm Sci 3:186–197

Debey LB, Pyenson ND (2013) Osteological correlates and phylogenetic analysis of deep diving in living and extinct pinnipeds: what good are big eyes? Mar Mamm Sci 29:48–83

Dehnhardt G (1990) Preliminary results from psychophysical studies on the tactile sensitivity in marine mammals. In: Thomas JA, Kastelein R (eds) Sensory abilities of cetaceans. Pleum Press, New York, pp 435–446

Dehnhardt G (1994) Tactile size discrimination by a California sea lion *(Zalophus californianus)* using its mystacial vibrissae. J Comp Physiol A 175:791–800

Dehnhardt G, Dücker G (1996) Tactual discrimination of size and shape by a California sea lion (Zalophus californianus). Anim Learn Behav 24(4):366–374

Dehnhardt G, Mauck B (2008) Mechanoreception in secondarily aquatic vertebrates. In: Thewissen JGM, Nummela S (eds) Sensory evolution on the threshold: adaptations in secondarily aquatic vertebrates. University of California Press, Berkeley, pp 295–314

Dehnhardt G, Mauck B, Bleckmann H (1998) Seal whiskers detect water movements. Nature 394:235–236

Dehnhardt G, Hanke W, Wieskotten S, Krüger Y, Miersch L (2014) Hydrodynamic perception in seals and sea lions. In: Bleckmann H (ed) Flow sensing in air and water. Springer, Berlin, pp 147–167

Dusenbery DB (1992) Sensory ecology—how organisms acquire and respond to information. W. H. Freeman, New York
Erbe C, Reichmuth C, Cunningham K, Lucke K, Dooling R (2016) Communication masking in marine mammals. a review and research strategy. Mar Pollut Bull 103:15–38
Fay RR (1988) Hearing in vertebrates: a psychophysics databook. Hill-Fay, Winnetka
Feldkamp SD, DeLong RL, Antonelis GA (1989) Diving patterns of California sea lions, *Zalophus californianus*. Can J Zool 67:872–883
Finneran JJ (2015) Noise-induced hearing loss in marine mammals: a review of temporary threshold shift studies from 1996 to 2015. J Acoust Soc Am 138:1702–1726
Friedl WA, Nachtigall PE, Moore PWB, Chun NKW (1990) Taste reception in the bottlenose dolphin *(Tursiops truncatus gilli)* and the California sea lion *(Zalophus californianus)*. In: Thomas JA, Kastelein RA (eds) Sensory abilities of cetaceans. Plenum, New York, pp 447–451
Gentry L (1967) Underwater auditory localization in the California sea lion (*Zalophus californianus*). J Aud Res 7:187-193
Gentry RL (1998) Behavior and ecology of the Northern fur seal. Princeton University Press, Princeton
Ghazanfar AA, Maier JX, Hoffman KL, Logothetis NK (2005) Multisensory integration of dynamic faces and voices in rhesus monkey auditory cortex. J Neurosci 25:5004–5012
Ginter CC, DeWitt TJ, Fish FE, Marshall CD (2012) Fused traditional and geometric morphometrics demonstrate pinniped whisker diversity. PLoS One 7:e34481
Gläser N, Wieskotten S, Otter C, Dehnhardt G, Hanke W (2011) Hydrodynamic trail following in a California sea lion *(Zalophus californianus)*. J Comp Physiol A 197:141–151
Griebel U, Schmid A (1992) Color vision in the California sea lion *(Zalophus californianus)*. Vision Res 32:477–482
Griebel U, Schmid A (1997) Brightness discrimination ability in the West Indian Manatee *(Trichechus manatus)*. J Exp Biol 200:1587–1592
Hanke W et al (2010) Harbor seal vibrissa morphology suppresses vortex-induced vibrations. J Exp Biol 213:2665–2672
Hanke W, Wieskotten S, Marshall CD, Dehnhardt G (2013) Hydrodynamic perception in true seals *(Phocidae)* and eared seals *(Otariidae)*. J Comp Physiol A 199:421–440
Heffner RS, Heffner HE (1992) Evolution of sound localization in mammals. In: Webster DB, Fay RR, Popper AN (eds) The evolutionary biology of hearing. Springer, New York, pp 691–715
Heinrich T, Ravignani A, Hanke FD (2020) Visual timing abilities of a harbour seal *(Phoca vitulina)* and a South African fur seal *(Arctocephalus pusillus pusillus)* for sub- and supra-second time intervals. Anim Cogn 23(5):851–859
Hemilä S, Nummela S, Berta A, Reuter T (2006) High-frequency hearing in phocid and otariid pinnipeds: an interpretation based on inertial and cochlear constraints (L). J Acoust Soc Am 120:3463-3466
Hobson ES (1966) Visual orientation and feeding in seals and sea lions. Nature 210:326–327
Holt MM, Schusterman RJ, Southall BL, Kastak D (2004) Localization of aerial broadband noise by pinnipeds. J Acoust Soc Am 115:2339–2345
Holt MM, Schusterman RJ, Kastak D, Southall BL (2005) Localization of aerial pure tones by pinnipeds. J Acoust Soc Am 118:3921–3926
Hyvärinen H, Palviainen A, Strandberg U, Holopainen IJ (2009) Aquatic environment and differentiation of vibrissae: comparison of sinus hair systems of ringed seal, otter and pole cat. Brain Behav Evol 74:268–279
Johnsen S (2017) Open questions: we don't really know anything, do we? Open questions in sensory biology. BMC Biol 15. https://doi.org/10.1186/s1291-017-0385-3
Johnson CS (1967) Sound detection thresholds in marine mammals. In: Tavolga WN (ed) Marine bioacoustics. Pergamon, New York, pp 247–260
Kastak D, Schusterman RJ (1998) Low-frequency amphibious hearing in pinnipeds: methods, measurements, noise, and ecology. J Acoust Soc Am 103:2216–2228
Kasumyan AO (2008) Sounds and sound production in fishes. J Ichthyol 48:981–1030

Kim SK, Amundin M, Laska M (2013) Olfactory discrimination ability of South African fur seals *(Arctocephalus pusillus)* for enantiomers. J Comp Physiol A 199:535–544
King JE (1983) Seals of the world. Cambridge University Press, Cambridge
Kishida T, Kubota S, Shirayama Y, Fukami H (2007) The olfactory receptor gene repertoires in secondary-adapted marine vertebrates: evidence for reduction of the functional proportions in cetaceans. Biol Lett 3:428–430
Krebs JR, Davies NB (1981) An introduction to behavioral ecology. Blackwell, Oxford
Kubota K (1968) Comparative anatomical and neurohistological observations on the tongue of the northern fur seal *(Callorhinus ursinus)*. Anat Rec 161:257–266
Kuznetzov VB (1990) Chemical sense of dolphins. In: Thomas JA, Kastelein RA (eds) Sensory abilities of cetaceans. Plenum, New York, pp 481–503
Ladygina TF, Popov VV, Supin AY (1985) Topical organization of somatic projections in the fur seal cerebral cortex. Neurophysiology 17:246–252
Ladygina TF, Popov VV, Supin AY (1992) Somatotopic projections in the cerebral cortex of the fur seal *(Callorhinus ursinus)*. Acad Sci Moskow 17:344–351
Landau D, Dawson WW (1970) The histology of retinas from the pinnipedia. Vision Res 10:691–702
Laska M, Svelander M, Amundin M (2008) Successful acquisition of an olfactory discrimintation paradigm by South African fur seals, *Arctocephalus pusillus*. Physiol Behav 93:1033–1038
Laska M, Lord E, Selin S, Amundin M (2010) Olfactory discrimination of aliphatic odorants in South African fur seals *(Arctocephalus pusillus)*. J Comp Psychol 124:187–193
Lavigne DM, Ronald K (1975a) Evidence of duplicity in the retina of the California sea lion *(Zalophus californianus)*. Comp Biochem Physiol 50A:65-70
Lavigne DM, Ronald K (1975b) Pinniped visual pigments. Comp Biochem Physiol 52:325–329
Levenson DH, Schusterman RJ (1997) Pupillometry in seals and sea lions: ecological implications. Can J Zool 75:2050–2057
Levenson DH, Schusterman RJ (1999) Dark adaptation and visual sensitivity in shallow and deep-diving pinnipeds. Mar Mamm Sci 15:1303–1313
Levenson DH, Ponganis PJ, Crognale MA, Deegan JF II, Dizon A, Jacobs GH (2006) Visual pigments of marine carnivores: pinnipeds, polar bear, and sea otter. J Comp Physiol A 192:833–843
Ling JK (1977) Vibrissae of marine mammals. In: Harrison RJ (ed) Functional anatomy of marine mammals. Academic, London, pp 387–415
Marshall CD, Amin H, Kovacs KM, Lydersen C (2006) Microstructure and innervation of the vibrissal follicle-sinus complex in the bearded seal, *Erignathus barbatus (Pinnipedia: Phocidae)*. Anat Rec 288A:13–25
Mass AM (1992) Retinal topography in the walrus *(Odobenus rosmarus divergence)* and fur seal *(Calorhinus ursinus)*. In: Thomas J, Kastelein R (eds) Marine mammal sensory systems. Plenum, New York, pp 119–135
Mass AM, Supin AY (1992) Peak density, size and regional distribution of ganglion cells in the retina of the fur seal *Callorhinus ursinus* Brain. Behav Evol 39:69–76
Mass AM, Supin AY (2005) Ganglion cell topography and retinal resolution of the Stellar sea lion *(Eumetobias jubatus)*. Aquat Mamm 31:393–402
Mass AM, Supin AY (2007) Adaptive features of aquatic mammals' eyes. Anat Rec 290:701–715
Mattson EE, Marshall CD (2016) Follicle microstructure and innervation vary between pinniped micro- and macrovibrissae. Brain Behav Evol 88:43–58
Mauck B, Eysel U, Dehnhardt G (2000) Selective heating of vibrissal follicles in seals *(Phoca vitulina)* and dolphins *(Sotalia fluviatilis guianensis)*. J Exp Biol 203:2125–2131
McGovern KA, Marshall CD, Davis RW (2015) Are vibrissae viable sensory structures for prey capture in Northern elephant seals, *Mirounga angustirostris*? Anat Rec 298:750–760
McHuron EA, Williams T, Costa DP, Reichmuth C (2020) Contrasting whisker growth dynamics within the phocid lineage. Mar Ecol Prog Ser 634:231–236

Miersch L et al (2011) Flow sensing in pinniped whiskers. Philos Trans R Soc Lond B Biol Sci 366:3077–3084
Miller EH (1975) A comparative study of facial expressions of two species of pinnipeds. Behaviour 53:268–284
Miller SN, Colitz CMH, Dubielzig RR (2010) Anatomy of the California sea lion globe. Vet Ophthalmol Suppl:63–71
Milne AO, Grant RA (2014) Characterization of whisker control in the California sea lion *(Zalophus californianus)* during complex, dynamic sensorimotor task. J Comp Physiol A 200:871–879
Moore PWB (1975) Underwater localization of click and pulsed pure tone signals by the California sea lion *(Zalophus californianus)*. J Acoust Soc Am 57:406–410
Moore PWB, Au WL (1975) Underwater localization of pulsed pure tones by the California sea lion *(Zalophus californianus)*. J Acoust Soc Am 58:721–727
Moore PWB, Schusterman RJ (1987) Audiometric assessment of northern fur seals, *Callorhinus ursinus*. Mar Mamm Sci 3:31–53
Mulsow J, Reichmuth C (2007) Electrophysiological assessment of temporal resolution in pinnipeds. Aquat Mamm 33:122–131
Mulsow J, Reichmuth C (2010) Psychophysical and electrophysiological aerial audiograms of a Steller sea lion *(Eumetopias jubatus)*. J Acoust Soc Am 127:2692–2701
Mulsow J, Finneran JJ, Houser DS (2011) California sea lion *(Zalophus californianus)* aerial hearing sensitivity measured using auditory steady-state response and psychophysical methods. J Acoust Soc Am 129:2298–2306
Mulsow J, Houser DS, Finneran JJ (2012) Underwater psychophysical audiogram of a young male California sea lion *(Zalophus californianus)*. J Acoust Soc Am 131:4182–4187
Nummela S (1995) Scaling of the mammalian middle ear. Hear Res 85:18–30
Nummela S (2008) Hearing in aquatic mammals. In: Thewissen JGM, Nummela S (eds) Sensory evolution on the threshold: adaptations in secondarily aquatic vertebrates. University of California Press, Berkeley, pp 211–224
Odend'hal S, Poulter TC (1966) Pressure regulation in the middle ear cavity of sea lions: a possible mechanism. Science 153:768–769
Oppermann D, Schramme J, Neumeyer C (2016) Rod-cone based color vision in seals under photopic conditions. Vision Res 125:30–40
Peichl L (1997) Die Augen der Säugetiere: unterschiedliche Blicke in unsere Welt. Biologie in unserer Zeit 27:96–105
Peichl L, Behrmann G, Kröger RHH (2001) For whales and seals the ocean is not blue: a visual pigment loss in marine mammals. Eur J Neurosci 13:1520–1528
Pfaffmann C, Bartoshuk LM, McBurney D (1971) Taste psychophysics. In: Beidler LM (ed) Handbook of sensory physiology. Springer, Berlin, pp 73–99
Pitcher BJ, Harcourt RG, Schaal B, Charrier I (2011) Social olfaction in marine mammals: wild female Australian sea lions can identify their pup's scent. Biol Lett 7:60–62
Pütter A (1903) Die Augen der Wassersäugethiere. Zoologische Jahrbuecher Abteilung fuer Anatomie und Ontogenie der Tiere 17:99–402
Rahmann H (1967) Die Sehschärfe bei Wirbeltieren Naturwissenschaftliche Rundschau 1:10–14
Reep RL, Finlay BL, Darlington RB (2006) The limbic system in mammalian brain evolution. Brain Behav Evol 70:57–70
Reichmuth C (2012) Psychophysical studies of auditory masking in marine mammals. In: Popper AN, Hawkins A (eds) The effects of noise on aquatic life. Springer, Berlin, pp 23–27
Reichmuth C, Southall BL (2012) Underwater hearing in California sea lion *(Zalophus californianus)*: expansion and interpretation of existing data. Mar Mamm Sci 28:358–363
Reichmuth C, Holt MM, Mulsow J, Sills JM, Southall BL (2013) Comparative assessment of amphibious hearing in pinnipeds. J Comp Physiol A 199:491–507
Reichmuth C, Sills JM, Ghoul A (2017) Psychophysical audiogram of a California sea lion listening for airborne tonal sounds in an acoustic chamber. Proc Meet Acoust 30:010001

Repenning CA (1972) Underwater hearing in seals: functional morphology. In: Harrison RJ (ed) Functional anatomy of marine mammals. Academic, London, pp 307–331
Sawyer EK, Turner EC, Kaas JH (2016) Somatosensory brainstem, thalamus, and cortex of the California sea lion *(Zalophus californianus)*. J Comp Neurol 52:1957–1975
Scholtyssek C, Dehnhardt G (2013) Brightness discrimination in the South African fur seal *(Arctocephalus pusillus)*. Vision Res 84:26–32
Scholtyssek C, Kelber A, Dehnhardt G (2008) Brightness discrimination in the harbor seal *(Phoca vitulina)*. Vision Res 48:96–103
Schusterman RJ (1972) Visual acuity in pinnipeds. In: Winn HE, Olla BL (eds) Behavior of marine animals, vol 2. Plenum, New York
Schusterman RJ (1974) Auditory sensitivity of a California sea lion to airborne sound. J Acoust Soc Am 56:1248–1251
Schusterman RJ, Balliet RF (1969) Underwater barking by male sea lions *(Zalophus californianus)*. Nature 222:1179–1181
Schusterman RJ, Balliet RF (1970) Visual acuity of the harbour seal and the Stellar sea lion under water. Nature 226:563–564
Schusterman RJ, Balliet RF (1971) Aerial and underwater visual acuity in the California sea lion *(Zalophus californianus)* as a function of luminance. Ann N Y Acad Sci 188:37–46
Schusterman RJ, Moore PW (1978) The upper limit of underwater auditory frequency discrimination in the California sea lion. J Acoust Soc Am 63:191–1595
Schusterman RJ, Balliet RF, Nixon J (1972) Underwater audiogram of the California sea lion by the conditioned vocalization technique. J Exp Anal Behav 17:339–350
Schusterman RJ, Kastak D, Levenson DH, Reichmuth CJ, Southall BL (2000) Why pinnipeds don't echolocate. J Acoust Soc Am 107:2256–2264
Sills JM, Southall BL, Reichmuth C (2015) Amphibious hearing in ringed seals *(Pusa hispida)*: underwater audiograms, aerial audiograms and critical ratio measurements. J Exp Biol 218:2250–2259
Sonntag CF (1923) The comparative anatomy of the tongues of the mammalia. VIII, Carnivora. Proc Zool Soc Lond 9:129–153
Southall BL, Schusterman RJ, Kastak D, Reichmuth Kastak C (2005) Reliability of underwater hearing thresholds in pinnipeds. J Acoust Soc Am 6:243–249
Southall BL et al (2019) Marine mammal noise exposure criteria: updated scientific recommendations for residual hearing effects. Aquat Mamm 45:125–232
Sprowls CD, Marshall CD (2019) Innervation patterns of mystacial vibrissae support active touch behaviors in California sea lions *(Zalophus californianus)*. J Morphol 280:1617–1627
Stephens RJ, Beebe IJ, Poulter TC (1973) Innervation of the vibrissae of the California sea lion, *Zalophus californianus*. Anat Rec 176:421–442
Stevens M (2013) Sensory ecology, behaviour, and evolution. Oxford University Press, Oxford
Tuckerman F (1890) On the gustatory organs of some of the mammalia. J Morphol 4:151–193
Turner EC, Sawyer EK, Kaas JH (2017) Optic nerve, superior colliculus, visual thalamus, and primary visual cortex of the northern elephant seal *(Mirounga angustirostris)* and California sea lion *(Zalophus californianus)*. J Comp Neurol 525(9):2109–2132
Wartzok D, Ketten DR (1999) Marine mammal sensory systems. In: Reynolds JE III, Rommel SA (eds) Biology of marine mammals. Smithsonian Institution Press, Washington, pp 117–175
West JA, Sivak JG, Murphy CJ, Kovacs KM (1991) A comparative study of the anatomy of the iris and ciliary body in aquatic mammals. Can J Zool 69:2594–2607
Wierucka K, Pitcher BJ, Harcourt R, Charrier I (2018) Multimodal mother-offspring recognition: the relative importance of sensory cues in a colonial mammal. Anim Behav 146:135–142
Wilson B, Batty RS, Dill LM (2004) Pacific and Atlantic herring produce burst pulse sounds. Biol Lett 271:S95–S97
Witte M, Hanke W, Wieskotten S, Miersch L, Brede M, Dehnhardt G, Leder A (2012) On the wake flow dynamics behind harbor seal vibrissae—a fluid mechanical explanation for an extraordinary capability. In: Tropea C, Bleckmann H (eds) Nature-inspired fluid mechanics. Springer, Berlin, pp 241–260

Chapter 17
The Mind of a Sea Lion

Peter Cook, Colleen Reichmuth, and Frederike D. Hanke

Abstract The mind of a sea lion links sensory information gathered from the surrounding environment to the decision operations, or cognitive processes, that enable behavioral responses based on judgement and experience. Here, we illustrate the sea lion's impressive capabilities for complex associative learning and highlight some of the unique and essential contributions sea lions have made to our understanding of problem solving, concept formation, and memory in non-human animals. We suggest that the persistence and behavioral flexibility that allow sea lions and fur seals to exploit unpredictable environments make them particularly well suited to performing in the behavioral laboratory, and therefore, highly valuable models for comparative studies of learning and intelligence.

Keywords Sea lions · Fur seals · Otariidae · Behavior · Cognition · Learning · Memory · Neuroscience · Neurobiology

Electronic Supplementary Material The online version of this chapter (https://doi.org/10.1007/978-3-030-59184-7_17) contains supplementary material, which is available to authorized users.

P. Cook
Department of Psychology, New College of Florida, Sarasota, FL, USA

C. Reichmuth (✉)
Long Marine Laboratory, Institute for Marine Sciences, University of California Santa Cruz, Santa Cruz, CA, USA
e-mail: coll@ucsc.edu

F. D. Hanke
Neuroethology, Institute for Biosciences, University of Rostock, Rostock, Germany

C. Campagna, R. Harcourt (eds.), *Ethology and Behavioral Ecology of Otariids and the Odobenid*, Ethology and Behavioral Ecology of Marine Mammals,
https://doi.org/10.1007/978-3-030-59184-7_17

17.1 Introduction

The otariidae are familiar to the public for high-energy participation in animal shows and their intelligence and persistence in exploiting human-made opportunities for fish predation. From these general impressions, a lay-person might sketch the sea lion's intellect as keen, and their personality as stubborn. The empirical evidence gathered over the past half century supports this impression. Given the small number of otariid research subjects featured in behavioral and cognitive sciences, there is a surprising wealth of careful, detailed studies; the majority of these feature one species, the California sea lion, *Zalophus californianus*.

Sea lions have demonstrated some cognitive and behavioral capabilities that are either unique or rarely observed in other non-human animals, including cetaceans and primates. Sea lions are behaviorally and cognitively flexible, with intense task focus and motivation, and therefore may play a disproportionately large role in comparative cognitive research. Schusterman et al. (2002) provided a comprehensive review of sea lion cognition. Here, we summarize research on perception and cognition and highlight recent findings, with particular attention to the cognitive world of the California sea lion. We also consider potentially rich avenues for future behavioral and neurobiological research, and begin with a history of behavioral research with California sea lions and other otariid carnivores.

17.2 Foundational Research

The sea lion's entry into behavioral science came largely as a response to similar research being conducted with fully aquatic odontocetes such as the bottlenose dolphin. Excitement regarding the cognitive and sensory capabilities of dolphins drove a wave of comparative research with marine mammals in the mid-twentieth century. Much of this work focused on the dolphins' remarkable ability to echolocate under water (Kellogg and Kohler 1952; Kellogg 1961). Some early researchers believed that other marine mammals might also share this biosonar capability, and a particular focus was placed on sea lions, which had been observed to navigate effectively in low-light underwater environments and produce pulsed trains of underwater sound (e.g., Poulter 1963). This interest led to federal funding in the United States and the subsequent establishment of a sea lion sensory/behavior research laboratory at the Stanford Research Institute in California (Fig. 17.1a). One of the young researchers tasked with examining echolocation in sea lions was Ronald Schusterman—a mentor to all three authors of this chapter—who would become instrumental in advancing behavioral and cognitive research with sea lions for more than 50 years (see Schusterman 1981, 2010; Schusterman et al. 2002; Fig. 17.1b). Much of his work was conducted at the Stanford Research Institute, the California State University of Hayward, and at the pinniped research facility he

Fig. 17.1 (**a**) The biosonar laboratory at Stanford Research Institute's Coyote Hills Laboratory in Fremont, California (**b**) Professor Ronald Schusterman with California sea lion *Rocky* at Long Marine Laboratory in Santa Cruz, California (**c**) California sea lion *Bibi* performing a stimulus discrimination task in which the size of the incorrect object was gradually increased

founded in 1985 at Long Marine Laboratory at the University of California Santa Cruz.

Schusterman was a behavioral scientist with training in psychophysics—the study of how physical stimuli relate to mental phenomena—and he grounded his work with sea lions in rigorous assessments of the animals' sensory capabilities (see Reichmuth 2018). By working with highly trained sea lion subjects eager to engage in carefully designed behavioral tasks in exchange for fish rewards, Schusterman and others discovered a great deal about the specific sensory capabilities and general learning strategies that sea lions could use to solve difficult problems (Schusterman 1981). They did not, however, find convincing evidence of 'echo-ranging' abilities despite an assortment of clever observations and experiments aimed at revealing such a sense (summarized in Schusterman et al. 2000a). Schusterman concluded that sea lions do not actively echolocate as cetaceans do but instead rely upon a combination of highly developed amphibious visual, auditory and tactile abilities (see Chap. 16 by Hanke et al.; Chaps. 14 and 15 by Charrier).

Schusterman's negative findings put an end to attempts to develop sea lions as subjects for biosonar and 'echo-ranging' research, but promoted sea lions as excellent models for general comparative studies of sensory biology and cognition. Healthy captive sea lions learned quickly, demonstrated high task focus, and were cooperative and enthusiastic participants in sensory and behavioral experiments. Following Schusterman's early work on sensory biology, he and other investigators continued to study sea lions in a range of behavioral and cognitive domains, including learning, memory, communication, language, and concept formation. Much of this work has been previously reviewed (see Schusterman et al. 2002; Schusterman and Kastak 2002). We highlight some of the most significant results before moving to more recent work.

Early research on sensory biology in sea lions led to a variety of clever experiments that revealed important learning abilities transcending specific sensory modalities. Following work on 'errorless' learning techniques in pigeons (Terrace 1963),

Schusterman began applying similar methods for sea lions, using 'fading' (gradually reducing the salience of a stimulus over successive trials or exposures) to demonstrate that sea lions could learn new sensory discriminations with few to no errors along the way (Schusterman 1966; Fig. 17.1c). Not only could sea lions learn to reliably select one stimulus over another in this manner, but they could learn to reverse their preferences, again and again, with minimal errors throughout the learning process (Schusterman and Thomas 1966). The same was later shown to be true in spatial learning tasks with sea lions, when predictable response locations were reversed (Beach and Pepper 1974). Such 'reversal learning,' in which an animal learns to switch its selection preference depending on reward contingencies, represents an impressive display of behavioral flexibility afforded by application of basic learning mechanisms. Schusterman also demonstrated efficient, flexible, and generalizable learning mechanisms through exploration of what has been termed 'learning sets,' which represent an animal's *learning to learn* ability. This ability involves learning to solve different problems of a similar type more quickly on repeated exposure (Harlow 1949). Schusterman charted this performance improvement effect in sea lions learning successive stimulus discrimination problems with arbitrary visual shapes (Schusterman and Thomas 1966), and the results supported his ongoing belief in the importance of using many example problems during training (e.g., Kastak and Schusterman 1994). By providing experimental animals with many training problems of a given type prior to testing a specific ability or process with novel problems, Schusterman believed an experimenter could reduce the potential effects of response strategy and idiosyncratic responses on cognitive performance. These and other studies were conducted during a time when animal learning studies were developing within behavioral frameworks from the emerging field of operant conditioning (Skinner 1938; Breland and Breland 1966). Schusterman's sea lion studies were pioneering demonstrations of fundamental cognitive abilities that paved the way for systematic evaluation of animal intelligence.

The 1970s and 1980s were golden years for research in animal cognition, particularly for work with primates and marine mammals. A major focus was to establish artificial communication frameworks that would enable human-animal informational exchanges. Language-learning studies with non-human animals were centered primarily on sign language and symbol use with primates, considered of great interest due to their close relationship to humans, as well as complex vocal production learners such as parrots and dolphins. A general consensus has since emerged that while animals are capable of learning arbitrary stimulus-symbol bindings—with some primates, parrots, and even dogs capable of learning hundreds of these word-like associations (Hilix and Rumbaugh 2013; Pilley and Hinzmann 2013)—non-human animals have not demonstrated an ability to use grammar or syntax in a systematic way. Some animals, however, including great apes and cetaceans, apparently have the capability to respond appropriately to syntactic elements of signals, specifically, the order or pattern of symbol or words (Kako 1999; Shanker and King 2002). This ability has been taken by some as evidence of convergent evolution of syntactically relevant cognitive capability (e.g., Fitch 2011).

Fig. 17.2 (**a**) California sea lion *Rocky* demonstrating application of the problem-solving concept of reflexivity or 'sameness' by matching unfamiliar but identical visual shapes, see also Supplementary Video 17.2 (**b**) California sea lion *Rio* demonstrating the formation of functional stimulus categories by matching shapes belonging to the same abstract grouping ('letters'), see also Supplementary Video 17.3

While sea lions lack complicated social communication systems, they have provided an excellent 'generalist' mammalian system with which to consider the true nature of certain cognitive abilities, including those related to 'artificial' language learning. Schusterman and his colleagues built upon their strong foundation of basic studies to show that sea lions could also perform well in highly complex instructional tasks. He trained several sea lions, including star pupil *Rocky*, to learn gestural cues for objects, their descriptive characteristics, and actions that could be performed with them. *Rocky* could respond appropriately to even novel assemblies of these cues (e.g., *large white cone, black small ball fetch*—meaning "*bring the small black ball to the large white cone*"), alter her behavior in response to changes in cue order, and even answer questions about the presence or absence of specific items in her environment (Gisiner and Schusterman 1992; see Supplementary Video 17.1). Schusterman and Gisiner (1997) eventually argued that these impressive capabilities could potentially be explained not via specific language- or meaning-related adaptations, but through complex associative learning. Whether receptive syntax can be fully explained by associative learning or not, the ability of sea lions to respond appropriately to meaning carried by novel combinations of 'word' order suggests that syntax-relevant abilities are spread very widely in the animal kingdom, and not just among close relatives to primates, or complex vocal production learners such as dolphins and parrots.

Following the era of animal language studies, Schusterman and colleagues shifted to exploring the mechanisms that underpin apparently complex cognitive behavior. These studies examined what can be considered as concept learning or 'rule-governed' learning. For example, trained sea lions were able to solve novel problems based on strategies learned from experience with similar examples solved by trial and error (Kastak and Schusterman 1992). These problem-solving strategies included using the process of elimination or 'exclusion' (Schusterman et al. 1993a), as well as rules of logic, including sameness matching or 'reflexivity' concept (Kastak and Schusterman 1994; Fig. 17.2a, see Supplementary Video

17.2), interchangeability of arbitrary cues or 'symmetry' concept, and the more complicated rules of transitive inference and stimulus equivalence (Schusterman and Kastak 1993). These logical *if...then* rules can be expressed in mathematical terms as: *if* A = A and B = B...*then* C = C (reflexivity), *if* A = B...*then* B = A (symmetry), if A = B and B = C...*then* A = C (transitivity), and finally A = B and B = C...*then* C = A (stimulus equivalence, which combines all three logical rules). In these experiments, the stimuli were arbitrary visual shapes, having no prior meaning to the sea lions. Examining the ability of sea lions to make these logical inferences with arbitrary items enabled, for the first time, a systematic analysis of the learning skills that support complex behavior, including some of the behavioral performances described in artificial language studies. These evaluations of concept learning have implications that extend beyond laboratory studies of animal cognition. For example, the concept of stimulus equivalence can be used to explain the complex behavior of wild sea lions, including multimodal recognition between mothers and their dependent pups (Schusterman et al. 1992; Schusterman and Kastak 1998), the 'dear enemy' phenomenon observed in male territorial sea lions, and even how sea lions and other animals classify individuals into functional categories such as 'friends' and 'foes' (Schusterman et al. 2000b). Each of these examples require individuals to use information about familiar stimuli to solve unfamiliar problems.

Learning experiments conducted with trained sea lions provided insight into how sea lions and other animals remember and recall information over both short and long periods of time. For example, in language-learning tasks, sea lion behavior is influenced by 'priming' from recent behavioral experiences (Schusterman et al. 1993b). In other stimulus association tasks, sea lions could hold information about visual shapes in working memory for at least two minutes without a decrement in problem-solving (Schusterman and Kastak 2002). On a much longer time scale, a highly trained California sea lion called *Rio* showed a memory ability exceeding that of any other animal who had been tested over multiple years. *Rio* had taken part in systematic studies of concept formation as a young adult, and later showed she was capable of remembering previously learned concepts—as opposed to memorizing specific solutions to familiar problems—for periods of 1–10 years (Reichmuth Kastak and Schusterman 2002). To date, *Rio's* remarkable memory performance remains among the best formal evidence concerning the longevity and complexity of animal memories.

It may seem intuitive that long-lived and sentient animals like sea lions must solve novel problems and remember essential information for extended periods of time. For example, sea lions and fur seals show astounding natal site fidelity, with females sometimes giving birth within a few meters of their pupping sites from prior years (see Chap. 1 by Gentry). Fur seals also show the ability to learn and remember the location of productive foraging sites, which they return to over successive at-sea trips (e.g., Bonadonna et al. 2001). In a social context, female otariids recognize the calls of their pups for multiple years, well beyond the period of maternal dependency, and pups seem equally able to remember the features of their mothers' voices (Chap. 14 by Charrier; Insley 2000; Mathevon et al. 2004; Pitcher et al. 2010).

Competing neighboring males can recognize and remember their territorial neighbors over multiple years, and treat these individuals differently than unfamiliar rivals (Chap. 1 by Gentry). Despite many examples of natural behavior that seem to call upon advanced cognitive abilities, there are few species for which complex behavior observed in nature can be linked to explicit learning and memory abilities demonstrated in controlled experiments. It is possible that behaviorally flexible animals such as otariids have greater pressure for associative learning or long-term memory than animals relying on instinct or less changeable circumstances. Thus, yet another aspect of otariids that sets them apart from other animals is our ability to view their behavior clearly from the standpoints of both natural problem-solving and cognitive capability (see Schusterman et al. 2002).

In the years since Schusterman et al. (2002) summarized the cognitive abilities of sea lions, otariids have contributed to a number of advances in comparative cognition. We focus in the next sections on more recent studies of otariid cognitive capabilities.

Box 17.1: Brain and Behavior

Recent research has expanded our understanding of otariid neurobiology and its potential relevance to behavior and cognition. For much of the history of sea lion research, these animals were not considered to have a nervous system predictive of high intelligence or advanced cognitive performance. This was on account of their Encephalization Quotient or 'EQ': a measure of how a species' ratio of brain size to body size compares to typical trends across species, which has been considered by some to be representative of a species' raw intellectual capabilities (Jerison 1977). Sea lion EQ has been measured close to 1 or 1.1 (Worthy and Hickie 1986), suggesting they have a typical brain-to-body ratio for a mammal, though note that due to sexual dimorphism, female sea lions have a higher EQ than males. The EQ measure for sea lions and other otariids is lower than for some other highly productive model species such as dolphins and non-human apes. However, as addressed below, contemporary neuroscientists no longer believe EQ to be a strong correlate to cognitive capability.

The impressive success of sea lions in cognitive tasks—some of which animals with much higher EQs failed or struggled to complete—has been attributed to the importance of simple associative mechanisms, broadly conserved across vertebrate species, rather than specializations of brain structures (Schusterman et al. 2002). Perhaps sea lions weren't necessarily smarter than other animals (or why the relatively low EQ?), but rather just highly motivated and allied with persistent trainers. This is a difficult interpretation to reconcile with contemporary knowledge of brain and behavior, which has cast doubt on the relevance of EQ in predicting species intelligence. EQ likely does not equal IQ, despite the fact that many species considered to be highly intelligent have

(continued)

Box 17.1 (continued)
EQ > 1. Some studies have found that overall brain size is more predictive of cognitive capability than EQ when assessed in closely related species (Deaner et al. 2007). Despite their unremarkable EQ, sea lions actually have large brains, between 300 and 400 cm^3 as measured in well over 40 California sea lions (Montie et al. 2010; Cook et al. 2015). In terms of absolute size, their brains are similar to those of chimpanzees,[1] and four to five times larger than the brain size of most dogs (Horschler et al. 2019). More important than size is likely total number of neurons, particularly cortical neurons, which vary greatly across species, and correlate only loosely with gross brain size. We do not have cell counts for sea lion brains, but large pinnipeds like walruses are estimated to have in the range of 4 billion cortical neurons, far higher than the numbers found in terrestrial carnivores.[2] Yet another perspective acknowledges the brain as a complex and dynamic network for information processing. The features of that network should be most predictive of how information is processed, rather than overall size (Buzsáki and Draguhn 2004; Sporns and Honey 2006). Indeed, neurobiological evidence suggests that brain network features, such as conduction speed and distance between processing areas may be most predictive of intelligence. Primates excel in these measures, with carnivores close behind (Roth and Dicke 2017), suggesting a link to social or ecological (predatory) problem solving.

In humans, there is growing evidence that the organization of the brain's connections—more than brain size or number of neurons—is perhaps most relevant to determining intelligence (Wu et al. 2013). Such analyses have yet to be conducted for sea lions, but, anecdotally, assessment of neural tractography suggests dense and varied cortico-cortical connection pathways (P. Cook, unpublished data—see Fig. 17.3). Different types of cognitive and behavioral tasks rely differentially on different brain structures and networks. Along these lines, and also anecdotally (P. Cook, unpublished data), sea lions have an apparently enlarged caudate nucleus compared to primates and many carnivores. This structure, found in both brain hemispheres, is predominantly featured in reward learning and motor learning (Brovelli et al. 2011)—and so may be related to the sea lions' often noted drive and motivation for task-based learning.

---

[1]The largest carnivores, like the polar bear (*Ursus maritimus*) or Steller sea lion (*Eumetopias jubatus*), have brain weights around 500 g and are expected to have at least a billion cortical neurons. Due to higher neuron packing density, carnivores have more cortical neurons than ungulates of the same brain size (Herculano-Houzel 2012; Kazu et al. 2014).

[2]African lions and domestic dogs have approximately 500 million cortical neurons (Jardim-Messeder et al. 2017). Humans have approximately 16 billion (Herculano-Houzel 2018).

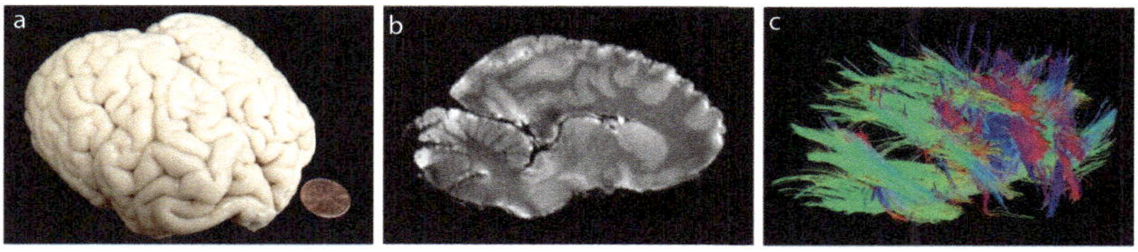

Fig. 17.3 Three views of a sea lion brain. (**a**) An extracted brain, viewed with coin for size reference (image provided by Eva Sawyer). The extensive convolutions of the sea lion brain are particularly notable. Whether the convolutions are functionally relevant remains to be determined. (**b**) A T2-weighted sagittal MRI slice of a post-mortem sea lion brain (image provided by Peter Cook). Traditional structural neuroimaging allows us to assess distribution of gray matter (cell bodies), distribution of white matter (connections between cell bodies), and to visualize and measure gross structures. The anterior (front) portion of the brain is oriented to the right of the image. Lighter tissue is gray matter (high in cell bodies) and darker tissue is white matter (myelinated axons, or communication channels between processing regions). The complexity of the cortical convolutions is apparent, even in this slice near sagittal midline. The extremely large caudate nucleus is visible near the front middle of the image, a large light gray oval. This structure, central to motor learning and reward, tends to be particularly large in canid carnivores such as the sea lion. (**c**) Deterministic fiber mapping of a post-mortem sea lion brain in sagittal view (image provided by Peter Cook). This type of image is produced from diffusion tensor imaging (DTI), a brain imaging protocol that maps large connective pathways in the brain, revealing characteristics of brain organization that are not accessible with typical structural MRI. Each colored thread is a representation of likely white matter tracts connecting processing regions. Tracts oriented front to back of the brain are shown in green (the brain stem, at the back bottom of the image contains a particularly visible green set of tracts). Tracts oriented right to left are shown in red (parts of the corpus callosum, connecting left and right hemispheres of the brain, are visible in the upper middle portion of the image in cross section). Tracts oriented up and down are shown in blue. The complex pattern of interwoven tracts is evident in sea lion cortex. Many of these represent cortico-cortical tracts connecting different cortical processing regions with each other. It is likely that complexity of cortico-cortical connections is correlated with breadth of cognitive capability

17.3 Concept Formation

After the rigorous study demonstrating flexible formation of concept classes by Schusterman and Kastak (1993), subsequent research demonstrated sea lions' facility with mental manipulation of abstract classes of visual shapes. Two sea lions were able to form large categories composed of alphanumeric symbols on the basis of their common function in a reversal learning paradigm, and later were able to link stimuli belonging to the same category in a matching task (Reichmuth Kastak et al. 2001; Fig. 17.2b, see Supplementary Video 17.3). The sea lion subjects subsequently expanded these concept classes of "letters" or "numbers" not merely through direct positive association between new stimuli and prior exemplars, but also through the logical process of exclusion (Reichmuth Kastak and Schusterman 2002). Here, sea lions were given new problems where a novel shape was presented as a possible match for a familiar class member along with a familiar shape that was not related to that class. Sea lions *Rocky* and *Rio* immediately selected the novel object by process of elimination. More notably, when the previously novel objects were used in follow-up transfer tests, the sea lions correctly paired them with every stimulus belonging to the same class as that of the initial match. In other words, having determined by the process of elimination that a novel stimulus must belong to a particular concept class, sea lions immediately integrated the new stimulus into the broader class with no further training required.

More recent follow-up work has shown that sea lion *Rio* was able to expand learned concept classes of visual shapes to include arbitrary sounds, suggesting a robust, cross-modal representational faculty. Lindemann-Biolsi and Reichmuth (2013) added novel sounds to the visual sets the sea lion had previously learned. After linking one sound and one shape through direct training, *Rio* demonstrated the ability to group the sounds with visual shapes from the same set appropriately, on the very first exposure of each stimulus combination. This level of functional categorization is among the most complex demonstrated thus far by non-human animals.

In research expanding earlier studies of a 'reflexivity' or 'sameness' concept—that is, matching items that are alike, sea lions have also been shown capable of visual 'oddity' learning—that is, matching items that are *not* alike (Hille et al. 2006). In this experiment, multiple visual shapes were presented, and the sea lion was rewarded for selecting the item that is not like the others. Following training, the sea lion was able to perform the oddity task with novel problems. Relatively few mammals have demonstrated the ability to perform stimulus matching of novel items based on the concepts of 'sameness' or 'difference'; in addition to humans, these include harbor seals (Mauck and Dehnhardt 2005; Scholtyssek et al. 2013), several monkey species, chimpanzees (*Pan troglodytes*), and a dolphin (*Tursiops truncatus)* (for review, see Katz et al. 2007).

Collectively, the ability of sea lions to acquire abstract concepts indicates a robust conceptual mechanism whereby sea lions maintain long-term, flexible representations of stimulus groupings that cross sensory boundaries. They can integrate new information into established stimulus classes, even when the new information is quite different in form from prior class exemplars. Why sea lions have been uniquely successful at demonstrating concept formation compared to other animals is not clear, as parallel data are still largely lacking for other species. Their success could be due in part to the need to respond rapidly and appropriately to complex stimuli, as sea lions are fast-moving, predatory animals that operate in three-dimensional underwater environments. They use an assortment of sensory cues in an adaptable manner as they move from shore to sea and back again (Chap. 16 by Hanke et al.). Their robust performance in cognitive studies may also be due to their persistence as experimental animals and their impressive task focus, and may be related to aspects of their neurobiology (see Box 17.1). Sea lions are clearly capable of a high degree of complex associative learning. Whether additional cognitive specializations come into play is at yet unknown.

17.4 Learning in the Vocal Domain

The ability of sea lions to relate novel sounds to classes of items suggests a degree of what is termed vocal 'comprehension' learning (Janik and Slater 1997; Lindemann et al. 2006). Learning to respond differentially to different sounds is one way in which learning may influence vocal communication in animals. Contextual control of sound production is also well developed in sea lions. For example, Schusterman's early studies of sound production (for review see Schusterman 2008), showed that sea lions could be readily trained to emit vocalizations and to inhibit them on cue; sea lions could also learn to vocalize in the presence of certain objects and respond vocally to sets of objects.

While some forms of vocal comprehension learning and vocal contextual learning are fairly common among animals, evidence for vocal production learning (learning to vary emitted sounds and learn new call types) is much less common. Sea lions have a relatively simple vocal repertoire that is linked to predictable situations and socio-emotional factors (e.g., Schusterman et al. 1966; Peterson and Bartholomew 1969; Chap. 14 by Charrier). Their calls are often stereotyped and repetitive. Even so, wild otariids can learn to recognize and respond to the calls of other individuals (Insley et al. 2003). Sea lions can also learn to modulate the production of calls both in captivity and on breeding rookeries in response to social cues (see Schusterman 2008). In terms of learning to produce sounds or calls that are not typically part of the vocal repertoire, there is less evidence for vocal plasticity. While trained sea lions appear to have limited ability to vary the complexity or "phonemic" characteristics of their calls, they do have substantial flexibility and control over call rate, which has apparent relevance for social interactions (Schusterman 1977). While sea lions maintain some long-term social relationships mediated by the exchange of vocal signals, they apparently do not possess the ability to learn entirely new vocalizations, as observed in phocids, walruses, and cetaceans (see Reichmuth and Casey 2014). It is unlikely that this limitation reflects upon otariids' general cognitive capacities. While vocal production learning is rare in the animal kingdom, and predominantly found in animals with relatively complex brains and flexible behavior (Jarvis 2006), it is virtually absent in non-human apes, and likely depends on specific social evolutionary pressures (Marler and Mitani 1988; Locke and Snow 1997; Owren et al. 2011). There is also conjecture that prior breathing and foraging adaptations may have subserved evolution of vocal learning in the pinnipeds (Reichmuth and Casey 2014; Schusterman 2008).

17.5 Rhythmic Behavior

Sensitivity to repetition rate during streams of redundant calls may be one reason sea lions have been shown so capable of perceiving and motorically matching a periodic beat in an auditory signal, also known as 'beat keeping' behavior (Repp and Su

2013). Specifically, exhaustive and progressive experiments with the trained sea lion *Ronan*[3] provided convincing evidence that she could (1) learn to bob her head in time to a rhythmic beat of a given tempo, (2) transfer that beat keeping ability to novel tempos, and (3) entrain her rhythmic movements to complex, musical stimuli (Cook et al. 2013).[4] This research has significant implications for the origins of what has been termed 'biomusicality' in animals (see Honing 2019). Prior to *Ronan's* unequivocal beat keeping performance, this entrainment capability had been observed only in humans and parrot-type birds—both of which are highly developed vocal production learners (Janik and Slater 1997). Some scientists had suggested that evolutionary changes to sensorimotor neural circuits in vocal production learners might also allow beat keeping to complex stimuli such as music (Patel et al. 2009a, b). However, the evidence for beat keeping in sea lions, including entrainment to dynamic musical stimuli at novel tempos, has challenged this theory (Rouse et al. 2016; Wilson and Cook 2016) as they are not considered to be vocal production learners. Once again, sea lions have played an outsize role in a contentious domain of comparative cognition.

We suspect that beat keeping capability in sea lions is probably due to strong general motor control and high motivation (Wilson and Cook 2016), as opposed to a unique neurobiological adaptation specifically supporting vocal motor control in species with complex vocal communication. The ability to judge temporal patterns is fundamental to animal behavioral ecology (DeCoursey 2004), and sensorimotor entrainment is unlikely to be restricted to rarified phylogenies, regardless of the current dearth of strong experimental evidence (Ravignani and Cook 2016). While relatively little is known about how animals perceive time, emerging evidence from a South African fur seal *Arctocephalus pusillus* suggests they can discriminate time intervals extremely well (Heinrich et al. 2020). Even for those species with fine-scale timing abilities, training animals to move in time to a beat is potentially difficult and may require highly capable research animals with dedicated trainers. Perhaps such 'high-performance' animals can more easily learn complicated sensorimotor tasks. Others have wondered whether sea lions, as relatives of other pinniped species with some evidence of vocal production learning, may have something like a vestigial vocal production learning brain circuit supporting their beat keeping (Patel 2014).

Resolving the evolutionary bases of rhythmic behavior in sea lions and other animals will require two parallel lines of evidence: behavioral data on beat keeping behavior in a range of other species (as called for in Ravignani and Cook 2016) and mapping the neural circuitry of pinniped brains. The latter is currently being explored (P. Cook unpublished data). Regardless, given the capability of sea lions to perceive and entrain to rhythmic sounds, future work in this domain should further probe these abilities with a wider range of acoustic stimuli, including the types of

---

[3] *Ronan* is a young female sea lion raised in captivity following stranding and rehabilitation; she was named in honor of Dr. Ron Schusterman.

[4] Video examples of *Ronan's* beat keeping performance can be found at https://youtu.be/6yS6qU_w3JQ and https://youtu.be/qwcR5LSsTUI.

heavily syncopated and metered musical stimuli that humans are able to resolve (Fitch 2013) and the rhythmic streams of repeated communicative calls that are relevant to sea lions in the wild (Schusterman 1977).

17.6 Temperament and Self-Regulation

The experimentally demonstrated cognitive capabilities of sea lions likely have something to do with temperament. Collectively, otariids are curious, eager, opportunistic, and gregarious animals whose lives are patterned by the ebb and flow of the seasonal, temperate environments where they live. As with most mammals, they demonstrate meaningful individual differences in temperament (Ciardelli et al. 2017)—which can be loosely defined as consistent behavioral differences in the way different individuals react to changes and challenges in the environment (Manteca and Deag 1993). Individual traits related to temperament that may contribute to learning abilities are explored by DeRango and Schwarz (Chap. 19). Well-developed learning abilities are also related to executive function and cognitive control—the cognitive processes involved in behavioral self-regulation. In humans, these processes are considered 'mental faculties' that help individuals to filter distractions, prioritize tasks, set and achieve goals, and control behavioral impulses.

There are several reasons that sea lions are interesting from the dual perspectives of temperament and executive function. As described in Box 17.1, otariids have large brains with complicated corticocortical connection patterns. They also show robust ability to manipulate abstract concepts and integrate and generalize concepts across sensory boundaries. These cognitive attributes suggest dense 'polysensory' cortical adaptations, which may be linked to enhanced fronto-parietal behavioral control mechanisms (Jonides et al. 1998). Sea lions appear to have a high capacity for motor inhibition, an ability which is fundamental to self-control and related to working memory (Roberts et al. 1994; Borella et al. 2008). By inhibiting or overriding behavioral impulses, sea lions can consider and select from a wider range of potential responses that may be informed by prior experience. Anecdotally, there are many instances in which individual sea lions appear to exhibit cognitive control. For example, adult breeding males chase juvenile males within their territories, but rarely if ever attack them, while adult competitors in the same areas face extreme aggressive consequences for intrusion (Gentry 1970). During development, sea lions may show role reversal during play with individuals taking turns in status and position during sparring (Chap. 20 by Llamazares-Martín and Palagi). In captivity, we observed that a young sea lion occasionally placed her open mouth on the hand or arm of a trainer, but did not bite or break the skin; this occurred predominantly when the animal was asked to endure extended close contact training sessions (P. Cook, unpublished data). One possible interpretation of these examples is that sea lions sometimes inhibit aggressive impulses—for example, beginning an aggressive signal (biting), by using top-down inhibitory processes to avoid completing the aggressive action. Empirical studies here are limited, but suggest meaningful capability for complex, top-down executive control mechanisms.

One study with trained California sea lions has explored self-control directly (Genty and Roeder 2006). In this case, sea lions showed apparently greater inhibitory control than that observed by non-human primates in similar studies. Four sea lions were trained on a reverse reward contingency task in which they were presented with two arrays of fish: one containing five fish, the other only one. To receive the five fish, a subject had to select the array containing only one, while selection of the larger array was reinforced with the smaller reward. Prior to training, the sea lions showed an unsurprising preference for the array containing five fish. With additional experience however, three of the sea lions learned to select the smaller array to receive the larger reward, and they learned the reversed-reward task almost perfectly in less than 200 trials. The sea lions' performance rivaled that previously observed by orangutans *Pongo pygmaeus*, and was better than that observed in other non-human primates. The authors speculated that the sea lions' performance might be related to their non-competitive foraging ecology, which might not require the same strength of competitive impulse to seize food when presented. Others have suggested their success on the task might be attributed to subtle features of experimental design (Beran 2018). Further study of inhibitory control in otariidae is warranted.

Self-control can be considered from the perspective of working memory, which relies on similar neural substrates as inhibitory control in humans and other mammals (Engle and Kane 2004; Singer et al. 2013). In addition to early work by Schusterman and colleagues with trained sea lions that showed durable memory (spanning at least 2 min) in delayed matching procedures (see Schusterman et al. 2002), newer findings have continued to probe a range of working-memory processes in sea lions. For example, Abramson et al. (2011) demonstrated a primate-like capability for numerosity (number) discriminations in South American sea lions *Otaria flavescens*. Discriminating between different values is believed to rely on complex working memory processes in humans and other animals (Botvinick and Watanabe 2007; Ditz and Nieder 2016). Other work indicates that sea lions have an apparently impressive capability for mental manipulation of visual representations. Trained sea lions are quite good at relating rotated shapes to their static counterparts in visual matching tasks (Mauck and Dehnhardt 1997; Stich et al. 2003). Further—as is the case with humans, but not with a number of other species tested—sea lions take longer to make their choice when the angle of disparity between the rotated and static shape is greater. In humans, similar performance has been attributed to 'mental rotation' of a simulated object until it meets (or does not meet) the orientation of the choice stimulus. This type of performance displays robust working memory, using cognitive control processes to manipulate sensory representations. Interestingly, sea lions differ from humans in that while their accuracy and speed of response vary by axis of rotation, performance is better and faster when objects are rotated around what has been termed the 'skew' axis (Stich et al. 2003), which has more relevance for underwater motion than terrestrial motion. This finding echoes an old suggestion of Schusterman and Thomas (1966): that visual processing in marine mammals adapted to move freely in three dimensions will function somewhat differently from that of land-based animals moving predominantly in two dimensions.

Some related studies support the view that sea lions are capable of some degree of mental transposition of visual information. Hill et al. (2015) showed that some sea lions have the ability to locate hidden objects viewed in a mirror's reflection. In a further demonstration of the ability to use cues to indirectly navigate the environment, Scheumann and Call (2004) showed that South African fur seals *Arctocephalus pusillus* could follow a graded range of pointing gestures from a human experimenter to obtain food rewards. Point following has been readily observed in domestic dogs cued by their trainers (Soproni et al. 2002) but has proven difficult to demonstrate in primates, leading some to speculate that the ability to follow gestural cues may be related to domestication by humans. However, the ease with which otariids demonstrate point-following in research and training contexts suggests that an evolutionary history entangled with humans is not required. It may be that the interpretation of pointing cues by sea lions is supported by early training history in human care, strong inhibitory control, low impulsiveness, or possibly a reliance on joint attention for traveling and hunting with conspecifics, but, here again, otariids have spoiled an evolutionary 'just-so' story regarding the uniqueness of certain capabilities in the animal kingdom.

While sea lions may be capable of using mirror reflections in practical ways, they have not shown clear evidence of self-focused or self-recognition behavior when presented with mirrors (Delfour and Marten 2001). Rather, they appear to treat their own reflection as a social stimulus (Schusterman 1966, 1967). Mirror 'self-recognition' is a controversial phenomenon that has been observed in some primates and dolphins (Gallup 1977; Reiss and Marino 2001), and has been suggested as an analog or at least a component of self-conception. Interestingly, but as yet anecdotally, exploratory examination of sea lion limbic connectivity via DTI imaging of the brain has shown remarkably sparse connectivity in a brain region called the cingulate cortex (P. Cook, unpublished data). This is one of the central regions involved in self-conception among human subjects, but the neurobiological basis of this ability has not been extensively studied in other species. Perhaps the lack of neural development in this part of the brain is reflective of a reduced capability for treating one's self image as an object of complex mental processes.

While a full range of higher cognitive faculties has yet to be explored in sea lions, these intriguing studies suggest that sea lions may be strong candidates for more rigorous experimental research.

17.7 Sea Lions as Natural Models for Studying Brain and Behavior

We have reviewed several laboratory studies, field observations, and anecdotes that demonstrate the short- and long-term memory capabilities of otariids. While lab studies are constrained by small sample size, and field studies may lack robust controls, a confluence of recent events has allowed in-depth study of sea lion memory capabilities and their neurobiological substrates with relatively large sample sizes. This unexpected research opportunity has arisen from what can only be considered an unfortunate series of events for some free-ranging California sea lions.

For decades, sea lions along the California coast have been stranding in distress, suffering disorientation and seizures. The mysterious cause of these striking and increasingly frequent natural events was eventually linked to exposure to domoic acid, a neurotoxin produced by invasive algae *Pseudonitzchia* spp. (Scholin et al. 2000). Domoic acid is a glutamate agonist that leads to overactivity in the medial temporal lobe of the brain—and, in many cases, chronic epilepsy and gross hippocampal damage (Silvagni et al. 2005). In humans, similar exposures of domoic acid accumulated through the ingestion of seafood can lead to amnesic shellfish poisoning, an irreversible condition that may involve permanent loss of experiential memory (Perl et al. 1990).

The large numbers of stranded sea lions entering rehabilitation facilities and the subset of animals that survived domoic acid exposure created a situation that allowed for follow-up neurobehavioral study. In vivo structural brain imaging revealed gross hippocampal damage in individuals who showed symptoms of exposure to the toxin (Montie et al. 2009). This finding motivated subsequent research that combined brain imaging and comprehensive behavioral assessments in the same individuals, in an effort to improve understanding of the behavioral consequences of brain damage, if present. Cook et al. (2013) systematically evaluated 30 wild sea lions that entered rehabilitation following stranding events. They discovered impairment in spatial working memory and reference memory tasks in sea lions that was correlated to the extent of damage to hippocampal structures, where most brain lesions were found. Notably, the greatest impairment was observed in animals with damage to the dorsal right hippocampus, a region which is specialized for spatial memory in humans. These sea lions showed additional evidence of disrupted functional connectivity (active communication) between the hippocampus and thalamus (Cook et al. 2013). Finally, in brains obtained from individuals that did not survive, there was evidence of white matter pathology in the fornix, a brain pathway that connects the hippocampus and mammillary bodies, both structures essential for supporting spatial and experiential memory across a range of species (Cook et al. 2018). As in humans, amnesic shellfish poisoning may be fatal to sea lions. In addition to brain damage and permanent memory loss, sea lions can also experience altered habituation and apparent sensitization to environmental stimuli (Cook et al. 2011, 2016), which has also been observed in rats exposed to these types of toxins in the laboratory (Zuloaga et al. 2016).

These findings with stranded sea lions are relevant to understanding neurotoxic environmental exposure in wild animals, and to treating affected sea lions. In addition, because sea lions (like humans) are large-brained, long-lived animals that may be exposed to marine toxins repeatedly, sea lions are also a promising neurobiological model for understanding low-dose effects of marine toxins in humans, including exploration of developmental effects following maternal exposure. While the circumstances surrounding the periodic exposure of sea lions, humans, and other animals to marine toxins remain a problem with few solutions, neurobehavioral data arising from these cases continues to broaden our comparative knowledge of epilepsy, the biology of memory, and behavioral regulation as well.

17.8 Conclusions

It is worth considering some of the evolutionary and ecological factors that may contribute to otariids' performances in laboratory behavioral science, along with the behavioral and physiological traits we have discussed. Their sensory faculty, which is highly developed across multiple senses and expressed cross-modally as they shift between land and water, provides them with incoming sensory streams rich with environmental information (see Chap. 16 by Hanke et al.). A large, densely packed brain with apparently complex patterns of cortico-cortical connectivity can support complex and flexible behavior across a range of contexts. Their well-developed sensory and cognitive abilities, combined with their agile, responsive, and muscular bodies, make them formidable predators on a wide variety of fish and cephalopods. Moreover, due to their high maneuverability in water, and their size and speed on land, they are capable of evading predation themselves. They are not overwhelmed on the raucous, densely crowded rookeries where they are often found, but are able to focus on attending to their own needs and the behavior of their offspring, neighbors, competitors, and mates. Despite decades of study, subtle aspects of their sociobiology continue to be revealed (e.g., Chap. 19 by DeRango and Schwarz; Wolf et al. 2007) and much remains to be learned. It seems likely that the ease with which they transition between the weightless, three-dimensional, aquatic realm and the terrestrial rocky shorelines poses specific challenges to their cognitive and neurobiological systems. Otariids have exploited productive, coastal waters as they have evolved and radiated from the north Pacific over the last 5 million years while at the same time developing an extraordinary degree of plasticity in their central-place foraging behavior (e.g., Bearzi 2006; Chap. 2 by Costa and Valenzuela Toro; Lowther et al. 2013; Staniland et al. 2010). This flexibility can also be seen in real time, as at least some species can manage the transition to living alongside humans (Chap. 18 by Schakner and Blumstein). Some of their success may be related to their apparent 'sangfroid'; their ability to cope with novel situations is clearly part of why they encounter humans so frequently, inhabit wharfs and jetties, interact with fisheries, and hunt fish at dams far upriver despite extensive efforts to dissuade them. Sea lions may be adapted to exploit unpredictable environments, and their persistence and drive may be temperamental tools to help them succeed in doing so.

To date, otariids have overperformed in the behavioral laboratory. As traditional behavioral comparative laboratories adapt for continued relevance in the twenty-first century, sea lions are well situated to continue contributing to our understanding of sensory biology, behavior, cognition, and neurobiology in long-lived, big-brained, and gregarious mammals. They excel in laboratory settings where careful, longitudinal work is required to establish an existence proof of a complex ability. Their spatiotemporal predictability, neophilia, and bravery make them relatively easy to study in the wild, in stark comparison to terrestrial carnivores. Their ability to learn rapidly across a wide range of contexts suits them for opportunistic assessment in zoos and aquaria, as well as rehabilitation settings. Sea lions also serve as a sentinel species, representing an accessible model to improve understanding of neurobehavioral impacts and challenges of rapid environmental change and near-shore habitat degradation. Sea lions and other otariids continue to demonstrate how much can be accomplished with careful training methods and a willing animal partner. Studies with other carnivores, including domestic dogs (e.g., Bensky et al. 2013; Miklósi 2014), are only starting to catch up to Schusterman's legacy of sea lion research.

References

Abramson JZ, Hernández-Lloreda V, Call J, Colmenares F (2011) Relative quantity judgments in South American sea lions *(Otaria flavescens)*. Anim Cogn 14(5):695

Beach FA III, Pepper RL (1974) Spatial habit reversal in two species of marine mammals. Psychol Rec 24:385–391

Bearzi M (2006) California sea lions use dolphins to locate food. J Mammal 87(3):606–617

Bensky MK, Gosling SD, Sinn DL (2013) The world from a dog's point of view: a review and synthesis of dog cognition research. In: Brockman HJ et al (eds) Advances in the study of behavior, vol 45. Academic, Oxford, pp 209–406

Beran M (2018) Self-control in animals and people. Academic, Cambridge

Bonadonna F, Lea M-A, Dehorter O, Guinet C (2001) Foraging ground fidelity and route-choice tactics of a marine predator: the Antarctic fur seal *Arctocephalus gazella*. Mar Ecol Prog Ser 223:287–297

Borella E, Carretti B, De Beni R (2008) Working memory and inhibition across the adult life-span. Acta Psychol 128(1):33–44

Botvinick M, Watanabe T (2007) From numerosity to ordinal rank: a gain-field model of serial order representation in cortical working memory. J Neurosci 27(32):8636–8642

Breland K, Breland M (1966) Animal behavior. Macmillan, New York

Brovelli A, Nazarian B, Meunier M, Boussaoud D (2011) Differential roles of caudate nucleus and putamen during instrumental learning. Neuroimage 57(4):1580–1590

Buzsáki G, Draguhn A (2004) Neuronal oscillations in cortical networks. Science 304 (5679):1926–1929

Ciardelli LE, Weiss A, Powell DM, Reiss D (2017) Personality dimensions of the captive California sea lion *(Zalophus californianus)*. J Comp Psychol 131(1):50

Cook P, Reichmuth C, Gulland F (2011) Rapid behavioural diagnosis of domoic acid toxicosis in California sea lions. Biol Lett 7(4):536–538

Cook P, Rouse A, Wilson M, Reichmuth C (2013) A California sea lion *(Zalophus californianus)* can keep the beat: motor entrainment to rhythmic auditory stimuli in a non vocal mimic. J Comp Psychol 127(4):412

Cook PF, Reichmuth C, Rouse AA, Libby LA, Dennison SE, Carmichael OT et al (2015) Algal toxin impairs sea lion memory and hippocampal connectivity, with implications for strandings. Science 350(6267):1545–1547

Cook PF, Reichmuth C, Rouse A, Dennison S, Van Bonn B, Gulland F (2016) Natural exposure to domoic acid causes behavioral perseveration in wild sea lions: Neural underpinnings and diagnostic application. Neurotoxicol Teratol 57:95–105

Cook PF, Berns GS, Colegrove K, Johnson S, Gulland F (2018) Postmortem DTI reveals altered hippocampal connectivity in wild sea lions diagnosed with chronic toxicosis from algal exposure. J Comp Neurol 526(2):216–228
Deaner RO, Isler K, Burkart J, Van Schaik C (2007) Overall brain size, and not encephalization quotient, best predicts cognitive ability across non-human primates. Brain Behav Evol 70 (2):115–124
DeCoursey PJ (2004) The behavioral ecology and evolution of biological timing systems. In: Dunlap JC, Loros JJ, DeCoursey PJ (eds) Chronobiology: biological timekeeping. Sinauer, Sunderland, pp 27–65
Delfour F, Marten K (2001) Mirror image processing in three marine mammal species: killer whales (Orcinus orca), false killer whales (Pseudorca crassidens) and California sea lions (Zalophus californianus). Behav Process *53*(3):181–190
Ditz HM, Nieder A (2016) Sensory and working memory representations of small and large numerosities in the crow endbrain. J Neurosci 36(47):12044–12052
Engle RW, Kane MJ (2004) Executive attention, working memory capacity, and a two-factor theory of cognitive control. Psychol Learn Motiv 44:145–200
Fitch W (2011) The evolution of syntax: an exaptationist perspective. Front Evol Neurosci 3:9
Fitch W (2013) Rhythmic cognition in humans and animals: distinguishing meter and pulse perception. Front Syst Neurosci 7:68
Gallup GG (1977) Self recognition in primates: A comparative approach to the bidirectional properties of consciousness. Am Psychol 32(5):329
Gentry RL (1970) Social behavior of the Steller sea lion. PhD thesis, University of California, Santa Cruz, CA
Genty E, Roeder JJ (2006) Self-control: why should sea lions, *Zalophus californianus*, perform better than primates? Anim Behav *72*(6):1241–1247
Gisiner R, Schusterman RJ (1992) Sequence, syntax, and semantics: Responses of a language-trained sea lion *(Zalophus californianus)* to novel sign combinations. J Comp Psychol 106 (1):78–91
Harlow HF (1949) The formation of learning sets. Psychol Rev 56:51–65
Heinrich T, Ravignani A, Hanke FH (2020) Visual timing abilities of a harbour seal (Phoca vitulina) and a South African fur seal (Arctocephalus pusillus pusillus) for suband supra-second time intervals. Animal Cognition. Advance online publication
Herculano-Houzel S (2012) The remarkable, yet not extraordinary, human brain as a scaled-up primate brain and its associated cost. Proc Nat Acad Sci 109(Suppl 1):10661–10668
Herculano-Houzel S (2018) Longevity and sexual maturity vary across species with number of cortical neurons, and humans are no exception. J Comp Neurol 527(10):1689–1705
Hilix WA, Rumbaugh D (2013) Animal bodies, human minds: Ape, dolphin, and parrot language skills. Springer, New York
Hill HM, Webber K, Kemery A, Garcia M, Kuczaj SA II (2015) Can sea lions (Zalophus californianus) use mirrors to locate an object? Int J Comp Psych 28(1):1–12
Hille P, Dehnhardt G, Mauck B (2006) An analysis of visual oddity concept learning in a California sea lion *(Zalophus californianus)*. Learn Behav 34(2):144–153
Honing H (2019) The Origins of Musicality. MIT Press, Cambridge, MA
Horschler DJ, Hare B, Call J, Kaminski J, Miklósi Á, MacLean EL (2019) Absolute brain size predicts dog breed differences in executive function. Anim Cogn 22(2):187–198
Insley SJ (2000) Long-term vocal recognition in the northern fur seal. Nature 406(6794):404
Insley S, Phillips AV, Charrier I (2003) A review of social recognition in pinnipeds. Aqut Mamm 29(2):181–201
Janik VM, Slater PJ (1997) Vocal learning in mammals. Adv Stud Behav 26:59–100
Jardim-Messeder D, Lambert K, Noctor S, Pestana FM, de Castro Leal EM, Bertelsen MF, Alagaili AN, Mohammad OB (2017) Dogs have the most neurons, though Reinot the largest brain: trade-off between body mass and number of neurons in the cerebral cortex of large carnivoran species. Front Neuroanat 11:118

Jarvis P (2006) Towards a comprehensive theory of human learning (vol 1). Psychology
Jerison HJ (1977) The theory of encephalization. Ann NY Acad Sci 299:146–160
Jonides J, Smith EE, Marshuetz C, Koeppe RA, Reuter-Lorenz PA (1998). Inhibition in verbal working memory revealed by brain activation. Proc Natl Acad Sci USA 95(14):8410–8413
Kako E (1999) Elements of syntax in the systems of three language-trained animals. Anim Learn Behav 27(1):1–14
Kastak DA, Schusterman RJ (1992) Comparative cognition in marine mammals: a clarification on match-to-sample tests. Mar Mamm Sci 8:414–417
Kastak D, Schusterman RJ (1994) Transfer of visual identity matching-to-sample in two California sea lions (*Zalophus californianus*). Anim Learn Behav 22(4):427–435
Kastak D, Schusterman RJ (2002) Changes in auditory sensitivity with depth in a free-diving California sea lion (*Zalophus californianus*). J Acoust Soc Am 112:329–333
Katz JS, Wright AA, Bodily KD (2007) Issues in the comparative cognition of abstract-concept learning. Comp Cogn Behav Rev 2:79–92
Kazu RS, Maldonado J, Mota B, Manger PR, Herculano-Houzel S (2014) Cellular scaling rules for the brain of Artiodactyla include a highly folded cortex with few neurons. Front Neuroanat 8:128
Kellog WN (1961) Porpoises and sonar. University of Chicago Press, Chicago
Kellogg WN, Kohler R (1952) Reactions of the porpoise to ultrasonic frequencies. Science:250–252
Lindemann KL, Reichmuth-Kastak C, Schusterman RJ (2006) The role of learning in the production and comprehension of auditory signals by pinnipeds. Aquat Mamm 32(4):483–490
Lindemann-Biolsi KL, Reichmuth C (2013) Cross-modal transitivity in a California sea lion (*Zalophus californianus*). Anim Cogn 17(4):879–890
Locke JL, Snow C (1997) Social influences on vocal learning in human and nonhuman primates. In: Snowdon CT, Hausberger M (eds) Social influences on vocal development. pp 274–292
Lowther AD, Harcourt RG, Page B, Goldsworthy SD (2013) Steady as he goes: at-sea movement of adult male Australian sea lions in a dynamic marine environment. PLoS One 8(9):e74348
Manteca X, Deag JM (1993) Individual differences in temperament of domestic animals: a review of methodology. Anim Welfare 2(3):247–268
Marler P, Mitani J (1988) Vocal communication in primates and birds: parallels and contrasts. In: Todt D, Goedeking P, Symmes D (eds) Primate vocal communication. Springer, Berlin, pp 3–14
Mathevon N, Charrier I, Aubin T (2004) A memory like a female fur seal: long-lasting recognition of pup's voice by mothers. An Acad Bras Cienc 76(2):237–241
Mauck B, Dehnhardt G (1997) Mental rotation in a California sea lion (*Zalophus californianus*). J Exp Biol 200:1309–1316
Mauck B, Dehnhardt G (2005) Identity concept formation during visual multiple-choice matching in a harbor seal (*Phoca vitulina*). Learn Behav 33:428–436
Miklósi Á (2014) Dog behaviour, evolution, and cognition. Oxford University Press, Oxford
Montie EW, Pussini N, Schneider GE, Battey TW, Dennison S, Barakos J, Gulland F (2009) Neuroanatomy and volumes of brain structures of a live California sea lion (*Zalophus californianus*) from magnetic resonance images. Anat Rec 292(10):1523–1547
Montie EW, Wheeler E, Pussini N, Battey TWK, Barakos J, Dennsion S, Colegrowe K, Gulland F (2010) Magnetic resonance imaging quality and volumes of brain structures from live and postmortem imaging of California sea lions with clinical signs of domoic acid toxicosis. Dis Aquat Organ 91:243–256
Owren MJ, Amoss RT, Rendall D (2011) Two organizing principles of vocal production: Implications for nonhuman and human primates. Am J Primatol 73(6):530–544
Patel AD (2014) The evolutionary biology of musical rhythm: was Darwin wrong? PLoS Biol 12 (3):e1001821
Patel AD, Iversen JR, Bregman M, Schulz I (2009a) Studying synchronization to a musical beat in nonhuman animals. Ann NY Acad Sci 1169(1):459–469

Patel AD, Iversen JR, Bregman MR, Schulz I (2009b) Experimental evidence for synchronization to a musical beat in a nonhuman animal. Curr Biol 19(10):827–830
Perl TM, Bédard L, Kosatsky T, Hockin JC, Todd EC, Remis RS (1990) An outbreak of toxic encephalopathy caused by eating mussels contaminated with domoic acid. N Engl J Med 322 (25):1775–1780
Peterson RS, Bartholomew GA (1969) Airborne vocal communication in the California sea lion, *Zalophus californianus*. Anim Behav 17:17–24
Pilley JW, Hinzmann H (2013) Chaser: Unlocking the genius of the dog who knows a thousand words. Houghton Mifflin Harcourt, New York
Pitcher BJ, Harcourt RG, Charrier I (2010) The memory remains: long-term vocal recognition in Australian sea lions. Anim Cogn 13(5):771–776
Poulter TC (1963) Sonar signals of the sea lion. Science 139:753–755
Ravignani A, Cook P (2016) The evolutionary biology of dance with frills. Curr Biol 26(19):R878–R879
Reichmuth C (2018) Ronald Schusterman. In: Vonk J, Schackelford TK (eds) Encyclopedia of animal cognition and behavior. Springer, Berlin. https://doi.org/10.1007/978-3-319-47829-6_940-1
Reichmuth C, Casey C (2014) Vocal learning in seals, sea lions, and walruses. Curr Opin Neurobiol 28:1–6
Reichmuth Kastak C, Schusterman RJ (2002) Long-term memory for concepts in a California sea lion. Anim Cogn 5:225–232
Reichmuth Kastak C, Schusterman RJ, Kastak D (2001) Equivalence classification by California sea lions using class-specific reinforcers. J Exp Anal Behav 76(2):131–158
Reiss D, Marino L (2001) Mirror self-recognition in the bottlenose dolphin: A case of cognitive convergence. Proc Natl Acad Sci USA 98(10):5937–5942
Repp BH, Su YH (2013) Sensorimotor synchronization: a review of recent research (2006–2012). Psychon Bull Rev 20(3):403–452
Roberts RJ, Hager LD, Heron C (1994) Prefrontal cognitive processes: Working memory and inhibition in the antisaccade task. J Exp Psychol Gen 123(4):374
Roth G, Dicke U (2017) Evolution of cognitive brains: mammals. In: Watanabe S, Hofman MA, Shimizu T (eds) Evolution of the brain, cognition, and emotion in vertebrates. Springer, Tokyo, pp 12–146
Rouse AA, Cook PF, Large EW, Reichmuth C (2016) Beat keeping in a sea lion as coupled oscillation: implications for comparative understanding of human rhythm. Front Neurosci 10:257
Scheumann M, Call J (2004) The use of experimenter-given cues by South African fur seals *(Arctocephalus pusillus)*. Anim Cogn 7(4):224–230
Scholin CA, Gulland F, Doucette GJ, Benson S, Busman M, Chavez FP et al (2000) Mortality of sea lions along the central California coast linked to a toxic diatom bloom. Nature 403(6765):80
Scholtyssek CA, Kelber A, Hanke FD, Dehnhardt G (2013) Same different concept formation in a harbor seal *(Phoca vitulina)*. Anim Cogn 16(6):915–925
Schusterman RJ (1966) Serial discrimination-reversal learning with and without errors by the California sea lion. J Exp Anal Behav 9(5):593–600
Schusterman RJ (1967) Perception and determinants of underwater vocalization in the California sea lion. In: Busnel RG (ed) Les Systemes Sonars Amimaux, Biologie et Bionique. Laboratoir de Physiologic Acoustique, Jouy-en-Josas, France
Schusterman RJ (1977) Temporal patterning in sea lion barking *(Zalophus californianus)*. Behav Biol 20(3):404–408
Schusterman RJ (1981) Behavioral capabilities of seals and sea lions: a review of their hearing, visual, learning and diving skills. Psychol Rec 31(2):125–143
Schusterman RJ (2008) Vocal learning in mammals with special emphasis on pinnipeds. In: Oller DK, Griebel U (eds) The evolution of communicative flexibility: complexity, creativity, and adaptability in human and animal communication. MIT Press, Cambridge, pp 41–70

Schusterman RJ (2010) Historical perspectives - pinniped psychobiology: the early years. Aquat Mamm 36(1):84–110
Schusterman RJ, Gisiner R (1997) Pinnipeds, porpoises and parsimony: Animal language research viewed from a bottom-up perspective. In: Mitchell RW, Thompson NS, Miles HL (eds) Anthropomorphism, anecdotes, and animals: the emperor's new clothes? State University of New York Press, Albany, NY, pp 370–382
Schusterman RJ, Kastak DA (1993) California sea lion *(Zalophus californianus)* is capable of forming equivalence relations. Psychol Rec 43:823–839
Schusterman RJ, Kastak D (1998) Functional equivalence in a California sea lion: relevance to social and communicative interactions. Anim Behav 55:1087–1095
Schusterman RJ, Kastak D (2002) Problem solving and memory. In: Hoelzel AR (ed) Marine mammals—an evolutionary approach. Blackwell, Oxford, pp 371–387
Schusterman RJ, Thomas T (1966) Shape discrimination and transfer in the California sea lion. Psychon Sci 5:21–22
Schusterman RJ, Gentry R, Schmook J (1966) Underwater vocalization by sea lions: Social and mirror stimuli. Science 154(3748):540–542
Schusterman RJ, Hanggi E, Gisiner R (1992) Acoustic signaling in mother-pup reunions, interspecies bonding, and affiliation by kinship in California sea lions *(Zalophus californianus)*. In: Thomas JA, Kastelein RA, Ay S (eds) Marine mammal sensory systems. Plenum, New York
Schusterman RJ, Gisiner R, Grimm BK, Hanggi EB (1993a) Behavior control by exclusion and attempts at establishing semanticity in marine mammals using match-to-sample paradigms. In: Roitblat H, Herman L, Nachtigall P (eds) Language and communication: comparative perspectives. Erlbaum, Hillsdale, NJ, pp 249–274
Schusterman RJ, Gisiner RG, Hanggi EB (1993b) Remembering in California sea lions: using priming cues to facilitate language-like performance. Anim Learn Behav 21:377–383
Schusterman RJ, Kastak D, Levenson D, Reichmuth CJ, Southall BL (2000a) Why pinnipeds don't echolocate. J Acoust Soc Am 107:2256–2264
Schusterman RJ, Reichmuth CJ, Kastak D (2000b) How animals classify friends and foes. Curr Dir Psychol Sci 9(1):1–6
Schusterman RJ, Reichmuth Kastak C, Kastak D (2002) The cognitive sea lion: meaning and memory in the lab and in nature. In: Bekoff M, Allen C, Burghardt G (eds) The Cognitive animal: empirical and theoretical perspectives on animal cognition. MIT Press, Cambridge, pp 217–228
Shanker SG, King BJ (2002) The emergence of a new paradigm in ape language research. Behav Brain Sci 25(5):605–620
Silvagni PA, Lowenstine LJ, Spraker T, Lipscomb TP, Gulland FMD (2005) Pathology of domoic acid toxicity in California sea lions (*Zalophus californianus*). Vet Pathol 42(2):184–191
Singer P, Hauser J, Lopez LL, Peleg-Raibstein D, Feldon J, Gargiulo PA, Yee BK (2013) Prepulse inhibition predicts working memory performance whilst startle habituation predicts spatial reference memory retention in C57BL/6 mice. Behav Brain Res 242:166–177
Skinner BF (1938) The behavior of organisms: an experimental analysis. Appleton Century Croft, New York
Soproni K, Miklósi Á, Topál J, Csányi V (2002) Dogs' *(Canis familaris)* responsiveness to human pointing gestures. J Comp Psychol 116(1):27–34
Sporns O, Honey CJ (2006) Small worlds inside big brains. Proc Natl Acad Sci USA 103 (51):19219–19220
Staniland IJ, Gales N, Warren NL, Robinson SL, Goldsworthy SD, Casper RM (2010) Geographical variation in the behaviour of a central place forager: Antarctic fur seals foraging in contrasting environments. Mar Biol 157:2383–2396
Stich KP, Dehnhardt G, Mauck B (2003) Mental rotation of perspective stimuli in a California sea lion *(Zalophus californianus)*. Brain Behav Evol 61:102–112
Terrace HS (1963) Errorless transfer of a discrimination across two continua. J Exp Anal Behav 6 (2):223–232

Wilson M, Cook PF (2016) Rhythmic entrainment: why humans want to, fireflies can't help it, pet birds try, and sea lions have to be bribed. Psychon B Rev 23(6):1647–1659

Wolf JB, Mawdsley D, Trillmich F, James R (2007) Social structure in a colonial mammal: unravelling hidden structural layers and their foundations by network analysis. Anim Behav 74(5):1293–1302

Worthy GA, Hickie JP (1986) Relative brain size in marine mammals. Am Nat 128(4):445–459

Wu K, Taki Y, Sato K, Hashizume H, Sassa Y, Takeuchi H et al (2013) Topological organization of functional brain networks in healthy children: differences in relation to age, sex, and intelligence. PLoS One 8(2):e55347

Zuloaga DG, Lahvis GP, Mills B, Pearce HL, Turner J, Raber J (2016) Fetal domoic acid exposure affects lateral amygdala neurons, diminishes social investigation and alters sensory-motor gating. Neurotoxicology 53:132–140

Chapter 18
The California Sea Lion: Thriving in a Human-Dominated World

Zachary A. Schakner and Daniel T. Blumstein

Abstract California sea lions offer an important case study of a species that has successfully adapted and thrived in a human-dominated world. The recovery of California sea lion populations over the past four decades is a conservation success story. Unfortunately, their recovery has put them in direct conflict with human activities resulting in new management challenges in the regions where they occur. Here, we review the role of learning in California sea lions and their capacity to tolerate and successfully capture prey from commercial or recreational fishing lines or salmon at dams. Learning underlies tolerance to novel human-related stimuli, locating novel foraging resources, and responding to environmental change. According to modern animal learning theory, there are basic mechanisms, or types of experiences underlying animal learning. The simplest learning process of habituation is non-associative because it involves an individual's experience with a single stimulus, whereas complex associative learning mechanisms elicit changes in behavior as a result of experience with two stimuli or stimulus and response. We focus on these fundamental associative and non-associative learning mechanisms in California sea lions which could be used to manage wildlife-human conflicts involving otariids. For instance, understanding what kinds of stimuli California sea lions respond to and learn from, or how social factors influence learning processes, are all important parameters that can be used by managers for modifying animal behavior. Lessons from both human-tolerant species and those in conflict could inform best practices for ensuring human-wildlife coexistence in a human-dominated environment will be enhanced with lessons from both human-tolerant species as well as those that do not do well with humans.

Z. A. Schakner (✉)
Office of Science and Technology, National Marine Fisheries Service (NMFS), National Oceanic and Atmospheric Administration (NOAA), Silver Spring, MD, USA
e-mail: zachary.schakner@noaa.gov

D. T. Blumstein
Department of Ecology & Evolutionary Biology, University of California Los Angeles, Los Angeles, CA, USA

C. Campagna, R. Harcourt (eds.), *Ethology and Behavioral Ecology of Otariids and the Odobenid*, Ethology and Behavioral Ecology of Marine Mammals,
https://doi.org/10.1007/978-3-030-59184-7_18

Keywords *Zalophus* · Animal learning · Coastal development · Human-animal interactions · Human-tolerant

18.1 Introduction

Relationships among otariids and humans have been intertwined for centuries. Interactions include exploitation through historical hunting of sea lions, competition over prey (Chap. 24 by Crespo; Betts et al. 2009; Braje and DeLong 2009; Lyman 2003; Hildebrandt and Jones 1992; Gerber and Hilborn 2001) and space for haulouts (Schakner et al. 2019). Most otariid populations experienced intense hunting in the nineteenth and twentieth centuries, which dramatically reduced population sizes. Protective management measures have led to the recovery of some, but not all populations (Chap. 24 by Crespo; Gerber and Hilborn 2001). As humans develop our coastal areas, new threats and challenges have arisen from species being forced to live in a human-dominated land and seascape. Human-induced rapid environmental change (HIREC; Sih 2013; Sih et al. 2016) shifts prey concentrations (exacerbated by intense fishing effort), creates novel stimuli from vessel traffic and reduces habitat on land from coastal development (see also Chap. 21 by Giardino et al.).

As aquatic mammals that forage at sea but rest, socialize, and pup on land, sea lions are exposed to urbanization, intense competition from fishing, shipping noise, and coastal development. Human-induced disturbances range from adverse exposure to stimuli such as vessel noise, reduction of natural food availability, to the creation of resources around fishing gear/pens or at dams. A fundamental question within conservation behavior is how some species respond adaptively to HIREC, yet others do not (Geffroy et al. 2015).

California sea lions (*Zalophus californianus*) in particular illustrate one of the most successful recoveries in the modern conservation era (Chap. 28 by Elorriaga et al.; Fig. 18.1; NOAA). California sea lion populations have grown considerably since the inception of the United States' Marine Mammal Protection Act of 1972 (MMPA 1972; Laake et al. 2018). The MMPA offers robust protections for marine mammals in US waters. The law has brought species back from the brink of extinction, and many populations have fully recovered since it was enacted.

For California sea lions, the capacity for learning, including learning from human trainers, which makes them popular in zoos, circuses, and aquaria, seemingly predisposes them to tolerate humans. This behavioral flexibility, and their capacity to behaviorally innovate and learn in a human-dominated land and seascape likely contributed to their success. In a rapidly changing environment, California sea lions have learned to exploit novel concentrations of prey, novel habitats, and even capitalize on human fishing effort. As an unintended consequence of their recovery, there are increasing competitive interactions between sea lions and humans. Along the west coast of the United States, expanding sea lion populations create management conflicts of consumption of endangered salmonid species, interactions with

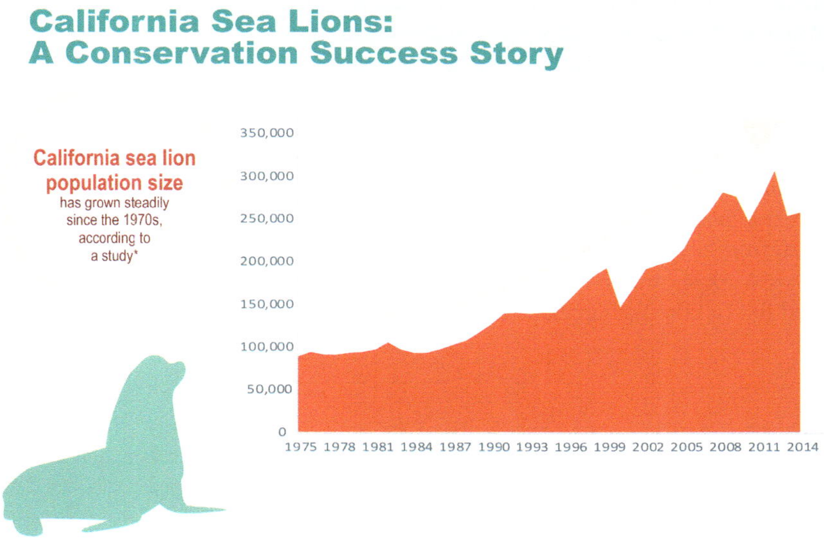

Fig. 18.1 Changes in California sea lion populations since the institution of the Marine Mammal Protection Act (source: Laake et al. 2018)

fisheries, and damage to docks and personal vessels (Jeffries and Scordino 1997; Marshall et al. 2015).

California sea lions thus offer an important case study into a species that has successfully adapted to living in a human-dominated world. Yet, California sea lion's success is in stark contrast with other USA west coast otariids such as some stocks of Steller sea lions (*Eumetopias jubatus*) which remain endangered (Atkinson et al. 2008; Chap. 23 by Trites). This is potentially because of the formers' ability to learn rapidly and socially transmit this knowledge, as well as being particularly behaviorally plastic with respect to habitat and diet. Such traits lead to the rapid tolerance of humans and human infrastructure.

18.2 Learning in a Changing World

Learning is a key aspect of behavior that may greatly enhance the survival of animals, especially in a dynamic environment. We and others have reviewed learning mechanisms elsewhere (Chap. 17 by Cook et al.; Greggor et al. 2019, Schakner et al. 2016), but believe it is worth focusing on fundamental mechanisms given the extensive comparative and experimental psychology literature. According to modern animal learning theory, there are three basic mechanisms, or types of experiences underlying animal learning. The simplest learning process, habituation, is non-associative because it involves an individual's experience with a single stimulus, whereas more complicated, associative learning mechanisms elicits behavioral change as a result of experience with two stimuli or stimulus and response. We believe that applying insights from the psychology literature provides mechanistic insights for a variety of human-wildlife conflict management interventions. For instance, understanding what kinds of stimuli animals respond to and learn from, as well as the reinforcement schedules necessary to sustain such learning, are all important parameters that can be used by managers for modifying animal behavior. We review these fundamental mechanisms derived from animal learning theory, and then discuss how these mechanisms facilitate California sea lion survival on the USA west coast, and then discuss how understanding these mechanisms may be applied to sea lion management issues.

18.3 Single Stimulus Learning: Habituation and Sensitization in a Changing World

Anthropogenic human disturbance provokes fear responses because many animals respond to humans as predators (Frid and Dill 2002; Lima and Dill 1990). This ecology of fear has wide-ranging negative effects on individual, population, species and ecosystem scales. However, some species are able to detect, recognize and learn to tolerate humans while some species do not (Blumstein 2014). Habituation, the simplest learning process, is *the* fundamental mechanism by which animals reduce their response to non-threatening stimuli.

Habituation is non-associative single-stimulus learning. It involves diminishing of behavioral responses after repeated exposure to stimuli (Groves and Thompson 1970; Schakner and Blumstein 2016a, b). In a human-dominated world, learning to ignore novel yet non-threatening cues via habituation versus overreacting to these cues via sensitization (the opposite of habituation) can determine which species can adapt to HIREC (Sih et al. 2016). Habituation involves stimulus specificity. That is to say, if the stimulus is changed in intensity or a novel stimulus is introduced after habituation has occurred, responsiveness is restored (Rankin et al. 2009). This stimulus specificity suggests that habituation filters innocuous stimuli from novel or significant stimuli (Rankin et al. 2009) by modulating responsiveness. In contrast to habituation, heightened/increasing responsiveness after repeated exposure is termed sensitization. The dual process theory suggests that the observed behavior after repeat exposure to a stimulus is the summation of two underlying processes habituation and sensitization (Groves and Thompson 1970; Schakner and Blumstein 2016a).

We expect that human disturbance is likely to impose the greatest fitness costs under two contexts; feeding and reproduction. California sea lions habituate to humans and anthropogenic disturbance in both contexts. This is evidenced by their use of human docks and buoys as haul-out sites and their generalized tolerance to being in close proximity to humans or fishing vessels. Even more, they are

Fig. 18.2 A growing California sea lion rookery on a man-made structure in California

remarkably tolerant of human disturbance at breeding grounds (Gerber and Hilborn 2001), and captive experiments have shown that even pregnant female sea lions habituate to the presence of human visitors (de Vere 2018). In some locations, California sea lions have begun breeding on human structures. For example, the Morro Bay breakwater was constructed by the Los Angeles District of the U.S. Army Corps of Engineers in the 1940s. Recently increasing numbers of California sea lions use it as a breeding ground and some even occupy it year round (Fig. 18.2). California sea lions illustrate the adaptive value in learning to adapt to a human-dominated environment (For an interesting parallel please also see the southern sea lion male haulouts described in Chap. 21 by Giardino.

In captivity, California sea lions rapidly habituate to humans (de Vere 2018), fishing stimuli (Bowles 2012), and novel sounds (Kastak and Schusterman 1983); and subsequent tolerance to novelty has been well documented for decades (Kastak and Schusterman 1983; Table 18.1). Perhaps most tellingly, this habituation response pathway is often in sharp contrast to other species of pinnipeds. For instance, in a series of individual playback experiments, California sea lions rapidly habituated to a novel, pulsed acoustic stimulus, whereas northern elephant seals *(Mirounga leonina)* had the opposite reaction, sensitization, subsequently hauling out away from and avoiding the playback pool altogether (Kastak and Schusterman 1983). These studies demonstrate important species-specific constraints on habituation to novel stimuli in pinnipeds.

Despite decades of laboratory experiments focused on the parameters influencing the underlying process of habituation in organisms such as humans, rats, and pigeons, less is known about the evolutionary ecology of habituation (Blumstein 2014, 2016). Among otariids, early observations of the differences in habituation go

Table 18.1 Early descriptions of otariid tolerance toward humans based upon experiments on captive animals

Species	Tameness[a]	Trainability[b]
California sea lion	Excellent	Excellent
Steller sea lion	Poor	Good
Southern sea lion	Poor	Good
Northern fur seal	Fair	Good
South African fur seal	Excellent	Excellent
Antarctic fur seal	Excellent	Unknown

Recreated from the original publication in Schusterman (1981)

[a]Tameness refers to an animal reliably accepting approach by strangers and handling by familiar trainers

[b]Trainability refers to the ease with which an animal can be brought under stimulus control by operant and Pavlovian procedures

back several decades. Schusterman suggested different responses to humans among a range of different pinnipeds including the same species at different locations, noting that "...all of these observations suggest that to some extent startle or flight reactions habituate at different rates in different species and in different populations as a function of age, sex, season, and time of day" (Schusterman 1981: 134).

There is extensive literature on the parameters that influence habituation in the laboratory, such as complexity or intensity of the stimulus, timing of stimulus presentation, etc. However, the parameters that influence habituation/sensitization in the wild are not well known, nor are the factors that determine habituation versus sensitization responses. Individual variation in habituation/sensitization is likely influenced by a variety of factors, including previous experience, sex, and personality (i.e., bold - shy continuum; Ellenberg et al. 2009, 2013; Fig. 18.3, see Chap. 19 by DeRango and Schwarz). Across populations, variation in habituation can be based upon a population's level of human disturbance or a sorting process by which sensitizers flee (Bejder et al. 2009; Fig. 18.3). Even less is known about the specific factors that underlie variations in species-specific habitation (Fig. 18.3), but there are likely selective advantages to sensitization or habituation depending on a species' life history.

A relatively new theory of habituation, Behavioral Homeostasis Theory (BHT), suggests that the primary determinants of an individual's response to a stimulus (i.e., sensitize or habituate) will involve signal detection and state/responsiveness (Eisenstein and Eisenstein 2006, 2012; Fig. 18.4). Whether habituation or sensitization occurs in response to repeated exposures is primarily a function of the individual's state of overall 'alertness/vigilance' at the time of the initial exposure. Behavioral homeostasis theory operates on the perspective that habituation/sensitization function as a mechanism for filtering novel stimuli. An alert individual will more rapidly detect, assess and recognize the novel stimulus, and thus respond with habituation. In contrast, an individual at a low level of alertness will respond with sensitization in order to increase responsivity and thus rapidly increase detection/assessment. According to the theory, how individuals respond is indicative of the stimulus significance to the organism at that point in time (Eisenstein and Eisenstein 2006, 2012).

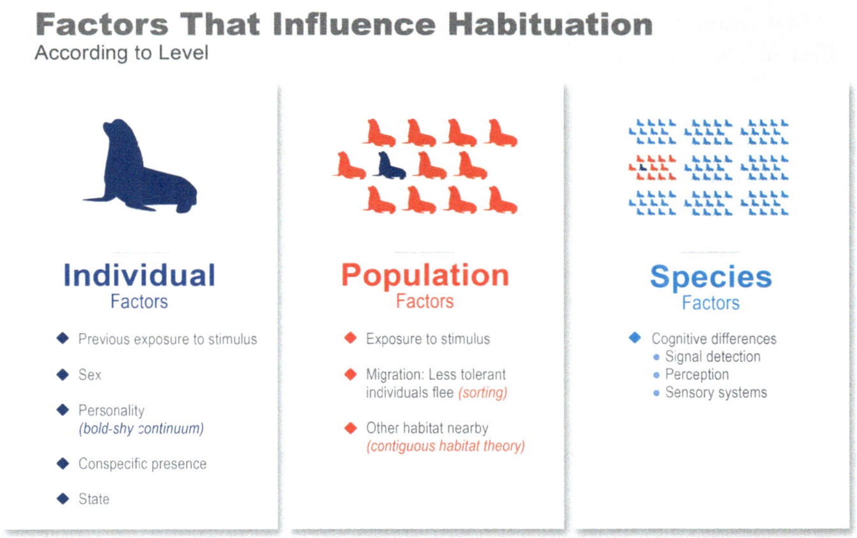

Fig. 18.3 Factors influencing habituation

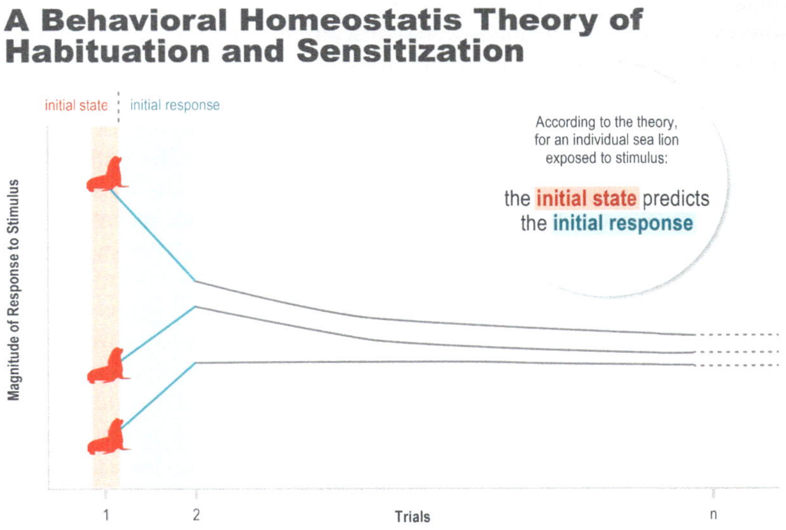

Fig. 18.4 Behavioral homeostasis theory (BHT) The state of individual at the moment of initial exposure influences habituation or sensitization. For California sea lions, state may be influenced by proximity to human or conspecifics

Behavioral homeostasis theory emphasizes the role at which an animal's state is prior/at the moment of initial exposure to a stimulus (Fig. 18.4). For most habituation studies in the wild, there is a minimal description of the initial state, and this shortcoming limits our ability to understand habituation in nature. For otariids and other pinnipeds that overlap with humans, individuals may be more often in a heightened state of awareness. This must be costly and according to BHT, the initial response to experiencing a novel stimuli will more often lead to habituation, because it helps the individual to return to baseline levels. Conversely, for those species that are less disturbed and thus less vigilant, the cost-effective response may be to be sensitized to novel stimuli. This view offers testable hypotheses for comparing the relative responsiveness among disturbed versus undisturbed populations, as well as furthers exploration into variation in individual responsiveness.

18.4 Management Implications of Habituation/ Sensitization

The goal of reducing habituation while simultaneously eliciting sensitization is the objective of many deterrents designed to reduce pinniped fishery interactions (Table 18.2, Götz and Janik 2013; Schakner et al. 2016a). In southern California, California sea lions haul out on and forage at bait receiver pens. In an attempt to deter individuals to reduce their interactions, we tested the responses of wild California sea lions to a variety of potentially aversive acoustic stimuli, including biologically relevant signals such as predator vocalizations, bee swarms, or dog barks. We found clear evidence of habituation to the biologically relevant signals as demonstrated by decreasing responsiveness to repeated exposures of the sounds (Fig. 18.5). However, responses to an acoustic startle device (Götz and Janik 2011; Schakner and Blumstein 2016b, Schakner et al. 2017a) were the opposite; sea lions sensitized to the sounds, as evidenced by increasing responsiveness. Signals that elicit sensitization may show some promise because they rapidly arouse the individual to a responsive state (Götz and Janik 2011).

Table 18.2 Direct versus indirect interactions

Types of pinniped fishery interactions		
Direct	Depredation: removal of fish from gear	Examples: California sea lion and sport fishery in California, South American Sea lion and bottom trawl fisheries, seal salmon farm predation
	Entanglement/bycatch: pinnipeds unintentionally captured	Example: Endangered New Zealand entanglement in Sub-Antarctic trawl fisheries,
Indirect	Competition for prey, displacement or broader ecosystem scale effects	Example: Alaska Steller Sea lion decline in part from overlap with fisheries, Northern Fur seal declining population at Pribilof Islands

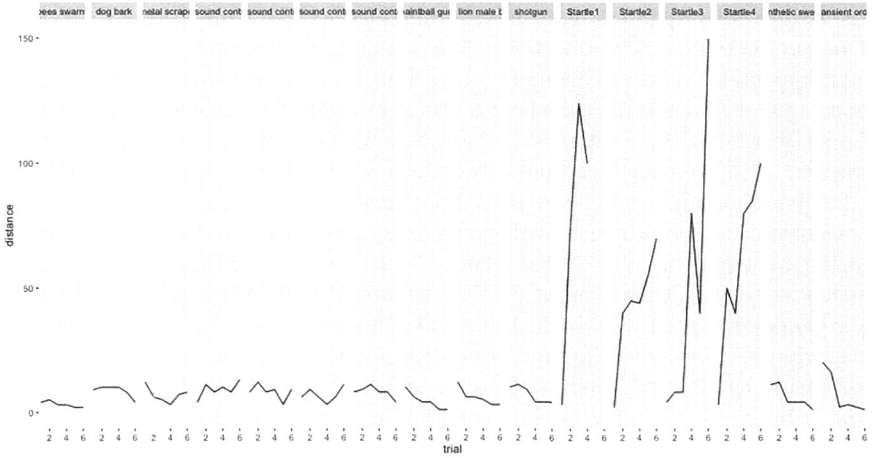

Fig. 18.5 Individual California sea lion responses (distance fled) during playback trials on bait barges (Schakner et al. 2016)

Habituation, Haul Outs and Humans California sea lion capacity to tolerate anthropogenic disturbances through habituation may enhance their survival in a human dominated environment, but this tolerance is a double-edged sword. Tolerating anthropogenic disturbance means that animals share spaces with humans, and for California sea lions this creates management issues. California sea lions haul out on almost any human structures and cause damage to docks and private vessels which create negative perceptions towards the animals and sometimes retaliatory actions (Scordino 2010). In addition, close proximity to humans increases the likelihood that sea lions may chase or bite humans, and these incidents are dangerous and may be reported widely in the news and social media (Schakner et al. 2019). Resolving conflicts over space is difficult and these conflicts are a lose-lose situation for all parties involved. Typically, most management actions occur after the animal has learned to associate humans with a favorable resource (haul-out space). Preventing the association from forming is ideal (Schakner et al. 2016), but we recognize that this is not always feasible. An understanding of associative learning processes is necessary to disrupt learning in the first place.

18.5 Associative Learning via Pavlovian and Instrumental Conditioning

Broadly, associative learning involves learning the relationship between stimuli and/or responses (Domjan 2005). The capacity to learn about the relationship between stimuli and/or responses is functional because it guides how an animal can adaptively respond to exogenous stimuli as well as anticipate future events (Domjan 2005; Shettleworth 2010). Individuals use associative learning to track environmental variation in a variety of contexts. In a variable world, individuals use associative learning to decide where to forage, what to forage on or where to avoid (Shettleworth 2010). Examples include associating sounds or context with the presence of a predator, taste cues associated with edible food, or honing a new extractive foraging technique to yield more prey.

California sea lions interact with nearly every commercial and recreational fishery on the west coast of the United States (Scordino 2010). Fishing or aquaculture creates novel concentrations of prey at a reduced cost by trapping fish in lines, nets or chumming. How individuals learn to remove fish from fisheries, referred to as 'depredation', is an open question, but likely involves a combination of Pavlovian and instrumental conditioning (Rescorla and Solomon 1967). Sounds associated with fishing vessels likely provide cues for individuals to locate the vessel. This Pavlovian conditioning process termed the 'dinner bell effect' has been observed in long term analysis of gillnet fishery, where pingers (sound emitters designed to deter porpoises) on gear appear to actually attract California sea lions (Carretta and Barlow 2011). In another putative example of the dinner-bell effect, sea lions may be able to cue in on sounds emitted on acoustic tags that are placed on salmonids to track their migration and this increases sea lion predation success (Bowles 2012; Cunningham et al. 2014). While plausible given the potential overlap of sea lion hearing thresholds and frequencies emitted from pingers, this dinner-bell effect requires more in situ observations and testing to determine whether sea lions are tracking tagged salmon.

In instrumental conditioning, the animal learns a relationship between operant behavior and the consequence of that behavior, and behavioral frequencies are adjusted accordingly (Thorndike and Bruce 1911; Domjan and Burkhard 1986). While Pavlovian conditioning through the dinner bell effect is likely involved in the localization of prey, instrumental conditioning is used to hone an individual's foraging technique to optimize foraging behaviors. Instrumental conditioning appears to underlie individual learning to extract prey from the fishing lines, nets, and aquaculture pens. In a series of experiments using underwater cameras attached to fishing lines from salmon trolling vessels, Woolery and Harvey (2005) tested whether the presence of captured fish increased sea lion interactions with salmon troll fishery. Interestingly, sea lions appeared to travel from line to line, irrespective of whether the line had hooked a salmon. It appears, therefore, that individuals iteratively visit all lines in the area to check for hooked prey.

18.6 Management Implications of Associative Learning

From first principles, we may not expect tactile or acoustic deterrents to work in a foraging context, but they are likely to work in terms of habitat selection. Garcia et al. (1955) showed that learning may be biased in such a way that it is easier for rats to learn to associate nausea with food avoidance and shock with location avoidance. If similar biases exist in California sea lions, harassment may work, but not by animals learning to avoid specific food items (e.g., salmon), but rather by learning specific locations. Conditioned taste aversion (CTA) was briefly tested on California sea lions in captive settings as well as predating salmon at Ballard Locks in Seattle, Washington USA. In captive settings, CSL that were nauseated from Lithium Chloride laced salmonids avoided those specific species, but not other salmonids. In the wild, emetic laced salmon were given to several individual CSL on fishing lines at Ballard. The individuals became nauseated from the laced fish (Gearin et al. 1988) but nevertheless continued to forage at similar rates as prior to getting sick. Interestingly, the individuals would not take laced fish on fishing lines from the researchers. Therefore, we predict that emetics are useful for specific foraging contexts, such as removal of fish from fishing lines or removal of baitfish from bait docks.

Our goal is to not distill animal behavior down to Pavlovian or instrumental conditioning, but instead, focus on these learning mechanisms given how much is known from decades of comparative psychological experimentation. We believe these underlying mechanisms give us tools to manipulate behavior, and to potentially reduce conflicts with otariids. It is our perspective that well-meaning managers (not necessarily well-equipped with a modern understanding of animal behaviors) can apply mechanistic understanding to minimize human/pinniped conflicts.

18.7 A Learning-Based Approach to Deterrents

Our acoustic deterrent work on California sea lions suggests that any single deterrent is unlikely to mitigate conflicts (Schakner et al. 2017a, b). We suggest that a more comprehensive approach to deterrence is necessary. We hope that by demonstrating the underlying learning processes that elicit responsiveness, or lack thereof, we can change the conversation from a one deterrent fits all approach/perspective, to a more comprehensive approach to deterrence.

We know of only one case of implementation of a deterrence program that successfully solved the conflict between sea lions and humans in the mid to long term. In Gold Beach, Oregon, the local port association and fishing community came together to fund a comprehensive approach to deter sea lions. The program consisted of a single full-time hazing vessel that used tactile deterrents, corresponding port activities to prevent sea lions from hauling out on docks, and banning the disposal of salmon carcasses into the water at docks (feeding attraction). The Port Authority installed pipe barricades and sprinklers with motion detectors on docks to discourage sea lions from hauling out. The full-time hazing vessel used seal bombs, cracker shells, rubber buckshot, and vessel pursuit on individuals in the area during peak sport fishing effort (Lottis 2007a, b, 2009). Such interventions may or may not be acceptable to local communities in other places. Once the sea lions fled outside the entrance bar at the mouth of the bay, the vessel used cracker shells and/or rubber buckshot to drive sea lions beyond the harbor jetties. This vessel operated daily during the fishery which covered a relatively small area and any sea lion observed was immediately targeted for hazing. This Gold Beach example highlights the necessity for community buy-in and demonstrates how a comprehensive approach can enhance mitigation of California sea lion port/fishery interactions.

Regardless of its real-world applicability to other locations, there are some lessons from Gold Beach for a comprehensive, learning-based deterrence scheme, especially from confined areas like bays, or ports that have sea lion issues.

1. Identify primary attractants to sea lions in the area and restrict them. In most cases, attractants can be food-related such as fishing discards or chum, open spaces for haul out, or live bait from bait receivers. Example mitigation measures include port activities to prevent hauling out on docks and eliminate the disposal of fishing discards into the water or live bait leakage from receivers.
2. Physical hazing to deter and drive out 'repeat offenders'. Once individuals have learned that the association between humans and food reinforcers has formed, management efforts rely on raising the cost to the individual depredator. For these repeat offenders, the association is difficult to extinguish, management efforts must rely on forming new negative associations or on decoupling the contingency between humans and reward. Hazing with physical deterrents appears to be the most effective at eliciting short term flight responses (Schakner et al. 2016). Additionally, emetics may be useful to prevent specific foraging behaviors, like removal of fish from fishing lines, bait docks, or specific salmonid species.
3. We predict that deterrent modalities should be tailored to the context. If the goal is to deter novel individuals from colonizing new habitat such as docks, then we recommend the use of acoustic deterrents or non-tactile efforts to repel novel individuals from colonizing or learning the association between human resources and the attractant. If the goal is to repel individuals from foraging contexts, we recommend the use of emetics to create negative associations with the novel foraging resources. New individuals are likely to be more neophobic since they have not yet learned the association between humans and food reward (or open space). We predict that they may be more likely to respond to conventional deterrent methods.

18.8 Social Learning Mechanisms

Social learning occurs as a result of interactions or observations with other individuals (conspecific or heterospecifics). There is evidence that asocial and social learning relies on the same underlying associative and non-associative mechanisms (Heyes 1994, 2012). From this perspective, Fig. 18.6 describes the overlap of associative/non-associative mechanisms in social learning and asocial learning. While the underlying asocial/social mechanism is similar, social learning has unique management implications. Social learning can function as a force multiplier, rapidly

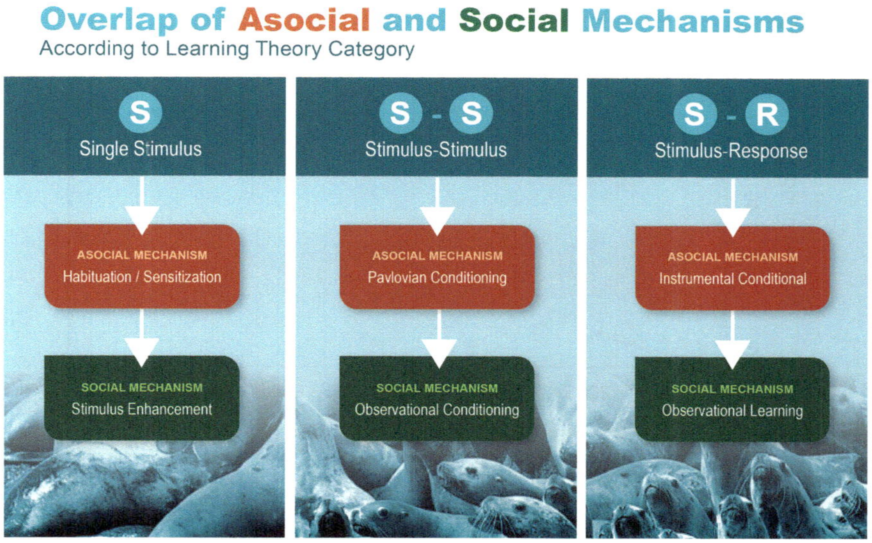

Fig. 18.6 Social and asocial learning mechanisms

spreading behaviors or knowledge through populations, almost like a disease (Hill et al. 2010; Schakner et al. 2016).

Social species with a life history that involves early parental care and overlap at breeding grounds have the opportunity to learn from conspecifics. Otariid mothers remain with their pups only during the first week or so following parturition (Costa 1991). This overlap of young otariid pups and juveniles with mothers in early life creates an opportunity for some vertical social transmission of foraging tactics, but there is little evidence for social transmission of foraging (Fowler et al. 2007; Leung et al. 2014). While studies have largely focused on breeding aggregations around rookeries, adult and subadult male social relationships are not well described. Yet animals do haul out together and this provides an opportunity to learn from others. Indeed, we know that male sea lions learn where to forage from their peers (Schakner et al. 2017a, b).

Little is known about social interactions among male otariids. Unlike females and pups, adult and subadult males range further than females, typically to forage and gain the mass that is required to successfully compete for mates at breeding grounds. Adult and subadult male otariids are the demographic group that more frequently interact with fisheries (Kirkwood et al. 2006). California sea lions, for example, migrate from breeding grounds off California to higher latitude regions in Oregon, Washington, and Alaska. This demographic group frequently interacts with salmonid fisheries (Scordino 2010; Weise and Harvey 2005) and their predation on endangered salmonids creates multiple challenges for endangered salmonid recovery. Whether males actively form social bonds to migrate and cooperatively forage is

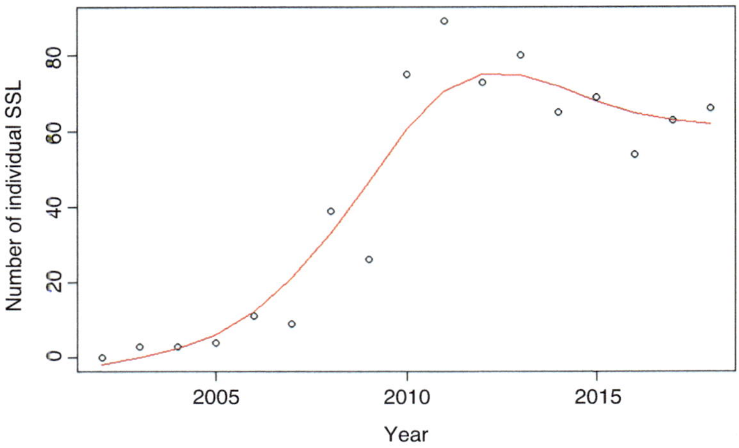

Fig. 18.7 Increasing numbers of Steller sea lions at Bonneville Dam (Tidwell et al. 2018)

not known, but we identified the occurrence of social networks at well-known haul-out sites in the Pacific Northwest of the United States (Schakner et al. 2017a, b).

The estuary of the Columbia River in the U.S. Pacific Northwest has several major California sea lion haul-out sites, with aggregations of tens to hundreds of migratory males during spring and summer months. We analyzed the haul-out patterns of males interacting at a large dock structure called the East Mooring Basin on the Columbia River. Social network analysis showed that social interactions were not patterned randomly; rather, individuals had preferential partners creating social structure. This social structure has implications for the transmission of knowledge of novel food sources/locations, which appears to be occurring at dams in the region. Foraging in the Columbia River estuary is not novel, but observations of sea lions foraging on bottlenecked salmon at upriver dams like Bonneville or Willamette (235 km upriver) from when (Bonneville was first built (in 1934 year) until the 1990s were rare. In an earlier study, we incorporated network-based diffusion analysis to demonstrate that knowledge of the novel food source is socially transmitted via California sea lion social networks (Schakner et al. 2017a, b). Not all individuals that haul out at the opening of the Columbia River estuary visit the dam, and not all the individuals that have discovered the dam are proficient foragers (Schakner et al. 2017a, b).

Increasingly, a sympatric otariid, the Steller sea lion (SSL), is hauling out at the East Mooring Basin and discovering bottlenecked salmon and steelhead at Bonneville or other upriver dams. The rapid, exponential increase (Fig. 18.7) in Steller sea lions at Bonneville suggests some form of social transmission because socially transmitted behaviors are expected to show accelerated diffusion through a population while traits acquired independently are not expected to spread as quickly (Smaldino et al. 2018, Rogers 1995; but see Hoppitt et al. 2010). The wave of increasing numbers of Steller sea lions at Bonneville mirrors the increasing numbers of California sea lions, suggesting heterospecific social transmission may underlie their foraging upriver. Interspecific interactions among these two species at haulouts occur elsewhere (Edgell and Demarchi 2012). The haul-out patterns and interactions among these species provide the ability to formally test whether Steller sea lions use heterospecific cues for foraging.

18.9 Management Implications of Social Learning

Social learning has unique consequences for management. Compared to associative (individual) learning mechanisms, socially transmitted behaviors can rapidly sweep through populations much like a disease. Borrowing insights from epidemiology, we and others have argued that timing is critical to stem the spread of socially transmitted behaviors (Schakner et al. 2017a, b; Snijders et al. 2017). Understanding the structure of the underlying social network is crucial for predicting transmission patterns. Among otariids, adult and subadult males appear to form less cohesive social connections compared to females and their dependent young, and there is greater potential for behaviors to spread more quickly. In contrast, vertical transmission (from mothers to dependent young) leads to a more conserved transmission pattern.

Our case study at the Columbia River and Bonneville dams underscores the importance of having individuals marked or otherwise distinguishable from each other. Managers from state and federal agencies have been collecting individual identification branding and photographs for over a decade. This information enables the tracking of social networks, social transmission, and individual foraging success. All lethal and non-lethal management activities for reducing salmon predation aim to reduce the number of individual sea lions at the dam, but not all individuals are successful foragers and some individuals may be more prone to transmission than others. Therefore, data from individually discriminable animals along with social networks provide managers a predictive tool to focus efforts on particularly successful foragers (i.e., super-spreaders). Since management interventions such as lethal removal or hazing are involved, focusing efforts on specific individuals is not only more effective but is also more ethical.

18.10 Conclusions

While we have focused on the role of learning in adapting to HIREC in California sea lions, these issues are global, involving many species of otariids and the ideas and strategies should apply widely among other representative species. For instance, South American sea lions, *Otaria byronia*, increasingly are bycaught in or depredate numerous fisheries, including gillnets, purse seines and trawl fisheries (e.g., Crespo et al. 1997; Sepúlveda et al. 2007). Additionally, conflicts have emerged when sea lions predate salmon farms in southern Chile, often resulting in retributive killings (Sepúlveda et al. 2015).

More generally, we view learning mechanisms as the levers that managers can use to adjust animal behavior and by modifying behavior, we have tools to solve human-wildlife conflicts. Decades of comparative psychology experimentation have described the underlying constraints of basic learning mechanisms. The parameters for learning and deterrents include stimulus intensity, presentation schedule, competing stimuli or state of organisms.

As humans rapidly change the environment, animals are challenged to behaviorally adapt through learning. California sea lions offer a glimpse into an organism that appears to not only adapt but succeeds along with human population expansion. Our ability to coexist with wildlife in an increasingly anthropogenically-driven world is enhanced with lessons from human-tolerant species and those that do not do well with humans. California sea lions offer a success story and with its successes come management challenges.

Acknowledgments Illustrations are by Jacqui Fenner, ECS Federal, under contract to the Office of Science and Technology, National Marine Fisheries Service, National Oceanic and Atmospheric Administration. We are indebted to Ken Franke and the Everingham Brothers Bait Co. for allowing testing to occur from bait docks. Our California sea lion empirical work was supported by a National Science Foundation Graduate Research Fellowship and a grant from the California Sportfishing Association.

References

Atkinson S, Demaster DP, Calkins DG (2008) Anthropogenic causes of the western Steller sea lion *Eumetopias jubatus* population decline and their threat to recovery. Mamm Rev 38:1–18

Bejder L, Samuels A, Whitehead H (2009) Impact assessment research: use and misuse of habituation, sensitisation and tolerance in describing wildlife responses to anthropogenic stimuli. Mar Ecol Prog Ser 395:177–185

Betts M, Maschner HDG, Lech V (2009) A 4500 year time-series of Otariid abundance on Sanak Island, Western Gulf of Alaska. In: Braje T, Rick T (eds) Sea dogs of the North Pacific: the archaeology and historical ecology of seals, sea lions, sea cows, and sea otters. University of California Press, Berkeley

Blumstein DT (2014) Attention, habituation, and antipredator behaviour: implications for urban birds. In Gill D, Brumm H (eds) Avian urban ecology: behavioural and physiological adaptations. Oxford University Press, Oxford, UK, pp 41–53

Blumstein DT (2016) Habituation and sensitization: new thoughts about old ideas. Anim Behav 120:255–262. https://doi.org/10.1016/j.anbehav.2016.05.012

Bowles A (2012) Behavioral responses and habituation of pinnipeds and small cetaceans to novel objects and simulated fishing gear with and without a pinger. Aquat Mamm 38(2):161–188

Braje TJ, DeLong RL (2009) Ancient sea mammals exploitation on the south coast of San Miguel Island. In: Damiani CC, Garcelon DK (eds) Proceedings of the 7th California Islands Symposium. Institute for Wildlife Studies, Arcata, CA, pp 43–52

Carretta JV, Barlow J (2011) Long-term effectiveness, failure rates, and "Dinner Bell" properties of acoustic pingers in a Gillnet fishery. Mar Technol Soc J 45:7–19. https://doi.org/10.4031/MTSJ.45.5.3

Costa DP (1991) Reproductive and foraging energetics of high latitude penguins, albatrosses and pinnipeds: implications for life history patterns. Am Zool 31:111–130

Crespo EA, Pedraza SN, Dans SL et al (1997) Direct and indirect effects of the highseas fisheries on the marine mammal populations in the northern and central Patagonian coast. J. NW Atlan Fish Sci 22:189–208

Cunningham KA, Hayes SA, Wargo Rub AM et al (2014) Auditory detection of ultrasonic coded transmitters by seals and sea lions. J Acoust Soc Am 135(4):1978–1985

de Vere AJ (2018) Visitor effects on a zoo population of California sea lions (Zalophus californianus) and harbor seals (Phoca vitulina). Zoo Biol 37: 162–170

Domjan M (2005) Pavlovian conditioning: a functional perspective. Ann Rev Psychol 56:179–206

Domjan M, Burkhard B (1986) The principles of learning & behavior. Brooks/Cole, Monterey

Edgell TC, Demarchi MW (2012) California and Steller sea lion use of a major winter haulout in the Salish Sea over 45 years. Mar Ecol Prog Ser 467:253–262

Eisenstein EM, Eisenstein D (2006) A behavioral homeostasis theory of habituation and sensitization: II. Further developments and predictions. Rev Neurosci 17(5):533–57

Eisenstein EM, Jonnalagedda DL, Smith JSM (2012) Some new speculative ideas about the "behavioral homeostasis theory" as to how the simple learned behaviors of habituation and sensitization improve organism survival throughout phylogeny. Commun Integr Biol 5 (3):233–239. https://doi.org/10.4161/cib.19480

Ellenberg U, Mattern T, Seddon PJ (2009) Habituation potential of yellow-eyed penguins depends on sex, character and previous experience with humans. Anim Behav 77:289–296

Ellenberg U, Mattern T, Seddon PJ (2013) Heart rate responses provide an objective evaluation of human disturbance stimuli in breeding birds. Conserv Physiol 1:cot013

Fowler SL, Costa DP, Arnould JPY (2007) Ontogeny of movements and foraging ranges in the Australian sea lion. Mar Mamm Sci 23:598–614

Frid A, Dill LM (2002) Human-caused disturbance stimuli as a form of predation risk. Conserv Ecol 6:11

Garcia J, Kimeldorf DJ, Koelling RA (1955) Conditioned aversion to saccharin resulting from exposure to gamma radiation. Science 122(3160):157–158

Gearin P, Pfeifer R, Jeffries SJ, DeLong SJ, Johnson MA (1988) Results of the 1986-87 California sea lion-steelhead trout predation control program at the Hiram M. Chittenden Locks. NWAFC Processed Rep. 88-30, 111

Geffroy B, Samia DSM, Bessa E et al (2015) How nature-based tourism might increase prey vulnerability to predators. Trends Ecol Evol 30:755–765

Gerber LR, Hilborn R (2001) Catastrophic events and recovery from low densities in populations of otariids: implications for risk of extinction. Mammal Rev 31:131–150

Götz T, Janik VM (2011) Repeated elicitation of the acoustic startle reflex leads to sensitisation in subsequent avoidance behaviour and induces fear conditioning. BMC Neurosci 12

Gotz T, Janik VM (2013) Acoustic deterrent devices to prevent pinniped depredation: efficiency, conservation concerns and possible solutions. Mar Ecol Prog Ser 492:285–302

Greggor AL, Trimmer PC, Barrett BJ et al (2019) Challenges of learning to escape evolutionary traps. Front Ecol and Evol 7:408

Groves PM, Thompson RF (1970) Habituation a dual process theory. Psychol Rev 77:419–450

Heyes CM (1994) Social-learning in animals – categories and mechanisms. Biol Rev 69:207–231

Heyes C (2012) What's social about social learning? J Comp Psychol 126:193–202

Hildebrandt W, Jones T (1992) Evolution of marine mammal hunting: a view from the California and Oregon Coasts. J Anthropol Archaeol 11:360–401

Hill AL, Rand DG, Nowak MA, Christakis NA (2010) Infectious disease modeling of social contagion in networks. PLoS Comput Biol 6:e1000968. https://doi.org/10.1371/journal.pcbi.1000968

Hoppitt W, Kandler, Kendal JR, Laland KN (2010) The effect of task structure on diffusion dynamics: Implications for diffusion curve and network-based analyses. Learn Behav 38:243–251

Jeffries SJ, Scordino J (1997) Efforts to protect a winter steelhead run from California sea lion predation at the Ballard Locks. and issues. In: Pinniped populations, eastern North Pacific: status, trends and issues. New England Aquarium, Boston, MA, and Monterey Bay Aquarium, Monterey, CA

Kastak D, Schusterman RJ (1983) Sensitization and habituation to underwater sound by captive pinnipeds. J Acoust Soc Am 99(4):2577–2603

Kirkwood R, Lynch M, Gales N, Dann P et al (2006) At-sea movements and habitat use of adult male Australian fur seals (Arctocephalus pusillus doriferus). Can J Zool 84(12):1781–1788. https://doi.org/10.1139/z06-164

Laake JL, Lowry MS, Delong RL et al (2018) Population growth and status of California sea lions. J Wild Manag 82:583–595. https://doi.org/10.1002/jwmg.21405

Leung ES, Chilvers BL, Moore AB et al (2014) Do yearling New Zealand sea lions (Phocarctos hookeri) learn foraging behavior from their mothers? Mar Mamm Sci 30(3):1220–1228

Lima SL, Dill LM (1990) Behavioral decisions made under the risk of predation – a review and prospectus. Can J Zool 68:619J

Lottis M (2007a) Field report - control of nuisance sea lions in the Rogue River estuary, 2006, Gold Beach, Oregon. Port of Gold Beach Report

Lottis M (2007b) Field report - control of nuisance sea lions in the Rogue River estuary, 2007. Port of Gold Beach Report

Lottis M (2009) Field report - control of nuisance sea lions in the Rogue River estuary, 2008. Port of Gold Beach Report

Lyman RL (2003) Pinneped behavior, foraging theory and the depression of metapopulations and nondepression of a local population on the southern Northwest Coast of North America. J Anthropol Archaeol 22:376–388

Marshall KN, Stier AC, Samhouri JF et al (2015) Conservation challenges of predator recovery. Conserv Lett. 9:70–79

Rankin CH, Abrams T, Barry RJ et al (2009) Habituation revisited: an updated and revised description of the behavioral characteristics of habituation. Neurobiol Learn Mem 92:135–138

Rescorla RA, Solomon RL (1967) Two-process learning theory: Relationships between Pavlovian conditioning and instrumental learning. Psychol Rev 74:151

Rogers E (1995) Diffusion of innovations, 4th edn. Free Press, New York

Schakner Z, Blumstein DT (2016a) Learning and conservation behavior: an introduction and overview. In: Conservation behavior: applying behavioral ecology to wildlife conservation and management. Cambridge University Press, Cambridge, pp 66–92

Schakner ZA, Blumstein DT (2016b) Unpublished Data

Schakner ZA, Buhnerkempe MG, Tennis MJ et al (2016) Epidemiological models to control the spread of information in marine mammals. Proc R Soc B 283:20162037

Schakner ZA, Götz T, Janik VM et al (2017a) Can fear conditioning repel California sea lions from fishing activities? Anim Conserv 20:425–432

Schakner ZA, Petelle MB, Tennis MJ et al (2017b) Social associations between California sea lions influence the use of a novel foraging ground. R Soc Open Sci 4:160820

Schakner ZA, Purdy C, Blumstein DT (2019) Contrasting attitudes and perceptions of California sea lions by recreational anglers and the media, Mar Pol 109

Schusterman RJ (1981) Behavioral capabilities of seals and sea lions: a review of their hearing, visual, learning and diving skills. Psychol Rec 31(2):125–143

Scordino J (2010) West Coast pinniped program investigations on California sea lion and Pacific harbor seal impacts on salmonids and other fishery resources. Pac States Mar Fish Comm Portland

Sepúlveda M, Pérez M, Sielfeld W et al (2007) Operational interaction between South American sea lions Otaria flavescens and artisanal (small-scale) fishing in Chile: Results from interview surveys and on-board observations. Fish Res 83:332–340

Sepúlveda M, Newsome SD, Pavez G et al (2015) Using satellite tracking and isotopic information to characterize the impact of South American sea lions on salmonid aquaculture in southern Chile. PLoS One. https://doi.org/10.1371/journal.pone.0134926

Shettleworth SJ (2010) Cognition, Evolution, and Behavior, 2nd edn. Oxford University Press, New York

Sih A (2013) Understanding variation in a responses to human-induced rapid environmental change: a conceptual overview. Anim Behav 85:1077–1088

Sih A, Trimmer PC, Ehlman SM (2016) A conceptual framework for understanding behavioral responses to HIREC. Curr Opin Behav Sci 12:109–114

Smaldino PE, Aplin LM, Farine DR (2018) Sigmoidal acquisition curves are good indicators of conformist transmission. Sci Rep 8:14015

Snijders L, Blumstein DT, Stanley C et al (2017) Animal social network theory can help wildlife conservation. Trends in Ecol Evol 32:567–577

Thorndike EL, Bruce D (1911) Animal intelligence: experimental studies. Macmillan, Lewiston

Tidwell KS, van der Leeuw BK, Magill LM et al (2018) Evaluation of pinniped predation on adult salmonids and other fish in the Bonneville dam tailrace, 2018. U.S. Army Corps of Engineers, Portland District Fisheries Field Unit, Cascade Locks, OR

Weise MJ, Harvey JT (2005) Impact of the California sea lion (*Zalophus californianus*) on salmon fisheries in Monterey Bay. Cal Fish Bull 685–696.

Woolery ES, Harvey JT (2005) Role of visual and acoustic cues in attracting California sea lion (*Zalophus californianus*) to commercial trolling vessel gear. MLML Progress Report

Chapter 19
Pinniped 'Personality', or Consistent Individual Differences

Eugene J. DeRango and Jonas F. L. Schwarz

Abstract Animal behaviorists now extensively recognize that individuals within a population differ and are consistent within structured behavioral traits related to boldness, exploration, aggression, social behavior and activity level, in aggregate defined as animal personalities. Animal personality traits pervade and influence the ecology, life history, and fitness of a diverse array of taxa, yet studies of otariids appear to be overwhelmingly underrepresented. This chapter gives an overview of the history and evolution of concepts of animal personality and then highlights studies that serve as important foundations for understanding the extent of personality within free-ranging otariids. We hypothesize on the development of personality in young pups, and how this is likely shaped by the early social environment and driven by maternal effects. We also consider that, especially within polygynous systems, personality traits such as boldness and aggression may influence reproductive success, and therefore fitness. We then transition to at-sea behavior and ponder how consistency may influence the acquisition of individual foraging strategies in a dynamic marine environment. Finally, we consider how knowledge of individual differences and behavioral plasticity may strengthen conservation efforts when considering how individuals may be limited in their resilience to environmental change.

Keywords Boldness · Individual variation · Intra-specific strategies · Personality · Plasticity

19.1 Introduction

> No one supposes that all individuals of the same species are cast in the same mould.
> Charles Darwin, On the Origin of Species

E. J. DeRango (✉) · J. F. L. Schwarz
Department of Animal Behaviour, Bielefeld University, Bielefeld, Germany

C. Campagna, R. Harcourt (eds.), *Ethology and Behavioral Ecology of Otariids and the Odobenid*, Ethology and Behavioral Ecology of Marine Mammals,
https://doi.org/10.1007/978-3-030-59184-7_19

Pinnipeds exhibit an extensive range of behaviors throughout their lives within both the terrestrial and aquatic environment. 'Highly social and gregarious.' 'Curious and intelligent.' 'Playful.' These phrases may be used to describe the complex and often enigmatic behavior of sea lions and fur seals. Using these descriptors, however, makes broad-brush generalizations, and may not leave room for recognizing the intricate variation in behavior between and within individuals. Pinniped researchers vividly enlighten us with countless stories from long-term studies of well-known individuals that may behave 'differently' from others within the colony. Maybe you have walked through a colony to perform a census, and one specific adult male fur seal always aggressively charged at you much more than others. Perhaps certain sea lion pups were obviously more docile than others and may even have seemed to study your habits as you undertook your research, eventually actively approaching you rather than fleeing with their neighbors. Or at the other end of the spectrum, when performing observations from afar, certain individuals may have seemed to trigger stampedes and appeared more 'nervous' or responsive to disturbances than others. In general, though, it does not take a marine mammalogist to recognize that individual animals differ. Even pet owners have long observed that animals display different 'personalities', with some more 'excitable', 'dominant', or 'more shy' than others (Mendl and Harcourt 2000, reviewed in Gartner 2015). This section focuses on consistent individual differences such as these, and how behaviorists examine the importance of variation in behavior from the mean in wildlife populations.

Individual animals differ in structured behavioral traits; and these traits are consistent within individuals across specific time periods and standardized contexts (reviewed in Roche et al. 2016). In this sense, consistency does not mean fixed, but that individuals within a given situation consistently vary with each other. If such variation involves multiple traits and is repeatable, it is defined as animal 'personalities'. Because these differences in behaviors define how individual animals may interact with and respond to their environment, personality differences can influence all aspects of an animal's niche, ecology and life history (Dingemanse et al. 2010; Dall and Griffith 2014). Some earlier studies of otariids occasionally acknowledged and attempted to capture individual variation for a variety of behaviors. For example, Harcourt (1993) found that male southern sea lions (*Otaria byronia*) in Peru, identified via distinct pelage patterns, differed in their aggression towards and predation of fur seals. Perhaps by design, the field of acoustics has often relied on individual differences in vocal cues to show how individuals recognize conspecifics within dense colonies in many otariids (Charrier, Chap. 14). Similarly, Steller sea lion (*Eumetopias jubatus*) mothers differed consistently in styles of maternal care, such as how much time they spent foraging vs. nursing dependent pups (Maniscalco et al. 2006). Therefore, it is likely that consistent individual differences in behavior and physiology greatly influence many facets of otariid life history (Fig. 19.1).

Fig. 19.1 Individual variation in personality traits is expressed early in life and may influence many aspects of life history in the otariids, such as pup social behavior, at-sea foraging strategies, and later reproductive success. It is unclear whether personalities remain stable across life stages. Illustrations credit, Janina Weissenborn

19.2 Recognizing Animal Personalities

19.2.1 Historic Perspectives

Recognition of personality is ingrained in us as humans, and we are sensitive and predisposed to the categorization of personality traits because it encompasses a huge aspect of our own social lives. Even at the dawn of modern biology, Charles Darwin, in one of his pivotal works *The Expression of Emotion in Man and Animals* in 1872, noted that behavioral traits in addition to morphology are subject to natural selection on the level of the individual. In the early 1900s, comparative psychologists created a Five Factor Model for human personality, which describes individual humans based on categories of shared consistent dimensions of behavior. These categories are 'openness to experience, conscientiousness, extraversion, agreeableness, and neuroticism' (Goldberg 1990). Unsurprisingly, researchers later borrowed this framework for familiar animal groups (reviewed in Gosling and John 1999). For example, a founder of modern animal psychology, Ivan Pavlov, categorized his dogs (*Canis lupus familiaris*) using terms such as 'excitable' or 'inhibited' and found differences in learning and conditioning based on these personality types (Pavlov 1906). Meredith Crawford applied the Five Factor Model to our closest of kin, the chimpanzee (*Pan troglodytes*), and found strong individual variation in several behavioral traits (Crawford 1938).

The Five Factor model is applied to many taxa today, and includes studies of marine mammals (reviewed in Frick et al. 2017). Two different studies have recognized individual consistency in personality dimensions for California sea lions (*Zalophus californianus*) under human care and condensed observed behaviors into traits such as 'extraversion', 'dominance', and 'reactivity' (Ciardelli et al. 2017; de Vere et al. 2017). More extroverted sea lions tended to score higher in training performance and learning when directed by their keepers (Ciardelli et al. 2017), and displayed more tactile and play behaviors towards conspecifics (de Vere et al. 2017). While these studies establish the existence of personality and offer relevant knowledge to improve welfare and management for housed pinnipeds (Frick et al. 2017) they do not allow us to judge the relevance of a behavior that functions under the strong selective pressures of the natural environment. Therefore, we focus on free-ranging individuals observed or manipulated within their natural environment.

19.2.2 Ecological Applications

As the field of animal personality grows more complex, there has been a huge push to focus on ecological and evolutionary concepts, such as how personality directs the adaptability and fitness of individuals within a population (Réale et al. 2007; Wolf and Weissing 2012). We describe individual variation in behavioral traits maintained through environmental gradients through time and/or specific life history contexts (Dingemanse et al. 2010). Table 19.1 gives a summary of the terminology encountered within the field.

Table 19.1 Common terminology encountered within the field of animal personality

Term	Definition	References
Behavioral syndrome	A correlated suite of behavioral traits (e.g. positive links between boldness and exploration) across contexts	Sih et al. (2004)
Behavioral type	Individuals classified and grouped by traits within behavioral syndromes (e.g. those that are bolder, more explorative)	Sih et al. (2004)
Context	A functional category (e.g. presence or absence of predators, conspecifics, maternal care)	Sih et al. (2004)
Situation	A given set of conditions within a specific time point or context (e.g. degree of predator exposure)	Sih et al. (2004)
Repeatability	A standard measurement of the proportion of phenotypic variance explained by differences between individuals	Bell et al. (2009) and Dingemanse et al. (2010)
Consistent	High repeatability of behavioural differences between and within individuals	Réale et al. (2007)
Behavioral plasticity	The ability to modify behavior as a result of exposure to changing or new stimuli within the environment	Dingemanse and Wolf (2013)
Behavioral reaction norm	The average response of an individual across environmental gradients (i.e. time, context)	Dingemanse et al. (2010)
Coping style	Behavioral and physiological responses towards stressors, which are consistent over time and characteristic to a certain group of individuals	Koolhaas et al. (1999)

An adjusted method describes animal personalities according to behavioral axes such as shyness–boldness, exploration, activity level, aggressiveness and sociability (Réale et al. 2007). Boldness is generally interpreted as willingness to take risks, while exploration usually describes the rate of movement within or inspection of unfamiliar situations. These traits often co-vary and interact within individuals, structured into 'behavioral syndromes' (Bergmüller 2010). Personality trait correlations have been linked to differences in multiple aspects of life history, such as social behavior toward conspecifics (Bergmüller and Taborsky 2010), reproductive success (Wolf and Weissing 2012), productivity (Biro and Stamps 2008), and overall survival (Réale et al. 2010). For example, a consistently aggressive individual may be less risk-adverse and show a reduced fear response towards predators, thus allowing it access to explore new habitats or foraging grounds despite increased dangers. Personality traits can shape an individual's reactivity towards its environment, how it forms routine behaviors, and how it changes its behavior in response to new or stressful experiences (Koolhaas et al. 1999; Coppens et al. 2010). Therefore, animal personalities are not always fixed, and we take this opportunity to explore how they may develop or change throughout life (Trillmich et al. 2018).

19.2.3 Modern Methodologies

The field of animal personality would not be as robust as it is today without the cornerstone of a consistent reliable methodology. Methods are designed to account for the ecology and natural behavior of the species and fall into broad- vs. narrow-sense definitions of animal personality. Broad-sense personality relies on performing behavioral observations without altering the animal's natural behavior (Roche et al. 2016). Observing behavior remotely can avoid potentially confounding effects of artificial situations and disturbance (Koski 2011), which may be useful in 'flighty' species of otariid seals. The narrow-sense approach consists of standardized behavioral assays and manipulation which elicit high-arousal states (Roche et al. 2016). Examples of tests include responses to novel objects, activity within open fields or arenas (either familiar or novel), or reactions to human approach or manual restraint (i.e. struggle or hand escape tests). The goal of these assays is to recreate some ecological trade-off that may influence consistent variation in individuals and correlations between personality traits in nature (Dall and Griffith 2014). For example, does an individual who is more docile when restrained following capture also show less anxiety or activity when placed within a novel environment? How do these two personality traits relate to its competitiveness for food resources in daily life, thus impacting individual fitness? To be considered consistent, behaviors must show high levels of statistical repeatability within the population, or limited variance through different contexts (Dall and Griffith 2014). Consequently, repeatability can robustly estimate how differences in behaviors develop or remain stable throughout an individual's lifespan.

19.3 Capturing Individual Differences in Pinnipeds

Pinnipeds occur in complex and often logistically challenging habitats, which presents hurdles to reliably describe individual differences. Even individual identification may prove challenging with free-ranging otariids, which often live in dense terrestrial aggregations. Individual identification can be achieved using well-established methods, such as bleach or dye marks, shaves, brands, flipper tags, or even distinctive natural scars and markings (Fig. 19.2). Otariid seals are generally wary of human presence, and depending on the species or location, may be difficult to access or approach. In these cases, observational studies may lend themselves to the study of consistent individual differences in a variety of behaviors. Researchers are often constrained to working within limited windows of time when animals haul out for breeding, rearing pups or molting. Luckily, many pinniped species show high degrees of breeding and natal site fidelity, and relocating individuals makes it possible to retest and determine stability of behavior across different life history contexts. Researchers should acknowledge potentially confounding effects within a field setting, such as the state of the animals before tests or observations, and take care to standardize conditions as much as possible.

Studies have provided insight to aspects of consistent individual differences in pinnipeds on land (Table 19.2), relying on methodologies tailored toward each species (Fig. 19.3), and they have demonstrated repeatability of behavior. Twiss

Fig. 19.2 Methods, such as shaves, dye or bleach marks (**a**), tags (**b**) or brands (**c**), and natural scars/markings (**d**) allow for identification of individuals within a given context or time period, which is crucial to determine the repeatability of behaviors Photo credit, E. DeRango (**a**, **b**, **d**), P. Robinson, NMML permit # 16087-02 (**c**)

Table 19.2 Current studies which measure consistent individual differences through behavioral manipulation and assays in pinniped species

Family, Species	Topic	References
Phocidae, grey seal	Standardized behavioral assay using a novel stimulus	Twiss et al. (2012a)
Phocidae, grey seal	Proactive-reactive coping strategies in response to disturbance in adults	Twiss et al. (2012b)
Phocidae, grey seal	Recognition of and aggression towards conspecifics in weanlings	Robinson et al. (2017)
Phocidae, grey seal	Boldness in adult females, consequences for offspring fitness	Bubac et al. (2018)
Otariidae, Galapagos sea lion	Standardized behavioral assays towards novel objects and human approach in pups	DeRango et al. (2019a)
Otariidae, Galapagos sea lion	Proactive-reactive coping strategies in response to disturbance (capture) in pups	DeRango et al. (2019b)

Fig. 19.3 Behavioral assays and standardized tests, such as novel object presentation (**a**), human approach (**b**), and struggle tests (**c**), are commonly used to elicit high-arousal states. Behavioral observations and recordings can also elucidate behavioral types in lieu of altering natural behavior (**d**). Photo credit: J. Schwarz (**a**, **d**) and E. DeRango (**b**, **c**), DPNG permit # 74-18

et al. (2012a) were the first to describe a reliable novel object test in pinnipeds that used a modified remote-controlled vehicle that could be driven towards the animals over rugged terrain. This test is convenient as it alleviates human presence and the test subject focuses exclusively on the novel object. DeRango et al. (2019a) later applied novel object tests with young Galápagos sea lion (*Zalophus wollebaeki*) pups. Compared to most other otariids, this species is unusually tame, which lends itself well to develop and test practical field methods. Experimenters constructed a simple, hand-held stimulus (a long pole attached to a randomized series of colored objects) to elicit approach and exploration behaviors. In colonies where animals can be accessed without causing large disturbances, human approach tests are valuable to elicit repeatable individual reaction types (Bubac et al. 2018; DeRango et al. 2019a). A human is considered the aversive stimulus and approaches the target in a standardized manner to measure its general response towards risk. If routine handling of animals is necessary and performed in a standardized way (such as for health assessments or monitoring), the intensity of struggle or escape after release can also give insights into how personality affects anxiety or aggressiveness towards an induced anthropogenic disturbance (DeRango et al. 2019b). Although relatively few, these studies serve as a basis to further explore whether personality may be ingrained into multiple aspects of otariid ecology.

The amphibious nature of pinnipeds produces a unique challenge. Some studies indirectly capture individual differences in foraging behavior by measuring stable isotope concentrations in the tissue of individuals. Stable isotope analysis does not necessarily give high-resolution information about diving behavior, but it can inform us about individual dietary specialization based on the trophic level and the habitat use (e.g. Elorriaga-Verplancken et al. 2013). Vibrissae grow slowly and create an archive of stable isotope values over months or even years (Crawford et al. 2008). For this reason, a single whisker sample makes it feasible to study the consistency of an individual's foraging strategy over time (Franco-Trecu et al., Chap. 29).

While stable isotope analysis can be a useful tool to study trophic levels and preferred habitats, diving behavior is a much more complex aspect of individual foraging strategies (du Dot and Guinet, Chap. 4). Attaching bio-loggers directly on an animal enables us to observe fine-scale diving behavior. In the wake of the digital revolution, technologies and analysis techniques have progressively improved along with the resolution of diving data that researchers can acquire. A single recorder can stay on an individual for an extended period, simultaneously collecting data such as dive depth, temperature, light level, location, three-dimensional acceleration, magnetism, and even record video (McGovern et al. 2019). While traditionally dives are divided into three segments (descent, bottom and ascent), better resolution and new analytical procedures, like the broken stick algorithm (Heerah et al. 2014), make the division into more segments possible, allowing accurate description of duration, depth and ranges of foraging events within individual dives through measures of velocities (du Dot and Guinet, Chap. 4). Such high-resolution dive data collated with GPS location and stable isotope analyses will make it easier to unfold small scale individual differences in foraging strategies.

19.4 Ontogeny of Pup Personality

19.4.1 Internal Mechanisms and External Stimuli

Mechanisms that spark the onset of personality traits are not always clear. Juvenile animals are influenced by stimuli from a variety of sources (Fig. 19.4). Genetic makeup, individual physiology, including hormone regulation, along with disparate exposure to stimuli (from the physical and social environment) are factors which strongly interact to promote differences in personality and behavioral plasticity (Stamps and Groothuis 2010; Trillmich et al. 2018). It can be difficult to disentangle these factors. Are young animals already fixed into predisposed personality types based on heredity and maternal conditions within the prenatal environment, and is personality later molded further by these same environmental effects? We explore how personality shapes the way pups navigate their early life environment and which internal and external factors may contribute to differences in personality traits.

19.4.2 Early Life Challenges

Sea lions and fur seals have a high-pressure job shortly after birth. First, pups must quickly bond with and be able to recognize their mothers (Charrier, Chap. 14), as the mother will soon leave and forage on her own before intermittently returning to

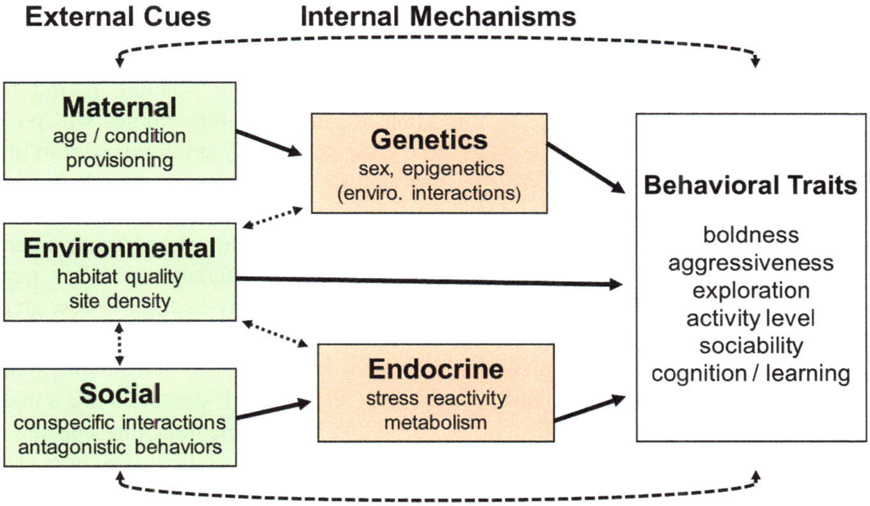

Fig. 19.4 The shaping of pup personality and the structure of behavioral traits consists of complex interactions between external cues, maternal characteristics and internal mechanisms, such as genetics and hormone expression

nurse (Sepulveda and Harcourt, Chap. 3). Pups initially lack swimming skills and are land-locked within an unpredictable environment that presents many risks to their well-being and survival. As pups begin to explore their terrestrial environments, they may encounter potentially dangerous situations due to conspecifics. For example, other adult females may reject and strongly attack un-related pups if they attempt to suckle (Harcourt 1992). As is often the case with Galapagos sea lions, older siblings may still be un-weaned and directly compete with neonates for suckling opportunities, which can lead to intense conflict and instances of aggression (Trillmich and Wolf 2008). In some species, adult males in the colony cause pup mortality, either inadvertently by trampling (Ryazanov, Chap. 10) or by intentionally preying upon pups (for instance, large sea lion species predation on fur seals) (Harcourt 1993). Additionally, otariid species have evolved under widely differing degrees of threats from land predators, such as shorebirds (e.g. Seguel et al. 2017) or large carnivores, or sometimes lack land predators completely (Riedman 1990). Furthermore, once pups develop swimming skills and wander into the ocean, they encounter a whole new series of risks in the form of aquatic predators or exposure to other anthropogenic stressors such as marine debris and fisheries. Therefore, there are likely huge differences in the timing and degree of risks during early life, both between individuals and entire populations for each otariid species.

Perhaps related to the strong presence of early life risks, individual differences in boldness exist as early as the first 2 months of life in Galapagos sea lion pups (DeRango et al. 2019a, b). Male pups, and those that were older within their cohort and in better body condition tended to be bolder (DeRango et al. 2019a) and habituated quicker to disturbance and manipulation (DeRango et al. 2019b). Female pups tended to be more anxious and may have perceived specific threats during tests as more aversive than males. Males tend to be the more aggressive sex, and for many species, boldness is linked to extroversion and aggression (Réale et al. 2007). Boldness was consistent within most individuals during the first 2–3 months and again after the first year, representing roughly half of the weaning period for Galapagos sea lions. Therefore, an important avenue for future studies will be to see how far pups will be constrained to these same personality structures later in life.

Boldness and exploration of the surrounding habitat were also linked in Galapagos sea lion pups. These characteristics increased with age in the first 2–3 months of life, implying that, as pups become more mobile, they may learn from and adapt to unique risky experiences. Intriguingly though, Galapagos sea lion pups which seemed to be less sensitive to stressors were less likely to explore new areas away from their site of birth (DeRango et al. 2019b). This is puzzling, as greater exploration rates are often argued to create great knowledge of surroundings and allow individuals to better respond to new risks, especially if predators are a main selective force (Rodríguez-Prieto et al. 2010). It is possible that this species is unusual, as pups are tame and have no need to worry about land predators. For Galapagos sea lions, familiarity with a limited environment may allow them to be more 'comfortable' when faced with novel or risky situations. It would be interesting to expand similar studies towards other species and determine how differences in the type and quantity of early life risks play a role in individual decision making and the expression of behavior.

19.4.3 Social Interactions and Play

Juvenile otariids are motivated to engage in play behavior, thus making up a large portion of the activity budget when not suckling (Llamazares-Martín and Palagi, Chap. 20). When mothers are away, pups either sleep or play. The traditional philosophy towards play suggests it is used as "preparation", or a way to develop motor skills that become useful during future opportunities (Bekoff and Allen 1998, see Llamazares-Martín and Palagi, Chap. 20). A small tide pool of pups biting, wrestling, or chewing on flippers could be thought of as a sort of kindergarten, where pups hone their skills for later foraging or reproductive interactions. Few studies, however, have addressed the intraspecific variation of play among juveniles and related this variation to aspects of personality (Fig. 19.5). For example, Gentry (1974) showed that male Steller sea lion pups engaged in more aggressive posturing, displaying, and biting than females. Personality may play a role in these behaviors, as bolder pups (which may be sex-biased) may be more aggressive and extroverted and therefore more likely to assert themselves in play behavior with conspecifics, prey items or even inanimate objects, such as rocks or plants. Play itself may also be risky, as pups may be more susceptible to predation (Harcourt 1991) or harassment (Campagna et al. 1992) when away from their mothers, and therefore rates of play

Fig. 19.5 Personality traits, such as boldness and aggression, along with site density and geography, likely influence social dynamics between pups, mothers, and other conspecifics. Photo credit, E. DeRango, DPNG permit # 74-18

may be higher in bolder individuals. If play is preparatory and incurs high costs, it would be relevant to determine if consistent differences in play behavior throughout development can directly translate to differences in adult risk-taking behavior or even foraging and reproductive success that may eventually benefit the individual. It may prove difficult to collect these data, as this may require constant annual monitoring of individuals across years. However, if the degree of play behavior in young pinnipeds is a stable and consistent trait this may result in life history consequences.

Another intriguing concept for the study of play is that it is often incited by novel or unpredictable stimuli provided by partners (Spinka et al. 2001). This "training for the unexpected hypothesis" therefore has direct implications for how pups will develop coping styles when responding to new environments or stressors. As a direct result of local conditions and abundance, population density varies strongly with habitat for individual sea lions and fur seals within a population. Therefore, individual pups will inevitably gather widely different experiences (amounts and qualities of stimuli) from their play partners depending on where they are born. Exposure to these cues can increase or reduce flexibility of behavior, and social learning depends on how pups integrate this information (Hoppitt and Laland 2008). For example, a pup born in a lone environment with little stimuli other than from its mother may become less adaptable towards new and stressful situations than pups who are born into large groups and constantly acquire information from their partners. It is then important to stress how individual differences in reactions towards conspecifics vary within the context of local density, and how this may influence social behavior and niches at an early age.

19.4.4 Maternal Effects

Even before pups are born, mothers may be setting the conditions which lead to individual differences during early life. The chosen birth site can play a huge role in stimuli that pups receive after they are born, and it may also impact the prenatal condition. Denser sites within colonies are often considered to have more social stressors, which impact chronic levels of stress and sex hormones in Antarctic fur seals (*Arctocephalus gazella*) (Meise et al. 2016). Cortisol and testosterone levels are linked between mother and pup, possibly indicating a genetic component but potentially also interactions between genetics and similar environments or maternal transmission. Early exposure to cortisol can alter the structure of personality traits in juvenile mammals (Guenther et al. 2018), which may be regulated by differences in energy levels and metabolism (Biro and Stamps 2010). Testosterone can also have profound behavioral and neuroendocrine effects on offspring, including aggression (Kaiser et al. 2003). Twiss et al. (2012b) found that bolder adult female gray seals (*Halichoerus grypus*) tended to live in higher density sites, suggesting that bolder individuals may be better equipped to deal with increased stressors in these environments. Whether this is due to interplay between behavior or physiology remains to be seen. Although there is minimal knowledge of the inheritance of personality in otariids (either through genetics, learned behavior or some combination of both), it would be advantageous for pups to express a similar personality as the mother if conditions are similar for both.

Maternal effects do not stop at birth. For the first few months of life, pups lack the skills to feed themselves and remain entirely nutritionally dependent on their mothers. Mothers provide support through nursing, which directly contributes to the energetic reserves and growth efficiency of dependent pups. The maintenance of good body condition has been linked to boldness (DeRango et al. 2019a) and activity levels and movement patterns in individual Galapagos sea lion pups (DeRango et al. 2019b). The level of maternal energy input will determine a pup's energy to explore surroundings or engage in extroverted behaviors, thus impacting its expression of personality. During dependency, maternal age, condition, and experience also may have strong impacts on how mothers grant social support to their pups. Shy Galapagos sea lion pups were less inclined to flee from a novel object when their mothers were present, and this effect became stronger in older and larger mothers (DeRango et al. 2019a). Shy, withdrawn animals are known to adapt their behavior in response to social cues (Ólafsdóttir and Magellan 2016); therefore, offspring may receive social support from mothers based on her state and characteristics related to her investment.

19.5 Reproductive Consequences

19.5.1 Life History Theory

From an evolutionary perspective, personality traits are maintained within individuals through a variety of life history trade-offs (Wolf et al. 2007). For example, reproductive behaviors are one of the strongest drivers of individual variation within a population. Some individuals may put more emphasis on future reproduction and be less risky in early life or in their current state so that they survive into later adulthood (Wolf et al. 2007). The life history trade-off hypothesis suggests that we do not see an 'optimal' personality type within populations because personality under one set of environmental conditions may be advantageous at times but maladaptive under others. In species where high levels of boldness, exploration and activity lead to higher food intake rates or the ability to give birth to more or better offspring, then this trait may lead to a selective advantage in the population (Biro and Stamps 2008). However, as previously noted, boldness may also lead to exposure to predators and an increase in mortality, so individuals adopt different strategies to account for these trade-offs. Personality traits and their physiological correlates, such as metabolism, growth rates and reproductive output, converge along a 'pace-of-life syndrome' continuum in response to environment conditions (Réale et al. 2010). Shy and thorough explorers may be 'slower', in that they live longer and reproduce later, while the opposite is true for aggressive and superficial explorers. Therefore, it is important to consider how these evolutionary models may explain the amount of individual variance in personality in the otariids.

19.5.2 Fortune Favors the Bold?

As long-lived species, otariids likely develop consistent differences in personality based on strategies related to current and future reproduction. Most otariids are polygynous where animals annually aggregate into densely settled habitats that promote a great deal of social conflict amongst individuals (Harcourt 1992). Males directly compete for territory and mates, and females compete for choice breeding areas to raise pups (Cassini, Chaps. 6 and 7). Sexual selection can produce large sex differences in personality based on competition between and within both sexes (Schuett et al. 2010). Although not an otariid, the gray seal is another polygynous pinniped where several aspects of consistent individual differences in reproductive behavior have been found. We will use these studies to consider how personality and proactivity within reproductive behaviors creates strong fitness consequences for reproductive adults and their offspring.

The first step for most reproductive males is establishing and controlling a territory, and eventually various-sized harems of females who settle on the territory. An early observational study demonstrated that adult male grey seals were consistent in the amount of time spent alert when defending territories across breeding seasons (Twiss and Franklin 2010). Later it was determined that consistency in aggression during breeding was associated less with dominance status and more related to the number of interactions with unfamiliar neighbors (Bishop et al. 2015). Adult males were aggressive to unfamiliar competitors more than familiars within their own territories. Twiss et al. (2012a) used their modified novel object test to show that adult gray seals also varied in taking risks. Some adult males consistently approached or showed aggression towards the object, while others fled. The authors hypothesized that these differences in aggression towards unfamiliar objects may be related to how males hold territories and defend females within a harem. Only very recently has it been shown that boldness can affect an animal's ability to hold resources during direct territory contests (Kaiser et al. 2019). Butterflies which were winners of territorial contests were more active and explorative, and intriguingly, losers became bolder in later contests. Boldness may then be advantageous if costs are not too high. For male otariids, high levels of boldness and aggression may increase access to females and increase fitness benefits in the short term, but it may also be associated with risks and trade-offs which increase mortality, such as severe wounds leading to deadly infection, increased energy expenditure and stress, or even overheating depending on the climate.

Adult female otariids have the dual role of defending and feeding themselves and their pups, which carries its own set of reproductive trade-offs. Females must frequently ward off unwanted advances from territorial males, and defend their pups from aggression from conspecifics. Adult female gray seals showed consistent degrees of vigilance in checking on their pups when confronted by a potentially aversive stimulus, a proxy for maternal attentiveness (Twiss et al. 2012a). Some adult female gray seals demonstrated behavioral plasticity and were more reactive towards disturbance. These females increased pup-checking behavior when disturbed rather than flee and separate from their pup. In a separate study, adult female gray seals maintained levels of boldness when approached by humans across several years during a long-term study (Bubac et al. 2018). Older (and likely more experienced) females were bolder. This fits life history theory and younger females were more risk adverse, consistent with the reproductive value of the pup for that life stage. Even within young females, bold mothers produced pups which were 2 kg heavier than those of shy mothers. Boldness—as it relates to maternal experience and the ability to alter behavior when assessing risks—is important for pup defense and potentially even fitness and survival.

19.6 Individual Foraging Strategies

19.6.1 Drivers of Intraspecific Variation

To comprehensively understand the mechanisms and consequences of individual differences in behavior of an amphibious predator, studying individuals exclusively on land excludes perhaps half of their lifetime. Foraging for otariids is almost exclusively done in the aquatic environment, where most of their predators are also found. Consequently, behavior while at sea has a huge influence on foraging success, predator avoidance, and ultimately fitness. Considering the high variability of individual behavior found on land and the complexity of the marine ecosystem, it is reasonable to also expect great intraspecific variation in the diving and foraging behavior of otariid species.

A strong driver of different specialized foraging strategies may be avoiding intraspecific competition. Adult females with dependent pups are restricted to foraging close to their colony, which can cause competition with other adult females within their colonies. Mothers can cope with this challenge by using only a narrow subset of the ecological niche, thereby reducing competition with their neighbors (Kuhn et al. 2006). Specialized foraging strategies could also increase familiarity with the feeding conditions (Elorriaga-Verplancken et al. 2013) hence increasing foraging success of animals with a specialized foraging strategy (Terraube et al. 2014). Different foraging strategies may also be adaptive in different environmental conditions. The ocean is under daily, seasonal and annual shifts in conditions, as well as unpredictable environmental changes such as the El Niño–Southern Oscillation event. Changes in sea surface temperature, the position of the thermocline (influencing the depth of prey organisms), in currents, in winds, etc. can have a huge effect on the distribution of prey organisms and the productivity of an area and thus the accessibility of prey (Soto et al. 2006). These effects can differ hugely between prey species and/or ecological regions. Changes in prey abundance will have strong implications for the success of different foraging strategies.

Studies that concentrate on the individual are still underrepresented, but they can give us insights into advantages and disadvantages that different strategies may have for individuals, thus providing tradeoffs with implications for individual fitness. Different foraging strategies might be adaptive within specific environmental conditions, even among individuals of the same species. The marine environment offers widely different habitats and prey, and thus, possibilities for foraging specialization. Stable isotope analysis in Antarctic fur seals revealed more pronounced individual foraging strategies in colonies with high intra-specific competition compared to colonies that lived in sympatry with subantarctic fur seals characterized by high inter-specific competition (Kernaléguen et al. 2015). Similarly, the Galapagos sea lion has a relatively small distribution, mostly restricted to less-productive waters surrounding the Galapagos Islands. However, different foraging strategies have been identified even within a colony (Villegas-Amtmann et al. 2008) or between colonies of different islands less than 80 km apart (Jeglinski et al. 2015; Páez-Rosas et al. 2017), likely driven by local prey availability. To understand the significance of foraging strategies on the ecology of a population, such differences on a subpopulation level should be accounted for. In lactating northern fur seals (*Callorhinus ursinus*) on St. Paul Island, stable isotope and dive data analysis revealed specialization of some females on small, energy-dense schooling fish of deeper waters, while other females preyed on larger, but less calorically rich prey on the bottom of shallower neritic waters (du Dot et al. 2018). Although females specializing in epipelagic foraging catch higher quality prey, they must swim longer distances to reach their foraging grounds (Costa and Toro, Chap. 2). Females specializing on the neritic waters can forage closer to the colony, allowing a higher frequency of nursing bouts, but they must settle for prey of lesser quality. The trade-off between the two strategies results in two alternative, stable strategies (du Dot et al. 2018).

19.6.2 Consistency: The Key to Success?

Consistency within foraging strategies, with only small degrees of flexibility, has been found in many species, such as northern gannets (*Morus bassanus*) (Patrick et al. 2015), minke whales (*Balaenoptera acutorostrata*) (Hoelzel et al. 1989), and cichlid fish (*Lepidiolamprologus profundicola*) (Kohda 1994). The ability of individuals to alternate foraging strategies seems to be rather uncommon in the otariids. For example, early studies of individual Australian sea lions (*Neophoca cinerea*) found strong multi-season consistency between the foraging sites used and the type of prey consumed (Lowther et al. 2011, 2013). Strong repeatability of foraging behaviors was recently also shown across different timescales in lactating California sea lion females (McHuron et al. 2018). While sea lions alter their behavior in response to changing environments, the prevailing strong persistence of foraging strategies in individuals suggests high degrees of consistency. This may allude to a

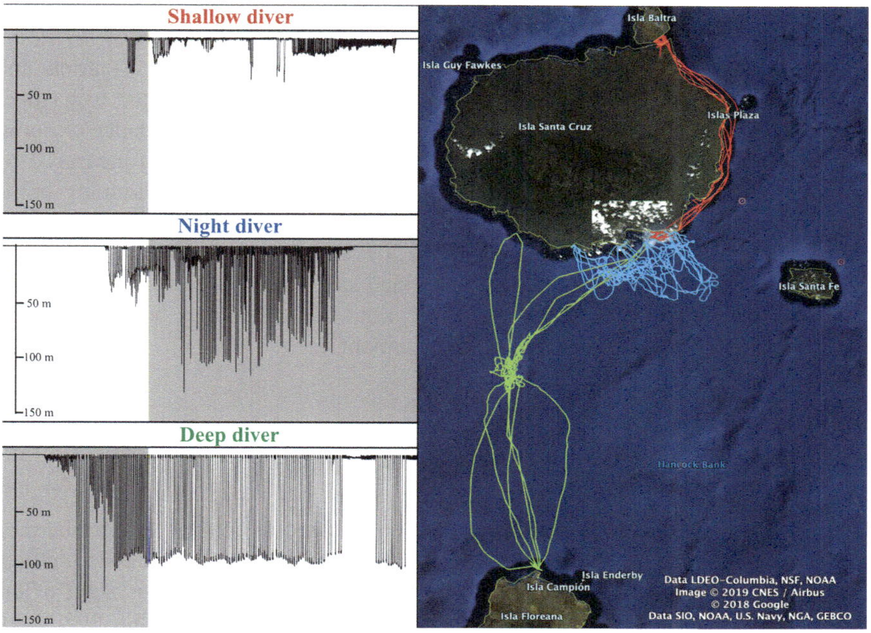

Fig. 19.6 Diving profiles (18 h) and GPS tracks (2 weeks) of three individual Galapagos sea lion mothers with three different foraging strategies. Gray background represents night hours. Strong variability in dive depth and duration (left) along with location (right) can be found

cost associated with high flexibility in foraging strategies. Individual stability might be the best strategy to manage uncertain environmental conditions (McHuron et al. 2018). Within a variable, dynamic environment, several foraging strategies can coexist in a population. Some individuals may have greater fitness depending on local and present conditions, and no single individual will outcompete the other strategies under all conditions. However, it is still unclear if strong fidelity for a certain foraging strategy or site may limit the ability of sea lions to adapt on a population level, if prey species decline or local conditions deteriorate (Baylis et al. 2015). An ongoing study in a colony of Galapagos sea lions aims to unravel the adaptability of individual foraging strategies. Resulting analysis of duration, depth and ranges of foraging events not only helps to identify individual foraging strategies even more reliably but provides great insight into the consequences of a certain strategy for an individual. Preliminary results show that different foraging strategies in lactating females were stable over several foraging trips (Fig. 19.6). However, the real value of this research lies in the opportunity to study the influence of different strategies on the body condition of the females and their offspring across years, and consequently over different environmental conditions (e.g. across El Niño and La Niña events).

Rarely have studies built a bridge between consistency in aquatic foraging strategies and personality on land, and none have been carried out in pinnipeds. A study of black-browed albatross (*Thalassarche melanophris*) found that bolder birds during a novel object test at the nest site foraged in shallower waters closer to the colony, while shyer animals foraged over deeper oceanic waters (Patrick and Weimerskirch 2014). Even fitness differences were found between the two strategies; the bold individuals fared better in years with higher prey abundance closer to shore. Similarly, bolder African penguins (*Spheniscus demersus*), as measured by human approach to their nest site, covered larger distances when swimming in the water column and followed a more sinuous path, thus allowing a greater exploration of their foraging sites (Traisnel and Pichegru 2019). Correlating at-sea behavior with measures of physiology and personality on land can reveal the mechanisms behind the acquisition of foraging strategies, and to what extent this may be driven by the diversity and dynamics of the marine environment.

19.7 Implications for Conservation

There has been a call to integrate concepts in behavioral ecology when planning conservation and management efforts for pinnipeds (Brakes and Dall 2016). Often, conservation focuses on large-scale population level trends and loses sight of the way that individuals cope with threats or respond to challenges within their environments. The ability to respond to human-induced changes in the environment is crucial for pinnipeds (and all species), and resiliency can have huge fitness consequences for individuals within a population (Dingemanse et al. 2004). The ability to discriminate between novel or risky stimuli can have direct implications for the development of personality (e.g. boldness or exploration), and also real-world consequences that may impact survival. By studying animal personality, its role in the ability to cope with changing environments and its influence on population dynamics and overall behavior, we increase our knowledge about animal personality in otariids in general, and might help to optimize conservation efforts.

Variation in personalities and behavioral plasticity are essential concepts to consider when making informed management decisions, as they may predict how individuals are flexible in coping with induced stressors or changing conditions within their environment (McDougall et al. 2006). Personality may even affect the structure and distribution of populations. For example, individuals select and use suitable habitat based on their personality type (Leclerc et al. 2016). Bold individuals are more likely to live in areas with a strong predator presence (Croft et al. 2009) or learn to explore new habitats faster (Gibelli and Dubois 2016). Individual sea lions react to human presence differently based on risk-taking behavior towards an adverse stimulus (DeRango et al. 2019a, b). Human disturbance can greatly affect natural behavior and haul-out patterns for Steller sea lions (Kucey 2005), and continued tourist presence affects the distribution of haul-out sites for southern sea lions (Túnez et al. 2008). It would prove informative to know if bolder or more aggressive individuals are better equipped to respond or habituate to these pressures, and potentially which individuals in colonies are responsible for causing mass group disturbances based on personality type. Using this same principle, it is also intriguing to speculate on how historic pressures may have altered the variability of personality within today's populations of sea lions and fur seals. Otariids, like many marine mammals, went through a recent period of over-exploitation and hunting, and populations of some species are still recovering. It has been shown in many other species that bolder, more explorative, or active individuals are more likely to be 'trapped' (e.g. captured) for research purposes (summarized in Michelangeli et al. 2015), potentially due to being less averse to humans or other risky situations. It is not unreasonable to wonder if this same principle holds true for past populations of otariids who were confronted with a new selective pressure of mass commercial hunts. Perhaps, bold, tame, and more visible sea lions or fur seals were more easily targeted for capture and culling. Although we have no baseline behavioral data, this thought experiment may give a useful perspective when considering how personality traits within otariids may cause differences in risk assessment towards humans (see Schakner and Blumstein, Chap. 18).

Individuals can also differ hugely in their susceptibility to anthropogenic threats within the marine environment. The study of at-sea behavior and individual foraging strategies can help predict the prevalence of interactions between anthropogenic factors, such as offshore wind farms, aquafarming or fisheries. In lactating female New Zealand sea lions (*Phocarctos hookeri*), for example, mesopelagic divers have a much greater overlap with the activity of the subantarctic squid trawl fishery than benthic divers (Chilvers and Wilkinson 2009). This results in direct competition with fisheries and an increased chance for mesopelagic divers to have a deadly encounter with a fishing vessel or entanglement. Obviously, there would be strong selection against individuals that follow this strategy, and could lead to the disappearance of this strategy and genetic diversity (Chilvers and Wilkinson 2009). Of great concern is also social transmission of novel behavior. Certain individual California sea lions within a social network influence others to exploit anthropogenically-influenced food sources (Schakner et al. 2017), thereby increasing risk of human-wildlife conflict. These studies would greatly benefit from knowing individual personality types, as this may help predict how animals disperse and use these newly learned behaviors. One proposed mechanism for how individuals learn from each other within populations is the concept of 'leaders and followers' (Harcourt et al. 2009), where bold individuals often take initiative and lead by example. This powerful idea could explain how individuals influence the foraging behavior of others. Better knowledge of individual foraging strategies can thus provide the needed information to reduce interactions between fisheries and sea lions to optimize conservation efforts.

19.8 Final Questions

Otariids are an underrepresented group in animal personality research; however, their diverse ecology and complex life histories should lead to interesting avenues for studies in personality in response to various contexts. We hope these examples lead to an improved emphasis on the ecological importance of personality within closely related but diverse sea lion and fur seals. We conclude with a set of remaining questions which can be used as a springboard for the ecological application of the personality concept in pinnipeds:

- How do maternal conditions, early life experiences, and local environments within a colony shape the ontogeny of personality in young pups?
- Does personality contribute to the development of social niches and competition among conspecifics?
- Considering trade-offs and costs of behavior, will variation in personality traits, such as boldness and aggression, lead to greater reproductive success?
- Do personality traits that are expressed on land impact at-sea behavior, such as dispersal and individual foraging strategies?
- Do individual foraging strategies play a role in individual or offspring adaptability to variation in marine conditions?
- How will different coping styles and degrees of behavioral plasticity affect an individual's ability to respond to dynamic conditions, diminishing resources or human disturbance?

Acknowledgments We extend our thanks to O. Krüger, F. Trillmich, and F. Zenth, along with editors C. Campagna and R. Harcourt, for thoughtful guidance on this chapter. We are also grateful to everyone who has contributed to our recent studies on personality in Galápagos sea lions, which inspired many of the ideas found here, mainly S. Kalberer, P. Piedrahita, D. Páez-Rosas, the Charles Darwin Foundation, and Galápagos National Park.

References

Baylis AMM, Orben RA, Arnould JPY et al (2015) Diving deeper into individual foraging specializations of a large marine predator, the southern sea lion. Oecologia 179:1053–1065

Bekoff M, Allen C (1998) Intentional communication and social play: how and why animals negotiate and agree to play. In: Bekoff M, Byers JA (eds) Animal play: evolutionary, comparative, and ecological perspectives. Cambridge University Press, Cambridge, pp 97–114

Bell AM, Hankison SJ, Laskowski KL (2009) The repeatability of behaviour: a meta-analysis. Anim Behav 77:771–783. https://doi.org/10.1016/j.anbehav.2008.12.022

Bergmüller R (2010) Animal personality and behavioural syndromes. In: Kappeler P (ed) Animal behaviour: evolution and mechanisms. Springer, Cham, pp 587–621

Bergmüller R, Taborsky M (2010) Animal personality due to social niche specialisation. Trends Ecol Evol 25:504–511

Biro PA, Stamps JA (2008) Are animal personality traits linked to life-history productivity? Trends Ecol Evol 23:361–368

Biro PA, Stamps JA (2010) Do consistent individual differences in metabolic rate promote consistent individual differences in behavior? Trends Ecol Evol 25:653–659. https://doi.org/10.1016/j.tree.2010.08.003
Bishop AM, Pomeroy P, Twiss SD (2015) Variability in individual rates of aggression in wild gray seals: fine-scale analysis reveals importance of social and spatial stability. Behav Ecol Sociobiol 69:1663–1675
Brakes P, Dall SRX (2016) Marine mammal behavior: a review of conservation implications. Front Mar Sci 3:87
Bubac CM, Coltman DW, Bowen WD et al (2018) Repeatability and reproductive consequences of boldness in female gray seals. Behav Ecol Sociobiol 72:100
Campagna C, Bisioli C, Quintana F et al (1992) Group breeding in sea lions: pups survive better in colonies. Anim Behav 43:541–548. https://doi.org/10.1016/S0003-3472(05)81014-0
Chilvers BL, Wilkinson IS (2009) Diverse foraging strategies in lactating New Zealand sea lions. Mar Ecol Prog Ser 378:299–308
Ciardelli LE, Weiss A, Powell DM, Reiss D (2017) Personality dimensions of the captive California sea lion (*Zalophus californianus*). J Comp Psychol 131:50–58. https://doi.org/10.1037/com0000054
Coppens CM, de Boer SF, Koolhaas JM (2010) Coping styles and behavioural flexibility: towards underlying mechanisms. Philos Trans R Soc Lond B Biol Sci 365:4021–4028
Crawford MP (1938) A behavior rating scale for young chimpanzees. J Comp Psychol 26:79
Crawford K, Mcdonald RA, Bearhop S (2008) Applications of stable isotope techniques to the ecology of mammals. Mammal Rev 38:87–107
Croft DP, Krause J, Darden SK et al (2009) Behavioural trait assortment in a social network: patterns and implications. Behav Ecol Sociobiol 63:1495–1503
Dall SRX, Griffith SC (2014) An empiricist guide to animal personality variation in ecology and evolution. Front Ecol Evol 2:3
De Vere AJ, Lilley MK, Highfill L (2017) Do pinnipeds have personality? Broad dimensions and contextual consistency of behavior in harbor seals (*Phoca vitulina*) and California sea lions (*Zalophus californianus*). Int J Comp Psychol 30:0–15
DeRango EJ, Schwarz JFL, Kalberer S et al (2019a) Intrinsic and maternal traits influence personality during early life in Galápagos sea lion, *Zalophus wollebaeki*, pups. Anim Behav 154:111–120
DeRango EJ, Schwarz JFL, Piedrahita P, Páez-Rosas D, Trillmich F, Krüger O (2019b) Intraspecific variation in boldness and exploration shapes behavioural responses to stress in Galápagos sea lion pups. Behav Ecol Sociobiol 73(12):166
Dingemanse NJ, Wolf M (2013) Between-individual differences in behavioural plasticity within populations: causes and consequences. Anim Behav 85:1031–1039
Dingemanse NJ, Both C, Drent PJ, Tinbergen JM (2004) Fitness consequences of avian personalities in a fluctuating environment. Proc R Soc Lond Ser B Biol Sci 271:847–852
Dingemanse NJ, Kazem AJN, Réale D, Wright J (2010) Behavioural reaction norms: animal personality meets individual plasticity. Trends Ecol Evol 25:81–89
Du Dot TJ, Trites AW, Arnould JPY et al (2018) Trade-offs between foraging efficiency and pup feeding rate of lactating northern fur seals in a declining population. Mar Ecol Prog Ser 600:207–222
Elorriaga-Verplancken F, Aurioles-Gamboa D, Newsome SD, Martínez-Díaz SF (2013) δ 15 N and δ 13 C values in dental collagen as a proxy for age-and sex-related variation in foraging strategies of California sea lions. Mar Biol 160:641–652
Frick EE, de Vere AJ, Kuczaj SA (2017) What do we want to know about personality in marine mammals? In: Personality in nonhuman animals. Springer, Cham, pp 237–253
Gartner MC (2015) Pet personality: a review. Pers Individ Differ 75:102–113
Gentry RL (1974) The development of social behavior through play in the Steller sea lion. Am Zool 14:391–403. https://doi.org/10.1093/icb/14.1.391

Gibelli J, Dubois F (2016) Does personality affect the ability of individuals to track and respond to changing conditions? Behav Ecol 28:arw137

Goldberg LR (1990) An alternative "description of personality": the big-five factor structure. J Pers Soc Psychol 59:1216

Gosling SD, John OP (1999) Personality dimensions in nonhuman animals: a cross-species review. Curr Dir Psychol Sci 8:69–75

Guenther A, Groothuis AGG, Krüger O, Goerlich-Jansson VC (2018) Cortisol during adolescence organises personality traits and behavioural syndromes. Horm Behav 103:129–139

Harcourt R (1991) Survivorship costs of play in the South American fur seal. Anim Behav 42 (3):509–511

Harcourt R (1992) Maternal aggression in the South American fur seal in Peru. Can J Zool 70:320–325

Harcourt R (1993) Individual variation in predation on fur seals by southern sea lions (*Otaria byronia*) in Peru. Can J Zool 71:1908–1911

Harcourt JL, Ang TZ, Sweetman G et al (2009) Social feedback and the emergence of leaders and followers. Curr Biol 19:248–252

Heerah K, Hindell M, Guinet C, Charrassin J-B (2014) A new method to quantify within dive foraging behaviour in marine predators. PLoS One 9:e99329

Hoelzel AR, Dorsey EM, Stern SJ (1989) The foraging specializations of individual minke whales. Anim Behav 38:786–794

Hoppitt W, Laland KN (2008) Social processes influencing learning in animals: a review of the evidence. Adv Study Behav 38:105–165

Jeglinski JWE, Wolf JBW, Werner C, Costa DP, Trillmich F (2015) Differences in foraging ecology align with genetically divergent ecotypes of a highly mobile marine top predator. Oecologia 179(4):1041–1052. https://doi.org/10.1007/s00442-015-3424-1

Kaiser S, Heemann K, Straub RH, Sachser N (2003) The social environment affects behaviour and androgens, but not cortisol in pregnant female Guinea pigs. Psychoneuroendocrinology 28 (1):67–83

Kaiser A, Merckx T, Van Dyck H (2019) Personality traits influence contest outcome, and vice versa, in a territorial butterfly. Sci Rep 9:2778

Kernaléguen L, Arnould JPY, Guinet C, Cherel Y (2015) Determinants of individual foraging specialization in large marine vertebrates, the Antarctic and subantarctic fur seals. J Anim Ecol 84:1081–1091

Kohda M (1994) Individual specialized foraging repertoires in the piscivorous cichlid fish *Lepidiolamprologus profundicola*. Anim Behav 48:1123–1131

Koolhaas JM, Korte SM, De Boer SF et al (1999) Coping styles in animals: current status in behavior and stress-physiology. Neurosci Biobehav Rev 23:925–935

Koski SE (2011) How to measure animal personality and why does it matter? Integrating the psychological and biological approaches to animal personality. In: From genes to animal behavior. Springer, Cham, pp 115–136

Kucey L (2005) Human disturbance and the hauling out behaviour of Steller sea lions (*Eumetopias jubatus*). University of British Columbia, Vancouver

Kuhn CE, Aurioles-Gamboa D, Weise MJ, Costa DP (2006) Oxygen stores of California sea lion pups: implications for diving ability. In: Trites AW, Atkinson SK, DeMaster DP, Fritz LW, Gelatt TS, Rea LD, Wynne KM (eds) Sea Lions world. Alaska Sea Grant College Program. University of Alaska, Fairbanks, pp 31–44

Leclerc M, Vander Wal E, Zedrosser A et al (2016) Quantifying consistent individual differences in habitat selection. Oecologia 180:697–705

Lowther AD, Harcourt RG, Hamer DJ, Goldsworthy SD (2011) Creatures of habit: foraging habitat fidelity of adult female Australian sea lions. Mar Ecol Prog Ser 443:249–263

Lowther AD, Harcourt RG, Page B, Goldsworthy SD (2013) Steady as he goes: at-sea movement of adult male Australian sea lions in a dynamic marine environment. PLoS One 8:e74348

Maniscalco JM, Parker P, Atkinson S (2006) Interseasonal and interannual measures of maternal care among individual Steller sea lions (*Eumetopias jubatus*). J Mammal 87:304–311
McDougall PT, Réale D, Sol D, Reader SM (2006) Wildlife conservation and animal temperament: causes and consequences of evolutionary change for captive, reintroduced, and wild populations. Anim Conserv 9:39–48
McGovern KA, Rodríguez DH, Lewis MN, Davis RW (2019) Diving classification and behavior of free-ranging female southern elephant seals based on three-dimensional movements and video-recorded observations. Mar Ecol Prog Ser 620:215–232
McHuron EA, Hazen E, Costa DP (2018) Constrained by consistency? Repeatability of foraging behavior at multiple timescales for a generalist marine predator. Mar Biol 165:122
Meise K, Von Engelhardt N, Forcada J, Hoffman JI (2016) Offspring hormones reflect the maternal prenatal social environment: potential for foetal programming? PLoS One 11:e0145352. https://doi.org/10.1371/journal.pone.0145352
Mendl M, Harcourt R (2000) Individuality in the domestic cat: origins, development and stability. In: Turner DC, Bateson P (eds) The domestic cat: the biology of its behaviour, 2nd edn. Cambridge University Press, Cambridge, pp 47–64
Michelangeli M, Wong BBM, Chapple DG (2015) It's a trap: sampling bias due to animal personality is not always inevitable. Behav Ecol 27:62–67
Olafsdottir GA, Magellan K (2016) Interactions between boldness, foraging performance and behavioural plasticity across social contexts. Behav Ecol Sociobiol 70(11):1879–1889
Páez-Rosas D, Villegas-Amtmann S, Costa D (2017) Intraspecific variation in feeding strategies of Galapagos sea lions: a case of trophic specialization. PLoS One 12:e0185165
Patrick SC, Weimerskirch H (2014) Personality, foraging and fitness consequences in a long-lived seabird. PLoS One 9:e87269
Patrick SC, Bearhop S, Bodey TW et al (2015) Individual seabirds show consistent foraging strategies in response to predictable fisheries discards. J Avian Biol 46:431–440
Pavlov IP (1906) The Huxley lecture on the scientific investigation of the psychical faculties or processes in higher mammals. Lancet 2:911–915
Réale D, Reader SM, Sol D et al (2007) Integrating animal temperament within ecology and evolution. Biol Rev 82:291–318. https://doi.org/10.1111/j.1469-185X.2007.00010.x
Réale D, Garant D, Humphries MM et al (2010) Personality and the emergence of the pace-of-life syndrome concept at the population level. Philos Trans R Soc Lond B Biol Sci 365:4051–4063
Riedman M (1990) The pinnipeds: seals, sea lions, and walruses. University of California Press, Berkeley
Robinson KJ, Pomeroy PP, Hazon N et al (2017) Individual size, sex and rearing environment impact on aggression in newly weaned seals. Mar Mammal Sci 33(2):621–629
Roche DG, Careau V, Binning SA (2016) Demystifying animal 'personality' (or not): why individual variation matters to experimental biologists. J Exp Biol 219:3832–3843
Rodríguez-Prieto I, Martín J, Fernández-Juricic E (2010) Individual variation in behavioural plasticity: direct and indirect effects of boldness, exploration and sociability on habituation to predators in lizards. Proc R Soc B Biol Sci 278:266–273
Schakner ZA, Petelle MB, Tennis MJ et al (2017) Social associations between California sea lions influence the use of a novel foraging ground. R Soc Open Sci 4:160820
Schuett W, Tregenza T, Dall SRX (2010) Sexual selection and animal personality. Biol Rev 85:217–246
Seguel M, Gottdenker N, Muñoz F, Montalva F, Perez-Venegas D, Pavés H (2017) Kelp and dolphin gulls cause perineal wounds in South American fur seal pups (*Arctocephalus australis*) at Guafo Island, Chilean Patagonia. R Soc Open Sci 4(7):170638
Sih A, Bell A, Johnson JC (2004) Behavioral syndromes: an ecological and evolutionary overview. Trends Ecol Evol 19:372–378
Soto KH, Trites AW, Arias-Schreiber M (2006) Changes in diet and maternal attendance of South American sea lions indicate changes in the marine environment and the abundance of prey. Mar Ecol Prog Ser 312:277–290. https://doi.org/10.3354/meps312277

Spinka M, Newberry RC, Bekoff M (2001) Mammalian play: training for the unexpected. Q Rev Biol 76:141–168. https://doi.org/10.1086/393866

Stamps J, Groothuis TGG (2010) The development of animal personality: relevance, concepts and perspectives. Biol Rev 85:301–325. https://doi.org/10.1111/j.1469-185X.2009.00103.x

Terraube J, Guixé D, Arroyo B (2014) Diet composition and foraging success in generalist predators: are specialist individuals better foragers? Basic Appl Ecol 15:616–624

Traisnel G, Pichegru L (2019) Boldness at the nest predicts foraging and diving behaviour of female but not male African penguins. Anim Behav 150:113–125

Trillmich F, Wolf JBW (2008) Parent–offspring and sibling conflict in Galápagos fur seals and sea lions. Behav Ecol Sociobiol 62:363–375

Trillmich F, Müller T, Müller C (2018) Understanding the evolution of personality requires the study of mechanisms behind the development and life history of personality traits. Biol Lett 14:20170740

Túnez JI, Cappozzo HL, Cassini MH (2008) Natural and anthropogenic factors associated with the distribution of South American sea lion along the Atlantic coast. Hydrobiologia 598:191–202

Twiss SD, Franklin J (2010) Individually consistent behavioural patterns in wild, breeding male grey seals (*Halichoerus grypus*). Aquat Mamm 36:234–238. https://doi.org/10.1578/AM.36.3.2010.234

Twiss SD, Cairns C, Culloch RM et al (2012a) Variation in female grey seal (*Halichoerus grypus*) reproductive performance correlates to proactive-reactive behavioural types. PLoS One 7: e49598. https://doi.org/10.1371/journal.pone.0049598

Twiss SD, Culloch R, Pomeroy PP (2012b) An in-field experimental test of pinniped behavioral types. Mar Mammal Sci 28:415–429. https://doi.org/10.1111/j.1748-7692.2011.00523.x

Villegas-Amtmann S, Costa DP, Tremblay Y et al (2008) Multiple foraging strategies in a marine apex predator, the Galapagos sea lion *Zalophus wollebaeki*. Mar Ecol Prog Ser 363:299–309. https://doi.org/10.3354/meps07457

Wolf M, Weissing FJ (2012) Animal personalities: consequences for ecology and evolution. Trends Ecol Evol 27:452–461

Wolf M, Van Doorn GS, Leimar O, Weissing FJ (2007) Life-history trade-offs favour the evolution of animal personalities. Nature 447:581

Chapter 20
Playing at the Edge of the Sea: A Comparative Analysis in Otariids and Odobenids

Clara Llamazares-Martín and Elisabetta Palagi

Abstract The role of play in the superfamily Otarioidea has not been widely explored. However, all play types found in other so-called "playful groups" (e.g., primates or canids) have been reported among sea lions, fur seals and walruses. Otariids and odobenids perform object play with animate and non-animate objects, such as kelp, rocks or seabirds. Locomotor-rotational play in otariids is mainly performed at sea, where animals exhibit complex and sophisticated movements such as pirouettes, torpedoes and somersaults. In the same family, play-fighting is the most common form of terrestrial social play, and is composed of competitive behavioral patterns that are exchanged between players. As in other mammalian species, in otariids play distribution follows a skewed bell-shape developmental curve, starting in infancy, peaking in juvenility and disappearing in adulthood. The distribution and modality of this behavior reflect the level of the sociality of a given species. In those species characterized by high levels of tolerance and maternal permissiveness, social play starts earlier in ontogeny. During play-fighting, play signals are displayed and exchanged between players to avoid escalating into aggression. In other mammalian taxa, the same signals have been reported and seem to serve the same purpose (e.g., Relaxed Open Mouth, "ROM"). Testing hypotheses on the evolution of play requires that many species with diverse habits (e.g., terrestrial, aquatic, arboreal) and kinds of sociality (e.g., despotic, tolerant) are studied using a comparative perspective. Herewith, the superfamily Otarioidea can be important to help understand this puzzling set of behaviors.

C. Llamazares-Martín
Ethology and Animal Welfare Section, Universidad Cardenal Herrera CEU, CEU Universities, Valencia, Spain

E. Palagi (✉)
Unit of Ethology, Department of Biology, University of Pisa, Pisa, Italy

Natural History Museum, University of Pisa, Calci, Italy
e-mail: elisabetta.palagi@unipi.it

C. Campagna, R. Harcourt (eds.), *Ethology and Behavioral Ecology of Otariids and the Odobenid*, Ethology and Behavioral Ecology of Marine Mammals,
https://doi.org/10.1007/978-3-030-59184-7_20

Keywords Play-fighting · Object play · Locomotor-rotational play · Play patterns · Cooperative/competitive sessions · Play signals · Facial expressions · Play in otariids · Play in walrus · Maternal styles

20.1 Introduction

Play is a widespread behavior across many mammals (Burghardt 2005; Fagen 1981) and the species belonging to the superfamily Otarioidea are no exception. In these species, play can be present from the very first phases of life and characterized by different types depending on the different species considered. For example, in otariids, pups begin to play with their mothers, and then expand their playful spheres to other group members, especially peers (see Farentinos 1971). Play between newborns and mothers includes gentle bites that pups direct to the vibrissae, face and neck of the mother. To reach the face of the mother, pups often climb over her back thus engaging in an energetically demanding activity. From infancy to juvenility, social play becomes more complex and challenging. Play-fighting, the most vigorous form of play, is typical of youngsters, which confront each other by engaging in chest pushing, neck fencing and biting. This type of playful activity requires an accurate control of movements and a sophisticated communicative system to avoid escalation into aggression. In many species of Otarioidea, social play is not the only type of play observed. Solitary play can also occur, and it can include both acrobatic motor actions and manipulative patterns (e.g., play with objects). Solitary play seems to be present in juvenile walruses (*Odobenus rosmarus*), which catch and repeatedly hit seabirds without eating them (Giljov et al. 2017), and in Australian and New Zealand sea lions (*Neophoca cinerea* and *Phocarctos hookeri*), which play with kelp by throwing it into the air and trying to grab it again (Marlow 1975).

All these authors described such activities as play, but the main problem to tackle the topic relies on the difficulty of finding an appropriate definition of play, which can be applied to all species.

Play behavioral patterns are recruited from different maintenance behaviors such as reproductive, aggressive or predatory activities. Moreover, the assemblage of the different behavioral patterns may vary according to the species' distinctive features, such as morphological traits, reproductive and social systems, and ecology. This high variability makes the operational definition of play difficult (Bekoff 2001; Palagi et al. 2016; Pellis and Pellis 1996). For this reason Burghardt (2005) set up five criteria to define play. The behavior (1) must depart from being completely functional in the form or context in which it appears; (2) should be rewarding and spontaneous; (3) must differ from its analogous functional behavior in form (exaggeration and rearrangement of motor actions) or in time (different ontogenetic stages); (4) must be performed repeatedly (absence of inhibition), but not stereotypically; and (5) must occur in the absence of heavy environmental or social stressors (e.g., food shortage, social conflicts or presence of a predator threat).

Play between pups and their mothers in otariids, is a good example summarizing all these criteria. Climbing on and biting the mother (1) is not functional (the behavioral pattern does not produce any kind of immediate benefit), (2) is spontaneous because the activity is initiated by the pup, (3) is limited to the first developmental stages, (4) is often repeated many times per bout and (5) occurs when energetic needs of pups are covered by mothers' milk.

Play can have both delayed (e.g., developing physical and cognitive skills, *motor training hypothesis*) and immediate benefits (e.g., creating and strengthening social bonds) (Byers and Walker 1995; Smith 1997). However, play has immediate costs in terms of time, energy and survival (Fagen 1993; Harcourt 1991a; Palagi 2007). The time spent in playful activities is subtracted from the time that could be spent in other essential activities (e.g., feeding, sleeping, mating). The percentage of time animals spend in playing differs across species in relation to their ontogenetic trajectories, personality, habits, diets and metabolic survival costs (DeRango and Schwarz, Chap. 19; Fagen 1993; Palagi 2018). In some species, play represents a strong metabolic investment deriving from the physical effort (e.g., chasing playmates, acrobatic movements) (Fagen 1981). In juvenile Assamese macaques (*Macaca assamensis*), for example, the investment in locomotor play leads to reduced growth in both sexes (Berghänel et al. 2015). This reduction is more accentuated in males that play more frequently than females and, consequently, grow less (Berghänel et al. 2015). This suggests that play is an important factor in the developmental process, since it has ontogenetic priority over the physical growth rates. Play also implies some risks. It increases the probability to be detected by predators especially when the attention of the subject is focused on the playmate and not on the surrounding environment, as occurs during play-fighting. Harcourt (1991a) quantified the number of attacks performed by southern sea lions (*Otaria byronia*) towards immature South American fur seals (*Arctocephalus australis*). The author reported 102 attacks from which 26 resulted in kills, and 85% of these kills occurred when immature pups were playing in shallow tidal pools. During their activity, players can be victim of injuries provoked by playmates, and by the difficulty in the execution of some motor sequences (Harcourt 1991a). Mild injuries can derive from falls and tumbles. During play-fighting, cheetah (*Acinonyx jubatus*) cubs often show signs of pain due to mild injuries that can cause temporary impairment in the subject (Caro 1995).

Play can be expressed in a variety of modalities. Solitary play can occur either by manipulating objects/body parts or performing locomotor-rotational actions. During locomotor-rotational play, an animal repeatedly performs body movements in an exaggerated way such as running or torpedoing. These can be punctuated by rotational maneuvers such as body twisting, or somersaulting (Wilson and Kleiman 1974). Putting themselves in unnatural body/head positions in relation to gravity produces a strong limitation in sensory perception (Palagi 2014; Špinka et al. 2001). This lack of physical and sensory control seems to be self-rewarding for the player, even though it involves a certain amount of risk (Pellis and Pellis 2009).

Object play differs from exploration at both functional and operational levels. When exploring, animals acquire information on their environment and familiarity with novel objects (Palagi 2014). In evolutionary terms, exploration is a *receptor* activity (i.e., activity providing information about the unknown environment) which could have been favored by natural selection because of its survival functions (e.g., locating food, escaping from predators, exploiting the environment in an effective way). During object play, objects are used to create novel, uncertain, and challenging situations, with which animals have to cope (Špinka et al. 2001). This is a self-rewarding activity that helps animals to assess and improve their motor and cognitive abilities (Palagi 2014). In this view, object play is an *effector* activity.

Therefore, *receptor* activity of object manipulation contrasts with *effector* activity of object play in the sense that the former is focused on increasing familiarity with the object and creating the conditions to use it for adaptive purposes (e.g., I can take a melon never seen before, explore it, break it and eat it). The latter acts on the object, which can already be familiar, to use it in a way that can be totally disentangled from its intrinsic characteristics (e.g., I can take the same melon, roll it and jump on it while trying to maintain my equilibrium).

When two or more individuals interact in a playful manner, play becomes social. The most common form of social play in toddlers and non-human animals is play-fighting, which is characterized by offensive and defensive patterns that are generally balanced (Fagen 1981; Norscia and Palagi 2016). Obviously, animal play can involve a mix of activities. Locomotor–rotational play can involve the use of objects (e.g., kelp, sticks, branches) and it can be carried out in a social way. It is often difficult to disentangle each type of play from the other, since they can be performed together in sophisticated sequences forming a single natural category. Even though no single, simple definition of play category is satisfactory, such categorization helps ethologists identify, quantify and describe, in a standardized way, the different playful activities of animals (Burghardt 2005).

Play has been largely investigated in particular mammalian groups, especially primates, rodents, and canids. However, it has been less studied in other mammalian taxa such as marine mammals, probably due to the difficulty in following elusive animals in their natural habitats (Burghardt 2005; Hill et al. 2017). This review deals with some aspects of solitary and social play of otariids and odobenids to take stock of knowledge regarding play in these species. Moreover, we aim to explore play using a comparative perspective. To explore if there are some homologous traits in the shaping of play behavior, we select species (e.g., harbor seals, *Phoca vitulina*) that are phylogenetically close to otariids and odobenids but that differ from them in their mating systems and maternal styles. At the same time, to understand whether some traits of play can be shaped through a process of evolutionary convergence, we look at those groups which are more phylogenetically distant from otariids and odobenids but that share similar mating systems and maternal styles (e.g., lowland gorillas, *Gorilla gorilla gorilla*; macaques, *Macaca* spp). Due to the paucity of studies focusing on play in otariids and odobenids, we also discuss anecdotal reports.

20.2 One Step Before We 'Playfully' Jump into Our Target Group

Demographic, natural and life history variables can influence play behavior (Ciani et al. 2012; Cordoni 2009; Cordoni and Palagi 2016; Palagi 2007). Play in adults is often inhibited, while play in juveniles is very conspicuous (Fagen 1981). On the other hand, in species showing sexual dimorphism and/or sex differences in adult roles, play is often more frequent in males especially when adult competition for mates is high (Byers and Walker 1995; Fagen 1993). Play variability is affected by social features typical of the different species.

Mating system and social organization define the level of tolerance and proximity of individuals and, in turn, can shape play across group members (Palagi 2018). In polygynous species, male competition tends to be strong, and this may lead to vigorous fights to monopolize females (Cassini, Chap. 7; Clutton-Brock and Huchard 2013; Emlen and Oring 1977). These aggressive adult tendencies may help explain the high levels of competition in juvenile play-fighting (Paquette 1994; Smith 1997). In this view, the most competitive players will likely have higher chances to gather high ranking status and, in turn, better reproductive opportunities in the future (Byers and Walker 1995). The so-called *motor training hypothesis* predicts that through play-fighting, juveniles acquire and develop those physical and cognitive skills that will be useful in the future. In species, such as lowland gorillas, which are sexually dimorphic, male-male competition is common and adult play is usually inhibited due to the risk that play-fighting entails (Cordoni et al. 2018). Nonetheless, adult play is frequent in tolerant species in which social relationships are not established and mediated through agonistic interactions and where the roles of males and females are well-balanced (e.g., bonobos, *Pan paniscus*, Palagi 2006; ring-tailed lemurs, *Lemur catta*, Palagi 2009).

Otariids and odobenids are sexually dimorphic (Boness 1991) and share mating systems ranging from female-defense (males directly herd their females) to resource-defense polygyny (males defend the territory to attract females) (e.g., Cassini, Chap. 6; Campagna and Le Boeuf 1988; Emlen and Oring 1977; Fay 1982; Le Boeuf 1991; Soto and Trites 2011). Except in Australian sea lions (McIntosh and Pitcher, Chap. 26), females are highly synchronous and during the breeding season they congregate in specific areas. In otariid females, agonistic interactions are higher once they give birth, as the space for nursing their pups on land is reduced (Harcourt 1992a; Wartzok 2012). Males of both the Otariidae and Odobenidae families strongly compete to monopolize females, and individuals engage in fights that range from agonistic displays to violent confrontations with injuries. Conflicts occur within and between sexes, with males displaying coercive behaviors to retain and mount females. Aggression towards pups has also been described in otariids (Ryazanov, Chap. 10; Campagna 2009). As the space for nursing is reduced, females display agonistic interactions towards alien pups in order to block their movements and may even kill them (Harcourt 1992b; Riedman 1990). In addition, infanticide by adult and juvenile males, although rare, has been reported in several species (e.g., Campagna et al. 1988; Higgins and Tedman 1990; Ryazanov, Chap. 10).

Taken together, all these features reveal the strong social competition of this superfamily, involving (in some species) individuals of all age- and sex-classes. Such high levels of competition make this taxon particularly interesting to hypothesize about the role of mating systems and social organizations in shaping the distribution of play in terms of frequency and modalities. In particular, a comparative approach across the species could shed light on the potential benefits of play-fighting in developing those motor and social skills necessary to compete in adulthood.

20.3 Different Maternal Styles Matter: The Ontogenetic Pathways of Social Play

20.3.1 Looking at the Mother: Mother-Pup Play

Maternal styles strongly affect the frequency and modality of infant social play in many mammalian species (Pellis and Pellis 2009). Therefore, it is mandatory to discuss play diversity in infants as a function of the role of mothers in determining play attitude across different developmental stages (Palagi 2018). Pinnipeds offer many possibilities to investigate the issue due to their extremely diverse maternal styles. There are three main maternal strategies that have been described for pinnipeds: the 'nursing aquatic strategy', the 'foraging cycle strategy' and the 'fasting strategy' (Boness and Bowen 1996). In the 'nursing aquatic strategy', characteristic of the single odobenid species, the walrus, *Odobenus rosmarus*, pups remain with their mothers wherever they go and are nursed on ice or land, and at sea (Fay 1982; Miller and Boness 1983). Lactation lasts 2–3 years, although pups start feeding autonomously at ~5 months of age (Boness and Bowen 1996). Otariids show the 'foraging cycle strategy'. Females fast during the 'perinatal period' and later alternate between long foraging trips to the sea and short attendance periods at the colony to nurse their pups, with the consequence that pups can remain alone for long periods between maternal visits (Sepulveda and Harcourt, Chap. 3). In these groups, lactation ranges from 4 months to 3 years. Most phocid species adopt the 'fasting strategy' with females who fast while not separating from their pups for the entire lactation, which ranges from 5 to 50 days. The remaining phocid species may adopt the "foraging cycle strategy" or both (Boness and Bowen 1996; Wheatley et al. 2008). It appears that the main difference across these three strategies is the amount of time that mothers spend with their pups until weaning, the energetic costs and the place used for nursing.

In the only species adopting the 'nursing aquatic strategy', the walrus, most social interactions between mothers and pups occur in the water and so they are particularly difficult to observe (Miller 1976); therefore, we cannot exclude that social play may be present in walruses. Otariid and phocid females are protective towards their offspring. They keep their pups in close proximity, preventing them from moving away or interacting with other individuals (Oftedal et al. 1987). This precludes pups from interacting with peers and limits their interactions, including social play, solely to their mothers (*social canalization*, Berman 1982). This phenomenon has also been described in macaque species with high levels of despotism and intraspecific competitiveness (Thierry 2004). Ciani et al. (2012) compared social play of immature subjects of two macaque species characterized by different levels of mother protectiveness and dominance hierarchy: the despotic Japanese macaque (*Macaca fuscata*) and the tolerant Tonkean macaque (*Macaca tonkeana*). The authors found that the former played at lower frequencies than the latter, thus suggesting a covariation between despotism and social play inhibition in immatures (*social canalization*).

In otariids, mother-pup play has been described in two species of sea lions (Australian sea lion, Marlow 1975; Steller sea lion, *Eumetopias jubatus*, Farentinos 1971), and one species of fur seal (New Zealand fur seal, *Arctocephalus forsteri*, McNab and Crawley 1975) (see Table 20.1). This type of play is always initiated by pups and may differ over the course of lactation, becoming more complex and sophisticated when pups get older. During the first days of life, pups of New Zealand fur seals have been seen to mouth their mother's vibrissae and pass them over their head and nose (Fig. 20.1). In the Steller and Australian sea lions, no play patterns have been reported in the first days of life; however, during the first weeks of lactation, pups have been observed to climb over the mother's neck, biting at her face and head while she is in a reclined position (Fig. 20.2) (Farentinos 1971; Marlow 1975). In a play-fighting bout, the pup takes the aggressive role and the mother is almost always passive (i.e., self-handicapping strategy). The pup directs bites at its mother's cephalic region and neck, and the mother responds giving it back gentle bites. The bout ends when the mother pushes away her pup with her snout and moves her head away (McNab and Crawley 1975) (Figs. 20.1 and 20.2).

20.3.2 Looking at Peers: Pup-Pup Play

When otariid mothers make feeding trips, they leave their offspring alone, leaving them free to interact with other pups. In some otariid species, pups gather in groups called "pods" in nearby areas where they spend most of their time resting and playing together (Farentinos 1971; Higgins and Gass 1993; McNab and Crawley 1975). The ontogenetic period in which inter-pup play begins differs across species and depends on the time mothers leave their pups to make their first feeding trip after the perinatal period. New Zealand adult sea lions seem to be more tolerant towards alien pups than adults of other otariid species (Marlow 1975). For this reason, mothers may not be so protective towards their offspring. They remain with their pups and defend them during the first 2–3 days after birth, but soon after they let them move freely. Pups of this species are precocious, and they move around and join other pups in large pods. Then they become more confident and play in small groups (polyadic play-fighting) at the edge of the sea at rock pools (Marlow 1975).

Table 20.1 Classification of play types found in the species of the superfamily Otar ioidea

Species	Object play	Self-object play	Play fighting/ Mock fighting	Pup-pup play	Mother-pup play	Juvenile-juvenile play	Adult-adult play	References
Zalophus californianus	V[1,2]	X[3]	X[4]	X[4]	Ø	Ø	Ø	[1]Riedman (1990), [2]Peterson and Bartholomew (1967), [3]R. Harcourt (pers. obs.), and [4]E. Palagi (pers. obs.)
Zalophus wollebaeki	V[1,2]	V[2]	V[2]	Ø	Ø	V[2]	Ø	[1]Eibl-Eibesfeldt (1955) and [2]Eibl-Eibesfeldt (1984)
Eumetopias jubatus	V[1]	V[1]	V[1,2]	V[1,2]	V[1]	V[1]	V[1]	[1]Farentinos (1971) and [2]Gentry (1974)
Otaria byronia	X[1]	X[2]	V[3,4]	Ø	Ø	V[3,4]	Ø	[1]C. Llamazares-Martín (pers. obs.), [2]R. Harcourt (pers. obs.), and [3,4]Llamazares-Martín et al. (2017a, b)
Neophoca cinerea	V[1]	X[2]	V[1]	V[1]	V[1]	V[1]	Ø	[1]Marlow (1975) and [2]R. Harcourt (pers. obs.)
Phocarctos hookeri	V	V	V	V	Ø	V	Ø	Marlow (1975)
Callorhinus ursinus	Ø	Ø	Ø	Ø	Ø	Ø	Ø	
Arctocephalus townsendi	V	Ø	Ø	Ø	Ø	Ø	Ø	Riedman (1990)
A. philippii	Ø	Ø	Ø	Ø	Ø	Ø	Ø	
A. galapagoensis	Ø	Ø	V	V	Ø	Ø	Ø	Arnold and Trillmich (1985)
A. australis	V[1]	X[2]	V[1]	V[1]	Ø	Ø	Ø	[1]Harcourt (1991b) and [2]Harcourt (pers. obs.)
A.forsteri	Ø	Ø	V	V	V	Ø	Ø	McNab and Crawley (1975)
A. gazella	V	Ø	Ø	Ø	Ø	Ø	Ø	Bonner and Hunter (1982)
A. tropicalis	Ø	Ø	Ø	Ø	Ø	Ø	Ø	
A. pusillus	V[1,2,3]	Ø	V[1]	Ø	Ø	V[1]	Ø	[1]Caudron (1995), [2]du Toit et al. (2004), and [3]Marks et al. (1997)

Odobenus rosmarus	V	Ø	Ø	Ø	Ø	Ø	Ø	Giljov et al. (2017)

Ø = no information available
X = reported but not published
V = play type reported in literature or studied

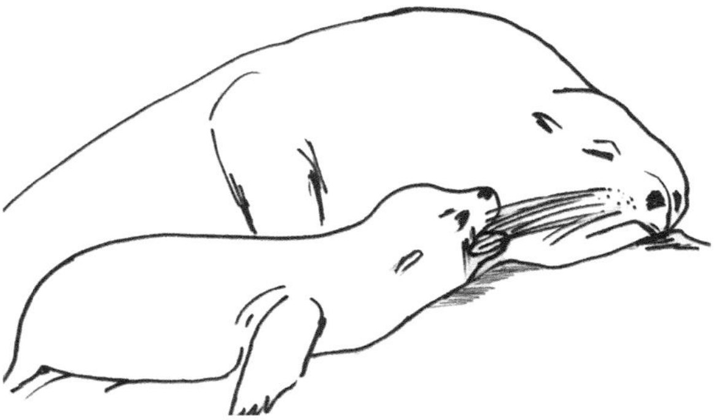

Fig. 20.1 Fur seal pup initiating a play session by biting its mother's vibrissae. (Drawing by Daniele Santerini)

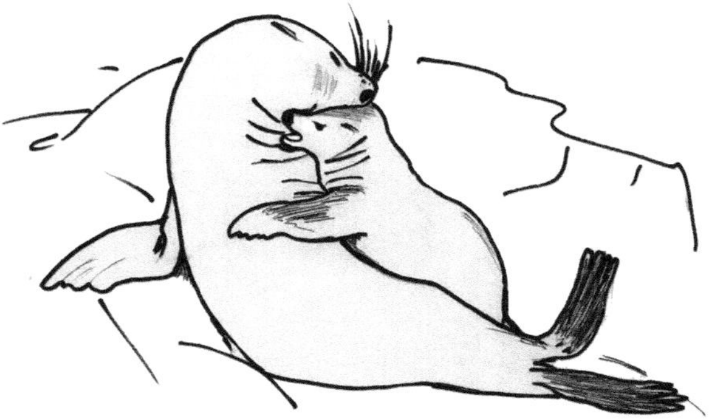

Fig. 20.2 Fur seal pup gently biting her mother's neck while climbing over its back. (Drawing by Daniele Santerini)

In contrast to New Zealand sea lions, Australian sea lion pups are less precocious. Mothers leave their pups for the first feeding trip only 14 days after birth. At this time, pups remain alone and the only form of play they engage in is object play. At 1 month of age they become less diffident and join pods where they play-fight together in small groups (Marlow 1975). Considering the differences between the two species in play developmental timing, it seems evident that maternal styles regarding mothers' protectiveness and foraging periods strongly influence pups' social interactions. This interpretation also finds support in Arnold and Trillmich's (1985) study of Galápagos fur seals (*Arctocephalus galapagoensis*). The authors highlighted the differences of pups' activities in the presence and absence of their mothers, and found a positive correlation between the amount of time spent in play-fighting and the amount of time they stayed separated from their mothers.

Play-fighting between pups is more complex than play-fighting in mother-pup dyads. For example, the number of players can increase, and up to four pups may join a play bout (Arnold and Trillmich 1985; Farentinos 1971; Gentry 1974). In play-fighting, pups face off with their chests in contact and bite one another on their necks and cephalic regions (Farentinos 1971) (Fig. 20.3). Bites may also be directed at other body areas such as fore and hind flippers (Arnold and Trillmich 1985). While directing bites, playmates try to dodge the other's bites by moving their heads from side to side, as the apparent goal is to bite without being bitten (Gentry 1974).

Fig. 20.3 California sea lion pups biting each others' mouths. (Photo by Elisabetta Palagi)

They sometimes grab the playmate's skin and vigorously shake their heads from side to side. Chest and neck pushing are also used as a tool to defeat the opponent. Short chases may also take place during play-fighting. All these patterns resemble the serious aggressive patterns that adult males engage in during the breeding season. In the Steller sea lion, play-fighting bouts can last more than 3–4 min, thus suggesting a strong energetic investment (Farentinos 1971).

Throughout the development pathway of pups, they become more agile and skillful, not only on land but also at sea. Time spent in underwater activities increases, and makes observing and monitoring pups more difficult. This is the reason why research on play behavior has only focused on early stages of pup development. The only study which shed light on behavioral play types during the entire lactation period was carried out by Harcourt (1991b) in the South American fur seal. In this species, pups do not engage in social play until about 1 month after birth. Terrestrial social play in pups was only prevalent between the 1st and the 4th month of age, immediately after there was a strong increase in social swimming play that tended to gradually decrease until the end of the first year of life. Terrestrial and aquatic social play not only differ in occurrence, but also in form. In social play on land, animals engage in play-fighting, while at sea they mainly engage in locomotor-rotational play (e.g., chasing, pirouetting, somersaulting). The adaptability of social play to different environments across ontogenetic stages clearly indicates the importance of this behavior in the achievement of skills indispensable in adult life (delayed benefits).

20.4 Towards Maternal Independence: Play Among Juveniles

A substantial amount of social play can be found among juvenile otariids, but it has only been systematically documented in five species of sea lions (Australian sea lion, Marlow 1975; Galápagos sea lion, *Zalophus wollebaeki*, Eibl-Eibesfeldt 1984; New Zealand sea lion, Marlow 1975; South American sea lion, Llamazares-Martín et al. 2017a; Steller sea lion, Farentinos 1971) and one species of fur seal

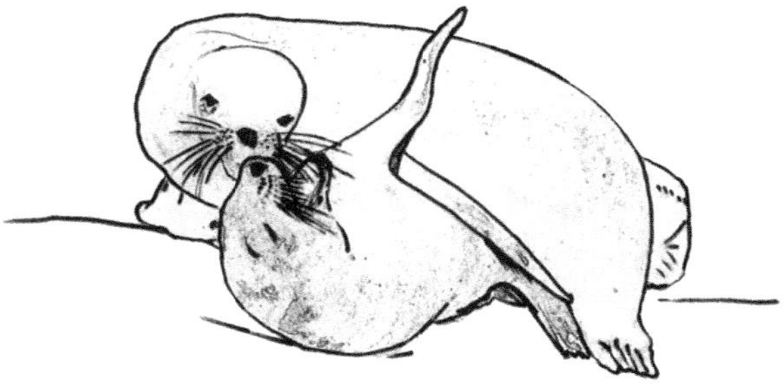

Fig. 20.4 Play-fighting in juvenile sea lions. The upper individual takes the dominant role by leaning on the belly of its playmate, thus restricting its movements. The lower individual performs a Relaxed Open Mouth (ROM). (Drawing by Daniele Santerini)

(South African fur seal, *Arctocephalus pusillus pusillus*, Caudron 1995) (see Table 20.1). Terrestrial play-fighting in juveniles does not strongly differ from that of pups, but juveniles appear to be more coordinated and precise in their motor actions. Juvenile play lasts longer than pup play. For example, in Steller sea lions some play sessions last over 1 h (Farentinos 1971). However, juvenile play involves a higher level of roughness. Play in older juveniles and non-breeding males includes more aggressive elements and, sometimes, it escalates into overt aggression (Farentinos 1971; Marlow 1975) and this is probably the reason why in breeding males play is totally inhibited. In Australian sea lions, heavier juveniles repeatedly push their lighter playmates backwards until the play partners manage to get away (Marlow 1975). Between matched partners, the more successful players are those whose competitive patterns are not dodged by the opponent (Fig. 20.4).

The play-fighting style of a given species is determined by its competitive nature and by the social benefits of play, being more imbalanced when it is useful for acquiring higher ranking positions, and well-balanced when it favors the creation and maintenance of social bonds (Pellis and Pellis 2016). In the southern sea lion, the only otariid species in which play-fighting style has been measured, play-fighting is well-balanced although it is rich in rough elements (Llamazares-Martín et al. 2017a). Contrary to expectations, in this competitive species with low levels of social tolerance, play-fighting seems to have an important role in establishing and maintaining social relationships and is predictive of the relationship quality of the subjects. Those animals that frequently engaged in play and, in particular, in well-balanced play sessions also shared strong affiliative relationships. The exchange of rough patterns during play-fighting may potentially lead to escalation towards aggression. In those species characterized by high aggressive tendencies and low levels of social tolerance, play-fighting can become even riskier. Moreover, since rough play may lead to aggression, southern sea lions avoid engaging in play involving more than two players (polyadic play). However, when polyadic play takes place, the sessions have similar durations compared to dyadic sessions, suggesting that polyadic sessions are also well-managed. Moreover, in this species the level of balance characterizing the session seems to have an effect on play duration (Llamazares-Martín et al. 2017a).

Male prevalence in play-fighting is especially important in juvenility, which is the last stage in which males have the opportunity to improve their fighting skills

Fig. 20.5 California sea lion pups reciprocally displaying a Relaxed Open Mouth (ROM). (Photo by Elisabetta Palagi)

through play before reaching the breeding phase. In southern sea lions, juvenile male dyads play at higher rates than female dyads, and the latter play at lower rates than mixed dyads (Llamazares-Martín et al. 2017a). Therefore, juvenile males of this species seem to motivate females to play. In the other three species of sea lions in which juvenile play has been reported, female play has not been observed (Farentinos 1971; Marlow 1975). A similar result has also been found in lowland gorillas (Palagi et al. 2007). Despite the large phylogenetic divergence between gorillas and the species of the superfamily Otarioidea, they share common social features that lead to similarities in the distribution and performance of social play. All these species show a marked sexual dimorphism and a polygynous mating system, in which males compete to monopolize and retain females (Watts 1991). As a whole, these findings suggest that similar social features concur in shaping similar play distribution in different species despite their phylogenetic distance.

Due to the risk of play-fighting in juveniles, sophisticated motor coordination and appropriate communication of the players' motivations are needed (Palagi et al. 2016). This effective communication is mediated by messages that are conveyed through play signals. These signals consist of gestures and facial expressions that can be borrowed from other functional contexts and performed in a repetitive or exaggerated manner (Palagi et al. 2016). By the performance of these signals, players point out the meaning of their subsequent actions, specifying that "what follows is only play", thus clarifying players' motivation (Bekoff 2001; Pellis and Pellis 1996). A congruent reciprocity of play signals between the players is required to prolong the playful interaction. Signal reciprocity means that the receiver has appropriately codified the meaning of the signal and, by reproducing the same pattern, the receiver expresses its agreement to prolong the playful interaction (Palagi et al. 2014; Palagi and Mancini 2011).

The Relaxed Open Mouth (ROM) is one of the most widespread play signals across different taxonomic groups (primates, e.g., Palagi et al. 2016; van Hooff and Preuschoft 2003; rodents, Panksepp and Burgdorf 2003; carnivorans, e.g., Cordoni et al. 2016; Poole 1978). It is a ritualized facial expression that derives from a biting action whose biting sequence has been inhibited (Fig. 20.5) (Palagi et al. 2014;

Fig. 20.6 California sea lion pups displaying a reversed Relaxed Open Mouth (ROM). (Photo by Tommaso Ragaini)

Tinbergen 1952). In otariids, it has only been anecdotally reported in the South African fur seal (Caudron 1995) and studied in the southern sea lion (Llamazares-Martín et al. 2017b). In the latter species, the Attempt to Bite pattern (PAB) was used as a control condition to test if ROM acts also as a play signal, since PAB shares very similar motor modules with ROM. The communicative nature of ROM is indicated as there is a significant difference in the frequency in which ROMs and PABs were followed by real bites, thus suggesting that ROM is not a preparation to bite. That ROMs were more frequently reciprocated than PABs reinforces this hypothesis. To reach communicative effectiveness, a play signal has to be reciprocated (Fig. 20.6). In the southern sea lion, the interactive nature of the signal is confirmed when its reciprocity prolongs the duration of the interaction. In the southern sea lion, ROM was equally performed during dyadic and polyadic play, but was more reciprocated in dyadic sessions. The similar frequency of ROM in polyadic and dyadic interactions suggests that the motivation to play does not depend on the number of players involved. However, to reach an effective communication, play signals must be reciprocated by the receiver. The reciprocity of ROM reaches its peak in dyadic play, when there are more face-to-face interactions between the two players. This type of visual engagement favors the detection of the signal and, therefore, increases the probabilities that it may be reciprocated (Llamazares-Martín et al. 2017b). Taken together, these results suggest that sea lions rely on visual signals for complex communication, since their reciprocity makes play more successful.

20.5 To Be Predator or Prey: The Role of Solitary Play in Developing Future Successful Strategies

In our review on play, solitary play cannot be ignored because it is pervasive in many terrestrial and aquatic species (Burghardt 2005). As already mentioned, it can be categorized as play with objects or locomotor-rotational play.

In pinnipeds, when animals are confident at sea, they may perform swimming solitary play. Most sea lion species have been seen bodysurfing (Eibl-Eibesfeldt 1984; R. Harcourt, personal observation), that is when they ride a moving wave with their bellies, which carries them towards the shore. This behavior is repeated over again with no apparent goal, but adults usually perform it during storms to land high on the rocks (R. Gentry, personal observation). Harcourt (1991b) reported that swimming solitary play patterns in South American fur seals resembled those that sea lions use when they escape from predators but did not provide a detailed description of the variety of patterns employed. Due to the paucity of studies on solitary play in this superfamily, we cannot conclude that these are the only species, which perform swimming solitary play. On the other hand, in the superfamily Phocoidea, which is the closest phylogenetic group to Otarioidea, the only species in which swimming solitary play has been described is the harbor seal (Renouf 1993; Renouf and Lawson 1986). In this species, swimming solitary play is composed of complex locomotor, rotational and acrobatic movements such as pirouettes, porpoises, or torpedoes. The striking similarity between locomotor-rotational patterns and anti-predatory movements in this and other mammal species suggests that the short- and long-term benefits obtained from engaging in these behaviors reside in the improvement of physical skills and motor coordination that would increase survival when animals must flee from predators (Špinka et al. 2001). Moreover, since all pinniped species share similar morphological adaptations to an aquatic life and have similar anti-predator tactics, it is reasonable to think that several members of the Otarioidea superfamily perform solitary acrobatic play during their juvenile period.

Solitary object play can be performed with both inanimate objects and prey. Animals can also use their own body parts, especially the extremities, in a playful manner. The body parts then become "external" objects that can provide stimuli for body exploration. In humans, for example, infants tickle their own feet and, sometimes, introduce their toes into their mouths and bite them (Rochat 1998). By playing with body parts at the beginning of their lives, animals, including humans, acquire the self-consciousness needed for developing self-recognition. Moreover, this peculiar play practice favors motor coordination and performance of movements (Rochat and Hespos 1997). In Steller sea lions, pups may play with their fore or hind flippers. In the first case, they wave their fore flippers in front of their faces while attempting to catch them. Once pups have grabbed their flippers, they begin to repeatedly bite them. In the second case, pups approach their hind flippers to their heads while they are sitting upright and bite them following the same sequence (Farentinos 1971). These rudimentary actions of grabbing and biting body parts provide infants with somatosensory and proprioceptive information that may help them to recognize

Fig. 20.7 Walrus performing object play with a ball. (Photo by Robert Harcourt)

those body parts as theirs (Blanke 2012). The identification of their own body as a different object from those found in the environment (*ecological self*, Rochat 1998), derives from the double perception of stimuli acquired when two parts of the body are in contact. It appears that the player coincides with the playmate.

The most common and rudimentary version of solitary object play has been described in five species of sea lions (Australian sea lions, Marlow 1975; California sea lions, *Zalophus californianus*, Peterson and Bartholomew 1967; Galápagos sea lions, Eibl-Eibesfeldt 1955, 1984; New Zealand sea lions, Marlow 1975; Steller sea lions, Farentinos 1971) and in two species of fur seals (South African fur seal, Caudron 1995; South American fur seal, Harcourt 1991b) (see Table 20.1). But it is also common in captive juvenile southern sea lions (C. Llamazares-Martín, personal observation) and Pacific walruses (*Odobenus rosmarus divergens*) (R. Harcourt, personal observation, Fig. 20.7). In this kind of play, animals manipulate (on land or at sea) inanimate objects such as rocks, feathers, sticks or pieces of kelp. When they play with a hard material object (e.g., rocks) they grab it with their mouths and bite it several times (Eibl-Eibesfeldt 1955; Peterson and Bartholomew 1967). When manipulating a soft or manageable material object (e.g., pieces of kelp), animals usually grab and hold it tightly, then vigorously shake their heads from side to side. The objects may be tossed into the air and immediately grabbed again (Farentinos 1971; Marlow 1975). These behaviors bring to mind those performed when preying on medium and big size animals, in which otariids firmly grab their prey by their teeth and quickly shake their heads from side to side (Hocking et al. 2016, Chap. 5).

In certain cases, animals display predatory maneuvers towards prey without feeding on them. This behavior has been defined as play since it loses its primary function, and because the patterns involved are strongly modified. The motor actions are slowed down, the predatory patterns are often interrupted and mixed with other behaviors and the sequences forming the session are frequently repeated without inhibition (Bonner and Hunter 1982; du Toit et al. 2004; Giljov et al. 2017). Such behavior has been reported in one species of sea lion (California sea lion, Riedman 1990), three species of fur seal (Antarctic fur seal, *Arctocephalus gazella*, Bonner and Hunter 1982; South African fur seal, du Toit et al. 2004; Marks et al. 1997;

Fig. 20.8 Failed attack from below performed by a juvenile walrus towards seabirds. (Drawing by Daniele Santerini)

Guadalupe fur seal, *Arctocephalus townsendi*, Riedman 1990) and the Pacific walrus (Giljov et al. 2017). Antarctic fur seals were seen to kill penguins tossing them into the air, but rarely eating them (Bonner and Hunter 1982). Du Toit et al. (2004) observed a similar behavior in South African fur seals towards seabirds just after a feeding bout. Marks et al. (1997), however, described some cases in which an adult fur seal seemed to teach several juveniles (1 and 2 years old) to prey on seabirds. Juveniles observed how the adult caught and degloved the animal and mimicked its behavior until they correctly performed it. The adult did not participate in those interactions, and in one of the cases, juveniles tossed the carcass into the air without killing the potential prey.

In the walrus, three main solitary playful interactions towards seabirds have been described (Giljov et al. 2017). While approaching a seabird from the surface, a walrus can shake its head from side to side and splash its face and tusks against the water. During this behavioral sequence there is never physical contact between the two animals. In a second form of play, the walrus approaches the bird and when quite close, lunges towards it trying to hit it with its tusks. In a third form, the attack from below, the walrus approaches from below the surface trying to hit the bird with its tusks and to catch it by mouth (Fig. 20.8). Despite the aggressive nature of the patterns described in the last two types of interaction, only 1 out of 51 birds was killed. This suggests that the behavior does not have any immediate benefits and is not functional for nutrition (see also Miller and Kochnev, Chap. 22).

Walruses may also interact with seabird carcasses both solitarily or together with other individuals (social play). In the latter, an animal carries the carcass in its mouth. It brings the carcass close to another individual's head and subsequently pulls the carcass backwards. This behavior results in a play bout that usually continues after the walruses lose interest in the "object". This peculiar pattern strongly resembles that observed in many primate and canid species that is often defined as "capture the flag" (Palagi et al. 2007). As in the case of the South African fur seal, juveniles may acquire these predatory techniques from adults or, as in the case of walruses, they may become skilled by directly interacting with the birds or their carcasses together with other juveniles or alone. Although seabird consumption may be opportunistic in otariids and odobenids, seabirds can be an important energy source during periods of food scarcity (Long and Gilbert 1997). In this perspective, playing with "potential prey" may be a scaffold to developing effective predatory strategies, thus increasing survival opportunities in the future.

20.6 Conclusion

This review underlines that play behavior is present in the superfamily Otarioidea. Play relates to different patterns of behavior, depending on the species. Infant play is directly related to maternal attendance styles, which affects the development of social, cognitive and physical skills of the pup. An accurate monitoring of mother-pup early interactions (mother permissiveness, time spent with the pup and stimulating actions by the mothers) and the ontogeny of pup play in a more complex social environment would contribute to filling gaps related to this complex behavior.

Studying play in the wild is challenging. That is why a captive environment offers the opportunity to gather useful information in these kinds of studies. Management in captive settings is relevant because it allows evaluating the development of behavior under controlled conditions. For instance, in zoos and aquaria, lactation can be changed according to management requirements and, in some cases, the mother-pup dyad may be isolated from the group for several months, thus limiting pups' social interactions just to their mothers. Mother-pup isolation produces an artificial and complete social canalization, which is present when the mother exercises control over the actions of the infant, thus limiting its social activity including play. Data collection carried out through the same observational methods on groups belonging to the same species living in structures with different kinds of management may provide opportunities to test hypotheses on the development of social and physical skills in the superfamily Otarioidea.

Another important point to test hypotheses on the evolution of play in a comparative manner is the number of species on which data should be collected. The selection should be done on the basis of their social features. To most efficiently allow comparisons, species differing in lactation time, level of dimorphism, degree of social tolerance, and ecological correlates should be followed and observed via similar methodological protocols.

There is little knowledge of how pinnipeds communicate their motivations to play. Understanding how communicative systems modulate social interactions between subjects is fundamental to clarify some aspects of social cognition of large-brained animals (Palagi et al. 2016). For this reason, further data on the patterns that act as play signals are needed. In exploring these questions, we should be particularly attentive to the different sensory modalities through which the play message can be conveyed. We cannot exclude the possibility that, in otariids, more than one cue may be integrated in a multi-modal system to improve the significance of the message (Wierucka et al. 2018). It is possible that acoustic, olfactory and visual cues are combined to optimize transmission and decoding of signals, thus making social play sessions more successful. Due to the sophisticated abilities required to manage a play session, many researchers consider play-fighting as a window into social cognition and a fertile field to explore cognitive skills.

Acknowledgements Supported by the Spanish Ministry of Economy and Competitiveness (PSI2011-29016-C02-02) and the Universidad Cardenal Herrera CEU (PRCEU-UCH PT-16-01, PSI2017-85922-P).

References

Arnold W, Trillmich F (1985) Time budget in Galapagos fur seal pups: the influence of the mother's presence and absence on pup activity and play. Behaviour 92:302–321

Bekoff M (2001) Social play behaviour. Cooperation, fairness, trust, and the evolution of morality. J Conscious Stud 8:81–90

Berghänel A, Schulke O, Ostner J (2015) Locomotor play drives motor skill acquisition at the expense of growth: a life history trade-off. Sci Adv 1(7):e1500451

Berman CM (1982) The ontogeny of social relationships with group companions among free-ranging infant rhesus monkeys I. Social networks and differentiation. Anim Behav 30:149–162

Blanke O (2012) Multisensory brain mechanisms of bodily self-consciousness. Nat Rev Neurosci 13:556–561

Boness DJ (1991) Determinants of mating systems in the Otariidae (Pinnipedia). In: Renouf D (ed) The behaviour of pinnipeds, 1st edn. Springer, Dordrecht, pp 1–44

Boness DJ, Bowen WD (1996) The evolution of maternal care in pinnipeds. Bioscience 46:645–654

Bonner WN, Hunter S (1982) Predatory interactions between Antarctic fur seals, macaroni penguins and giant petrels. Br Antarct Surv B 56:75–79

Burghardt GM (2005) The genesis of animal play: testing the limits. MIT Press, Cambridge

Byers JA, Walker C (1995) Refining the motor training hypothesis for the evolution of play. Am Nat 146:25–40

Campagna C (2009) Aggressive behavior, intraspecific. In: Perrin WF, Bürsig B, Thewhiseen JGM (eds) Encyclopedia of marine mammals, 2nd edn. Academic Press, San Diego, pp 18–22

Campagna C, Le Boeuf BJ (1988) Reproductive behaviour of southern sea lions. Behaviour 104:233–261

Campagna C, Le Boeuf BJ, Cappozzo HL (1988) Pup abduction and infanticide in southern sea lions. Behaviour 107:44–960

Caro TM (1995) Short-term costs and correlates of play in cheetahs. Anim Behav 49:333–345

Caudron AK (1995) Social behaviour of Cape fur seals *Arctocephalus pusillus pusillus* in captivity. Aquat Mamm 21:7–17
Ciani F, Dall'Olio S, Stanyon SR et al (2012) Social tolerance and adult play in *macaque* societies: a comparison with different human cultures. Anim Behav 84:1313–1322
Clutton-Brock TH, Huchard E (2013) Social competition and selection in males and females. Philos Trans R Soc B 368:20130074
Cordoni G (2009) Social play in captive wolves (*Canis lupus*): not only an immature affair. Behaviour 146:1363–1385
Cordoni G, Palagi E (2016) Aggression and hierarchical steepness inhibit social play in adult wolves. Behaviour 153:749–766
Cordoni G, Nicotra V, Palagi E (2016) Unveiling the "secret" of play in dogs (*Canis lupus familiaris*): asymmetry and signals. J Comp Psychol 130:278–287
Cordoni G, Norscia I, Bobbio M et al (2018) Differences in play can illuminate differences in affiliation: a comparative study on chimpanzees and gorillas. PLoS One 13:e0193096
Du Toit MP, Bartlett A, Bester MN et al (2004) Seabird predation by individual seals at Ichaboe Island, Namibia. S Afr J Wildl Res 34:45–54
Eibl-Eibesfeldt I (1955) Ethologische Studien am Galapagos-Seelöwen, *Zalophus californianus wollebaeki* Swersten. Z Tierpsychol 12:286–303
Eibl-Eibesfeldt I (1984) The natural history of the Galapagos sea lion (*Zalophus californianus wollebaeki*, Sirvertsen). In: Perry R (ed) Key environments Galapagos, 1st edn. Pergamon Press, Oxford, pp 207–214
Emlen ST, Oring LW (1977) Ecology, sexual selection, and the evolution of mating systems. Science 197:215–223
Fagen R (1981) Animal play behavior. Oxford University Press, New York
Fagen R (1993) Primate juveniles and primate play. In: Pereira ME, Fairbanks LA (eds) Juvenile primates, 2nd edn. Oxford University Press, New York, pp 182–196
Farentinos RC (1971) Some observations on the play behavior of the Steller sea lion (*Eumetopias jubata*). Z Tierpsychol 28:428–438
Fay FH (1982) Ecology and biology of the Pacific walrus, *Odobenus rosmarus divergens*. United States Department of the Interior Fish and Wildlife Service, Washington, DC
Gentry RL (1974) The development of social behavior through play in the Steller sea lion. Am Zool 14:391–403
Giljov A, Karenina K, Kochnev A (2017) Prey or play: interactions between walruses and seabirds. Acta Ethol 20:47–57
Harcourt R (1991a) Survivorship costs of play in the South American fur seal. Anim Behav 42:509–511
Harcourt R (1991b) The development of play in the South American fur seal. Ethology 88:191–202
Harcourt R (1992a) Maternal aggression in the South American fur seal in Peru. Can J Zool 70:320–325
Harcourt R (1992b) Factors affecting mortality in the South American fur seal (*Arctocephalus australis*) in Peru: density-related effects and predation. J Zool 226:259–270
Higgins LV, Gass L (1993) Birth to weaning: parturition, duration of lactation, and attendance cycles of Australian sea lions (*Neophoca cinerea*). Can J Zool 71:2047–2055
Higgins LV, Tedman RA (1990) Effect of attacks by male Australian sea lions, *Neophoca cinerea*, on mortality of pups. J Mammal 71:617–619
Hill HM, Dietrich S, Cappiello B (2017) Learning to play: a review and theoretical investigation of the developmental mechanisms and functions of cetacean play. Learn Behav 45:335–354
Hocking DP, Fitzgerald EM, Salverson M et al (2016) Prey capture and processing behaviors vary with prey size and shape in Australian and subantarctic fur seals. Mar Mammal Sci 32:568–587
Le Boeuf BJ (1991) Pinniped mating systems on land, ice and in the water: emphasis on the phocidae. In: Renouf D (ed) The behaviour of pinnipeds, 1st edn. Springer, Dordrecht, pp 45–65

Llamazares-Martín C, Scopa C, Guillén-Salazar F et al (2017a) Strong competition does not always predict play asymmetry: the case of South American sea lions (*Otaria flavescens*). Ethology 123:1–13

Llamazares-Martín C, Scopa C, Guillén-Salazar F et al (2017b) Relaxed Open Mouth reciprocity favours playful contacts in South American sea lions (*Otaria flavescens*). Behav Process 140:87–95

Long DJ, Gilbert L (1997) California sea lion predation on chicks of the Common Murre (León Marino de California (*Zalophus californianus*) Depredando Sobre Pichones de Uria aalgae). J Field Ornithol 68:152–154

Marks MM, Brooke RK, Gildenhuys AM (1997) Cape fur seal *Arctocephalus pusillus* predation on Cape cormorants *Phalacrocorax capensis* and other birds at Dyer Island, South Africa. Mar Ornithol 25:9–12

Marlow BJ (1975) The comparative behaviour of the Australasian sea lions *Neophoca cinerea* and *Phocarctos hookeri* (Pinnipedia: Otariidae). Mammalia 39:159–230

McNab AG, Crawley MC (1975) Mother and pup behaviour of the New Zealand fur seal, *Arctocephalus forsteri* (Lesson). Mauri Ora 3:77–88

Miller EH (1976) Walrus ethology. II. Herd structure and activity budgets of summering males. Can J Zool 54:704–715

Miller EH, Boness DJ (1983) Summer behavior of Atlantic walruses *Odobenus rosmarus rosmarus* (L.) at Coats Island, N. W. T. (Canada). Z Saugetierkd 48:298–313

Norscia I, Palagi E (2016) The missing lemur link: an ancestral step in the evolution of human behaviour. Cambridge University Press, Cambridge

Oftedal OT, Boness DJ, Tedman RA (1987) The behavior, physiology, and anatomy of lactation in the pinnipedia. Curr Mammal 1:175–245

Palagi E (2006) Social play in bonobos (*Pan paniscus*) and chimpanzees (*Pan troglodytes*): implications for natural social systems and interindividual relationships. Am J Phys Anthropol 129:418–426

Palagi E (2007) Play at work: revisiting data focusing on chimpanzees (*Pan troglodytes*). Am J Phys Anthropol 85:63–81

Palagi E (2009) Adult play fighting and potential role of tail signals in ringtailed lemurs (*Lemur catta*). J Comp Psychol 123:1–9

Palagi E (2014) Playing alone and playing with others: a lesson from animals. In: Coplan RJ, Bowkereds J (eds) The handbook of solitude: psychological, 1st edn. Wiley, Chichester, pp 463–482

Palagi E (2018) Not just for fun! social play as a springboard for adult social competence in human and non-human primates. Behav Ecol Sociobiol 72:90. https://doi.org/10.1007/s00265-018-2506-6

Palagi E, Mancini G (2011) Playing with the face: playful facial "chattering" and signal modulation in a monkey species (*Theropithecus gelada*). J Comp Psychol 125:11–21

Palagi E, Antonacci D, Cordoni G (2007) Fine-tuning of social play by juvenile lowland gorillas (*Gorilla gorilla gorilla*). Dev Psychobiol 49:433–445

Palagi E, Norscia I, Spada G (2014) Relaxed open mouth as a playful signal in wild ring-tailed lemurs. Am J Primatol 76:1074–1083

Palagi E, Burghardt GM, Smuts B et al (2016) Rough-and-tumble play as a window on animal communication. Biol Rev 91:311–327

Panksepp J, Burgdorf J (2003) "Laughing" rats and the evolutionary antecedents of human joy? Physiol Behav 79:533–547

Paquette D (1994) Fighting and playfighting in captive adolescent chimpanzees. Aggress Behav 20:49–65

Pellis SM, Pellis VC (1996) On knowing it's only play: the role of play signals in play fighting. Aggress Violent Behav 1:249–268

Pellis SM, Pellis VC (2009) The playful brain: venturing to the limits of neuroscience. Oneworld, Oxford

Pellis SM, Pellis VC (2016) Play fighting in Visayan warty pigs (*Sus cebifrons*): insights on restraint and reciprocity in the maintenance of play. Behaviour 153:727–747
Peterson RS, Bartholomew GA (1967) The natural history and behaviour of the California sea lion. Am Soc Mammal Special Publication No. 1:79
Poole TB (1978) An analysis of social play in polecats (Mustelidae) with comments on the form and evolutionary history of the open mouth play face. Anim Behav 26:36–49
Renouf D (1993) Play in a captive colony of harbour seals (*Phoca vitulina*): constrained by time or by energy? J Zool 231:351–363
Renouf D, Lawson JW (1986) Play in harbour seals (*Phoca vitulina*). J Zool 208:73–82
Riedman M (1990) The pinnipeds: seals, sea lions, and walruses. University of California Press, Berkeley
Rochat P (1998) Self-perception and action in infancy. Exp Brain Res 123:102–109
Rochat P, Hespos SJ (1997) Differential rooting response by neonates: evidence for an early sense of self. Early Dev Parent 6:105–112
Smith PK (1997) Play-fighting and real fighting: perspectives on their relationship. In: Schmitt A, Atzwanger K, Grammar K et al (eds) New aspects of human ethology, 1st edn. Plenum Press, New York, pp 47–64
Soto KH, Trites AW (2011) South American sea lions in Peru have a lek-like mating system. Mar Mammal Sci 27:306–333
Špinka M, Newberry RC, Bekoff M (2001) Mammalian play: training for the unexpected. Q Rev Biol 76:141–168
Thierry B (2004) Social epigenesis. In: Thierry B, Singh M, Kaumanns W (eds) *Macaque Societies:* a model for the study of social organization, 1st edn. Cambridge University Press, Cambridge, pp 267–290
Tinbergen N (1952) The curious behaviour of the stickleback. Sci Am 187:22–26
van Hooff JARAM, Preuschoft S (2003) Laughter and smiling: the intertwining of nature and culture. In: de Waal FBM, Tyack PL (eds) Animal social complexity, 1st edn. Harvard University Press, Cambridge, pp 260–287
Wartzok D (2012) Physiology of behaviour in pinnipeds. In: Renouf D (ed) The behaviour of pinnipeds, 1st edn. Springer Science and Business Media, Bristol, pp 236–299
Watts DP (1991) Mountain gorilla reproduction and sexual behavior. Am J Primatol 24:211–225
Wheatley KE, Bradshaw CJA, Harcourt RG, Hindell MA (2008) Feast or famine: evidence for mixed capital-income breeding strategies in the Weddell seal. Oecologia 155:11–20
Wierucka K, Pitcher BJ, Harcourt R, Charrier I (2018) Multimodal mother-offspring recognition: the relative importance of sensory cues in a colonial mammal. Anim Behav 146:135–142
Wilson S, Kleiman DG (1974) Eliciting play: a comparative study. Am Zool 14:341–370

Part V
Diving Deeper into Species

Chapter 21
Responding to Human Influence: Southern Sea Lion Males Adapt to Harbor Habitats

Diego H. Rodríguez, Gisela V. Giardino, María A. Mandiola, Joaquín C. M. Gana, Marta C. De León, Julián Bastida, Sergio G. Morón, Pablo Denuncio, and Ricardo O. Bastida

Abstract Male-only hauling grounds occur in close association to breeding colonies in several otariids. The occasional use of human-built infrastructure by these groups is also common. We observed permanent male aggregations of southern sea lions, *Otaria flavescens*, in urban contexts, such as two of the main fishing harbors in Argentina and investigated how they changed over 35 years. These haul-outs are located approximately 700 km from Uruguayan and northern Patagonian breeding grounds. Males in urban colonies have functional connections with both populations. A predictable pattern of dispersion during the breeding season, returning for winter, indicate site fidelity. Harbor habitats present ever changing stimuli, threats and challenges to habituated sea lions. Sequential, rapid and spontaneous changes in the location of the aggregations may be a response to quick learning and socially transmitted knowledge. Food provision via consumption of fish offal—complemented with natural foraging—provide support to these aggregations over prolonged periods. The presence of hundreds of sea lions in harbors represents a management challenge. Fidelity to resting areas and rapid habituation to human presence makes it difficult to design strategies that can deter sea lions from intensive use of critical sectors of port activity. Interaction with fishing gear results in frequent entanglements. The risk of transmission of diseases from and to dogs and other domestic animals is a major potential problem for individuals in unnatural aggregations.

D. H. Rodríguez (✉) · G. V. Giardino · M. A. Mandiola · J. C. M. Gana · M. C. De León · P. Denuncio · R. O. Bastida
Instituto de Investigaciones Marinas y Costeras (IIMyC), Facultad de Ciencias Exactas y Naturales, Universidad Nacional de Mar del Plata – CONICET, Mar del Plata, Argentina

J. Bastida
Instituto Nacional de Investigación y Desarrollo Pesquero (INIDEP), Mar del Plata, Argentina

S. G. Morón
Fundación Mundo Marino, San Clemente del Tuyú, Argentina

C. Campagna, R. Harcourt (eds.), *Ethology and Behavioral Ecology of Otariids and the Odobenid*, Ethology and Behavioral Ecology of Marine Mammals,
https://doi.org/10.1007/978-3-030-59184-7_21

Keywords Southern sea lions · South American sea lion · *Otaria flavescens* · Urban rookeries · Harbor habitats

21.1 Introduction

Male-only hauling grounds are frequently recorded in close association with larger breeding colonies in several otariid species, including Steller sea lions *Eumetopias jubatus* (Belkin 1966; Harestad and Fisher 1975; Perlov 1980; Raum Suryan et al. 2002), New Zealand sea lions *Phocarctos hookeri* (Hawke 1986; Beentjes 1989; Robertson et al. 2006), California sea lions *Zalophus californianus* (Riedman 1990), South American fur seals *Arctocephalus australis* (Vaz Ferreira 1956, 1960), New Zealand fur seals *A.forsteri* (Johnstone and Davis 1987; Bradshaw et al. 1999) and subantarctic fur seals *A. tropicalis* (Bester 1982). Absence of male parental investment, exclusion of males from breeding opportunities, delayed sexual maturation, and avoidance of local prey competition with breeding females maybe some of the drivers for the establishment of all-male haul-outs. Although some of these aggregations may ultimately change in social composition and become breeding sites (Grandi et al. 2008), the functions of non-breeding male aggregations are still not clearly understood.

Male South American or southern sea lions (*Otaria flavescens*; see Rodriguez and Bastida 1993) form aggregations that vary in social structure ranging from peripheral groups of subadult and adult males around central breeding areas, through mixed sex juvenile aggregations, to winter aggregations of juvenile, subadult and adult males fully integrated with young females or mother and pups (Carrara 1952; Vaz Ferreira 1982; Lewis and Ximenez 1983; Crespo and Pedraza 1991; Dans et al. 1996; Grandi et al. 2008). Both seasonal and permanent male southern sea lions haul-out sites distant from the reproductive colonies of Uruguay and Patagonia have been recorded on coastal islands (Ilha dos Lobos, Brazil; Isla Trinidad, Argentina), sand banks (Banco Culebra, Argentina), breakwaters (Molhe Leste da Barra do Rio Grande, Brazil) and inside harbors (Mar del Plata and Quequén, Argentina, Fig. 21.1).

The occasional use of marine infrastructure (offshore buoys, marinas, breakwaters, pontcons, docks, piers, oil rigs) by otariids is pervasive throughout their geographic range, but is rarely referred to in the scientific literature. Reported examples include South African fur seals (*A. pusillus pusillus*), which are frequently reported near fishing facilities in Cape Town Harbor, South Africa (Shaughnessy and Chapman 1984), Australian fur seals (*A. pusillus doriferus*) that gather on oil-rig struts in Bass Strait (Kirkwood and Goldsworthy 2013), New Zealand fur seals and Australian sea lions (*Neophoca cinerea*) recorded on the breakwaters in Adelaide Harbor, Australia (Shaughnessy et al. 2018), southern sea lions in Ilha dos Lobos and Molhe Leste (Brazil; Rosas et al. 1994; Curi Estima 2002; Silva 2004; Pavanato et al. 2013; Procksch et al. 2020) and California sea lions seen in Ballard Locks (Fraker and Mate 1999), Pier 39 in San Francisco harbor and Columbia River system in the West coast of the United States.

Fig. 21.1 Current male harbor haul outs in Mar del Plata (**a**) and Quequén (**b**). Photo credits: (**a**) Diego Rodríguez; (**b**) Gisela Giardino

These groupings in human-made habitats is sometimes limited by seasonal movements, while in other cases the seals are permanently present. The existence of easily accessible and non-restricted open spaces for hauling out maybe the primary attractant, in combination with the disposal of fishing discards that provides an abundant and predictable food source. Moreover, coastal urban infrastructure may in some cases reduce the potential activity of predators, as sharks and killer whales would avoid highly disturbed and often polluted waters.

Harbors, as novel and dynamic human-made habitats, present new stimuli, threats and challenges to sea lions. However, the stability and site fidelity of sea lions hauling out in harbors strongly suggest that these non-natural places provide new habitats. Habituation to human presence (*sensu* Schakner and Blumstein, Chap. 18) is likely to be a key element in the formation this type of aggregation. These unisexual groups may provide an important social and ecological context for non-breeding males, whether seasonally or prior to sexual and social maturity.

We here describe the behavior of southern sea lions in two permanent male haul-outs in harbors (MHH) in northern Argentina. The first site was established by a small group of southern sea lions decades ago at the Mar del Plata harbor whereas the second example is more recent and was established at Puerto Quequén in the early 1990s (Rodríguez 1996; Giardino 2006). These haul-outs are nearly equidistant (ca. 700 km) to the breeding colonies of Uruguay and northern Patagonia. Both groups may be spatially and socially isolated, yet they do have strong functional connections with both breeding populations (Giardino et al. 2016).

Long-term monitoring of MHH in northern Argentina provides an opportunity to observe social behavior, and describe the composition, preferences and influence of food provisioning in human-made habitats. We use this information to discuss management issues arising from urban aggregations and outline their potential application to otariid groups occurring in harbors elsewhere. The interaction between humans and sea lions in urban contexts has relevance to the conservation of the species as well as to safety and public health considerations.

21.2 Origin of the Harbor Colonies

The rocky coast of Mar del Plata was inhabited by extensive southern sea lion colonies, which disappeared during the second half of the nineteenth century due to rapid urban development and to the indirect effect of pinniped over-exploitation in other areas of the south-western Atlantic (Rodriguez and Bastida 1998). Mar del Plata harbor was built between 1914–1918 and occasional sea lion vagrants were recorded during the first half of the twentieth century. By the late 1960s a small group pf circa 80–100 sea lions were established on a small internal breakwater of a local Yacht Club (Fig. 21.2), and a few of them used the docks and boats as resting places (Baldás et al. 1987; Rodríguez 1990).

During the 1970s and early 1980s, solitary sea lions were occasionally sighted inside Puerto Quequén, but it was only between 1988 and 1990 when upwards of

Fig. 21.2 Different haul out areas in harbors. Docks (concrete blocks), breakwaters, vessel decks and sandy beaches. Original Mar del Plata settlement in black and white. Photo credits: (**a**) Ricardo Bastida; (**b**) Diego Rodríguez; (**c**) Joaquin Gana; (**d**) Gisela Giardino; (**e**) Joaquin Gana; (**f**) Julián Bastida

50–80 sea lions hauled out on the docks (Westergaard 2001). During these years, a group of coastal boats from Mar del Plata were docked temporarily in Puerto Quequén during the anchovy (*Engraulis anchoita*) fishery season, and sea lions frequently interacted with these midwater trawl operations. Although the origin of the sea lions was uncertain, it seems probable that were animals associated with the boats from Mar del Plata.

21.3 Age Structure

The southern sea lion harbor haul-outs were composed of males of different ages, predominantly juveniles and subadults. Yearlings or two-year old animals were born in either Uruguay or Patagonia and moved to MHH after weaning or during their second year of life. They were found in relatively low numbers (Table 21.1). Seventy percent of marked juveniles (3–4 years of age) recorded at MHHs were born in the Uruguayan colonies (Latin American Network of Pinniped Marking—http://marcacionpinnipedos.blogspot.com/-, unpublished data). The presence of this age class varied, but juveniles persisted throughout the year.

Subadult males (5–6 years old) were also a frequent component of MHHs. Males become fully sexually and socially mature at about 9 years (Grandi et al. 2010). The presence of adults in harbors is therefore likely to be influenced by the breeding cycle. The relatively low numbers of adults suggests that some may reside, while others only visit the MHH sporadically.

The dominance of subadult and adult males in harbor colonies seems to be consistent across otariids. Southern sea lions in Southern Brazil (Ilha dos Lobos and Molhe Leste; Rosas et al. 1994; Curi Estima 2002; Silva 2004; Pavanato et al. 2013; Procksch et al. 2020) and Isla Trinidad (Argentina; Petracci et al. 2010), South African fur seals in Cape Town Harbor (Shaughnessy and Chapman 1984) and New Zealand fur seals and Australian sea lions at the Outer Adelaide Harbor (Shaughnessy et al. 2018) have the same composition and dynamics. Even in those harbors where only a few vagrant southern sea lions patrol the docks in search for fish discharges (i.e., Valvidia, Rawson, Comodoro Rivadavia, Punta del Este), they are exclusively subadult or adult males.

21.4 Long Term Trends in Abundance

Although sea lions are present in both MHH all year around since colony formation, long-term monitoring has confirmed marked changes in the number of animals hauled out. The Mar del Plata MHH, after an initial settlement of less than a hundred animals, experienced a noticeable increase in abundance in 1985–1987, reaching average values above 600 animals during the following decade (Rodríguez 1990; Fig. 21.3). A gradual decline in the mean number of animals was then recorded, and once again recovering to historical values from 2010 on. In contrast, Puerto Quequén MHH numbered only approximately one hundred animals for the first 10 years,

Table 21.1 Mean (± SD) percentage of each age class concentrated in Puerto Mar del Plata (1984–2019) and Puerto Quequén (1996–2019)

Age Class	Mar del Plata	Quequén
Annuals	4.7 ± 3.5	5.8 ± 4.2
Juveniles	45.4 ± 9.7	30.2 ± 9.3
Subadults	45.1 ± 5.4	53.6 ± 9.7
Adults	5.3 ± 4.7	10.4 ± 10.0

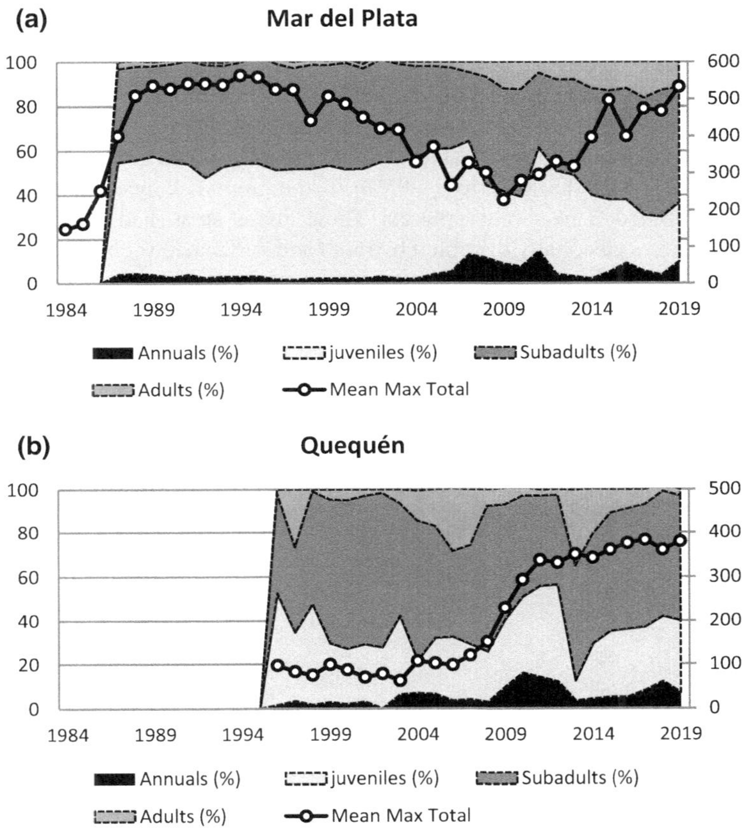

Fig. 21.3 Historical fluctuations of age composition at Mar del Plata (**a**) and Quequén (**b**) HMHs. Left axis indicates the mean annual percentage of sea lions of each age class, and the right axis the mean annual number of sea lions hauled out

followed by a noticeable increase in abundance to a mean of ca 350–400 sea lions (Fig. 21.3).

Southern sea lions were bleach marked and sightings in both MHH confirmed a flux of animals between haul-outs (Giardino 2014). Around 14% of the sea lions hauled out alternatively at both MHHs, with a greater exchange of younger animals. Giardino et al. (2017) estimated that the effective number of animals forming both MHH would at least triple the number of animals that are hauled out at any one time. The ratio between the number of animals observed inside the harbor and the total estimate remained relatively constant, ranging between two and four times more individuals estimated than counted. An estimate of total SSL hauled out in 2019 at both MHHs was *ca.* 900 animals, producing an estimated total number of about 2850 animals.

21.5 Habitat Use and Internal Distribution

Southern sea lions in the harbors of Mar del Plata and Quequén will use all substrate types present within each port, including docking and loading areas, internal and external beaches, breakwaters, boats and floating pontoons. As a rule, there is a main haul-out area, where the majority (>85%) of the animals concentrate, plus small groups distributed across other places. These main areas had long periods of permanent use, occasionally disrupted by rapid and spontaneous changes in location and substrate. MHHs went through major changes in main haul-out areas at least twice in the last 30 years (Fig. 21.4). When numbers were low (<250 animals), docks were actively used for resting, generating conflicts with harbor operations. Animals then moved to sandy beaches, with a concomitant rapid increase in

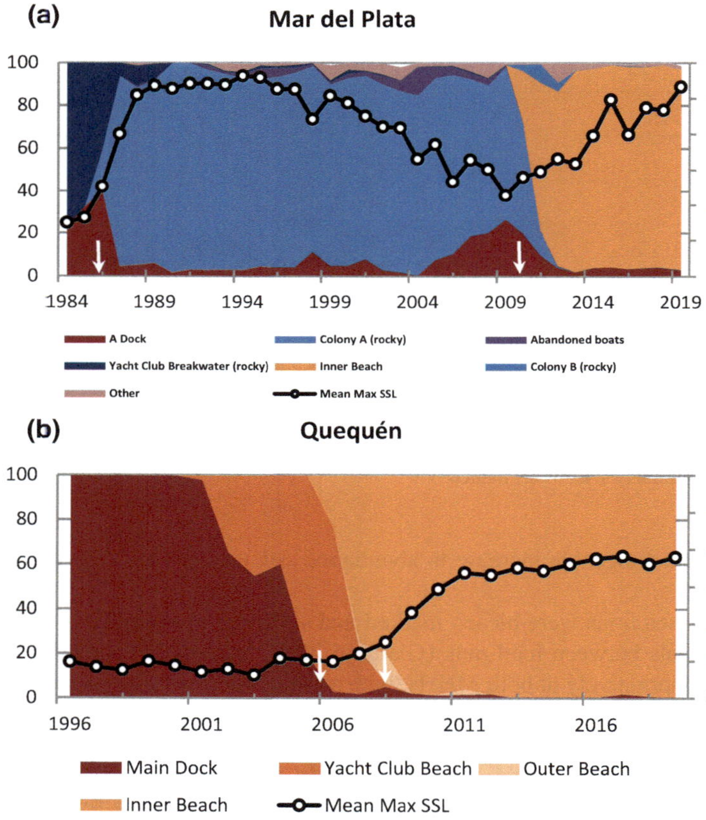

Fig. 21.4 Historical distribution of southern sea lions in Mar del Plata (**a**) and Quequén (**b**) harbors by area and substrate type. White arrows indicate rapid relocalization of the main haulout area. Left axis indicate the mean annual percentage of sea lions in each area, and the right axis the mean annual number of sea lions hauled out

Table 21.2 Internal haul out areas in harbors

Area	Type of surface	Restrictions to sea lions	Potential sea lion density	Human disturbance	Age/size limitations
Breakwaters	Rocky	Restricted by topography	Low	Low	No
Docks	Flat + vertical	Restricted by tide	High	High	Yes (Young animals at low tide)
Boats	Flat	Restricted by boat operation	Low	High–low	Yes (operative boats)
Beaches	Flat	Unrestricted	High	Low	No

abundance attributed to low human disturbance, no topographic restriction, and no size (age) access limitations (Fig. 21.2; Table 21.2).

Stability of the main haul-out area was influenced by high fidelity to specific resting sites. Observations of bleach-marked animals revealed that most sea lions haul out in only one or two specific places, regardless of age (Mandiola et al. 2012).

21.6 Food Provisioning and Foraging

Sea lions in harbors show a combination of natural foraging and fishing discard utilization. Thirty prey items were identified from scats from Mar del Plata and Quequén MHHs. Bony fishes were the most frequent prey (Frequency of Occurrence; FO = 92%), followed by cartilaginous fish (FO = 33%) and cephalopods (FO = 17%). Most of the identified prey (70.8%) was common to both MHHs. The Index of Relative Importance (IRI) indicated that the most important bony fishes in the diet of animals from Mar del Plata were Argentine hake (*Merluccius hubbsi*), anchovy (*Engraulis anchoita*) and striped weakfish (*Cynoscion guatucupa*). At Puerto Quequén, the most common species were raneya (*Raneya fluminensis*) and the rough scad (*Trachurus lathami*), followed by striped weakfish and snapper. In Mar del Plata, the presence of squid (*Illex argentinus*) was also very important.

The diets in both MHH differed in terms of the importance of commercial prey, and the potential provision of food from discharging fleets (Table 21.3). Animals from the Mar del Pata harbor had a diet primarily composed of juveniles of commercial species (IRI = 72.9%), while at Puerto Quequén, sea lion diet was mainly based on a single non-commercial species (Raneya; IRI = 65.9%). Southern sea lions from Mar de Plata also consume fish offal, while in Puerto Quequén natural foraging was dominant. The only two consumed species that included items of commercial size (anchovy and rough scad) had similar importance (Mar del Plata = 19.1%; Puerto Quequén = 14.8%), mainly because the bulk landing favors the fish fall that is used by the sea lions. A component of natural foraging should not be ruled out completely as it is present in natural diets, albeit of low importance (IRI <5%) (Koen Alonso et al. 2000; Romero et al. 2011).

Table 21.3 Sea lion diet in HMH. Prey item are expressed in Index of Relative Importance (IRI %), and the landings of each prey item in the different harbors, expressed as a percentage of the total catch (1989–2019)

Species	Type	Prey size	Mar del Plata (IRI%)	Quequén (IRI%)	Mar del Plata (% catch)	Quequén (% catch)
Merluccius hubbsi (hake)	Commercial	Non commercial	59.8	0.5	52.8 ± 8.5	22.6 ± 28.8
Cysnoscion guatucupa (striped weakfish)	Commercial	Non commercial	12.8	4.8	3.0 ± 1.0	2.6 ± 2.4
Xistreuris rasile (Sole)	Commercial	Non commercial	0.3	4.0	1.5 ± 0.7	8.2 ± 6.0
Umbrina canosai (Argentine croaker)	Commercial	Non commercial	0.1	4.7	0.4 ± 0.3	0.7 ± 0.8
Engraulis anchoita (anchovy)	Commercial	Commercial	17.7	0.2	4.7 ± 1.8	9.6 ± 14.2
Trachurus lathami (Rough scad)	Commercial	Commercial	1.4	14.6	0.04 ± 0.04	0.1 ± 0.3
Raneya fluminensis (raneya)	Non commercial	–	0.8	65.9	–	–
Illex argentinus (long-finned squid)	Commercial	Commercial	99.2	30.3	10.5 ± 6.1	13.2 ± 19.7

21.7 Functional Connections with Breeding Colonies

MHHs are focal concentrations that are socially and spatially segregated (*sensu* Conradt 1998) during the non-breeding season, but during the breeding season there is a consistent displacement of juvenile, subadult and adult males to both Uruguay (~600 km north) and Patagonia (~700 km south) breeding rookeries (Giardino et al. 2016; Fig. 21.5).

Male departure from MHHs is finely tuned to the establishment of territorial male positions in Uruguay (late November) and Patagonia (early December; Vaz Ferreira 1982; Lewis and Ximenez 1983; Campagna 1985). Breeding success was confirmed from MHH animals, as nearly half of them held harems or single females in both peripheral and central positions on the beach (*sensu* Campagna and Le Boeuf 1988).

Sixty-five percent of the trips to the breeding colonies were roundtrips, returning to both MHH after the breeding season. Sea lion trips were relatively consistent in

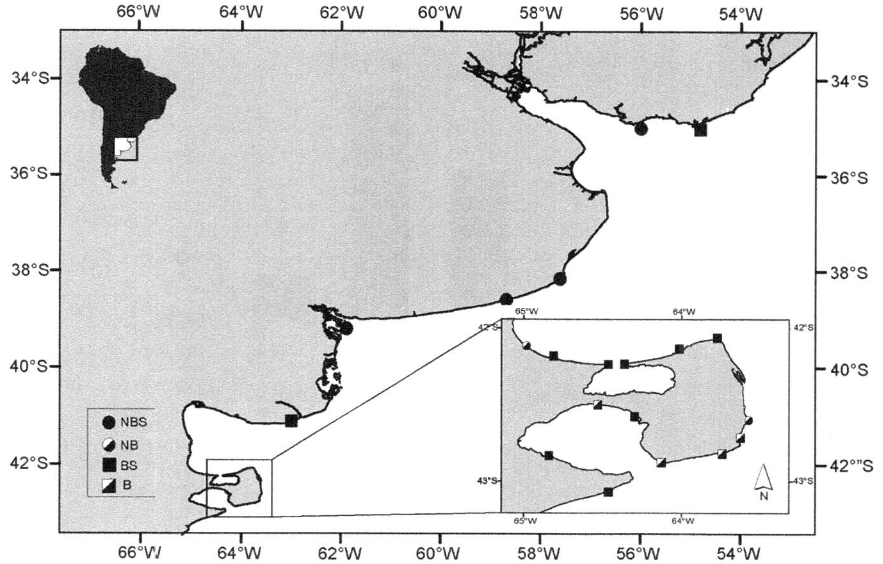

Fig. 21.5 Location of southern sea lion non breeding rookeries at Puerto Quequén (PQ) and Puerto Mar del Plata (PM) in northern Argentina. References: NBS = Non breeding rookeries with resightings of marked animals from PQ-PM. NB = Non breeding rookeries with no resightings. BS = Breeding rookeries with resightings of marked animals from PQ-PM. B = breeding rookeries with no resightings. (after Giardino et al. 2016)

duration (72 ± 26.7 days; range 14–161; n = 325), with no significant effect of age and haul-out location (Giardino et al. 2016). Round trip duration tended to decrease throughout the breeding season, with the longest trips occurring when animals departed from the MHH earlier in the breeding season (Fig. 21.6). On some occasions, the round trip was preceded by a pre-breeding trip in July and August to places where small, and young animal gathered.

No significant effect of age, haul-out or destination rookery was found, showing that the summer flux of animals to the breeding colonies is not restricted to adult sea lions. Multiyear (2–3 years) sightings of a small group of re-marked sea lions confirmed that animals perform roundtrips every breeding season and return to their original wintering MHH. Considering that the number of individuals responsible for the maintenance of the genetic diversity and the evolutionary potential or *effective population size*—of SSLs in the northern Patagonia was reportedly around 9000 animals in 2009 (Grandi et al. 2012), the ca 3000 sea lions wintering in harbor colonies of northern Argentina would be an important pool of Patagonian adult males.

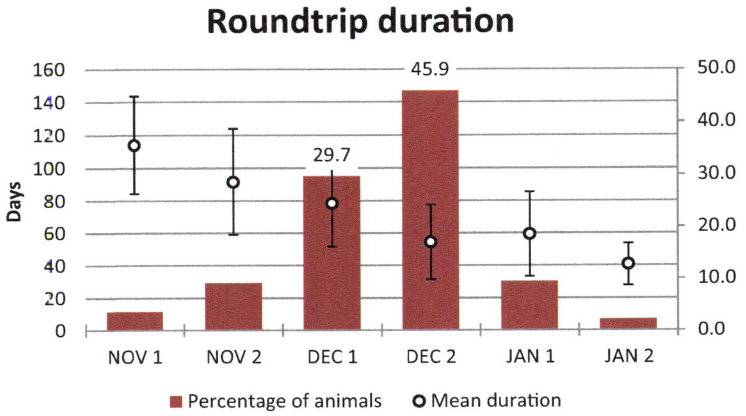

Fig. 21.6 Roundtrip duration and the frequency of abandonment of bleach-marked sea lions in both HMH

Fig. 21.7 High tolerance of sea lions in HMH to human presence. Note placing children, or indeed any naïve person in such close proximity to any wild otariid is considered inappropriate and potentially very dangerous. Photo credit: Julián Bastida

21.8 Social Structure and Individual Association

The existence of stable and abundant male harbor haul-outs in the southern sea lion suggests behavioral plasticity, tolerance of humans, and exploitation of human infrastructure, as well as the ability to learn rapidly, and to socially transmit, this knowledge (Schackner and Blumstein, Chap. 18). The contrast in response to human approach distance in natural areas (ca 25 m; Pavez et al. 2011, 2015) and the lack of response in MHHs—particularly in docks (Fig. 21.7)—emphasizes the importance of habituation in MHH.

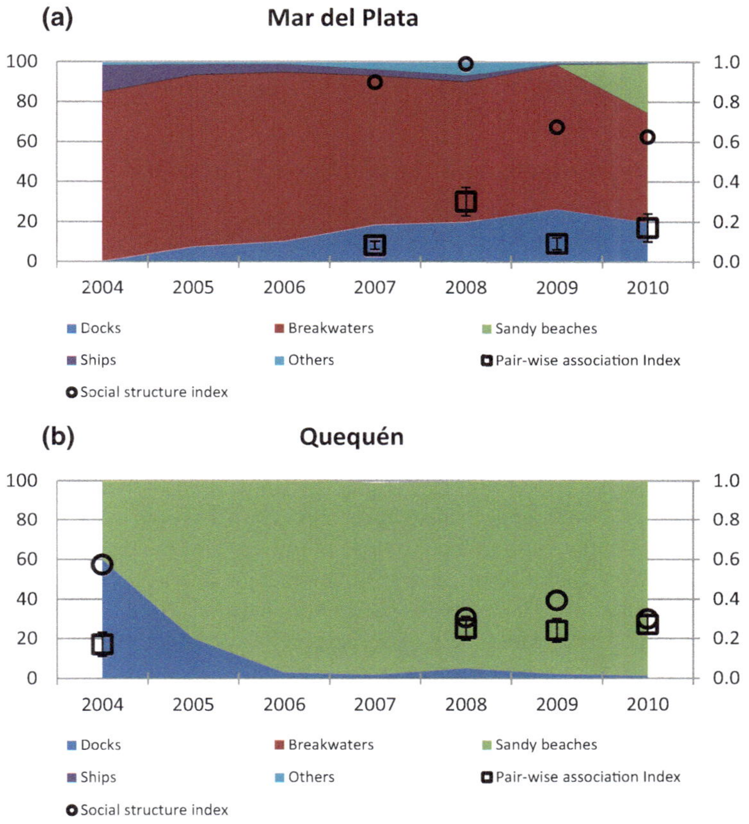

Fig. 21.8 Pairwise association index and social structure index in Mar del Plata (**a**) and Quequén (**b**) HMHs. Left axis indicate the mean annual percentage of sea lions in each area, and the right axis the pair-wise association and social structure indexes

Giardino (2014) conducted a study of association patterns at both MHHs. Pair association occurred if marked individuals were recorded in the same place (sandy beaches, rocky breakwaters or concrete block docks) on the same day. In Mar del Plata many sea lion pairs spent more time together (associated) than expected by chance, resulting in a heterogeneous social structure (*sensu* Whitehead 2008a, b, 2009). At Puerto Quequén, conversely, random association between individuals were found, suggesting homogeneous social structure.

In both MHHs, animals formed mixed-age groups, showing no preference to form age-specific aggregations. No significant differences were found in the association between or within age classes in any of the study years (Giardino 2014). The highest heterogeneity in the social structure (Social differentiation indexes; $S > 0.5$) at both MHHs were found when the animals were mainly concentrated on docks and breakwaters and may be the result of habitat preference or site fidelity rather than genuine social behaviors. Rather homogeneous social groupings (S < 0.4) were found when the main haul-out areas were beaches (Fig. 21.8).

21.9 Activity Budgets and Agonistic Interactions

In MHH southern sea lions, the predominant activity was *Laying down* (following Beentjes 1989), followed by *sitting up* (or "alert"), *grooming*, *thermoregulation*, *social interactions* and *locomotion* (Westergaard 2001; Gana 2016). Regardless of close proximity to humans and the frequency of human activity, the allocation of time in resting behaviors (> 85%) is similar to other species of otariids in natural colonies, with values greater than 60% (New Zealand sea lion: Hawke 1986; Beentjes 1989; New Zealand fur seal: Stirling 1971; Crawley et al. 1977; Johnstone and Davis 1987; Guadalupe fur seal: Pierson 1978; Steller sea lion: Sandegren 1976; subantarctic fur seal Bester 1982; Bester and Rossouw 1994). Under frequent noise exposure—e.g. vessel refloating—there was a significant decrease in the time spent on laying down, and an increase in locomotion, sitting up (alert) and social interactions (Westergaard 2001). Reduced onshore resting behavior due to anthropogenic noise has also been seen in the Australian fur seal (Tripovich et al. 2012; Back et al. 2018).

In breeding contexts, most male-male aggression is associated with defending positions on the beach where the females settle (Campagna and Le Boeuf 1988; see Campagna et al., Chap. 9). The more experienced they are, the longer they can maintain their dominance over a group of females, leading to breeding success. Social interactions in pre-reproductive males may contribute to the development of communication and motor skills, and social abilities to cope with the complexity of an adult life (Harestad and Fisher 1975).

Social interactions between males in MHH are frequent and involve all age classes. Although most of the interactions occur between animals of the same age class, over 30% include sea lions of different age class (Rodríguez 1990; Konigheim 1999). Juvenile and subadults are involved in most social interactions, a similar trend observed in non-breeding colonies of the Steller sea lion and are thought to establish hierarchy relationships with their neighbors and to develop new skills (Harestad and Fisher 1975). Imitation maybe important in one to two year old sea lions, as they seldom interact within the same age class, but actively socialize with senior classes.

A repertoire of 13 types of interactions was recorded from more than 2000 individual interactions observed at MHH (Rodríguez 1990; Konigheim 1999). Only juveniles and subadults executed them all, with fewer performed by the adult males and annuals. A similar pattern of activity—Subadults> juveniles> adults> annual—was found in the Steller sea lion in non-breeding contexts (Harestad and Fisher 1975).

Campagna and Le Boeuf (1988) characterized agonistic interactions between adult males as a combination of threat vocalizations, attack and retreat displays, and fights. Physical contact was restricted to fights and resulted in severe injuries to contestants. The avoidance of direct contact reduces physical damage and saves energy, essential to maintaining harems and enhancing reproductive success. In MHH the proportion of interactions without physical contact increased in frequency in older males. As sea lions grow, they are potentially more dangerous and able to injure opponents more seriously. In southern sea lions less than 10% of agonistic interactions between adult males are violent fights (Campagna and Le Boeuf 1988).

Although no territoriality is performed in MHHs, the attack and retreat display described by Campagna and Le Boeuf was the second most common interaction (ca 20%; Rodríguez 1990; Konigheim 1999). These agonistic interactions occur mainly between sea lions of the same age class (83%) and are predominantly executed by subadults. The low frequency of these interactions between adult males may be indicative of low testosterone levels in non-reproductive contexts. Lower fecal testosterone was found in non-territorial New Zealand fur seal males and associated with a reduction in mean frequency of aggressive intra-sexual interactions (Negro et al. 2010).

21.10 Management Challenges in Harbor Habitats

The presence of hundreds of sea lions inside a harbor creates a full set of positive, negative, dangerous, or simply bothersome interactions that makes management a challenge. Harbor male haul-outs were spontaneously formed and have existed for decades, but no scientifically based management plans have been proposed and educational resources are to date minimal. While these colonies are a great tourist attraction, being must-see self-guided visits in both Mar del Plata and Quequén, sea lions have had adverse impacts upon management. Some types of infrastructure, mainly those made of wood or plastic, have been broken by the weight of large numbers of resting animals. Sea lions' tendency to haul out in groups means that they are often perceived as nuisances by the fishery community as they crowd docks and/or create unpleasant pungent odors. Periods of conflict became associated with the intense use of fishing docks and boat decks as resting areas. Fortuitously, these do decline when sea lions move to other resting areas (Bastida and Rodríguez 2002).

In Mar del Plata, the presence of plastic waste (remains of nets, plastic straps) causes injury or death of many sea lions. A local NGO (Fundación Fauna Argentina) is dedicated to cutting the plastic remains, but this is a long-lasting conflict that is difficult to solve without improved waste management policies by the harbor authorities. The presence of contaminants in sea lion tissues has also been confirmed (Fossi et al. 1997a, b; Marsili et al. 1997; Junín et al. 1998).

The dense aggregation of sea lions and close proximity and interactions with people and dogs creates an opportunity for zoonotic disease transmission and long-term disease maintenance. Seal tuberculosis is present in otariids from Argentina (Bernardelli et al. 1996; Zumarraga et al. 1999; Cousins et al. 2003), and both seal tuberculosis and avian influenza have been reported in Uruguayan fur seal and sea lion colonies (Arbiza et al. 2012). Conflicts with domestic animals have been reported for the southern sea lion (Osman and Pavés 2007), and zoonotic disease transmission from contact between marine mammals and humans includes infections of marine mammals with zoonotic agents, gastritis and localized skin infections in attending veterinarians (Hunt et al. 2008). In both MHHs visitors are often less than 20–30 m from the animals (Mar del Plata harbor) or in direct contact (Puerto Quequén; Figs. 21.1 and 21.2), so they should be considered—and managed—for potential cross infection and as a threat to wildlife, domestic animals and public health.

The age structure, social context and group dynamics of MHHs can facilitate or challenge the application of some management tools. As a non-breeding group, there is no sensitive pupping season that would restrict the visits or port activities, which in turn is favored by the high levels of sea lion habituation. The annual presence facilitates educational tourism planning, as it is a permanent self-guided activity promoted at local level, and only restricted by weather conditions. The high gregariousness and the presence of hundreds of animals in an urban area is a great attraction, taking into account that Mar del Plata and Quequén are very important sun and beach destinations and both harbors are frequently visited by tourists.

High fidelity to resting areas and habituation to human presence makes it difficult to design strategies to avoid intensive use of critical sectors for port activity, and to date the only deterrent strategy has been to set restriction fences. Rapid and spontaneous changes in internal distribution may lead to use of new areas where sea lions are not welcome. An intervention occurred in Mar del Plata harbor in 2004, with a new "rookery" built but never colonized The surface area at the main sea lion haul-out at that time was reduced to force the sea lions to colonize the new "rookery", but instead the animals moved to the docks and resurrected a conflict that had been resolved in the mid-1980s (Bastida and Rodríguez 2002). Paradoxically, the new "rookery" produced sedimentation and expanded an existing small beach, which was quickly colonized in 2010 by a small group of sea lions from Puerto Quequén and in a few weeks ca 90% of the animals moved to this beach as the main resting area (Fig. 21.1).

21.11 Conclusions

Southern sea lions show marked plasticity and fidelity in the use of harbor facilities as haul-outs, emphasizing the importance of habituation to man-made novel habitats. Sequential, rapid and spontaneous changes in the location of aggregations could be a related to the ability to learn rapidly and socially transmit knowledge. Food provision via consumption of fish offal, complemented by natural foraging, could also be a driver to support aggregations of sea lions.

Harbor male haul-outs in northern Argentina are important to functionally connect distinct female genetic breeding stocks from Uruguay and Patagonia with gene flow maintained by male movements (Szapkievich et al. 1999; Túnez et al. 2007, 2010; Feijoo et al. 2011). The predictable pattern of summer dispersion of sea lions and later returning to the original MHH, indicate a high degree of non-breeding site fidelity.

Social interactions between males in MHH are frequent and involve all age classes, including agonistic interactions similar to adult male competitions in breeding colonies but mainly performed by subadult males. The proportion of interactions with physical contact between contenders clearly decreases with age.

The presence of several hundred sea lions inside a harbor makes management a challenge. High fidelity to resting areas and habituation to human presence make it difficult to design strategies to deter sea lions from using critical sectors of port activity. MHH are a must-see tourist attraction with no sensitive pupping season that would restrict the visits or port activities, but plastic entanglement and potential cross infection between sea lions, domestic animals and humans should be managed carefully.

Acknowledgements We would like to acknowledge many colleagues and volunteers for helping in many aspects of the study of the male harbor haul-outs of Mar del Plata and Quequén for more than 30 years. Harbor administrators and coastal guards (M.Bruno, M. Merlini, A.Zufriategui, M. Goicoechea, E.Cerfoglio), lifeguards and rangers (Lopez, G., Leiva, F., Gerardi, E., Bracco, E) and volunteers students (Asselborn, Y., Bagnato, R., Biagiotti, C., Cardozo, M., Castro, A., Chaijale, N., Delli Antoni, M., Fernandez, M., Fernandez de Landa, G., Friedman, I., Gallo, M., Konigheim, B., Leonardi, C., Magrini, C., Mercado, A., Perez Salles, S., Olmedo, L., Piscirelli, J., Pizzorno Buceta, JM., Recayte, MC., Sal Moyano, MP., Temperoni, B., Teryda, N., Vulcano, G., Westergaard, G.) helped in maintaining the monthly census and monitoring. Drs. Cappozzo, HL., Dassis, M., Quse, V., and Elisssamburu, A. provided invaluable ideas for several analysis. Grants were provided by Universidad Nacional de Mar del Plata and CONICET.

Figure 21.5 is reprinted by permission from Springer Nature Customer Service Centre GmbH: Springer Nature, Mammalian Biology, Giardino, G.V., Mandiola, M.A., Bastida, J. et al. Travel for sex: Long-range breeding dispersal and winter haulout fidelity in southern sea lion males. Mamm Biol 81, 89–95 (2016). https://doi.org/10.1016/j.mambio.2014.12.003.

References

Arbiza J, Blanc A, Castro-Ramos M et al (2012) Uruguayan Pinnipeds (*Arctocephalus australis* and *Otaria flavescens*): evidence of influenza virus and Mycobacterium pinnipedii infections. New Approaches to the Study of Marine Mammals. In Tech Open, Rijeka, Croatia, 248, pp 151–182

Back JJ, Hoskins AJ, Kirkwood R et al (2018) Behavioral responses of Australian fur seals to boat approaches at a breeding colony. Nat Conserv 31:35

Baldás MI, Bastida RO, Castello HP (1987) Características generales de un apostadero de lobos marinos de un pelo, *Otaria flavescens*, del puerto de Mar del Plata. Resumos da 2da Reuniao de Trabalho de Especialistas em Mamíferos Aquáticos de América do Sul (Río, 1986), p111

Bastida R, Rodríguez D (2002) Puerto de Mar del Plata: Pacto con Lobos. Vida Silvestre 82:15–19

Beentjes MP (1989) Haul out patterns, site fidelity and activity budgets of male Hooker's sea lions (*Phocarctos hookeri*) on the New Zealand mainland. Mar Mamm Sci 5:281–297

Belkin AN (1966) Summer distribution, stocks, prospects for commercial utilization and certain features of the biology of sea lions inhabiting the Kurile Islands. Iavestiya Tikhookeanskogo N-I (TINRO) 58:69–95

Bernardelli A, Bastida R, Loureiro J et al (1996) Tuberculosis in sea lions and fur seals from the south-western Atlantic coast. Rev Sci Tech 15:985–1005

Bester MN (1982) Distribution, habitat selection and colony types of the Amsterdam Island fur seal *Arctocephalus tropicalis* at Gough Island. J Zool 196:217–231

Bester MN, Rossouw GJ (1994) Time budgets and activity patterns of sub-Antarctic fur seals at Gough Island. Afr Zool 29:168–174

Bradshaw CJ, Lalas C, Perriman L et al (1999) Seasonal oscillation in shore attendance and transience of New Zealand fur seals. Can J Zool 77:814–823

Campagna C (1985) The breeding cycle of the Southern sea lion, *Otaria byronia*. Mar Mamm Sci 1:210–218

Campagna C, Le Boeuf BJ (1988) Reproductive behaviour of southern sea lions. Behaviour 104:233–261

Carrara IS (1952) Lobos marinos, pingüinos y guaneras del litoral marítimo e islas adyacentes de la República Argentina. Cátedra de Higiene e Industrias, Facultad de Ciencias Veterinarias, Universidad Nacional de La Plata. La Plata, Argentina

Conradt L (1998) Measuring the degree of sexual segregation in group-living animals. J Anim Ecol 67:217–226

Cousins DV, Bastida R, Cataldi A et al (2003) Tuberculosis in seals caused by a novel member of the *Mycobacterium tuberculosis* complex: *Mycobacterium pinnipedii* sp. nov. Int J Syst Evol Microbiol 53:1305–1314

Crawley MC, Stark JD, Dodgshun PS (1977) Activity budgets of New Zealand fur seals *Arctocephalus forsteri* during the breeding season. N Z J Mar Freshw Res 11:777–788

Crespo EA, Pedraza SN (1991) Estado actual y tendencia de la población de lobos marinos de un pelo (*Otaria flavescens*) en el litoral norpatagónico. Ecol Aust 1:87–95

Curi Estima S (2002) O leão-marinho, *Otaria flavescens* (Shaw, 1800) (Pinnipedia, Otariidae) no estuário da Lagoa dos Patos. Tese de Bachelerado em Ecologia, Universidade Católica De Pelotas

Dans SL, Crespo EA, Pedraza SN et al (1996) Estructura y tendencia de los apostaderos de lobos marinos de un pelo (*Otaria flavescens*) en el norte de la Patagonia. Informes Técnicos del Plan de Manejo Integrado de la Zona Costera Patagónica 13:1–21

Feijoo M, Lessa EP, Loizaga de Castro R et al (2011) Mitochondrial and microsatellite assessment of population structure of South American sea lion (*Otaria flavescens*) in the Southwestern Atlantic Ocean. Mar Biol 158:1857–1867

Fossi M, Marsili L, Junin M et al (1997a) Use of nondestructive biomarkers and residue analysis to assess the health status of endangered species of pinnipeds in the south-west Atlantic. Mar Pollut Bull 34:157–162

Fossi M, Savelli C, Marsili L et al (1997b) Skin biopsy as a nondestructive tool for the toxicological assessment of endangered populations of pinnipeds: preliminary results on mixed function oxidase in *Otaria flavescens*. Chemosphere 35:1623–1635

Fraker M, Mate B (1999) Seals, sea lions, and salmon in the Pacific Northwest. Conservation and management of marine mammals. Smithsonian Inst. Press, Washington, DC, pp 99–119

Gana JCM (2016). Presupuestos de actividad de lobos marinos de un pelo (*Otaria flavescens*) del Puerto Mar del Plata: Variaciones individuales y ontogenéticas. Tesis de Grado (B.Sci.Hons thesis) Universidad Nacional de Mar del Plata (Argentina)

Giardino GV (2006) Patrones de ingreso y egreso y estimación del tamaño de la colonia de lobos marinos de un pelo de Puerto Quequén. Tesis de grado (B.Sci.Hons thesis). Universidad Nacional de Mar del Plata

Giardino G (2014) Estructura y dinámica de las colonias de lobos marinos de un pelo de la Provincia de Buenos Aires, y su relación con pesquerías de la región. Tesis Doctoral (PhD thesis). Universidad Nacional de Mar del Plata

Giardino GV, Mandiola MA, Bastida J (2016) Travel for sex: long range breeding dispersal and winter haulout fidelity in southern sea lion males. Mammal Biol – Zeitschrift für Säugetierkunde 81:89–95

Giardino G, Bastida J, Mandiola MA et al (2017) Estimated population size of two South American sea lion male haulouts from the northern coast of Argentina. Mammalia 81:197–202

Grandi MF, Dans SL, Crespo EA (2008) Social composition and spatial distribution of colonies in an expanding population of South American sea lions. J Mammal 89:1218–1228

Grandi MF, Dans SL, García NA et al (2010) Growth and age at sexual maturity of South American sea lions. Mamm Biol 75(5):427–436

Grandi MF, de Oliveira LR, Dans SL et al (2012) A hunted population in recovery: effective population size for South American sea lions from Patagonia. Anim Biol 62:433–450

Harestad AS, Fisher HD (1975) Social behavior in a non-pupping colony of Steller sea lions (*Eumetopias jubatus*). Can J Zool 53:1596–1613

Hawke DJ (1986) Observations of Hooker's sea lion, *Phocarctos hookeri*, at a hauling ground on Otago Peninsula, New Zealand. N Z J Mar Freshw Res 20:333–337

Hunt TD, Ziccardi MH, Gulland FM et al (2008) Health risks for marine mammal workers. Dis Aquat Org 81:81–92

Johnstone RM, Davis LS (1987) Activity in a non-breeding colony of New Zealand fur seals, *Arctocephalus forsteri*. N Z J Mar Freshw Res 21:153–155

Junín M, Castello HP, Hollmann P (1998) Efectos de la contaminación crónica por hidrocarburos en el puerto de Mar del Plata (Argentina), sobre una colonia de lobo marino de un pelo, *Otaria flavescens*. Acta Oceanográfica del Pacífico. INOCAR, Ecuador 9:71–78

Kirkwood R, Goldsworthy S (2013) Fur seals and sea lions. CSIRO Publishing

Koen Alonso M, Crespo EA, Pedraza SN et al (2000) Food habits of the South American sea lion, *Otaria flavescens*, off Patagonia, Argentina. Fish Bull 98:250–263

Konigheim BS (1999) Comportamiento social de la colonia de lobos marinos de un pelo del Puerto de Mar del Plata. Tesina de Grano (B.Sci.Hons. thesis). Universidad Nacional de Cordoba (Argentina)

Lewis M, Ximenez I (1983) Dinámica de la población de *Otaria flavescens* en el área de Península Valdés zonas adyacentes (Segunda parte). Contribución No. 79, Centro Nacional Patagónico pp. 21

Mandiola MA, Giardino GV, Bastida R et al (2012) Patrones de asociación individual en la colonia de lobos marinos de un pelo del Puerto de Mar del Plata (Argentina). In: Crespo, E. A., Oliva, D., Dans, S., and Sepúlveda, M. Estado de situación del lobo marino común en su área de distribución. Current situation of the South American sea lion along the distribution range. Universidad de Valparaíso, Chile

Marsili L, Fossi M, Casini S et al (1997) Fingerprint of polycyclic aromatic hydrocarbons in two populations of southern sea lions (*Otaria flavescens*). Chemosphere 34:759–770

Negro SS, Caudron AK, Dubois M et al (2010) Correlation between male social status, testosterone levels, and parasitism in a dimorphic polygynous mammal. PLoS One 5(9):e12507

Osman LP, Pavés H (2007) Conflictos entre animales domésticos, el lobo fino de Juan Fernández (*Arctocephalus philippii*) y el lobo marino común (*Otaria flavescens*). Comisión permanente del Pacífico Sur-CPPS Programa de las Naciones Unidas para el Medio Ambiente-PNUMA:65

Pavanato H, Silva K, Estima S, Monteiro D et al (2013) Occupancy dynamics of South American sea lions in Brazilian haul-outs. Braz J Biol 73:855–862

Pavez G, Muñoz L, Inostroza P et al (2011) Behavioral response of South American sea lion *Otaria flavescens* to tourist disturbance during the breeding season. Rev Biol Mar Oceanogr 46:135–140

Pavez G, Muñoz L, Barilari F et al (2015) Variation in behavioral responses of the South American sea lion to tourism disturbance: implications for tourism management. Mar Mamm Sci 31:427–439

Perlov AS (1980) Age composition and sex ratio in Steller sea lions, *Eumetopias jubatus* (Otariidae) on Kurile Island rookeries. Zoologiske Zhurnal 49:1545–1553

Petracci PF, Sotelo M, Massola V et al (2010) Actualización sobre el estado del apostadero de lobo marino de un pelo sudamericano (*Otaria flavescens*) en la isla Trinidad, estuario de Bahía Blanca, Argentina. Mastozoología Neotropical 17:175–182

Pierson MO (1978) A study of the population dynamics and breeding behavior of the Guadalupe fur seal, *Arctocephalus townsendi*. PhD thesis. University of California, Santa Cruz

Procksch N, Grandi MF, Ott PH, Groch K, Flores PA, Zagonel M et al (2020) The northernmost haulout site of South American sea lions and fur seals in the western South Atlantic. Sci Rep 10 (1):1–15

Raum Suryan LK, Pitcher KW, Calkins DG et al (2002) Dispersal, rookery fidelity, and metapopulation structure of Steller sea lions (*Eumetopias jubatus*) in an increasing and decreasing population in Alaska. Mar Mamm Sci 18:746–764

Riedman M (1990) The pinnipeds: Seals, sea lions and walruses. University of California Press

Robertson BC, Chilvers BL, Duignan PJ (2006) Dispersal of breeding, adult male *Phocarctos hookeri*: implications for disease transmission, population management and species recovery. Biol Conserv 127:227–236

Rodríguez DH (1990) Aspectos biológicos, ecológicos e históricos de la colonia de lobos marinos de un pelo, *Otaria flavescens* (Shaw,1800), el puerto Mar del Plata. Tesis de Grado (B.Sci.Hons thesis). Universidad Nacional de Mar del Plata (Argentina)

Rodríguez DH (1996) Biología y Ecología de los Pinnípedos del Sector Bonaerense. Tesis Doctoral (PhD thesis) Universidad Nacional de Mar del Plata (Argentina)

Rodriguez DH, Bastida RO (1993) The southern sea lion, *Otaria byronia* or *Otaria flavescens*? Mar Mamm Sci 9:372–381

Rodriguez D, Bastida R (1998) Four hundred years in the history of pinniped colonies around Mar del Plata, Argentina. Aquat Conserv Mar Freshwat Ecosyst 8(6):721–735

Romero MA, Dans S, González R et al (2011) Solapamiento trófico entre el lobo marino de un pelo *Otaria flavescens* y la pesquería de arrastre demersal del golfo San Matías, Patagonia, Argentina. Lat Am J Aquat Res 39:344–358

Rosas FWC, Pinedo MC, Marmontel M et al (1994) Seasonal movements of the South American sea lion (*Otaria flavescens* Shaw) off the Rio Grande do Sul coast, Brazil. Mammalia 58:51–59

Sandegren FE (1976) Courtship display, agonistic behavior and social dynamics in the Steller sea lion (*Eumetopias jubatus*). Behaviour LVII:159–172

Shaughnessy P, Chapman P (1984) Commensal Cape fur seals in Cape Town docks. S Afr J Mar Sci 2:81–91

Shaughnessy PD, Bossley M, Nicholls A (2018) Fur seals and sea lions (family Otariidae) on the breakwaters at Adelaide's Outer Harbor, South Australia. Aust Mammal 40:157–161

Silva KG (2004) Os pinípedes no Brasil: ocorrências, estimativas populacionais e conservação (Doctoral dissertation, Ph. D. Thesis. Fundação Universidade Federal do Rio Grande, Rio Grande, Brazil)

Stirling I (1971) Studies on the behavior of the South Australian fur seal (*Arctocephalus forsteri*). Annual cycles, postures and calls, and adult males during the breeding season. Aust J Zool 19:243–266

Szapkievich VA, Cappozzo HL, Crespo EA et al (1999) Genetics relatedness in two southern sea lion (*Otaria flavescens*) rookeries in the southwestern Atlantic. Mamm Biol 64:1–5

Tripovich JS, Hall-Aspland S, Charrier I et al (2012) The behavioural response of Australian fur seals to motor boat noise. PLoS One 7(5):e37228

Túnez JI, Centrón D, Cappozzo HL et al (2007) Geographic distribution and diversity of mitochondrial DNA haplotypes in South American sea lions (*Otaria flavescens*) and fur seals (*Arctocephalus australis*). Mamm Biol 72:193–203

Túnez J, Cappozzo H, Nardelli M et al (2010) Population genetic structure and historical population dynamics of the South American sea lion, *Otaria flavescens*, in north-central Patagonia. Genetica 138:831–841

Vaz Ferreira R (1956) Etologia terrestre de *Arctocephalus australis* (Zimmermann) ("lobo fino") en las islas uruguayas. Trabajos sobre Islas de Lobos y Lobos Marinos (SOYP, Uruguay) 2:22

Vaz Ferreira R (1960) Islas de Lobos y lobos marinos en Uruguay. Boletín Informativo del Departamento Científico Técnico del SOYP (Uruguay) 1:19–25

Vaz Ferreira R (1982) *Otaria flavescens* (Shaw), South American sea lion. FAO Mammal Seas IV:477–496

Westergaard GG (2001) El apostadero de lobos marinos de un pelo de Puerto Quequén: estructura social, dinámica y presupuestos de actividad. Tesis de Grado (B.Sci.Hons., thesis). Universidad Nacional de Mar del Plata (Argentina)
Whitehead H (2008a) Precision and power in the analysis of social structure using associations. Anim Behav 75:1093–1099
Whitehead H (2008b) Analyzing animal societies: quantitative methods for vertebrate social analysis. University of Chicago Press
Whitehead H (2009) SOCPROG programs: analysing animal social structures. Behav Ecol Sociobiol 63:765–778
Zumarraga M, Bernardell A, Bastida R et al (1999) Molecular characterization of mycobacteria isolated from seals. Microbiology 145:2519–2526

Chapter 22
Ethology and Behavioral Ecology of the Walrus (*Odobenus rosmarus*), with Emphasis on Communication and Social Behavior

Edward H. Miller and Anatoly A. Kochnev

Abstract Large tusks characterize the extant walrus (*Odobenus rosmarus*) and its extinct relatives. Those socially selected organs vary between the walrus and fossil relatives, intraspecifically, and between the sexes. Tusks are used in innumerable interactions on land and in water, including fights during rut. "Play fighting" appears even in young calves that lack tusks. Complex sounds resembling those of rutting males underwater and at the water surface occur throughout the year; some are produced by young males. Short-range graded communication (acoustic; tactile; chemical) is important but has scarcely been investigated. Underwater communication within traveling or feeding groups is likely to occur, but also has not been investigated. Specialized integumentary "bosses" on the chests and necks of adult males probably function in optical signaling. Knowledge of movements, diving, feeding, rhythms, time-activity budgets, and effects of weather on behavior has increased greatly; little information is available on associated finer-scale behavioral structure. Field observations on benthic feeding and seabird predation have revealed previously unknown and ecologically interesting behaviors. Walruses are the most gregarious species of pinniped and are almost always in groups in the water and on land or ice, and in extensive body contact with one another. Gregariousness enables huddling for warmth and cultural transmission of information. Many anecdotes from over more than a century suggest more complex social structure than usually

---

Electronic Supplementary Material The online version of this chapter (https://doi.org/10.1007/978-3-030-59184-7_22) contains supplementary material, which is available to authorized users.

E. H. Miller (✉)
Biology Department, Memorial University of Newfoundland, St. John's, NL, Canada
e-mail: tmiller@mun.ca

A. A. Kochnev
Institute of Biological Problems of the North, Far East Branch, Russian Academy of Sciences, Magadan, Russia

Beringia National Park, Provideniya, Russia

C. Campagna, R. Harcourt (eds.), *Ethology and Behavioral Ecology of Otariids and the Odobenid*, Ethology and Behavioral Ecology of Marine Mammals,
https://doi.org/10.1007/978-3-030-59184-7_22

assumed, and the species expresses extensive social play that continues into adulthood. In light of those traits, plus the species' high intelligence and longevity, low reproductive rate, and site fidelity, it seems timely to investigate cultural aspects of the walrus social system.

Keywords Acoustic communication · Feeding behavior · Gregariousness · Maternal behavior · *Odobenus rosmarus* · Optical communication · Play behavior · Social structure · Tactile communication · Walrus · Weapon

22.1 Introduction

Explorers and scientists have reported on characteristics, natural history, and behavior of the walrus (*Odobenus rosmarus*) over centuries. They did so in the scientific contexts of their times, so early descriptions and illustrations have been succeeded by different and increasingly modern and more complete treatments. Before he traveled to the Pribilof Islands, Alaska, Elliott (1881) felt that he would learn little new about the walrus because of the extensive published material that he had consulted. He wrote (1881: 93; Fig. 22.1a)[1]:

> When, therefore, looking for the first time upon the walrus of Bering sea (*sic*), judge of my astonishment as I beheld the animal before me. It was a new species; it was a new creature, or all that had been written by five hundred authors in regard to the appearance and behavior of its Atlantic cousin had been in error.

Knowledge of the walrus has grown enormously since early scientific studies (Chapsky 1936; Belopolsky 1939; Freiman 1941; Nikulin 1941; Vibe 1950; Fay 1955; Mansfield 1958; Loughrey 1959), and through new forms of instrumentation, methods of analysis, and quantitative and modeling techniques. In addition, previously unknown behaviors are being observed in field and laboratory settings and contribute to increasingly comprehensive knowledge of the species' behavioral ecology.

We first discuss the iconic tusks of the walrus, which function mainly as a social organ. Tusk characteristics conform to widespread patterns of socially-selected structures (Sect. 22.2). We comment on behavioral description (Sect. 22.3) and elaborate on this topic in Supplementary Material III. Display behavior of rutting males has been described reasonably well and serves as an introduction to the subject of communication (Sect. 22.4). Most communication is subtle and involves much less conspicuous or stereotyped behavior than that of rutting males; examples are a specialization of the integument for passive optical signaling, and tactile communication (Sect. 22.5; the latter topic is also treated in Sect. 22.6). The extraordinary gregariousness and sociability of walruses are treated in two sections, on

[1]Elliott himself made many errors of observation and interpretation, *e.g.* in regard to food habits (Fay 1982).

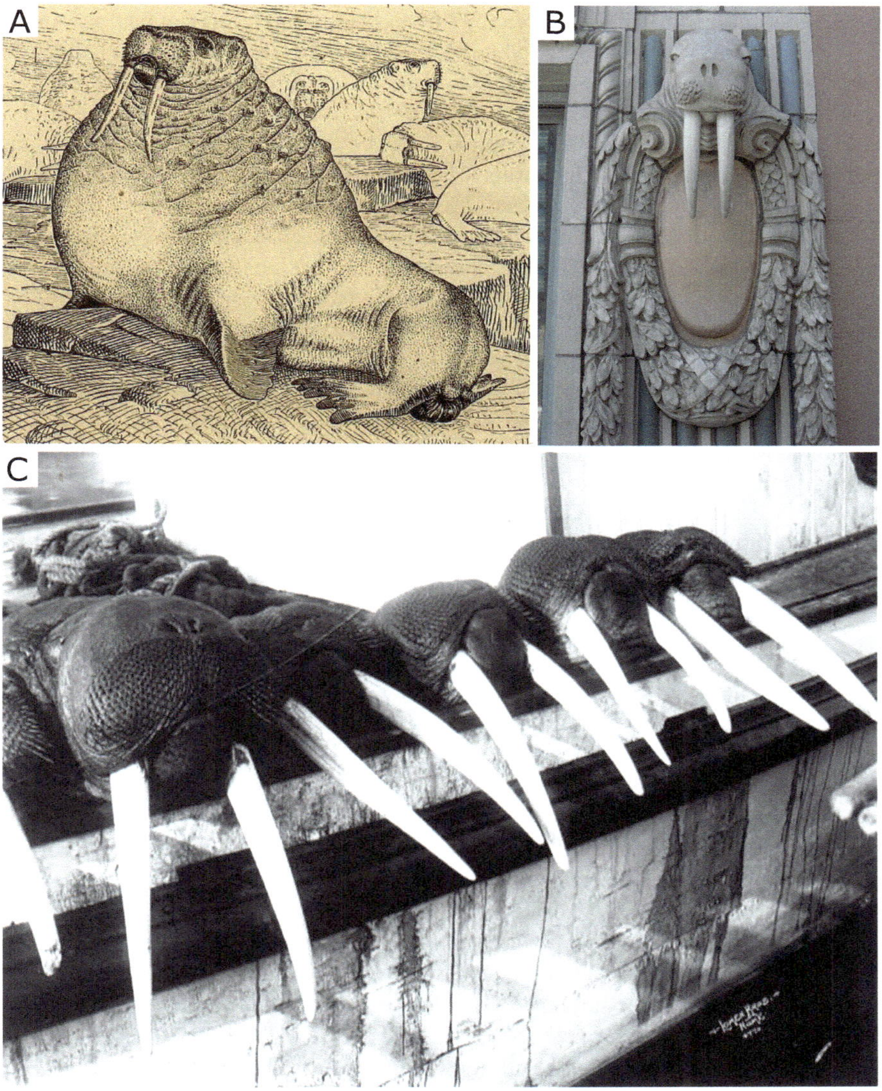

Fig. 22.1 (**a**) Henry Wood Elliott studied the walrus (*Odobenus rosmarus*) at the Pribilof Islands, Alaska. His drawing (**a**) clearly shows the largely naked skin, and large bumps on the neck and shoulder, that characterize adult males. (**b**) The walrus and its tusks are an icon for exploration and other activities in the Arctic. This image shows one of 27 walrus heads that decorate the outside of the Arctic Building (now the Arctic Club Hotel) in Seattle, Washington (the original ivory tusks were replaced with plastic copies in the 1940s). (**c**) Walruses were often killed just for their tusks. These heads were from Pacific walruses killed near Nome, Alaska, sometime between 1894 and 1904. **a**, After figure preceding p. 95 in Elliott (1881); **b**, Wikimedia Commons; **c**, University of Washington Libraries, Special Collections (PH Coll 328.228a; negative number UW12474)

gregariousness and thigmotaxis (Sect. 22.6) and herds and terrestrial haulouts (Sect. 22.7); the social system is discussed in Sect. 22.8. Technical and analytical advances have greatly improved knowledge of walrus behavior over large spatiotemporal scales; we discuss some aspects of movements and diving in Sect. 22.9. The walrus is probably the most specialized pinniped species in terms of feeding ecology and behavior, and in morphological traits related to its specialized trophic habits (Sect. 22.10). The concluding part is Sect. 22.11, in which we make some suggestions about research priorities, with an emphasis on social behavior. We encourage investigators to include descriptive anecdotes, illustrations, and personal impressions in their observations of walrus behavior. Knowledge of behavior contributes in many ways to fields other than science, such as welfare of captive animals; we comment on some such contributions.

In this chapter we emphasize communication and social behavior, and stress the need for information over multiple scales, from motor patterns to social structure. Behavioral studies over small spatiotemporal scales and that pertain to swimming, diving, and feeding are also desirable (*e.g.* locomotory patterns used in travel or diving). We include citations, descriptions, and data from older accounts, particularly from Soviet/Russian research, to emphasize the importance of natural history observations and subjective impressions of experienced observers, in building a broad understanding of the behavioral ecology of walruses.

22.2 Walrus Tusks as Social Organs

Tusks (greatly enlarged upper canines) are the most striking visible trait of the living walrus. The importance of tusks as a trade item led to targeted voyages to the Old World Arctic as early as the ninth century, and may have precipitated European settlement of both Iceland (c. 850–75 CE) and Greenland (c. 980–90 CE; Frei et al. 2015; Dectot 2018). Tusks were mainly shipped as raw tusks to a few centers where they were worked (*e.g.* Norway, Great Britain, and France; Dectot 2018). Carvings found their way throughout Europe, and even to the Middle East and China (Dectot 2018). Some walrus populations were over-exploited in and even extirpated from some places (*e.g.* Iceland; Dectot 2018; Keighley et al. 2019b; Barrett et al. 2020), or exploited over a long time (*e.g.* from eighth to the 20th centuries around the North Water polynya; Gotfredsen et al. 2018). The generation time of the walrus is 15–20 years (Lowry 2016; COSEWIC 2017), so it is possible that evolutionary changes in behavior have occurred in some populations due to exploitation. Here we describe some uses of tusks, the primary one being as a structure for display and striking in agonistic encounters. Like many other socially-selected[2] structures, tusks differ interspecifically, vary much within the species, and differ between the sexes.

[2]Social selection: "an alternative conceptual framework ... in which sexual selection is one component of a more general form of selection resulting from all social interactions" (Tobias et al. 2012: 2274; Lyon and Montgomerie 2012; West-Eberhard 1983, 2014).

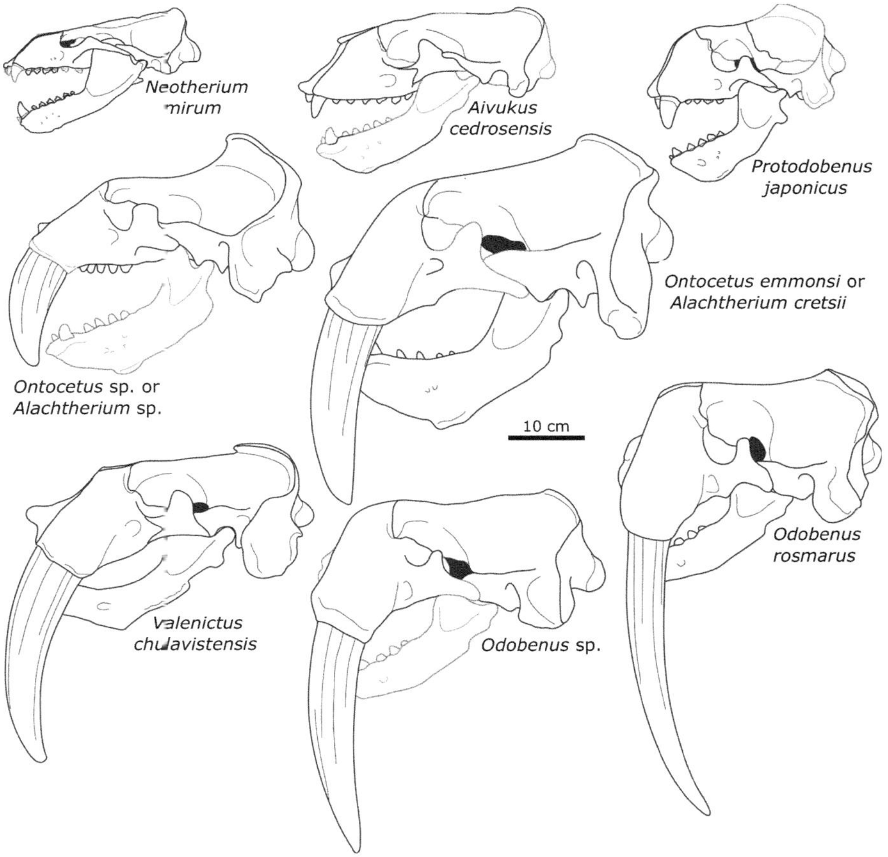

Fig. 22.2 The upper canine teeth varied in size and shape over the evolution of the Odobenidae. Skulls and tusks of some Miocene and Pliocene taxa, plus Pleistocene *Odobenus rosmarus*, are illustrated here. For divergence dates and recorded presence in the fossil record, see Berta et al. (2018). After illustration by R.W. Boessenecker (*in litt.*, 27 March 2020)

Throughout most of the clade's evolutionary history, and in most species, the dentition of Odobenidae was similar to that of their sister clade, the Otariidae. Extreme enlargement of the upper canines evolved independently several times and is a synapomorphy for the clade to which *Odobenus rosmarus* belongs (Odobenini; Boessenecker 2017; Berta et al. 2018; Magallanes et al. 2018; Biewer et al. 2020; Fig. 22.2).

The origins and current adaptive functions of social organs vary across and even within species, and functions almost certainly have changed and diversified even just within the Odobeninae (Stankowich 2011; Cabrera and Stankowich 2020; O'Brien 2019). A dominant old view was that tusks are used for rooting molluscs out of the sea floor, an interpretation that was unsound for multiple reasons (Fay 1981, 1982: 134–135, 1985). Some uses of tusks are as an aid to hauling out or in-air locomotion

Fig. 22.3 The large tusks (upper canines) of the walrus (*Odobenus rosmarus*) serve various non-social functions. (**a**) Tusks assist in locomotion on ice or land; here a thin ailing adult female is using her tusks as a fifth limb, pulling her body up the beach. (**b**) A male resting on land uses his tusks as props. (**c**) An adult male asleep in a breathing hole, holding himself in place by the tusks. (**d**) Most of the males (foreground) in this mixed herd hold themselves in place with their tusks against the ice. (**e**, **f**) Adult walruses threaten polar bears (*Ursus maritimus*) approaching them on sea ice and land. Credits: **a**, **b**, **d**, **f**, Anatoly A. Kochnev; **c**, Eliezer Gurarie; **e**, Svetlana Artemyeva

(probably a major cause of fractures, especially for hauling out on land), when animals on land or ice are resting on their ventral surface, to chop through ice, or to abrade the edges of ice at breathing holes by swinging the head rapidly from side to side (Fig. 22.3a, b; Tomilin and Kibal'chich 1975; Fay 1982: 136–137). Walruses commonly hook the tusks on ice while resting or sleeping in the water, when they serve both as prop and anchor (Fay 1982; Fig. 22.3c, d).

Tusks also serve as weapons in defense against predators, especially the polar bear (*Ursus maritimus*; Fig. 22.3e–h), which they sometimes kill (Freuchen 1921; Krylov 1971; Kiliaan and Stirling 1978; Calvert and Stirling 1990; Kochnev 2001; Thiemann et al. 2008), and the killer whale (*Orcinus orca*; Freuchen 1921; Nikulin 1941; Fay 1982; Born et al. 1995; Kochnev et al. 2008b; Kryukova et al. 2012; Kryukova 2016a; see further). Finally, walruses use the tusks to kill seals for food, and seabirds for food or in play (Freuchen 1935; Lowry and Fay 1984; Seymour et al. 2014; Giljov et al. 2017; see further).

Fig. 22.4 Tusks are used extensively in almost all agonistic and most play interactions between walruses (*Odobenus rosmarus*). (**a**) Typical mutual tusk display between two adult males who are in upright postures on land. The tusks are approximately horizontal in an absolute frame of reference (see text); the oblique orientations of the animals enable them to see one another (Kastelein et al. 1993: Fig. 29B). (**b**) Dominant animal (on the right) orienting his tusks directly toward the head region of the other walrus, who is on his back and stretching away. Here a partner-wise frame of reference needs to be included in the description. (**c**) Dominant walrus (on the left) raises his tusks only slightly while oriented more-or-less directly at the other walrus, who has his tusks raised defensively while placing his right foreflipper against the dominant animal, rearing away, and barking with mystacial vibrissae erected. (**d**) Play-fighting between two young males; note the similarity of the postures to those in **a**. Sources: **a**, **b**, Pacific walrus, Lee Rentz; **c**, Atlantic walrus, Claude Lester; **d**, Laptev walrus, Anatoly A. Kochnev

Bud Fay had studied walruses for about 15 years when he read a landmark publication on the adaptive (behavioral) significance of horn-like structures in mammals (Geist 1966a), which shifted his thoughts about the functional significance of tusks. When he and EHM first met in 1971 he suggested that walrus tusks, like horn-like structures, may function as social organs. They investigated this possibility in a study of males at Round Island, Alaska, over two summers and found strong support for the hypothesis, even though their study took place well outside the period of rut: walruses brandish and strike one another with their tusks in a myriad of agonistic encounters whose outcomes are determined largely by tusk size and intactness, plus body size (Miller 1975a; Tomilin and Kibal'chich 1975; Salter 1979a; Krushinskaya and Lisitsyna 1983; Taggart 1987; Fig. 22.4). In addition, walruses exhibit formalized offensive, defensive, and submissive behavior in relation to the tusks during such encounters (Miller 1975a, b, 1985; Tomilin and Kibal'chich 1975; Krushinskaya and Lisitsyna 1983). The tusks are used similarly and invariably in innumerable agonistic encounters involving both sexes and all ages that take place on land or ice and in the water throughout year, ranging from play-fighting by very young animals to escalated fights between breeding adult males in the water during rut (Miller 1975a, 1985; Ray and Watkins 1975; Fay et al. 1984b; Fig. 22.4). Play-fighting and agonistic encounters appear very early in development, even before tusks are erupted: in agonistic interactions between captive calves only a few months old, one individual struck the others "quite forcibly with the mandibular region" (Brown 1963: 15), and "the games of pups on ice consist in hitting and pushing one another with their heads, necks, and chests" (Krushinskaya and Lisitsyna 1983: 277; Supplementary Material I). Rutting males often strike the tusks forcefully against those of the opponent, which likely contributes to tusk breakage (Fay 1982). Sjare and Stirling (1996: 903) described a fight between adult male Atlantic walruses:

"Most mature males had scars and superficial bleeding wounds on their neck and shoulders, but surprisingly few fights ... were observed and only one of these involved an attending male. In late April 1988, a silent mature satellite male ... closely approached a female herd member. This action incited an intense fight with the attending male ... One male reared out of the water to about midchest height and attempted to drive his tusks into the neck region of the other male. After striking one another two or three times the pair rolled around at or near the surface of the water ... it appeared as though one male was trying to remain on top. The fight lasted about 2.5 min and when it was over the former attending male ... became silent and left the area alone, while the satellite male ... took over his position and sang[3] for at least the rest of the afternoon. Both males had fresh wounds prior to the fight, indicating they had recently been involved in other aggressive interactions (not necessarily with each other)."

The primary functions of walrus tusks thus are social: in intraspecific offense and defense, and in optical signaling during agonistic interactions. In addition, they are visually obvious and hence informative to other individuals even when the bearer is not engaged in overt interactions. Fay (1982: 138) suggested that the shift from piscivory to molluscivory in walrus evolution emancipated the upper canines and they thereafter evolved principally as social organs; convergent evolution of tusks (enlarged upper incisors) in the bottom-feeding tusked odontocete *Odobenocetops* (de Muizon 2018) strengthens this inference (see also Born 2005; Magallanes et al. 2018).

Tusk size and shape vary over multiple levels: across species and populations, and between the sexes (Emlen 2008, 2014). For example, compared with *Ontocetus*, the tusks of *Odobenus rosmarus* are relatively thicker laterally than anteroposteriorly, and are more weakly curved (Fig. 22.5). Traits of tusks also vary across different walrus populations, being "much longer and thicker ... less incurved and more convergent" in Pacific than in Atlantic walruses (Allen 1880: 158; Boeskorov et al. 2018; McLeod et al. 2014; Fig. 22.5b).[4] Finally, tusks vary substantially within populations (Fay 1982; Miller 1991; MacCracken and Benter

[3]"Song" refers to the acoustic displays of rutting males at and below the water surface (Sect. 22.4).

[4]Atlantic and Pacific walruses diverged from one another nearly 1 mya (Andersen et al. 2017).

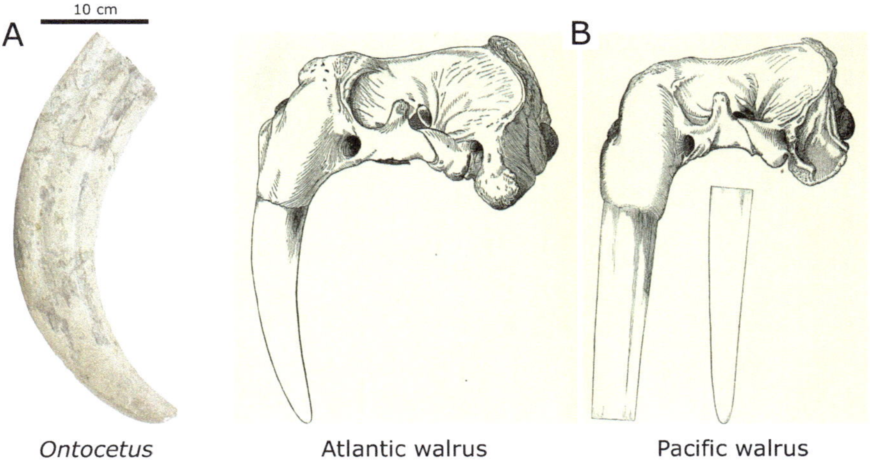

Fig. 22.5 Socially-selected structures like walrus tusks typically vary substantially among related species and across conspecific populations. (**a**) Tusk of Pleistocene specimen of the odobenid *Ontocetus emmonsi* (sister to *Valenictus* + *Odobenus*; see Fig. 22.2), which differs in basal thickness and curvature from the modern walrus *Odobenus rosmarus* (see panel **b** and Fig. 22.6). (**b**) Tusks of Atlantic and Pacific walruses differ in curvature and thickness. All images are shown in left lateral view; the 10-cm scale applies to *Ontocetus* only. (**a**) After Fig. 2 of Boessenecker et al. (2018); (**b**) After Figs. 15 and 14 (respectively) of Allen (1880)

2016). Fractures occur disproportionately in males with curved, highly divergent tusks (Fay 1982). Female walruses also possess tusks.

Sexual differences in social organs are common (Stankowich and Caro 2009), and walrus tusks conform to that trend (Nikulin 1941; Fay 1982; Fay et al. 1984a; Sjare and Stirling 1996; Kryukova 2019; Fig. 22.6). Adult males are 13–19% longer in body length and slightly larger (18–22%) in tusk length; however their tusks are much thicker: >40% on average (Supplementary Material II). Eberhard et al. (2018) and O'Brien et al. (2018) predicted that the size of social organs should generally scale positively on body size; this has not been investigated in the walrus.

Many explanations have been put forward to explain the presence of weapons or ornaments in female mammals, but none applies well to the walrus (Estes 1974, 1991, 2011; Stankowich and Caro 2009). Finally, female walruses protect their offspring from other walruses in densely packed herds through interactions that include tusk displays and strikes (Salter 1979a; Krushinskaya and Lisitsyna 1983; Miller and Boness 1983; Taggart 1987; Trillmich 1996). Females with longer tusks are dominant over females with shorter tusks and they display tusks to each other just like males, *e.g.* as females try to haul out and join a group (L.T. Quakenbush, *in litt.*, 23 April 2020).

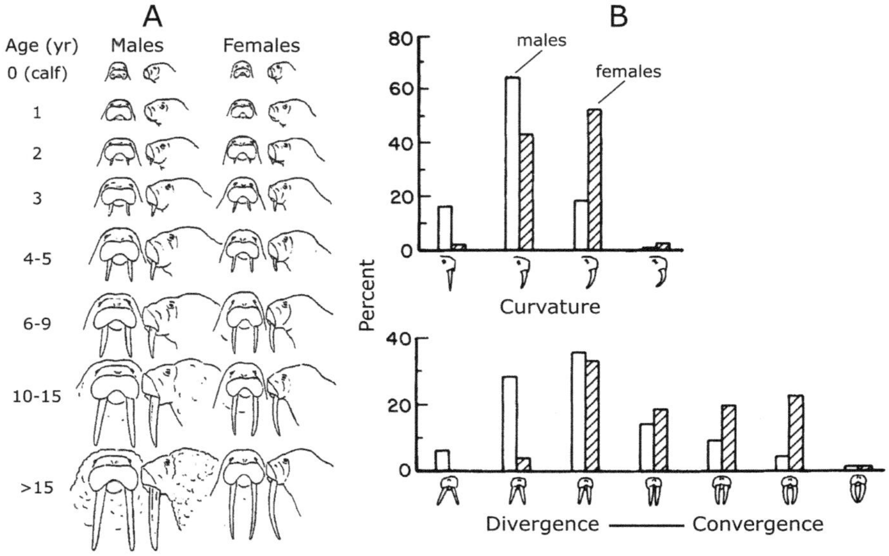

Fig. 22.6 Tusks of the walrus (*Odobenus rosmarus*) differ between the sexes in many ways apart from just size, *e.g.*: male tusks are less curved in lateral aspect (**a**; **b** top) and more divergent in frontal aspect (**a**; **b** bottom). **a**, After Fig. 2 of Fay et al. (1984a); **b**, After Figure 76 of Fay (1982)

22.3 Notes on Behavioral Description

Behavior can be described in many ways, according to the purposes of a study and the kind of behavior being described. We make three points here (see also Supplementary Material III). First, frames of reference (*e.g.* absolute; body-wise; partner-wise) need to be explicit even for descriptions of fairly simple motor patterns (Figs. 22.4, 22.7, 22.8). Second, animals rarely do only one thing at any time (*e.g.* sleeping animals at a haulout also are breathing and in physical contact with others). Third, even seemingly simple kinds of behavior like “sleep” may not be unitary, as they can have different underlying states (Fig. 22.9).

22.4 Walrus Breeding Behavior and Male Displays

The mating systems of pinnipeds have long fascinated biologists. The mating system of the walrus has been of particular interest, due to the species’ sister-group relationship with but completely different ecology from otariids (Fay 1982; Krushinskaya and Lisitsyna 1983; Fay et al. 1984b). However, direct observations of breeding animals were impossible because of the species’ remote breeding sites and the mid to late winter breeding period (Fay et al. 1984b; Born 2003; Sjare et al. 2003). The first observations of breeding behavior by scientists were made in the

Fig. 22.7 The description of different kinds of motor behavior of walrus (*Odobenus rosmarus*) should be explicit with respect to the frame of reference used: the animals are orienting to external structures/stimuli in panels **a** and **b**, and these externalities should be used as points of reference for descriptions: **a**, Male Atlantic walrus resting against a log; **b**, Feeding male Atlantic walrus turning in the water while maintaining contact with the sea floor. Body-wise and partner-wise frames of reference are appropriate for describing motor patterns in **c** and **d**, respectively: **c**, Atlantic walrus scratching with a hind flipper; **d**, Juvenile (likely a 3–4-year-old male) Laptev walrus orients face (with nostrils closed) to investigate newly arriving calf that lolls in the water as it rubs itself on the bottom; the calf's mother (on the right) watches attentively. In all scenarios, descriptions should include details such as medium and posture. Photographers: **a**, Nick Garbutt; **b**, Phil Nicklin/ National Geographic Society; **c**, Russell Millner/Alamy Stock Photo; **d**, Anatoly A. Kochnev

Bering Sea in March 1972 (Ray and Watkins 1975; Fay et al. 1984b). Those observations were opportunistic and short-term. Subsequently, Ian Stirling began a research program on breeding Atlantic walruses in the Canadian High Arctic (Stirling et al. 1987; Sjare 1993; Sjare and Stirling 1991, 1996; Sjare et al. 2003); this remains as the most important study of walrus breeding behavior to date (Supplementary Material IV).

Observations suggest a lek-like mating system in which large adult males display in water and compete with one another, through threats and fights, to be close to females in mixed herds on ice. Males appear to show site fidelity across breeding seasons (Sjare and Stirling 1996, Freitas et al. 2009). Males may stay by a particular group for several days at a time; a dominance-based system among males may prevail. Dominant males close to mixed herds may display for lengthy periods at and below the water surface (see further). Females occasionally enter the water and engage in facial and extensive body contact with males at and below the water surface, including mounting of males, before possible underwater copulation (McCord 2016). Below we focus on male behavior (Schevill et al. 1966; Ray and Watkins 1975; Fay 1982; Stirling et al. 1983, 1987; Sjare 1993; Verboom & Kastelein 1995; Sjare and Stirling 1996; Nowicki et al. 1997; Sjare et al. 2003; Denes 2014).

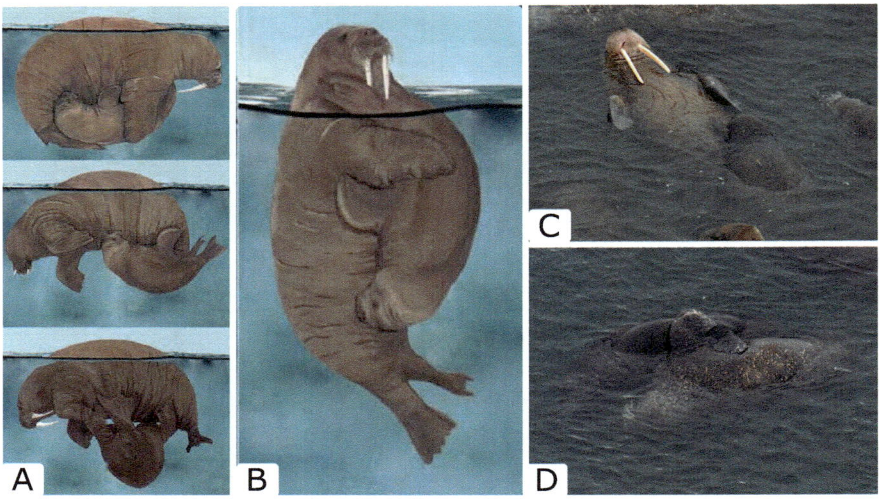

Fig. 22.8 The appropriateness of using a body-wise frame of reference for describing social interactions is exemplified by calf walruses (*Odobenus rosmarus*) nursing underwater. They do so in varied ways, determined by the mother's position, *e.g.* roughly horizontally below the mother (**a**), or approximately vertically upside-down (**b**). They also nurse at the water surface with the female on her back (**c**) or side (**d**). **a**, **b**, after Figs. 22.2 and 22.1 of Kryukova (2017); **c**, **d**, Anatoly A. Kochnev

Rutting adult male Atlantic walruses attending mixed herds dive repeatedly and give diverse sounds between dives at the water surface and under the water (at-surface time ~ 1–2 min long; dives ~4–6 min long; Fig. 22.10); see also Charrier (2021a). At the surface, males produce in-air *whistles* and underwater *knocks* and *taps* with the face submerged; during dives they give *knocks* and *taps* plus a *diving vocalization*, a *bell* sound (occasionally given by itself; Sjare 1993, X. Mouy *in litt.* 6 January 2020), and a *strum* sound.[5] Thus they produce six kinds of sound,[6] and these can be structured as compound sounds in various ways: through repetition of a single sound type, combinations of different sound types, or both (*e.g.* two *knocks* plus a *bell* sound); more complex sequences also occur. The *bell* sound is audible in air. The first example in Fig. 22.10b shows two non-harmonically related frequencies differing by ~300 Hz, and both frequencies differ from the single frequency of the second *bell* sound. Schevill et al. (1966: 104) made similar observations, and implied that two sound sources were responsible for examples like the first pulse in Fig. 22.10b (sometimes "there are two 'bells' involved"; see also Fig. 5d of Stirling et al. 1983). This is in keeping with the presence of paired pharyngeal pouches in

[5]In a captive study, *knocks* and *bells* also were given in air during the rut (Hughes et al. 2011).

[6]Correspondence between some sounds mentioned by Schevill et al. (1966; *e.g.* "rasp"), Gehnrich (1984; *e.g.* "splutter"), and Schusterman and Reichmuth (2008; *e.g.* "moan") with those named by Sjare and colleagues is not clear.

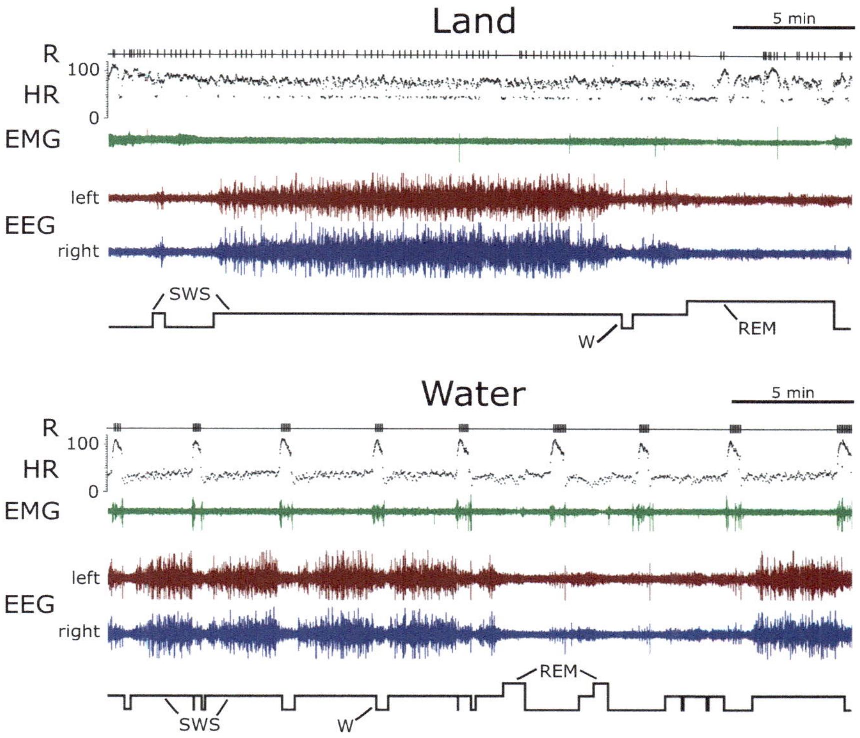

Fig. 22.9 Even seemingly simple categories of behavior in walrus (*Odobenus rosmarus*) like "sleep" can be physiologically complex and can differ physiologically between land and water. In this figure, polygrams at the bottom of each panel show wakefulness (W), slow-wave sleep (SWS), and rapid-eye-movement sleep (REM). Other abbreviations: EEG, electroencephalogram of left and right hemispheres; EMG, electromyogram of neck muscles; HR, instantaneous heart rate (beats-min^{-1}); and R, respiratory acts (breaths). After Fig. 25.8 of Lyamin and Siegel (2019; see also Lyamin et al. 2012)

male walruses, which are presumed to be responsible for the *bell* sound (Fay 1960, 1982; Schevill et al. 1966; Tyack and Miller 2002).[7] Nuptial sounds of rutting male walruses have no counterpart in other species of the Otarioidea, although contrasting airborne and underwater sound repertoires typify ice-breeding pinnipeds (Van Opzeeland and Miksis-Olds 2011).

The size of pharyngeal pouches increases with age in male walruses, and pouches are sometimes present in females (one of four females examined by Fay 1960); this accounts for the production of *bell* sounds by some captive females (Schusterman and Reichmuth 2008). Pharyngeal pouch distension and production of *bell* sounds

[7]McCord (2016: 18) noted that the "neck area" of a captive male expanded "concurrent with a vocal (*sic*) resembling the scraping of a washboard followed by hitting a gong" .

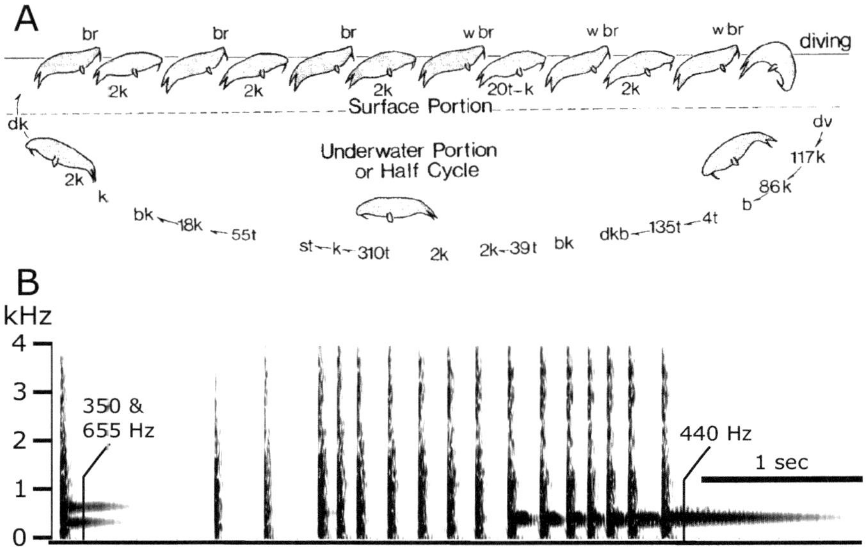

Fig. 22.10 Dominant adult male walruses (*Odobenus rosmarus*) position themselves in the water near mixed herds on ice, where they perform repeated displays at and below the water surface. (**a**) This diagram shows a male at the water surface, alternating breaths (br) and some airborne *whistle* sounds (w), with other kinds of phonation with the head below water: *knock* (k) and *tap* (t). They produce a *diving vocalization* (dv) when they begin a dive, and *bell* sounds (b), and *strum* sounds (st), plus *knocks* and *taps* alone, repeated, or in combination with other sound types (other abbreviations: *double-knock*, dk; *double-knock bell*, dkb) (The *double-knock bell* was later renamed, because it contains three pulses plus a tonal component (associated with the terminal pulse; Sjare 1993: 59)). (**b**) A typical part of the surface and underwater acoustic displays ("*bell knock coda*": I. Stirling *in litt.* 2 January 2020), showing two *bell* sounds, each beginning with a pulse; the second longer *bell* sound also appears to be refreshed (hence lengthened) at each successive pulse. Finally, (**a**) the pulses differ in frequency and number of frequency bands, and (**b**) the two frequency bands in the first *bell* sound are not harmonically related. A, From Fig. 1 of Stirling et al. (1987); (**b**) Spectrogram prepared from Macaulay Library (Cornell University) recording #53276 (Becky Sjare, recordist)

occur commonly in males outside the breeding season, for example close to summer haulouts (Fay 1960; Miller 1975a; Kibal'chich 1978; Krushinskaya and Lisitsyna 1983; Mouy et al. 2012; Rideout et al. 2013), and in captivity (Gehnrich 1984). Pharyngeal pouches also have another function, as males often rest/sleep suspended in the water with pouch(es) inflated (Vibe 1950; Bel'kovich and Yablokov 1961; Born 2005; Kryukova 2016b; see further).

Sjare and Stirling (1996) and Sjare et al. (2003) analyzed structure and variation of male "song", and Denes (2014) detailed fine structure and temporal organization. Some principal findings of Sjare and colleagues were: song sequences range from stereotyped to variable (as noted); songs are individually distinctive, even across years; and song structure varies with social context (*e.g.* males displaying alone *vs.* adjacent to a mixed herd on ice). In addition, many continuous display/singing performances lasted for long periods; for 24 observations of performances that were at least 8 h long, median duration = 13.5 h and the maximum was 81 h (data from Fig. 4 of Sjare et al. 2003). Therefore, adult males compete physically and through optical displays to establish positions close to adult females; they produce complicated and striking acoustic displays at such positions; and they expend considerable time and energy in the displays (though dive durations do not exceed the aerobic limit; Nowicki et al. 1997).

Kibal'chich (1978) stated that the intense *knocks* of male walruses (as young as 3–4 years old) are produced by the lower jaw striking the upper; preliminary results of research by Larsen and Reichmuth (2012) support this idea; publication of their study should further illuminate mechanisms of sound production. The *knock* sound is of high amplitude (177–186 dB; Hughes et al. 2011; Mouy et al. 2012; Denes et al. 2015). Therefore these and other display sounds can carry over considerable distances, which led Stirling et al. (1983) to propose their use in surveys and monitoring. Many techniques have been developed since then (*e.g.* passive acoustic monitoring) and have been used to detect or track walruses even outside the breeding period (Mouy et al. 2012; Denes 2014; Miksis-Olds et al. 2016; Marcoux et al. 2017; Chou et al. 2020; Fig. 22.11); at some sites, songs/song elements are not restricted to the period of rut (*e.g.* the A2W Bering Strait monitoring station of Chou et al. 2020; "a" in Part 2 of Fig. 22.11).

The rutting display of male walruses is a fairly discrete type of display. Display[8] behavior, and behavior more generally, is rarely so simple to describe or enumerate (Supplementary Material V). For example, the airborne vocal repertoire of the walrus is unremarkable in structure, with much acoustic variation and intergradation, as is typical of short-range displays (Kibal'chich and Lisitsyna 1979; Krushinskaya and Lisitsyna 1983; Miller 1985, 1991; Kastelein et al. 1995; Charrier et al. 2010, 2011, 2021a). The high gregariousness of walruses enables other forms of important short-range communication as well.

22.5 Optical and Tactile Communication[9]

Here we discuss integumentary specializations of the walrus for protection against tusk strikes and optical signaling. We then address some general aspects of communication, much of which relies on general and unspecialized sources of information.

[8]In the field of animal communication, the term "display" conventionally refers to acts (often complex and stereotyped) that have evolved through the evolutionary process of ritualization. Overly restrictive use of the term seems undesirable however, and perhaps it is preferable to simply apply the term to acts that "appear to serve a signal or communication function, and to have been evolutionarily 'designed' to that end" (Beer 1977: 156). See Appendix V.

[9]See also Charrier (2021b).

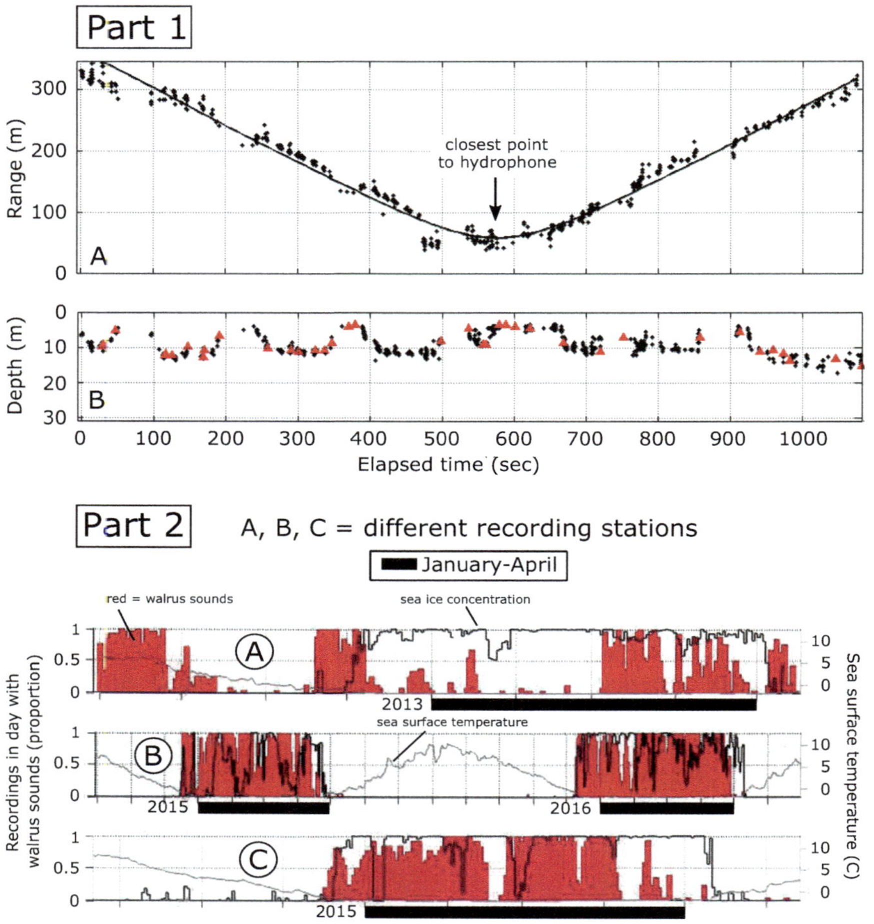

Fig. 22.11 Walruses (*Odobenus rosmarus*) produce underwater sounds throughout the year, not just during the rut. Part 1, movement track and depth of a walrus in the eastern Chukchi Sea, tracked by its display sounds (8 August 2007, hence well outside the rut). The symmetrical track in panel **a** shows that a single animal swam in a straight-line trajectory past the hydrophone. The animal was closest (~60 m) to the hydrophone at about 500–600 s from the beginning of the record and its horizontal velocity was ~0.6 ms^{-1}. During its movement, it uttered both *bell* sounds and *knocks* (red and black symbols, respectively, in panel **b**); *knocks* were uttered at a rate of ~75 min^{-1}. Part 2, long-term records of walrus sounds detected at three recording stations near Gambell and Savoonga (St. Lawrence Island, Alaska), and in the Bering Strait (**a**, **b**, and **c**, respectively). The black bars cover the period of January through April, which includes the period of rut. Note the regional differences in the circannual patterns of sounds. Part 1, After Fig. 9 of Mouy et al. (2012); Part 2, After Figs. 3, 4, and 5 of Chou et al. (2020)

Adaptive thickening of parts of the integument occurs in males of many species of mammal that use behavior like ramming or structures like horns in male-male combat. Thickened areas occur in body areas where contact is normally made in combat. These thickened areas, or *dermal shields*, show recurrent seasonal thickening in some species (Geist 1966b; Sokolov 1982; Jarman 1989; Shadwick et al. 1992); this has not been investigated in the walrus. Dermal shields in adult male elephant seals (*Mirounga*) have been attributed to scarring (Bartholomew 1952; Laws 1953), but seem likely to also have a genetic contribution. In the Pacific walrus, the integument is thickest on the neck and shoulders, thickens with age, and is thicker in males than females (Fay 1982). Integument on the neck and shoulders of mature males is additionally thickened by rounded bosses that are about 1 cm thicker (sometimes more) than the surrounding skin; these bosses increase in number, thickness, and diameter with age and are present in all males (Fay 1982; Figs. 22.12, 22.13c). The bosses have long been regarded as a secondary sexual character unique to males (Chapsky 1936; Nikulin 1941; Freiman 1941; Mansfield 1958; Pedersen 1962; Burns 1965, citations in Fay 1982). Their possible communicative significance is summarized well by Fay (1982: 49):

"The thick, nearly hairless neck of the adult male, with its pale, "lumpy" skin contrasts markedly with the more slender, browner, and smoother neck of the adult female and of immature animals of either sex. Hence, it probably serves also as an important visual cue to other walruses of the sex and maturity of the individual."

Finally, the view that bosses are a secondary sexual character is supported by geographic variation in the structure: they are more strongly developed in the Pacific walrus than in the Atlantic form (Pedersen 1962; Figs. 22.12a, c, e–g, 22.13c). Bosses, like tusks, are informative about sex, sexual maturity, and age of the bearer – general classes of information that are all important in social interactions. Other kinds of information are also available from unspecialized integumentary traits.

Some integumentary traits that are not selected for as signals *per se* provide socially useful information about age or sex available to other animals. At birth, calves are very dark, but this changes even in the first several weeks after birth (Burns 1965; Fay 1982: 44); do walruses behave differently toward calves of different ages, based in part on such characteristics? Old males can be nearly hairless and differ in appearance from younger males of the same size that retain pelage. The appearance of old hairless males can change strikingly due to peripheral blood flow, as individuals change from a pale appearance due to peripheral ischemia when in or just emerging from cold water, to a ruddy "sunburnt" appearance after resting in warm temperatures for some time; is such information available from hairlessness informative and hence useful to other males—do males appraise relative age partly by such appearance, *e.g.* when approached by or interacting with hairless mature males? (Burns 1965; Fay 1982: 44–45; Miller 1991; Fig. 22.13a–c)? Information about vision of the species is needed to assess this possibility, but walruses could detect such a difference simply by brightness, even if the species cannot distinguish color (F.D. Hanke, *in litt.* 12 March 2020).

Tactile behavior has been very poorly described in pinnipeds generally (Miller 1991), and in walruses has been described only incidentally and in general terms (Kastelein et al. 2015). This is ironic because body contact is so pronounced in the species, and ranges from brief to lengthy structured touching between interacting individuals, to protracted contact of much of the body surface between adjacent

Fig. 22.12 Sexual differences in the walrus (*Odobenus rosmarus*) are expressed in many attributes, including tusk size and shape (Fig. 22.6), and external appearance. Sexual differences in appearance include the greater size and more massive neck and shoulders of old adult males (**a**, **c**), and in the "more slender, browner, and smoother neck of the adult female and of immature animals" (Fay 1982: 49; **b**, **d**). Raised "bosses" on the chest, neck, and shoulders of old males are generally more strongly developed in the Pacific walrus (**a**) than in the Atlantic form (**c**), but extremes are similar (**a**, G; see also Fig. 22.13c). The bosses grow over years; examples from a few small to many large bosses, and increasing loss of epidermal pigment, are shown in panels **e–f–g** (Barents Sea). (**a**) Pacific walrus, Klaus Steinkamp/Alamy Stock Photo; (**b**) Pacific walrus, Sarah A. Sonsthagen, U.S. Geological Survey; (**c**) Atlantic walrus, Andy Williams/Muench Workshops; (**d**) Atlantic walrus, agefotostock/Alamy Stock Photo; (**e**, **f**, **g**) Atlantic walrus, Anatoly A. Kochnev

animals in herds on land or ice. Examples of patterned touching are: in agonistic interactions, when one or both (but especially the subordinate) animals place a foreflipper against one another, to maintain control and distance (Miller 1975a; Fig. 22.4c); when calves are on the back of the mother or the mother and other adults, in rest or having climbed there to avoid disturbance (Fig. 22.14a); in nursing between offspring and mother on ice or land, at the water surface, or under the water

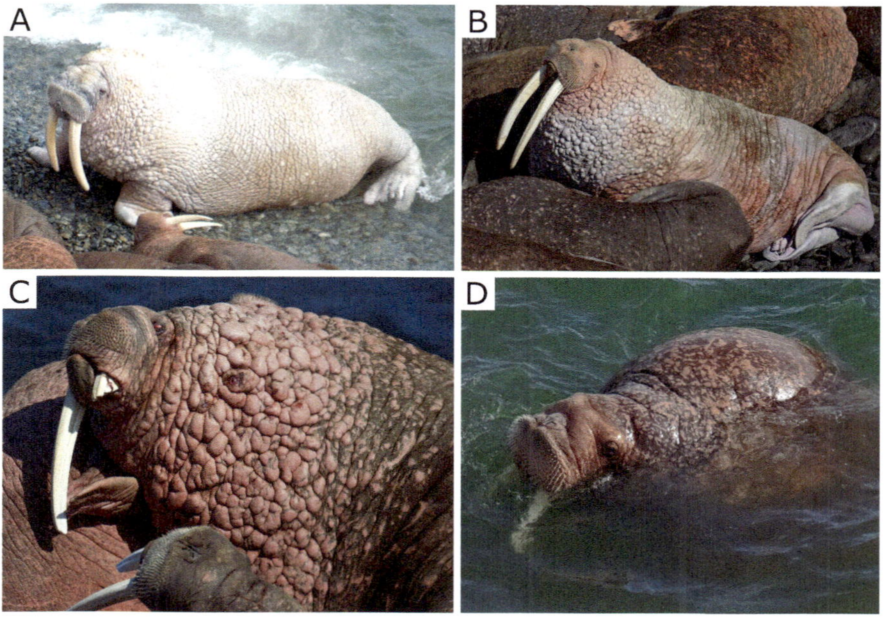

Fig. 22.13 Adult male walruses (*Odobenus rosmarus*) are nearly hairless. (**a**) In the summer, when they initially haul out, the skin is strikingly pale due to ischemia. This gradually (**b**) changes to a completely ruddy appearance coloration after they have been hauled out for some time (**c**). Therefore information about both physiological state and age is available to other individuals from the appearance of the skin of old males. (**d**) Near summer haulouts, male walruses spend much time in the water producing *knock*, *bell* and other sounds with a pharyngeal pouch inflated. (**a**) Hal Brindley/halbrindley.com; (**b–d**) Lee Rentz

(Fig. 22.8); and in nuzzling, when an animal may place its mystacial pad against another animal (often this is mutual), with or without open nostrils (it occurs frequently under the water, for example; Figs. 22.7d, 22.12b, d, 22.14b–d).[10] Body contact or touching occur in innumerable other circumstances as well (*e.g.* Fig. 22.14e). Touching behavior has been investigated extensively in humans and other primates, and many of the questions asked in research on human tactile ("haptic") communication could be adapted for studying this behavior in walruses; for example (Thayer 1986: 13):

> "Consider ... the stimulus qualities of touch which can vary and alter the meaning of a touch: duration, frequency, intensity, breadth, continuity, rhythm, and sequence. Finally, consider the body parts involved, the settings in which touch occurs, the relation of touch to

---

[10]Fay (1982: 53) noted large sweat glands "in the anterior surface of the snout ... [and] ... associated with groups of small hairs in the spaces between the mystacial vibrissae", and suggested that these might be important in chemical signaling. The mystacial vibrissae of walrus are numerous, mobile, and highly sensitive (Yablokov and Klevezal' 1964; Ling 1977; Kastelein et al. 1989, 1990, 1991a; Kastelein & van Gaalen 1988; Milne et al. 2020).

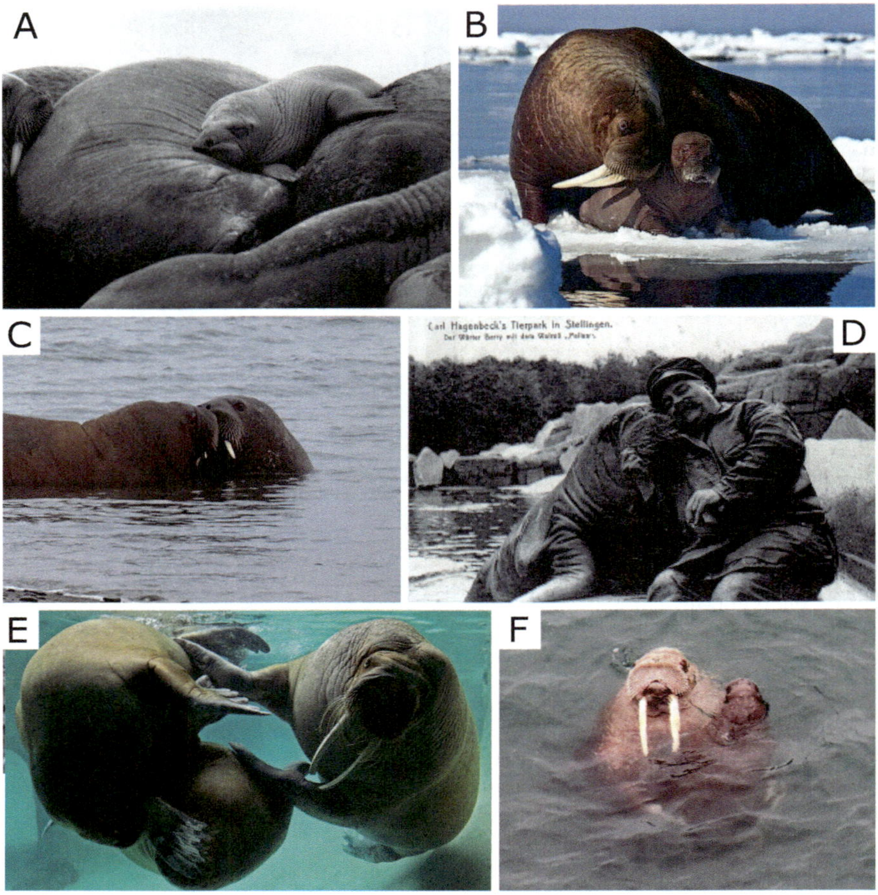

Fig. 22.14 Patterned touching takes many forms in the walrus (*Odobenus rosmarus*), as when a calf rest on top of its mother or other herd members (**a**); or when walruses nuzzle others (**b**, female nuzzles her calf; **c**, face-to-face nuzzling between two young males (note open nostrils of animal on the right)). Captive walruses also nuzzle and otherwise touch humans with whom they have social bonds (**d**). Walruses touch one another extensively under the water (**e**), and mothers and calves commonly are in extensive body contact with one another (**f**). **a**, After Figure 4.38 of Miller (1991); **b**, Rodney Ungwiluk Jr.; **c**, Anatoly A. Kochnev; **d**, Archiv Hagenbeck, Hamburg (see Svanberg 2010); **e**, The Seattle Times; **f**, After Karenina et al. (2017)

other communication signals, who initiates touch, whether touch is reciprocated, whether an expected touch is omitted, how a touch is responded to, and the relationship and roles of the individuals involved."

Fig. 22.15 Walruses (*Odobenus rosmarus*) are gregarious throughout the annual cycle and are invariably in body contact with other group members when hauled out. (**a**) Small group on sea ice, Chukchi Sea. (**b**) Large herd in the process of hauling out near Point Lay, Alaska, September 2014. Note the discreteness of the herd, due to animals seeking close proximity and contact with one another. (**c**) Group of males floating in the water with pharyngeal pouches inflated (Vankarem Cape, Chukchi Sea, Russia; August 2011). **a**, After Fig. 10 of Beatty et al. (2019); **b**, Photographer, Corey Accardo; **c**, Photographer, Natalia V. Kryukova

22.6 Gregariousness and Thigmotaxis

Characteristics, behavior, and adaptive significance of gregariousness in pinnipeds vary greatly. The walrus can be considered as the most gregarious pinniped for several reasons: group sizes can be enormous; walruses are gregarious throughout the year; and walruses are invariably in groups, whether hauled out on land or ice, resting in the water, traveling, or feeding (Pedersen 1962; Fay and Ray 1968; Fay 1981, 1982, 1985; Fig. 22.15). It is important to have baseline information about herd behavior and characteristics because, in addition to inter- and intra-annual variations due to ecological factors, many aspects of walrus herds and their ecology are being affected increasingly both directly and indirectly by loss of sea ice due to climate change (predicted effects are not necessarily negative; Born 2005).

An obvious possible benefit of group living in the walrus pertains to locating patchily distributed food resources. Learning must be an important factor in facilitating the return to areas where walruses have fed in previous years (Born 2005); cultural knowledge and transmission may thus contribute substantially to this behavior.

Another possible advantage of gregariousness and group living is detection of predators.[11] In the case of polar bears, walruses at a terrestrial haulout in herds of a hundred or more may respond by fleeing when they detect a bear about 60 m away, or closer (Laptev walruses, and perhaps Atlantic walruses, are much calmer in the presence of polar bears than are Pacific walruses). Most walruses do not detect the bear directly, but instead respond to the behavior of other walruses and flee in response.[12] Reactions to killer whales differ. Walruses detect carnivorous *Orca* up to several km away, presumably by the whales' sounds. In response, all walruses swim quickly toward one another, resulting in a dense group. Killer whales do not approach such large groups and prefer to pursue small groups or single individuals. Hence the gregariousness of walruses facilitates predator detection in the case of polar bears, and protection (through predator discouragement) in the case of killer whales. Photographs of the reaction of walruses to killer whales are in Sect. 22.11.

Gregariousness also enables body contact that seems likely to afford thermal benefits.[13] These are likely to be very important in the walrus, as they are substantial even in the low-latitude California sea lion (*Zalophus californianus*; Liwanag et al. 2014; Gilbert et al. 2010). Thermal benefits to body contact must be greatest in the winter and especially important for very young walruses, which have a thin insulative blubber layer (Liwanag et al. 2012). Walruses can regulate behaviorally how much surface and which body surfaces are exposed to the air *via* adjustments of body posture and positions of the flippers (Fay and Ray 1968). Additionally, most walruses in groups on ice or land are in extensive body contact with one another and, in cold weather, most calves are "brooded ... against the mother's breast, between her forelimbs, and so completely concealed and sheltered that [a calf's] presence was

[11]The following notes are based on observations of AAK, who saw >100 walrus-polar bear interactions over nearly four decades.

[12]The behavior of fleeing because other herd members are doing so is a generalized response that occurs in many circumstances, *e.g.* "disturbances can be caused by as little as a raven [*Corvus corax*] cawing" (M. Winfree in Robards and Garlich-Miller 2013: 26). Of course, a raven's call in turn could be a reaction to the presence of a polar bear.

[13]Advantages to being in a central position in a herd are suggested by Fig. 2c of Lydersen et al. (2012).

not detected until the mother became alarmed and began to flee" (Fay and Ray 1968: 4); in this position about half of the calf's body surface is against its mother. In addition, as noted, calves often rest atop the mother or other walruses, which provides a warmer surface than the ice (Fig. 22.14a). Walruses huddle extensively throughout the year, indicating a strong social component to the behavior. Huddling behavior (and touching behavior generally) of the walrus is important energetically and socially, so merits detailed study (Miller 1991; Gilbert et al. 2010).

Regardless of the adaptive advantages to gregariousness and (separately) thigmotaxis, the affinity of walruses for one another's close company is apparent from the facts that they (a) are rarely alone, and (b) form closely packed groups (regardless of whether the groups are small or large; Krushinskaya and Lisitsyna 1983). These behavioral tendencies have been well described by the walrus biologist Tony Fischbach[14]: "walruses love to be next to each other"; and "as soon as they're concerned about something their first response is to turn to their companions and ... sniff them and nudge them". Over larger distances, walruses are strongly attracted to and move toward distant airborne sounds from herds,[15] which can carry considerable distances (Nyholm 1975); *e.g.* (Tomilin and Kibal'chich 1975: 5):

"... walruses resting on ice floes roar from time to time; this roar resounds through the air and attracts the attention of other walruses. The roar may be far enough away as to be still inaudible to the human ear, but the animals have perceived the call and head for the meeting points, orientating themselves by the direction from which the roar came. It may be that roars emitted from an ice floe are propagated through the water and thus perceived by those animals that are in the water. Upon hearing the call, feeding walruses interrupt their activities and head toward the assembly point (hunters note this direction when they are searching for walrus assemblages). Upon arrival at the floe from which the roar came, the new arrivals climb out of the water and join the others. If there is insufficient room on that ice floe, they climb onto the next one. Gregariousness is accompanied by synchronized behavior in groups (small or large) in the water as they travel or feed. In addition to just mothers and their dependent young, synchronous feeding occurs in larger groups of females and calves (Fay 1982). Tomilin and Kibal'chich (1975) observed small feeding groups of 2–3 animals at Wrangel Island diving and surfacing synchronously; for traveling groups in Greenland, Freuchen (1935: 6; 1921) noted that "whether there are many or only a few they always dive and surface together".

Gregariousness in general, and the habit of forming large gatherings on land at traditional sites in some parts of the range, made walruses vulnerable to hunting by some indigenous populations; the impact of exploitation varied greatly

---

[14]https://www.youtube.com/watch?v=pF-aNYhCr8k

[15]Many observers have commented on the great distances over which airborne sounds of walruses can be transmitted, especially in calm conditions and fog (*e.g.* Pedersen 1962). Krushinskaya and Lisitsyna (1983: 264) remarked that even an injured or seriously ill individual may remain separate from herds "but ... always stays within earshot of the herd's sound signals"; Kryukova and Ivanov (2012) made similar observations.

geographically, and also was affected by which other marine mammals were being hunted and which hunting technology was used (Drew et al. 2016; Hill 2011). The same traits predisposed the species to overexploitation and even extinction in certain areas after "discovery" by Europeans, who extirpated the distinctive populations of Iceland and that of Québec (Gulf of St. Lawrence) plus the Maritime Provinces of Canada (Fay et al. 1984a; Chugunkov 1991; Dyke et al. 1999; McLeod et al. 2014; Lydersen 2018; Keighley et al. 2019b; Barrett et al. 2020).

22.7 Herds and Terrestrial Haulouts

Size and composition of walrus groups vary generally according to substrate (smallest in the water, larger on ice, and largest on land; Fay 1982, 1985), size of area available,[16] and stage of the annual cycle. The species mainly associates with moving pack ice over continental shelves where they feed. When ice is not available they will haul out on land (though this tendency may be overstated in the literature; Born 2005). Most populations are migratory, in accord with seasonal changes in the distribution of ice. For much of the year, adult males and mixed herds containing females and dependent young are segregated from one another (Fay 1982, 1985; Gjertz and Wiig 1994; Born et al. 1997; Lydersen et al. 2008; Monson et al. 2013). Group size and composition are important structural elements of a species' social system (see further), and more information about herds on ice and in the water is desirable.

Sites of terrestrial haulout are well documented throughout most of their range (Fischbach et al. 2016; Lindqvist et al. 2016; Heide-Jørgensen et al. 2017; Semenova et al. 2019). A comprehensive review of historical and recent records of the Pacific walrus in Russia and the United States for the period 1852–2016 provides a valuable resource for long-term monitoring (Fischbach et al. 2016). Some haulout sites have been in use since they were first encountered by Europeans,[17] whereas others have formed sporadically in recent decades at times when sea ice was unavailable. When the herds contain many females with calves, substantial mortality from crushing can occur. This has been observed since the 1960s in the Russian part of the Chukchi Sea (Kochnev 2002) and, with the reduction of ice cover since the 1990s, happens almost annually (AAK, unpubl.).

Large mixed herds of Pacific walruses are also forming on land more frequently in the absence of sea ice in the eastern part of the range. Monson et al. (2013) studied size, composition, and age-class assortment in mixed herds (mainly females and young animals) at haulouts in late summer. Walruses were not distributed randomly within the herds: dependent young in year classes 0, 1, and 2 tended to be closest to

[16]See photographs of confined beaches of different sizes in Kryukova et al. (2019).

[17]Use of traditional sites and the size of aggregations were strongly affected by European exploitation (Fay et al. 1984a).

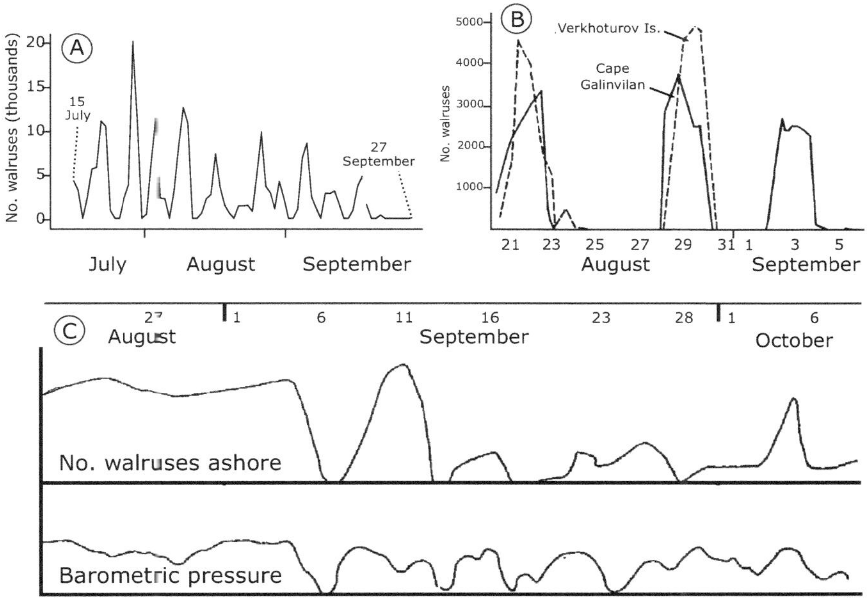

Fig. 22.16 Herds of walruses (*Odobenus rosmarus*) at terrestrial haulouts often display pronounced and rapid changes in numbers (**a**) and apparent synchrony across haulout sites in patterns of their presence and absence (**b**). Many early observations tied the presence or absence of walruses on land to weather, subjectively and through simple correlational analyses (**c**). **a**, Erratic changes in numbers at the Meyechkyn haulout site in the Gulf of Anadyr, Russia, in 1999. **b**, Nearly simultaneous changes in numbers hauled out at two Russian haulout sites that are separated by about 80 km; daily counts were made at each site over the period 21 August to 4 September 1987. **c**, Concurrent temporal trends in the number of walruses ashore and several meteorological measures at a haulout on Arakamchechen Island, Chukotska, for the period 20 August–10 September 1977. **a**, After Fig. 3 of Smirnov et al. (2000); **b**, After Fig. 2 of Semenov et al. (1988); **c**, After Fig. 2 of Kibal'chich (1988)

the water, a trend that strengthened over time as herds remained ashore. Socially subordinate walruses are over-represented peripherally in herds of summering male walruses (Miller 1976); non-random sex and age assortment in mixed summering herds that include males also has been noted in the Atlantic walrus, with females and dependent young closest to the water (Loughrey 1959; Krushinskaya and Lisitsyna 1983; Miller and Boness 1983); Vishnevskaya and Bychkov (1985, 1990) observed the same pattern, and tied it to the behavior of the females: mothers with first- or second-year calves did not attempt to enter the herd, but stayed at its periphery. In his 1977 study at Cape Kagynin, Kibal'chich (1978) noted that sick animals mainly occupied positions at the side of the haulout, and Freuchen (1935: 238) noted complete segregation of "bulls" and "cows" at a terrestrial haulout. Finally, at Russian haulouts, females and calves are more common on rocky sites, and males are more common on sandy beaches (Kochnev 2013).

Walrus aggregations on land often form and dissolve suddenly and erratically, a pattern that sometimes seems synchronous even across fairly distant haulout sites (Yablokov and Bel'kovich 1963; Vishnevskaya and Bychkov 1985, 1990; Taggart 1987; Chugunkov 1991; Kochnev 2001; Lydersen et al. 2008; Kryukova et al. 2019; Fig. 22.16a).[18] The most parsimonious explanation of synchrony at different nearby sites is that walruses react similarly to common meteorological or sea conditions. Many early observers noted that weather affects presence on land (Belopolsky 1939; Nikulin 1941). Yablokov and Bel'kovich (1963), Kryukova (2012), and Kryukova et al. (2019) stressed the influence of wind; Kibal'chich (1978, 1988), Vishnevskaya and Bychkov (1985, 1990), and Hills (1992) noted the importance of air pressure, with departure associated with falling pressure and return to land occurring with increasing pressure (Fig. 22.16b). More recently, rigorous quantitative assessments of the relationships of haulout behavior to weather have identified wind speed and air temperature as important influences on haulout behavior, in addition to a diel pattern (Born and Knutsen 1997; Smirnov et al. 2000; Lydersen et al. 2008; Udevitz et al. 2009; Hamilton et al. 2015; Jay et al. 2017).[19] The patterns are not uniform around the year, as temperature is an important factor mainly in winter, when mature males also spend much time in the water during the period of breeding (Hamilton et al. 2015). Haulout patterns and activity budgets also vary regionally and inter-annually due in part to ice availability and location (Garde et al. 2018). Behavioral observations at haulout sites are indispensable for understanding higher-level attributes of herds.

Aggregations on land appear to be established by adults who are initially cautious and halting in hauling out; the presence of some walruses on land appears to embolden others (Nikulin 1947; Chugunkov 1991; Supplementary Material VI). In mixed herds with only a small proportion of males, the males haul out first and females with young haul out several hours later (Kochnev 1999). Nikulin (1941: 43, see also Salter 1979b; Fay et al. 1984a) remarked on how animals become calmer over time after hauling out:

> "After they get out on the firm ground, walrus behave very restlessly for the first few hours and it is very easy to frighten them at this time. But after lying for certain time they react very little or not at all to the approach of men or any foreign object"

Comparable information for the formation of herds on ice appears to be unavailable.

Individual walruses show site fidelity to particular haulout sites within and between years (Born and Knutsen 1997; Born et al. 2005; Jay and Hills 2005; Lydersen and Kovacs 2014). There is some evidence that individuals tagged at the

[18] In her detailed study, conducted for five months in each of five successive years, Hills (1992) observed no synchrony between two haulout sites separated by about 100 km.

[19] The animals' behavior is not directly observable in such studies, so "haulout" must be defined operationally. For example, Hamilton et al. (2015) defined the onset and termination of haulout as how long a sensor was continuously dry ($\geq$15 min) or wet ($\geq$40 s), respectively. Operational definitions vary across studies, and direct observations also would probably result in slightly different findings.

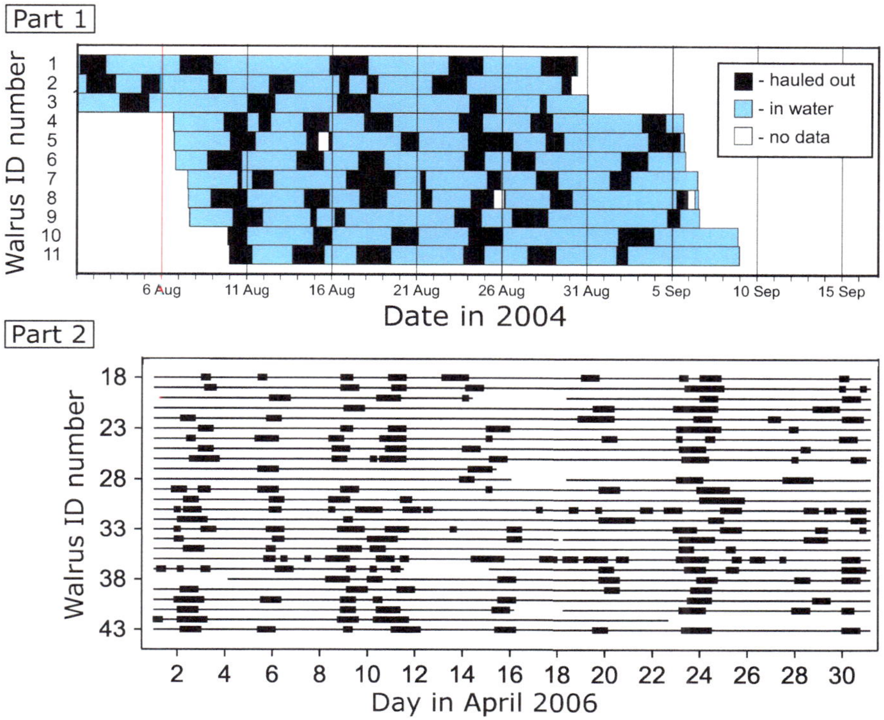

Fig. 22.17 Walruses (*Odobenus rosmarus*) in herds exhibit synchrony in patterns of haulout on land (male Atlantic walruses; Part 1) and ice (female Pacific walruses; Part 2). The patterns may be explained largely as responses to weather, but social factors likely also contribute. Assessment of synchrony depends on temporal scale, which differs between the two parts of this figure. Part 1, after Fig. 3 of Lydersen et al. (2008); Part 2, after Fig. 2 of Udevitz et al. (2009)

same time at a haulout site are synchronous in their at-sea/on-land activities thereafter (Fig. 22.17). This pattern could be explained most parsimoniously by similar behavioral reactions of individual animals to the same weather and sea conditions, but a social component may be involved and should be investigated, especially in small populations.

22.8 Social System

Walrus society can be considered within the framework of Kappeler (2019), who suggested that the term "social system" be standardized to include four core elements: mating system, social organization, social structure, and care system. We have adapted his treatment of the latter three elements for this part of the discussion:

Fig. 22.18 Females and their offspring are the core social unit in the walrus (*Odobenus rosmarus*), and remain together for years during and after weaning (**a**). For much of the year they associate in mixed herds containing other females and calves, plus young males (**b**). Multiple females and their calves sometimes associate closely, enabling calves to play among themselves (**c**). Photographs are of Pacific walruses. **a**, **b**, Sarah A. Sonsthagen, U.S. Geological Survey. **c**, Anatoly A. Kochnev

Social organization refers to three basic features of a social unit. Two of these are (i) *group size* and (ii) *group composition* (age; sex; reproductive status); many data on these features exist and more can be obtained for the walrus. The core social unit of walruses is the mother-calf pair, as they are inseparable throughout lactation and remain together for years thereafter (females longer than males: Fay 1981, 1982; Fig. 22.18). Fay (1982: 138) stated that "After [weaning] young males stay for 2 or 3 years longer with the adult females before joining all-male herds; the young females tend to stay with the adult females continuously". Genetic analysis has refined this picture: analysis of samples from animals >2 years old in mixed herds disclosed slightly higher mean relatedness among herd members than expected by chance, and suggested that 3-year-old juveniles may associate and travel with the mother whereas offspring ≥4 years of age are wholly independent (Beatty et al. 2020). The long period of association between females and calves (especially female calves), coupled with strong individual recognition, could enable future "preferential interactions" between them (Trillmich 1996: 566).

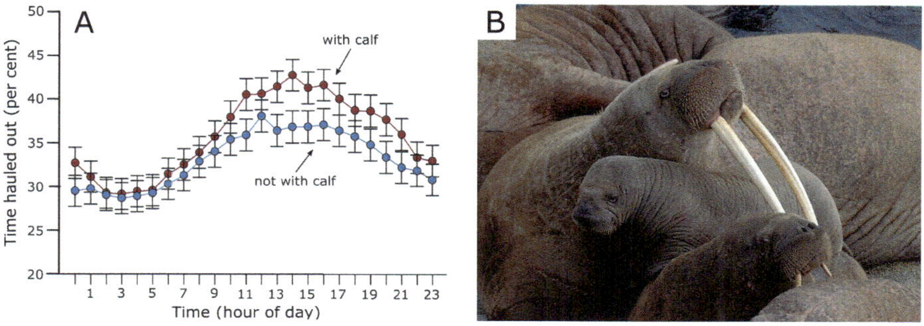

Fig. 22.19 Costs of motherhood: (**a**) female walruses (*Odobenus rosmarus*) with calves spend more time hauled out on ice (and hence not feeding) than do females without calves. Data for the graph are from tagged females in the Alaskan portion of the Chukchi Sea, June–September 2013–2015. Error bars represent 95% confidence intervals. (**b**) Females and calves often assume positions that could be interpreted as protective of the calf. **a**, After Figure 24 of Quakenbush et al. (2016); **b**, Anatoly A. Kochnev

Female walruses and their offspring travel and presumably feed together over their long period of association. Indications of high costs of walrus maternal care over this time include: observations of mothers defending or carrying their young; females with calves hauling out longer than females without calves; long female-calf bond; and high energetic costs of lactation (Kochnev et al. 2008a; Citta et al. 2014; Noren et al. 2016; Quakenbush et al. 2016; Fig. 22.19); see also Sepúlveda and Harcourt (2021).

Costs to female Pacific walruses with calves presumably are increasing due to global warming, because more feeding is taking place at greater depths (Kochnev 2004). Vishnevskaya and Bychkov (1985, 1990) noted that walruses with young-of-the-year hauled out on land and returned to the sea more often than did other adults, and interpreted this as reflecting their higher energetic requirements; they also observed that females with young-of-the-year occurred disproportionately in small mixed groups. Trillmich (1996) cautioned that energetic measures reflect maternal expenditure, not maternal investment; those measures are not static but vary greatly throughout the long reproductive lifetimes of female walruses. Female Pacific walruses are predicted to rest and forage less due to future changes in the distribution and amount of sea ice, which could have negative effects on multiple aspects of reproduction if females cannot replenish their endogenous reserves through winter feeding (Udevitz et al. 2017).

For these reasons, mixed herds outside the breeding period may be structured partly on (iii) *genetic relationships* (Beatty et al. 2019, 2020).[20] These three features

[20]Fay (1981: 16) felt that this probably was not the case: "Apparently, the individual does not necessarily associate with the same retinue of companions and must re-affirm his or her social status continually in each new situation." In contrast, Krushinskaya and Lisitsyna (1983: 267, 271) assumed that groups of adult females and young animals (up to the age of 7 years) consisted of related individuals, but commented for mature males that "ties of kinship are scarcely possible".

provide a purely structural description that does not refer to or imply anything about how individuals in groups interact with each other. In addition, different age and sex classes of walruses are not distributed evenly across the range at any season. Such influences affect the composition of and hence the social dynamics within herds.

2. *Social structure* concerns the nature of social relationships within a group, as reflected in interactions among individuals (*e.g.* frequency or nature of social interactions). Characteristics of this element are usually positively related to high longevity and slow reproductive rate, as in the walrus. In walruses, the kind and frequency of interactions are influenced by age, sex, body size, and the female-offspring bond. Krushinskaya and Lisitsyna (1983) stressed the strength and nature of the female-calf bond, with young calves in essentially continuous body contact with their mothers from birth, whether hauled out or in the water, and with extensive vocal communication between them. Is walrus society also shaped by individual familiarity or kinship apart from females and their offspring? Does it reflect dominance relationships that emerge from repeated interactions involving animals familiar with one another in a group? Apart from females and their offspring, do individuals preferentially associate with other individuals?

3. *Care structure* refers to all forms of care. It includes maternal care but also care provided by kin or unrelated individuals, including seemingly altruistic behavior. For example, group defense against a polar bear has been observed (Stirling 1984, plus citations in Taggart 1987). This element too is often positively related to life-history traits of the species. Information on care is anecdotal but consistent: some reports suggest that young calves may be attended by other adult females when the mother is feeding; and many published descriptive accounts suggest that aid-giving is common. For example, many observers have noted that older walruses will attack hunters, especially when other group members are threatened or injured; Burns (1965) interpreted many (not all) of such observations as animals returning to young animals in response to their vocalizations (Supplementary Material I, VII; Bel'kovich and Yablokov 1961; Krushinskaya and Lisitsyna 1983). Care-giving and "mutual assistance" have been noted mainly in mixed herds of females and juveniles (Krushinskaya and Lisitsyna 1983); females were not seen to exhibit aggression toward first-year pups in herds dominated by that age class (Vishnevskaya and Bychkov 1990).

Behavioral development is intimately bound with social life in mammals, especially in group-living long-lived species like the walrus. It has not been studied specifically in the walrus, though diverse play behavior occurs throughout the year over all ages and in both sexes (see also Llamazares-Martín and Palagi 2021). It includes individual and social play with seabirds (wounded or and dead); play-fighting on land, ice, or in the water; mounting and apparent copulation; and diverse contact and rolling behavior (Miller 1975a; Vishnevskaya and Bychkov 1990; Giljov et al. 2017; Figs. 22.20, 22.21). Sometimes several females and their calves associate for some time, with the females moving together while monitoring the calves who play with one another (Fig. 22.18c). The most extensive descriptions of play are in Krushinskaya and Lisitsyna (1983; Supplementary Material I); here is an excerpt (p. 278):

Fig. 22.20 Pacific walruses (*Odobenus rosmarus*) actively pursue, eat, and play with seabirds. (**a**) Young walrus (probably a 3–4-year-old male) pursuing a juvenile pelagic cormorant (*Phalacrocorax pelagicus*). (**b**) Male walrus (6–9 years old) about to strike a juvenile black-legged kittiwake (*Rissa tridactyla*) after upending it by rising quickly beneath it from below the water surface. (**c**) Mixed-age group of walruses approaching a juvenile glaucous gull (*Larus hyperboreus*), with a juvenile male in the lead. (**d**) Juvenile walrus "offers" the carcass of a juvenile black-legged kittiwake to an adult male (foreground), perhaps as an invitation to play. After Figs. 22.1, 22.2, and 22.3 of Giljov et al. (2017)

Fig. 22.21 Play in the walrus (*Odobenus rosmarus*). (**a**) Playing between two young males, when they were generally in extensive body contact while twisting and turning at and below the water surface. (**b**) Multiple males form a "water carousel" (see text); note also one male mounting another—slightly above and to the left of the carousel. Photographs, Anatoly A. Kochnev

> "Males of different ages very often play in the water next to a costal rookery. The game is usually played by two, or less frequently, three animals, which can be of the same size, or one larger than the other. The playing pair surface simultaneously, and the animals touch with their vibrissae, then they dive, also simultaneously, and underwater one can see the walruses turning, so that a ring is formed in which the snout of one animal almost touches the rear flippers of the other. . ."

It is apparent that the social system of the walrus is poorly known and patchily known. Some information on social structure can come from focused investigations on particular topics like tactile communication or inter-individual associations (*e.g.* between individual animals at summer haulouts). Key information that will illuminate social processes can come from jointly investigating genetic structure, age/sex composition, and stability of mixed herds over the species' range and through the annual cycle. Observations on social behavior mentioned above and in previous sections, and in Supplementary Material VII, are strikingly similar to some features of odontocete social systems, *e.g.*: fission-fusion dynamics; cooperative behavior; and aid-giving (Gowans 2019; McHugh 2019). Structure and dynamics of odontocete societies therefore may provide some direction to future research on walruses.

22.9 Movements and Diving: Descriptions Across Scales

Walruses travel great distances, feed on the sea floor in frigid water, and spend much time in areas that are remote or impossible to visit. For these reasons, researchers have tried to deploy instrumentation attached to walruses for many years. The first successful attachment of tags to walruses appears to have been in 1954; none of the 115 tags was recovered (Mansfield 1958; see Born et al. 1995: Fig. 14). Soviet biologists attached tags to 500 Pacific walruses in 1961 (Krylov 1965); none of those tags had been recovered as of 1976 (Krylov 1965: translator's note). Later workers deployed diverse sophisticated instruments and used other techniques such as passive acoustic monitoring to document movements and diving. The investigations have yielded rich information on movements (sometimes over very long distances); travel time; diving; influences of weather on behavior; time-activity budgets; and within and between-year fidelity to terrestrial haulout sites, feeding grounds, and winter breeding areas; and other topics (Born and Knutsen 1992; Hills 1992; Wiig et al. 1993; Mouy et al. 2012; Rideout et al. 2013; Dietz et al. 2014; Marcoux et al. 2017; Semenova et al. 2019; Chou et al. 2020; *e.g.* Fig. 22.17). Most investigations of this kind have operated over large scales, so it is useful to consider smaller-scale studies that ask different questions, and provide complementary information that increases understanding across scales (Krupnik and Ray 2007; Schneider 2009; Robards and Garlich-Miller 2013; Kiszka et al. 2015). Swimming speed can serve as an example.

Many estimates of swimming speed from tracking studies are derived from long records (*e.g.*: 3.1 km h^{-1}; Born and Knutsen 1992: 281[21]). Heide-Jørgensen et al. (2017) gave estimates of 2.3 and 1.9 km h^{-1} for records of $\leq$5 h and over a longer period (up to 24 h), respectively; for a 20-min sample of a walrus moving in a straight line, Mouy et al. (2012) estimated speed at <1 km h^{-1} for 12- and 18-min records of walruses swimming in a straight line. Averages from long records will usually be lower than from short records, because animals rarely travel in straight lines for long periods, and they engage in other kinds of behavior than just traveling. Estimates of swimming velocity through direct observation are much greater: Loughrey (1959) estimated maximal swimming speed as 11–12 km h^{-1}; and Fay (1981: 14; 1982) estimated "normal cruising speed" at about 7 (up to 10) and "maximal 'sprint' speed" as $\geq$35 km h^{-1}. More important than the quantitative differences, the differing estimates for the different scales inform about different behavioral processes and differ in biological significance. Large-scale estimates may be useful in energetics analysis, bioenergetics modeling, or in relating travel velocity to sea-ice conditions; estimates at small scales can inform about whether swimming walruses can escape from a polar bear or killer whale, for example. Diving behavior is another example of scale differences.

Walruses dive when they travel to and feed on the sea floor. Most investigations of diving have concerned feeding and have interpreted dive characteristics in that light. Lowther et al. (2015) analyzed dive data for adult males in summer and winter, and noted an increase in shallow dives in the latter period. They interpreted the trend as an effect of rutting behavior, when males engage in many shallow dives during their displays. So-called "surface dives" are not observed directly so must be inferred from attached instruments and defined operationally. Wiig et al. (1993) considered "surface" to be $\leq$2 m from the water surface and "bottom" as $\geq$85% of a dive's maximal depth; Garde et al. (2018) considered a dive to occur at >2 m; Lowther et al. (2015) recognized dives as submergences >4 m deep; and Jay et al. (2001) modified a criterion for maximal depth used by others. Analysis of coarse behavioral categories is also necessarily applied to data on rates of ascent and descent, and on various properties of feeding dives (*e.g.* do quantitatively defined classes of dives used in different studies mirror natural classes of behavior?). Observations on motor patterns of walruses while swimming at the surface, in non-feeding dives, and during ascent and ascent, would illuminate all these matters. Basic knowledge of anatomy and motor patterns is available as a starting point for such investigations.

Anatomical traits associated with and motor patterns used by walruses swimming at and just below the water surface have been well described (Ray 1963; Fay 1981; English 1975; Gordon 1981, 1982; Pierce et al. 2011). Hind limbs are the major source of propulsion; the forelimbs are used mainly for maneuvering, and also as hydroplanes, to initiate dives, *etc.* (Gordon 1981, 1982). Motor patterns and basic quantitative aspects of locomotion during sustained traveling, or during ascent or descent from dives, have not been described in detail, although Levermann et al.

[21]The estimate of 3.08 is the grand mean of the seven individual means in their Table 6.

(2003: third page) made direct observations of feeding walruses and pointedly stated that "walruses dived directly to the bottom to feed, and when finished they went straight to the surface for air" (the motor behavior used, such as position or movements of flippers, *etc.*, remain undescribed; more descriptive details of behavior from that publication are in Supplementary Material VIII). Other aspects of motor patterns are important for understanding energetic costs: do the form of locomotion and energetic costs differ between large bulky adult males and more slender adult females, or between females with and without calves, or over development from calf to adulthood? Parenthetically, information on contextual details is invaluable in reporting on directly observed behavior, such as whether animals are alone or in groups (and if so, what is the group size?), whether a female is alone or diving with her offspring, and the age of her offspring.[22]

22.10 Feeding

Abrasion patterns on tusks and mystacial vibrissae, and the strongly keratinized dorsal margin of the mystacial pad, suggested to Fay (1982 and earlier) that walruses root in the sea floor for their invertebrate prey. Sam Stoker, a Ph.D. student at the University of Alaska, made the first direct observations of feeding furrows in 1972 from an icebreaker-supported mini-submarine[23]; furrow characteristics supported Fay's inference (Ray et al. 2006). Further studies have provided more details about the structure and meaning of feeding signs.[24] As feeding walruses move forward on the sea floor (see further) they encounter prey, which they excavate in particular ways according to the type of prey that they encounter, and often leave identifying shells beside the excavation pits (Oliver et al. 1983; Nelson and Johnson 1987; Fig. 22.22a).

Detailed knowledge of the structure of feeding pits and how they are formed, coupled with knowledge of which prey species that were excavated, have enabled the use of excavations to be used as proxies for the prey that were eaten. The information has been used at different scales of analysis: Nelson et al. (1984, 1987) could resolve furrows but not pits; Bornhold et al. (2005: 296) could resolve furrows and pits, but not different kinds of pits. Nelson et al. (1987) described variation in the tracks of furrows and noted that the persistence (hence detectability) of furrows over time likely is affected by bottom-current water velocity and grain size of the substrate. Details of feeding behavior have been refined over time.

---

[22]Some authors have suggested that very young calves cannot dive with their mothers (*e.g.* Loughrey 1959), but even very young calves dive with (and are carried by) their mothers (Krushinskaya and Lisitsyna 1983). Calves have accelerated development of physiological traits that are important to diving at a young age (Noren et al. 2015; Noren and Edwards 2020).

[23]This was the same cruise on which the first observations on rutting behavior were made (Ray and Watkins 1975).

[24]Such behavioral traces are termed *ichnofossils* in paleontology.

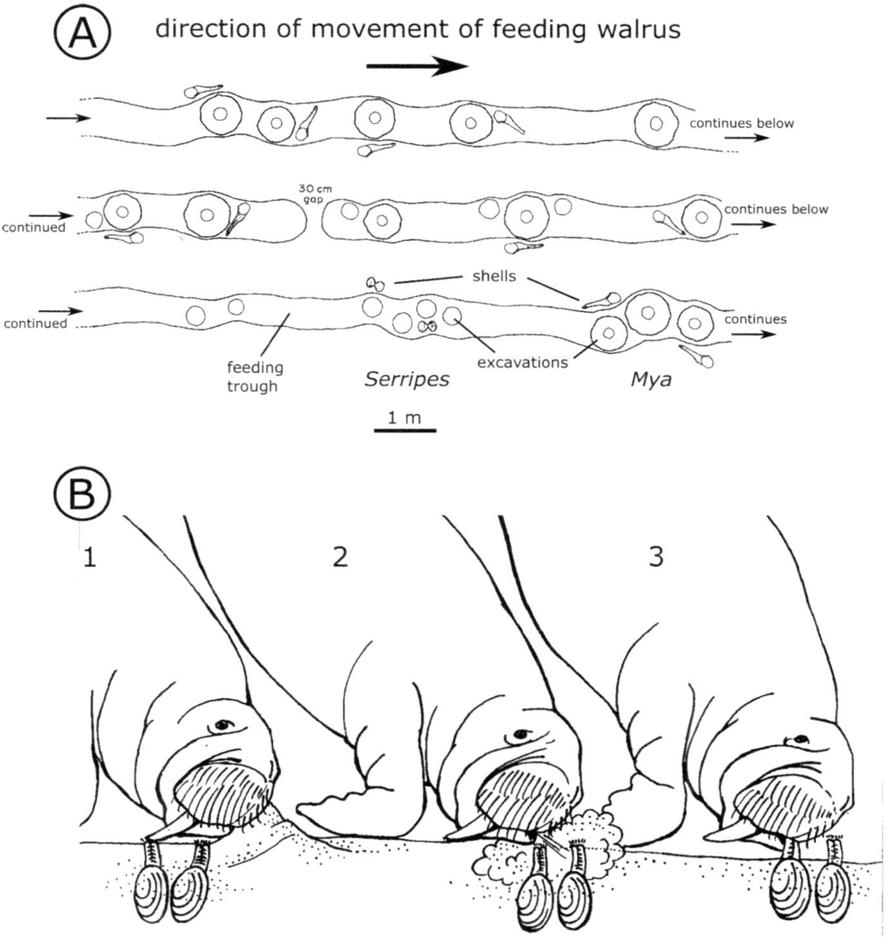

Fig. 22.22 (**a**) Walruses (*Odobenus rosmarus*) leave distinctive signs of feeding on the sea floor. They root in the substrate and, when they encounter prey, excavate them, and then suck out the soft parts and discard the shells. The excavations differ across prey types (here two are depicted: *Serripes* and *Mya*). (**b**) Sequence of feeding behavior by a captive walrus: 1, rooting in the substrate; 2, jetting with water to expose the food items (*Mya arenaria*); and 3, sucking the siphons from the *Mya* and ingesting them. Placement of the foreflippers on the substrate while feeding may have been related to the shallow placement of the prey items in the experiment; in the wild, prey are deeper in the substrate and flippers are needed to disperse overlying sediment (Levermann et al. 2003). **a**, After Fig. 3 of Oliver et al. (1983); **b**, After Fig. 2 of Kastelein and Mosterd (1989)

Vibe (1950: 34) offered the suggestion that walruses suck out the soft parts of molluscs and eject the shells: "[maybe] the Walrus takes the mussel into its mouth or between the lips, sucking out the firm portions, whereupon the rest of the mussel is spit out again . . . It is the general opinion among the Polar Eskimos that the Walrus in this way sucks out the mussel, spitting out the shells." The first direct observations of feeding behavior were made on a captive animal (Kastelein and Mosterd 1989: 3–4; Fig. 22.22b):

> "The animals showed the following behaviour. They positioned themselves on their front flippers on the sand, with their body at an angle of 30–90 degrees with the bottom surface . . . They slowly moved snout-first along the bottom, exploring it with their eyes open. The animals rooted in the sand with the upper edge of the snout and then made a pumping motion with the mouth cavity. This created a strong current in the water and stirred up the sand . . . The [food items] were found and excavated. After inspection with the vibrissae (which was difficult to see because it was done quickly between the snout and the substrate) the flesh was sucked from the shells. Afterwards the shells were dropped."

The mechanics of suction include both fine and gross control of the tongue, which is used like a piston (Fay 1982; Kastelein and Gerrits 1990; Kastelein et al. 1991b, 1994). Walruses have a highly vaulted palate and greatly enlarged facial and labial muscles that are used to minimize the size of the mouth opening by (a) pursing the rostral lips and (b) occluding the lateral gape; this helps to maintain subambient (*i.e.* suction) pressure in the mouth (Marshall and Pyenson 2019; Horning et al. 2017). Hydraulic jetting is also used by the bearded seal (*Erignathus barbatus*), and suction is used widely in Phocidae and Otariidae (Marshall et al. 2008; Marshall 2016; Marshall et al. 2014, 2015; Kienle and Berta 2016, 2018; Kienle et al. 2018, 2019, references in Hocking et al. 2021).

Knowledge of how walruses feed has advanced a lot through direct observations and video analysis of 32 bottom sequences of 12 wild walruses by Levermann et al. (2003): walruses kept their eyes open and usually faced into the bottom current, perhaps so they could see most clearly, as they moved forward (walruses may detect some food items visually, such as *Mya* siphons protruding above the surface; Oliver et al. 1983; see Kastelein et al. 1993: Fig. 29D). The animals moved a foreflipper (usually the right one) to displace water that then removed sediment from an area in front of the head. Finally, the animals jetted water into the sediment to expose prey items, and rooted through the sediment with the muzzle, confirming the observations of Kastelein and Mosterd (1989) and the long-held view of Fay. Lateralization has also been noted in social interactions and during interactions with seabirds (Giljov et al. 2017, 2018; Karenina et al. 2017; Karenina and Giljov 2018; Fig. 22.14f).

22.11 Concluding Comments

Quantitative analyses of allometry and sexual differences in tusk size and shape would be informative. Many such analyses exist for socially-selected structures in other animals. Further, what are the quantitative relationships among tusk characteristics like length, thickness, curvature, *etc.*, and how are those related in turn to tusk strength and resistance to breakage, and to "condition" or" quality" of the bearer?

Socially-selected breeding displays in animals typically vary substantially across populations. Considering the substantial phenotypic differences between Pacific and

Fig. 22.23 Poorly understood aspects of walrus (*Odobenus rosmarus*) behavior range from individual behavior like nuzzling, to group behavior. **a**, Young male nuzzling a rock as it hauls out. **b**, Two sexually mature males, one 6–9 and the other 10–15 year old; (note the bosses) investigate a walrus carcass with the mystacial vibrissae (and smell?). **c**, Pacific walruses can detect the presence of carnivorous killer whales (*Orcinus orca*) up to several km away, presumably by the whales' underwater sounds. In response, all walruses swim quickly toward one another, resulting in a dense group (see also Kryukova 2016a; Kryukova et al. 2010). Killer whales do not approach such large groups and prefer to pursue small groups or single individuals (**d**). All photographs: Anatoly A. Kochnev

Atlantic walruses, and their long period of divergence, differences in breeding displays likely also occur. More information on displays of the Pacific walrus, more populations of the Atlantic walrus, and of the Laptev walrus, is desirable.

Walruses communicate extensively, mainly over short distances. Communication and the social system are intertwined, and knowledge about them will deepen our understanding of gregariousness and many other distinctive features of walrus biology. To date, only coarse patterns of the structure of conspicuous signals ("displays") in a narrow range of situations have been studied. The broader structure of and variation in communicative signals in all modalities need to be described, and patterns of interaction documented. The program should include ontogeny of communicative signals. A striking gap in knowledge concerns communication within groups that are traveling or feeding, a situation in which some odontocete species engage in extensive tactile and acoustic communication (Norris 1991). The need for studies on behavior in relation to body contact and touching is noted above. Those studies should include special attention to the structure and contextual use of the extremely common behavior of social nuzzling, and to its sensory and behavioral significance. As noted, nuzzling occurs in air and under the water, so is not always associated with olfaction. However airborne nuzzling appears to be usually associated with olfaction, judging by the open nostrils of nuzzling animals. Olfaction begs to be studied, considering the species' sensitivity to and alarm in reaction to smelling foreign odors (*e.g.* humans; smoke; Fay et al. 1984a), the common behavior of smelling conspecific animals, smelling the substrate when hauling out on land (Fig. 22.23), and in interactions between females and their calves and between individuals of other classes.

Understanding walrus society more fully will require knowledge about the presence and nature of inter-individual associations over time, both kin and non-kin, and about possible genetic structure and relationships within herds (Beatty et al. 2019, 2020; O'Corry-Crowe et al. 2020). Certainly the social bonds between females and their calves can be broken (and of course calves killed) in stampedes in large terrestrial aggregations, which are increasing in occurrence and size especially in Alaska and Russia. Much mortality can be experienced in such circumstances, and young animals are affected disproportionately directly through physical injuries, and indirectly through predation by polar bears on young animals (Tomilin and Kibal'chich 1975; Born et al. 1995; Kochnev 2002, 2004; Garlich-Miller et al. 2011; Jay et al. 2011; Udevitz et al. 2013; citations in Fischbach et al. 2016; Øren 2017). Separation of females and their calves also can result from stronger storms caused by reduction in the amount of and later seasonal formation of sea ice (Kochnev 2004). The importance of obtaining information from multiple haulout sites and over time is obvious from inter-annual and geographic fluctuations in and increasing loss of sea ice (Fischbach et al. 2016; Descamps et al. 2017; Kochnev 2019), and in the varied causes of mortality across different sites (Kibal'chich 1978; Gol'tsev 1968; Kochnev 2002, 2004).

Basic knowledge of individual and social behavior is relevant and important directly and indirectly to many aspects of walrus captive welfare, biology, and conservation. For example, through studies on anatomy and behavioral elements of feeding, we can understand better how walruses find, obtain, and consume their invertebrate prey, and what signs they leave behind when they do so (Kastelein and Mosterd 1989; Kastelein et al. 1991a, b, 1994). Improved conditions for captive animals have been developed using results from those studies plus increased knowledge of wild walruses (Kastelein and Wiepkema 1989; Kastelein et al. 1989, 2007, 1991b). Similarly, Fernandez and Timberlake (2019), recognizing the great amount of time that wild walruses spend foraging, and the highly specialized behavior that they use in feeding, developed methods to ameliorate behavioral stereotypies of captives such as repetitive swimming and sucking.

Understanding the social system and social culture of walruses is vital to conservation of the species. Genetic investigations like that of Beatty et al. (2019, 2020) will help to reveal the bases for herd structure, group, and care-giving behavior. As part of advancing knowledge in this area and in many other facets of walrus biology, it is important to incorporate natural history information and hence to train students in how to gather and report on such information (Estes and Gilbert 1978; Graham et al. 2011; Able 2016; Barrows et al. 2016); observer observations and impressions over the past century, despite seemingly casual interpretive references to "family groups" *etc.*, have been invaluable in building an overall picture of walrus society that must be real in part (*e.g.* in respect to care-giving behavior) but needs to be built further through more critical observations. Only anecdotal evidence about behavioral development of any sort—individual or social—is available, yet behavioral development is a topic of great importance for deeper understanding of feeding and social structure, for example.

Knowledge of behavior is useful in diverse other kinds of studies. For example, information about time-activity budgets, haulout patterns, and influences of weather is vital for assessing survey data; for survey data as recent as 2006, the authors of one study could claim that it was "the first Pacific walrus survey to account for the proportion of the population in the water during the survey" (Speckman et al. 2011: 545). Subsequent studies on instrumented walruses have provided finer detail about activity budgets, time at and below the surface effects of weather, *etc*.

Assessment of effects due to human disturbance needs adequate understanding of many aspects of behavior, ranging from sensory biology to social structure: individuality ("personality"; de Vere et al. 2017; DeRango and Schwarz 2021); mechanics of feeding behavior; site fidelity; social structure; and walrus cultural traditions (Brakes and Dall 2016). Of course, the relevance of selected behaviors to a problem needs to be articulated, especially with respect to management problems (Greggor et al. 2016). Some forms of human disturbance occur directly at the hands of researchers, as in handling and marking or, over a longer period, effects of attached instruments (Rosen et al. 2017). For example, it is long been accepted that walrus tusks function as "runners" when feeding on the sea floor; this can account for patterns of wear on the anterior surface (Fay 1982). In addition, walruses appear to be lateralized and to favor the right flipper while feeding (Levermann et al. 2003), which may explain the generally greater wear on right tusks (MacCracken and Benter 2016; B. L. Benter *in litt.* 15 January 2020). What are the effects on feeding behavior and energy budgets of walruses resulting from instruments attached to tusks? Various kinds of instrument have been attached to walruses (Dietz et al. 2014). Horning et al. (2017; 2019) review some general concerns about instrumentation in their consideration of devices that are implanted or attached externally.

Species that have evolved in a setting of recurrent and predictable selective forces may lack sufficient behavioral flexibility to adjust to novel challenges over short, non-evolutionary time scales, or to rare, severe environmental events in general (Trillmich 1993; Kappeler et al. 2013). Knowledge of the walrus's evolutionary history and adaptive traits therefore can inform about its resilience to anthropogenic climate change and other future challenges (Estes 1979; Born 2005); different populations of the species may differ in their responses (Born 2005). Ecological research and conservation concerns for the walrus presently address mainly small to moderate spatiotemporal scales, which is appropriate for investigating certain topics (*e.g.* effects of human disturbance; use of terrestrial haulouts in response to receding ice). Studies of behavior can contribute to planning and decision-making over multiple scales and can usually benefit by being integrated across different disciplines (Greggor et al. 2016; Keighley et al. 2019a).

Early observations on and interpretations of movements, diving, and feeding were based on direct observations or inference. Technical advances have led to a spectacular radiation of instrumentation and analyses that have illuminated our knowledge and understanding of the behavioral ecology of walruses. However, knowledge of those subjects is mainly disconnected from knowledge of the behavior that underlies the patterns. For example, different motor patterns have different energetic costs, and those costs likely differ over ontogeny and whether a walrus is alone or moving in a group (Trillmich 1996).

Acknowledgments Many people have helped us in many ways with this chapter. First, we thank Claudio Campagna and Rob Harcourt for their great patience and trust, as we pushed them repeatedly to limits on deadlines. Second, we thank many colleagues for providing reviews, clarification, data, feedback, figures, or information: Corey Accordo, Sveta Artemyeva, Rebecca Baker, Bill Beatty, Brad Benter, Pierre-Paul Bitton, Bobby Boessenecker, Sarah Boessenecker, Hal Brindley, Casey Clark, Emily Chou, Justin Crawford, Olga Filatova, Nick Garbutt, Andrey Giljov, Klaus Gille, Eliezer Gurarie, Frederike Hanke, Dick Harington, Lauri Jemison, Karina Karenina, Xénia Keighley, Kit Kovacs, Natalia Kryukova, Claude Lester Oleg Lyamin, Christian Lydersen, Russell Millner, Xavier Mouy, Phil Nicklin, Jim Parham, Lori Quakenbush, Lee Rentz, Colleen Reichmuth, Beth Sinclair, Sarah Sonsthagen, Klaus Steinkemp, Ian Stirling (whom we thank also for letting us include his reflections in Supplementary Material IV), Ingvar Svanberg, Rodney Ungwiluk Jr., Carolyn Walsh, and Andy Williams. Publications like this one are burdened with innumerable details, and we thank the following people who helped us to deal with them: Hal Dringenberg, Tony Fischbach, Electra Georgiades, Aren Gunderson, Chadwick Jay, Brendan Kelly, Sarah Kienle, Sonja Kromann, Christian Lydersen, Shawn Noren, Link Olsen, Morten Olson, Colleen Reichmuth, Celia Rozen, Jerry Siegel, Ted Stankowich, Garry Stenson, and Øystein Wiig. Our close colleague Kit Kovacs got us involved in this project; we thank her for trusting us to deliver. EHM thanks his ever-supportive wife Margarita and her mother Leo, for giving him the time and peace of mind to write substantial parts of this chapter while on 2019 Christmas vacation with our Chilean family in Punta Arenas. He acknowledges financial support from the Natural Sciences and Engineering Research Council of Canada over many years. AAK thanks many institutions and foundations for assistance and support during his 37 years of field research in the Russian Arctic. First, he thanks the Okhotskrybvod Marine Hunting Inspection, Wrangel Island State Reserve, and Pacific Research Fisheries Center (Chukotka branch), where he worked. He has received financial support from the World Wildlife Fund, Wildlife Conservation Society, Pacific Environment, U.S. Geological Survey, and U.S. Fish and Wildlife Service. AAK also warmly thanks those who helped him in the field, especially the native hunters of Chukotka. Finally, he thanks his sister Svetlana and her loving family, for providing a place for a quiet life and a space to work on this chapter when he was forced to be away from his home and family during the COVID-19 epidemic.

References[25]

Able KW (2016) Natural history: an approach whose time has come, passed, and needs to be resurrected. ICES J Mar Sci 73:2150–2155

[25]English translations of cited Soviet/Russian publications not in Canadian translation series are in the National Marine Mammal Laboratory Library (https://www.librarything.com/venue/93367/National-Marine-Mammal-Laboratory-Library-Alaska-Fisheries-Science-Center) or ARLIS, the library of the Alaska Resources Library & Information Services (https://www.arlis.org/), or both. The citation format that we used lists the year of publication followed by the year of translation (*e.g.* "Nikulin PG (1941 (1953))"; our in-text citations use the year of publication. All other cited works are available via internet searches.

Allen JA (1880) History of North American pinnipeds. A monograph of the walruses, sea-lions, sea-bears and seals of North America. Department of the Interior, U.S. Geological and Geographical Survey of the Territories, Miscellaneous Publications 18. Government Printing Office, Washington, DC

Andersen LW, Jacobsen MW, Lydersen C et al (2017) Walruses (*Odobenus rosmarus rosmarus*) in the Pechora Sea in the context of contemporary population structure of Northeast Atlantic walruses. Biol J Linn Soc 122:897–915

Barrett JH, Boessenkoold S, Kneale CJ et al (2020) Ecological globalisation, serial depletion and the medieval trade of walrus rostra. Quat Sci Rev 229:106122

Barrows CW, Murphy-Mariscal ML, Hernandez RR (2016) At a crossroads: the nature of natural history in the twenty-first century. BioScience 66:592–599

Bartholomew GA (1952) Reproductive and social behavior of the northern elephant seal. Univ Calif Publ Zool 47:369–472

Beatty WS, Lemons PR, Sethi SA et al (2019) Estimating Pacific walrus abundance and demographic rates from genetic mark-recapture. U.S. Department of the Interior, Bureau of Ocean Energy Management, Alaska OCS Region, OCS Study BOEM 2019–059

Beatty WS, Lemons PR, Sethi SA et al (2020) Panmixia in a sea-ice associated marine mammal: evaluating genetic structure of the Pacific walrus (*Odobenus rosmarus divergens*) at multiple spatial scales. J Mammal 101:755–765

Beer CG (1977) What is a display? Am Zool 17:155–165

Bel'kovich VM, Yablokov AV (1961) Among the walruses. Priroda 1961(3):50–56

Belopolsky LO (1939 (1959)) On the migrations and ecology of reproduction of the Pacific walrus (*Odobaenus rosmarus divergens* Illiger). Fish Res Board Can Transl Ser 246

Berta A, Churchill A, Boessenecker RW (2018) The origin and evolutionary biology of pinnipeds: seals, sea lions, and walruses. Annu Rev Earth Planet Sci 46:203–228

Biewer JN, Velez-Juarbe J, Parham JF (2020) Insights on the dental evolution of walruses based on new fossil specimens from California. J Vertebr Paleontol 40:e1833896

Boeskorov GG, Tikhonov AN, Protopopov AV et al (2018) New records of Holocene polar bear and walrus (Carnivora) in the Russian Arctic. Russ J Theriol 17:68–77

Boessenecker RW (2017) A new early Pliocene record of the toothless walrus *Valenictus* (Carnivora, Odobenidae) from the Purisima Formation of northern California. PaleoBios 34: Ucmp_paleobios_35289

Boessenecker SJ, Boessenecker R, Geisler JH (2018) Youngest record of the extinct walrus *Ontocetus emmonsi* from the Early Pleistocene of South Carolina and a review of North Atlantic walrus biochronology. Acta Palaeontol Pol 63:279–286

Born EW (2003) Reproduction in male Atlantic walruses (*Odobenus rosmarus rosmarus*) from the north water (N Baffin Bay). Mar Mamm Sci 19:819–831

Born EW (2005) An assessment of the effects of hunting and climate on walruses in Greenland. Ph. D. thesis, Natural History Museum, University of Oslo, Oslo, Norway

Born EW, Knutsen LØ (1992) Satellite-linked radio tracking of Atlantic walruses (*Odobenus rosmarus rosmarus*) in northeastern Greenland, 1989–1991. Z Säugetierkd 57:275–287

Born EW, Knutsen LØ (1997) Haul-out and diving activity of male Atlantic walruses (*Odobenus rosmarus rosmarus*) in NE Greenland. J Zool 243:381–396

Born EW, Gjertz, Reeves RR (1995) Population assessment of Atlantic walrus (*Odobenus rosmarus rosmarus* L.). Norsk Polarinstitutt, Meddelelser Nr. 138

Born EW, Dietz R, Heide-Jørgensen MP, Knutsen LØ (1997) Historical and present distribution, abundance and exploitation of Atlantic walruses (*Odobenus rosmarus rosmarus* L.) in eastern Greenland. Meddel Grønland Biosci 46

Born EW, Acquarone M, Knutsen LØ et al (2005) Homing behaviour in an Atlantic walrus (*Odobenus rosmarus rosmarus*). Aquat Mamm 31:23–33

Bornhold BD, Jay CV, McConnaughey R et al (2005) Walrus foraging marks on the seafloor in Bristol Bay, Alaska: a reconnaissance survey. Geo-Mar Lett 25:293–299

Brakes P, Dall SRX (2016) Marine mammal behavior: a review of conservation implications. Front Mar Sci 3:87
Brown DH (1963) The health problems of walrus calves, and remarks on their general progress in captivity. Int Zoo Yearb 4:13–22 + two tables
Burns JJ (1965) The walrus in Alaska. Its ecology and management. Alaska Department of Fish & Game, Juneau, AK
Cabrera D, Stankowich T (2020) Stabbing slinkers: tusk evolution among artiodactyls. J Mamm Evol 27:265–272
Calvert W, Stirling I (1990) Interactions between polar bears and overwintering walruses in the central Canadian High Arctic. Int Conf Bear Res and Manage 8:351–356
Chapsky KK (1936) The walrus of the Kara Sea. Trans Arctic Inst 67:1–124
Charrier I (2021a) Vocal communication in otariids and odobenids. Chapter 14, this volume
Charrier I (2021b) Non-vocal communication in otariids and odobenids: involvement of visual and olfactory cues in their social life. Chapter 15, this volume
Charrier I, Aubin T, Mathevon N (2010) Mother-calf vocal communication in Atlantic walrus: a first field experimental study. Anim Cogn 13:471–482
Charrier I, Burlet A, Aubin T (2011) Social vocal communication in captive Pacific walruses *Odobenus rosmarus divergens*. Mamm Biol 76:622–627
Chou E, Antunes R, Sardelis S et al (2020) Seasonal variation in Arctic marine mammal acoustic detection in the northern Bering Sea. Mar Mamm Sci 36:522–547
Chugunkov DI (1991 (1992)) Distribution and population dynamics of walruses on Verkhoturov Is. in the summer of 1990. In: Popov LA (compiler) Research papers on marine mammals of the North Pacific in 1989–1990. Environment Canada, Translation OOENV-2691, pp 21–30
Citta JJ, Quakenbush LT, Taras BD (2014) Estimation of calf:cow ratios of Pacific walruses for use in population modeling and monitoring. Mar Mamm Sci 30:20–43
COSEWIC (2017) COSEWIC assessment and status report on the Atlantic Walrus *Odobenus rosmarus rosmarus*, High Arctic population, Central-Low Arctic population and Nova Scotia-Newfoundland-Gulf of St. Lawrence population in Canada. Committee on the Status of Endangered Wildlife in Canada, Ottawa, ON
de Vere AJ, Lilley MK, Highfill L (2017) Do pinnipeds have personality? Broad dimensions and contextual consistency of behavior in harbor seals (*Phoca vitulina*) and California sea lions (*Zalophus californianus*). Int J Comp Psychol 30:1–16
Dectot X (2018) When ivory came from the seas. On some traits of the trade of raw and carved sea-mammal ivories in the Middle Ages. Anthropozoologica 53:159–174
Denes SL (2014) Ocean environmental effects on walrus communication. Ph.D. thesis, The Pennsylvania State University, State College, PA
Denes SL, Miksis-Olds SL, Mellinger D et al (2015) Comparison of managed care and wild walrus source characteristics. J Acoust Soc Am 137:2196
DeRango EJ, Schwarz JFL (2021) Pinniped 'personality', or consistent individual differences. Chapter 19, this volume
Descamps S, Aars J, Fuglei E et al (2017) Climate change impacts on wildlife in a High Arctic archipelago – Svalbard, Norway. Glob Chang Biol 23:490–502
Dietz R, Born EW, Stewart REA et al (2014) Movements of walruses (*Odobenus rosmarus*) between central west Greenland and southeast Baffin Island, 2005–2008. In: Stewart REA, Kovacs K, Acquarone M (eds) Walrus of the North Atlantic. The North Atlantic Marine Mammal Commission Scientific Publications, vol. 9. Trømso, Norway, pp 53–74
Drew J, López EH, Gill L et al (2016) Collateral damage to marine and terrestrial ecosystems from Yankee whaling in the 19th century. Ecol Evol 6:8181–8192
Dyke AS, Hooper J, Harington RC et al (1999) The late Wisconsinan and Holocene record of walrus (*Odobenus rosmarus*) from North America: a review with new data from Arctic and Atlantic Canada. Arctic 52:160–181
Eberhard WG, Rodríguez RL, Huber BA et al (2018) Sexual selection and static allometry: the importance of function. Q Rev Biol 93:207–250

Elliott HW (1881) The seal-islands of Alaska. Government Printing Office, Washington, DC
Emlen DJ (2008) The evolution of animal weapons. Annu Rev Ecol Syst 39:387–413
Emlen DJ (2014) Animal weapons: the evolution of battle. Henry Holt and Company, LLC, New York
English AW (1975) Functional anatomy of the forelimb in pinnipeds. Ph.D. thesis, University of Illinois at the Medical Center, Chicago, IL
Estes JA (1979) Exploitation of marine mammals: *r*-selection of *K*-strategists? J Fish Res Board Can 36:1009–1017
Estes JA, Gilbert JA (1978) Evaluation of an aerial survey of Pacific walruses (*Odobenus rosmarus divergens*). J Fish Res Board Can 35:1130–1140
Estes RD (1974) Social organization of the African Bovidae. In: Geist V, Walther F (eds) The behaviour of ungulates and its relation to management. The papers of an international symposium held at the University of Calgary, Alberta, Canada, 2–5 November 1971, vol. 1. International Union for Conservation of Nature and Natural Resources, Morges, Switzerland, pp 166–205
Estes RD (1991) The significance of horns and other male secondary sexual characters in female bovids. Appl Anim Behav Sci 29:403–451
Estes RD (2011) Evolution of horns in female bovids – a functional perspective. Gnusletter 29:18–20
Fay FH (1955) The Pacific walrus (*Odobenus rosmarus divergens*): spatial ecology, life history, and population. Ph.D. thesis, University of British Columbia, Vancouver, BC
Fay FH (1960) Structure and function of the pharyngeal pouches of the walrus (*Odobenus rosmarus* L.). Mammalia 24:361–371
Fay FH (1981) Walrus *Odobenus rosmarus* (Linnaeus, 1758). In: Ridgway SH, Harrison RJ (eds) Handbook of marine mammals, vol. 1: sea lions, fur seals and sea otter. Academic, New York, pp 1–23
Fay FH (1982) Ecology and biology of the Pacific walrus, *Odobenus rosmarus divergens* Illiger. N Am Fauna 74:1–279
Fay FH (1985) *Odobenus rosmarus*. Mamm Species 238:1–7
Fay FH, Ray C (1968) Influence of climate on the distribution of walruses, *Odobenus rosmarus* (Linnaeus). I Evidence from thermoregulatory behavior Zoologica 53:1–18
Fay FH, Kelly BP, Gehnrich PH et al (1984a) Modern populations, migrations, demography, trophics, and historical status of the Pacific walrus. U.S. Department of Commerce, National Oceanic and Atmospheric Administration, Outer Continental Shelf Environmental Assessment Program. Final Report 37:231–376
Fay FH, Ray GC, Kibal'chich AA (1984b) Time and location of mating and associated behavior of the Pacific walrus, *Odobenus rosmarus divergens* Illiger. In: Fay FH, Fedoseev GA (eds) Soviet-American cooperative research on marine mammals. Volume 1 – pinnipeds. NOAA Technical Report NMFS 12. U.S. Department of Commerce, National Oceanic and Atmospheric Administration, National Marine Fisheries Service, pp 89–100
Fernandez EJ, Timberlake W (2019) Foraging devices as enrichment in captive walruses (*Odobenus rosmarus*). Behav Process 168:103934
Fischbach AS, Kochnev AA, Garlich-Miller JL et al (2016) Pacific walrus coastal haulout database, 1852–2016. Background report: U.S. Geological Survey Open-File Report 2016–1108, 27 pp, https://pubs.er.usgs.gov/publication/ofr20161108
Frei KM, Coutu AN, Smiarowski K et al (2015) Was it for walrus? Viking Age settlement and medieval walrus ivory trade in Iceland and Greenland. World Archaeol 47:439–466
Freiman SIu (1941 (1954)) Biology of the Chukchi walrus. Izvestiia TNIRO 20:3–20
Freitas C, Kovacs KM, Ims RA et al (2009) Deep into the ice: over-wintering and habitat selection in male Atlantic walruses. Mar Ecol Prog Ser 375:247–261
Freuchen P (1921 (1973)) The occurrence and migrations of the walrus near the west coast of Greenland. Fish Res Board Can Transl Ser 2383

Freuchen P (1935) Mammals. Part II. Field notes and biological observations. Report of the Fifth Thule Expedition 1921–24. Volume II, Numbers 4–5, pp 68–278
Garde E, Jung-Madsen S, Ditlevsen S et al (2018) Diving behavior of the Atlantic walrus in High Arctic Greenland and Canada. J Exp Mar Biol Ecol 500:89–99
Garlich-Miller J, MacCracken JG, Snyder J et al (2011) Status review of the Pacific walrus (*Odobenus rosmarus divergens*). U.S. Fish and Wildlife Service, Anchorage, AK
Gehnrich PH (1984) Nutritional and behavioral aspects of reproduction in walruses. M.S. thesis. University of Alaska, Fairbanks, AK
Geist V (1966a) The evolution of horn-like organs. Behaviour 27:175–214
Geist V (1966b) On fighting injuries and dermal shields of mountain goats. J Wildl Manag 31:192–194
Gilbert G, McCafferty D, Le Maho Y et al (2010) One for all and all for one: the energetic benefits of huddling in endotherms. Biol Rev 85:545–569
Giljov A, Karenina K, Kochnev A (2017) Prey or play: interactions between walruses and seabirds. Acta Ethol 20:47–57
Giljov A, Karenina K, Malashichev Y (2018) Facing each other: mammal mothers and infants prefer the position favouring right hemisphere processing. Biol Lett 14:20170707
Gjertz I, Wiig Ø (1994) Past and present distribution of walruses in Svalbard. Arctic 47:34–42
Gol'tsev VN (1968 (1971)) Dynamics of coastal walrus rookeries in connection with distribution and numbers of walruses. In: Arsen'ev VA, Panin KI (eds) Pinnipeds of the North Pacific. Israel Program for Scientific Translations, Jerusalem, pp 201–212
Gordon KR (1981) Locomotor behaviour of the walrus (*Odobenus*). J Zool 195:349–367
Gordon KR (1982) Mechanics of the limbs of the walrus (*Odobenus rosmarus*) and the California sea lion (*Zalophus californianus*). J Morphol 175:73–90
Gotfredsen AB, Appelt M, Hastrup K (2018) Walrus history around the North Water: human–animal relations in a long-term perspective. Ambio 47(suppl. 2):193–212
Gowans S (2019) Grouping behaviors of dolphins and other toothed whales. In: Würsig B (ed) Ethology and behavioral ecology of odontocetes. Springer, Basel, pp 3–24
Graham MH, Parker J, Dayton PK (eds) (2011) The essential naturalist. Timeless readings in natural history. University of Chicago Press, Chicago, IL
Greggor AL, Berger-Tal O, Blumstein DT et al (2016) Research priorities from animal behaviour for maximising conservation progress. Trends Ecol Evol 31:953–964
Hamilton CD, Kovacs KM, Lydersen C (2015) Year-round haul-out behaviour of male walruses *Odobenus rosmarus* in the northern Barents Sea. Mar Ecol Prog Ser 519:251–263
Heide-Jørgensen MP, Flora J, Andersen AO, Stewart REA et al (2017) Walrus movements in Smith Sound: a Canada – Greenland shared stock. Arctic 70:308–318
Hill E (2011) The historical ecology of walrus exploitation in the North Pacific. In: Braje TJ, Rick TC (eds) Human impacts on seals, sea lions, and sea otters: integrating archaeology and ecology in the Northeast Pacific. University of California Press, Oakland, CA, pp 41–64
Hills S (1992) The effect of spatial and temporal variability on population assessment of Pacific walruses. Ph.D. thesis, University of Maine, Orono, ME
Hocking DP, Park T, Rule JP, Marx FG (2021) Prey capture and processing in fur seals, sea lions and walruses. Chapter 5, this volume
Horning M, Haulena M, Tuomi PA et al (2017) Best practice recommendations for the use of fully implanted telemetry devices in pinnipeds. Anim Biotelemetry 5:13
Horning M, Andrews RD, Bishop AM et al (2019) Best practice recommendations for the use of external telemetry devices on pinnipeds. Anim Biotelem 7:20
Hughes WR, Reichmuth C, Mulsow JL et al (2011) Source characteristics of the underwater knocking displays of a male Pacific walrus (*Odobenus rosmarus divergens*). J Acoust Soc Am 129:2506
Jarman PJ (1989) On being thick-skinned: dermal shields in large mammalian herbivores. Biol J Linn Soc 36:169–191

Jay CV, Hills S (2005) Movements of walruses radio-tagged in Bristol Bay, Alaska. Arctic 58:192–202

Jay CV, Farley SD, Garner GW (2001) Summer diving behavior of male walruses in Bristol Bay, Alaska. Mar Mamm Sci 17:617–631

Jay CV, Marcot BG, Douglas DC (2011) Projected status of the Pacific walrus (*Odobenus rosmarus divergens*) in the twenty-first century. Polar Biol 34:1065–1084

Jay CV, Taylor RL, Fischbach AS, Udevitz MS et al (2017) Walrus haul-out and in water (*sic*) activity levels relative to sea ice availability in the Chukchi Sea. J Mammal 98:386–396

Kappeler PM (2019) A framework for studying social complexity. Behav Ecol Sociobiol 73:13

Kappeler PM, Barrett L, Blumstein DT et al (2013) Constraints and flexibility in mammalian social behaviour: introduction and synthesis. Philos Trans R Soc Lond B 368:20120337

Karenina K, Giljov A (2018) Mother and offspring lateralized social behavior across mammalian species. In: Forrester GS, Hopkins WD, Hudry K, Lindell A (eds) Cerebral lateralization and cognition: evolutionary and developmental investigations of behavioral biases. Progress in Brain Research 238, pp 115–141

Karenina K, Giljov A, Ingram J et al (2017) Lateralization of mother-infant interactions in a diverse range of mammal species. Nat Ecol Evol 1:0030

Kastelein RA, Gerrits NM (1990) The anatomy of the walrus head (*Odobenus rosmarus*). Part 1: the skull. Aquat Mamm 16:101–119

Kastelein RA, Mosterd P (1989) The excavation technique for molluscs of Pacific walruses (*Odobenus rosmarus divergens*) under controlled conditions. Aquat Mamm 15:3–5

Kastelein RA, van Gaalen MA (1988) The sensitivity of the vibrissae of a Pacific walrus (*Odobenus rosmarus divergens*) part I. Aquat Mamm 14:123–133

Kastelein RA, Wiepkema PR (1989) A digging trough as occupational therapy for Pacific walrusses (*sic*) (*Odobenus rosmarus divergens*) in human care. Aquat Mamm 15:9–17

Kastelein RA, Wiepkema PR, Slegtenhorst (1989) The use of molluscs to occupy Pacific walruses (*Odobenus rosmarus divergens*) in human care. Aquat Mamm 15:6–8

Kastelein RA, Stevens S, Mosterd P (1990) The tactile sensitivity of the mystacial vibrissae of a Pacific walrus (*Odobenus rosmarus* divergens). Part 2: masking. Aquat Mamm 16:78–87

Kastelein RA, Gerrits NM, Dubbeldam JL (1991a) The anatomy of the walrus head (*Odobenus rosmarus*). Part 2: description of the muscles and of their role in feeding and haul-out behaviour. Aquat Mamm 17:156–180

Kastelein RA, Paasse M, Klinkhamer P et al (1991b) Food dispensers as occupational therapy for the walrus *Odobenus rosmarus divergens* at the Harderwijk Marine Mammal Park. Int Zoo Yearb 30:207–212

Kastelein RA, Zweypfenning RCVJ, Spekreijse H et al (1993) The anatomy of the walrus head (*Odobenus rosmarus*). Part 3: the eyes and their function in walrus ecology. Aquat Mamm 19:61–92

Kastelein RA, Muller M, Terlow A (1994) Oral suction of a Pacific walrus (*Odobenus rosmarus divergens*) in air and under water. Z Säugetierkd 59:105–115

Kastelein RA, Postma J, Verboom WC (1995) Airborne vocalizations of Pacific walrus pups (*Odobenus rosmarus divergens*). In: Kastelein RA, Thomas JA, Nachtigall PE (eds) Sensory systems of aquatic mammals. De Spil Publishers, Woerden, pp 265–285

Kastelein RA, Jennings N, Postma J (2007) Feeding enrichment methods for Pacific walrus calves. Zoo Biol 26:175–186

Kastelein RA, van den Belt I, Jennings N et al (2015) Behavior and body mass changes of a mother and calf Pacific walrus (*Odobenus rosmarus divergens*) during the suckling period. Zoo Biol 34:9–19

Keighley X, Olsen MT, Jordan P (2019a) Integrating cultural and biological perspectives on long-term human-walrus (*Odobenus rosmarus rosmarus*) interactions across the North Atlantic. Quat Res First View:1–21. https://doi.org/10.1017/qua.2018.150

Keighley X, Pálsson S, Einarsson BF et al (2019b) Disappearance of Icelandic walruses coincided with Norse settlement. Mol Biol Evol 36:2656–2667

Kibal'chich AA (1978 (1990)) Observations on Arakamchechen of a herd of walruses. In: Agarkov GB (ed) Marine mammals: abstracts of reports, VII All-Union Conference, Simferopol. Tsneiteirkh, Moscow, Russia, pp 148–149
Kibal'chich AA (1988 (1990)) Data on the biology of the Pacific walrus (expedition of the "Zakharovo" hunting and fishing vessel in 1985). In: Chernysheva NS (ed) Scientific research on sea mammals of the northern part of the Pacific Ocean in 1986–1987. Can Transl Fish Aquat Sci 5506, pp 116–132
Kibal'chich AA, Lisitsyna TYu (1979) [Some acoustical signals of the Pacific walrus]. Zool Zhurnal 58:1247–1249 (in Russian)
Kienle SS, Berta A (2016) The better to eat you with: the comparative feeding morphology of phocid seals (Pinnipedia, Phocidae). J Anat 228:396–413
Kienle SS, Berta A (2018) The evolution of feeding strategies in phocid seals (Pinnipedia, Phocidae). J Vertebr Paleontol 38:6
Kienle SS, Hermann-Sorensen H, Costa DP et al (2018) Comparative feeding strategies and kinematics in phocid seals: suction without specialized skull morphology. J Exp Biol 221: jeb179424
Kienle SS, Cacanindin A, Kendall T et al (2019) Hawaiian monk seals exhibit behavioral flexibility when targeting prey of different size and shape. J Exp Biol 222:jeb194985
Kiliaan HPL, Stirling I (1978) Observations on overwintering walruses in the eastern Canadian High Arctic. J Mammal 59:197–120
Kiszka JJ, Heithaus MR, Wirsing AJ (2015) Behavioural drivers of the ecological roles and importance of marine mammals. Mar Ecol Prog Ser 523:s267–s281
Kochnev AA (1999) [Pacific walrus in coastal waters of Wrangel Island (1991–1994). 2. Sex and age composition] (in Russian). Izvestiya TINRO 126:465–471
Kochnev AA (2001) [Pacific walruses in the area of Wrangel and Herald islands and their protection] (in Russian). In: Pavlov DS, Bychkov VA (eds) The walrus: mode of the species. Nauka, Moscow, pp 180–205
Kochnev AA (2002) The factors of the Pacific walrus mortality at the coastal haulouts at the Wrangel Island. In: Aristov AA, Bel'kovich VM, Zemskii VA, Vladimirov VA, Smelova IV (eds) Marine mammals (results of research conducted in 1995–1998). Collection of articles. Catran, NIP MORE, Moscow, Russia, pp 191–215
Kochnev AA (2004) Warming of eastern Arctic and present status of the Pacific walrus (*Odobenus rosmarus divergens*) population. In: Belkovich VM (ed) Marine mammals of the Holarctic, third international conference, Koktebel, Crimea, Ukraine, 11–17 October 2004, pp 284–288
Kochnev AA (2013) Sex/age composition of Pacific walruses on the Russian Federation haul-outs. In: Robards M, Garlich-Miller J (eds) A workshop on assessing Pacific walrus population attributes from coastal haul-outs. USFWS Administrative Report, R7/MMM 13-1. Marine Mammals Management, U.S. Fish and Wildlife Service, Anchorage, AK, pp 17–18
Kochnev AA (2019) Phenology of the Pacific walrus (*Odobenus rosmarus divergens*) in coastal waters of Wrangel Island: the impact of the sea ice dynamics. Biol Bull 46:1156–1164
Kochnev AA, Fischbach AS, Jay CV, Speckman SG (2008a) Satellite radio-tracking of Pacific walruses (*Odobenus rosmarus divergens*) in the Chukchi Sea. In: Marine mammals of the Holarctic, Fifth International Conference, 14–18 October, Odessa, Ukraine, vol 5, pp 263–267
Kochnev AA, Kryukova NV, Pereverzev AA et al (2008b) Coastal haulouts of the Pacific walruses (*Odobenus rosmarus divergens*) in Anadyr Gulf (Bering Sea), 2007. In: Marine mammals of the Holarctic, fifth international conference. Odessa, Ukraine, October 14–18, vol. 5. Astroprint, Odessa, Ukraine, pp 267–272
Krupnik I, Ray GC (2007) Pacific walruses, indigenous hunters, and climate change: bridging scientific and indigenous knowledge. Deep-Sea Res II 54:2946–2957
Krushinskaya NL, Lisitsyna TY (1983) Behavior of marine mammals. Nauka Press, Moscow
Krylov VI (1965) First marking of the Pacific walrus. Fur Ind 10:24–25

Krylov VI (1971 (1974)) The feeding of the Pacific walrus (*Odobaenus rosmarus divergens* Ill.). In: Mil'chenko ES (ed) Research on marine mammals. Fish Marine Serv (Canada) Transl Ser 3185, pp 177–189

Kryukova NV (2012) Use of waters near coastal haulouts by Pacific walruses (*Odobenus rosmarus divergens*) in Chukotka. In: Marine mammals of the Holarctic, seventh international conference, September 24–28, 2012, vol. 1. Suzdal, Russia, pp 338–343

Kryukova NV (2016a) Killer whales hunting for walruses near the Village of Enmelen in August, 2006. In: Alaska Marine Science Symposium, 25–29 January 2016, Anchorage Alaska, p 146

Kryukova NV (2016b) [Wild walrus rest at the water surface] (in Russian). In: Theriofauna of Russia and Adjacent Territories Tenth Congress of Russian Theriological Society RAS, 1--5 February 2016, Moscow. KMK Scientific Press, Moscow, p 203

Kryukova NV (2017) Suckling behaviour in wild Pacific walruses (*Odobenus rosmarus divergens*). In: Marine mammals of the Holarctic, ninth international conference, 31 October–05 November 2016, vol 1. Astrakhan, Russia, pp 238–242

Kryukova NV (2019) [Know me, if you can: opportunities of the identification of walrus.] Priroda 2019(8):20–29 (in Russian)

Kryukova NV, Ivanov DI (2012) Rest of single walruses on the shore. In: Fifth all-Russian conference of animal behavior, 20–23 November 2012, Moscow, Russia, p 105

Kryukova NV, Ivanov DI, Pereverzev AA (2010) Observations on walruses (*Odobenus rosmarus divergens*) in the area of their haulout on the Retkyn Spit. In: Marine mammals of the Holarctic, sixth international conference, 11–15 October 2010, Kaliningrad, Russia, pp 291–296

Kryukova NV, Kruchenkova EP, Ivanov DI (2012) Killer whales (*Orcinus orca*) hunting for walruses (*Odobenus rosmarus divergens*) near Retkyn Spit, Chukotka. Biol Bull 39:768–778

Kryukova NV, Krupin IL, Burkanov VN (2019) Pacific walrus (*Odobenus rosmarus*) survey results near Cape Inchoun (Chukchi Sea) haulout, 2017. In: Marine mammals of the Holarctic, ninth international conference, 29 October–2 November 2018, vol. 1. Arkhangelsk, Russia, pp 154–162

Larsen ON, Reichmuth C (2012) Proximal mechanisms for sound production in male Pacific walruses. Can Acoust 40:139

Laws RM (1953) The elephant seal (*Mirounga leonina* Linn.) I. Growth and age. FIDS Sci Rep 8

Levermann N, Galatius A, Ehlme G et al (2003) Feeding behaviour of free-ranging walruses with notes on apparent dextrality of flipper use. BMC Ecol 3(9)

Lindqvist C, Roy T, Lydersen C et al (2016) Genetic diversity of historical Atlantic walruses (*Odobenus rosmarus rosmarus*) from Bjørnøya and Håøya (Tusenøyane), Svalbard, Norway. BMC Res Notes 9:112

Ling JK (1977) Vibrissae of marine mammals. In: Harrison RJ (ed) Functional anatomy of marine mammals, vol 3. Academic, London, pp 387–415

Liwanag HEM, Berta A, Costa DP et al (2012) Morphological and thermal properties of mammalian insulation: the evolutionary transition to blubber in pinnipeds. Biol J Linn Soc 107:774–787

Liwanag HEM, Oraze J, Costa DP et al (2014) Thermal benefits of aggregation in a large marine endotherm: huddling in California sea lions. J Zool 293:152–159

Llamazares-Martín C, Palagi E (2021) Playing at the edge of the sea: a comparative analysis in otariids and odobenids. Chapter 20, this volume

Loughrey AG (1959) Preliminary investigation of the Atlantic walrus *Odobenus rosmarus rosmarus* (Linnaeus). Can Wildl Serv Wildl Manage Bull Ser 1 no. 14

Lowry L (2016) *Odobenus rosmarus*. The IUCN red list of threatened species 2016: e. T15106A45228501. https://doi.org/10.2305/IUCN.UK.2016-1.RLTS.T15106A45228501.en

Lowry LF, Fay FH (1984) Seal eating by walruses in the Bering and Chukchi seas. Polar Biol 3:11–18

Lowther AD, Kovacs KM, Griffiths D et al (2015) Identification of motivational state in adult male Atlantic walruses inferred from changes in movement and diving behavior. Mar Mamm Sci 31:1291–1313

Lyamin OI, Siegel JM (2019) Sleep in aquatic mammals. In: Dringenberg HC (ed) Handbook of sleep research. Academic Press/Elsevier, San Diego, CA, pp 375–393
Lyamin OI, Kosenko PO, Vyssotski AL et al (2012) Study of sleep in a walrus. Dokl Biol Sci 444:188–191
Lydersen C (2018) Walrus. In: Würsig B, Thewissen JGM, Kovacs KM (eds) Encyclopedia of marine mammals, 3rd edn. Academic, New York, pp 1045–1048
Lydersen K, Kovacs KM (2014) Walrus *Odobenus rosmarus* research in Svalbard, Norway, 2000-2010. In: Stewart REA, Kovacs K, Acquarone M (eds) Walrus of the North Atlantic. The North Atlantic Marine Mammal Commission Scientific Publications, vol. 9. Trømso, Norway, pp 175–190
Lydersen C, Aars J, Kovacs KM (2008) Estimating the number of walruses in Svalbard from aerial surveys and behavioural data from satellite telemetry. Arctic 61:119–128
Lydersen C, Chernook VI, Glazov DM et al (2012) Aerial survey of Atlantic walruses (*Odobenus rosmarus rosmarus*) in the Pechora Sea, August 2011. Polar Biol 35:1555–1562
Lyon BE, Montgomerie R (2012) Sexual selection is a form of social selection. Philos Trans R Soc B 367:2266–2273
MacCracken JG, Benter RB (2016) Trend in Pacific walrus (*Odobenus rosmarus divergens*) tusk asymmetry, 1990–2014. Mar Mamm Sci 32:588–601
Magallanes I, Parham JF, Santos G-P et al (2018) A new tuskless walrus from the Miocene of Orange County, California, with comments on the diversity and taxonomy of odobenids. Peer J 6:e5708
Mansfield AW (1958) The biology of the Atlantic walrus *Odobenus rosmarus rosmarus* (Linnaeus) in the eastern Canadian Arctic. Ph.D. thesis, McGill University, Montreal, QC
Marcoux M, Ferguson SH, Roy N et al (2017) Seasonal marine mammal occurrence detected from passive acoustic monitoring in Scott Inlet, Nunavut, Canada. Polar Biol 40:1127–1138
Marshall CD (2016) Morphology of the bearded seal (*Erignathus barbatus*) muscular-vibrissal complex: a functional model for phocid subambient pressure generation. Anat Rec 299:1043–1053
Marshall CD, Pyenson ND (2019) Feeding in aquatic mammals: an evolutionary and functional approach. In: Bels V, Whishaw IQ (eds) Feeding in vertebrates. Springer, Basel, pp 743–785
Marshall CD, Kovacs K, Lydersen C (2008) Feeding kinematics, suction, and hydraulic jetting capabilities in bearded seals (*Erignathus barbatus*). J Exp Biol 211:699–708
Marshall CD, Wieskotten S, Hanke W et al (2014) Feeding kinematics, suction, and hydraulic jetting performance of harbor seals (*Phoca vitulina*). PLoS One 9:e86710
Marshall CD, Rosen DAS, Trites AW (2015) Feeding kinematics and performance of basal otariid pinnipeds, Steller sea lions and northern fur seals: implications for the evolution of mammalian feeding. J Exp Biol 218:3229–3240
McCord (2016) Mating behaviors exhibited by a captive male Pacific walrus (*Odobenus rosmarus divergens*). Honors Thesis, The University of Southern Mississippi, Hattiesburg, MS
McHugh K (2019) Odontocete social strategies and tactics along and inshore. In: Würsig B (ed) Ethology and behavioral ecology of odontocetes. Springer, Basel, pp 165–182
McLeod BA, Frasier TA, Lucas Z (2014) Assessment of the extirpated Maritimes walrus using morphological and Ancient DNA analysis. PLoS One 9(6):e99569
Miksis-Olds JL, Van Opzeeland IC, Van Parijs S et al (2016) Pinniped sounds in the polar oceans. In: Au WW, Lammers MO (eds) Listening in the ocean. New discoveries and insights on marine life from autonomous passive acoustic recorders, Springer, Science+Business Media LLC, New York, pp 257–308
Miller EH (1975a) Walrus ethology. I. The social role of tusks and applications of multidimensional scaling. Can J Zool 53:590–613
Miller EH (1975b) A comparative study of facial expressions of two species of pinnipeds. Behaviour 53:268–284
Miller EH (1976) Walrus ethology. II. Herd structure and activity budgets of summering males. Can J Zool 54:704–715

Miller EH (1985) Airborne acoustic communication in the walrus *Odobenus rosmarus*. Natl Geogr Res 1:124–145
Miller EH (1991) Communication in pinnipeds, with special reference to non-acoustic signalling. In: Renouf D (ed) The behaviour of pinnipeds. Chapman and Hall, London, England, pp 128–235
Miller EH, Boness DJ (1983) Summer behavior of Atlantic walruses *Odobenus rosmarus rosmarus* (L.) at Coats Island, N. W. T. (Canada). Z Säugetierkd 48:298–313
Milne AO, Smith C, Orton LD et al (2020) Pinnipeds orient and control their whiskers: a study on Pacific walrus, California sea lion and harbor seal. J Comp Physiol A 206:441–451
Monson DH, Udevitz MS, Jay CV (2013) Estimating age ratios and size of Pacific walrus herds on coastal haulouts using video imaging. PLoS One 8(7):e69806
Mouy X, Hannay D, Zykov M et al (2012) Tracking of Pacific walruses in the Chukchi Sea using a single hydrophone. J Acoust Soc Am 131:1349–1358
Muizon D (2018) *Odobenocetops*. In: Würsig B, Thewissen JGM, Kovacs KM (eds) Encyclopedia of marine mammals, 3rd edn. Elsevier, Amsterdam, pp 654–656
Nelson CH, Johnson KR (1987) Whales and walruses as tillers of the sea floor. Sci Am 256 (2):112–118
Nelson CH, Phillips RL, McRea Jr J et al (1984) Gray whale and Pacific walrus benthic feeding grounds and sea floor interaction in the Chukchi Sea. U.S. Department of the Interior, Minerals Management Service, Technical Report IA, No. 141571, Anchorage, AK
Nelson CH, Johnson KR, Barber JH Jr (1987) Gray whale and walrus feeding excavation on the Bering Shelf, Alaska. J Sediment Petrol 57:419–430
Nikulin PG (1941 (1953)) Chukotsky walrus. Izvestiia TNIRO 20:21–59
Nikulin PG (1947 (1957)) Biological characteristics of the shore aggregations of the walrus in the Chukotka Peninsula. Fish Res Board Can Transl Ser 115
Noren SR, Edwards MS (2020) Mature hematology in neonatal walruses supports diving alongside their mothers. Mar Mamm Sci 36:595–611
Noren SR, Jay CV, Burns JM et al (2015) Rapid maturation of the muscle biochemistry that supports diving in Pacific walruses (*Odobenus rosmarus divergens*). J Exp Biol 218:3319–3329
Noren SR, Udevitz MS, Jay CV (2016) Sex-specific energetics of Pacific walruses (*Odobenus rosmarus divergens*) during the nursing interval. Physiol Biochem Zool 89:93–109
Norris KS (1991) Dolphin days. The life & times of the spinner dolphin. Avon Books, New York
Nowicki SN, Stirling I, Sjare B (1997) Duration of stereotyped underwater vocal displays by male Atlantic walruses in relation to aerobic dive limit. Mar Mamm Sci 13:566–575
Nyholm ES (1975) Observations on the walrus (*Odobenus rosmarus* L.) in Spitsbergen in 1971–1972. Ann Zool Fenn 12:193–196
O'Brien DM (2019) Static scaling and the evolution of extreme canine size in a saber-toothed cat (*Smilodon fatalis*). Integr Comp Biol icz054. https://doi.org/10.1093/icb/icz054
O'Brien DM, Allen CE, Van Kleeck MJ et al (2018) On the evolution of extreme structures: static scaling and the function of sexually selected signals. Anim Behav 144:95–108
O'Corry-Crowe G, Suydam R, Quakenbush L et al (2020) Group structure and kinship in beluga whale societies. Sci Rep 10:11462
Oliver JS, Slattery PN, O'Connor EF et al (1983) Walrus, *Odobenus rosmarus*, feeding in the Bering Sea: a benthic perspective. Fish Bull 81:501–512
Øren K (2017) Haul-out behaviour of walruses (*Odobenus rosmarus*) monitored by camera stations – potential impacts of tourist visitations. M.Sc. thesis, The Arctic University of Norway, Tromsø, Norway
Pedersen A (1962) Das Walross. Die Neue Brehm-Bücherei, Heft 306. A. Ziemsen, Wittenberg, Germany
Pierce SE, Clack JA, Hutchinson JR (2011) Comparative axial morphology in pinnipeds and its correlation with aquatic locomotory behaviour. J Anat 219:502–514

Quakenbush LT, Crawford JA, Citta JJ et al (2016) Pinniped movements and foraging: village-based walrus habitat use studies in the Chukchi Sea. U.S. Department of the Interior, Bureau of Ocean Energy Management, Alaska OCS Region, OCS Study BOEM 2016-053

Ray C (1963) Locomotion in pinnipeds. Nat Hist 72(3):10–21

Ray GC, Watkins WA (1975) Social function of underwater sounds in the walrus *Odobenus rosmarus*. In: Ronald K, Mansfield AW (eds) Biology of the seal: proceedings of a symposium held in Guelph 14–17 August 1972. Rapports et procès-verbaux des reunions, Conseil International pour l'Exploration de la Mer 169, pp 524–526

Ray GC, McCormick-Ray J, Berg P et al (2006) Pacific walrus: benthic bioturbator of Beringia. J Exp Mar Biol Ecol 330:403–419

Rideout BP, Dosso SE, Hannay DE (2013) Underwater passive acoustic localization of Pacific walruses in the northeastern Chukchi Sea. J Acoust Soc Am 134:2534

Robards M, Garlich-Miller J (eds) (2013) A Workshop on assessing Pacific walrus population attributes from coastal haul-outs. USFWS Administrative Report, R7/MMM 13-1. Marine Mammals Management, U.S. Fish and Wildlife Service, Anchorage, AK

Rosen DAS, Gerlinsky CG, Trites AW (2017) Telemetry tags increase the costs of swimming in northern fur seals, *Callorhinus ursinus*. Mar Mamm Sci 34:385–402

Salter RE (1979a) Observations on social behaviour of Atlantic walruses (*Odobenus rosmarus* (L.)) during terrestrial haul-out. Can J Zool 58:461–463

Salter RE (1979b) Site utilization, activity budgets, and disturbance responses of Atlantic walruses during terrestrial haul-out. Can J Zool 57:1169–1180

Schevill WE, Watkins WA, Ray C (1966) Analysis of underwater *Odobenus* calls with remarks on the development and function of the pharyngeal pouches. Zoologica 51:103–106 + 5 plates

Schneider DG (2009) Quantitative ecology: measurement, models, and scaling, 2nd edn. Elsevier, Amsterdam

Schusterman RJ, Reichmuth (2008) Novel sound production through contingency learning in the Pacific walrus (*Odobenus rosmarus divergens*). Anim Cogn 11:319–327

Semenov AR, Burkanov VN, Mashagin SA (1988 (1990)) Walrus hauling-out grounds on the Kamchatka Peninsula. In: Chernysheva NS (ed) Scientific research on sea mammals of the northern part of the Pacific Ocean in 1986–1987. Can Transl Fish Aquat Sci 5506, pp 90–96

Semenova V, Boltunov A, Nikiforov V (2019) Key habitats and movement patterns of Pechora Sea walruses studied using satellite telemetry. Polar Biol 42:1763–1774

Sepúlveda M, Harcourt R (2021) Maternal behavior in otariids and the walrus. Chapter 3, this volume

Seymour J, Horstmann-Dehn L, Wooller MJ (2014) Proportion of higher trophic-level prey in the diet of Pacific walruses (*Odobenus rosmarus divergens*). Polar Biol 37:941–952

Shadwick RE, Russell AP, Lauff RF (1992) The structure and mechanical design of rhinoceros dermal armour. Philos Trans R Soc Lond B 337:419–428

Sjare BL (1993) Vocalizations and breeding behaviour of Atlantic walruses in the Canadian High Arctic. Ph.D. thesis, University of Alberta, Edmonton, Alberta

Sjare B, Stirling I (1991) I hear you knocking. Nat Hist 100(3):60–63

Sjare B, Stirling I (1996) The breeding behavior of Atlantic walruses, *Odobenus rosmarus rosmarus*, in the Canadian High Arctic. Can J Zool 74:897–911

Sjare B, Stirling I, Spencer C (2003) Structural variation in the songs of Atlantic walruses breeding in the Canadian High Arctic. Aquat Mammal 29:297–318

Smirnov G, Kochnev A, Kompantseva Y et al (2000) Environmental monitoring of coastal walrus haul-outs in the Gulf of Anadyr, 1999. U.S. Fish and Wildlife Service Marine Mammals Management, Anchorage, AK

Sokolov VE (1982) Mammal skin. University of California Press, Berkeley, CA

Speckman SG, Chernook VI, Burn DM, Udevitz MS, Kochnev AA, Vasilev A, Jay CV (2011) Results and evaluation of a survey to estimate Pacific walrus population size, 2006. Mar Mamm Sci 27:514–553

Stankowich T (2011) Armed and dangerous: predicting the presence and function of defensive weaponry in mammals. Adapt Behav 20:32–43

Stankowich T, Caro T (2009) Evolution of weaponry in female bovids. Proc Royal Soc B 276:4329–4334

Stirling I (1984) A group threat display given by walruses to a polar bear. J Mammal 65:352–353

Stirling I, Calvert W, Cleator H (1983) Underwater vocalizations as a tool for studying the distribution and relative abundance of wintering pinnipeds in the High Arctic. Arctic 36:262–274

Stirling I, Calvert W, Spencer C (1987) Evidence of stereotyped underwater vocalizations of male Atlantic walruses (*Odobenus rosmarus rosmarus*). Can J Zool 65:2311–2321

Svanberg I (2010) Walruses (*Odobenus rosmarus*) in captivity. Svenska Linnésällskapets årsskrift (Yearbook of the Swedish Linnaeus Society) 2010:119–136

Taggart SJ (1987) Grouping behavior of Pacific walruses (*Odobenus rosmarus divergens* Illiger), an evolutionary perspective. Ph.D. thesis, University of California, Santa Cruz, CA

Thayer S (1986) History and strategies of research on social touch. J Nonverbal Behav 10:12–28

Thiemann GW, Iverson SJ, Stirling I (2008) Polar bear diets and arctic marine food webs: insights from fatty acid analysis. Ecol Monogr 78:591–613

Tobias JA, Montgomerie R, Lyon BE (2012) The evolution of female ornaments and weaponry: social selection, sexual selection and ecological competition. Philos Trans R Soc Lond B 367:2274–2293

Tomilin AG, Kibal'chich AA (1975 (1976)) Walruses of the region of Wrangel Island. Fish Marine Serv (Canada) Transl Ser 3721

Trillmich F (1993) Influence of rare ecological events on pinniped social structure and population dynamics. In: Boyd IL (ed) Marine mammals: advances in behavioural and population biology. Symposia of the Zoological Society of London, no. 66, pp 95–114

Trillmich F (1996) Parental investment in pinnipeds. In: Rosenblatt JS, Snowdon CT (eds), Parental care: evolution, mechanisms, and adaptive significance. Advances in the study of behavior, vol 25. Academic. New York, pp 533–577

Tyack P, Miller EH (2002) Vocal anatomy, acoustic communication and echolocation. In: Hoelzel AR (ed) Marine mammal biology: an evolutionary approach. Blackwell Science, Oxford, pp 142–184

Udevitz MS, Jay CV, Fischbach AS et al (2009) Modeling haul-out behavior of walruses in Bering Sea ice. Can J Zool 87:1111–1128

Udevitz MS, Taylor RL, Garlich-Miller JL et al (2013) Potential population-level effects of increased haulout-related mortality of Pacific walrus calves. Polar Biol 36:291–298

Udevitz MS, Jay CV, Taylor RL et al (2017) Forecasting consequences of changing sea ice availability for Pacific walruses. Ecosphere 8(11):e02014

Van Opzeeland IC, Miksis-Olds JL (2011) Acoustic ecology of pinnipeds in polar habitats. In: Eder DL (ed) Aquatic animals. Nova Science, Hauppauge NY, pp 1–52

Verboom WC, Kastelein RA (1995) Rutting whistles of a male Pacific walrus (*Odobenus rosmarus divergens*). In: Kastelein RA, Thomas JA, Nachtigall PE (eds) Sensory systems of aquatic mammals. De Spil, Woerden, The Netherlands, pp 287–298

Vibe C (1950) The marine mammals and the marine fauna in the Thule District (northwest Greenland) with observations on ice conditions in 1939–41. Medd Grønland 150(6)

Vishnevskaya TYu, Bychkov VA (1985 (1986)) A mixed herd of the Laptev walrus in M. Pronchishchev Bay and prospects for its protection. In: Flint VE (ed) Ecological aspects of protection of the beasts of the world. Ministry of Agriculture, Moscow, pp 3–14

Vishnevskaya TYu, Bychkov VA (1990 (1993)) The Laptev walrus. Can Transl Fish Aquat Sci 5581

West-Eberhard MJ (1983) Sexual selection, social competition, and speciation. Q Rev Biol 58:155–183

West-Eberhard MJ (2014) Darwin's forgotten idea: the social essence of sexual selection. Neurosci Biobehav Rev 46:501–508

Wiig Ø, Gjertz I, Griffiths D et al (1993) Diving patterns of an Atlantic walrus *Odobenus rosmarus rosmarus* near Svalbard. Polar Biol 13:71–72

Yablokov AV, Bel'kovich VM (1963) Observations on walruses on hauling grounds in the Gulf of Anadyr and the Chukchi Sea. Krayevedcheskiye Zapiski (Magadan Oblast Executive, Magadan) 4:156–174

Yablokov AV, Klevezal GA (1964) Vibrissae of whales and seals, their distribution, structure, and significance. In: Kleinenberg SE (ed) Morphological features of aquatic mammals. Akademii Nauk SSSR, Moscow, Russia, pp 48–81. Fish Res Board Can Transl Ser 1335

Chapter 23
Behavioral Insights into the Decline and Natural History of Steller Sea Lions

Andrew W. Trites

Abstract Two competing hypotheses were proposed to explain why Steller sea lions had declined in the Gulf of Alaska, Bering Sea and Aleutian Islands. One of the theories was that young sea lions were starving because fisheries had reduced the abundance of groundfish—the overfishing hypothesis. The other was that these low-fat species of fish had increased in abundance as the sea lion population declined following the 1976–1977 oceanic regime shift, and were compromising sea lion reproductive and survival rates—the junk-food hypothesis. Behavioral ecologists tested these hypotheses by comparing sea lion behaviors in the declining region (Gulf of Alaska and Aleutian Islands) with sea lion behaviors in an increasing region (Southeast Alaska) to determine whether the populations exhibited behavioral differences consistent with food shortages. These studies involved comparing dive depths, dive durations, time spent foraging, and time spent nursing by regions and seasons. Research also focused on weaning—a critical life-history stage—to determine when and how it occurs. Collectively, these observations and measures of behavioral responses revealed that most dependent young begin supplementing their milk diet with fish between April and May, and wean just before the start of the upcoming June breeding season. However, the proportion of young sea lions that wean at 1, 2 or 3 years of age appears to vary by year due to regional and temporal differences in the quantity and quality of prey available to them once weaned. None of the behavioral studies of adult and juvenile Steller sea lions supported the overfishing hypothesis—but were, instead, consistent with the junk-food hypothesis. It appears that lactating females that consume large amounts of low-energy fish (such as walleye pollock and Pacific cod) have a high probability of miscarriage, and will keep their dependent young for an extra one or two years—thereby causing birth rates and population size to decline. In contrast, lactating females that consume larger amounts of fattier fish (such as sand lance and Pacific herring) can successfully wean a pup every year. Plasticity in age at weaning appears to be an

A. W. Trites (✉)
Marine Mammal Research Unit, Institute for the Oceans and Fisheries, University of British Columbia, Vancouver, B.C., Canada
e-mail: a.trites@oceans.ubc.ca

C. Campagna, R. Harcourt (eds.), *Ethology and Behavioral Ecology of Otariids and the Odobenid*, Ethology and Behavioral Ecology of Marine Mammals,
https://doi.org/10.1007/978-3-030-59184-7_23

evolutionary adaptation to natural shifts in community prey structure in the North Pacific Ocean—and is an adaptation that successfully slows population declines of Steller sea lions until the ocean shifts to an alternative state containing greater proportions of energy-rich fish that allows sea lion numbers to increase again.

Keywords *Eumetopias jubatus* · Life history · Population decline · Endangered · Maternal strategies · Phenology · Weaning · Abortions · Breeding · Junk food · Overfishing · Regime shifts

23.1 Introduction

In 1980, biologists reported seeing fewer Steller sea lions than expected at breeding sites in the Aleutian Islands (Braham et al. 1980). Subsequent surveys confirmed the sea lion population was in steep decline (Merrick et al. 1987; Trites and Larkin 1996)—which in turn led the United States to list Steller sea lions as *threatened* and later as *endangered* (NMFS 1992). In 1995, the North American population of Steller sea lions was split into two populations—an Eastern (*threatened*) and Western (*endangered*) population (Fig. 23.1; NMFS 1995). The dividing line at 144° W longitude (east of Prince William Sound) corresponded to a break in the genetic and breeding-site distributions of Steller sea lions, as well as a difference in population trajectories (Fig. 23.2; Bickham et al. 1998; Trites and Larkin 1996).

Since the two Steller sea lion populations were listed, the Eastern population (ranging from southeast Alaska to California) has increased at a rate of ~5% per year (Fig. 23.2)—and was delisted (NMFS 2013). Animals permanently emigrating from the Western population also augmented sea lion numbers in Southeast Alaska (Jemison et al. 2013, 2018; O'Corry-Crowe et al. 2014; Hastings et al. 2020). In contrast, the Western population has remained *endangered.* It has continued to decline in the Aleutian Islands, but has increased in the Gulf of Alaska. Overall, the Western population numbers ~20% of its peak 1970s' abundance (Fig. 23.2).

Once Steller sea lions were listed under the US Endangered Species Act, the US government restricted fisheries and initiated an unprecedented research effort in Alaska to uncover the underlying cause of the population decline (Mansfield and Haas 2006; NMFS 2008; Cheever and Riley 2019). Much of the initial scientific research focused on overfishing and fisheries-related mortality—the leading theory to explain the decline of Steller sea lions in the Gulf of Alaska and Aleutian Islands (Trites and Larkin 1992; Springer 1992; Atkinson et al. 2008; Hui et al. 2015; Conn et al. 2014). Major fisheries targeting species that dominated sea lion diets (i.e., walleye pollock, Pacific cod, and Atka mackerel) had expanded as Steller sea lions declined (Alverson 1992). Fisheries were therefore thought to have reduced the abundance and accessibility of prey, and impeded the ability of sea lions to meet their daily energy needs (Trites and Donnelly 2003).

Steller sea lions in the declining population were hypothesized to be nutritionally stressed (Trites and Donnelly 2003). Put more simply, adults were presumed to be no

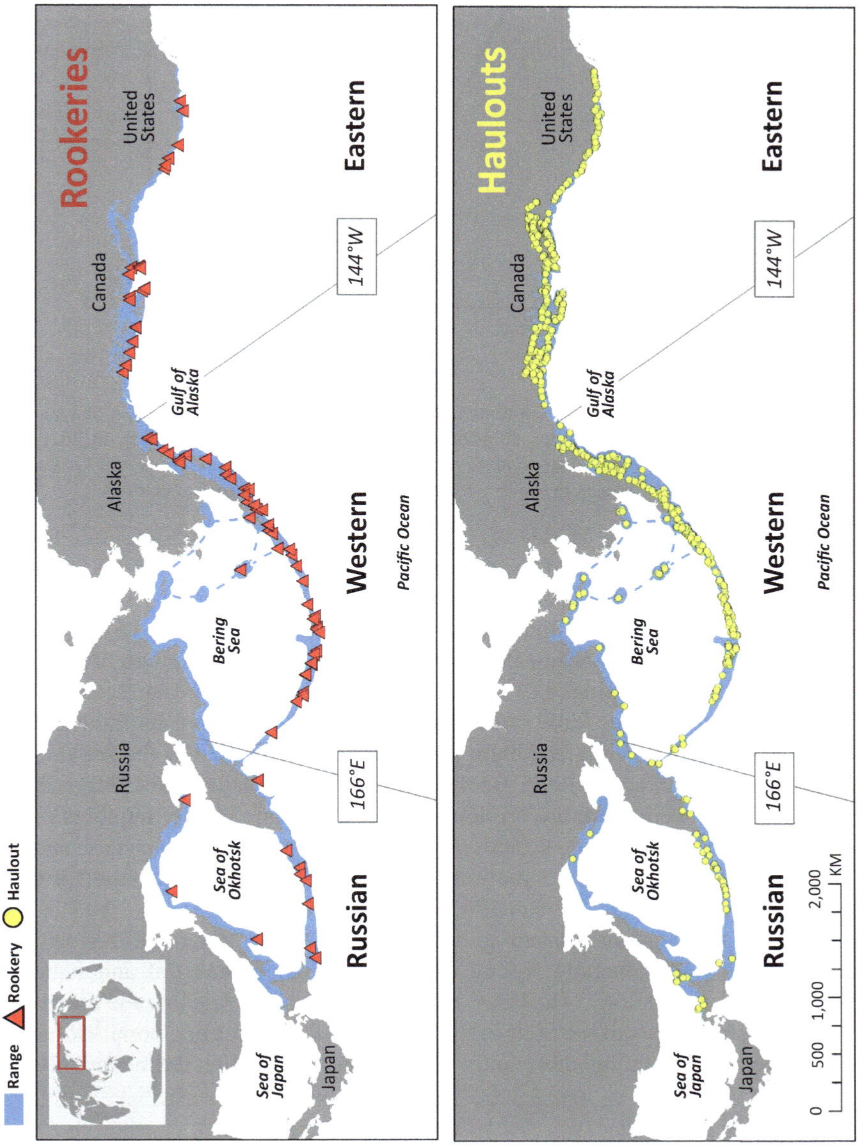

Fig. 23.1 Locations of rookeries and haulouts used by the Russian, Western and Eastern populations of Steller sea lions in 2020. The shaded range reflects the tendency for Steller sea lions to stay within ~30 km of haulouts and rookeries, and to feed over shelf areas. Dashed lines indicate presumed inter-site movements.

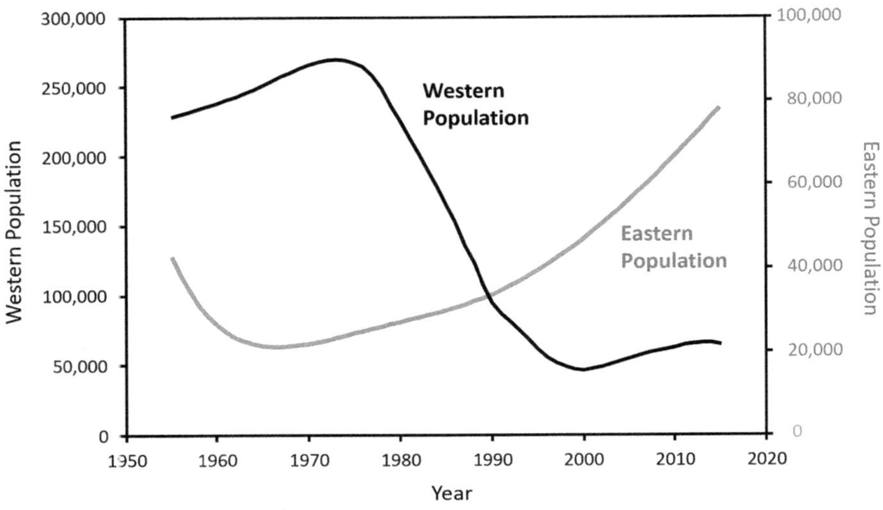

Fig. 23.2 Estimated numbers of Steller sea lions in the Western (Gulf of Alaska, Bering Sea and Aleutian Islands) and Eastern populations (California to Southeast Alaska) from 1956 to 2018. Numbers were reconstructed from census counts of pups and non-pups as per Trites and Larkin (1996). Note the three-fold difference in scales used to plot the two population trajectories

longer getting enough to eat, and their young were thought to be starving after weaning. Some speculated that fisheries were out-competing sea lions by removing too many cod, pollock, and Atka mackerel—the over-fishing hypothesis (Atkinson et al. 2008). However, others speculated that rather than not finding enough of these fish, the sea lions were eating too many of them—the junk-food or nutri-lite hypothesis (Rosen and Trites 2000; Alverson 1992). Stock assessments indicated that these species of commercially caught groundfish had increased as the sea lion population declined (Alverson 1992; Bakkala et al. 1987; Bulatov 2014; Merrick 1997). The sea lions were therefore presumed to be consuming greater numbers of fish that had less lipid and fewer calories compared to the forage fishes such as sand lance and smelts that were a greater portion of sea lion diets prior to the population decline (Merrick et al. 1997; Alverson 1992; Winship and Trites 2003).

Much of the research undertaken to test whether young were starving and whether sea lions were generally having difficulty obtaining prey fell under the umbrella of behavioral ecology. The behavioral studies involved tracking the movements and diving behaviors of sea lions in the increasing and decreasing populations. It also involved documenting diets and watching and recording the attendance

Fig. 23.1 (continued) Note that the range of the Russian population once extended further south to the Toyama Gulf in Japan, and along the Asian coast of the Sea of Japan to the Korean Peninsula (Map produced by S. Agbayani and A.W. Trites based on Loughlin et al. 1992; Trites and Larkin 1996; Burkanov and Loughlin 2005; Olesiuk 2018; Gregr and Trites 2008; Jemison et al. 2018; Hui et al. 2015)

Fig. 23.3 An observation blind used by Milette and Trites (2003) to record perinatal periods and duration of foraging trips of Steller sea lions breeding on Lowrie Island (Forester Island complex) in Southeast Alaska (Photo by A.W. Trites)

patterns of sea lions at their breeding (rookery) and resting (haulout) sites from blinds (Fig. 23.3) and through remotely operated cameras. Interactions between mothers and their dependent young were particularly noteworthy to determine the weaning process and the perilous time of year when young sea lions might be dying. Collectively, behavioral studies yielded critical pieces of new information about the life history of Steller sea lions. They also revealed much about the plasticity of sea lion behavior and the ability of Steller sea lions to adapt to changes in their environment. More importantly, behavioral observations provided significant insights needed to resolve why Steller sea lions had declined.

23.2 Natural History

Georg Wilhelm Steller was the first to describe Steller sea lions (Steller 1751). During June 1742, while shipwrecked on Bering Island, Steller noted that the sea lions were polygamous (with each male having 2–4 females), and that females gave birth on land to a single pup at the beginning of June (Fig. 23.4). Pups, he said, bleated like sheep, and the females bellowed like cows. He also observed that bulls ate little or nothing during June and July—and that they generally preyed on fish. However, Steller reported little about sea lions beyond the breeding season, other than they only used certain areas during winter and summer that were "rocky and near precipices".

Today, the range of Steller sea lions is known to cross the Pacific Rim from California through Alaska and Russia, down to northern Japan (Fig. 23.1). They

Fig. 23.4 A Steller sea lion bull in July with pups, a harem of adult females—and a dependent juvenile (upper right corner) (Photo by A.W. Trites)

have been grouped into three populations or stocks—Russian, Western and Eastern (Baker et al. 2005). Despite their huge range, Steller sea lions only breed at about 100 sites (rookeries, Figs. 23.1 and 23.5) and rest at over 600 other sites (winter and year-round haulouts, Figs. 23.1 and 23.6) (Loughlin et al. 1992; Trites and Larkin 1996; Burkanov and Loughlin 2005; Olesiuk 2018). As noted by Georg Steller, the terrestrial sites used by Steller sea lions tend to be on relatively steep barren rocks and wave-cut outcrops associated with sheer drop offs (Ban and Trites 2007). Most of their terrestrial sites have been continuously used for centuries (Lyman 1988; Steller 1751) and tend to be exposed to ocean swells, with limited protection (Bigg 1985). While some haulouts are on steep cobblestone beaches, most are characteristically on solid stone from which the sea lions can quickly enter deep waters. The consistency with which they have repeatedly used so few sites over centuries (Lyman 1988) suggests that Steller sea lions have a fairly narrow set of criteria for haulout selection throughout their range—from California to Japan.

Indigenous hunters likely played a significant role over the past 10,000 years in shaping the distribution and natural history of Steller sea lions and other hunted species (Maschner et al. 2014; Suraci et al. 2019; Gaynor et al. 2019). Hunting may, for example, explain the tendency for rookeries and haulouts to be far from land in areas with challenging topographies and currents for people to navigate (Lyman 1989). However, reduced consumption of Steller sea lions by Indigenous people in recent times combined with the legal protection of sea lions from disturbance and commercial hunting may be leading to new rookeries and haulouts being established in the growing Eastern population that are more easily accessed by people (Fig. 23.1; Pitcher et al. 2007; Olesiuk 2011; Cammen et al. 2019).

It was almost 200 years after Georg Steller first reported his observations that a second paper was published on Steller sea lions. In it, Scheffer (1945) provided information on the timing of births, and reported the weights and measurements of

Fig. 23.5 A Steller sea lion rookery in Southeast Alaska in June showing typical substrate and the distribution of pups and adult females across the territories of four bulls (Photo by A.W. Trites)

11 young sea lions shot at a small sea lion rookery. He also described how mothers attempted to carry their young to safety by biting onto the skin fold of the neck or rump of their pups. He further noted how the young followed their mothers into the water.

A more complete account of the timing of births, reproductive behaviors, and the attendance patterns of bulls and adult female Steller sea lions was obtained in the 1960s and 1970s (Orr and Poulter 1967; Mathisen et al. 1962; Sandegren 1976; Gentry 1974). However, none of the behavioral observations documented weaning

Fig. 23.6 A Steller sea lion haulout in British Columbia in August showing typical substrate and the presence of mostly mature and immature females, as well as a few bulls and dependent juveniles. The two pups (near the water edge, left side) would have swum over 100 km from the closest rookery with their mothers (Photo by A.W. Trites)

behavior or provided baseline measurements that could be used to determine why the Western population of Steller sea lions was declining.

Variations in attendance patterns of pinnipeds (e.g., time spent at sea, and time spent nursing) are generally accepted to reflect variation in the availability of prey near breeding areas (Trillmich and Ono 1991; Hood and Ono 1997). For example, when the availability of prey declined during the 1982–83 El Niño, sea lions and fur seals from California to South America spent more time at sea searching for prey—and less time on land nursing their pups—because prey were difficult to obtain (Trillmich and Ono 1991). The perinatal period (time between giving birth and making the first feeding trip) was also shortened; and pups were underweight, and died at higher rates. These measures of maternal behavior were therefore considered useful proxies for inferring whether prey availability was reduced or adequate to support Steller sea lions (Hood and Ono 1997).

The challenge in using behavioral studies to resolve whether there was a prey shortage is that there were no baseline of historical behavioral data for comparison. Normal nursing times and normal foraging times were unknown for Steller sea lions. As a result, the control population for comparison with the declining Western population became the increasing Eastern population of Steller sea lions in Southeast Alaska, where sea lions were presumed to have more prey available to them. As such, sea lions in the declining population were predicted to compensate for reduced prey by making longer feeding trips and spend less time nursing their pups compared with sea lions in the increasing population. Results, however, proved counter-intuitive.

23.3 Breeding Behavior

Steller sea lions are the largest of the five species of sea lions, and are the fourth largest pinniped (after elephant seals and walrus). At birth, male pups are about 4% longer and 10% heavier than females—with male pups measuring 0.98 m and weighing 22 kg, and females averaging 0.94 m and 20 kg (Winship et al. 2001; Brandon et al. 2005). However, when fully grown and holding a territory (Figs. 23.4 and 23.7), males are 2.5 times heavier than adult females (~700 kg vs ~275 kg), but just 1.3 times longer (3.0 m males vs 2.3 m female). Breeding males will average ~ 700 kg at their peak body weight, although weights as heavy as 910 kg have been recorded in the wild (Fig. 23.7) (Winship et al. 2001) and some captive raised Steller sea lions have exceeded 1000 kg.

Aerial photographs taken in 2008 suggest that adult sea lions in the Eastern population were shorter than in the Western population (Sweeney et al. 2015). This difference in body size may reflect an inherent morphological difference between the two sea lion populations (Sweeney et al. 2015). However, it may also reflect relative differences in the quantity and quality of prey available to the two populations. Reduced food availability could have stunted body growth in the increasing region—while better feeding conditions augmented body growth in the declining population further north.

Steller sea lions grow quickly. By the time females have turned 4 years old, many are sexually mature and have attained ~90% of their asymptotic lengths (Winship et al. 2001). Most will ovulate for the first time between the ages of 3 and 5 years old (range 2–8 years), and will, on average, give birth for the first time when they are 5 years old (Pitcher and Calkins 1981; Perlov 1971). Female body mass, however, does not reach 90% of its asymptotic mass until the females are 13 years old. In contrast, males attain 90% of their final body length when much younger—at ~8 years old (Winship et al. 2001)—which coincides with the start of the male growth spurt in body mass. Similar to females, males become sexually mature between 3–8 years old (Pitcher and Calkins 1981; Perlov 1971). However, most males lack the social and physical maturity to defend a territory until they are 9 years or older (Fig. 23.7). Average age of ~90% of territorial bulls is 9–13 years old (range 6–15 years) (Thorsteinson and Lensink 1962).

Males have a much shorter life expectancy than females. About 40% of male pups die on average during their first year (range 27–60%), compared with an average death rate of 30% for female pups (range 29–44%) (Hastings et al. 2011; Pendleton et al. 2016; Maniscalco et al. 2015). Death rates can vary considerably between years and breeding sites. Of the sea lions that survive to their first birthdays, average age at death (life expectancy) is 4 years old for males (range 2–9 years) and 10 years old for females (range 2.5–13 years; calculated from life tables in: Pendleton et al. 2006;

Fig. 23.7 A scarred Steller sea lion bull weighing over 700 kg among young females (Photo by A.W. Trites)

Hastings et al. 2011; Maniscalco et al. 2015). The longest Steller sea lions might live (i.e., longevity—defined as the oldest age attained by $<1\%$ of the population) is about 16 years old for males and 23 years old for females.

Large body size provides males with an advantage over other bulls while breeding, but carries a metabolic cost outside of the breeding season. Thus, territorial bulls lose mass through the breeding season and into the fall before fattening up again from November to March (Winship et al. 2001). Females also experience seasonal growth and fattening between November and March, and grow little during the breeding season (May–July) (Winship et al. 2001).

Acquiring a territory is a prolonged process. Most bulls begin their tenure a year earlier by becoming a post-season territorial male (i.e., by establishing a territory on unused periphery areas of the rookery after copulation activity has ended). They will hold their empty territory until they can safely move onto the territory abandoned by a breeding bull (Gisiner 1985).

Most males that arrive at the start of the breeding season to reclaim past territories are considered dear-enemies of each other—having gotten to know each other through past interactions. New males that try to fight their way onto a rookery to establish a breeding territory have low success (<15% obtain a territory) compared to the post-season territorial males (Gisiner 1985). Bulls typically hold a territory for ~2 years (range 1–8 years) (Gisiner 1985; Parker and Maniscalco 2014).

Most combative challenges between competing bulls start as chest-to-chest pushing matches—much like sumo wrestlers—but can escalate into bloody fights. Fighting involves hard bites and shaking around the neck, shoulders, face, and edges of the front flippers (Fig. 23.7) (Gentry 1970; Gisiner 1985). The weaker of the two will either be pushed out or will take a submission posture while backing away and fleeing. These fights usually occur only with males that cross territorial boundaries or with unknown bulls approaching from the sea. For the most part, bulls engage in little fighting and spend most of their time making territorial displays towards neighboring territorial males (i.e., by rushing towards the territory boundary and falling forward on their bellies with head raised and mouth open towards the neighboring male) (Fig. 23.8) (Gentry 1970). However, territorial bulls have noticeably fewer boundary displays with bulls they know than with unfamiliar neighbors (Gisiner 1985).

The breeding season begins in early May with the arrival of bulls that held territories in past years (Gisiner 1985; Thorsteinson and Lensink 1962; Pitcher and Calkins 1981). They will wait for the first pregnant females to arrive a couple of weeks later. An average bull will maintain a territory for ~6 weeks (range 3–10 weeks) without eating (Gentry 1970)—and most will begin leaving their territories around mid-July when mating is complete and post-season males begin securing peripheral territories (Gisiner 1985; Parker and Maniscalco 2014).

The average territorial bull copulates ~7 times in a breeding season (range 0–38 times) (Gisiner 1985). However, as many as two-thirds of territorial bulls are thought to not copulate successfully in their first year (Parker and Maniscalco 2014). Those that are most successful tend to have used the rookery for 3 or more previous years, and have held territories that were near the water edge and center of the rookery with favored access to and from the water (Gisiner 1985; Parker and Maniscalco 2014). Another successful strategy for maximizing reproductive success is to occupy peripheral territories where fewer births occur, but a longer tenure (7–8 years) can be had (Parker and Maniscalco 2014). Lifetime reproductive success is thereby a combination of tenure duration and territory location.

Within a week of arriving on shore (starting in mid-May), females will give birth to a single pup (Fig. 23.9). Those first to arrive and give birth appear to be ~10–13 years of age, with progressively younger and older females giving birth as the season progresses (Hastings and Jemison 2016). Pups born earlier tend to be marginally heavier than later born pups, with male pups being born about a day earlier on average than female pups (Maniscalco and Parker 2017).

Fig. 23.8 Steller sea lion bulls making territorial displays towards each other. Note the scars on the face, chest and flippers from fights that involved hard bites and shaking. Note also the difference in color and texture of dry and wet pelage (Photo by A.W. Trites)

This parabolic relationship between age of mothers and her date of parturition (Hastings and Jemison 2016) is consistent with the non-linear relationship between age of northern fur seal mothers and the size of fetuses they carry (Trites 1991). Fetuses of northern fur seals increase in size as mothers age until the females are 10–13 years, after which adult females continue to growth, but the sizes of their fetuses decline (Trites 1991). This suggests that timing of birth in Steller sea lions reflects the size of a fetus at term, with larger fetuses being delivered earlier on average than smaller fetuses. Male pups are heavier on average than female pups (Brandon et al. 2005; Merrick et al. 1995), which suggests that differences in body size may also explain the apparent difference in timing of male and female births.

Nearly all births occur between May 15 and July 15 (Pitcher et al. 2001; Kuhn et al. 2017b). Mean date of birth is earliest in Southeast Alaska (June 4) and becomes progressively later further north and further south (by as much as over 2 weeks later in California) (Pitcher et al. 2001). Almost 90% of pups are born within a ~3-week window (Pitcher et al. 2001).

As parturition approaches, most pregnant females (~80–85%) go to the rookery of their birth in the Eastern population, and a slightly smaller proportion (~75%) return to their natal rookeries in the Western population (Hastings et al. 2017; Raum-Suryan et al. 2002). Females choose a location within a bull's territory, and tend to return to the same areas in subsequent years (Hastings et al. 2017)—with some females pupping within about 6 m of where they pupped the previous year (Parker et al. 2008).

Following birth, mothers remain with their pups for an average of 8–11 days (range 3–13 days) before leaving for their first feeding trip (the perinatal period; Sandegren 1970; Hood and Ono 1997; Milette and Trites 2003; Maniscalco et al.

Fig. 23.9 A nursing Steller sea lion pup. Females have four or six functional teats, such as this female with six teats (Photo by A.W. Trites)

2006). Typically, first time mothers (i.e., primiparous females aged 3–6 years) return to sea to feed before multiparous females—due perhaps to having a smaller body size with less body reserves than multiparous females to produce milk for their pups (Maniscalco et al. 2006; Hastings and Jemison 2016).

Copulations typically occur at the end of the perinatal period before females leave their pups for the first time (Sandegren 1970; Gentry 1970). Males rarely hinder female movement across territories for more than a few minutes, and general rely on olfactory senses in most pre-copulatory interactions to detect estrus (>65%) (Gisiner 1985). In other cases, females may initiate sexual solicitation (<10%), or males may spontaneously mount females with no pre-copulatory interactions (<25% of copulations) (Gisiner 1985).

Once the egg is fertilized in June and early July, the blastocyst will remain dormant until implantation in the fall (~3–3.5 months later between late September and mid-October) (Pitcher and Calkins 1981). Based on the distribution of births (occurring as early as late May and as late as the end of June), gestation is between 8–9 months, and likely averages ~8.5 months.

Some females are seen nursing juveniles during the breeding season at their natal rookeries (~40% of mothers with juveniles), at other rookeries (~20%), and at haulouts (~40%) (Hastings et al. 2017). Some of these females returning to rookeries with juveniles in tow will give birth, and will have to choose between keeping their pup and continuing to nurse their juvenile. Females nursing juveniles at haulouts will breed with territorial males (which occurs ~1–2 weeks earlier than on rookeries; Trites, unpubl. data). Thus, females nursing juveniles during the breeding season have mating options (Hastings et al. 2017).

When pups are young (~2 weeks), females spend ~1.2 days with their pups on land and 0.8 days at sea (Milette and Trites 2003). As the pup ages, mothers spend more time at sea and less time on shore, such that by ~7 weeks old, females are spending ~0.5 days with their pups and 1.1 days at sea (Milette and Trites 2003). The increased time spent feeding is presumed to reflect the time needed to increase milk production to meet the increased energetic needs of the growing pup.

Pups will enter the water for the first time when 2–4 weeks old (Sandegren 1970), and will move with their mothers to haulouts when 2–3 months old (Fig. 23.6; Calkins and Pitcher 1982). Mothers with pups may start dispersing from the rookeries as early as the last week of June, with significant increases in numbers leaving during the last two weeks of July (Kuhn et al. 2017a; Hastings and Jemison 2016). Some females will call their pups into the water, while others will drag and carry them by the scruffs of their necks, and block their attempts to climb back onto the rookery.

Dependent pups generally stay within 500 km of where they are born, although a few as young as 5 months have been seen with their mothers at haulouts further away (Raum-Suryan et al. 2002). Juveniles (1–3 years old) disperse much further than pups (up to ~2000 km), with males being resighted further on average from their natal rookeries, and at older ages, than females (Raum-Suryan et al. 2002; Fuller 2012).

Steller sea lions rely on a thick layer of blubber to insulate themselves from the cold water. However, while on land, their fur provides some additional protection from convective heat-loss to air and wind. Steller sea lion fur is thick, short and coarse when dry—and lies flat against the skin when wet to reduce drag while swimming (Fig. 23.8). However, the main function of the fur is to protect the skin from abrasion against rocks and barnacles. As such, the hairs become damaged over time, and are replaced each year during an annual molt (Daniel 2003).

It takes ~45 days for a Steller sea lion to shed its old hair (Daniel 2003). The first age-class to molt are the juveniles (starting in June), which are followed by adult females (August), and bulls and pups (October) (Daniel 2003). The molt progresses over the body surface much like an unravelling sweater. Pups will lose their dark chocolate-colored coats—and older sea lions will attain a fresh tan to golden-brown colored coat, which darkens to chocolate brown on their flippers and underside (Fig. 23.9). However, once wet, the pup appears black while the coats of older age-classes look dim- or dark-gray (Figs. 23.4 and 23.8).

23.4 Behavioral Proxies of Environmental Conditions

Changes in the abundance of prey available to fur seals and sea lions are known to affect the duration of perinatal periods, and the time that mothers spend foraging and nursing their pups (e.g., Ono et al. 1987; Trillmich and Ono 1991; Boyd et al. 1994; McCafferty et al. 1998; Goldsworthy 2006). Dive depths and duration of dives can also be used to infer differences in the abundance and distribution of prey accessible to different populations of pinnipeds (e.g., Chilvers 2018). Collectively, these measures of maternal investment and foraging efficiency translate into differences in growth and weights of pups (e.g., Kirkman et al. 2002; Goldsworthy 2006; Jeanniard-du-Dot et al. 2017; Trillmich 1990; Trillmich and Limberger 1985; Roux 1997)—and can be used to assess the relative abundance of prey available to fur seals and sea lions.

Behavioral studies of Steller sea lions in the 1990s set out to compare the behavioral ecology of sea lions in the declining and increasing populations. Under the assumption that the Western population in the Gulf of Alaska was declining because of a shortage of prey, nursing females were predicted to have (1) short perinatal periods, (2) spend more time searching for prey, (3) have greater difficulty finding and capturing prey, and (4) spend less time nursing their pups (Milette and Trites 2003; Trites and Donnelly 2003). It was further predicted that pups in the declining population should be (5) smaller during the breeding season than pups in the increasing population (Trites and Donnelly 2003). However, none of the behavioral studies yielded results consistent with Steller sea lions in the declining population having difficulty accessing prey. Rather they indicated that the population having trouble finding food was the increasing population in Southeast Alaska.

Counter to predictions, Steller sea lions spent more time with their pups where the population was declining than in the areas where the population was growing. Mean perinatal periods (time between birth and first feeding trip) of females observed at rookeries were significantly longer where they were declining (9.9 vs 7.9 d), and the length of their foraging trips was also shorter on average (19.5 vs 24.9 h) (Milette and Trites 2003). In addition, females in the declining population spent more time with their pups between feeding trips (27.0 vs 22.6 h) (Milette and Trites 2003). Satellite-tracked lactating females were also found to make shorter feeding trips in the area of decline (Seguam Island) than in the area of growth (Forrester Island) (Andrews et al. 2002). These females from the declining population generally made single bouts of uninterrupted dives to capture prey, while those from the increasing population made multiple bouts separated by time travelling and resting (Andrews 2004). These tracked sea lions were also 5-times faster at capturing their first prey in the declining area than in the increasing area (based on stomach temperature records) and ingested it twice as fast (Andrews et al. 2002). Thus, sea lions in the declining area appeared to find and consume prey faster, and spend more time nursing their pups compared with sea lions in the increasing population.

Body weights of pups taken throughout the Steller sea lion's range also supported the conclusion that lactating Steller sea lions in the declining regions were not having difficulty obtaining prey during summer. Instead of being undersized, pups grew faster in the declining areas than in the increasing areas (0.41 vs 0.25 kg/day), and required half as much milk (Adams 2000). Pups (2–6 weeks old) were heavier on average at declining rookeries (Aleutian Islands and Gulf of Alaska) compared to pups at stable and increasing rookeries (Southeast Alaska and Oregon) (Merrick et al. 1995; Rea 1995). Older pups (7–9 months old) were also similarly heavier and in better condition in the Aleutian Islands (with 36% total body lipid) and Gulf of Alaska (30% body fat) than in Southeast Alaska (28% body fat) where the population was increasing (Rea et al. 2016). No one expected Steller sea lion pups from the increasing population to be leaner than pups from the declining population. Nor did anyone expect pups in the declining area to be heavier in the 1980s and 1990s than during 1965–1975—prior to when the population declined (Merrick et al. 1995).

The most notable change that occurred in the Gulf of Alaska, Bering Sea and Aleutian Islands pre- and post-decline was a dramatic shift in ocean climate conditions, ecosystem community structure, and sea lion diets. Prior to the decline, fat-rich forage fishes such as smelts and sand lance made up a significant portion of Steller sea lions diets in the Gulf of Alaska—to the exclusion of walleye pollock (Mathisen et al. 1962; Thorsteinson and Lensink 1962). However, pollock dominated the post-decline diet (a low-lipid fish not reported in the earlier diet studies) (Sinclair et al. 2013; Sinclair and Zeppelin 2002). This dietary shift in the Gulf of Alaska corresponded with an oceanic regime shift in 1976–1977 (Hare and Mantua 2000; Benson and Trites 2002) that altered the structure of fish communities (Anderson and Piatt 1999; Trites et al. 1999) and the relative abundances of dominate prey species available to Steller sea lions (Alverson 1992; Trites et al. 2007b; Bakkala et al. 1987; Bulatov 2014).

Regional populations of Steller sea lions in the Western population that experienced the most rapid declines consumed diets dominated by energy-poor species (Merrick et al. 1997; Winship and Trites 2003) that consisted primarily of pollock in the Gulf of Alaska, and Atka mackerel (a hexagramid) and Pacific cod in the Aleutian Islands (Sinclair et al. 2013; Tollit et al. 2017; Sinclair and Zeppelin 2002). In contrast, the growing population of sea lions in Southeast Alaska were consuming a more energy-rich and diverse diet dominated by gadids, forage fish and salmon (Trites et al. 2007a) that appears to be closer to what was once consumed by sea lions in the Gulf of Alaska prior to the population decline.

As the Western population of Steller sea lions rapidly declined through the late 1970s and 1980s (Fig. 23.2), Atka mackerel increased in the Aleutian Islands, and walleye pollock increased in the Gulf of Alaska and Bering Sea (Bakkala et al. 1987; Bulatov 2014; Merrick 1997). These two prey species appear to have been very abundant and available to lactating females during summer and fall throughout the period of sea lion decline based on the ease with which mothers seemed to capture them, the relative shortness of their feeding trips, time spent nursing pups, and increased weights of pups (from birth—9 months of age). Thus, the high abundance of pollock and Atka mackerel are consistent with the observed maternal behaviors and increased sizes of pups—and suggest that lactating females in the declining population were not food limited. Rather, they appear to have had abundant prey that allowed them to nurse bigger pups. However, the high abundance of these prey species may have placed a nutritional limitation on the ability of pups to successfully wean as shown by behavioral observations of the weaning process and the frequencies of abortions.

23.5 Weaning Behavior

Comparing the age composition of the Steller sea lion population in the 1970s and 1980s revealed significantly fewer juvenile sea lions among the animals present in the 1980s (York 1994). This relatively small number of juveniles in the 1980s led many to conclude that the decline of sea lions in the Gulf of Alaska was due to an increase in juvenile mortality (Merrick and Loughlin 1997; Alaska Sea Grant 1993; DeMaster and Atkinson 2002). One line of logic held that the missing juveniles had starved and died after weaning because fisheries had reduced prey biomass and altered the distribution of fish consumed by sea lions. It was theorized that young sea lions were food-limited because their poor diving abilities prevented them from catching prey that had become less accessible (Merrick and Loughlin 1997). However, there was no evidence that fisheries had reduced the biomass or changed the distribution of prey available to Steller sea lions. Nor were there observations of starving juveniles to support the hypothesis that young sea lions were nutritionally stressed due to consuming less prey. Research was therefore initiated to find malnourished juveniles and document the poorly understood weaning process that potentially held the key to understanding why trajectories of the Western and Eastern populations of Steller sea lions differed.

It has been generally accepted that most Steller sea lions wean during their first year, with some young staying with their mothers for as long as 3 years (Pitcher and Calkins 1981). Less certain, however, was whether Steller sea lions wean from mid-April to late-May before the start of the next breeding season (Pitcher and Calkins 1981; Raum-Suryan et al. 2004), or whether they wean much earlier between November and March (Merrick and Loughlin 1997; Loughlin et al. 2003). The extent to which immature sea lions supplemented their milk diet with fish (and when this might occur) was equally uncertain (Kuhn et al. 2017a; Raum-Suryan et al. 2004; Rehberg and Burns 2008). Nor was it understood why some sea lions should continue to suckle beyond their first year.

Attempts to document the weaning process began with field observations of Steller sea lions at a winter haulout (late January to end of March) when young were expected to transition from milk to independent foraging (Trites and Porter 2002). During this time, mothers were observed coming and going from their haulout, while dependent pups (8–10 months) and yearlings (20–22 months) played in the intertidal zone or made short independent trips. No immature sea lions were observed leaving and returning with their mothers on their feeding trips. However, some dependent young were seen playing with fish and octopus near their haulout—although none of these sea lions were weaned, and none were observed swallowing the prey in their mouths (Trites and Porter 2002).

A few pups at the winter haulout appeared to have lost their mothers and were starving (Porter and Trites 2004). Some of these pups tried to steal milk from other mothers, but were rebuffed with bites when discovered by the females. Curiously, none of the starving pups switched to eating fish that similarly sized pups had been observed to be readily capable of catching.

The tenacity of the starving pups to acquire milk rather than catch fish was similar to what occurred when bottle-raised Steller sea lions pups were weaned at the Vancouver Aquarium in the early 1990s (Trites, unpubl. data). Under the mistaken belief that wild pups weaned during winter, the captive-raised pups were cut off of milk and offered live fish during January when 8 months-old. All of the hand-raised pups were adept at catching and killing the fish they were offered, and played with them ceaselessly until only bits of flesh hung from the skeletons. However, none of these 8-month-old pups—which were no longer receiving milk—would consume any of the fish they killed.

Subsequent attempts to switch the trained pups from milk to solid food (after giving up on them eating live fish) involved putting fish into their mouths, and holding their mouths closed until they swallowed (Trites, unpubl. data). Initial attempts at force-feeding resulted in the unweaned pups coughing the fish back up. With time and patience, however, the pups accepted whole fish. This experience in trying to hand-wean Steller sea lion pups—and the subsequent lack of field observations that weaning occurs during winter (Trites and Porter 2002)—pointed to a flawed conceptual understanding of weaning and the factors that influence it. Neither weaning nor transitioning to live food occurs during winter (Jan–Mar).

Additional teams of researchers sent to observe sea lions behavior at haulouts during spring and summer (Apr–Aug) concluded that most sea lions begin supplementing their milk diet with fish between April and May, and wean before the start of the breeding season when 1 or 2 years old (Trites et al. 2006). These conclusions were drawn from the marked decline in the proportions of time that immature sea lions spent suckling (Trites et al. 2006), and are consistent with an independent study of vibrissae from known-aged immature sea lions between November and April that looked for trends in stable isotopes indicative of fish consumption (Rehberg et al. 2018).

As with their reduced time spent suckling, the depths and durations of dives of immature sea lions also change significantly towards the end of the first and second years (May and June) when pups are 11–12 months old, and yearlings are 22--24 months old (Loughlin et al. 2003; Rehberg and Burns 2008). The diving behaviors of the dependent pups and yearlings (recorded by satellite dive-depth recorder tags) become more adult-like in May and June. Maximum daily dive depths and dive durations also increase around their first and second birthdays (Pitcher et al. 2005). Similarly, round-trip distances and durations of trips made by tracked pups and yearlings show notable changes from April to June (Raum-Suryan et al. 2004). Collectively, all of these studies of diving and movement behaviors suggest that pups and yearlings undertake a gradual process of supplementing milk with solid food 2–3 months before the oncoming breeding season when they are most likely abruptly weaned.

Mothers of dependent pups and yearlings make significantly longer foraging trips during winter than in spring and summer (Trites et al. 2006; Merrick and Loughlin 1997). These lactating females spend about 2 days away from the haulout during winter, and 1 day away during spring (Trites et al. 2006). Lactating females also make longer trips if nursing a yearling than if supporting a pup (Trites et al. 2006).

However, no difference was noted in the attendance patterns of lactating females between regions during winter and spring (i.e., between increasing and decreasing populations) (Trites et al. 2006). This suggests that lactating sea lions observed at haulouts in the 1990s did not have more difficulty capturing prey during winter, spring and summer in the area of decline compared to where sea lions were increasing.

Male and female pups and yearlings do not appear to follow the same weaning schedule (Trites et al. 2006). During the 1990s, about half of the female pups observed in Southeast Alaska weaned at 1 year, and the other half continued suckling until 2 years old. In contrast, most males weaned at 2 years—with a small number weaning at 3 years, and one male continuing to suckle at 4 years of age (Trites et al. 2006).

This difference between the age at which young males and females wean presumably reflects their different size-based nutritional needs and their likelihood of surviving and successfully reproducing when independent. Weaning (at 1 to 3 years of age; Pitcher and Calkins 1981; Trites et al. 2006) appears to occur at the start of summer when survival is likely optimal, and pregnant mothers must return to rookeries to give birth and mate. However, the proportion of young sea lions that wean at 1, 2 or 3 years old likely varies by year due to regional and temporal differences in the available quantity and quality of prey needed to ensure the survival of recently weaned individuals.

Some biologists have assumed that the limited diving abilities of pups make them nutritionally dependent on their mothers during winter (Rehberg and Burns 2008). However, this explanation for why pups do not wean earlier is unlikely given that yearlings have the diving abilities of adults, but also continue to depend on their mothers for nutrition. Dependent pups and yearlings are also all capable of catching prey as shown by field observations (Porter and Trites 2004). The fact that pups and so many yearlings remain with their mothers rather than wean implies that they obtain greater energy (and benefit) from milk than from fish to support rapid body growth and high daily energy needs (Winship et al. 2002; Trites et al. 2006).

Feeding trials and bioenergetics models show that growth of young sea lions (from birth to 2 years) can be achieved on a diet of milk (Winship et al. 2002; Rosen and Trites 2000, 2004; Rosen 2009). Similarly, older pups and yearlings can grow on a diet of high-fat fish (such as herring, sardines, and sand lance). However, young Steller sea lions can lose body mass on a diet of only low-fat fish (such as cod and pollock) because they become full before meeting their daily energy needs (Rosen and Trites 2000, 2004; Rosen 2009). Older and larger sea lions, however, have lower relative energy requirements and can do equally well on high- or low-fat fish—they simply have to eat more of the low-fat fish to compensate for the caloric difference.

Consuming fish at the expense of not having room for milk—or simply replacing milk with fish—would most likely lead to the death of a pup in the wild. The behavior of the captive-raised Steller sea lion pups when presented with fish further suggests that pups evolved the skills to hunt and kill by at least 8 months, but not to swallow because they cannot assimilate sufficient energy from a stomach full of fish to meet their daily energy requirements. This simple physiological process can explain the observations made of starving and healthy pup behaviors in the wild

(Porter and Trites 2004). It can also explain why so many sea lions stay an extra year with their mothers when milk is an option.

The community structure of the ecosystem in which Steller sea lions evolved periodically shifts between one dominated by low-energy gadids and one dominated by high-energy forage fishes (Trites et al. 1999, 2007b; Maschner et al. 2014; Benson and Trites 2002; Mantua et al. 1997; Alverson 1992; Anderson and Piatt 1999; Hare and Mantua 2000). Such an ecosystem shift is consistent with the observed shift in Steller sea lion diet from forage fishes to gadids following the 1976–1977 oceanic regime shift (Mathisen et al. 1962; Thorsteinson and Lensink 1962; Sinclair et al. 2013; Sinclair and Zeppelin 2002). However, pups weaning at 1 year of age are unlikely to survive on a gadid-dominated diet, but could do so on a diet dominated by energy-rich forage fish (Rosen and Trites 2000, 2004; Rosen 2009). Young sea lions simply do not have the stomach capacity to process enough gadids to meet their daily needs. Thus, Steller sea lions should stay an extra year with their mothers until they are large enough (approaching 2 years of age) to continue normal growth on a diet dominated by gadids. This plasticity in age of weaning displayed by Steller sea lions is likely an evolutionary adaptation to natural shifts in ocean productivity in the North Pacific.

The thought that an animal can get full before it has consumed enough to meet its daily needs (or that energy density of fish matters to apex marine predators) is accepted by seabird biologists (e.g., Romano et al. 2006; Grémillet et al. 2008; Osterblom et al. 2008; Whitfield 2008; Jodice et al. 2006), but is rejected by some marine mammalogists (e.g., Fritz and Hinckley 2005; Calkins et al. 2013). Arguments raised to dispute the notion that quality of prey matters to young sea lions range from an obfuscation of facts to failure to recognize that feeding studies purported to disprove that young animals require high-caloric food are actually consistent with this simple premise. They also fail to recognize that methodically weaning pups on a fixed schedule as done by most other pinniped species is a poor evolutionary strategy for mothers to follow if there is little likelihood of offspring surviving to pass on their mother's genes during years when low-energy prey species will dominate yearling diets.

If quality of prey and stomach capacity were inconsequential to young Steller sea lion, males should wean earlier than females—thereby allowing young males to gorge on low-energy prey species and free themselves of the 1–2 day waits for mothers to return to haulouts to nurse them. However, the opposite occurs—males tend to wean later than females. This suggest that the sporadic delivery of milk is significantly more advantageous to males than daily consumption of fish to ensure that males attain sufficient body sizes to successfully hold a territory and breed. Larger sizes at weaning likely ensure higher breeding success as adults.

Feeding trials, mathematical models, and field observations of Steller sea lions all suggest that age at weaning is a function of the energy densities of fish available to young sea lions. They further suggest that age at weaning should be later in the declining sea lion population because the average energy densities of fish available to weaning sea lions is significantly less than the energetic densities of prey available to sea lions weaning in the increasing population (Trites et al. 2007b; Winship and

Trites 2003). Such an increase in age at weaning would increase the inter-annual interval between females having subsequent pups (from 1 to 2 or 3 years)—thereby causing birth rates and the size of the population to fall.

While age at weaning can be linked to quality of prey available to weaned individuals, determining who initiates the weaning process is less obvious—as is the question of how Steller sea lions assess the likelihood that the prey base at weaning will support recently weaned individuals. Possible answers lie with the high frequency of abortions and stillbirths among Steller sea lions.

23.6 Abortions and Stillbirths

Collectively, the behavioral observations of Steller sea lions made at rookeries and summer haulouts indicate that most mature females breed each year either on rookeries following the birth of a pup, or on haulouts if they are not pregnant and are nursing a pup or yearling. First time breeders may breed on either rookeries or haulouts, and some non-pregnant females with dependent young may return to rookeries instead of haulouts to breed. Thus, females have mating choices and opportunities to ensure maximum numbers become pregnant each year.

Having sex, becoming pregnant, and carrying a fetus to term comes at a small energetic cost compared to the cost of producing milk (Winship et al. 2002). It therefore makes sense for all females to become pregnant each year and give birth 12 months later if they can successfully wean their offspring. It would also make evolutionary sense to cease pregnancies during years when the prey base is insufficient to support weaned offspring—thereby ensuring that pups and yearlings approaching weaning remain with their mothers for another year until they have grown big enough to thrive on a low-energy diet.

One means by which pregnancies might be terminated is through a hormonal process associated with lactating and consuming a high-protein low-fat diet that triggers spontaneous abortions midway through fetal development. Such a cause-and-effect scenario is consistent with observing aborted fetuses at haulouts during winters (Fig. 23.10; Marcotte 2006; Gentry 1970; Mate 1973; Calkins and Pitcher 1982), and with seasonal trends in pregnancy rates observed during the 1970s and 1980s before and during the Western population decline (Pitcher et al. 1998; Pitcher and Calkins 1981).

Examining the reproductive tracts of Steller sea lions shot in the Gulf of Alaska showed that all sexually mature females are pregnant during early gestation (Pitcher et al. 1998). However, nearly half may lose their fetus before the start of the breeding season as occurred during the start of the population decline (1980s) when only 55% of sexually mature females remained pregnant in late gestation (Pitcher and Calkins 1981). Conversely, a greater percentage (67%) remained pregnant at this stage in the 1970s prior to the population decline (Pitcher and Calkins 1981).

The decline in pregnancy rates that occurs as the fetus develops is consistent with the discovery of fetuses on Steller sea lion haulouts from December through mid-May (Fig. 23.10; Marcotte 2006; Gentry 1970; Mate 1973; Calkins and Pitcher

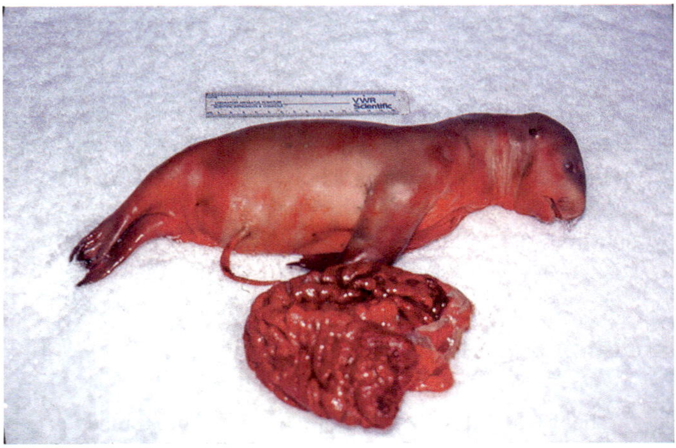

Fig. 23.10 An aborted Steller sea lion fetus found on a haulout during January. Photographed on snow with a 15 cm ruler for scale (Photo by A.W. Trites)

1982). The highest frequency of abortions occur from January to March during the second trimester of pregnancy when the fetus is about 2–5 months developed. Fetuses collected and tested for disease known to cause abortions in mammals have not revealed underdevelopment or pathogen-related causes (Esquible et al. 2019; Burek et al. 2005). There is no medical explanation for the high frequency of abortions among Steller sea lions.

Most of the females aborting fetuses are likely still nursing pups and yearlings. During the 1980s, pregnancy rates during late gestation were just 30% for lactating females compared with 84% for non-lactating females (Pitcher et al. 1998). It is equally noteworthy that pregnancy rates of lactating females were twice as high during late gestation in the 1970s when Steller sea lions were presumed to consume a more energy-rich diet than in the 1980s (i.e., 63% in the 1970s vs 30% pregnancy rates in the 1980s) (Pitcher et al. 1998).

The small number of lactating females that return to give birth on rookeries with a pup or yearling in tow face inevitable squabbles with juveniles pushing newborns away or attempting to suckle at the same time. Females do not appear to continue supporting both, and ultimately appear to choose juveniles over newborns.

From a strategic point of view, it makes sense that all lactating females should become pregnant, and that their pregnancies should continue if their pup or yearling dies, or conditions prove favorable for successful weaning. Under favorable conditions, females could wean a pup as frequently as once a year. However, should conditions not favor successful weaning, aborted fetuses ensure that mothers can continue putting all of their energy into their pups for a second year or longer. Thus, abortions in sea lions may be an evolutionary selected process that simultaneously maximizes reproductive potential and increases the likelihood that newborns survive and pass on a mother's genes. The high frequency of abortions among Steller sea lions is most likely an innate hormonal response associated with lactation and lipid content of prey—and not a function of female choice or body condition.

23.7 Conclusions

Behavioral ecology can help resolve conservation conflicts and aid species at risk—and has proven to be a powerful means to test whether the behaviors of the declining population of Steller sea lions were consistent with the predictions of overfishing or over-abundance of low-quality prey. However, behavioral studies require context from physiological ecology and a solid understanding of the natural history of the species at risk to attain their full power.

So much of what has been learned about the behavioral ecology of Steller sea lions has come from graduate thesis research (e.g., Sandegren 1970; Gentry 1970; Harestad 1973; Gisiner 1985; Kucey 2005; Marcotte 2006; Milette 1999; Parker 2006; Porter 1997; Smith 1988; Higgins 1984; Vazquez 2013; Scordino 2006; Fuller 2012; Merrick 1987; Daniel 2003; Brandon 2000; Keech 2008). These university-based studies represent tens of thousands of hours of observations by students and their assistants in California, Oregon, British Columbia and Alaska—and are the backbone of much that has been learned about Steller sea lion behavior. Each of the behavioral studies contributed a small piece of knowledge needed to bring the puzzling natural history of Steller sea lions into focus. Without these core behavior studies, it would not have been possible to make behavioral inferences about food availability or causes of population declines and recoveries.

Behavioral ecology has provided significant insights into the decline and natural history of Steller sea lions. It has revealed that Steller sea lions wean when about 12, 24 or 36 months old, and that they start transitioning to solid food 2–3 months before they wean. The exact age at weaning appears to depend on whether their mothers remain pregnant in the last trimester of fetal development. It is unknown, however, if late-term pregnant-females discourage their pups and yearlings from suckling, or if some other progressive change initiates the weaning process. In all likelihood, the final stage of weaning is probably abrupt, and may well occur when their pregnant mothers leave them at the haulout and return alone to their natal rookeries to give birth.

All sexually mature females appear to breed each year on rookeries and haulouts, regardless of whether they have recently given birth or are still nursing a 1- or 2-year old. Most non-lactating females carry their fetuses to term. However, up to half of lactating females may lose their fetuses during the second trimester of pregnancy. Females that miscarry will likely continue nursing their pup or yearling for another year. Rates of abortions may therefore be related to lactation and the extent to which mothers consume species that are low in fat and high in protein.

Behavioral observations revealed regional differences in feeding behaviors of Steller sea lions that were inconsistent with there being less prey available to sea lions in the declining population. Instead, maternal attendance patterns and measures of foraging success indicate that more prey were available to the declining population than to the increasing population. However, the greater abundance of prey available to the declining population was of lower average quality (i.e., energy density) than the prey available to the increasing population. Sea lions beyond the age of 1.5 years appear to have no trouble meeting their daily energy requirements by increasing daily amounts of food consumed. However, the same cannot be said for yearlings that do not yet have sufficient stomach capacity.

Low quality prey likely explains the increased rate of abortions that occurred during the decline, which in turn allowed females to keep their dependent young longer and increase the interval between having subsequent pups. This longer period of dependency would increase body size and enhance reproductive fitness, but may come at the cost of increased risk of predation on dependent young by killer whales—a significant source of death (Barrett-Lennard et al. 1995; Williams et al. 2004; Guénette et al. 2007; Matkin et al. 2007).

The plasticity observed in age of weaning of Steller sea lions is likely an evolutionary adaptation to survive the reorganization of fish communities known to follow ocean-climate regime shifts in the North Pacific Ocean (Trites et al. 2007b). Extending age at weaning is one means by which population declines would slow until the ocean shifts naturally back to a state that supports a more energy-rich prey field and greater numbers of Steller sea lions. Shifts in the abundance of walleye pollock, Pacific cod and Atka mackerel (such as followed the 1976–1977 oceanic regime shift) have occurred repeatedly over millennia—as have shifts in sea lion numbers (Trites et al. 2007b; Maschner et al. 2014). Recovery of Steller sea lions requires another regime shift to increase the energy density of prey available to pups and yearlings—and may indeed be underway in the Gulf of Alaska and Eastern Aleutians where recent increases in sea lion numbers have been observed.

Acknowledgements I am grateful to the many biologists, graduate students and field assistants whose thousands of hours of behavioral observations have made it possible to decode the natural history of Steller sea lions and resolve a perplexing ecological mystery. I also thank Selina Agbayani for GIS support, David Rosen, Kate Colson and Claudio Campagna for comments and feedback, and Vladimir Burkanov and Chad Nordstrom for providing updated locations of currently used rookeries and haulouts in Russia and Canada.

References

Adams TC (2000) Foraging differences and early maternal investment in adult female Alaskan Steller sea lions (*Eumetopias jubatus*). PhD thesis, Texas A&M University, Galveston

Alaska Sea Grant (1993) Is it food? Addressing marine mammal and sea bird declines. Fairbanks, AK

Alverson DL (1992) A review of commercial fisheries and the Steller sea lion (*Eumetopias jubatus*): the conflict arena. Rev Aquat Sci 6:203–256

Anderson PJ, Piatt JF (1999) Community reorganization in the Gulf of Alaska following ocean climate regime shift. Mar Ecol Prog Ser 189:117–123

Andrews RD (2004) The population decline of Steller sea lions: testing the nutritional stress hypothesis. In: Gordon M, Bartol S (eds) Experimental approaches to conservation biology. University of California Press, Berkeley, CA, pp 132–146

Andrews RD, Calkins DG, Davis RW, Norcross BL, Peijnenberg K, Trites AW (2002) Foraging behavior and energetics of adult female Steller sea lions. In: DeMaster D, Atkinson S (eds) Steller sea lion decline: is it food II? University of Alaska Fairbanks, Alaska Sea Grant College Program (Report No. AK-SG-02-02), Fairbanks, AK, pp 19–22

Atkinson S, Demaster DP, Calkins DG (2008) Anthropogenic causes of the western Steller sea lion *Eumetopias jubatus* population decline and their threat to recovery. Mamm Rev 38:1–18

Baker AR, Loughlin TR, Burkanov V, Matson CW, Trujillo RG, Calkins DG, Wickliffe JK, Bickham JW (2005) Variation of mitochondrial control region sequences of Steller sea lions: the three-stock hypothesis. J Mammal 86:1075–1084

Bakkala RG, Wespestad VG, Low L-L (1987) Historical trends in abundance and current condition of walleye pollock in the eastern Bering Sea. Fish Res 5:199–215

Ban S, Trites AW (2007) Quantification of terrestrial haul-out and rookery characteristics of Steller sea lions. Mar Mammal Sci 23:496–507

Barrett-Lennard LG, Heise K, Saulitis E, Ellis G, Matkin C (1995) The impact of killer whale predation on Steller sea lion populations in British Columbia and Alaska. University of British Columbia, Fisheries Centre, 2204 Main Mall, Vancouver, BC V6T 1Z4

Benson AJ, Trites AW (2002) Ecological effects of regime shifts in the Bering Sea and eastern North Pacific Ocean. Fish Fish 3:95–113

Bickham JW, Loughlin TR, Calkins DG, Wickliffe JK, Patton JC (1998) Genetic variability and population decline in Steller sea lions from the Gulf of Alaska. J Mammal 79:1390–1395

Bigg MA (1985) Status of the Steller sea lion (*Eumetopias jubatus*) and California sea lion (*Zalophus californianus*) in British Columbia. Can Spec Publ Fish Aquat Sci 77:1–20

Boyd IL, Arnould JPY, Barton T, Croxall JP (1994) Foraging behaviour of Antarctic fur seals during periods of contrasting prey abundance. J Anim Ecol 63:703–713

Braham HW, Everitt RD, DJ R (1980) Northern sea lion population decline in the eastern Aleutian Islands. J Wildl Manag 44:25–33

Brandon EAA (2000) Maternal investment in Steller sea lions in Alaska. Ph.D. thesis, Texas A&M University, Galveston, Texas, USA

Brandon EAA, Calkins DG, Loughlin TR, Davis RW (2005) Neonatal growth of Steller sea lion (*Eumetopias jubatus*) pups in Alaska. Fish Bull 103:246–257

Bulatov OA (2014) Walleye pollock: global overview. Fish Sci 80:109–116

Burek KA, Gulland FMD, Sheffield G, Beckmen KB, Keyes E, Spraker TR, Smith AW, Skilling DE, Evermann JF, Stott JL, Saliki JT, Trites AW (2005) Infectious disease and the decline of Steller sea lions (*Eumetopias jubatus*) in Alaska, USA: Insights from serologic data. J Wildl Dis 41:512–524

Burkanov VN, Loughlin TR (2005) Distribution and abundance of Steller sea lions, *Eumetopias jubatus*, on the Asian coast, 1720's–2005. Mar Fish Rev 67(2):1–62

Calkins DG, Pitcher KW (1982) Population assessment, ecology and trophic relationships of Steller sea lions in the Gulf of Alaska. In: Environmental assessment of the Alaskan continental shelf: final reports of principal investigators, vol 19. U.S. Dept. Commer., NOAA, Juneau, Alaska, pp 445–546

Calkins DG, Atkinson S, Mellish J-A, Waite JN, Carpenter JR (2013) The pollock paradox: juvenile Steller sea lions experience rapid growth on pollock diets in fall and spring. J Exp Mar Biol Ecol 441:55–61

Cammen KM, Rasher DB, Steneck RS (2019) Predator recovery, shifting baselines, and the adaptive management challenges they create. Ecosphere 10:e02579

Cheever E, Riley T (2019) Steller sea lion thirty year review. Bibliography. US Department of Commerce, NOAA Central Library, Silver Spring, Maryland

Chilvers BL (2018) Preliminary assessment of the foraging behaviour and population dynamics of a cryptic population of the endangered New Zealand sea lion. N Z J Ecol 42:48–57

Conn PB, Johnson DS, Fritz LW, Fadely BS (2014) Examining the utility of fishery and survey data to detect prey removal effects on Steller sea lions (*Eumetopias jubatus*). Can J Fish Aquat Sci 71:1229–1242

Daniel RG (2003) The timing of moulting in wild and captive Steller sea lions (*Eumetopias jubatus*). MSc thesis, University of British Columbia, Vancouver, BC

DeMaster D, Atkinson S (eds) (2002) Steller sea lion decline: Is it food II? University of Alaska Fairbanks, Alaska Sea Grant College Program (Report No. AK-SG-02-02), Fairbanks, AK

Esquible JA, Burek-Huntington K, Atkinson S, Klink AC, Bortz E, Goldstein TA, Beckmen K, Pabilonia K, Tiller R (2019) Pathological findings and survey for pathogens associated with reproductive failure in perinatal Steller sea lions *Eumetopias jubatus*. Dis Aquat Org 137:131–144

Fritz LW, Hinckley S (2005) A critical review of the regime shift-"junk food"-nutritional stress hypothesis for the decline of the western stock of Steller sea lion. Mar Mammal Sci 21:476–518

Fuller AR (2012) Spatial and temporal distribution, haulout use and movement patters of Steller sea lions (*Eumetopias jubatus*) in northern California. MSc thesis, Humboldt State University, Arcata, CA

Gaynor KM, Brown JS, Middleton AD, Power ME, Brashares JS (2019) Landscapes of fear: spatial patterns of risk perception and response. Trends Ecol Evol 34:355–368

Gentry RL (1970) Social behavior of the Steller sea lion. PhD thesis, University of California, Santa Cruz, CA

Gentry RL (1974) The development of social behavior through play in the Steller sea lion. Am Zool 14:391–403

Gisiner RC (1985) Male territorial and reproductive behaviour in the Steller sea lion, *Eumetopias jubatus*. PhD thesis, University of California, Santa Cruz, CA

Goldsworthy SD (2006) Maternal strategies of the New Zealand fur seal: evidence for interannual variability in provisioning and pup growth strategies. Aust J Zool 54:31–44

Gregr EJ, Trites AW (2008) A novel presence-only validation technique for improved Steller sea lion *Eumetopias jubatus* critical habitat descriptions. Mar Ecol Prog Ser 365:247–261

Grémillet D, Pichegru L, Kuntz G, Woakes AG, Wilkinson S, Crawford RJM, Ryan PG (2008) A junk-food hypothesis for gannets feeding on fishery waste. Proc R Soc Lond Ser B Biol Sci 275:1149–1156

Guénette S, Heymans SJJ, Christensen V, Trites AW (2007) Ecosystem models of the Aleutian Islands and Southeast Alaska show that Steller sea lions are impacted by killer whale predation when sea lion numbers are low. In: Piatt JF, Gende SM (eds) Proceedings of the fourth Glacier bay science symposium. U.S. Geological Survey, Juneau, Alaska, pp 150–154

Hare SR, Mantua NJ (2000) Empirical evidence for North Pacific regime shifts in 1977 and 1989. Prog Oceanogr 47:103–146

Harestad AS (1973) Social behaviour in a non-pupping colony of Steller sea-lion (*Eumetopias jubatus*). MSc thesis, University of British Columbia, Vancouver, BC

Hastings KK, Jemison LA (2016) Age-specific variation in timing of parturition in Steller sea lions at Forrester Island Complex, Alaska. Mar Mammal Sci 32:777–785

Hastings KK, Jemison LA, Gelatt TS, Laake J, Pendleton G, King JC, Trites AW, Pitcher KW (2011) Cohort effects and spatial variation in age-specific survival of Steller sea lions from southeastern Alaska. Ecosphere 2:111

Hastings KK, Jemison LA, Pendleton GW, Raum-Suryan KL, Pitcher KW (2017) Natal and breeding philopatry of female Steller sea lions in southeastern Alaska. PLoS One 12:e0176840

Hastings KK, Rehberg MJ, O'Corry-Crowe GM, Pendleton GW, Jemison LA, Gelatt TS (2020) Demographic consequences and characteristics of recent population mixing and colonization in Steller sea lions, *Eumetopias jubatus*. J Mammal 101:107–120

Higgins L (1984) Maternal behavior and attendance patterns of the Steller sea lion in California. MSc thesis, University of California, Santa Cruz, CA

Hood WR, Ono KA (1997) Variation in maternal attendance patterns and pup behaviour in a declining population of Steller sea lions (*Eumetopias jubatus*). Can J Zool 75:1241–1246

Hui TCY, Gryba R, Gregr EJ, Trites AW (2015) Assessment of competition between fisheries and Steller sea lions in Alaska based on estimated prey biomass, fisheries removals and predator foraging behaviour. PLoS One 10(5):e0123786

Jeanniard-du-Dot T, Trites AW, Arnould JPY, Guinet C (2017) Reproductive success is energetically linked to foraging efficiency in Antarctic fur seals. PLoS One 12:e0174001

Jemison LA, Pendleton GW, Fritz LW, Hastings KK, Maniscalco JM, Trites AW, Gelatt TS (2013) Inter-population movements of Steller sea lions in Alaska with implications for population separation. PLoS One 8(8):e70167

Jemison LA, Pendleton GW, Hastings KK, Maniscalco JM, Fritz LW (2018) Spatial distribution, movements, and geographic range of Steller sea lions (*Eumetopias jubatus*) in Alaska. PLoS One 13:e0208093

Jodice PGR, Roby DD, Turco KR, Suryan RM, Irons DB, Piatt JF, Shultz MT, Roseneau DG, Kettle AB, Anthony JA (2006) Assessing the nutritional stress hypothesis: relative influence of diet quantity and quality on seabird productivity. Mar Ecol Prog Ser 325:267–279

Keech AL (2008) Fecal triiodothyronine assay validation using captive Steller sea lions (*Eumetopias jubatus*) and subsequent application to free-ranging populations to examine nutritional stress. MSc thesis, University of British Columbia, Vancouver, BC

Kirkman SP, Bester MN, Hofmeyr GJG, Pistorius PA, Makhado AB (2002) Pup growth and maternal attendance patterns in Subantarctic fur seals. Afr Zool 37:13–19

Kucey L (2005) Human disturbance and the hauling out behaviour of Steller sea lions (*Eumetopias jubatus*). MSc thesis, University of British Columbia, Vancouver, BC

Kuhn CE, Chumbley K, Fritz L, Johnson D (2017a) Estimating dispersal rates of Steller sea lion (*Eumetopias jubatus*) mother-pup pairs from a natal rookery using mark-resight data. PLoS One 12:e0189061

Kuhn CE, Chumbley K, Johnson D, Fritz L (2017b) A re-examination of the timing of pupping for Steller sea lions *Eumetopias jubatus* breeding on two islands in Alaska. Endang Spec Res 32:213–222

Loughlin TR, Perlov AS, Vladimirov VA (1992) Range-wide survey and estimation of total number of Steller sea lions in 1989. Mar Mammal Sci 8:220–239

Loughlin TR, Sterling JT, Merrick RL, Sease JL, York AE (2003) Diving behavior of immature Steller sea lions (*Eumetopias jubatus*). Fish Bull 101:566–582

Lyman RL (1988) Zoogeography of Oregon coast marine mammals: the last 3000 years. Mar Mammal Sci 4:247–264

Lyman RL (1989) Seal and sea lion hunting: a zooarchaeological study from the southern Northwest Coast of North America. J Anthro Archaeol 8:68–99

Maniscalco JM, Parker P (2017) Maternal and offspring effects on the timing of parturition in western Steller sea lions (*Eumetopias jubatus*). Can J Zool 96:333–339

Maniscalco JM, Parker P, Atkinson S (2006) Interseasonal and interannual measures of maternal care among individual Steller sea lions (*Eumetopias jubatus*). J Mammal 87:304–311

Maniscalco JM, Springer AM, Adkison MD, Parker P (2015) Population trend and elasticities of vital rates for Steller sea lions (*Eumetopias jubatus*) in the eastern Gulf of Alaska: a new life-history table analysis. PLoS One 10:e0140982

Mansfield B, Haas J (2006) Scale framing of scientific uncertainty in controversy over the endangered Steller sea lion. Environ Polit 15:78–94

Mantua NJ, Hare SR, Zhang Y, Wallace JM, Francis RC (1997) A Pacific interdecadal climate oscillation with impacts on salmon production. Bull Am Meteorol Soc 78:1069–1079

Marcotte ML (2006) Steller Watch: timing of weaning and seasonal patterns in numbers and activities of Steller sea lions at a year-round haulout site in Southeast Alaska. MSc thesis, University of British Columbia, Vancouver, BC

Maschner HDG, Trites AW, Reedy-Maschner KL, Betts M (2014) The decline of Steller sea lions (*Eumetopias jubatus*) in the North Pacific: insights from indigenous people, ethnohistoric records and archaeological data. Fish Fish 15:634–660

Mate BR (1973) Population kinetics and related ecology of the northern sea lion, *Eumetopias jubatus*, and the California sea lion, *Zalophus californianus*, along the Oregon coast. PhD thesis, University of Oregon, Eugene, OR

Mathisen OA, Baade RT, Lopp RJ (1962) Breeding habits, growth and stomach contents of the Steller sea lion in Alaska. J Mammal 43:469–477

Matkin CO, Barrett-Lennard LG, Yurk H, Ellifrit D, Trites AW (2007) Ecotypic variation and predatory behavior among killer whales (*Orcinus orca*) off the eastern Aleutian Islands, Alaska. Fish Bull 105:74–87

McCafferty DJ, Boyd IL, Walker TR, Taylor RI (1998) Foraging responses of Antarctic fur seals to changes in the marine environment. Mar Ecol Prog Ser 166:285–299

Merrick RL (1987) Behavioral and demographic characteristics of northern sea lion rookeries. MSc thesis, Oregon State University, Corvallis, OR

Merrick RL (1997) Current and historical roles of apex predators in the Bering Sea ecosystem. J Northwest Atl Fish Sci:343–356

Merrick RL, Loughlin TR (1997) Foraging behavior of adult female and young-of-year Steller sea lions in Alaskan waters. Can J Zool 75:776–786

Merrick RL, Loughlin TR, Calkins DG (1987) Decline in abundance of the northern sea lion, *Eumetopias jubatus*, in Alaska, 1956–86. Fish Bull US 85:351–365

Merrick RL, Brown R, Calkins DG, Loughlin TR (1995) A comparison of Steller sea lion, *Eumetopias jubatus*, pup masses between rookeries with increasing and decreasing populations. Fish Bull 93:735–758

Merrick RL, Chumbley MK, Byrd GV (1997) Diet diversity of Steller sea lions (*Eumetopias jubatus*) and their population decline in Alaska: a potential relationship. Can J Fish Aquat Sci 54:1342–1348

Milette LL (1999) Behaviour of lactating Steller sea lions (*Eumetopias jubatus*) during the breeding season: A comparison between a declining and stable population in Alaska. MSc thesis, University of British Columbia, Vancouver, BC

Milette LL, Trites AW (2003) Maternal attendance patterns of Steller sea lions (*Eumetopias jubatus*) from a stable and a declining population in Alaska. Can J Zool 81:340–348

NMFS [National Marine Fisheries Service] (1992) Recovery plan for the Steller sea lion (*Eumetopias jubatus*). Steller Sea Lion Recovery Team, Silver Spring, MD

NMFS [National Marine Fisheries Service] (1995) Steller sea lions; reclassification. Fed Regist 60 (192):51968–51978

NMFS [National Marine Fisheries Service] (2008) Recovery plan for the Steller sea lion. Eastern and western distinct population segments (*Eumetopias jubatus*). Revision. US National Marine Fisheries Service, Silver Spring, MD

NMFS [National Marine Fisheries Service] (2013) Endangered and threatened species; delisting of the eastern distinct population segment of Steller sea lion under the Endangered Species Act. Federal Register Parts 223 and 224(78):66140

O'Corry-Crowe G, Gelatt T, Rea L, Bonin C, Rehberg M (2014) Crossing to safety: dispersal, colonization and mate choice in evolutionarily distinct populations of Steller sea lions, *Eumetopias jubatus*. Mol Ecol 23:5415–5434

Olesiuk PF (2011) Abundance of Steller sea lions (*Eumetopias jubatus*) in British Columbia. Can Sci Adv Sec Res Doc 2010(000):1–43

Olesiuk PF (2018) Recent trends in abundance of Steller sea lions (*Eumetopias jubatus*) in British Columbia. Can Sci Adv Sec Res Doc 2018(006):1–72

Ono KA, Boness DJ, Oftedal OT (1987) The effect of a natural environmental disturbance on maternal investment and pup behavior in the California sea lion. Behav Ecol Sociobiol 21:109–118

Orr RT, Poulter TC (1967) Some observations on reproduction, growth, and social behavior in the Steller sea lion. Proc Calif Acad Sci 35:193–226

Osterblom H, Olsson O, Blenckner T, Furness RW (2008) Junk-food in marine ecosystems. Oikos 117:967–977

Parker P (2006) Maternal attendance and pupping site fidelity of Steller sea lions in Alaska. MSc thesis, San José State University, San Jose, CA
Parker P, Maniscalco JM (2014) A long-term study reveals multiple reproductive behavior strategies among territorial adult male Steller sea lions (*Eumetopias jubatus*). Can J Zool 92:405–415
Parker P, Harvey JT, Maniscalco JM, Atkinson S (2008) Pupping-site fidelity among individual Steller sea lions (*Eumetopias jubatus*) at Chiswell Island, Alaska. Can J Zool 86:826–833
Pendleton GW, Pitcher KW, Fritz LW, York AE, Raum-Suryan KL, Loughlin TR, Calkins DG, Hastings KK, Gelatt TS (2006) Survival of Steller sea lions in Alaska: a comparison of increasing and decreasing populations. Can J Zool 84:1163–1172
Pendleton GW, Hastings KK, Rea LD, Jemison LA, O'Corry-Crowe GM, Beckmen KB (2016) Short-term survival of Steller sea lion (*Eumetopias jubatus*) pups: investigating the effect of health status on survival. Mar Mammal Sci 32:931–944
Perlov AS (1971) On the age of sexual maturation in sea lions. VNIRO Proc 80:174–189
Pitcher KW, Calkins DG (1981) Reproductive biology of Steller sea lions in the Gulf of Alaska. J Mammal 62:599–605
Pitcher KW, Calkins DG, Pendleton GW (1998) Reproductive performance of female Steller sea lions: an energetics based reproductive strategy? Can J Zool 76:2075–2083
Pitcher KW, Burkanov VN, Calkins DG, Le Boeuf BJ, Mamaev EG, Merrick RL, Pendleton GW (2001) Spatial and temporal variation in the timing of births of Steller sea lions. J Mammal 82:1047–1053
Pitcher KW, Rehberg MJ, Pendleton GW, Raum-Suryan KL, Gelatt TS, Swain UG, Sigler MF (2005) Ontogeny of dive performance in pup and juvenile Steller sea lions in Alaska. Can J Zool 83:1214–1231
Pitcher KW, Olesiuk PF, Brown RF, Lowry MS, Jeffries SJ, Sease JL, Perryman WL, Stinchcomb CE, Lowry LF (2007) Abundance and distribution of the eastern North Pacific Steller sea lion (*Eumetopias jubatus*) population. Fish Bull 105:102–115
Porter B (1997) Winter ecology of Steller sea lions (*Eumetopias jubatus*) in Alaska. MSc thesis, University of British Columbia, Vancouver, BC
Porter BT, Trites AW (2004) Suckling attempts during winter by two non-filial Steller sea lion pups (*Eumetopias jubatus*). Mammalia 68:23–26
Raum-Suryan KL, Pitcher KW, Calkins DG, Sease JL, Loughlin TR (2002) Dispersal, rookery fidelity, and metapopulation structure of Steller sea lions (*Eumetopias jubatus*) in an increasing and a decreasing population in Alaska. Mar Mammal Sci 18:746–764
Raum-Suryan KL, Rehberg MJ, Pendleton GW, Pitcher KW, Gelatt TS (2004) Development of dispersal, movement patterns, and haul-out use by pup and juvenile Steller sea lions (*Eumetopias jubatus*) in Alaska. Mar Mammal Sci 20:823–850
Rea LD (1995) Prolonged fasting in pinnipeds. PhD thesis, University of Alaska, Fairbanks, AK
Rea LD, Fadely BS, Farley SD, Avery JP, Dunlap-Harding WS, Stegall VK, Eischens CAB, Gelatt TS, Pitcher KW (2016) Comparing total body lipid content of young-of-the-year Steller sea lions among regions of contrasting population trends. Mar Mammal Sci 32:1200–1218
Rehberg MJ, Burns JM (2008) Differences in diving and swimming behavior of pup and juvenile Steller sea lions (*Eumetopias jubatus*) in Alaska. Can J Zool 86:539–553
Rehberg M, Rea L, Eischens C (2018) Overwintering Steller sea lion (*Eumetopias jubatus*) pup growth and behavior prior to weaning. Can J Zool 96:97–106
Romano MD, Piatt JF, Roby DD (2006) Testing the junk-food hypothesis on marine birds: effects of prey type on growth and development. Waterbirds 29:407–414
Rosen DAS (2009) Steller sea lions *Eumetopias jubatus* and nutritional stress: evidence from captive studies. Mammal Rev 39:284–306
Rosen DAS, Trites AW (2000) Pollock and the decline of Steller sea lions: testing the junk-food hypothesis. Can J Zool 78:1243–1258
Rosen DAS, Trites AW (2004) Satiation and compensation for short-term changes in food quality and availability in young Steller sea lions (*Eumetopias jubatus*). Can J Zool 82:1061–1069

Roux J-P (ed) (1997) Proceedings of the international workshop on research and management of Cape Fur Seals in Namibia. Ministry of Fisheries and Marine Resources of Namibia, Swakopmund, Namibia, 24–26 June 1997

Sandegren FE (1970) Breeding and maternal behavior of the Steller sea lion (*Eumetopias jubatus*) in Alaska. MSc thesis, University of Alaska, Fairbanks, AK

Sandegren FE (1976) Courtship display, agonistic behavior and social dynamics in the Steller sea lion. Behaviour 57:136–158

Scheffer VB (1945) Growth and behavior of young sea lions. J Wildl Manag 26:390–392

Scordino J (2006) Steller sea lions (*Eumetopias jubatus*) of Oregon and northern California: seasonal haulout abundance patterns, movements of marked juveniles, and effects of hot-Iron branding on apparent survival of pups at Rogue Reef. MSc thesis, Oregon State University, Corvalis, OR

Sinclair EH, Zeppelin TK (2002) Seasonal and spatial differences in diet in the western stock of Steller sea lions (*Eumetopias jubatus*). J Mammal 83:973–990

Sinclair EH, Johnson DS, Zeppelin TK, Gelatt TS (2013) Decadal variation in the diet of western stock Steller sea lions (*Eumetopias jubatus*). US Dep Commer, NOAA Tech Memo NMFSAFSC-248

Smith LN (1988) The influence of rookery terrain on population structure, territorial behavior, and breeding success of Steller sea lions in the Gulf of Alaska. MSc thesis, University of Alaska, Fairbanks, AK

Springer AM (1992) A review: walleye pollock in the North Pacific—how much difference do they really make? Fish Oceanogr 1:80–96

Steller GW (1751) De bestiis marinis, or, the beasts of the sea (1751). Translated by J.E. Miller and W. Miller (2005). Faculty Publications, University of Nebraska, Lincoln

Suraci JP, Clinchy M, Zanette LY, Wilmers CC (2019) Fear of humans as apex predators has landscape-scale impacts from mountain lions to mice. Ecol Lett 22:1578–1586

Sweeney KL, Shertzer KW, Fritz LW, Read AJ (2015) A novel approach to compare pinniped populations across a broad geographic range. Can J Fish Aquat Sci 72:175–185

Thorsteinson FV, Lensink CJ (1962) Biological observations of Steller sea lions taken during an experimental harvest. J Wildl Manag 26:353–359

Tollit DJ, Fritz L, Joy R, Miller K, Schulze A, Thomason J, Walker W, Zeppelin T, Gelatt T (2017) Diet of endangered Steller sea lions in the Aleutian Islands: new insights from DNA detections and bio-energetic reconstructions. Can J Zool 95:853–868

Trillmich F (1990) The behavioral ecology of maternal effort in fur seals and sea lions. Behaviour 114:3–20

Trillmich F, Limberger D (1985) Drastic effects of El Niño on Galapagos pinnipeds. Oecologia 67:19–22

Trillmich F, Ono K (eds) (1991) Pinnipeds and El Niño: responses to environmental stress. Springer, New York

Trites AW (1991) Fetal growth of northern fur seals: life history strategy and sources of variation. Can J Zool 69:2608–2617

Trites AW, Donnelly CP (2003) The decline of Steller sea lions *Eumetopias jubatus* in Alaska: a review of the nutritional stress hypothesis. Mammal Rev 33:3–28

Trites AW, Larkin PA (1992) The status of Steller sea lion populations and the development of fisheries in the Gulf of Alaska and Aleutian Islands. Report to the Pacific States Marine Fisheries Commission pusuant to National Oceanic and Atmospheric Administration award no. NA17FD0177. Available from Marine Mammal Research Unit, Institute for the Oceans and Fisheries, University of British Columbia, Vancouver, BC

Trites AW, Larkin PA (1996) Changes in the abundance of Steller sea lions (*Eumetopias jubatus*) in Alaska from 1956 to 1992: How many were there? Aquat Mamm 22:153–166

Trites AW, Porter BT (2002) Attendance patterns of Steller sea lions (*Eumetopias jubatus*) and their young during winter. J Zool (Lond) 256:547–556

Trites AW, Livingston PA, Mackinson S, Vasconcellos MC, Springer AM, Pauly D (1999) Ecosystem change and the decline of marine mammals in the eastern Bering Sea: testing the ecosystem shift and commercial whaling hypotheses. Fish Centre Res Reports 7(1):1–106

Trites AW, Porter BP, Deecke VB, Coombs AP, Marcotte ML, Rosen DAS (2006) Insights into the timing of weaning and the attendance patterns of lactating Steller sea lions (*Eumetopias jubatus*) in Alaska during winter, spring, and summer. Aquat Mamm 32:85–97

Trites AW, Calkins DG, Winship AJ (2007a) Diets of Steller sea lions (*Eumetopias jubatus*) in Southeast Alaska, 1993–1999. Fish Bull 105:234–248

Trites AW, Miller AJ, Maschner HDG, Alexander MA, Bograd SJ, Calder JA, Capotondi A, Coyle KO, Lorenzo ED, Finney BP, Gregr EJ, Grosch CE, Hare SR, Hunt GL, Jahncke J, Kachel NB, Kim H-J, Ladd C, Mantua NJ, Marzban C, Maslowski W, Mendelssohn R, Neilson DJ, Okkonen SR, Overland JE, Reedy-Maschner KL, Royer TC, Schwing FB, Wang JXL, Winship AJ (2007b) Bottom-up forcing and the decline of Steller sea lions (*Eumetopias jubatus*) in Alaska: assessing the ocean climate hypothesis. Fish Oceanogr 16:46–67

Vazquez NA (2013) Spatial and temporal distribution of juvenile Steller sea lion (*Eumetopias jubatus*) predation events in the Gulf of Alaska. MSc thesis, Oregon State University, Corvalis, OR

Whitfield J (2008) Does 'junk food' threaten marine predators in northern seas? Science 322:1786–1787

Williams TM, Estes JA, Doak DF, Springer AM (2004) Killer appetites: assessing the role of predators in ecological communities. Ecology 85:3373–3384

Winship AJ, Trites AW (2003) Prey consumption of Steller sea lions (*Eumetopias jubatus*) off Alaska: How much prey do they require? Fish Bull 101:147–167

Winship AJ, Trites AW, Calkins DG (2001) Growth in body size of the Steller sea lion (*Eumetopias jubatus*). J Mammal 82:500–519

Winship AJ, Trites AW, Rosen DAS (2002) A bioenergetic model for estimating the food requirements of Steller sea lions (*Eumetopias jubatus*) in Alaska, USA. Mar Ecol Prog Ser 229:291–312

York AE (1994) The population dynamics of northern sea lions, 1975–1985. Mar Mammal Sci 10:38–51

Chapter 24
Exploitation and Recovery of the South American Sea Lion in the Southwestern Atlantic

Enrique A. Crespo

Abstract The South American sea lion, *Otaria flavescens,* is one of the most conspicuous marine mammals along the coasts of the Atlantic and Pacific Oceans. It has a long history of exploitation throughout its range. In the southwestern Atlantic it was heavily exploited throughout the first half of the twentieth century. As a K-strategist subject to exploitation this population fell to very low levels. After protection was instigated, recovery was at first very slow. At the time of writing, the population size is about one third of the original size. The opportunistic and plastic behavior of the South American sea lion together with a high level of juvenile survival resulted in a faster rate of population recovery in the most recent decades. Possible reasons for this improved survivorship may include an increase in the availability of food resources, prey-switching, and a decrease in exogenous causes of mortality.

Keywords South American sea lion · *Otaria flavescens* · Exploitation · Recovery · Recolonization · Survivorship · Feeding behavior · Female behavior · Age structure · Food webs

24.1 Introduction

Many marine mammals have been subject to intense human exploitation (Bonner 1982). Their entire natural history has been strongly affected by this process, but at the same time, many have been provided time to recover post-exploitation. Thus, they provide an opportunity to explore links between population dynamics and demographic parameters, foraging and reproductive behavior, under the influence of interactions with humans. The purpose of this chapter is to explore these links for

E. A. Crespo (✉)
Marine Mammal Laboratory, Centro Nacional Patagónico (CONICET), Puerto Madryn, Chubut, Argentina
e-mail: kike@cenpat-conicet.gob.ar

C. Campagna, R. Harcourt (eds.), *Ethology and Behavioral Ecology of Otariids and the Odobenid*, Ethology and Behavioral Ecology of Marine Mammals,
https://doi.org/10.1007/978-3-030-59184-7_24

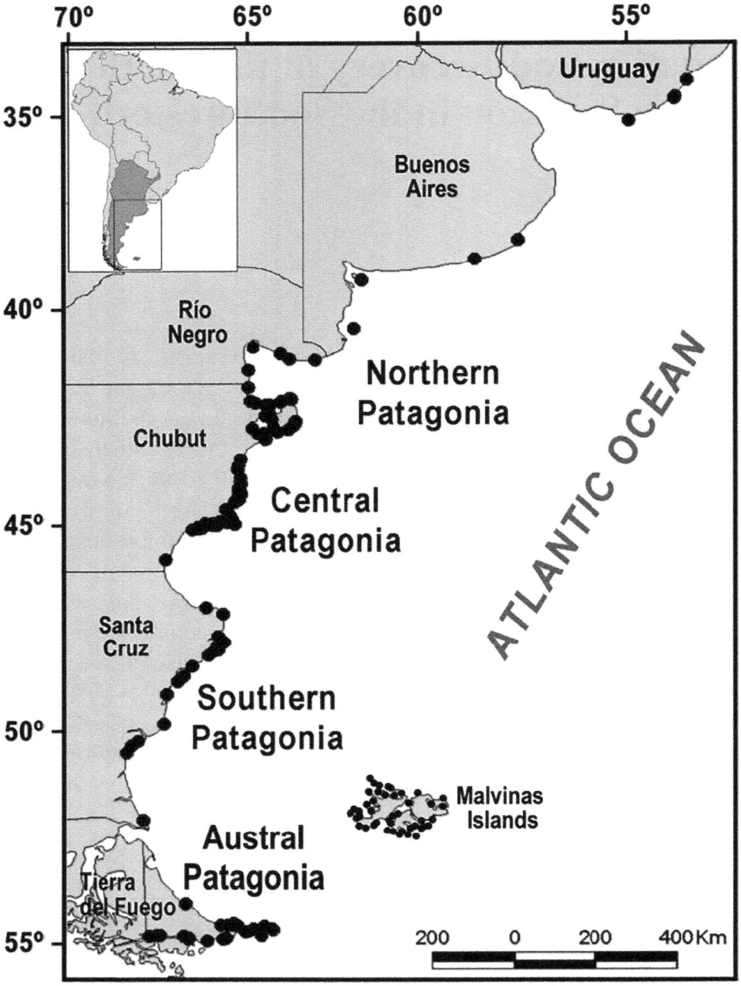

Fig. 24.1 Distribution of South American sea lion rookeries along the SW Atlantic

the South American sea lion, *Otaria flavescens*, a prime example of a sea lion in the southern hemisphere that suffered dramatic reduction in population size due to harvesting, but that since protection, has partially recovered over its entire range (Fig. 24.1).

Many species of aquatic mammals have been driven to the edge of extinction due to indiscriminate hunting, poorly planned harvesting or both. Several species have since recovered, or are well on the way to recovery, simply due to the removal of the cause of decline i.e. when hunting stopped (Best 1993). For other species, recovery has been slow or did not occur (see Carroll et al. 2014). It has been argued that social disruption and/or interspecific competition may impede the recovery process (see discussion in Clapham and Brownell 1996). This raises questions concerning the mechanisms dictating recovery, and the extent to which density dependence, and other environmental effects, may influence population recovery. From the ecosystem management perspective, even resilient ecosystems may take decades to recover after major impacts, and there is no certainty that they will return to pre-exploitation conditions. While it is possible to eliminate or mitigate negative impacts within an ecosystem, it is not always possible to predict the trajectory of the response or its endpoint (Crespo and Hall 2001).

In general, aquatic marine mammals are close to the top of food webs and most are considered to be K-strategists (Caughley 1977; Estes 1979; McLaren and Smith 1985). These characteristics include large body size, long life-spans, low reproductive rates, parental care of the offspring, delayed reproduction, juvenile survival less predictable than that of the adults, and predictable adult survival rates. In some cases they appear to show density-dependent responses (Fowler 1981, 1984; Doidge et al. 1984; Harcourt 1992; Evans and Stirling 2001; Crespo et al. 2018). Large mammals like the aquatic ones may be placed at one of the extremes of the r-K continuum; at the other end, we find rodents and insectivores (Lockyer 1984; Evans 1987; Crespo and Hall 2001).

There are several characteristics that made aquatic mammals a valuable resource for humans, in particular the semi-aquatic seals and sea lions. First, the protein and fat contained in aquatic mammals comes in large and concentrated "packages," compared to other harvested animals. Aquatic mammals are rich in fat, which provide a high-calorific value diet and—in recent human history but not applicable anymore today—other valuable resources such as oils which could be burned for light and warmth (Bonner 1982). Many semi-aquatic mammals are also predictable regarding the location of colonies and haul-out places, as well as seasonal migration trajectories. Finally, marine mammals with an amphibious life cycle, such as fur seals and sea lions, are particularly vulnerable when ashore because their adaptations for aquatic life hinder their movements on land, i.e., they are relatively slow and clumsy.

These features were exploited by both ancient and modern humans. A prime example is the South American sea lion which inhabits the coasts of the Atlantic and Pacific Oceans in South America, from Ilha de Lobos in Torres, southern Brazil, (23° S), to the southern tip of the American continent, and from there to the north to Zorritos in Peru (4° S) (Vaz-Ferreira 1982). In Brazil, they have been observed up to the latitude of Rio de Janeiro. In Uruguay, they occupy and breed on islands and islets, the most important being Isla de Lobos, off the coast of Punta del Este and Cabo Polonio (Vaz-Ferreira 1950, 1952).

The species has a long history of exploitation across most of its distribution, from prehistorical times (Zenteno et al. 2015) to the modern era which peaked during the 1930's and the 1950's along the southwestern Atlantic. The southwestern Atlantic harvest was meticulously documented by governments. South American sea lions are social and have a predictable annual cycle, alternating periods on land and at sea (Campagna 1985). As a result it is highly predictable where they haul out and breed. These behavioral features have made them vulnerable to hunters. In contrast to other southern hemisphere sea lion species, South American sea lions were relatively abundant, widely distributed on islands and the mainland in the Atlantic and Pacific coasts of South America. Large colonies of many thousands of animals have been found throughout their range, both breeding and wintering aggregations (Vaz-Ferreira 1982). When not foraging at sea, females spend considerable time on land attending their pups (Campagna 1985). As the species is highly polygynous, a large proportion of the males remains in the vicinity of the breeding colonies, making them easy targets for hunters. It is not surprising that South American sea lions were exposed to intense killing and that this impacted their populations.

While their behavior on land facilitated their demise, the behavior at sea has been central to explaining their recovery. I will overview the demography of the species as it relates to its behavior, addressing first the history of exploitation and then that of recovery. South American sea lions are high trophic level predators in a temperate ecosystem with a trophic web that may be simplified to a few abundant keystone species components. Behavior at sea is driven by their highly plastic foraging behavior (Koen-Alonso et al. 2000), pursuing both pelagic and benthic prey, but also learning rapidly to take advantage of fish caught in gillnets, loglines, and bottom trawls nets. Movements at sea have been generally described by satellite tracking of individuals (Campagna et al. 2001; Riet-Sapriza et al. 2013; Baylis et al. 2015, 2017). Feeding behavior differs between males and females, with males usually displaying more pelagic behavior and a more restricted diet compared to females who are more coastal with a more diverse diet (Koen Alonso et al. 2000).

24.2 History of Exploitation

24.2.1 Pre-commercial Period: 7000 BP Until the Colonization of America

Prior to European colonization, subsistence hunting with a variable degree of intensity occurred along the entire Atlantic coast (Bayón and Politis 2014; Orquera and Piana 1999; Borella 2014; Gómez Otero 2007; Castilho and Simões-Lopes 2008). This hunting pressure was considered low given past population abundance (Schiavini 1992; Orquera and Piana 1999; Zangrando et al. 2014). South American fur seals *Arctocephalus australis* were apparently preferred to sea lions, possibly due to their smaller size and the luxuriance of their pelt. However, no record exists of the extent nor rate of this harvest.

24.2.2 Colonization by Europeans

In Uruguay, after the "western" discovery of the Río de la Plata by Juan Díaz de Solís, and subsequent to his death, the crew killed and transported a cargo of sea

Table 24.1 South American sea lion leather and oil produced in Argentina from 1931 to 1960 (Data from Godoy 1963 in Crespo 1988)

Period	Total leathers	Kg Oil	Kg/leather	Leathers exported
1931/36	32,911	232,800	7.07	
1937/40	230,339	764,373	3.32	185,554
1941/45	153,971	697,628	4.53	102,199
1946/50	57,371	266,964	4.65	21,576
1951/55	10,728	112,850	10.52	
1956/60	8138	89,000	10.94	
Totals	493,438	2,163,615		309,329

lions and fur seals that were sold in the Seville market (Vaz-Ferreira 1982). Beginning in 1520, Europeans used to stock up on sea lion meat for long ocean voyages. According to Kellogg (1942), (cited by Vaz-Ferreira 1982), Simón de Alcazaba killed sea lions off the coast of Chubut in 1535, Sir Francis Drake killed 200 specimens in Puerto Deseado in 1577, and similar events were recorded on the coasts of Uruguay, Chile and Perú. According to Vaz-Ferreira (1982) the meat was used for human consumption, and the preferred parts were the heart, brain, liver and tongue. However while the flesh of the sea lion was consumed in the days of sailing, this practice soon ceased about 1600.

Sea lion and fur seal leather was used by the aboriginal peoples of Tierra del Fuego for the construction of slings. Furs were used also as coats to cover their backs and shoulders (Weddell 1825). Exploitation of sea lions along the southwestern Atlantic coast of Argentina was exercised locally by European settlers who produced leather from the pelts for the manufacture of goods such as harnesses, reins, etc., at least until the beginning of the twentieth century (Crespo 1988). It is possible that similar events occurred on the coasts of the Province of Buenos Aires, where a reduction in the number of breeding sites and haul-outs occurred (Valette 1928), and along the entire Patagonian coast. In continental Argentina and in Tierra del Fuego, there were also permit holders who hunted from 1920 until 1960 (Table 24.1) (Godoy 1963).

Exploitation was carried out in a rudimentary way (Fig. 24.2); skin was salted for preservation, and subcutaneous fat was passed through boilers and then filtered to obtain oil. The rest of the flesh was wasted. Skin of pups was used to manufacture fur and leather goods; that of the adults was destined to saddlery. Oil was used in the tannery industry (López Arregui and González Regalado 1940). In Uruguay, the flesh was used in the production of meat meal (Ximénez Pers. Comm.) or as a supplementary food for ermine, mink, or fish farms, as it was in Alaska with the Northern Fur Seal (Bekiashev and Serebriakov (1981).

Holders of licenses to harvest sea lions exploited settlements from the Province of Buenos Aires to Tierra del Fuego (almost all the rookeries). The largest takes were on the settlements of Península Valdés (Table 24.2) and Tierra del Fuego. The Islands and Islets of Chubut and Santa Cruz, perhaps because of their difficult access, and the precarious exploitation from the beginning and middle of the

Fig. 24.2 Harvesting sea lions in Península Valdés rookeries (source: Family Machinea)

twentieth century, apparently remained unexploited despite the large number of animals there, or if exploitation did occur it was not recorded (Carrara 1952; Godoy 1963).

Despite the overall utilitarian attitudes that operated at the beginning and middle of the twentieth century in Argentina, Valette (1928), recognized very early on the need to protect areas for their repopulation. López Arregui and González Regalado (1940) also made a series of recommendations to improve and optimize the industry, highlighting its precariousness and wastefulness, which was limited only to

Table 24.2 Exploitation of South American sea lion rookeries of Península Valdés (Data from Godoy 1963 in Crespo 1988)

Period	Total Leathers	Kg of oil
1921	1887	
1937/40	84,665	401,810
1941/45	71,103	226,230
1946/50	24,738	154,210
1951/55	10,728	112,850
1956/60	8138	89,000
Totals	201,259	984,100

preparing leathers and turning fat into oil, disregarding the full use of animals. However, they also grossly overestimated the abundance of sea lions in Argentina (1,500,000 individuals) and their supposed consumption of fish. If they had been correct their estimates would imply that all of the schools of anchovies, sardines, squid and shrimp would have disappeared from large areas (López Arregui and González Regalado op. cit.).

The irrational nature of the exploitation severely reduced stocks. There is evidence from the chronicles of early sealers of complete extermination of some rookeries. For example, more than 10,000 males were reported killed per season between 1923 and 1926 from an assumed population of 60,000 animals in southern Chubut (Kirn 1986). While it is possible that both the abundance estimate and rate of slaughter were somewhat overestimated, the latter clearly lacked control. Throughout the twentieth century, maritime hunting was regulated. Hunting of fur seals was banned and sea lion hunting was regulated in 1953 (Godoy 1963).

Sea lion breeding groups, according to twentieth century pre-sealing data, tended to be large, in the order of many thousand animals, with several groups clustered within a few hundreds kilometers of coast. In Península Valdés and nearby areas there were four large colonies of several thousand animals distributed less than 100 kilometers apart. Sealers exploited these features and targeted reproductive animals. Sea lion females give birth to one pup per year of reproduction, this relatively low rate, plus the fact that pups and mothers were also hunted, contributed to the quick demise of populations.

But, post World War 2 there occurred a rather abrupt change in attitudes about the killing of marine mammals. From the 1970's, South American sea lions became a tourism attraction in Argentine Patagonia, together with southern elephant seals (*Mirounga leonina*), fur seals, southern right whales (*Eubaelana australis*), penguins and other seabirds. Some breeding areas were declared provincial wildlife reserves. Finally, in the twenty-first century, the National Parks of Argentina became involved in the protection of the species.

24.3 History of Recovery

Population recovery relates to expansion of colonies but also colonization of new places. Describing how this occurs is one purpose of this section. Providing demographic data is the second.

24.3.1 *Demographics*

The lowest numbers of the sea lion population due to intensive over-exploitation occurred in northern Argentine Patagonia, with less than 10,000 individuals counted during the 1960's, with a rate of increase not different from zero (Crespo 1988; Crespo et al. 2012). The original population size in the same area was estimated as more than 160,000 individuals (Koen Alonso et al. 1999). The rate of increase was slightly positive after 1990 (Crespo and Pedraza 1991), reaching a rate of 5.6% a few years later (Dans et al. 2004; Grandi et al. 2015, 2016). More recently, using a Bayesian state-space modeling framework, the original population size of northern and central Argentine Patagonia was estimated to be between 300,000 and 400,000 animals (Romero et al. 2017).

Given the high rates of exploitation and new estimates of the pre-exploitation population, this raises the question of what are the behavioral mechanisms that influence recovery. Several mechanisms may contribute including: (a) cleavage of juveniles from central breeding areas, leading to new colonies; (b) increased survivorship of juveniles; (c) availability of feeding resources and (d) plastic behavior to take advantage of multiple food sources.

24.3.2 *New Colonies*

Typical breeding rookeries are composed of around 80% females and pups, with the remainder made up of adult and subadult males and juveniles of both sexes. During the 1980's, along with the overall population increase, a proliferation of new colonies with a different social structure to that of central breeding areas began to appear (Crespo 1988; Grandi et al. 2008). From time to time, new groups, composed essentially of juveniles, split and settled a few kilometers away from traditional breeding colonies. These groups differ in a number of demographic parameters. Long-established colonies have lower, constant rates of increase compared to new ones, and it appears that the different age, size and social structure of these now colonies influence individual female breeding success. Long-established, "traditional" breeding colonies remain constant in social composition and breeding behavior for decades. By comparison, new colonies grow with the addition of new juveniles every year. After a few years, the juvenile females become sexually mature

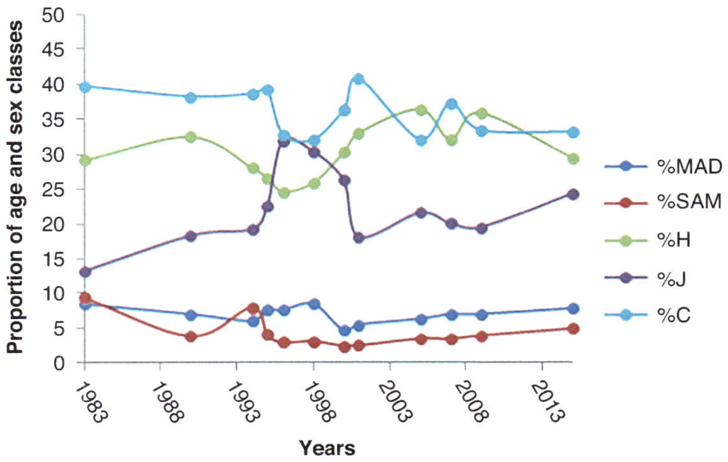

Fig. 24.3 Change in the age structure of the population through time

and reproduce, but in a social composition that is different from old breeding colonies. Juveniles of both sexes prevail in new groups, and newborn pups are exposed to frequent contact with sub-adult males. These new groups have higher rates of increase when compared with old breeding colonies, and these rates are explained by an important migration component, of juveniles moving from old colonies to new ones. This, in turn, is likely to impact the growth rate of the population as a whole. The rate of increase of new colonies is always more than 10%, reaching almost 20% with immigration. By contrast, rates of increase in old traditional colonies were never higher than 3–4%. This colonization mechanism suggests that potential areas for population expansion are likely to be close to areas where growing colonies already exist (Grandi et al. 2008).

Despite the higher growth rates, pups born in new colonies appear more at risk. Pup mortality ranges from 4–6% in new colonies, a significantly higher rate than the less than 2% in traditional colonies (Svendsen et al. 2012). This may be explained by higher potential for infanticide by solitary or subadult males in the new colonies (Campagna et al. 1988) and/or to higher proportions of younger females with less experience in nursing pups. Critically, social structure of the colony is important to survivorship of the pups.

Commercial harvesting of sea lions from 1920 to 1960 decimated population abundance. Population recovery was not immediate after hunting ceased in 1962. The northern Patagonian stock was fairly stable until 1989 or at least no increase was detected. After 2000, the population has grown at a constant rate of 5.6%/annum. Along with this growth, there has been an increase in the juvenile fraction and other changes in the social composition of colonies (Fig. 24.3), which likely relate to changes in population vital rates.

24.3.3 Female Behavior

Colony formation is related to dispersal, philopatry, conspecific attraction, available suitable habitat, proximity and availability of food resources, and reproductive success (Grandi et al. 2018). Grandi and colleagues analyzed whether females show natal fidelity at small geographic scales (between colonies of the same breeding area) in the context of a recovering population. Given the existence of philopatry in South American sea lion females, this study expected to find that: (a) maternal lineages within a colony will be more similar to each other than lineages in other colonies, and (b) geographically closer colonies will be more genetically related than distant ones. They found that female sea lions display different strategies when they choose where to breed: some were born and stayed in one particular colony, others moved to nearby colonies while some moved to new juvenile settlements within the study area. Recolonization of traditional areas may be influenced by weak female philopatry, attenuated and/or interacting with other processes such as site fidelity to nearby feeding grounds, breeding success, dispersal and terrestrial habitat selection (Grandi et al. 2018).

To analyze changes in survivorship patterns of sea lions through time, a comparison of survivorship curves of males and females was obtained from life tables constructed in two periods with different population trends: 1981–1987 (stationary, Crespo 1988) and 2000–2008 (recovering, Grandi 2010; Grandi et al. 2016). The results showed that age-specific survivorship increased during the past three decades for both males and females. The survivorship curves were more a Type II curve during the 80's and transformed gradually to a Type I curve in the 2000's, more typical of a K-strategist population. The process may continue to the present day. All demographics, juveniles, adult males and adult females from the recent period show higher survival rates than those from the 1980's. In fact, survivorship of mature males increased by 10%, and for mature females by 20% (see Fig. 24.4).

This improvement in survivorship is very likely to be one of the essential factors that drove population recovery in the most recent decades. The possible reasons for this change may include an increase in the availability of food resources, a decrease in exogenous mortality causes, or a combination of both factors.

Sea lions feed on at least 41 prey species, including fishes, cephalopods, crustaceans and gastropods. However, only a few prey species dominate in the diet (Koen Alonso et al. 2000), including Argentine hake (*Merluccius hubbsi*), red octopus (*Enteroctopus megalocyathus*), Argentine shortfin squid (*Illex argentinus*), 'raneya' (*Raneya brasiliensis*), Patagonian squid (*Loligo gahi*), and Argentine anchovy (*Engraulis anchoita*). Sex differences in diet are evident, but not between geographic areas or season. Females are benthic feeders, whereas males feed mostly on demersal-pelagic species. Sex differences are explained by feeding grounds, home ranges, and perhaps constraints in the feeding behavior to capture or handle prey. Females feed in more coastal and shallower waters than males, who move more offshore and also may follow fishing vessels. Sex differences in diet may have influenced recovery, as 60% of the male diet is Argentine hake, increasing their vulnerability to hake depletion by overfishing. Males are also part of the incidental

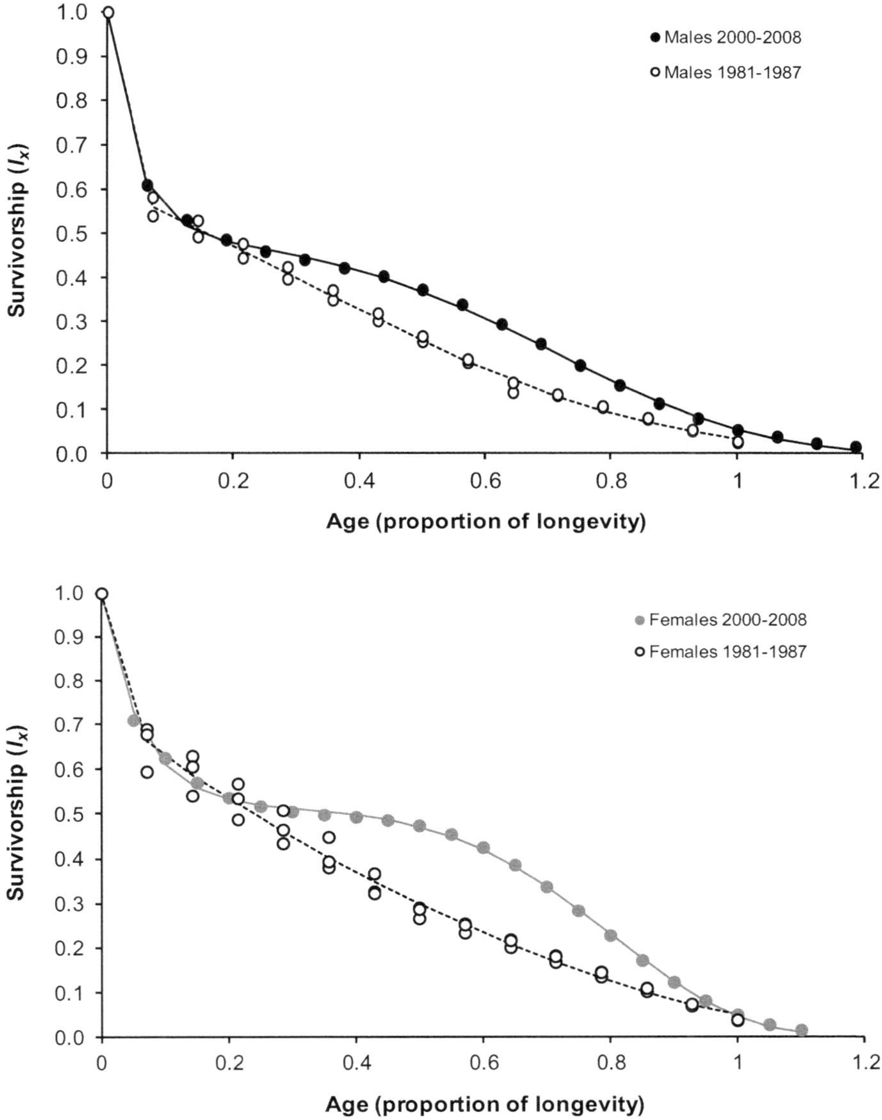

Fig. 24.4 Survivorship of males (left) and females (right) in the 80's (open circles) and the 2000's (closed circles) (After Grandi et al. 2016)

catch in trawlers, probably due to their habit of following trawlers to take fish from the nets. Females are less dependent on any single prey.

The effect of sealing and industrial fisheries on the carbon and nitrogen isotopic footprint in sea lions from northern Patagonia indicate that specimens killed in the

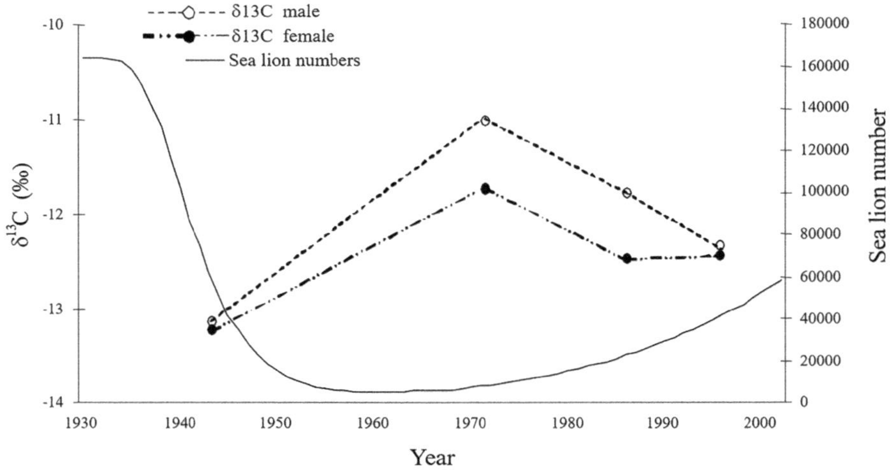

Fig. 24.5 SASL population in Northern Patagonia and median δ13C in skull bone in four contrasting periods (after Drago et al. 2009)

earlier years of hunting showed a pelagic signature, and then the δ13C of males and females increased from the 1940s to the 1970s, declining again after that period (Fig. 24.5) (Drago et al. 2009). As benthic prey off northern Patagonia are more enriched in δ13C than pelagic prey, this suggests an increase in consumption of benthic coastal prey in the 1970s. The period 1940–1970 was a time when sea lion abundance was at its lowest due to commercial hunting. The increased consumption of pelagic prey since the 1970s is coincident with sea lion population recovery.

The main isotopic signal during the 1970's was identified as coming from red octopus (*Enteroctopus megalocyathus*), while the isotopic signal in early years before and more recently was identified as common hake (*Merluccius hubbsi*) (Fig. 24.5) (Drago et al. 2009). It appears that the opportunistic behavior of sea lions allows them to switch prey depending on the size of the population. High density may force sea lions to move offshore for pelagic prey, low density allows them to feed inshore on coastal prey. During the period 1970–2000, Argentine hake was predated by sea lions while at the same time the bottom trawling fishery was depleting this most important of Argentina's fishing resources. Reinforced intraspecific competition and massive discards of pelagic fish may have contributed to the observed dietary shift. The ability of sea lions to shift prey in different situations may have been key in their population recovery.

Recently, Jarma et al. (2019) compared sea lion feeding habits over a small spatial scale, specifically two gulfs, Golfo San Matías and Golfo Nuevo. These gulfs vary in their physiography, oceanography, and anthropic pressures. Sea lions exhibited different patterns of prey-resource exploitation in the two gulfs, in particular with respect to the dominant species, the ecological and zoological groups, and the size of the prey consumed. This flexibility in sea lion diet underscores the value of local-scale studies to enable complete characterization of the diet and an understanding of different population contexts over the overall distribution range (Jarma et al. 2019).

24.4 Conclusions: Old Problems—New Problems

An understanding of the underlying processes and comprehensive history of population growth after a harvest-driven depletion is necessary to assess the long-term effectiveness of management and conservation strategies. The South American sea lion is the most conspicuous marine mammal along the South American coasts, and it was heavily exploited. Many of its colonies were decimated during the early twentieth century but now are showing clear recovery. The most important elements of recovery are diverse and we are beginning to understand how the population responded at low densities, how human-induced mortality interplayed with natural mechanisms, and how density-dependence regulates population growth. The observed population trajectory shows a non-linear relationship with density, with the population now recovering with a maximum increase rate of 5.6%. However, 50 years after hunting ceased, the population is still only at 35% of its pre-exploitation abundance (Romero et al. 2017).

Sealing for leather and oil ceased many years ago, but anthropogenic stressors remain. Interactions between pinnipeds and fisheries include damage to fishing gear, loss of fish, and injury and death of sea lions. Ecological interactions involve competition, predation and parasite transmission. Conflicts occur in all the areas in which colonies are near fishing zones, due to overlap in resources and the areas used by sea lions and fisheries (Crespo et al. 2012).

Since the development of the hake trawling fishery in Patagonia, during the 1970's, sea lions have drowned in nets. Their opportunistic behavior of taking advantage of new sources of food makes them vulnerable. Sea lions interact with all types of fisheries, but are not killed in all of them (Crespo et al. 1997, 2012). They take the bait or fish caught in longlines and gillnets, but are rarely caught in that gear. In Patagonian hake and shrimp trawls, mortality is roughly between one and two percent of population size, which is today compensated by a higher rate of increase. Sea lions are still increasing and have not yet reached carrying capacity. Given the huge amount of biomass taken by the trawling and jigging fisheries, in which the main targets are also prey of sea lions, future carrying capacity may be very much lower than the original one.

A recent analysis of food webs across the past 7000 years considered three top predators, sea lions, fur seals and Magellanic penguins (*Spheniscus magellanicus*). Ancient food webs were shorter, more redundant and overlapped more than current ones, both in northern-central and in southern Patagonia. These surprising results may be best explained by the huge impact of sealing on pinnipeds during the fur trade period, indicating that in modern food webs there has been a release from intraspecific competition and a shift towards larger and higher trophic level prey

(Saporiti et al. 2014). The three top predators overlapped more in the past. This, in turn, has led to longer and less overlapping food webs with fur seals and sea lions well below the original carrying capacity.

24.5 Future Scenarios

The present chapter describes the history of exploitation and recovery of one of the most conspicuous top predators of the southwestern Atlantic. It also attempts to explain the ecological and ethological mechanisms that allowed the species to recover. However, which are the potential scenarios that the sea lions may expect in an ecosystem in continuous change? South American sea lions did not reach a stable population size, nevertheless, they will reach one soon and lower compared to the pre-exploitation population. The huge amount of biomass extracted by the fishery will remain for future decades. Therefore, in order to keep safe this and other top predators there is a need to reach agreement of an equilibrium point in which the fishery catches may coexist with sustainable abundances of every and all top predators in the ocean.

Acknowledgments The author thanks CONICET, ANPCyT, Amnéville Zoo and UNP for supporting him and his students for almost 38 years of research and conservation. To his colleagues from LAMAMA for sharing ideas, field work and papers throughout these years. To Néstor A. García and Florencia Grandi for their help drawing figures.

References

Baylis AMM, Orben RA, Arnould JPY et al (2015) Diving deeper into individual foraging specializations of a large marine predator, the South American sea lion. Oecology 179:1053–1065

Baylis AMM, Orben RA, Costa DP et al (2017) Habitat use and spatial fidelity of male South American sea lions during the nonbreeding period. Ecol Evol 7:3992–4002

Bayón C, Politis G (2014) The Inter-Tidal Zone Site of La Olla: early–middle holocene human adaptation on the Pampean Coast of Argentina. In: Evans AM, Flatman JC, Flemming NC (eds) Prehistoric archaeology on the continental shelf: a global review. Springer, New York, p 115130

Bekiashev KA, Serebriakov VV (1981) North Pacific Fur Seal Commission (NPFSC). In: International marine organizations. Springer, Dordrecht. https://doi.org/10.1007/978-94-009-8261-1_28

Best PB (1993) Increase rates in severely depleted stocks of baleen whales. Int Comm Explor Sea J Mar Sci 50:169–186

Bonner WN (1982) Seals and man: a study of interactions. Univ Wash Press 170pp

Borella F (2014) Zooarchaeological evidence of Otariids in continental coast of Patagonia, Argentina: Old and new perspectives. En Neotropical and Caribbean Aquatic Mammals. Perspective from archaeology and conservation biology, Muñoz AS, Götz CM, Roca ER (eds). Nova Science: New York; 135–160

Campagna C (1985) The breeding cycle of the South American sea lion, *Otaria byronia*. Mar Mamm Sci 1:210–218
Campagna C, Le Boeuf BJ, Cappozzo HL (1988) Pup abduction and infanticide in Southern Sea Lions. Behavior 107(1/2):44–60
Campagna C, Werner R, Karesh W et al (2001) Movements and location at sea of South American sea lions (*Otaria flavescens*). J Zool Lond 257:205–220
Carrara IS (1952) Lobos marinos, pingüinos y guaneras de las costas del litoral marítimo e islas adyacentes de la República Argentina. Fac Cs Vet, Univ Nac La Plata
Carroll G, Hedley S, Bannister J et al (2014) No evidence for recovery in the population of sperm whale bulls off Western Australia, 30 years post-whaling. Endanger Species Res 24:33–43
Castilho PV, Simões-Lopes PC (2008) Mammals in archaeological sites on the southern coast of Brazil. Rev Mus Arqu e Etnol, São Paulo 18:101–113
Caughley G (1977) Analysis of vertebrate populations. Wiley, Chichester
Clapham PJ, Brownell RL Jr (1996) The potential for interspecific competition in baleen whales. Rep Int Whal Comm 46:361–367
Crespo EA (1988) Dinámica poblacional del lobo marino del sur *Otaria flavescens* (Shaw, 1800), en el norte del litoral patagónico. PhD thesis dissertation, Fac Cien Exact Nat, Univ Nac Buenos Aires, Argentina
Crespo EA, Hall MA (2001) Interactions between aquatic mammals and humans in the context of ecosystem management, Chapter 13. In: Evans PGH, Raga JA (eds) Marine mammals: biology and conservation. Kluwer Academic/Plenumm, pp 463–490
Crespo EA, Pedraza SN (1991) Estado actual y tendencia de la población de lobos marinos de un pelo (*Otaria flavescens*) en el litoral norpatagónico. Ecol Aust 2(1):87–95
Crespo EA, Pedraza SN, Dans SL et al (1997) Direct and indirect effects of the highseas fisheries on the marine mammal populations in the northern and central Patagonian Coast. J Northw Atl Fish Sci 22:189–207
Crespo EA, Oliva D, Dans SL et al (eds) (2012) Estado de situación del lobo marino común en su área de distribución. Universidad de Valparaíso, Valparaiso, Chile
Crespo EA, Pedraza SN, Dans SL et al (2018) Southwestern Atlantic southern right whales *Eubalaena australis* still growing but at a decelerated speed. Mar Mamm Sci. https://doi.org/10.1111/mms.12526
Dans SL, Crespo EA, Pedraza SN et al (2004) Recovery of the South American sea lion population in northern Patagonia. Can J Fish Aquat Sci 61:1681–1690
Doidge DW, Croxall JP, Baker JR (1984) Density–dependent pup mortality in the Antarctic fur seal *Arctocephalus gazella* at South Georgia. J Zool 202:449–460
Drago M, Crespo EA, Aguilar A et al (2009) Historic change of the diet of the South American sea lion in Patagonia as revealed by isotopic analysis. Mar Ecol Prog Ser 384:273–286
Estes JA (1979) Exploitation of marine mammals: r-selection of K-strategists? J Fish Res Bd Can 36:1009–1017
Evans PGH (1987) The Natural History of Whales and Dolphins. Academic Press/Facts on File, London
Evans PGH, Stirling I (2001) Life history strategies of marine mammals. In: Evans PGH, Raga JA (eds) Marine mammals: biology and conservation. Kluwer Academic/Plenum, pp 7–62
Fowler CW (1981) Density dependence as related to life history strategies. Ecology 62:602–610
Fowler CW (1984) Density dependence in cetacean populations. Rep Int Whal Comm Special Issue 6:373–379
Godoy JC (1963) Evaluación de los Recursos Naturales de la Argentina (Primera Etapa) Tomo VIII. Fauna Silvestre. Vol 1. Consejo Federal de Inversiones
Gómez Otero J (2007) Dieta, uso del espacio y evolución en sociedades cazadoras-recolectoras de la costa centro-septentrional de Patagonia durante el Holoceno medio y tardío. PhD thesis, Universidad de Buenos Aires, Argentina

Grandi MF (2010) Dinámica poblacional y cambios estructurales en una población en crecimiento de lobo marino común, *Otaria flavescens*, del litoral norpatagónico. PhD thesis, Universidad Nacional del Comahue, Argentina
Grandi MF, Dans SL, Crespo EA (2008) Changes in social composition and spatial distribution of colonies in an expanding population of South American sea lions, *Otaria flavescens*. J Mammal 89(5):1218–1228
Grandi MF, Dans SL, Crespo EA (2015) The recovery process of a population is not always the same: the case of *Otaria flavescens*. Mar Biol Res J 11(3):225–235
Grandi MF, Dans SL, Crespo EA (2016) Improvement in survivorship: the key for population recovery? Zool Stud 55:9. https://doi.org/10.6620/ZS.2016.55-09
Grandi MF, Loizaga R, Teran E et al (2018) Is recolonization pattern related to female philopatry? An insight into a colonially breeding mammal. Mamm Biol 89:21–29
Harcourt R (1992) Factors affecting mortality in the South American fur seal (*Arctocephalus australis*) in Peru: density-related effects and predation. J Zool Lond 226:259–270
Jarma DA, Romero MA, García NA et al (2019) Small-scale variation in the diet of South American Sea lion (*Otaria flavescens*) in northern Patagonia (Argentina). Reg Stud Mar Sci 28:1–11
Kellogg R (1942) Tertiary, quaternary, and recent marine mammals of South America and the West Indies. Proc 8th Am Sci Cong Wash 3:445–473
Kirn W (1986) Un poco de historia: Cuando en la lobería había lobos. Patagonia Agropecuaria Revista de la Sociedad Rural de Comodoro Rivadavia 8:27–29
Koen-Alonso M, Pedraza SN, Crespo EA et al (1999) A multispecies approach to the dynamics of some high trophic predators of the northern and central Patagonia marine community. 14th Ann Conf Europ Cet Soc, p 47
Koen-Alonso M, Crespo EA, Pedraza SN et al (2000) Feeding habits of the South American sea lion *Otaria flavescens* of Patagonia. Fish Bull 97(2):250–263
Lockyer C (1984) Review of baleen whale (Mysticeti) reproduction and implications for management. Rep Int Whal Comm Special Issue 6:27–50
Lopez Arregui E, Gonzalez Regalado T (1940) Lobos marinos en la Argentina. Su aprovechamiento integral. Boletín de Fomento Ganadero, Ministerio de Agricultura de la Nación, p 18:19
McLaren I, Smith TG (1985) Population ecology of seals: retrospective and prospective views. Mar Mamm Sci 1:54–83
Orquera LA, Piana EL (1999) Arqueología de la región del canal Beagle (Tierra del Fuego, República Argentina). Sociedad Argentina de Antropología, Buenos Aires
Riet-Sapriza FG, Costa DP, Franco-Trecu V et al (2013) Foraging behavior of lactating South American sea lions, *Otaria flavescens* and spatial-resource overlap with the Uruguayan fisheries. Deep-Sea Res II 88–89:106–119
Romero MA, Grandi MF, Koen-Alonso M et al (2017) Analyzing the natural population growth of a large marine mammal after a depletive harvest. Sci Rep 7:5271. https://doi.org/10.1038/s41598-017-05577-6
Saporiti F, Bearhop S, Silva L et al (2014) Longer and less overlapped food webs in anthropogenically disturbed marine ecosystems: confirmations from the past. PLoS One 9(7):e103132
Schiavini ACM (1992) Hombres y lobos marinos en la Sudamérica Austral. En Huellas en la arena, sombras en el mar. Los mamíferos marinos de la Argentina y la Antártida, Lichter AA (ed). Ed Terra Nova: Buenos Aires; 250–254
Svendsen G, Dans SL, Crespo EA et al (2012) Mortalidad de crías asociada a cambios en la estructura social y al aumento poblacional. In: Crespo E, Oliva D, Dans S, Sepúlveda M (eds) Estado de situación del lobo marino común en su área de distribución. Ed Univ Valparaíso, Valparaíso, Chile
Valette L (1928) El lobo marino de un pelo. Consultas Dirección General de Agricultura, Ganadería e Industrias (Provincia de Buenos Aires), Circ Nro 97:6
Vaz-Ferreira R (1950) Observaciones sobre las Islas de lobos. Rev Fac Hum Cs Univ Urug 5:145–176

Vaz-Ferreira R (1952) Observaciones sobre las Islas de Torres y de Castillo Grande. Revista de la Facultad de Humanidades y Ciencias. Universidad de Montevideo, Uruguay 9:237–258

Vaz-Ferreira R (1982) *Otaria flavescens* (Shaw), South American sea lion. Mamm Seas, FAO, Fish Ser 5:477–495

Weddell J (1825) A Voyage towards the South Pole performed in the years 1822–1824 by James Weddell Master in the Royal Navy, London 1825, Loneman, Hurst, Rees, Orme, Brown y Green

Zangrando AF, Panarello H, Piana EL (2014) Zooarchaeological and stable isotopic assessments on pinniped–human relations in the Beagle Channel (Tierra del Fuego, Southern South America). Int J Osteoarchaeol 24:231–244

Zenteno L, Borella F, Otero JG et al (2015) Shifting niches of marine predators due to human exploitation: the diet of the South American sea lion (*Otaria flavescens*) since the late Holocene as a case study. Paleobiology 41:387–401

Chapter 25
Living on the Edge, the New Zealand Sea Lion

B. Louise Chilvers

Abstract The New Zealand sea lion (*Phocarctos hookeri*) is an endangered and declining species restricted in distribution by past human exploitation to southern New Zealand and its subantarctic islands. This chapter investigates how the behavior of New Zealand sea lions is influencing this species' recovery and recolonization back to mainland New Zealand. New Zealand sea lions show high levels of philopatry to their birth colonies. The colonization of new breeding habitats rarely occurs when philopatry is strong for females of a species; however, some female New Zealand sea lions have dispersed more than 700 km to start breeding far from their birth place. What behaviors, in both male and female New Zealand sea lions, encouraged or allowed this recolonization to occur and how do these behaviors affect current population parameters?

Keywords New Zealand sea lion · *Phocarctos hookeri* · Behavior · Recolonization · Philopatry · Population parameters

25.1 Introduction

New Zealand sea lions are the world's least abundant sea lion species (Chilvers and Meyer 2017). Due to subsistence hunting and then commercial sealing in the late nineteenth century, the species population and range shrank from New Zealand wide (a mostly oceanic area of over 3,000,000 km^2), to their now remaining population restricted predominantly to the New Zealand subantarctic islands. They are an endangered, declining, socially complex polygamous species that are relatively long lived with marked sexual dimorphism and protracted maternal care and lactation. They are apex predators who display individual specialized foraging behaviours. In this chapter, I investigate how the behavior of New Zealand sea lions is

B. L. Chilvers (✉)
Wildbase, School of Veterinarian Science, Massey University, Palmerston North, New Zealand
e-mail: b.l.chilvers@massey.ac.nz

C. Campagna, R. Harcourt (eds.), *Ethology and Behavioral Ecology of Otariids and the Odobenid*, Ethology and Behavioral Ecology of Marine Mammals,
https://doi.org/10.1007/978-3-030-59184-7_25

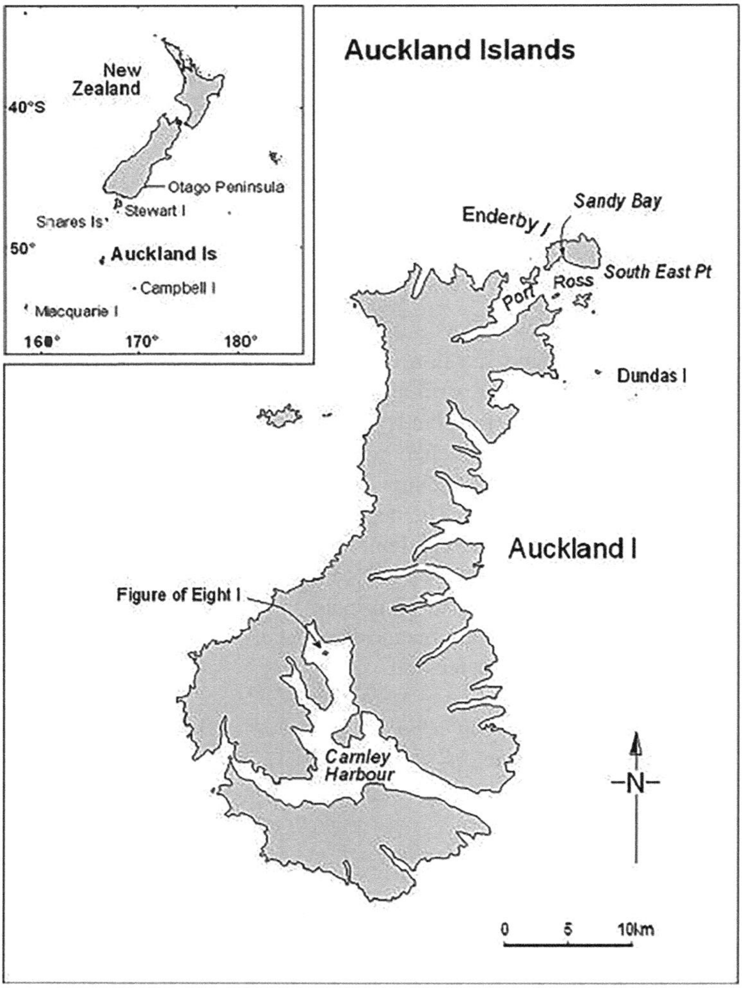

Fig. 25.1 Auckland Islands showing the main breeding areas for New Zealand sea lions *Phocarctos hookeri*: Sandy Bay and South East Point, Enderby Island, Dundas Island and Figure of Eight Island, Carnley Harbour. Inset: New Zealand's subantarctic area showing the Otago Peninsula, Stewart Island/Rakiura, Snares Island, Auckland Islands, Campbell Island/Motu Ihupuku and Macquarie Island (Australia), areas where New Zealand sea lions are known to occur

influencing the species' recovery, its recolonization back to the mainland, and its overall population trends.

New Zealand sea lions at the Auckland Islands (Fig. 25.1), their largest subantarctic colony, are well studied in regard to population dynamics, foraging ecology, genetics and causes of mortality. However, despite their predominantly-isolated distribution relative to areas that humans inhabit, they remain endangered, declining, with low reproductive rates, lower than expected female survival, and unique restless movement behaviors of mothers with pups.

25.2 Background

25.2.1 Distribution

Historically, New Zealand sea lions breed throughout New Zealand and its subantarctic islands from latitude 35° S to 52° S. However, due to Maori subsistence hunting and then European commercial sealing, the species were extirpated from mainland New Zealand and are now confined to a small, roughly triangular area centered on the Auckland Islands (Fig. 25.1, 50°50′S, 166°E, Childerhouse and Gales 1998). The range of New Zealand sea lions extends northwards from the Auckland Islands to the southern half of the South Island (to 46°S), south-westwards to Australia's' Macquarie Island (54°′0′S, 158°50′E), and eastwards to Campbell Island (52°′2′S, 169°08′E), predominantly all in the New Zealand subantarctics (Fig. 25.1). As of 2018, there are now four breeding areas recognized for the species: (1) Enderby and Dundas Islands, Northern Auckland Islands (equating to 66% of all pups born); (2) Figure of Eight Island, Carnley Harbour, southern Auckland Islands (2% of pups born); (3) Campbell Island (30% of pups born); and (4) Stewart Island, southern New Zealand (2% pups born). Since 1992, there have also been small numbers of pups born regularly (≤10) on the Otago Peninsula, South Island, New Zealand (Fig. 25.1).

25.2.2 Population Trends and Conservation Status

New Zealand sea lions are the only seal species endemic to New Zealand (Chilvers 2015). Their total abundance in 2015 (based on pup production of 2316 pups) was estimated to be 11,767 (95% confidence interval: 10,790–12,923), the lowest population size of any sea lion species (Chilvers and Meyer 2017). This population estimate is a decrease of 27% since 1998 with the decline driven by the significant decline of 48% at the Auckland Islands, an area that was once their breeding stronghold. New Zealand sea lions are listed as Endangered by IUCN Red List (Chilvers 2015) and Nationally Critical under the New Zealand Threat Classification List (Baker et al. 2016).

The current decline of New Zealand sea lions is believed to be driven by multiple factors. The most significant human impact on this species is the direct and indirect effects of trawl fisheries around the Auckland and Campbell Islands, in New Zealand's subantarctics (Robertson and Chilvers 2011). An estimated 1571 sea lions have died in the Auckland Island trawl fishery between 1995 and 2015, many of which are adult females. Additional to the fisheries mortality-caused decline at the Auckland Islands, New Zealand sea lions have also been affected by disease outbreaks causing high early mortality in young pups (Robertson and Chilvers 2011).

Away from the subantarctics, individuals are just starting to repopulate Stewart Island and South Island, New Zealand; however, they still occur only in small numbers. With this return to the mainland come additional threats common for more human populated areas. New Zealand sea lions have been killed by being hit by cars and trains, shot, and pups killed by domestic dogs (Chilvers 2017a).

So, how does the behavior of New Zealand sea lions affect this species' susceptibility to their currently known impacts and how could it help or hinder their recovery and recolonization back to the New Zealand mainland where they belong? There are several topics that could be investigated as to why the behavior of New Zealand sea lions may be restricting the current population and hinder their recovery. In this chapter, I start by looking at female breeding behaviors, particularly how philopatry to their places of birth restricts female dispersion for recolonization. Second, I investigate the behavior that, once females give birth, they continually move with young pups throughout lactation. This movement pattern may relate to New Zealand sea lion mothers protecting themselves and pups from male harassment, moving away from parasite loads known to exist in colonial breeding areas, or access to foraging areas needed and used by lactating females. Specific hypotheses from these behaviors would be that:

1. The colonization process of new breeding habitats rarely occurs when philopatry is strong in females;
2. Females move away from breeding colonies with young pups to protect pups and themselves from male-male fighting and female harassment within harems;
3. Females move away from breeding colonies with young pups to move away from parasite burdens likely in the area of the breeding harems and;
4. Females move with young and growing pups to reach better or more easily accessed foraging areas.

25.3 Site Fidelity, Philopatry and Recolonization Behavior

Like most otariids, New Zealand sea lions show high levels of philopatric behaviors to their birth colonies (Chilvers and Wilkinson 2008). One of the management goals for the species has been to encourage and establish new breeding areas away from the New Zealand subantarctic regions. However, a better understanding of the recolonization processes for otariids is needed to appreciate factors influencing individual dispersal. This is particularly true for a species that has been isolated from its next nearest breeding area (distance between the Auckland Islands and lower New Zealand mainland) by a minimum of 460 km.

Data from resights of female New Zealand sea lions marked as pups from the northern Auckland Island breeding area show high site fidelity and philopatry to their natal sites. For females, non-natal resightings and places they have given birth away from their own birthplace has predominantly been restricted to locations within the breeding areas where they were born (an area within ~10 km^2). In contrast, males disperse more widely, traveling distances up to 700 km away from their place of birth to areas on mainland New Zealand (Chilvers and Wilkinson 2008).

These data support the hypothesis that colonization of new breeding habitats rarely occurs when philopatry is strong for females, particularly when a population is at a low density or declining such as for New Zealand sea lions at the Auckland Islands. However, some female New Zealand sea lions have broken these behaviors and dispersed more than 700 km to start breeding far from home. One female New Zealand sea lion, affectionately dubbed 'Mum', appeared at the Taieri River mouth just south of the Otago Peninsula, New Zealand, in 1993 and gave birth to the first sea lion pup recorded on mainland New Zealand in more than 150 years. Mum, who had been tagged as a pup at the Auckland Islands, remained based around Otago's coastline for nearly two decades and produced 11 pups, with six females. Mum died in 2010 at the age of 24 and now her daughters, grand-daughters, great-grand-daughters and great, great grand-daughters make up the female population of New Zealand sea lions on the Otago Peninsula (https://sealiontrust.org.nz/). Since Mum, there have been a few other examples of females bucking the trend and moving from the Auckland Islands to Stewart Island, or lower South Island, New Zealand, away from their natal sites; however, the numbers are very low, but this does raise the question; what encourages these females to break the mold and move?

25.4 Restless Maternal Behavior

At the subantarctic breeding areas, New Zealand sea lions show polygynous colonial breeding behavior. Pup births are highly synchronized with all births concentrated within a single month and 69% of births within a two-week period centered around 26 December (Chilvers 2012). Pregnant females arrive ashore at breeding areas/beaches in early December, usually coming ashore two days prior to giving birth. Once their pup is 7 to 10 days old, females make regular foraging trips averaging 2.8 days (range 1–4 days), returning to suckle their pups for a day between trips (Chilvers et al. 2006a). Lactation is presumed to last about 10 months but pups/yearlings have been recorded to stay with their mothers after weaning (Leung et al. 2014). At about one month after mean pupping date, breeding harems start to break up and females and young pups start to move, off the breeding islands or up to 1.5 km inland and 400 m above sea level to areas of grass, herb fields and forests in small groups or solitarily (Augé et al. 2009). This behavior to move appears to continue throughout lactation, not stopping after the first movement away from the pups' birth site (Augé et al. 2012). Within the Auckland Island group, after late

January, Dundas Island females swim with their month-old pups to Sandy Bay, Enderby Island (a ~ 10 km swim), and to other locations on the main Auckland Islands, near Dundas Island (minimum distance ~2 km). For Enderby Island females, they either move inland away from the breeding beaches or swim to nearby islands in Ross Harbour or the main Auckland Island (distances 1 to 7 km). None have ever been recorded swimming to Dundas Island. For the newer, smaller breeding areas, females also move, but this time it is away from their assumed solitary secluded birth locations (no two or more females have been recorded giving birth at the same location for these areas—however, for Stewart Island, no one has actually investigated to know), to more communal areas with other mother and pup pairs. No female, at any location (subantarctic or New Zealand main islands), has been recorded to either (a) stay at the birth location of their pup; or (b) move to a beach or rocky coastal location; it is always inland.

It is not uncommon for sea lions to move continuously or swim far during lactation. It has never been investigated why female New Zealand sea lions move/swim for considerable distances, particularly with such small pups. For the Dundas Islands females, the swim to Sandy Bay, Enderby Island is often at the cost of the pups' life, with at least one pup per season dying soon after arriving from the swim or being half carried/nosed dead to shore by its mother (pers. obs). And that would not account for any pups that may not make the 10 km swim given cold subantarctic sea conditions, currents, tides and sharks. The reasons for these movements may be similar to southern sea lion (*Otaria flavescens*) or Steller sea lion (*Eumetopias jubata*) females moving with 2 to 3 month old pups, which is either to move to areas closer to foraging grounds or to move from highly exposed breeding rookeries to predominantly more sheltered haul-outs habitats further up inlets (Raum-Suran et al. 2004).

The breeding habitat of the New Zealand sea lion is diverse. The large breeding colonies, such as the Enderby and Dundas Island colonies at the Auckland Islands, usually form on sandy beaches. However, for Campbell Island and Figure of Eight Island at the Auckland islands, colonies are on solid rock shorelines or scrub/bush areas (Fig. 25.2). For the newer, smaller breeding sites of Stewart Island and the Otago Peninsula, the reverse happens, with solitary females giving birth in secluded bush or forested areas, away from shorelines and other sea lions. Then, when the pups are older, females move their pups to more communal areas with other mother and pup pairs; however, they are still sheltered by forests and areas with sheltered water (streams, ponds or protected sea areas) nearby where pups play and swim with their cohort (Chilvers 2017a).

Females moving young pups off Dundas Island would support a hypothesis that female New Zealand sea lions move to relocate their pups from highly exposed breeding areas to less exposed locations. Dundas Island is a small (2 km^2), low lying island (highest point 12 m above sea level), covered only in grasses and mega herbs. It is in an exposed location east of the main Auckland Islands, not in a harbor, and has no islands to the south for protection from high seas or winds. It is a very exposed island in summer, so is likely to be dangerous in winter, evidence of which can be seen from large 2–3 m logs which have been thrown up to 30–40 m above the

Fig. 25.2 New Zealand sea lion breeding harem in scrub, Figure of Eight Island, Auckland Islands

high tide mark, presumably during winter storms (pers. obs). However, this reason for moving would not explain why Enderby Island, Stewart Island or Otago Peninsula females move, as all of these breeding locations are already sheltered and well above sea level protected areas.

For all areas, dispersal away from harems or areas where females have pupped may be to move away from parasite loads in the sand and soil in harems or birthing locations. For example, hookworm burdens and bacterial epidemics are known in both New Zealand and California sea lion colonies (i.e. hookworm burdens; Micheal et al. 2016) and moving from breeding areas and harems that have high numbers of individuals in small areas can lower the risk of spread of parasites and bacteria. Bacteria epidemic is a particular problem for New Zealand sea lions in recent years with the Auckland Island colonies affected with significant acute bacterial epizootics (see more detail below, Wilkinson et al. 2006).

Harassment, accidental and intentional killing of pups and females by adult and subadult males have been documented for New Zealand and other sea lion species (Wilkinson et al. 2000; Cassini and Fernandez-Juricic 2003; Chilvers et al. 2005a), and moving could be a females' way of increasing protection for herself and her pup by moving away from harem locations and males. For the populations that move from isolated birth locations to more communal areas, possible reasons could include that: giving birth in isolation is protection from male harassment, as found in southern sea lions when females are single and in low densities, it is safer for females to be isolated and hopefully not be found and harassed, rather than be in small groups but most likely be out-numbered by males (Cassini and Fernandez-Juricic 2003). This scenario would be a possibility for female New Zealand sea lions as males outnumber females easily 25 to 1 in the Otago and Stewart Island areas (Chilvers 2017a). The second motivation for moving when pups are a month old (and importantly females are no longer thought to be in estrus) is moving back into social protective groupings, so that as the mothers start their longer foraging trips, the pups have other pups for warmth, protection and socializing; and with more females in a group, it is more likely for a female to be present with the pups, possibly providing some protection from males.

Let's look in more depth at these possible behavioral reasons for moving and how it may help or hinder species recolonization, recovery and population parameters.

25.4.1 Female Movement and Male Harassment: Dimorphism and Polygamous Breeding

Hypothesis: Females move away from breeding colonies with young pups to protect pups and themselves from male-male fighting within harems. Like all otariids, New Zealand sea lions are sexually dimorphic with adult male New Zealand sea lions being three times the size of females at 240–350 cm in length and 320–450 kg in weight, compared with adult females at 180–200 cm in length and 85–160 kg in weight (Fig. 25.3). The other predominant feature about New Zealand sea lion sexual dimorphism is that they are also strikingly different in color, with adult males dark blackish-brown all over, with a large, thick, coarse mane over their shoulders, while adult females are slender white to creamy grey (Fig. 25.3). The only other otariid with this color dimorphism is the Australian sea lion (*Neophoca cinerea*). This sexual dimorphism, both in color and size, is seen in New Zealand sea lions from birth. Male pups are born significantly heavier than females (males 10.6 kg, females 9.7 kg) with the natal pelage of pups ranging from pale cream for female pups to dark brown and black for males (Chilvers et al. 2006a). At approximately 5–6 months old, pups molt to their adult coloration, although light colored 1 to 2 year old males can be mistaken for females.

As mentioned above, New Zealand sea lions show polygynous breeding behaviour, with highly synchronized births (Chilvers 2012). During the breeding season, adult males can establish territories on the breeding beaches several weeks before females return, claiming and defending areas of about 5 m^2. Challenges from peripheral males are frequent throughout the breeding season but particularly in early periods of harem establishment lead to fierce fights and often all of harem brawls (Fig. 25.4a). This fighting can cause injuries and can cause the death of up to 5 females in every 1000 breeding females. In New Zealand sea lion harems and trauma is responsible for 35% of early pup deaths (Chilvers et al. 2005a, b; Fig. 25.4b). Male sea lions use ritualized posturing, bluff charges, and oblique stares

Fig. 25.3 Adult male and female New Zealand sea lion showing sexual dimorphism in size and colour

when faced with challengers, only physically biting each other in less than 5% of interactions. Peripheral males attempt to secure females physically and can cause significant injuries and sometimes death (Chilvers et al. 2005a). Adult male New Zealand sea lions breed and hold territories on multiple subantarctic breeding islands within one season, meaning that male-male fighting, territorial establishment and female harassment is continual throughout the breeding season (Chilvers et al. 2005a; Robertson et al. 2006). This continual male-male fighting during the breeding season would be a good reason for females to move away from breeding areas as soon as they can for protection for themselves and their pups.

Behaviorally, females protect themselves in harems by aggregating in groups with fluid compositions on the beach away from the surf line for both protection and warmth (Augé et al. 2012). Group compositions change predominantly with breeding status of females, i.e. after giving birth females move into groups where other mothers are located rather than groups of pregnant or non-reproductive females. Females have no discernible social hierarchy within their groups and prefer to lie in

A

B

Fig. 25.4 (**a**) Male New Zealand sea lion fights including almost the entire harem. (**b**) New Zealand sea lion pups accidentally being sat on by adult male sea lion, who is oblivious to the pup even being there. The male sitting on the pup is fighting with the male sitting in front of the female, and the two males further back are also fighting. Note the size difference between adult male and pup

contact with each other. Within harems, females are tolerant of young pups (except when they attempt stealing milk). Females about to give birth separate themselves slightly from others, become agitated, and flip sand over themselves to keep cool. Females and pups identify one another by smell and by sound (each makes a distinctive call quickly learned by the other). Female New Zealand sea lions display the discrete communal behavior of backing out of groups to urinate or defecate;

however, will not move away from the group when needing to regurgitate squid/octopus beaks and other indigestible hard parts, usually vomiting over the backs of the groups they are huddled in. The behavior of defecating away from groups may be a behavior to lower parasite loads directly under female groups, but vomiting over each other might be just as bad.

There has been behavioral research on female New Zealand sea lions to assess the factors they prefer for breeding locations and what attracts them to areas (Augé et al. 2012; Augé and Chilvers 2010). Overall, the most suitable terrestrial habitat configuration for a breeding aggregation of New Zealand sea lions appears to be a sandy beach, with a wide area above high tide and moderate intertidal zone (for breeding), backed with vegetated sand dunes and forest on primarily flat terrain (for later dispersion; Augé et al. 2012). There has also been behavioral research to investigate the visual cues that female sea lions look for when coming to shore. This behavioral experiment used a dummy female sea lion made from white canvas and taxidermy female sea lion decoys, (taxidermy females died from drowning in the fisheries around the Auckland Islands and bodies were returned for necropsy) to identify what females were visually attracted to. The research showed that female New Zealand sea lions were attracted to decoys by their color and shape, as is expected by their behavior, as other females are likely to offer protection and warmth. This technique of using female colored and shaped decoys could be trialed to direct recolonizing females to suitable pupping sites on mainland New Zealand.

There are data to support the hypothesis that female New Zealand sea lions travel with young pups to move away from males, which can cause their death and that of their pups. However, this does not explain why they keep moving throughout lactation.

25.4.2 Female Movement, Parasites and Disease

Hypothesis: Females move away from breeding colonies with young pups to move away from parasite burdens likely in the area of the breeding harems. For New Zealand sea lions, the causes of pup mortality are well-researched. Post-mortem examinations of pups found dead at Sandy Bay, Enderby Island show the primary causes of death in the first month of life were trauma (35%), bacterial infections (24%), hookworm infection (13%), starvation (13%), and stillbirth (4%). For most pups, more than one diagnosis was recorded meaning a combination of these factors were the likely cause of death (Castinel et al. 2007). Many of these causes of death are due to the species colonizing polygamous breeding behaviors. Adult and sub-adult males are responsible for pup deaths (trauma) during the breeding season, usually accidental while males are fighting over territories or defending females. However, such deaths have also been recorded as deliberate aggression towards pups and females (Fig. 25.4b, Wilkinson et al. 2000; Chilvers et al. 2005a).

New Zealand sea lions have been affected by two bacterial outbreak epizootics at the Auckland Islands, resulting in the death of 53, 32 and 21% of pups in the first month of their lives in the 1998, 2002 and 2003 seasons, respectively (Wilkinson et al. 2006). Pinniped colonial breeding behavior makes species susceptible to such outbreaks as large numbers of animals in close proximity means diseases can easily travel from one individual to the next (Kovacs et al. 2012). For many of the pups that died during these epizootics, and for pups with high hookworm loads, their deaths are usually the result of trauma. This is because trauma deaths are usually accidental, with pups not getting out of the way of fighting males quickly enough. Pups impacted by parasite loads (suffering conditions such as anaemia), or being slowed down by bacteria infections, have slower response reactions than healthy pups and this means they are less likely to be able to move fast enough from fighting males (Fig. 25.4b). These data support the hypothesis that the colonial breeding behavior of New Zealand sea lions impacts pup mortality and may be attractive for females to breed alone or move soon after a pup's birth. However, this still does not give reason for continual movement throughout lactation.

The epizootic events at the Auckland Islands have not been linked to known naturally occurring or anthropogenic influences, and there is also no apparent link between the timing of events and environmental factors, such as El Niño/Southern Oscillation (ENSO) cycles. Although the bacterial epizootic can cause high early pup mortality in the year they occur, population modeling shows that these pup deaths do not lead to changes in the overall mortality rate of pups in 1 year. Rather, the data suggest that disease-related mortality is compensatory to other sources of mortality, with the bacteria killing off potentially smaller weaker pups earlier than would normally be expected (or reported) but over the first year of life having no overall effect on the New Zealand sea lion 1-year mortality rates (Meyer et al. 2015).

25.4.3 Female Movement and Foraging Behavior

Hypothesis: Females move with young and growing pups to reach better or easier accessed foraging areas. Like all otariids, lactating New Zealand sea lions are central place foragers, foraging in areas close to where their pups are based so they can return to regularly suckle their pups. New Zealand sea lions are considered generalist opportunistic feeders, targeting varied prey species based on what is available in their foraging environment, introducing variation in their diet through time following changes in prey availability and with changes in environmental factors such as the southern oscillation (Meynier et al. 2009, 2010). Lactating female New Zealand sea lions are the deepest diving eared seal (Gales and Mattlin 1997; Chilvers et al. 2006b). Mean dive depth is 131 m with the deepest dive recorded 686 m. For the Auckland Island population, dive durations average 4 min with mean maximum dive length over 10 min. Auckland Island females dive continuously while at sea. Over 68% of their dives exceed their calculated aerobic dive limit, suggesting that female New Zealand sea lions are operating near their physiological maxima. Female New Zealand sea lions at the Auckland Islands have two distinct dive profile types or foraging ecotypes, a benthic diving profile (benthic or bottom foraging) and a deeper, more varied mesopelagic (mid-water) diving profile. These different diving behaviors also correlate to different foraging locations, with benthic foragers diving north-east of the Auckland Islands over the continental shelf, while the mesopelagic divers forage north-west in deeper waters. The foraging location of the mesopelagic diving females results in a much higher overlap with fisheries interactions and likely mortality than the benthic divers (Chilvers 2008a). Females only forage using one of these foraging types and consistently use this foraging behavior within and between years regardless of environmental differences across years (Chilvers and Wilkinson 2009). Stable isotope signatures (δ13C and δ15N) from female New Zealand sea lions determined that the isotopic composition of serum and whiskers reflects the differences between the two sea lion foraging types of benthic and mesopelagic (Chilvers 2017b). Individual lifetime chronology of foraging strategy through stable isotope analysis of female New Zealand sea lion whiskers revealed that individuals do not change their foraging strategies within and across years, showing either constant benthic or mesopelagic foraging ecotypes regardless of age or reproductive status.

In conjunction with their extreme diving behavior, female New Zealand sea lions also have extensive foraging trips, covering huge distances in a single trip (Chilvers et al. 2005b). Foraging trips last an average of 66 h with a mean return travel distance per trip of 423 km (max recorded = 1087 km) and up to 102 km direct line distance away from the breeding area (range 33 to 188 km). New Zealand sea lion foraging location distances are at the upper end of observed otariid foraging limits. For comparison, southern sea lions, make return travel foraging trips of 206 ± 117 km, yet have mean direct line distance to colony distances of only 103 km (Campagna et al. 2001), compared with New Zealand sea lions' 423 km. Combine this with their extreme diving behavior and it adds to the hypothesis that female New Zealand sea lions are operating at or near their physiological maximum in the Auckland Island area (Chilvers et al. 2006b; Augé et al. 2011a).

For Otago female New Zealand sea lions, foraging behavior is significantly different. In 2008 to 2010, all New Zealand sea lion females (n = 13, aged 2–14 years) born on the Otago Peninsula were equipped with time–depth recorders and satellite tags (Augé et al. 2011b). The mean dive depth for these females was 20 m and mean dive duration was 1.8 min, some of the lowest values reported for otariids. Otago female New Zealand sea lions did not exhibit two distinct diving specialization as reported at the Auckland Islands and foraged only 40 km from their breeding locations (Augé et al. 2011a). Otago adult females exceeded calculated aerobic dive limits in 7% of dives compared with 68% at the Auckland Islands. The contrasting differences in diving behavior between Otago and the Auckland Islands suggest that Otago represents a better marine habitat for New Zealand sea lions, with food easily accessible to animals of all ages. This would appear to be a good reason to move away from the Auckland Islands for a female New Zealand sea lion raising a pup.

Foraging data do not support the hypothesis that female movements with young pups are towards better foraging areas, as their site fidelity to foraging areas at the Auckland Islands is constant within and between years. However, given the differences in foraging effort needed between the Auckland Islands and Otago area foraging, it would be a good reason to move and start pupping further north, back into their original range, for those that make the move.

The consequences of extreme foraging behavior in New Zealand sea lions at the Auckland Islands appears to have significant impacts on population parameters. New Zealand sea lion survival estimates differ significantly by sex and age class. After 1 year of age, male New Zealand sea lions have higher survival rates than females, including as adults. Male survival between 4 and 15 years is estimated at 98%, while female survival for the same age class is estimated to be 88–90% (Chilvers and MacKenzie 2010; Meyer et al. 2015). This difference in survival between the sexes is a critical problem for New Zealand sea lions, as even a small change in adult female survival significantly affects population trends for long-lived, polygamous mammals (Meyer et al. 2015). These differences in survival rates are related to direct and indirect impacts from the trawl fishery around the Auckland Islands, as it is predominantly only the area where mesopelagic diving female New Zealand sea lions forage that overlaps with the fisheries, and no adult male has ever been reported killed in the fishery (Chilvers 2008b; Chilvers and MacKenzie 2010; Meyer et al. 2017).

The indirect consequences of female New Zealand sea lion extreme foraging behavior at the Auckland Islands is thought to be shown in their low breeding rates compared to Otago female New Zealand sea lions and other sea lion species. New Zealand sea lions at the Auckland Islands have the lowest population growth rate reported for any sea lion species. This low growth rate is a results of females having a low reproductive rate (0.67), late recruitment of females into the breeding population (50% of breeders do not start breeding until 6 years of age although capable of breeding at 4 years), and a high percentage of non-breeding females (approximately 27% of females who survive to 3 years of age will never breed; Chilvers and MacKenzie 2010). Concurrently, the composition of New Zealand sea lion milk has a mean lipid content of only 21.3% ± 8.1%, with milk fat content of otariids directly related to prey intake (Riet-Sapriza et al. 2012). Overall, the quality of milk of the New Zealand sea lion at the Auckland Islands is considered relatively low in fats compared with New Zealand sea lions on mainland New Zealand and other sea lion species. For comparison, the lipid content of New Zealand sea lion milk from Otago females is 40.7 ± 2.9% (n = 11), and for Stewart Island females 35.7 ± 2.7% (n = 11, Chilvers 2017a, b). Its fat content is lower than for California or southern sea lions but similar to Australian sea lions (Riet-Sapriza et al. 2012). Again, compared with the females in the Otago population, the extreme foraging behavior and consequent apparent low reproductive ability of Auckland Island New Zealand sea lions indicates their population is constrained by their behavior. This is consistent with a population that is occupying a marginal foraging environment, or for an area impacted by resource competition that they have been restricted to due to past human harvest (Chilvers et al. 2006b; Augé et al. 2011a).

25.5 Conclusions

The New Zealand sea lion is the only endemic seal species in New Zealand. Their total population size is the lowest of any sea lion species, with population decline continuing to occur in the Auckland Islands, their largest breeding area (Chilvers and Meyer 2017). The decline of New Zealand sea lions is thought to be driven by multiple factors, with the largest human impact the direct and indirect fisheries interactions occurring with Auckland Island females. New Zealand sea lion past exploitation has resulted in this restricted range to an area where females are having to undertake extreme diving behaviors, resulting in low reproductive ability and population resilience. New Zealand sea lions are naturally making a small come-back on mainland New Zealand, their pre-sealing territory, and the animals breeding there are doing well, with earlier breeding, higher pupping rate per female, relaxed foraging behavior, larger female body weight and higher milk fat content for pups (Augé et al. 2011b). Behaviorally, New Zealand sea lions are similar to other sea lion species except for the restless movement pattern of mothers with young pups, which appears to at the very least be a mechanism to reduce pup mortality from trauma caused by males in breeding areas, and potentially for lowering parasite and bacteria loads. As the recolonization process hopefully continues for New Zealand sea lions back to the New Zealand mainland, it will be interesting to observe how, or if, these foraging and maternal behaviors change as more females and possibly a breeding population establishes on the New Zealand main islands. For the New Zealand sea lion to successfully recolonize and return to the New Zealand mainland, habitat protection and public education are needed to attempt to lessen anthropogenic mortalities. Humans living on the New Zealand mainland need to get used to the behavior of big "on-land" marine mammal predators, something that has not occurred in recent New Zealand history while humans have occupied the land.

References

Augé AA, Chilvers BL (2010) Behavioural responses and attraction of New Zealand sea lions to on-land female decoys. NZ J Zool 37:107–119

Augé AA, Chilvers BL, Moore AB et al (2009) Aggregation and dispersion of female New Zealand sea lions at the Sandy Bay breeding colony, Auckland Islands: how unusual is their spatial behaviour? Behaviour 146:1287–1311. https://doi.org/10.1163/15683909x427687

Augé AA, Chilvers BL, Moore AB et al (2011a) Foraging behaviour indicates marginal habitat for New Zealand sea lions: remnant vs recolonizing. MEPS 432:247–256

Augé AA, Chilvers BL, Davis LS et al (2011b) In the shallow end: diving behaviour of recolonising female New Zealand sea lions *(Phocarctos hookeri)* around the Otago Peninsula. Can J Zool 89:1195–1205

Augé AA, Chilvers BL, Moore AB et al (2012) On-land habitat preferences of female New Zealand sea lions at Sandy Bay, Auckland Islands. Mar Mammal Sci 28:620–637

Baker CS, Chilvers BL, Childerhouse S et al (2016) Conservation status of New Zealand marine mammals, 2013. New Zealand Threat Classification Series 14. Department of Conservation, Wellington. 18 p

Campagna C, Werner R, Karesh W et al (2001) Movements and locations at sea of South American sea lions (*Otaria flavescens*). J Zool Lond 257:205–220

Cassini MH, Fernandez-Juricic E (2003) Costs and benefits of joining South American sea lion breeding groups: testing the assumptions of a model of female breeding dispersion. Can J Zool 81:1154–1160

Castinel A, Duignan PJ, Pomroy WE et al (2007) A survey of neonatal mortality in New Zealand sea lions *Phocarctos hookeri* at Sandy Bay Beach on Enderby Island from 1998 to 2005. J Wildl Dis 43:461–474

Childerhouse S, Gales N (1998) Historical and modern distribution and abundance of the New Zealand sea lion, *Phocarctos hookeri*. NZ J Zool 25:1–16

Chilvers BL (2008a) Foraging behaviour site fidelity of lactating New Zealand sea lions, *Phocarctos hookeri*. J Zool Lond 276:28–36

Chilvers BL (2008b) New Zealand sea lions, *Phocarctos hookeri* and squid trawl fisheries: bycatch problems and management options. ESR 5:193–204. https://doi.org/10.3354/esr00086

Chilvers BL (2012) Using life-history traits of New Zealand sea lions, Auckland Islands, to clarify potential cause of decline. J Zool (Lond) 287:240–249

Chilvers BL (2015) *Phocarctos hookeri*. The IUCN Red List of Threatened Species. Version 2015.2, www.iucnredlist.org, downloaded on November 3rd 2015

Chilvers BL (2017a) Preliminary assessment of the foraging behaviour and population dynamics of a cryptic population of the endangered New Zealand sea lion. NZ J Ecol 42:48–57

Chilvers BL (2017b) Stable isotope signatures of whisker and blood serum confirm foraging strategies for female New Zealand sea lions derived from telemetry. Can J Zool 95:955–963

Chilvers BL, MacKenzie D (2010) Age and sex specific survival estimates incorporating tag loss for New Zealand sea lions, *Phocarctos hookeri*. J Mammal 91:758–767

Chilvers BL, Meyer S (2017) Conservation needs for the endangered New Zealand sea lion, *Phocarctos hookeri*. Aquat Conserv Mari Freshwat Ecosyst 27:846–855. https://doi.org/10.1002/aqc.2742

Chilvers BL, Wilkinson IS (2008) Philopatry and site fidelity of New Zealand sea lions, *Phocarctos hookeri*. Wildl Res 35:463–470

Chilvers BL, Wilkinson IS (2009) Divers foraging strategies in lactating New Zealand sea lions. MEPS 378:299–308

Chilvers BL, Robertson BC, Wilkinson IS et al (2005a) Male harassment of female New Zealand sea lions, *Phocarctos hookeri*: mortality, injury, and harassment avoidance. Can J Zool 83:642–648

Chilvers BL, Wilkinson IS, Duignan PJ et al (2005b) Identifying the distribution of summer foraging areas for lactating New Zealand sea lions, *Phocarctos hookeri*. MEPS 304:235–247

Chilvers BL, Robertson BC, Wilkinson IS et al (2006a) Growth and survival of New Zealand sea lions, *Phocarctos hookeri*: birth to 3 months. Polar Biol 30:459–469

Chilvers BL, Wilkinson IS, Duignan PJ et al (2006b) Diving to extremes: are New Zealand sea lions (*Phocarctos hookeri*) pushing their limits in a marginal habitat? J Zool Lond 269:233–241

Gales NJ, Mattlin RH (1997) Summer diving behaviour of lactating New Zealand sea lions, *Phocarctos hookeri*. Can J Zool 75:1695–1706

Kovacs KM, Aguilar A, Aurioles D et al (2012) Global threats to pinnipeds. Mar Mamm Sci 28:414–436

Leung ES, Chilvers BL, Nakagawa S et al (2014) Do yearling New Zealand sea lions (*Phocarctos hookeri*) learn foraging behaviour from their mothers? Mar Mamm Sci 30:1220–1228. https://doi.org/10.1111/mms.12087

Meyer S, Robertson BC, Chilvers BL et al (2015) Population dynamics reveal conservation priorities of the threatened New Zealand sea lion *Phocarctos hookeri*. Mar Biol 162:1587–1596

Meyer S, Robertson BC, Chilvers BL et al (2017) Marine mammal population decline linked to obscured by-catch. PNAS. www.pnas.org/cgi/doi/10.1073/pnas.1703165114

Meynier L, MacKenzie DDS, Duignan PJ et al (2009) Variability in the diet of New Zealand sea lion (*Phocarctos hookeri*) at the Auckland Islands, New Zealand. Mar Mamm Sci 25:302–326

Meynier L, Morel PCH, Chilvers BL et al (2010) Quantitative fatty acid signature analysis on New Zealand Sea lions: model sensitivity and diet estimates. J Mammal 91:1484–1495

Micheal S, Chilvers BL, Roe W et al (2016) Long term survival and reproductive success of New Zealand sea lions (*Phocarctos hookeri*) treated with ivermectin as pups. Wildl Res 42 (8):660. https://doi.org/10.1071/WR15120

Raum-Suran KL, Rehberg MJ, Pendleton GW et al (2004) Development of dispersal, movement patterns and haul-out use by pup and juvenile Steller sea lions (*Eumetopias jubatus*) in Alaska. Mar Mamm Sci 20:823–850

Riet-Sapriza FG, Duignan PJ, Chilvers BL et al (2012) Interannual and individual variation in milk composition of New Zealand sea lions (*Phocarctos hookeri*). J Mamm 93:1006–1016

Robertson BC, Chilvers BL (2011) The population decline of the New Zealand sea lion *Phocarctos hookeri*: a review of possible causes. Mamm Rev 41:253–275. https://doi.org/10.1111/j.1365-2907.2011.00186.x

Robertson BC, Chilvers BL, Duignan PJ et al (2006) Dispersal of breeding, adult male *Phocarctos hookeri*: implications for disease transmission, population management and species recovery. Biol Conserv 127:227–236

Wilkinson IS, Childerhouse S, Duiganan PJ et al (2000) Infanticide and cannibalism in the New Zealand sea lion, *Phocarctos hookeri*. Mar Mamm Sci 16:494–500

Wilkinson IS, Duignan PJ, Grinberg A et al (2006) *Klebsiella pneumonia* epidemics: possible impact on New Zealand sea lion recruitment. In: Trites AW, DeMaster DP, Fritz LW, Gelatt LD, Rea LD, Wynne KM (eds) Sea Lions of the World. Lowell Wakefield Fisheries Symposium, Anchorage, Alaska, p 455–471

Chapter 26
The Enigmatic Life History of the Australian Sea Lion

Rebecca R. McIntosh and Benjamin J. Pitcher

Abstract Like most sea lions, Australian sea lions are benthic foragers with prolonged lactation, extensive parental care, complex social interactions, mild-moderate polygyny, are relatively long-lived and near apex predators. They have been well studied in regards to population dynamics, foraging ecology, parent-offspring behavior and social communication; in fact, they are probably the best understood of all the otariids in regards to mother-offspring recognition and mating strategies. Yet they remain an enigmatic beast with a unique, asynchronous aseasonal reproductive cycle, the evolution of which is still not fully understood. In this chapter we compare Australian sea lions to other otariids and the odobenid, describe the detailed understanding of their life history traits and explore putative hypotheses as to how their unique reproductive cycle evolved.

Keywords *Neophoca cinerea* · Aseasonal life history · Endemic · Population decline · Endangered · Benthic · Maternal strategies · Phenology

---

Electronic supplementary material The online version of this chapter (https://doi.org/10.1007/978-3-030-59184-7_26) contains supplementary material, which is available to authorized users.

R. R. McIntosh (✉)
Conservation Department, Phillip Island Nature Parks, Cowes, Victoria, Australia

Department of Biological Sciences, Macquarie University, Macquarie Park, NSW, Australia
e-mail: rmcintosh@penguins.org.au

B. J. Pitcher
Department of Biological Sciences, Macquarie University, Macquarie Park, NSW, Australia

Taronga Conservation Society Australia, Mosman, NSW, Australia

C. Campagna, R. Harcourt (eds.), *Ethology and Behavioral Ecology of Otariids and the Odobenid*, Ethology and Behavioral Ecology of Marine Mammals,
https://doi.org/10.1007/978-3-030-59184-7_26

26.1 Genetics and Biogeography

Similar to most regions, the fur seals and sea lions of Australia were overexploited and locally extirpated by the sealing industry at many locations during the early 1800s (Warneke 1982; Ling 2002). The Australian sea lion (*Neophoca cincerea*) is Australia's only endemic pinniped. Although the pre-exploitation distribution of breeding colonies is not fully known, its range is thought to have extended east to the Furneaux, Kent and Anser Groups of islands in Bass Strait, and included the islands around Albany and Perth in Western Australia (Warneke 1982; Gales et al. 1994; Ling 1999). The current range is restricted to South Australia and small populations in Western Australia (Fig. 26.1). Breeding is thought to occur at 28 sites in Western Australia and 48 in South Australia (Shaughnessy et al. 2011; DSEWPaC 2013). There is geographic isolation between the South Australian, south coast of Western Australia and west coast of Western Australia populations, and possibly some external differences indicative of genetic differentiation (such as the 'Snoopy nose' in the Western Australian population see Fig. 26.2). Today, 93% of the breeding population occurs in the Great Australian Bight (Gales and Costa 1997; Goldsworthy et al. 2017b). The Australian sea lion has not recovered, unlike the sympatric fur seals, and is currently in decline (Goldsworthy et al. 2017b). Despite

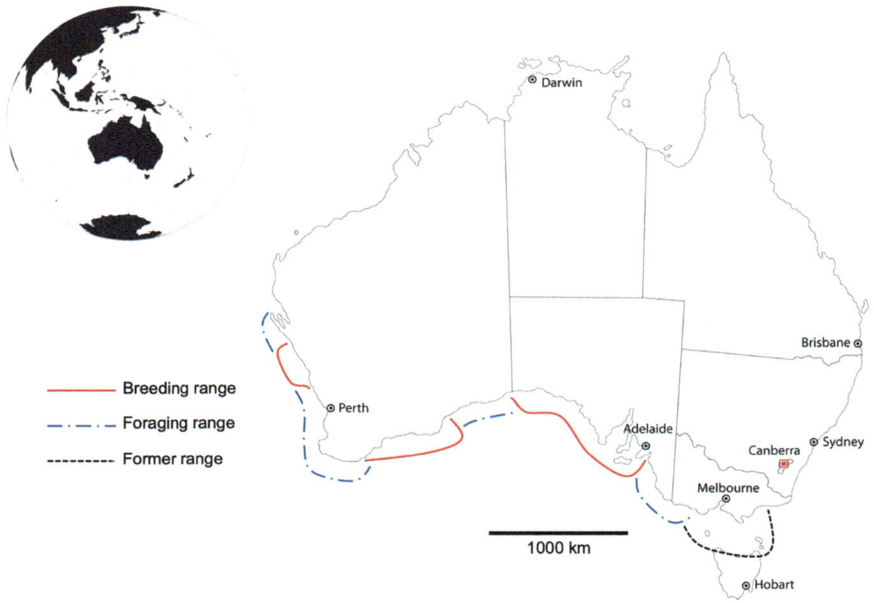

Fig. 26.1 The distribution of the Australian sea lion. The current breeding range is highlighted in red and extends from the east of Kangaroo Island, South Australia, west to the Houtman Abrolhos in Western Australia. The species formerly occurred in Bass Strait but was extirpated in the early 1800s

Fig. 26.2 A female from Olive Island, South Australia (**a**), with a typical nose is contrasted with a female from Beagle Island, Western Australia (**b**), where extended black coloration around the nose is more common. It is not known if this is a genetic difference, or the result of a different foraging strategy or substrate. Photos R. Harcourt

an extremely small population size with only an estimated 2800 pups born per breeding cycle (Goldsworthy et al. 2017b), genetic diversity is not exceptional in the Australian sea lion, being similar or slightly lower than other pinnipeds (Ahonen et al. 2016). While there is little doubt that the Australian sea lion was extirpated in Bass Strait of south-eastern Australia, low numbers of skins from sealing records compared to other species of the region suggest that the Australian sea lion has always been characterized by a small population size (Ling 1999).

While some mammals exhibit year-round breeding, few have broken the constraint of annual reproduction (Burthe et al. 2011). The Australian sea lion has an unusual life history with high maternal investment, low fecundity and extreme philopatry. In contrast to any other large carnivore, Australian sea lion breeding seasons are asynchronous across their range; even colonies that are separated by only a few kilometers may breed at different times for varying lengths of time; with a mean breeding interval of 18 months. This breeding asynchrony is thought to reinforce female natal site fidelity and consequently makes populations highly vulnerable to anthropogenic impacts. Should a colony be extirpated, it is unlikely to be recolonized.

Australian sea lions have the highest level of population subdivision of any marine mammal, with breeding sites being associated with a single matriline (Campbell et al. 2008; Lowther et al. 2012). The level of genetic differentiation at relatively short spatial scales (<40 km) is higher than that detected in other sea lions (e.g., southern sea lion (*Otaria flavescens*) (Feijoo et al. 2011), Californian sea lion (*Zalophus californianus*) (González-Suárez et al. 2009)). Genetic diversity measures are significantly higher in the larger South Australian colonies than in the smaller Western Australian colonies and this may result from a discontinuity in their geographic range. Long-term isolation from other colonies, genetic drift, and possibly founder effects could explain this low genetic diversity (Ahonen et al. 2016).

In common with other mammals, pinnipeds typically display male-biased dispersal. One could expect that the breeding asynchrony would benefit males by facilitating them to breed across multiple locations throughout the 18-month period; however, similar to females, males also have restricted dispersal and do not appear to be exploiting this opportunity (Ahonen et al. 2016). Within those constraints, when animals disperse between close colonies (<110 km) there is a tendency for males to move further than females (Ahonen et al. 2016) and it is possible that even this short dispersal difference may overcome the philopatry with panmictic behavior (Lowther et al. 2012). However, it is equally possible that many breeding colonies may not receive sufficient levels of new genetic material to buffer against risks associated with inbreeding and/or environmental change (Ahonen et al. 2016).

26.2 Conservation Status

The Australian sea lion is currently listed as Endangered on the IUCN Red List of Threatened Species and Vulnerable under the Australian Environmental Protection and Biodiversity Conservation Act. The most recent population assessment identified a population decline of up to 4.1% per breeding season between 2004 and 2017 (Goldsworthy et al. 2017b). Most breeding colonies for the species are difficult to access, being on remote offshore islands or at the base of high cliffs. Seal Bay Conservation Park on Kangaroo Island, South Australia, is the exception, being a tourist attraction (Fig. 26.3) and unusual in that it can be monitored daily. It is also one of the largest breeding colonies and the only one that has been continuously

Fig. 26.3 Tourists participating in a guided tour onto the beach at Seal Bay Conservation Park. Access to the colony is restricted and National Parks officers regulate approach distances. Photo B. Pitcher, Kangaroo Island

monitored for some 30 years. At Seal Bay, pup production has declined over eight breeding seasons (2004 to 2017–18) by 2.2% per breeding season (Goldsworthy et al. 2019). Fisheries interactions, disease and high pup mortality are all contributing factors, and vulnerability to environmental change is a looming long-term threat (Goldsworthy and Page 2007; McIntosh and Kennedy 2013; Schumann et al. 2013; Marcus et al. 2015a; Ladds et al. 2017; Fulham et al. 2018). Commercial gill-net exclusion zones and gear changes have been implemented around breeding colonies to try to reduce bycatch (Goldsworthy et al. 2010); however, this population decline continues and the behavior of the Australian sea lion compounds the extinction risk (see below).

26.3 Life History

26.3.1 Breeding Cycle and Demography

The 18-month breeding and lactation cycle and delayed primiparity of the Australian sea lion result in low lifetime reproductive output and a long generation time (12.6 years) compared to other pinnipeds (Goldsworthy et al. 2017b). Embryonic diapause is the same at 3–4 months for all otariids. Australian sea lions are the only pinniped to have an extended post-implantation period of 14 months (Gales et al. 1997) and most females (83%) pup every breeding season (Goldsworthy et al. 2019). As far as we know, there are only two mammals that break the rules of either annual cycling or being non-seasonal and polyestrous (all-year breeding). These species are the Australian sea lion and the two-toed sloths (*Choloepus hoffmanni* and *C. didactylus*). The sloths have an aseasonal breeding cycle with a 10-month gestation, an inter-birth interval of 15–16 months or more and a post-partum estrus of 10-days, but unlike the Australian sea lion, they are able to breed all-year and have no marked breeding season (Taube et al. 2001).

On average, the pupping season of the Australian sea lion is considerably longer than that of other otariids (King 1983). For Australian sea lions, 90% of births occur over 124 days (McIntosh et al. 2012), other sea lion species give birth over 20–60 days, except the Galapagos sea lion (*Zalophus wollebaeki*) which births over 122–280 days (Trillmich 1986; Boyd 1991). The Juan Fernandez (*Arctocephalus philippi*), Galapagos fur seal (*A. galapagoensis*) and walrus (*Odobenus rosmarus*) also have prolonged birth periods (70–90 days) (Fay 1982; Trillmich 1986; Torres 1987; Francis et al. 1998; Born 2001), but not as extended as the Australian sea lion. The Galapagos fur seal and walrus also have a long lactation period (630–680 days), but birth every 2 years and have therefore not broken with an annual cycle like the Australian sea lion (Boyd 1991; Kovacs and Lavigne 1992; Atkinson 1997). Further, the Australian sea lion breeding cycle is asynchronous between colonies, with the pupping season of individual colonies occurring at various times of the year (Gales et al. 1992; Goldsworthy et al. 2017b). The combination of this extended breeding cycle and asynchrony between colonies means that across the population breeding occurs at multiple times during a year; and over time, within a colony breeding will occur during all seasons of the year.

Seal Bay has become the major source of demographic information for the species because since 2002–03 nearly all pups born each season have been individually marked using 23 mm TIRIS ™ Passive Integrated Transponder (PIT) tags (McIntosh et al. 2012; Goldsworthy et al. 2016). Resight records provide a valuable resource for demographic and behavioral studies with the oldest tagged individuals being 17 years of age in 2020.

During the 2016 and 2017–18 breeding seasons, female recruitment to the breeding population occurred at age 6 years (Goldsworthy et al. 2017a, 2019). This supports findings from cross-sectional aging of a post-canine tooth from lactating females at 12 sites that defined primiparity at age 5 years (range 3.8–6.1 years) and the average age of lactating females to be 11 years (McIntosh 2007). From known-aged individuals at Seal Bay, the peak birth rate (0.70) occurs at 10.5 years of age (Goldsworthy et al. 2019).

A consistent eastward geographic shift has been identified in birthing areas at Seal Bay (Goldsworthy et al. 2019), and consequently more births are occurring in the Main Beach area that is visited by tourists than in the prohibited areas. Previously, most breeding occurred away from the Main Beach that has a higher density of sea lions and fewer caves and rocks that provide shelter to pups. The breeding areas became designated protected areas in the 1980's to protect females and pups from human interference. The Australian sea lion is more habituated to human approach in visited areas at Seal Bay, with reduced responses to controlled approach trials in tourist areas (Lovasza et al. 2008). At Carnac Island, Western Australia, a haul-out for itinerant individuals near Perth, the sea lions display a high level of tolerance to the unmanaged shared space. But only to a point: direct interactions including items thrown at the sea lions, physical contact and ball games (for example), lead to aggressive sea lion responses and movement away from the people (Orsini et al. 2006). There are many personal accounts of researchers chased in the field and rare attacks on people recreating in the vicinity of sea lions. Australian sea lions can be highly aggressive when protecting pups (not necessarily their own), during capture for research, and at times just because the people are present. Attacks can be triggered by pup vocalizations or by observing something that takes the animal's interest.

At an individual level, females show variation in maternal care behavior. For example: when pups are learning to swim some cows may drag them in, or lead them to the water's edge and call patiently from the water, some females wait until the pup is ready, while others have been observed to allow the pup to ride on their back. Females may be highly aggressive and protective of pups, both their own and others; they may also be aggressive towards pups that are not their own; and then there are females that are astonishingly relaxed and tolerant to conspecifics, people and disturbance. One female whose pup had died, stayed ashore and guarded the body for a week, while another was observed walking the dunes day after day, for ten days, calling for her pup that was known to be dead (R McIntosh Pers. Obs, see Fig. 26.4, Charrier Chap. 14). Females have been observed to move with pups

Fig. 26.4 Mothers are known to guard the bodies of stillborn or dead pups in a similar way to how they would typically guard a pup. Photo B. Pitcher, Lewis Island, South Australia

Fig. 26.5 A male guarding an individual female with a newborn pup. Photo B. Pitcher, Olive Island, South Australia

between locations and swim along the coast of Kangaroo Island with their pups beside them (R McIntosh Pers. Obs). Two 100 day old pups marked on Dangerous Reef in South Australia were seen 20 km away on the nearby colony of English Island (Lowther and Goldsworthy 2011b) and pups marked at Seal Bay have been recorded ~30 km away at the Seal Slide, Kangaroo Island, South Australia (R. McIntosh Pers. Comm.).

Females generally exhibit birth-site fidelity for parturition (Higgins and Tedman 1990) and males exhibit serial polygyny, with individual mate guarding occurring in sites where females are more dispersed such as Seal Bay (Fig. 26.5) and harem-like behavior (Fig. 26.6) including up to four females at sites with higher densities such as Dangerous Reef (Marlow 1975; Higgins 1990), or potentially lek-like behavior at Beagle Island, similar to southern sea lions (Soto and Trites 2011; R. Harcourt Pers. Comm.). Male tenure is related to female presence, while fasting periods are plastic. Males rest between mate-guarding periods and leave the colony for 1–3 days, presumably to feed and maintain condition during the extended breeding season. Males fight for access to females and surrender with a specific high-pitched vocalization and retreat; not surprisingly, more time spent in the colony results in more copulations (Higgins 1990).

Australian sea lion milk has relatively low lipid content (25.4%) compared to pinnipeds with shorter lactation periods (Kretzmann et al. 1991; Gales et al. 1996). Otariid females that make relatively short trips to sea produce milk with lower fat content than those species which make foraging trips of longer duration (Kretzmann et al. 1991). The approximately 18-month lactation period enables juvenile Australian sea lions to wean gradually while learning foraging techniques and how

Fig. 26.6 These three mothers have chosen the shelter of a group of boulders to give birth to their pups. This provides the opportunity for a single male to mate-guard a group of females. Photo B. Pitcher, Olive Island

to exploit local resources (Lowther and Goldsworthy 2012). There is evidence that this leads to higher survival rates of weaned juveniles compared to other fur seals and sea lions (McIntosh et al. 2013). Pups typically wean over a period of 3-–6 months before the new pup is born, but it is uncertain how this is regulated and it is not always successful (Higgins and Gass 1993; Lowther and Goldsworthy 2016). Similar to most otariids, female Australian sea lions (29%) may feed juvenile offspring after the new pup is born, which can sometimes lead to sibling competition and older siblings outcompeting the smaller and vulnerable pup (Higgins and Gass 1993; McIntosh and Kennedy 2013). Australian sea lions also exhibit variability in the expression of maternal care between and within breeding sites. Maternal lipid content, time spent onshore and pup growth rates vary between breeding sites (Lowther and Goldsworthy 2011b).

Other life history parameters are similar to other sea lions: females have a single estrous approximately ten days after parturition (Higgins and Gass 1993). Australian sea lions are long-lived with typical otariid growth rates: based on annual growth layers in teeth on average, wild females live to 24 years and males to 21 years (McIntosh 2007). The oldest known Australian sea lion was a female born at Kamogawa Sea World, Japan in September 1988 that lived for 30 years and 9 months. She was the offspring of a pair of sea lions that were part of a group sent by the South Australian Government to be exhibited at the Expo '75, the 1975 World's Fair in Okinawa, Japan (E. Katsumata Pers. Comm.).

26.3.2 Diet and Foraging Behavior: Near Apex Predator

The diet of the Australian sea lion is predominantly benthic but also diverse and opportunistic, with octopus, squid and fish the main prey items (Gales and Cheal 1992; McIntosh et al. 2006; Peters et al. 2015). Foraging is restricted to continental shelf waters, with average distances travelled by juveniles, adult females and adult males being 118, 190 and 340 km respectively (Kirkwood and Goldsworthy 2013). Males are typically wider ranging and exploit a broader range of habitats than females (Goldsworthy et al. 2009). Studies of adult females indicate that there may be some spatial variability in diet associated with localized prey availability and individual foraging strategies (Baylis et al. 2009; Lowther et al. 2013a; Peters et al. 2015). Australian sea lions are capable of occasional long-range movements with males moving up to 500 km, as far as Deen Maar Island (Lady Julia Percy Island) and Cape Bridgewater in western Victoria—part of their putative former range (Vincent Antony, Seals By Sea Tours of Cape Bridgewater, Pers. Obs.).

Australian sea lions have short trip durations compared to other species and do not rest at sea, perhaps to reduce vulnerability to shark predation (Kirkwood and Goldsworthy 2013). They dive continuously and the depth of the water column seems to dictate the time spent at sea, with more time spent at sea when the water is shallower (Goldsworthy et al. 2009). Females express two foraging strategies: inshore (8–20 m depth) vs offshore (40–60 m depth): with heavier and larger individuals exploiting offshore environments (Lowther and Goldsworthy 2011a; Kirkwood and Goldsworthy 2013). Individuals utilize a high diversity of foraging habitats and behaviors: underwater video cameras attached to the back of sea lions show females pursuing octopus across sandy seafloors, sitting and waiting to ambush prey such as leatherjackets on rocky reefs, and searching under ledges and pushing over rocks to find prey (Hunt 2015).

Australian sea lions work at their physiological maximum while exploiting benthic habitats. Females and juveniles dive continuously, maximize bottom time and regularly exceed their calculated aerobic dive limits (Fowler et al. 2006), a typical trait of benthic foragers (Costa et al. 2004; du Dot and Guinet Chap. 4). Pups develop foraging skills during the extended lactation period, and at some sites may attain adult foraging ranges by 6 months of age (Lowther and Goldsworthy 2012). However, at Seal Bay, 23 month-old juveniles only attained 40% of the home range exploited by adult females (Fowler et al. 2007a), 22 month old juveniles accessed 60% of adult female muscle oxygen stores (Fowler et al. 2007b), and dive development was slow, achieving 62% of adult female depth by 22 months (Fowler et al. 2006). This indicates that there may be spatial variability in the development of foraging ability or spatial variability in environmental conditions, habitat and prey availability.

Male Australian sea lions repeatedly use the same region at sea for foraging, irrespective of oceanographic conditions or seasons. This may indicate a lack of ability to adapt to localized change or a preference for a particular habitat or prey (Lowther et al. 2013b). Such behaviors contrast with the diverse foraging strategies of many other mammals that are more fluid and adapt to localized change (Croll et al. 1998; Viswanathan et al. 1999). Perhaps the extended breeding seasons and unpredictable nature of the asynchronous breeding strategy forces males to remain at locations with continuous access to known resources and minimize extended journeys to explore alternative locations (Ahonen et al. 2016).

26.4 Parent-Offspring Behavior and Social Communication

26.4.1 Maternal Behavior

Like other pinnipeds, all parental care is provided by female Australian sea lions. However, unlike most other species, maternal care is quite protracted and lasts for 15 to 18 months after birth, or as long as 40 months if a female doesn't raise offspring in consecutive breeding seasons (Higgins and Gass 1993). Extended maternal care in Australian sea lions may moderate energetic demands of reproduction by spreading it over greater time, allowing pups to develop foraging skills prior to independence.

Females typically move to suitable pupping sites within a colony about two days prior to giving birth (Higgins and Gass 1993). Pupping sites are selected for shelter such as vegetation or a rocky niche (Fig. 26.7) (Marlow 1975), and mothers often return very close to the same area over multiple breeding seasons (Higgins and Gass 1993). A female gives birth to a single pup and stays onshore with it for a perinatal

Fig. 26.7 Mothers typically select sheltered locations, such as vegetation or rocky areas to give birth. This mother is sniffing and calling to her newborn pup. Maternal recognition of offspring develops very quickly after birth. Photo B. Pitcher, Olive Island

attendance between 6 to 12 days, after which she commences regular foraging trips (Higgins and Gass 1993; Pitcher et al. 2010). Typically, females come into estrus and mate 1–2 days before the end of the perinatal attendance period (Higgins and Gass 1993; Pitcher et al. 2010). In contrast to subantarctic fur seals (*A. tropicalis*), where mothers appear to time departure after the pup recognizes her voice (Charrier et al. 2001), duration of perinatal attendance in Australian sea lions appears more related to timing of estrus, because pups do not recognize their mother's voice before she commences foraging trips (Pitcher et al. 2009).

At the end of perinatal attendance, mothers commence regular foraging trips. Lactating females alternate foraging at sea for about 2 days, with nursing the pup in the colony for 1–2 (Higgins and Gass 1993; Lowther and Goldsworthy 2011b). Unlike other otariids that increase foraging trip duration as lactation progresses (du Dot and Guinet Chap. 4), foraging trip lengths of female Australian sea lions do not tend to increase (Lowther and Goldsworthy 2011b). For otariids, foraging trips can lengthen in response to increasing energetic demands on females of pups and gestation, as well as seasonal variation of prey availability (Gentry and Kooyman 1986; Harcourt et al. 2002, see also du Dot and Guinet this volume). By contrast, Lowther and Goldsworthy (2011b) suggest that energetic requirements of lactation in Australian sea lions can be met across all seasons without the need to vary foraging strategies, and that breeding and lactation are not linked to seasonal abundance of prey.

Pups less than 1-month old stay at the birth site and do not roam during maternal foraging trips (Fig. 26.8) (Higgins and Gass 1993). After this time, females often move pups to other areas of the colony and by three months old pups move about the colony and interact with others while mothers are at sea (Marlow 1975; Higgins and Gass 1993). Pups make short forays into the sea and improve swimming ability until around three months old, when they are competent swimmers and spend extended periods in the water (Marlow 1975). They gradually travel further away from the colony and begin foraging trips at around 6 months of age (Fowler et al. 2007a; Lowther and Goldsworthy 2012). By 10 months age, pups occupy similar foraging habitat to adult females (Lowther and Goldsworthy 2012). While pups do not appear to accompany mothers on foraging trips, they may learn foraging locations and techniques from other non-parental individuals through social learning (Lowther and Goldsworthy 2012; Lowther et al. 2012).

26.4.2 Alloparental Care

Alloparental care, i.e. care of a conspecific young by an individual other than a genetic parent (Wilson 1975), is observed at varying frequencies in pinnipeds, from very rarely in subantarctic fur seals (Georges et al. 1999) and New Zealand fur seals (*Arctocephalus forsteri*) (Haase 2006), to more commonly (up to 11%) in Antarctic fur seals (*A. gazella*) (Gemmell 2003). The nursing of young by non-mothers is unusual in Australian sea lions (Pitcher et al. 2011a) with events rarely lasting for

Fig. 26.8 Young pups typically remain in shelter close to their birth site while mothers are at sea foraging. Photo R. McIntosh, Kangaroo Island

longer than one day (Higgins 1990), but 13% of pups were reported to have allosuckled during one unusual season at Seal Bay. While it was originally thought to occur only when a female's own offspring had died (Marlow 1972), it has also been observed in the days prior to a female giving birth to her own pup, and while she is nursing her own pup of similar age (Pitcher et al. 2011a).

Pitcher et al. (2011a) observed one instance of allosuckling lasting for longer than a day. A pup, approximately 6 months old, regularly suckled from two adult females for at least 2 weeks, while neither female suckled another pup. Asynchrony in foraging trips of the females allowed the pup to suckle for up to six days, and this pup was larger than other pups in its cohort. Both females responded to playbacks of the pup's calls by researchers. During another observed allosuckling event at Seal Bay, a female nursing two pups (Fig. 26.9) moved towards another female, the females vocalized and nuzzled, then one pup moved from the first female to the second female, and both settled to suckle one pup (R. McIntosh Pers. Comm.).

A number of hypotheses have been put forward as to why allosuckling occurs. Misdirected maternal care appears to be the most likely explanation for this behavior (Roulin 2002; Acevedo et al. 2016). Observations of females kidnapping pups prior to giving birth, or after stillbirth of their own, suggest that allosuckling in these cases occurs before recognition of a female's own pup has developed (Pitcher et al. 2010).

Fig. 26.9 While rare, allosuckling is occasionally observed in Australian sea lions. This mother accepted nursing two pups. Photo R. McIntosh, Kangaroo Island

Milk evacuation and kin selection are also potential explanations for allosuckling that include benefits to the female (Roulin 2002). Females may seek to evacuate surplus milk that is not consumed by their own offspring, and so tolerate suckling from a non-filial young. This may be beneficial in the absence of the female's own pup, to reduce pressure in the mammary tissue and avoid infection (Lee 1987; O'Brien and Robinson 1991). Females may preferentially allow related, non-filial, kin to allosuckle, increasing their own indirect fitness. Australian sea lions have strong female natal site fidelity and a high level of relatedness within colonies (Campbell et al. 2008), suggesting that the likelihood of nursing a related pup may be higher than in other otariids, although the relatedness of individuals observed in allosuckling events is not known. Pups may benefit from allosuckling through increased energy intake and subsequent growth rates (Pitcher et al. 2011a; Acevedo et al. 2016), as well as by obtaining more diverse specific immune compounds from several females, although they may also be at increased risk of pathogen transmission (Roulin and Heeb 1999; Marcus et al. 2014). Despite potential benefits to both pups and females, allosuckling is typically a rare behavior in Australian sea lions, suggesting that the costs of maternal effort outweigh the benefits. Once established, mother-pup recognition is generally robust, reducing the likelihood of misdirected maternal care. Further, mothers are typically highly aggressive to non-filial young that attempt to suckle (Marlow 1975), increasing the cost of gaining additional resources.

26.4.3 Communication and Recognition

The communication and recognition systems of Australian sea lions are among the most studied of all pinnipeds (See Charrier Chaps. 14 and 15). These studies have examined intrasexual vocal communication of males and the multimodal communication system involved in mother-offspring recognition. Male Australian sea lions produce three different call types (Gwilliam et al. 2008). The barking call, a short noisy sound that resembles a loud click, is the most common and can be made for several minutes. The bleating call is primarily produced by young males during agonistic interactions and potentially by adult males as a sign of submission during fights. The female-like call is rarely heard and appears to be given in response to female or pup vocalizations, however its function has not been determined (Fig. 26.10a and Video S1) (Gwilliam et al. 2008). The acoustic structure of male barks contains both an individual and colony signature (Gwilliam et al. 2008; Ahonen et al. 2014) that is detectable by males (Attard et al. 2010).

Like other otariids, male Australian sea lions produce a distinctive odor during the breeding season (Marlow 1975). Kirkwood and Goldsworthy (2013) suggest that otariid males use olfaction to communicate status with other males and that male odor is used as a cue by females when selecting mates. Similarly, mature male Australian sea lions have a large blonde patch of hair extending from the top of the head down the nape of the neck which contrasts against the darker brown or black of the body fur. It is likely that this fur patch acts as a cue to other males and/or females, providing information about the condition of the male (Fig. 26.10b).

Australian sea lions have a strong selective pressure placed on the mother-offspring recognition and reunion process. Colonial breeding raises the potential

Fig. 26.10 Adult male Australian sea lions have a distinctive blonde patch that covers the head and extends down the neck. It is likely that this patch signals a male's condition. (**a**) Male at Dangerous Reef calling like a female then moving off in direction of pup response. In the video content, Peter Shaughnessy is standing to the right of the male. Still and Video R. McIntosh. (**b**) Many contests between males are played out through vocalizations and posturing, but can escalate to intense physical fights where they attempt to throw their rivals. Photo B. Pitcher, Olive Island

for confusion among individuals, and mothers and pups are frequently separated over the course of the extended lactation period and must successfully reunite to ensure offspring survival (See also Charrier Chap. 14)). Mother-offspring communication and recognition involves multiple sensory cues, including vocalizations, olfactory cues, visual and spatial information. Immediately after birth, mothers and pups begin calling and sniffing each other (Marlow 1975; Higgins and Gass 1993). Both mothers and pups produce individually stereotyped vocalizations (Charrier and Harcourt 2006), and mothers learn to recognize their offspring's voice within 48 h post-partum (Pitcher et al. 2010). By contrast, pups do not learn to recognize their mother's voice before the end of the perinatal attendance period (Pitcher et al. 2009). For the first few weeks after the perinatal attendance period, pups do not move from the natal site and mother-offspring reunions are driven by the mother (Pitcher et al. 2009). Pup recognition of maternal calls is functional before two months of age, coinciding with when pups become more mobile in the colony (Pitcher et al. 2009, 2010). At this stage, pups begin more actively participating in reunions and respond more selectively to only their mother's calls (Fig. 26.11). The individual signatures of both mother and pup vocalizations use a combination of amplitude and frequency parameters making calls relatively resistant to degradation and noise in the colony, allowing individuals to be recognized over long distances (Charrier et al. 2009; Pitcher et al. 2012). Pups retain the ability to recognize their mother's voice for at least 2 years after weaning, highlighting the strength of learning and the importance of vocal recognition in offspring survival (Pitcher et al. 2010).

Visual cues such as morphological changes in pelage color pattern, and body size, are also important for mother-offspring reunions. Pups under two months of age tend to be black. They become brown between 2 and 4 months of age, and finally molt into the silver-beige coloration seen in adult females after around four months (Wierucka et al. 2017). Unlike other otariids where pups within a colony are born within a short time and are at similar stages of development, the extended pupping season of Australian sea lions results in pups of multiple age classes being present in the colony at the same time. Adult females are able to differentiate between pup age-classes based on visual characteristics and show a preference towards the age class that matches that of their own pup (Wierucka et al. 2017). Similarly, pups are able to differentiate between models of adult females and pups based solely on visual cues; however, they appear to rely more on vocal cues during reunions (Wierucka et al. 2018a).

Observations of mother-offspring reunions in several species of otariids indicate that mothers almost always make naso-nasal contact with pups during reunions and accept or reject pups based on olfactory cues (Insley et al. 2003). Australian sea lion mothers often return to their last suckling site and make naso-nasal and/or naso-flank contact with a pup before accepting or rejecting it without any vocal production (Pitcher et al. 2009). Mothers differentiate between the odor of their own pup and that of another (Pitcher et al. 2011b), although it is not known how soon after birth they learn their pup's odor. Analysis of the chemical compounds present on sea lion bodies indicate they have a general body odor that can be used in recognition (Wierucka et al. 2019b). It is likely that mothers learn the scent of their individual

Fig. 26.11 A pup, approximately six months old, calls to its mother during a reunion. Photo B. Pitcher, Kangaroo Island

offspring rather than recognizing them based on similarity with their own odor, as scents of mothers and offspring can be as similar as scents of other individuals (Wierucka et al. 2019a). While olfactory cues alone are capable of facilitating mother-offspring reunions, mothers rely more heavily on pup calls than scent when there is ambiguity in the combination of cues (Wierucka et al. 2018b).

Wierucka et al. (2018b) proposes that the hierarchy of cues used by mothers and pups during recognition and reunions is explained by the active space of the cues, and the costs and benefits to individuals of obtaining the information. They suggest that because females are aggressive towards non-filial young (Marlow 1972; Higgins and Tedman 1990; Gales et al. 1994), pups must rely on longer range vocal cues to recognize mothers and avoid injury from approaching unrelated females (Wierucka et al. 2018b). In contrast, females do not risk injury from pups, and can therefore approach closer and utilize long and short-range cues such as olfactory information.

26.5 Population Regulation and Limits to Recovery

High levels of genetic differentiation between colonies informs us that females breed where they were born, limiting the formation of new colonies. This increases the vulnerability and value of every population because colony extirpation is expected to be final. Population size is regulated by a combination of natural and anthropogenic factors affecting survival and/or birth rates. Fisheries interactions have strong regional effects and at Seal Bay, births have reduced while adult survival is high (Goldsworthy et al. 2019). Unfortunately, there is little information on birth and survival for other sites. Mitigating threats is complex and measuring the success of management is delayed because of the long generation time of the sea lion, taking up to 12 years before population trends may improve. Regular and reliable surveys of pup numbers are difficult to obtain from remote breeding sites where births are extended, asynchronous and unpredictable.

26.5.1 Predation, Neonatal Mortality and Disease

Natural predators include large sharks, in particular the white shark (*Carcarodon carcharias*) (Walker and Ling 1981; Shaughnessy et al. 2007). The level of predation is unknown and difficult to quantify because many sea lions die at sea. At Seal Bay between 5 to 25 individuals had shark inflicted injuries per year over a five year period with most observed attacks occurring in January on adult females and juveniles (Shaughnessy et al. 2007) coincident with the annual breeding season of the sympatric New Zealand fur seal (Bruce and Bradford 2015) (Fig. 26.12).

Similar to most otariids, young pups are highly vulnerable and suffer higher mortality than other age classes. However, average pup mortality is high for the Australian sea lion (mean 27.1%, from nine of the past 11 seasons at Seal Bay) and oscillates; being higher in summer/autumn and lower in winter/spring seasons (Goldsworthy et al. 2019). Pups of both sexes, younger than six weeks old are most affected, dying from trauma inflicted by conspecifics and disease caused by infection from the hookworm parasite *Uncinaria sanguinis* (Higgins and Tedman 1990; McIntosh and Kennedy 2013; Marcus et al. 2014, 2015a).

Fig. 26.12 Survivors of shark attacks often bear the scars and missing limbs that give evidence to their encounters. Photos R. McIntosh & B. Pitcher, Kangaroo Island and Dangerous Reef

Hookworm is passed to the pups during their first drink of milk (Marcus et al. 2014). Infection may cause reduced body condition and blood loss or anemia (Marcus et al. 2015a). Northern fur seals (*Callorhinus ursinus*), California sea lions and New Zealand sea lions (*Phocarctos hookeri*) are also affected by hookworm. Hookworm infection has been treated successfully for New Zealand sea lions using a drug called ivermectin (Lyons et al. 2005; Castinel et al. 2007; Chilvers et al. 2009; DeLong et al. 2009; Michael et al. 2015), and is likely to improve survival rates for Australian sea lion pups (Marcus et al. 2015b).

Biosecurity is important for protecting vulnerable populations. An Australian sea lion pup has tested positive for antimicrobial resistant *E. coli* at Seal Bay, suggesting that this environment is affected by human pollution (Fulham et al. 2018). The dispersal of human-associated bacteria into the marine environment can change the natural gut microbiome or introduce disease. This occurred in New Zealand when an epizootic associated with *K. pneumoniae* infection caused high pup mortality over two breeding seasons for New Zealand sea lion pups (Wilkinson et al. 2006; Leung et al. 2014). Tuberculosis has also been found in wild Australian sea lions in Western Australia (Cousins et al. 1993). Given the high levels of tourism at Seal Bay, poor biosecurity is a risk for Australian sea lions at this site.

26.5.2 Fisheries and Human Interactions Including Marine Plastic Entanglement

Fisheries interactions are difficult to quantify and difficult to avoid for marine predators such as Australian sea lions: the two main fisheries known to affect this species have been the gill-net and rock lobster fisheries (Shaughnessy et al. 2003; Goldsworthy and Page 2007; Kirkwood and Goldsworthy 2013). The gill-net fishery for sharks and finfish in South Australia provided a significant level of bycatch of Australian sea lions. An observer study estimated that 200 sea lions were drowned annually (Hamer et al. 2013). These findings resulted in spatial closures of the fishery around breeding colonies (most 4–10 nm), increased observer coverage and trigger limits that close a fishery if sea lions are caught (Goldsworthy et al. 2010; Kirkwood and Goldsworthy 2013). It has almost been a generation period for the Australian sea lions since this mitigation and we hope to see the positive effects of this management on coming population trends. The drowning of Australian sea lion pups in lobster pots has been largely solved using a simple measure. A vertical spike was introduced to the center of the pot that allowed lobsters to enter, but not sea lions; virtually eliminating sea lion deaths in this fishery (Campbell et al. 2008; Kirkwood and Goldsworthy 2013).

Direct killing by humans has been problematic. From 1980 to 1996, 179 stranded otariids were examined in Western Australia; 47 of these were Australian sea lions and 16 of these deaths were attributed to gunshot wounds (Mawson and Coughran 1999).

It is expected that this practice has reduced with improved education and changes to Australian gun laws in 1996.

26.6 Evolution of a Unique Reproductive Cycle

26.6.1 Breeding Cycle

The Australian sea lion's breeding biology is unique. It is the only reported instance of a mammal that does not breed opportunistically, does not ovulate spontaneously, and breeds supra-annually, independent of time and season (Bradshaw 2003). Like many sea lions, an adult female Australian sea lion typically faces the energetic demands of caring for a dependent young while pregnant with the next.

In the remainder of the chapter we explore hypotheses on how extended gestation and asynchrony became established. We begin with ideas that are no longer supported and move through novel ideas that may have merit.

26.6.2 Breeding Asynchrony and Nutrient Poor Environment

Gales et al. (1994) and Campbell et al. (2008) hypothesized that foraging site fidelity and breeding asynchrony between colonies was the driver selecting for adult female natal site philopatry. It was hypothesized that the nutrient poor environment resulted in extended maternal care (Gales and Costa 1997); however the Australian sea lion is sympatric with a large population of New Zealand fur seals (Shaughnessy et al. 1994) and one of the largest fishing fleets in Australia. We now understand that South Australia is characterized by nutrient rich, cold water upwelling and is highly productive (Middleton et al. 2007), putting this hypothesis to rest. The Australian sea lion and New Zealand fur seal exhibit different foraging and breeding strategies, and may exploit niche separation to avoid competition.

26.6.3 The Family Farm and Matrilineal Specialization

The family farm hypothesis considered whether cows would teach their offspring of local foraging patches and resources, and these offspring would inherit the patch. This is a hypothesis of localized adaptation (Goldsworthy et al. 2009) and has also been described as individual niche width specialization and matrilineal population structure with foraging specialization occurring across generations. Lowther et al. (2012) identified that behavioral specialization was not maintained along matrilines, but that foraging specialization occurs within discrete fine-scale foraging areas and habitats at the individual level. This in turn may limit the dispersive capacity of adult female Australian sea lions which in turn drives population structure and associated breeding asynchrony (Lowther et al. 2012). However, this alone would not drive asynchrony between colonies. Several species have fragmented distributions yet maintain synchrony in breeding as seasonality is controlled through some mechanism, such as photoperiod (see below). Isolation of populations is not sufficient to cause asynchrony without another cause; however, it may reinforce asymmetry once it occurs.

26.6.4 Prime Number Theory

Pupping asynchrony and staggered weaning of pups may limit the development of learned behavior by long-lived predators such as the white shark. Aseasonal breeding may also reduce targeted predation from white sharks compared to synchronized New Zealand fur seals. It has been hypothesized, with much controversy, that when species (e.g. cicadas) breed on a prime number cycle greater than one year, the ability for a predator to evolve to benefit from eating that species is mathematically improbable (Cox and Carlton 1998; Goles et al. 2001; Webb 2001; Koenig and Liebhold 2005). Sharks target fur seal breeding colonies during pupping and in particular, weaning periods (Bruce 2005; Martin et al. 2005; Bertillson-Friedman 2006). Little is known regarding the prevalence of Australian sea lions in the diet of white sharks. However, at Seal Bay, white shark attacks on Australian sea lions occur more frequently during summer, coinciding with the New Zealand fur seal breeding season (Shaughnessy et al. 2007).

Juvenile and female Australian sea lions have pale ventral pelage and darker dorsal pelage (Fig. 26.13), and spend the majority of foraging time on the benthos (Fowler et al. 2006). Adult females invest much time in gestation and lactation, which may enable pups to develop skills and be less naïve at weaning. Perhaps these features have evolved to in response to predation. It is possible that the sympatric otariid species are exhibiting alternative strategies such that the New Zealand fur seal swamps its predators during synchronized breeding and weaning while the Australian sea lion has a strategy to avoid its predators through diffusion, lack of predictability and cryptic benthic behavior (McIntosh 2007). However, for highly philopatric sea lions, the breeding and haul-out sites remain targets for predators.

26.6.5 Decoupling Embryonic Diapause from Photoperiod and Reducing Costs of Reproduction

Typically, pinnipeds exhibit short seasonal pupping periods (King 1983; Boyd 1991). In most species the entire gestation period is about one year, with an approximately three-month period of embryonic diapause followed by around nine

Fig. 26.13 Juvenile and female (right) Australian sea lions have pale ventral pelage and darker grey dorsal pelage, in contrast to dark brown adult males (left). Photo R. McIntosh, Kangaroo Island

months of active gestation (Bonner 1994). Embryonic diapause may have evolved to prolong gestation and ensure that young are born during optimal seasons to increase survival (Boyd 1991). Within many species of pinnipeds the duration of diapause is quite consistent (e.g. York and Scheffer 1997) and the end of diapause is controlled by photoperiod (Temte 1994; Atkinson 1997). Species with longer pupping periods, such as the monk seal (*Monachus* spp.) and Galapagos sea lion, tend to occur at lower latitudes (Kenyon 1981; Trillmich 1986; Temte 1994). Given the very long pupping period and aseasonality of the Australian sea lion breeding cycle it is apparent that, unlike other pinniped species, the duration and timing of embryonic diapause is not linked to photoperiod.

Long-lived species with low reproductive outputs face a trade-off between current reproductive effort, future survival and future breeding attempts (Stearns 1992; McNamara and Houston 1996). They may therefore adjust breeding effort by skipping reproductive events rather than reproducing at a cost to survival (Hamel et al. 2010). Female southern elephant seals (*Mirounga leonina*) (Desprez et al. 2017), Weddell seals (*Leptonychotes weddellii*) (Chambert et al. 2015), walrus (Kovacs and Lavigne 1992), Galapagos sea lions (Boyd 1991; Kovacs and Lavigne 1992; Atkinson 1997), Antarctic fur seals (Boyd et al. 1995), and male king penguins (*Aptenodytes patagonicus*) (Le Bohec et al. 2007) display intermittent breeding in response to reproductive costs. Individuals in these species skip reproductive events to lower reproductive costs, yet as a population, continue to breed annually. Perhaps through the decoupling of embryonic diapause from photoperiod, Australian sea lions have been able to achieve a similar reduction in reproductive costs by extending the length of the breeding cycle, and thus spreading energetic costs over a longer period. At Seal Bay, 83% of breeding females breed every season and 17% breed every second season (Goldsworthy et al. 2019). Australian sea lion milk contains lower lipid content than other otariids (Gales et al. 1996) and foraging duration does not vary across lactation (Lowther and Goldsworthy 2011b), suggesting that energetic costs are more dispersed than in other species.

Once decoupling of the breeding cycle and photoperiod occurred, perhaps drift in the timing of breeding caused the asynchrony of breeding between colonies. The pupping season is shorter than the average in colonies of smaller abundance, simply because fewer pups are born (Gales et al. 1992). Limited dispersal of females between colonies would then reinforce the asynchrony, driven by foraging specialization (Lowther et al. 2012).

26.7 The Future of Australian Sea Lions

The Australian sea lion is unique amongst mammals and one of the rarest pinniped species. Its unusual breeding cycle likely evolved to reduce the costs of reproduction and/or predation. The extended maternal care allows mothers to invest heavily in the growth and survival of their offspring. However, these highly maternal characteristics limit their capacity to recover from past exploitation, recent fishing pressures and the current, rapid changes in their environment. An Australian sea lion female has fewer pups over her lifetime than annually breeding otariids and the extreme philopatry places every colony at risk of extirpation. We are observing continued declines across the population, including at Seal Bay. This site is highly valued for tourism and connecting people with the species; it is also one of the largest and most researched breeding sites and therefore critical for their protection. The effects of any conservation efforts will take several generations to be reflected in population size, making it difficult to assess the effectiveness of mitigation. Meanwhile the small size of many colonies continues to make them vulnerable to stochastic events.

The Australian sea lion is a wonderful subject for study and theoretical enquiry. We have been privileged to get to know them so well and spend so much time in their world; yet even so, many of their secrets evade us. One thing for certain is that the world would be the poorer for the loss of this charming and beautiful creature, and we need to do all we can to preserve these isolated and vulnerable island populations.

References

Acevedo J, Torres D, Aguayo-Lobo A (2016) Offspring kidnapping with subsequent shared nursing in Antarctic fur seals. Polar Biol 39:1225–1232

Ahonen H, Stow A, Harcourt RG, Charrier I (2014) Adult male Australian sea lion barking calls reveal clear geographical variations. Anim Behav 97:229–239

Ahonen H, Lowther AD, Harcourt RG et al (2016) The limits of dispersal: fine scale spatial genetic structure in Australian sea lions. Front Mar Sci 3. https://doi.org/10.3389/fmars.2016.00065

Atkinson S (1997) Reproductive biology of seals. Rev Reprod 2:175–194

Attard MRG, Pitcher BJ, Charrier I et al (2010) Vocal discrimination in mate guarding male Australian sea lions: familiarity breeds contempt. J Ethol 116:704–712
Baylis AMM, Hamer DJ, Nichols PD (2009) Assessing the use of milk fatty acids to infer the diet of the Australian sea lion (*Neophoca cinerea*). Wildl Res 36:169–176
Bertillson-Friedman P (2006) Distribution and frequencies of shark-inflicted injuries to the endangered Hawaiian monk seal (*Monachus schauinslandi*). J Zool 268:361–368
Bonner N (1994) Seals and sea lions of the world. Blandford, London
Born EW (2001) Reproduction in female Atlantic walruses (*Odobenus rosmarus rosmarus*) from north-west Greenland. J Zool 255:165–174
Boyd IL (1991) Environmental and physiological factors controlling the reproductive cycles of pinnipeds. Can J Zool 69:1135–1148
Boyd IL, Croxall JP, Lunn NJ et al (1995) Population demography of Antarctic fur seals: the costs of reproduction and implications for life-histories. J Anim Ecol 64:505–518
Bradshaw D (2003) Vertebrate ecophysiology: An introduction to its principles and applications. Cambridge University Press, Cambridge
Bruce BD (2005) Site fidelity, residence times and home range patterns of white sharks around pinniped colonies. CSIRO Marine and Atmospheric Research, Hobart. Final Report to the Australian Government Department of the Environment and Heritage Hobart
Bruce B, Bradford B (2015) Segregation or aggregation? Sex-specific patterns in the seasonal occurrence of white sharks *Carcharodon carcharias* at the Neptune Islands, South Australia. J Fish Biol 87:1355–1370
Burthe S, Butler A, Searle KR et al (2011) Demographic consequences of increased winter births in a large aseasonally breeding mammal (*Bos taurus*) in response to climate change. J Anim Ecol 80:1134–1144
Campbell RA, Gales N, Lento GM et al (2008) Islands in the sea: extreme female natal site fidelity in the Australian sea lion, *Neophoca cinerea*. Biol Lett 4:139–142
Castinel A, Duignan PJ, Pomroy WE et al (2007) Neonatal mortality in New Zealand sea lions (*Phocarctos hookeri*) at Sandy Bay, Enderby Island, Auckland islands from 1998–2005. J Wildl Dis 43:461–474
Chambert T, Rotella JJ, Garrott RA (2015) Female Weddell seals show flexible strategies of colony attendance related to varying environmental conditions. Ecology 96:479–488
Charrier I, Harcourt RG (2006) Individual vocal identity in mother and pup Australian sea lions (*Neophoca cinerea*). J Mammal 87:929–938
Charrier I, Mathevon N, Jouventin P (2001) Mother's voice recognition by seal pups. Newborns need to learn their mother's call before she can take off on a fishing trip. Nature 412:873
Charrier I, Pitcher BJ, Harcourt RG (2009) Vocal recognition of mothers by Australian sea lion pups: individual signature and environmental constraints. Anim Behav 78:1127–1134
Chilvers BL, Duignan PJ, Robertson BC et al (2009) Effects of hookworms (*Uncinaria* sp.) on the early growth and survival of New Zealand sea lion (*Phocarctos hookeri)* pups. Polar Biol 32:295–302
Costa DP, Kuhn CE, Weise MJ et al (2004) When does physiology limit the foraging behavior of freely diving mammals? Int Congr 1275:359–366
Cousins DV, Williams SN, Reuter R et al (1993) Tuberculosis in wild seals and characterization of the seal bacillus. Aust Vet J 70:92–97
Cox RT, Carlton CE (1998) A commentary on prime numbers and life cycles of periodical cicadas. Am Nat 152:162–164
Croll DA, Tershy BR, Hewitt RP et al (1998) An integrated approach to the foraging ecology of marine birds and mammals. Deep-Sea Res 45:1353–1371
DeLong RL, Orr AJ, Jenkinson RS et al (2009) Treatment of northern fur seal (*Callorhinus ursinus*) pups with ivermectin reduces hookworm-induced mortality. Mar Mamm Sci 25:944–948
Desprez M, Gimenez O, McMahon CR et al (2017) Optimizing lifetime reproductive output: intermittent breeding as a tactic for females in a long-lived, multiparous mammal. J Anim Ecol 87:199–211

DSEWPaC (2013) Recovery plan for the Australian sea lion (*Neophoca cinerea*). Commonwealth of Australia
Fay FH (1982) Ecology and biology of the Pacific walrus, *Odobenus rosmarus divergens* Illiger. North Am Fauna 74:279
Feijoo ME, Lessa P, de Castro RL et al (2011) Mitochondrial and microsatellite assessment of population structure of South American sea lion (*Otaria flavescens*) in the Southwestern Atlantic Ocean. Mar Biol 158:1857–1867
Fowler SL, Costa DP, Arnould JPY et al (2006) Ontogeny of diving behavior in the Australian sea lion: trials of adolescence in a late bloomer. J Anim Ecol 75:358–367
Fowler SL, Costa DP, Arnould JPY (2007a) Ontogeny of movements and foraging ranges in the Australian sea lion. Mar Mamm Sci 23:598–614
Fowler SL, Costa DP, Arnould JPY et al (2007b) Ontogeny of oxygen stores and physiological diving capability in Australian sea lions. Funct Ecol 21:922–935
Francis J, Boness D, Ochoa-Acuna H (1998) A protracted foraging and attendance cycle in female Juan Fernandez fur seals. Mar Mamm Sci 14:552–574
Fulham M, Power M, Gray R (2018) Comparative ecology of *Escherichia coli* in endangered Australian sea lion (*Neophoca cinerea*) pups. Infect Genet Evol 62:262–269
Gales N, Cheal AJ (1992) Estimating diet composition of the Australian sea lion (*Neophoca cinerea)* from scat analysis: an unreliable technique. Wildl Res 19:447–456
Gales NJ, Costa DP (1997) The Australian sea lion: a review of an unusual life history. In: Hindell MA, Kemper C (eds) Marine mammal research in the southern hemisphere, Vol 1. Status, ecology and medicine. Surrey Beatty and Sons, Chipping Norton, Sydney, pp 78–87
Gales NJ, Cheal AJ, Pobar GJ et al (1992) Breeding biology and movements of Australian sea-lions, *Neophoca cinerea*, off the west coast of Western Australia. Wildl Res 19:405–416
Gales NJ, Shaughnessy PD, Dennis TE (1994) Distribution, abundance and breeding cycle of the Australian sea lion *Neophoca cinerea* (Mammalia: Pinnipedia). J Zool 234:353–370
Gales NJ, Costa DP, Kretzmann M (1996) Proximate composition of Australian sea lion milk throughout the entire supra-annual lactation period. Aust J Zool 44:651–657
Gales NJ, Williamson P, Higgins LV et al (1997) Evidence for a prolonged postimplantation period in the Australian sea lion (*Neophoca cinerea*). J Reprod Fertil 111:159–163
Gemmell NJ (2003) Kin Selection may influence fostering behavior in Antarctic fur seals (*Arctocephalus gazella*). Proc R Soc B Biol Sci 270:2033–2037
Gentry RL, Kooyman GL (1986) Fur Seals: maternal strategies on land and at sea. Princeton University Press, Princeton
Georges JY, Sevot X, Guinet C (1999) Fostering in a subantarctic fur seal. Mammalia 63:384–388
Goldsworthy SD, Page B (2007) A risk-assessment approach to evaluating the significance of seal bycatch in two Australian fisheries. Biol Conserv 139:269–285
Goldsworthy SD, McKenzie J, Shaughnessy PD et al (2009) An Update of the Report: Understanding the Impediments to the Growth of Australian Sea Lion Populations. SARDI Aquatic Sciences Publication Number F2008/000847-1. SARDI Research Report Series No. 356, South Australian Research and Development Institute, Adelaide
Goldsworthy S, Page B, Shaughnessy P et al (2010) Mitigating seal interactions in the SRLF and the gillnet sector SESSF in South Australia. Report to the Fisheries Research and Development Institute. SARDI Publication No. F2009/000613-1. SARDI Research Report Series No. 405. South Australian Research and Development Institute (Aquatic Sciences), Adelaide
Goldsworthy SD, Bailleul F, Shaughnessy PD et al (2016) Monitoring of pinniped populations on Kangaroo Island: 2015/16. Report to the Department of Environment, Water and Natural Resources, SARDI Publication No. F2014/000332-3, South Australian Research and Development Institute (Aquatic Sciences), Adelaide
Goldsworthy SD, Bailleul F, Shaughnessy PD et al (2017a) Monitoring of Seal Bay and other pinniped populations on Kangaroo Island: 2016/17. Report to the Department of Environment, Water and Natural Resources, SARDI Publication No. F2014/000322-4. SARDI Research

Report Series No. 951, South Australian Research and Development Institute (Aquatic Sciences), Adelaide
Goldsworthy SD, Mackay AI, Bilgmann K et al (2017b) Status, distribution and abundance of iconic species and apex predators in the Great Australian Bight. Final Report GABRP Project 4.1. Great Australian Bight Research Program, GABRP Research Report Series Number 15, South Australian Research and Development Institute (Aquatic Sciences), Adelaide
Goldsworthy SD, Shaughnessy PD, Smart J et al (2019) Monitoring of Seal Bay and other pinniped populations on Kangaroo Island: 2017/2018. SARDI Publication No. F2014/000322-5. SARDI Research Report Series No. 1018. Report to the Department of Environment and Water. South Australian Research and Development Institute (Aquatic Sciences), Adelaide
Goles E, Schulz O, Markus M (2001) Prime number selection of cycles in a predator-prey model. Complexity 6:33–38
González-Suárez M, Flatz R, Aurioles-Gamboa D et al (2009) Isolation by distance among California sea lion populations in Mexico: redefining management stocks. Mol Ecol 18:1088–1099
Gwilliam J, Charrier I, Harcourt RG (2008) Vocal identity and species recognition in male Australian sea lions, *Neophoca cinerea*. J Exp Biol 211:2288–2295
Haase T (2006) Suckling an additional pup to weaning in the New Zealand fur seal. Mar Mamm Sci 23:176–183
Hamel S, Gaillard JM, Yoccoz NG et al (2010) Fitness costs of reproduction depend on life speed: empirical evidence from mammalian populations. Ecol Lett 13:915–935
Hamer DJ, Goldsworthy SD, Costa DP et al (2013) The endangered Australian sea lion extensively overlaps with and regularly becomes by-catch in demersal shark gill-nets in South Australian shelf waters. Biol Conserv 157:386–400
Harcourt RG, Bradshaw CJA, Dickson K et al (2002) Foraging ecology of a generalist predator, the female New Zealand fur seal. Mar Ecol Prog Ser 227:11–24
Higgins LV (1990) Reproductive behavior and maternal investment of Australian sea lions. PhD Thesis, University of California, Santa Cruz
Higgins LV, Gass L (1993) Birth to weaning: Parturition, duration of lactation, and attendance cycles of Australian sea lions (*Neophoca cinerea*). Can J Zool 71:2047–2055
Higgins LV, Tedman RA (1990) Effect of attacks by male Australian sea lions, *Neophoca cinerea*, on mortality of pups. J Mammal 71:617–619
Hunt R (2015) Secrets of the Australian sea lion revealed. Australian Geographic, Sydney
Insley SJ, Phillips AV, Charrier I (2003) A review of social recognition in pinnipeds. Aquat Mamm 29:181–201
Kenyon KW (1981) Monk seals *Monachus* Flemming, 1882. In: Ridgway SH, Harrison RJ (eds) Handbook of marine mammals. 2. Seals. Academic Press, New York, pp 195–220
King JE (1983) Seals of the world, 2nd edn. Oxford University Press, London
Kirkwood RW, Goldsworthy SD (2013) Fur Seals and Sea Lions. CSIRO, Melbourne
Koenig WD, Liebhold AM (2005) Effects of periodical cicada emergences on abundance and synchrony of avian populations. Ecology 86:1873–1882
Kovacs KM, Lavigne DM (1992) Maternal investment in otariid seals and walruses. Can J Zool 70:1953–1964
Kretzmann MB, Costa DP, Higgins LV et al (1991) Milk composition of Australian sea lions *Neophoca Cinerea*, variability in lipid content. Can J Zool 69:2556–2561
Ladds MA, Slip DJ, Harcourt RG (2017) Intrinsic and extrinsic influences on standard metabolic rates of three species of Australian otariid. Conserv Physiol 5:cow074
Le Bohec C, Gauthier-Clerc M, Gremillet D et al (2007) Population dynamics in a long-lived seabird: I. Impact of breeding activity on survival and breeding probability in unbanded king penguins. J Anim Ecol 76:1149–1160
Lee PC (1987) Allomothering among African elephants. Anim Behav 35:278–291
Leung ES, Chilvers BL, Moore AB et al (2014) Do yearling New Zealand sea lions (*Phocarctos hookeri*) learn foraging behavior from their mothers? Mar Mamm Sci 30:1220–1228

Ling JK (1999) Exploitation of fur seals and sea lions from Australian, New Zealand and adjacent subantarctic islands during the eighteenth, nineteenth and twentieth centuries. Aust Zool 31:323–350

Ling JK (2002) Impact of colonial sealing on seal stocks around Australia, New Zealand and subantarctic islands between 150 and 170 degrees east. Aust Mammal 24:117–126

Lovasza T, Croft D, Banks P (2008) Establishing tourism guidelines for viewing Australian Sea Lions *Neophoca cinerea* at Seal Bay Conservation Park, South Australia. In: Lunney D, Munn A, Meikle W (eds) Too close for comfort: contentious issues in human-wildlife encounters. Royal Zoological Society of New South Wales, Mosman, NSW, pp 225–232

Lowther AD, Goldsworthy SD (2011a) Detecting alternate foraging ecotypes in Australian sea lion (*Neophoca cinerea*) colonies using stable isotope analysis. Mar Mamm Sci 27:567–586

Lowther AD, Goldsworthy SD (2011b) Maternal strategies of the Australian sea lion (*Neophoca cinerea*) at Dangerous Reef, South Australia. Aust J Zool 59:54–62

Lowther AD, Goldsworthy SD (2012) Head start: Australian sea lion pups gain experience of adult foraging grounds before weaning. Mar Biol 159:2687–2696

Lowther AD, Goldsworthy SD (2016) When were the weaners weaned? Identifying the onset of Australian sea lion nutritional independence. J Mammal 97:1304–1311

Lowther AD, Harcourt RG, Goldsworthy SD et al (2012) Population structure of adult female Australian sea lions is driven by fine-scale foraging site fidelity. Anim Behav 83:691–701

Lowther AD, Harcourt RG, Goldsworthy SD (2013a) Regional variation in trophic ecology of adult female Australian sea lions inferred from stable isotopes in whiskers. Wildl Res 40:303–311

Lowther AD, Harcourt RG, Page B et al (2013b) Steady as he goes: at-sea movement of adult male Australian sea lions in a dynamic marine environment. PLoS One 8. https://doi.org/10.1371/journal.pone.0074348

Lyons ET, DeLong RL, Spraker TR et al (2005) Seasonal prevalence and intensity of hookworms (*Uncinaria* spp.) in California sea lion (*Zalophus californianus*) pups born in 2002 on San Miguel Island, California. Parasitol Res 96:127–132

Marcus AD, Higgins DP, Gray R (2014) Epidemiology of hookworm (*Uncinaria sanguinis*) infection in free-ranging Australian sea lion (*Neophoca cinerea*) pups. Parasitol Res 113:3341–3353

Marcus AD, Higgins DP, Gray R (2015a) Health assessment of free-ranging endangered Australian sea lion (*Neophoca cinerea*) pups: Effect of hematophagous parasites on hematological parameters. Comp Biochem Physiol A 184:132–143

Marcus AD, Higgins DP, Gray R (2015b) Ivermectin treatment of free-ranging endangered Australian sea lion (*Neophoca cinerea*) pups: effect on hookworm and lice infection status, hematological parameters, growth, and survival. Parasitol Res 114:2743–2755

Marlow BJ (1972) Pup abduction in the Australian sea lion, *Neophoca cinerea*. Mammalia 36:161–165

Marlow BJ (1975) The comparative behavior of the Australasian sea lions *Neophoca cinerea* and *Phocarctos hookeri* (Pinnipedia: Otariidae). Mammalia 39:159–230

Martin RA, Hammerschlag N, Collier RS, Fallows C (2005) Predatory behavior of white sharks (*Carcharodon carcharias*) at Seal Island, South Africa. J Mar Biol Assoc UK 85:1121–1135

Mawson PR, Coughran DK (1999) Records of sick, injured and dead pinnipeds in Western Australia 1980-1996. J R Soc West Aust 82:121–128

McIntosh RR (2007) Life history and population demographics of the Australian sea lion. La Trobe University, Melbourne

McIntosh RR, Kennedy CW (2013) Morphology, sex ratio and cause of death in Australian sea lion (*Neophoca cinerea*) pups. Aust Mammal 35:93

McIntosh RR, Page B, Goldsworthy SD (2006) Dietary analysis of regurgitates and stomach samples from free-living Australian sea lions. Wildl Res 33:661–669

McIntosh RR, Goldsworthy SD, Shaughnessy PD et al (2012) Estimating pup production in a mammal with an extended and aseasonal breeding season, the Australian sea lion (*Neophoca cinerea*). Wildl Res 39:137–148

McIntosh RR, Arthur AD, Dennis T et al (2013) Survival estimates for the Australian sea lion: Negative correlation of sea surface temperature with cohort survival to weaning. Mar Mamm Sci 29:84–108

McNamara JM, Houston AI (1996) State-dependent life histories. Nature 380:215–221

Michael SA, Chilvers BL, Roe WD et al (2015) Long-term survival and reproductive success of New Zealand sea lions (*Phocarctos hookeri*) treated with ivermectin as pups. Wildl Res 42:660–667

Middleton JF, Arthur C, van Ruth P et al (2007) El Niño effects and upwelling off South Australia. J Phys Oceanogr 37:2458–2477

O'Brien TG, Robinson JG (1991) Allomaternal care by female wedge-capped capuchin monkeys: effects of age, rank and relatedness. Behavior 119:30–50

Orsini J-P, Shaughnessy PD, Newsome D (2006) Impacts of human visitors on Australian sea lions (*Neophoca cinerea*) at Carnac Island, Western Australia: implications for tourism management. Tourism Mar Environ 3:101–115

Peters KJ, Ophelkeller K, Bott NJ et al (2015) Fine-scale diet of the Australian sea lion (*Neophoca cinerea*) using DNA-based analysis of feces. Mar Ecol 36:347–367

Pitcher BJ, Ahonen H, Harcourt RG et al (2009) Delayed onset of vocal recognition in Australian sea lion pups (*Neophoca cinerea*). Naturwissenschaften 96:901–909

Pitcher BJ, Harcourt RG, Charrier I (2010) Rapid onset of maternal vocal recognition in a colonially breeding mammal, the Australian sea lion. PLoS One 5:e12195

Pitcher BJ, Ahonen H, Charrier I et al (2011a) Allosuckling behavior in the Australian sea lion (*Neophoca cinerea*): an updated understanding. Mar Mamm Sci 27:881–888

Pitcher BJ, Harcourt RG, Schaal B et al (2011b) Social olfaction in marine mammals: wild female Australian sea lions can identify their pup's scent. Biol Lett 7:60–62

Pitcher BJ, Harcourt RG, Charrier I (2012) Individual identity encoding and environmental constraints in vocal recognition of pups by Australian sea lion mothers. Anim Behav 83:681–690

Roulin A (2002) Why do lactating females nurse alien offspring? A review of hypotheses and empirical evidence. Anim Behav 63:201–208

Roulin A, Heeb P (1999) The immunological function of allosuckling. Ecol Lett 2:319–324

Schumann N, Gales NJ, Harcourt RG et al (2013) Impacts of climate change on Australian marine mammals. Aust J Zool 61:146

Shaughnessy PD, Gales NJ, Dennis TE et al (1994) Distribution and abundance of New Zealand fur seals, *Arctocephalus forsteri*, in South Australia and Western Australia. Wildl Res 21:667–695

Shaughnessy PD, Kirkwood RJ, Cawthorn M et al (2003) Pinnipeds, cetaceans and fisheries in Australia: a review of operational interactions. In: Gales NJ, Hindell MA, Kirkwood R (eds) Marine mammals: fisheries, tourism and management issues. CSIRO, Melbourne, pp 136–152

Shaughnessy PD, Dennis TE, Berris M (2007) Predation on Australian sea lions *Neophoca cinerea* by white sharks *Carcharodon carcharias* in South Australia. Aust Mammal 29:69–75

Shaughnessy PD, Goldsworthy SD, Hamer DJ et al (2011) Australian sea lions *Neophoca cinerea* at colonies in South Australia: distribution and abundance, 2004 to 2008. Endanger Species Res 13:87–98

Soto KH, Trites AW (2011) South American sea lions in Peru have a lek-like mating system. Mar Mammal Sci 27:306–333

Stearns SC (1992) Evolution of life histories. Oxford University Press, Oxford

Taube E, Keravec J, Vie J-C, Duplantier J-M (2001) Reproductive biology and postnatal development in sloths, *Bradypus* and *Choloepus*: review with original data from the field (French Guiana) and from captivity. Mammal Rev 31:173–188

Temte JL (1994) Photoperiod control of birth timing in the harbour seal (*Phoca vitulina*). J Zool 233:369–384

Torres DN (1987) Juan Fernandez fur seal, *Arctocephalus phillipii*. NOAA Technical Report Number 51, Cambridge

Trillmich F (1986) Attendance behavior of Galapagos sea lions. In: Gentry RL, Kooyman GL (eds) Fur Seals. Maternal strategies on land and at sea. Princeton University Press, Princeton, pp 198–208

Viswanathan GM, Buldyrev SV, Havlin S et al (1999) Optimizing the success of random searches. Nature 401:911–914

Walker GE, Ling JK (1981) Australian sea lion *Neophoca cinerea* (Péron, 1816). In: Ridgway SH, Harrison RJ (eds) Handbook of marine mammals Volume 1; the walrus, sea lions, fur seals and sea otter. Academic, London, pp 99–118

Warneke RM (1982) The distribution and abundance of seals in the Australasian region, with summaries of biology and current research. In: Mammals in the seas, FAO fisheries series No.5, vol 4. FAO, Rome, pp 431–475

Webb GF (2001) The prime number cicada problem. Disc Cont Dyn Sys 1:387–399

Wierucka K, Pitcher BJ, Harcourt R et al (2017) The role of visual cues in mother–pup reunions in a colonially breeding mammal. Biol Lett 13:20170444. https://doi.org/10.1098/rsbl.2017.0444

Wierucka K, Charrier I, Harcourt R et al (2018a) Visual cues do not enhance sea lion pups' response to multimodal maternal cues. Sci Rep 8:9845. https://doi.org/10.1038/s41598-018-28171-w

Wierucka K, Pitcher BJ, Harcourt R et al (2018b) Multimodal mother–offspring recognition: the relative importance of sensory cues in a colonial mammal. Anim Behav 146:135–142

Wierucka K, Barthes N, Harcourt R et al (2019a) Chemical fingerprints suggest direct familiarization rather than phenotype matching during olfactory recognition in Australian sea lions (*Neophoca cinerea*). J Exp Mar Biol Ecol 517:49–53

Wierucka K, Barthes N, Pitcher BJ et al (2019b) Chemical profiles of integumentary and glandular substrates in Australian sea lion pups (*Neophoca cinerea*). Chem Senses 44:205–214

Wilkinson IS, Duignan PJ, Castinel A et al (2006) *Klebsiella pneumoniae* epidemics: possible impact on New Zealand sea lion recruitment. In: Trites AW, Atkinson SK, DeMaster DP et al (eds) Sea Lions of the World. Alaska Sea Grant College Program. University of Alaska, Fairbanks, pp 385–403

Wilson EO (1975) Sociobiology. Belknap Press, Harvard

York AE, Scheffer VB (1997) Timing of implantation in the northern fur seal, *Callorhinus ursinus*. J Mammal 78:675–683

Chapter 27
Australian Fur Seal: Adapting to Coexist in a Shared Ecosystem

Roger J. Kirkwood and Rebecca R. McIntosh

Abstract The Australian fur seal (*Arctocephalus pusillus doriferus*) resulted from an emigration by South African fur seals (*A. p. pusillus*) across the Indian Ocean 18,000 to 12,000 years ago. The sub-species look and behave similarly. The Australian fur seal has a smaller range, smaller population density and smaller total population probably because waters around south-eastern Australia are less productive than coastal waters off Southern Africa. The Australian fur seals' distribution and habitat uses are likely influenced by interactions with New Zealand fur seals (*A. forsteri*) and Australian sea lions (*Neophoca cinerea*), which had prior occupation of southern Australian waters. Australian fur seals are primarily benthic foragers and adaptable generalist predators. Their life history traits are comparable to other otariids, being linked to an annual reproductive cycle, a territorial defense/polygynous mating system, and extreme sexual dimorphism—males are twice the size of females. Conservation issues facing Australian fur seals are comparable to those facing other otariids, and include competition from fisheries, entanglement in marine debris, exposure to pollutants, and disturbance on land.

Keywords *Arctocephalus pusillus doriferus* · Anthropogenic impacts · Benthic · Generalist predator · Interspecific competition · Population recovery

R. J. Kirkwood (✉)
SARDI Aquatic Sciences, Adelaide, Australia
e-mail: roger.kirkwood@sa.gov.au

R. R. McIntosh
Research Department, Phillip Island Nature Parks, Ventnor, Australia
e-mail: rmcintosh@penguins.org.au

C. Campagna, R. Harcourt (eds.), *Ethology and Behavioral Ecology of Otariids and the Odobenid*, Ethology and Behavioral Ecology of Marine Mammals,
https://doi.org/10.1007/978-3-030-59184-7_27

27.1 Establishment and Distribution

Australian fur seals (*Arctocephalus pusillus doriferus*) are the most recently established otariids. Around the end of the last ice-age, 18,000 to 12,000 years ago, an open-ocean crossing by South African fur seals (*A. p. pusillus*) led to their colonization of the southern Australian coast (King 1983). By comparison, the majority of otariids have been established within their ranges for one million years or more. Modern otariids evolved from the ancestral family Enaliarctidae which evolved in the Northern Pacific region circa 11 to 8 mya (Mitchell and Tedford 1973). About 8 to 6 mya, there was an ancestral crossing into the Southern Hemisphere. There, between 6.1 and 3.4 mya, genetic lines were founded to the extant sea lion genera (*Neophoca*, *Eumetopias*, *Zalophus*, *Otaria* and *Phocarctos*) and, 4.3 mya, to the South African fur seal of southern Africa (Berta and Sumich 1999). Other extant southern hemisphere fur seals probably derived from a rapid radiation within the past one million years (Berta and Sumich 1999). It is not known where on the Australian coast South African fur seals first colonized. Their arrival coincided with a period when sea levels rose approximately 100 m, so original colony sites undoubtedly were inundated. Hunting by Aboriginal peoples, typically by highly skilled women, also would have influenced the distribution of Australian fur seal colonies (Russell 2007; Stockton 1982). By the time European sealers arrived in the early 1800s, Australian fur seals predominantly occupied offshore islands in Bass Strait, between Victoria and Tasmania (Cumpston 1973). Their range extended around Tasmania and along the South Australian coast. This was within the broader ranges of Australian sea lions (*Neophoca cinerea*) and New Zealand fur seals (*Arctocephalus forsteri*), with which the new arrivals had to compete. The land habitats and prey species of all three seals overlap (Kirkwood and Goldsworthy 2013).

At first, Australian fur seals may have been attracted to colony sites of New Zealand fur seals, then steadily, though unconsciously, evicted them. New Zealand fur seals are smaller than Australian fur seals and intolerant of being touched when resting ashore, whereas Australian fur seals are highly thigmotactic, they like to be in direct contact with other seals when resting ashore (Goldsworthy et al. 1997; Miller 1974). At sites currently occupied by both Australian and New Zealand fur seals, the Australian fur seals tend to breed in the more comfortable, broader and flatter areas, whereas New Zealand fur seals breed in tumbled and often sharp, boulder areas (Goldsworthy et al. 1997). The gradual usurping of land by Australian fur seals is evident in recent occupations of sites in South Australia (Shaughnessy et al. 2010).

During the 1800s, sealers, including stolen Aboriginal women, harvested fur seals across southern Australia, with the biggest harvests in Bass Strait and around Kangaroo Island (Ling 1999). The sealers referred to the hair seal (sea lion), the black fur seal (New Zealand fur seal) and the brown fur seal (Australian fur seal) (Cumpston 1973). All seals of all ages were harvested, and seals were eliminated from many sites with minimal documentation of which species was taken (Ling

2002). Based on available records, though, prior to harvesting an estimated 100,000 to 200,000 Australian fur seals occupied >23 sites, mostly in Bass Strait (Warneke and Shaughnessy 1985). By the mid-1800s, the total population of Australian fur seals was likely <10,000. Although the seal-fur export industry ended during the 1820s, sealing for local consumption (fur, meat and oil) continued in Bass Strait up to 1923, ending when government regulation prohibited harvesting in the summer breeding period (Warneke and Shaughnessy 1985). Thereafter, recovery by the seals was gradual, being limited by sporadic harvesting, occasional culls and lethal interactions with establishing fisheries (Arnould et al. 2003). After receiving protected wildlife status in the 1970s, the population size increased (Kirkwood et al. 2005; Pemberton and Kirkwood 1994; Shaughnessy et al. 2002). By the early 2000s, Australian fur seals had reoccupied most of their former range, including an expansion up the east Australian coast (Kirkwood et al. 2010; McIntosh et al. 2015). Bass Strait has remained the centre for breeding, however, with almost 80% of Australian fur seal pups born at sites adjacent to the Victorian coast of Bass Strait (Kirkwood et al. 2005, 2010).

Following a peak in 2007 of an estimated 21,000 pups born (Kirkwood et al. 2010), there was an apparent decline in pup production of about 4.2% per annum to 2013 (McIntosh et al. 2018), which continued through to 2018 (McIntosh unpublished data, Phillip Island Nature Parks). Declines were greatest at the largest colonies in Bass Strait, i.e. Seal Rocks and Lady Julia Percy Island. Density-dependent factors, such as reduced food availability or disease could have contributed to the declines, though more information is needed to better understand the population changes. In South Australia and the New South Wales coast, Australian fur seal numbers are low, but have continued to increase despite the declines in Bass Strait.

Most sites occupied by Australian fur seals are small, rocky, offshore islands (Warneke and Shaughnessy 1985). The seals' preference appears to be for rocky substrates—including boulder, cobble and gravel beaches, although tussock vegetation close to the coast is also used. The seals also access large coastal caves. Apart from inside such a cave at Cape Bridgewater, Victoria, Australian fur seals have not taken to sandy beaches, unlike South African fur seals which have some of their largest colonies on sandy beaches.

The largest concentration of Australian fur seals away from Bass Strait is in southern Tasmanian waters, around the islands of Maatsuyker, Walker and Pedra Branca (Pemberton and Kirkwood 1994). Relatively few pups are born at these sites and their occupation is undoubtedly because they are close to productive feeding waters. In addition, growth in seal numbers at haul-outs along the east and west coasts of Tasmania may have been stimulated by the establishment since the 1980s of marine salmon farms in southern Tasmania (McIntosh et al. 2018). Predominantly male Australian fur seals have been attracted to these, typically with an economic cost to the farms and poor outcomes for individual fur seals (Robinson et al. 2008).

Australian fur seals rest at >50 haul-out sites across their range, including on man-made structures such as oil-rig struts and navigational structures. Compared with other fur seals, when on land, Australian fur seals are particularly skittish

(Kirkwood and Goldsworthy 2013). They stampede into the water if disturbed by an unfamiliar sight, sound or smell (Back 2010). Offshore rocky islands and isolated peninsulas afford the most secure resting locations.

At sea, Australian fur seals travel almost exclusively over shelf waters (Arnould and Kirkwood 2008; Kirkwood and Arnould 2011; Kirkwood et al. 2006; Salton et al. 2019). Their range extends from the Great Australian Bight in the west, around Tasmania and to northern New South Wales waters.

27.2 Population Genetics

Australian fur seals are almost identical to South African fur seals in appearance and behavior, but have slight morphological differences (King 1983). Accordingly, they are considered to be the same species (*Arctocephalus pusillus*) with >8000 km separating the two subspecies.

All fur seal stocks were subjected to over-exploitation by sealers during the 1700s and 1800s and, therefore, have passed through recent genetic bottlenecks (Bonin, Chap. 11). Due to their relatively recent founding by a small number of individuals, Australian fur seals have an even more limited genetic diversity than other fur seals (Lancaster et al. 2010) (Table 27.1). They also exhibit minimal subpopulation structure. Using microsatellite loci (on nuclear DNA and mtDNA) analysis of skin tissue collected from pups at nine separate colonies between 1992 and 2002, Lancaster et al. (2010) revealed no between-colony differences in allelic diversity or microsatellite heterozygosity and no differences in haplotype diversity within the mtDNA control region. Gene flow among Australian fur seal colonies appears to be substantial and facilitated by both sexes, indicating that the Australian fur seal population should be considered a single, genetically panmictic unit.

Table 27.1 Comparison of genetic diversity (mean ± SD) within the Australian fur seal and other fur seal species, indicating its relatively low levels, potentially reflecting recent establishment (Lancaster et al. 2010)

Arctocephalus sp	Fur seal	Alleles per locus	Heterozygosity (H_O)
A. pusillus doriferus	Australian	8.0 ± 3.1	0.58 ± 0.26
A. australis	South American	9.0 ± 4.6	–
A. gazella	Antarctic	15.0 ± 7.9	0.71 ± 0.17
A. forsteri	New Zealand	16.8 ± 5.9	0.82 ± 0.06
A. tropicalis	Subantarctic	17.2 ± 7.5	0.82 ± 0.08

27.3 Description

Australian (and South African) fur seals are distinguishable from other southern hemisphere fur seals by their 'sea-lion-like' characteristics (Berta and Churchill 2011). They rest in contact with conspecifics as do sea lions (thigmotaxis; Fig. 27.1), and are almost twice the size of other fur seals; females and males weigh up to 120 and 350 kg, respectively (compared with 50 and 150 kg, respectively, for other female and male fur seals; Fig. 27.2) (Arnould and Warneke 2002). They are fur seals, however, possessing the thick underfur that distinguishes fur seals from sea lions.

All seals undergo a regular molt to renew their coats. In otariid seals, each hair is replaced individually (Ling 1970). A typical molt commences around the face, then moves down the spine and around the body. For most seals, molting is an annual process with the peak in hair turnover occurring during late summer and autumn. In fur seals, the molt is gradual so the seals remain waterproof and do not need to fast. At the peak, though, individuals generally remain ashore for longer periods than usual—which likely hastens the shedding of hair (King 1983)—and at this time seal sites are filled with wind-blown hair (Kirkwood and Goldsworthy 2013). Adult females nursing pups are likely to maintain pup support and forgo the comfort of the extended shore periods. As a consequence, their molt is generally more protracted than the molt of males (King 1983).

Like other otariids, Australian fur seal pups vary in color depending on their age, wetness and duration ashore (Kirkwood and Goldsworthy 2013) (Fig. 27.3). They

Fig. 27.1 Australian fur seals are thigmotactic, they like to lie in contact with one-another (RK)

Fig. 27.2 A large Australian fur seal male (possibly >300 kg) lying with some juveniles (photo by Tony Mitchell)

are born with black guard hairs on their back and flanks, a grey belly and a dense coat of light-cream underfur. The pelage develops on the neonate and grows quickly from birth with guard hairs reaching 1–2 cm in length. Pup guard hairs are softer than hairs of older individuals and clump when they emerge from the water. Over several weeks, their hair fades to a dark grey with the belly and chest hairs taking on a fawn coloration.

Fur seals possess acute sensors of smell and hearing (Wartzok and Ketten 1999), and mothers and pups recognize each other by their unique calls and odors (Charrier Chaps. 14 and 15). Calls of Australian fur seals are at lower frequencies than those of other fur seals (Stirling and Warneke 1971). On return to a colony, the adult female calls for her pup (Fig. 27.4). Generally, the sound is similar to that of a sheep but it can sound like the fog-horn of a ship (Stirling and Warneke 1971). On hearing a call comparable to their mother, pups call back and run forward, they are less discriminating than the female (see Charrier Chap. 14). If a pup's call approximates that of her pup, the female will sniff the pup. It takes just one brief sniff to determine definitively if it is her pup or not (see Charrier Chap. 15).

In February, Australian fur seal pups molt into a silver-grey pelage (Fig. 27.3). Color-wise, they are then indistinguishable from juveniles. Even size is not indicative of age as some pups can be larger than yearlings. These age groups can however be distinguished by other features and behaviors. Pups tend to gallivant with enthusiasm and a lack of coordination while juveniles are more muscular and 'serious'. Also, compared to juveniles, pups' noses look shorter and they can have a 'chubby' appearance, because they are nursed rather than having to hunt for fish (although some yearlings are also nursed). Definitive distinction between pups and juveniles is evident in their teeth (Kirkwood and Goldsworthy 2013). The lower

Fig. 27.3 Australian fur seal pups: (1) Newborn, (2) Black coat fading, (3) Early moult, (4) Late moult, (5) fully moulted (RK)

canines of pups are about the same size as their incisors and post-canines, whereas those of juveniles are at least twice the length of their other teeth.

Juveniles are 1 to 3-year-old seals of either sex. They are much the same color as the adult females—silver-grey to brown, with lighter hair on the face, throat and

Fig. 27.4 An adult female Australian fur seal arrives at the colony and calls for her pup (RK)

chest, and with a light-reddish-brown underfur (Fig. 27.5). Individuals in poorer condition can have an overall brownish appearance, which may indicate a disruption or delay to a normal molting regimen. Unlike pups, juveniles may travel hundreds of kilometers from their natal sites, and rest at other colonies and haul-outs, so may be encountered across the species' range (Warneke 1975; Salton et al. 2019).

Adult females are a light greyish-brown with a pale-grey chest, brown belly, and reddish-brown underfur (Fig. 27.6). When wet, they appear uniform grey, with the paler chest (Warneke and Shaughnessy 1985). The head is smaller and narrower than that of the male, with no obvious brow, and they have a narrower neck and chest. Subadult and adult male Australian fur seals are similar in color to the adult females but a little darker. Adult males have a 'mane' of up to 5 cm long hair on the nape, chest and shoulders (Fig. 27.7). The nape may be lighter in color in older animals and resemble the 'wig' of adult male Australian sea lions.

27.4 Life-History

Most otariid seals share similar life history characteristics, including annual, synchronous breeding in spring and summer, sexual maturity at 3–6 years of age and longevity of approximately 20 years (Boyd et al. 1999). Mortality is greatest (up to 70%) in the first year of life, then declines exponentially, with fewer than 1% of individuals achieving ages beyond 18 years (Beauplet et al. 2006; Gibbens and

Fig. 27.5 Juvenile Australian fur seals resting together (RK)

Fig. 27.6 A female Australian fur seal reunited with her pup looks for a location to rest and suckle (RK)

Fig. 27.7 An adult male Australian fur seal (RK)

Arnould 2009a). In Australian fur seals, maximum recorded longevity in the wild is 21 years for females and 19 years for males (Warneke and Shaughnessy 1985).

Most deaths probably result from starvation, although predators such as large sharks and killer whales (*Orcinus orca*) account for some deaths. Starvation may be precipitated by a shortage of prey, old age, accidental injury or disease (Lynch et al. 2011a, b, c; Woods et al. 1995).

27.4.1 Territory Establishment and Defence

Similar to most otariids, Australian fur seals exhibit a 'resource defense polygyny' mating system (Boyd et al. 1999). Single males establish and defend territories where multiple females gather to pup (Fig. 27.8). The piece of land, rather than the females themselves, is the resource defended by the males. Males may even establish intertidal territories and remain in them during high tides, ducking under as the waves wash over, while the females resort to higher ground (Cane 1999). Territories of Australian fur seals typically contain five to nine (but up to 16) females during the breeding period. In Australian fur seals, average territory size is less than 60 m^2 (Warneke and Shaughnessy 1985).

Male Australian fur seals mature at 3–6 years of age. But do not attain sufficient size to hold a territory until they are 8–13 years old (Warneke and Shaughnessy 1985). Then their breeding careers average just 2 years (maximum 6) before being terminated by younger and stronger individuals. The potential territory-holding males are termed bulls. Males that are too young, too old or otherwise incapable of establishing a breeding territory are evicted from the breeding areas during the breeding period. These may gather to rest in non-breeding areas, 'bachelor parks'. Large non-territorial males patrol the near-shore waters where they try to intercept and mate with departing females. They also make occasional sorties into breeding areas, attempting to secure territories from territory holders.

Adult male Australian fur seals, like other otariids such as the New Zealand fur seal (Troy 1997), return to breeding sites year-round (Kirkwood et al. 2006). They rest within territories that they aim to secure during the breeding period, and actively evict other males from them. Their presence at the colony outside the breeding season advertises to both potential challengers and females that 'this space will be occupied during the breeding period'. Occasionally, mating is observed outside the generally synchronized breeding period (personal observations). Out-of-season mating tends to involve subadult and adult males that appear too small to hold a breeding territory, and smaller sized females, the latter potentially mating for the first time.

Adult males at the colony may provide a defensive role against predation, albeit inadvertently. On one occasion at Seal Rocks in Victoria, passengers on a tour vessel observed five large adult males mobbing a white shark (*Carcharadon carcharias*) (Kirkwood and Dickie 2005). This event occurred on 7 March 2003, outside of the breeding season. The males focussed on the sharks' head and averted its approach towards shore. South African fur seals have also been seen to mob a great white shark (Stewardson and Brett 2000). Such behavior could be altruistic but more likely is motivated by direct benefits to the individual, such as: increasing aggression, gaining confidence, showcasing prowess or increasing knowledge of a predator (Kirkwood and Dickie 2005).

Spermatogenesis in fur seals commences 3–4 months prior to their breeding season and continues to the end of the breeding period (Stewardson et al. 1998). Commencement of spermatogenesis cues the adult males to prepare for breeding. Adult male Australian fur seals tend to go on extended foraging trips to gain energy

Fig. 27.8 During the breeding period, adult male Australian fur seals vigorously defend their territory. (RK)

reserves prior to the breeding season (Kirkwood et al. 2006). The synchronized breeding cycle of fur seals involves territorial establishment by bulls in spring or early summer and then territory occupation for 30–50 days, up to an extreme of

Fig. 27.9 On a hot and muggy day during the breeding period, this adult male has abandoned his territory to go for a cooling swim (RK)

70 days (Boyd et al. 1999). During territory occupation, the males do not feed and lose a considerable amount of weight. In Australian fur seals, territory establishment starts in late October. Earliest arriving and biggest males tend to occupy the better breeding territories, which are around the high-tide mark so they can cool off if required, as well as on flat terrain (Warneke and Shaughnessy 1985).

Territorial defense behaviors include vocal threats, huffing and snorting, ritualized posturing, bluff charges, snaps and bites, and rare fierce battles (Tripovich et al. 2005). The threat calls of Australian fur seal males are at a lower frequency than those of other fur seals and do not include the high-pitched 'uff-uff-uff' sounds made by other fur seals (Stirling and Warneke 1971). Most ferocious battles occur during territory establishment. Usually, fights last only a few minutes but they may last longer than 15 min (Cane 1999). Savage wounds can be inflicted that can be life threatening. Contests end when one combatant flees, or vocalizes submissively and backs away. Excessive effort in territory defense is avoided as the males are fasting and so need to conserve energy.

During very hot days, the territorial bulls utilize many strategies to avoid heat stress (Garlepp et al. 2014). They minimize movement, pant, seek cooler microhabitats within their territory, flipper wave, hold their flippers in water and, as a last resort, temporarily leave the territory for a cooling swim (Fig. 27.9). On a hot day, even mating appears too arduous for territory-holding males. Individuals may try to ignore females that attempt to initiate mating.

27.4.2 Gestation, Parturition and Breeding

Like other female otariids, Australian fur seal females maximize reproductive output, by suckling one pup while being pregnant with the next. Gestation is characterized by a period of embryonic diapause, a delayed implantation following fertilization (Boyd et al. 1999). Durations of diapause (3–4 months) and active (placental) gestation (8–9 months) enable the birth of the pup 12 months after mating. This strategy allows seals to give birth and then breed during the one period ashore.

Recorded pregnancy rates for Australian fur seal females have been 70–90% at mid-term, then 45–65% at birth (Gibbens et al. 2010). This suggests Australian fur seals have a high rate of fetal mortality during later stages of gestation. Mid-aged females (8–13 years of age) are better able to sustain a pregnancy through to parturition than younger or older females.

As for most mammals, aborted fetuses, premature births and still-born births are common (Fig. 27.10). Aborted pups can be found at Australian fur seal colonies, as well as occasionally at non-breeding haul-out sites and very occasionally on random beaches away from colonies from June onwards (personal observations). Their frequency increases as the breeding period approaches. Premature births usually occur within one month of the breeding period in November but a newborn Australian fur seal pup has been recorded at Seal Rocks (Victoria) in late August. Pups born this early are unlikely to survive. Mycoplasma infection has been linked to some pre-term abortions (Lynch et al. 2011c).

The great majority of pups are born during a 5-week period between early November and mid-December, with a peak in late November/early December (Warneke and Shaughnessy 1985). The date of the peak varies slightly between years. In years when females can attain better body condition in winter, median pupping dates are earlier by four days and a greater proportion of females carry their pup to term (Gibbens and Arnould 2009b).

Females exhibit fidelity to a colony, which is likely to be the site of their birth (Warneke and Shaughnessy 1985), but can change sites (Kirkwood and Arnould 2011)—possibly stimulated to do so by disturbance or poor success at a site. Near-term females start to arrive at colonies shortly after the prime territories have been claimed by the bulls. Each female gives birth to a single pup within 48 h of arrival.

Pups are most frequently born in the head-first presentation, although breech (tail-first) births are common (Gibbens et al. 2010). The amniotic sac is usually broken during birth and the female often assists its removal from the pup. At colonies in southern Australia, the placenta is quickly detected and cleaned up by silver, kelp and Pacific gulls (*Larus* spp). Hence during the breeding period, avid flocking and calling of gulls is a reliable sign of a pup birth. Females appear to mourn dead fetuses

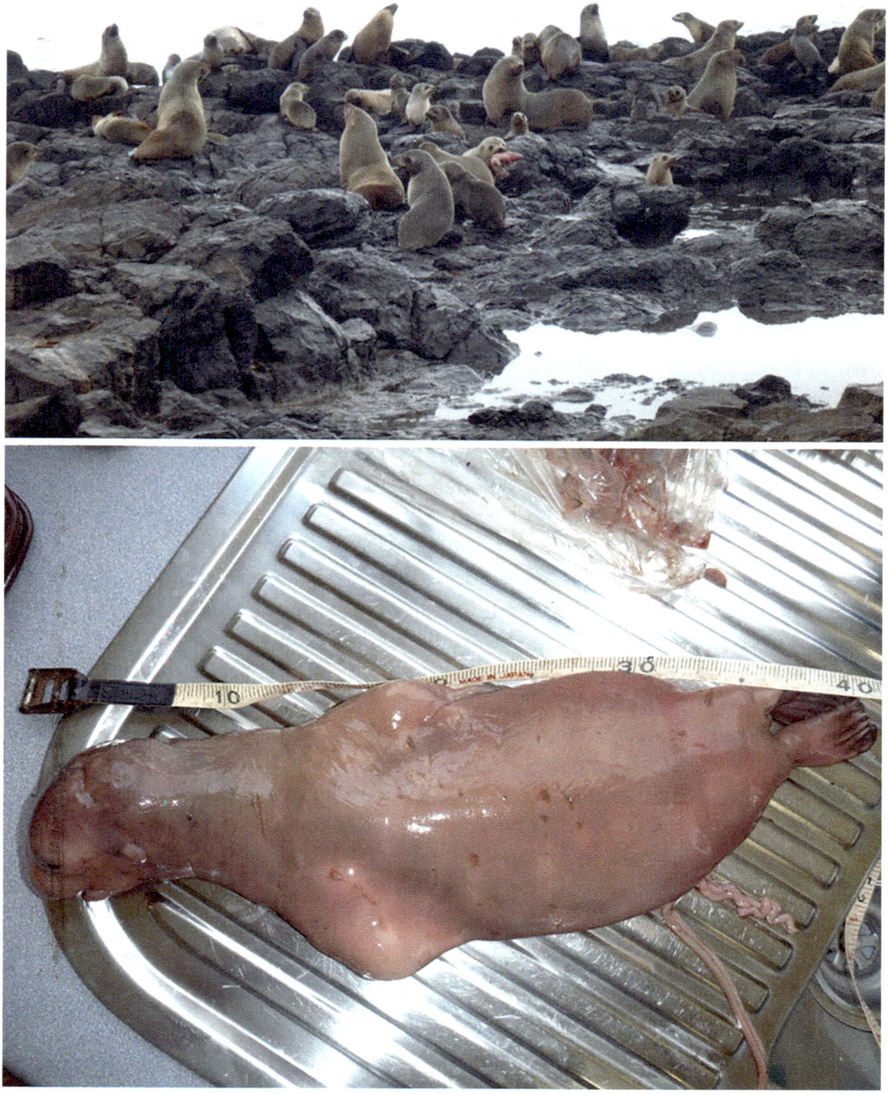

Fig. 27.10 In July, a female Australian fur seal carries her premature foetus around the colony. After several hours mourning, she abandoning the foetus, and it was collected by researchers. (RK)

or still-born pups, nuzzling them, carrying them around, and defending them against gulls and passing seals (personal observations). This mourning behavior is common in otariids (see McIntosh and Pitcher, Chap. 26).

Female otariids usually come into estrus 5–10 days after giving birth (Boyd et al. 1999). Males keep track of the receptive state of females in their territory, occasionally sniffing them and the ground on which they sit (personal observations). Mating is often initiated by the female and mostly occurs during the cooler hours of the day and at night. Each copulation lasts for 10–30 min. After mating, females usually depart the colony to forage at sea. As they leave the colony, they may mate with patrolling bachelor males.

27.4.3 Maternal Attendance and Lactation

In all otariids, females remain with their pup for 5–10 days post-partum, after which they alternate between foraging trips to sea and attendance bouts ashore to nurse their pup and rest (Sepulveda and Harcourt Chap. 3; Costa and Toro Chap. 2. This cycling termed 'maternal attendance behavior' continues until pups are weaned. Female Australian fur seals continue to suckle pups for 6–9 months, remaining at the colony for up to 2 days at a time between trips away that normally last 3–7 days, but can be >20 days (Arnould and Hindell 2002; Kirkwood and Arnould 2011). Throughout lactation, the bond between mother and pup is maintained by mutual recognition of calls and odor.

As with all seals (Baker 1984; Harcourt 1992a), pups have high rates of mortality in their first weeks—due to starvation (usually because they are offspring of young/inexperienced mothers), injury, drowning, and through diseases. The numbers succumbing each year varies enormously, depending on variation in weather conditions, prey distributions, and other location-dependent factors.

Black pups readily enter the water within days of birth, but endeavor to remain in pools or less than a few meters from the shore (personal observations). Mothers encourage their pups into the water when they are learning to swim. Once at Lady Julia Percy Island, a female was observed encouraging her month-old, black pup, to swim with her up to 100 m offshore. A large breaking wave caught the pair: the mother dived deep while the pup 'body-surfed' all the way to shore. For several minutes, the mother frantically searched and called for her pup before discovering it back on the rocks. If pups are washed away from colonies, they may not have the swimming ability to return. During summer, dead and dying individuals often come ashore on public beaches down-current from colonies and can attract considerable human attention.

Australian fur seal females normally suckle only their own pup, which is common to all fur seals (Fig. 27.11). Should another pup approach and attempt to obtain milk, it may be bitten and roughly flung away (Harcourt 1992b). As a pup grows, it may assist in aggressively deterring other pups that seek milk from its mother. Often though, Australian fur seals are seen suckling juveniles. On average, >20% of female Australian fur seals suckling at a colony are suckling juvenile seals, 1–3 years old (Hume et al. 2001). This can be deliberate support to a previous year's pup, or inadvertent or deliberate alloparenting—suckling of an opportunistic seal that is

Fig. 27.11 Australian fur seal female suckles her pup (top, RK) and a pup plus a juvenile at the same time (both, one or neither may be her progeny, photo by Vincent Antony)

not related. Females may alloparent if their fetus does not survive to term or their pup dies prematurely. In Australian fur seals, 45–65% of adult females give birth while 80–90% of adult females are lactating (Gibbens et al. 2010). Some pups and juveniles become adept at sneaking milk from more than one female and can get exceedingly chubby (personal observations). Others are poor at sneaking a suckle; they wake or irritate the females and are quickly expelled. Further behaviors observed include a juvenile vigorously evicting a pup at the teat (and this was tolerated by the female) (Hume et al. 2001), and a female simultaneously suckling a pup and a juvenile (Fig. 27.11).

Seal milk is richer in lipid (fat, up to 50%) and protein (10–18%) than the milk of terrestrial mammals, and can be transferred quickly to the pup (Cane et al. 2005; Costa and Valenzuela-Toro Chap. 2). Australian fur seal milk averages 42% lipid (fat) and 10% protein (Arnould and Hindell 2002). Milk composition varies throughout lactation, with lipid levels increasing from 30% early on to 50% later in lactation, before decreasing to 45% prior to weaning. Protein content increases only slightly during lactation from 10 to 12%. Average milk consumption by Australian fur seal pups is 557 g/day, ranging from 400 g/day shortly after birth to 675 g/day at 210 days of age (Arnould and Hindell 2002).

27.4.4 Pup Growth

Male fur seal pups are generally heavier than females at birth, grow faster and are heavier at weaning (Guinet et al. 1999; Trillmich 1986). Observations of pup behavior suggest there is no difference in the periods of nursing of the sexes or quality of milk delivered (Arnould and Hindell 2002; Lunn and Arnould 1997). One theory as to why males could grow faster is that they may suck harder. More likely though is that different growth of pups may be influenced by the types of body tissues being laid down. Males tend to direct more energy towards lean tissue (muscle) development whereas females direct more to fat (Arnould et al. 1996). Muscle is heavier than fat, so males may gain more mass than females for the same milk energy received. Growth rates of pups decline markedly as weaning is approached.

The average birth mass of female and male Australian fur seal pups is 7.2 and 8.3 kg, respectively. Average growth rates from birth to weaning are 53 and 62 g/day (Arnould and Hindell 2002). Pups double their birth mass after approximately 135 days, which is longer than in most fur seals. Mass at weaning (about 300 days) is about 23 and 27 kg for females and males, respectively. Most pups wean between September and October, when adult females depart for extended foraging trips to prepare for their next birth.

Growing pups are very playful and curious (Llamazares-Martin and Palagi, Chap. 20). They regularly play with each other and objects, including seaweed, marine plastic debris, and even a kelp gull (*Larus dominicanus*) chick that had fallen out of a nest (like a squeaky toy) (Fig. 27.12). Their curiosity includes investigations of researchers and their equipment. At times individuals investigate the research hut on Seal Rocks, scuffling around the feet of researchers, if they sit quietly enough (Fig. 27.13).

Fig. 27.12 Australian fur seal pups at play with: (1) Each other, (2) Kelp, (3) A gull chick (RK)

Fig. 27.13 A curious Australian fur seal pup enters the research hut (RK)

27.4.5 *Pup Transition to Independence*

During the period of dependence (6–9 months), Australian fur seal pups must stay at their colony of birth (Arnould and Hindell 2002; Spence-Bailey et al. 2007). A pup risks losing contact with the site, and maternal support, if it ventures too far away. Indeed, a pup that ventures >50 m from shore, quickly returns.

Pups less than 1–2 months old spend <8% of their time in the water (Spence-Bailey et al. 2007). After molting in January–February, at 2 months of age, their time in the water increases to >27%. From July, Australian fur seal pups start supplementing their milk diet with self-caught fish (Warneke and Shaughnessy 1985). The oxygen-carrying capacity of the blood of Australian fur seal pups steadily increases through time, reaching 71% of adult levels prior to weaning (Spence-Bailey et al. 2007). Just prior to weaning, they can dive to >60 m depth, although dive durations are short, typically under 1 min. Australian fur seal juveniles have higher swimming metabolic rates than adults, which limits their dive capacity, but are similar to other Australian otariids of similar age (Ladds et al. 2017a).

As with most animals, the transition from dependence to independence is associated with high rates of mortality. Pups improve their chance of survival after weaning by attaining the greatest possible weight and learning to catch prey while still suckling. When mothers are at sea, pups hone their swimming, diving and foraging skills: much of the in-water 'play' of seal pups is directed at developing these skills (see Llamazares-Martín and Palagi, Chap. 20). Even so, weanlings are initially naive of much of their prey, and where to locate it.

Weaning is a poorly understood process in seals. Some Australian fur seal pups may achieve independence as early as August but for the majority, weaning occurs in September and October. Pups appear to wean themselves by departing the colony on progressively longer foraging trips (Spence-Bailey et al. 2007). Their mothers may continue to return to the colony and search for their pups several times after their pup has weaned, a behavior that has been recognized for New Zealand fur seals (Haase 2004). In October, whether or not their pup has obviously weaned, all pregnant females will go on an extended foraging trip prior to returning to colony to give birth to their next pup (Kirkwood and Arnould 2011).

27.5 Foraging Ecology

27.5.1 Ranges

The first information on ranges of individual Australian fur seals came in the 1960s and 70s. Then, Australian fur seals flipper-tagged as pups on Seal Rocks were recovered up to 700 km away (Warneke 1975). Since the early 1990s, foraging ranges of individual seals from several colonies have been assessed using geolocation and satellite tracking devices. Research has focused on lactating females, a critical component of the population, whose ranges are restricted as they need to return to the colony regularly to suckle their pup. This obligation facilitates the recapture of individuals to recover instruments. The females from colonies in northern Bass Strait foraged principally in central Bass Strait but could range up to 300 km from their colony (Arnould and Hindell 2001; Arnould and Kirkwood 2008; Hoskins and Arnould 2013). From colonies situated at the edges of Bass Strait, the females tended to forage on the continental shelf and closer to their colonies (Arnould and Kirkwood 2008; Kirkwood and Arnould 2011).

While most research has been on female foraging, there are some data on males. In 1993, an adult male Australian fur seal that had frequently been trapped at a fish farm in southern Tasmania was released with a geolocator device 450 km away in Bass Strait and was re-trapped 15 days later back at the fish farm (Hindell and Pemberton 1997). The seal had used shelf waters off eastern Tasmania and rested at several known seal haul-outs along its route. In 1999–2001, satellite-tracked adult males from Seal Rocks foraged in western Bass Strait (Kirkwood et al. 2002, 2006). Several of the seals also travelled down the west coast of Tasmania to forage in southern Tasmanian waters, 500 km from Seal Rocks, and one foraged west of the Eyre Peninsula (South Australia), 1200 km from Seal Rocks. Male Australian fur seals tracked from Montague Island in NSW swam as far as 400 km north to Newcastle, NSW, and over 1000 km south to Southwest Tasmania (Rob Harcourt, personal communication).

Juvenile seals tracked from Lady Julia Percy and Seal Rocks display similar ranges to adult females (Salton et al. 2019). Adult females are restricted from more distant foraging by their need to return to the colony to suckle their pup, while juveniles are possibly constrained from more distant foraging by their small body size and naivety of prey distributions. Juveniles still being suckled also will want to remain near the colony to maximize the chance of encountering a female returning with milk for them. The juveniles spend about half their time in the water (36–69%) (Ladds et al. 2018), which is less than that of breeding females (75%) (Arnould and Hindell 2001), suggesting they don't need to forage as frequently as do adult females supporting pups.

Australian fur seals forage mostly in association with the sea floor. This behavior is more sea lion-like and contrasts with the mid-water (pelagic) foraging of other fur seals, including the conspecific South African fur seal and sympatric New Zealand fur seal. Australian fur seals also pursue prey pelagically, though, and their diet includes many pelagic-schooling prey (prey species are discussed below). Generally, Australian fur seals remain over the continental shelf. Occasionally, individuals leave the shelf, however. Seals are regularly sighted around trawl fishing operations both on and off the shelf around southern Australia (Hamer and Goldsworthy 2006). Also, several tracked seals have ventured off the shelf, including one adult male that spent several weeks in water depths of over 1000 m to the south-west of Tasmania (Kirkwood and Goldsworthy 2013).

27.5.2 *Foraging Trip Strategies*

Australian fur seals depart from shore at all times of the day, although there is a peak in departures in the few hours leading up to sunrise. On sunrise, following the early departures, is the most peaceful time of the day at a colony, with most of those still present sleeping deeply (Fig. 27.14).

On leaving the colony, the seals immediately commence diving to the bottom and transit directly out to feeding areas (Kirkwood and Arnould 2011). Presumably, they will feed opportunistically if they encounter prey during the transit but most outbound journeys involve continuous travel. After a day or more of transit, the seals arrive in a preferred foraging area and remain there for five days or more before returning to land, usually the site from where they departed. While some seals repeatedly return to uniquely preferred 'hot spots' within their ranges, others alternate between two or more focal areas and a few seals appear to 'roam' almost randomly during foraging trips (Hoskins et al. 2015). Routes can weave or traverse in circular patterns. Factors influencing changes in trip time and foraging area could be weather patterns, competition from other seals, encounters with predators, or encounters/non-encounters with prey concentrations. If foraging takes an individual some distance from the site it departed from, it may rest at an alternative site closer to where it is feeding; even lactating females do this (Kirkwood and Arnould 2011; Salton et al. 2019).

Fig. 27.14 Sunrise at an Australian fur seal colony, a peaceful time of day with much sleeping (RK)

Foraging trip durations by female Australian fur seals during lactation range from less than one day to more than 20 days (Arnould and Hindell 2001; Kirkwood and Arnould 2011). Trip lengths vary considerably within and between individuals and over the lactation period, but are usually shorter when the pups are smallest, i.e. 3- or 4-day trips in summer, and longer when the pups are bigger, i.e. 6- or 7-day trips in winter. The increasing durations of foraging trips occurs in response to the increasing demands of the growing pup, as well as of the fetus. Longer trips are possible because of the increased fasting capacity and supplementary feeding of the larger pup. Attendance durations range from under one to over eight days (averaging approximately 1.7 days) and remain relatively constant in duration throughout lactation (Arnould and Hindell 2001; Kirkwood and Arnould 2011).

27.5.3 Diving Behavior

Most Australian fur seals forage predominantly during daylight hours, although some individuals are mainly nocturnal hunters and a few forage without apparent regard for time of day (Dixon 2011; Knox et al. 2017). Bouts of diving can continue for up to 36 h at a time (Arnould and Hindell 2001). The majority of dives are

U-shaped, which is indicative of direct descent, travel along a near-flat substrate (or waiting in ambush), then direct ascent. This behavior is unusual among fur seals, although it is common among sea lions. Footage from Crittercam cameras attached to females at Kanowna Island confirm that the seals travel along the bottom on most dives, and have recorded individuals following man-made structures, such as power-cable trenches (Arnould et al. 2015).

Mean dive durations of Australian fur seals range from 2 to 4 min, with individual dives lasting up to 10 min (Arnould and Hindell 2001). The majority of dives achieve depths <80 m. The maximum depth recorded has been just over 160 m; however, cameras fitted to trawl-nets have recorded Australian fur seals at depths of 190 m (Hamer and Goldsworthy 2006).

27.5.4 Diet

The types of prey hunted by Australian fur seals differ to those of the conspecific South African fur seals. South African fur seals are primarily pelagic foragers, which is common among fur seals, whereas the Australian fur seal is a bentho-pelagic forager, which is more sea-lion like. The different foraging habits of these similar sub-species is likely to be an adaptation made by Australian fur seals following their arrival in southern Australian waters, where pelagic resources are scarcer than in the more productive waters off southern Africa. In Australian waters, the New Zealand fur seals already present could have further reduced pelagic prey stocks, steering the new arrivals toward a more benthic foraging niche (Table 27.2).

Australian fur seal diet has undoubtedly changed due to human influences. In the past 200 years, humans have modified the ecosystems which Australian fur seals inhabit, substantially reducing the availability of some species and increasing others. During European colonization of Australia, the seals were harvested to near extinction, then commercial fisheries reduced some prey and provided access to others (such as by bringing deep-water species to the surface in trawl nets). A specific example of altering prey availabilities was the broad-scale mortality of the ubiquitous sardine (*Sardinops sagax neopilchardus*), caused by a herpes virus that was likely introduced via contaminated feed to cage-reared tuna in South Australia during the mid-1990s (Griffin et al. 1997). This altered trophic structures across

Table 27.2 Foraging habitats and key prey groups of the three Australian otariids

	Australian fur seal	New Zealand fur seal	Australian sea lion
Habitat	On shelf, local and distant	On and off shelf, local and distant	On shelf, local
Foraging	Principally benthic, also demersal & epipelagic	Principally epipelagic and reef associated	Principally benthic
Prey	Fish, cephalopods	Fish, cephalopods, seabirds	Fish, cephalopods, crustaceans

southern Australia. Sardine stocks near the center of the outbreak (in South Australia) recovered quickly, but were slower to recover in Bass Strait (Gaughan et al. 2000). They appeared in the diet of Australian fur seals at Seal Rocks in 2009 (Kliska 2015), signaling a recovery in Bass Strait 15 years after the virus outbreak.

Research in the past 100 years reveals the adaptability of Australian fur seals to changing prey resources. During the 1920s and 1940s at colonies in northern Bass Strait, Australian fur seals shot for research purposes contained in their stomachs remains of barracouta (*Thyrsites atun*), cephalopods and southern rock lobster (*Jasus edwardsii*) (Lewis 1930; McNally and Lynch 1954). Diet monitoring since the 1980s has found barracouta and cephalopods present but few rock lobsters. Potentially, rock lobster harvesting since the 1940s has reduced their availability to the seals, or the few seals sampled in the 1920s and 1940s provided a biased representation of the seals' diet. Since the 1980s, Australian fur seal diet has been investigated primarily through analysis of prey remains in seal scats and regurgitates. Between the 1980s and 2000 at breeding and haul-out sites around Tasmania, the main prey of Australia fur seals appeared to be redbait (*Emmelichthys nitidus*), jack mackerel (*Trachurus declivis*), leatherjackets (family Monacanthidae), barracouta, red cod (*Pseudophycis bachus*) and Gould's squid (*Nototodarus gouldi*) (Gales and Pemberton 1994; Gales et al. 1993; Hume et al. 2004) (Fig. 27.15).

During 2001–02 at a non-breeding site in South Australia, redbait was the dominant prey of adult male Australian fur seals (Page et al. 2005). Between 1997–2000, at breeding sites in north and eastern Bass Strait, four prey dominated diets: redbait, jack mackerel, red cod and Gould's squid (Littnan et al. 2007). Between 1998 and 2014, routine sampling from Seal Rocks in northern Bass Strait revealed multi-year changes in the diet (Kirkwood and Goldsworthy 2013; Kirkwood et al. 2008; Kliska 2015). In the late 1990s, red cod (*Pseudophycis bachus*) and barracouta were the most abundant prey, then in the early 2000s, redbait were important—at times constituting >80% of the diet. After 2006, jack mackerel and leatherjackets rose to dominance, and in 2013 and 2014 gurnard (family Triglidae) took over. The temporal dietary changes were reflected in stable isotope ratios in Australian fur seal tissues at nearby Kanowna Island, and correlated with changes in body condition of the seals (Arnould et al. 2011). Such multi-year dietary shifts point to frequent ecosystem changes in Bass Strait, possibly related to large-scale oceanographic processes.

DNA based analysis of prey remains in seal scats have improved species recognition, and greatly extended the prey known to be taken by the seals (Deagle et al. 2009; Hardy et al. 2017). At breeding sites in northern Bass Strait during the 2007/08 summer, redbait and jack mackerel were the main prey (Deagle et al. 2009), while in 2014 at small breeding and haul-out sites on the NSW coast, where Australian fur seals were colonizing, leatherjackets (notably ocean jacket, *Nelusetta ayraudi*) and jack mackerel dominated (Hardy et al. 2017).

The size of prey eaten by Australian fur seals ranges from 10–80 cm in length, with the great majority of identified prey being <20 cm in size (Kirkwood et al. 2008; Kliska 2015; Page et al. 2005). Smaller prey may be captured using suction alone and swallowed whole, while larger prey are caught using raptorial biting, then

Fig. 27.15 Fur seals process large prey at the surface: (1) Australian fur seal eating a horseshoe leatherjacket (*Meuschenia hippocrepis*) and (2) New Zealand fur seal eating a Maori octopus (*Octopus maorum*), a method shared by both species (Hocking et al. 2016). (3) Multiple Australian fur seals feeding off schooling redbait (*Emmelichthys nitidus*). Photos by Vincent Antony

may be taken to the surface and vigorously shaken to tear off chunks sufficiently small to be swallowed (Hocking et al. 2016, Chap. 5) (Fig. 27.15).

In summary, Australian fur seals are generalist predators with >70 species of fish and cephalopod species recorded in their diet. The importance of particular prey varies, but the principal ones consumed include bentho-pelagic schooling species, such as jack mackerel, redbait, barracouta and Gould's squid, and benthic species, such as leatherjackets, gurnards, red cod and flathead (family Platycephalidae). Their diets are similar to those of Australian sea lions and New Zealand fur seals, the two other otariids in southern Australian waters. Notable differences are that the Australian sea lions frequently eat crustaceans (such as lobster, *Jasus edwardsii*) (McIntosh et al. 2006) and New Zealand fur seals eat seabirds (such as little penguins, *Eudyptula minor*) (Page et al. 2005). Crustaceans and seabirds are rarely recorded in Australian fur seal dietary analysis (Deagle et al. 2009; Hume et al. 2004; Kirkwood et al. 2008). An exception is the DNA of slipper lobsters (*Crenarctus crenatus*) in four of 17 scats collected at a haul-out on the east coast of Australia in Jan-Apr 2014 (Hardy et al. 2017).

27.6 Conservation

Human caused changes and cyclic oceanographic processes will continue to influence ecosystem structure and prey availability to Australian fur seals. Human-influenced climate change is increasing sea and air temperatures around south-eastern Australia, and further warming in coming decades is inevitable (Hobday and Pecl 2014). On the east coast of Tasmania, for example, an increase of over 1 °C has been recorded between the 1940s and 2000s (Ridgway 2007).

Modeling of impacts of climate change on marine communities predicts a range of possible scenarios for the short and long term. Under climate-change scenarios, low relief sites are predicted to experience increased ocean inundation that could reduce pupping areas for Australian fur seals at many of their larger breeding sites, forcing dispersal and colonization of new sites (McLean et al. 2018).

Australian fur seals could have physiological capacity to adapt to increases of several degrees in water and air temperatures (Ladds et al. 2017b), but their behaviors are likely to change. Weather and sea conditions influence the seals' resting and suckling behaviors. High air temperatures (above 25 °C) reduce the seal's tolerance to remain out of the water (Garlepp et al. 2014). Therefore, increased air temperatures and more frequent hot days could steadily reduce the time seals can spend resting ashore and suckling pups, and may alter occupation rates of hotter sites. Molt is also an important time for the seals, when their metabolic rates increase and they could be less tolerant of warmer conditions, in and out of the water (Ladds et al. 2017b). Conversely, high wind/wave action motivates seals to remain ashore while storms persist (Garlepp et al. 2014), thereby restricting their foraging ability. These results demonstrate how changes in climate could alter a seal's ability to remain ashore, to rest or breed, and its ability to forage effectively, thus driving changes in population status and range.

Other human activities that are influencing Australian fur seal survival and distribution include interactions with fishing operations, entanglement in marine debris, exposure to toxins and disturbances, particularly at breeding sites during the summer breeding period. Fishing activities alter prey abundances, changing ecosystem structures, and can directly result in seal mortalities (Goldsworthy et al. 2003; Hamer and Goldsworthy 2006). The South East Trawl Fishery, for example, catches on average 720 Australian fur seals per year (one seal per 50 shots), and wet boat sector of the blue grenadier fishery catches one seal per 12 shots (one seal per 33 shots drowns) (Stewardson and Knuckey 2005). Entanglement and death in marine debris is a problem for Australian fur seals across their range (McIntosh et al. 2015; Pemberton et al. 1992). At one colony, Seal Rocks, it has been estimated that 302 seals are entangled each year—with the majority of entangling material originating from fisheries operations (McIntosh et al. 2015).

Australian fur seal pup production is declining at several of the larger colonies and the declines could be related to exposure to toxins (Lynch et al. 2012; Taylor et al. 2018). Elevated concentrations of persistent organic pollutants (including polychlorinated dibenzo-p-dioxins and dibenzofurans, dioxin-like polychlorinated biphenyls, polybrominated diphenyl ethers and perfluoro-octane sulfonate/ perfluoro-octanoic acid) were evident in juvenile fur seals sampled on Lady Julia Percy Island that were suffering alopecia, a critical issue for an animal that relies on its fur for thermoregulation (Taylor et al. 2018). Exposure to increasing levels of disturbance by boat traffic could also reduce pup survival at colonies. Close approaches and rapid approaches by vessels can have a number of consequences: disturbing resting seals, increasing aggression between seals, increasing stress, reducing suckling times, and causing seals to stampede into the water (Back et al. 2018; Tripovich et al. 2012).

A further conservation issue is that while human related impacts on Australian fur seals are relatively easy to recognize, it is considerably more difficult to change the detrimental processes. Invariably there are complicated processes and economic costs required to instigate the change.

27.7 Conclusion

Australian fur seals have similar behavioral and physical characteristics to other fur seals. They exhibit sexual dimorphism, a territory-based polygynous mating system, and a strong link to an annual breeding cycle. They frequently exhibit play behavior and, while being aquatic mammals, spend a significant proportion of their time resting on land. They also possess many sea lion-like characteristics. They are twice the size of other fur seals, thigmotactic, and have a predominantly benthic foraging strategy. They also possess unique characteristics. For example, they are extremely flighty, stampeding to the water if an unusual sound, sight or smell is detected. They do not tolerate close approaches by humans. They have one of the smallest ranges and smallest populations of any fur seal. It is also unusual for an otariid to share its range with two other otariids, the New Zealand fur seal and the Australian sea lion. As a late arrival on the scene, the Australian fur seal has adapted its strategies to fit in with habitats available to it in the presence of these seals.

Harvesting of Australian fur seals by Aboriginal peoples probably influenced the seals' distribution—for example, by preventing large colonies from establishing on mainland coasts—but was unlikely to have had much impact on the species or offshore ecosystems where the seals foraged. The impact of colonial sealers, however, was huge. While of benefit to European colonization, the rapid removal of key predators undoubtedly and suddenly altered marine ecosystem structure in southern Australia. This happened prior to any understanding by the newcomers of species abundances and trophic interactions in the region. For 200 years, the seals failed to recover from this over-harvesting. Following legislative protection in the 1970s, Australian fur seal populations potentially re-approached pre-harvest levels by the early 2000s. Since then, though, there has been a decline for reasons that are unknown. With their population centres alongside the coast of Victoria, which is densely populated by humans relative to other Australian coasts where seals live, the Australian fur seal—more than other seal species in Australia—must coexist with humans. It will be interesting to see how the population fairs through the twenty-first century.

References

Arnould JPY, Hindell MA (2001) Dive behavior, foraging locations, and maternal-attendance patterns of Australian fur seals (*Arctocephalus pusillus doriferus*). Can J Zool 79:35–48

Arnould JPY, Hindell MA (2002) Milk consumption, body composition and pre-weaning growth rates of Australian fur seal (*Arctocephalus pusillus doriferus*) pups. J Zool 256:351–359

Arnould JPY, Kirkwood R (2008) Habitat selection by female Australian fur seals (*Arctocephalus pusillus doriferus*). Aquat Conserv Mar Freshwat Ecosyst 17:S53–S67

Arnould JPY, Warneke RM (2002) Growth and condition in Australian fur seals (*Arctocephalus pusillus doriferus*) (Carnivora: Pinnipedia). Aust J Zool 50:53–66

Arnould JPY, Boyd IL, Socha DG (1996) Milk consumption and growth efficiency in Antarctic fur seal (*Arctocephalus gazella*) pups. Can J Zool 74:254–266

Arnould JPY, Boyd IL, Warneke RM (2003) The historical dynamics of the Australian fur seal population: evidence of regulation by man? Can J Zool 81:1–9

Arnould JPY, Cherel Y, Gibbens J et al (2011) Stable isotopes reveal inter-annual and inter-individual variation in the diet of female Australian fur seals. Mar Ecol Prog Ser 422:291–302

Arnould JPY, Monk J, Ierodiaconou D et al (2015) Use of anthropogenic sea floor structures by Australian fur seals: potential positive ecological impacts of marine industrial development? PLoS One 10:e0130581. https://doi.org/10.1371/journal.pone.0130581

Back JJ (2010) Behavioural responses of Australian fur seals to boat-based ecotourism. Deakin University

Back JJ, Hoskins AJ, Kirkwood R et al (2018) Behavioral responses of Australian fur seals to boat approaches at a breeding colony. Nat Conserv 31:35–52

Baker JR (1984) Mortality and morbidity in grey seal pups (*Halichoerus grypus*) – studies on its causes, effects of environment, the nature and sources of infectious agents and the immunological status of pups. J Zool 203:23–48

Beauplet G, Barbraud C, Dabin W et al (2006) Age-specific survival and reproductive performances in fur seals: evidence of senescence and individual quality. Oikos 112:430–441

Berta A, Churchill M (2011) Pinniped taxonomy: review of currently recognized species and subspecies, and evidence used for their description. Mammal Rev 42:207–234

Berta A, Sumich JL (1999) Marine mammals: evolutionary biology. Academic, San Diego

Boyd IL, Lockyer C, Marsh HD (1999) Reproduction in marine mammals. In: Reynolds JE, Rommel SA (eds) Biology of marine mammals. Melbourne University Press, Melbourne, pp 218–286

Cane K (1999) Aspects of the behavior of male Australian fur seals *Arctocephalus pusillus doriferus* during the breeding season. University of Melbourne, Honours

Cane KN, Arnould JPY, Nicholas KR (2005) Characterisation of proteins in the milk of fur seals. Comp Biochem Physiol B 141:111–120

Cumpston JS (1973) First visitors to Bass Strait. vol 7. Roebuck Society. Union Offset Company PL, Fyshwick, ACT

Deagle BE, Kirkwood R, Jarman SN (2009) Evaluating trophic links by pyrosequencing prey DNA in faeces: a population level dietary study on Australian fur seals. Mol Ecol 18:2022–2038. https://doi.org/10.1111/j.1365-294X.2009.04158.x

Dixon V (2011) Foraging behavior of juvenile Australian fur seals (*Arctocephalus pusillus doriferus*). Deakin University

Gales R, Pemberton D (1994) Diet of the Australian fur seal in Tasmania. Aust J Mar Freshwat Res 45:653–664

Gales R, Pemberton D, Lu CC et al (1993) Cephalopod diet of the Australian fur seal: variation due to location, season and sample type. Aust J Mar Freshwat Res 44:657–671

Garlepp L, Logan M, Kirkwood R (2014) Behavioral responses of Australian fur seals (*Arctocephalus pusillus doriferus*) to environmental variations. Mar Mamm Sci 30:978–993. https://doi.org/10.1111/mms.12094

Gaughan DJ, Mitchell RW, Blight SJ (2000) Impact of mortality, possibly due to herpesvirus, on pilchard *Sardinops sagax* stocks along the south coast of Western Australia in 1998–99. Mar Freshw Res 51:601–612

Gibbens J, Arnould JPY (2009a) Age-specific growth, survival and population dynamics of female Australian fur seals. Can J Zool 87:902–911

Gibbens J, Arnould JPY (2009b) Interannual variation in breeding chronology and pup production of Australian fur seals. Mar Mamm Sci 25:573–587

Gibbens J, Parry LJ, Arnould JPY (2010) Influences on fecundity in Australian fur seals (*Arctocephalus pusillus doriferus*). J Mammal 91:510–518

Goldsworthy SD, Pemberton D, Warneke RM (1997) Field identification of Australian and New Zealand fur seals, *Arctocephalus* spp., based in external characters. In: Hindell M, Kemper C (eds) Marine mammal research in the southern hemisphere: status, ecology and medicine, vol 1. Surrey Beatty & Sons, Chipping Norton, NSW, pp 63–71

Goldsworthy SD, Bulman C, He X et al (2003) Trophic interactions between marine mammals and Australian fisheries: an ecosystem approach. In: Gales N, Hindell M, Kirkwood R (eds) Marine mammals: fisheries, tourism and management issues. CSIRO, Melbourne, pp 62–99

Griffin DA, Thompson PA, Bax NJ et al (1997) The 1995 mass mortality of pilchard: no role found for physical or biological oceanographic factors in Australia. Aust J Mar Freshwat Res 48:27–42

Guinet C, Goldsworthy SD, Robinson S (1999) Sex differences in mass loss rate and growth efficiency in Antarctic fur seal (*Arctocephalus gazella*) pups at Macquarie Island. Behav Ecol Sociobiol 46:157–163

Haase T (2004) The determinants of weaning in the New Zealand fur seal. La Trobe University

Hamer DJ, Goldsworthy SD (2006) Seal-fishery operational interactions: identifying the environmental and operational aspects of a trawl fishery that contributes to by-catch and mortality of Australian fur seals (*Arctocephalus pusillus doriferus*). Biol Conserv 130:517–529

Harcourt R (1992a) Factors affecting early mortality in the South American fur seal (*Arctocephalus australis*) in Peru: density-related effects and predation. J Zool 226:259–270

Harcourt R (1992b) Maternal aggression in the South American fur seal in Peru. Can J Zool 70:320–325

Hardy N, Berry T, Kelaher BP et al (2017) Assessing the trophic ecology of top predators across a recolonisation frontier using DNA metabarcoding of diets. Mar Ecol Prog Ser 573:237–254

Hindell MA, Pemberton D (1997) Successful use of a translocation program to investigate diving behavior in a male Australian fur seal, *Arctocephalus pusillus doriferus*. Mar Mamm Sci 13:219–228

Hobday AJ, Pecl GT (2014) Identification of global marine hotspots: sentinels for change and vanguards for adaptation action. Rev Fish Biol Fish 24:415–425. https://doi.org/10.1007/s11160-013-9326-6

Hocking DP, Fitzgerald EMG, Salverson M et al (2016) Prey capture and processing behaviors vary with prey size and shape in Australian and subantarctic fur seals. Mar Mamm Sci 32:568–587. https://doi.org/10.1111/mms.12285

Hoskins AJ, Arnould JPY (2013) Temporal allocation of foraging effort in female Australian fur seals (*Arctocephalus pusillus doriferus*). PLoS One 8:e79484. https://doi.org/10.1371/journal.pone.0079484

Hoskins AJ, Costa DP, Arnould JPY (2015) Utilisation of intensive foraging zones by female Australian fur seals. PLoS One 10:e0117997. https://doi.org/10.1371/journal.pone.0117997

Hume F, Arnould JPY, Kirkwood R et al (2001) Extended maternal dependence by juvenile Australian fur seals (*Arctocephalus pusillus doriferus*). Aust Mammal 23:67–70

Hume F, Hindell MA, Pemberton D et al (2004) Spatial and temporal variation in the diet of a high trophic level predator, the Australian fur seal (*Arctocephalus pusillus doriferus*). Mar Biol 144:407–415

King JE (1983) Seals of the world, 2nd edn. Oxford University Press, London

Kirkwood R, Arnould JPY (2011) Foraging trip strategies and habitat use during late pup rearing by lactating Australian fur seals. Aust J Zool 59:216–226

Kirkwood R, Dickie J (2005) Mobbing of a great white shark (*Carcharodon carcharias*) by adult male Australian fur seals (*Arctocephalus pusillus doriferus*). Mar Mamm Sci 21:336–339

Kirkwood R, Goldsworthy S (2013) Fur Seals and Sea Lions. CSIRO, Melbourne

Kirkwood R, Gales N, Lynch M et al (2002) Satellite tracker deployments on adult, male Australian fur seals *Arctocephalus pusillus doriferus*: methods and preliminary results. Aust Mammal 24:73–83

Kirkwood R et al (2005) Pup production and population trends of the Australian fur seal *Arctocephalus pusillus doriferus*. Mar Mamm Sci 21:260–282

Kirkwood R, Lynch M, Gales N et al (2006) Foraging strategies of an adult, male otariid: the Australian fur seal (*Arctocephalus pusillus doriferus*). Can J Zool 84:1781–1788

Kirkwood R, Hume F, Hindell M (2008) Sea temperature variations mediate annual changes in the diet of Australian fur seals in Bass Strait. Mar Ecol Prog Ser 369:297–309

Kirkwood R, Pemberton D, Gales R et al (2010) Continued population recovery by Australian fur seals. Mar Freshw Res 61:695–701

Kliska K (2015) Environmental correlates of temporal variation in the diet of Australian fur seals. Macquarie University

Knox TC, Baylis AMM, Arnould JPY (2017) Habitat use and diving behavior of male Australian fur seals. Mar Ecol Prog Ser 566:243–256

Ladds MA, Slip DJ, Harcourt R (2017a) Swimming metabolic rates vary by sex and development stage, but not by species, in three species of Australian otariid seals. J Comp Physiol B 187:503–516

Ladds MA, Slip DJ, Harcourt RG (2017b) Intrinsic and extrinsic influences on standard metabolic rates of three species of Australian otariid. Conserv Physiol 5. https://doi.org/10.1093/conphys/cow074

Ladds MA, Salton M, Hocking DP et al (2018) Using accelerometers to develop time-energy budgets of wild fur seals from captive surrogates. Peer J 6:e5814. https://doi.org/10.7717/peerj.5814

Lancaster ML, Arnould JPY, Kirkwood R (2010) Genetic status of an endemic marine mammal, the Australian fur seal, following historical harvesting. Anim Conserv 13:247–255

Lewis F (1930) Seals on the Victorian coast and their feeding habits. Aust Museum Magaz 4:39–44

Ling JK (1970) Pelage and molting in wild mammals with special reference to aquatic forms. Q Rev Biol 45:16–54

Ling JK (1999) Exploitation of fur seals and sea lions from Australian, New Zealand and adjacent subantarctic islands during the eighteenth, nineteenth and twentieth centuries. Aust Zool 31:323–350

Ling JK (2002) Impact of colonial sealing on seal stocks around Australia, New Zealand and subantarctic islands between 150 and 170 degrees east. Aust Mammal 24:117–126

Littnan CL, Arnould JPY, Harcourt RG (2007) Effect of proximity to the shelf edge on the diet of female Australian fur seals. Mar Ecol Prog Ser 338:257–267

Lunn NJ, Arnould JPY (1997) Maternal investment in Antarctic fur seals: evidence for equality in the sexes? Behav Ecol Sociobiol 40:351–362
Lynch M, Duignan PJ, Taylor T et al (2011a) Epidemiology of *Brucella* infection in Australian fur seals. J Wildl Dis 47:352–363
Lynch M, Nielsen O, Duignan PJ et al (2011b) Serological survey for potential pathogens and assessment of disease risk in Australian fur seals. J Wildl Dis 47:555–565
Lynch M, Taylor T, Duignan PJ et al (2011c) Mycoplasmas in Australian fur seals *Arctocephalus pusillus doriferus*: identification and association with abortion. J Vet Diagn Invest 23:1123–1130
Lynch M et al (2012) Characterization and causality investigations of an alopecia syndrome in Australian fur seals (*Arctocephalus pusillus doriferus*). J Mammal 93:504–513
McIntosh RR, Page B, Goldsworthy SD (2006) Dietary analysis of regurgitates and stomach samples from free-living Australian sea lions. Wildl Res 33:661–669
McIntosh RR, Kirkwood R, Sutherland DR et al (2015) Drivers and annual estimates of marine wildlife entanglement rates: a long-term case study with Australian fur seals. Mar Pollut Bull 101:716–725. https://doi.org/10.1016/j.marpolbul.2015.10.007
McIntosh RR et al (2018) Understanding meta-population trends of the Australian fur seal, with insights for adaptive monitoring. PLoS One 13:e0200253. https://doi.org/10.1371/journal.pone.0200253
McLean LJ, George S, Ierodiaconou D et al (2018) Impact of rising sea levels on Australian fur seals. Peer J 6:e5786. https://doi.org/10.7717/peerj.5786
McNally J, Lynch DD (1954) Notes on the food of Victorian seals. Office of the Director of Fisheries and Game, Melbourne
Miller EH (1974) Social behavior between adult male and female New Zealand fur seals, *Arctocephalus forsteri* (Lesson), during the breeding season. Aust J Zool 22:155–173
Mitchell E, Tedford RH (1973) The Enaliarctinae: a new group of extinct aquatic carnivora and consideration of the origin of the Otariidae. Bull Am Mus Nat Hist 151:202–284
Page B, McKenzie J, Goldsworthy SD (2005) Dietary resource partitioning among sympatric New Zealand and Australian fur seals. Mar Ecol Prog Ser 293:283–302
Pemberton D, Kirkwood RJ (1994) Pup production and distribution of the Australian fur seal, *Arctocephalus pusillus doriferus*, in Tasmania. Wildl Res 21:341–352
Pemberton D, Brothers NP, Kirkwood R (1992) Entanglement of Australian fur seals in man-made debris in Tasmanian waters. Wildl Res 19:151–159
Ridgway KR (2007) Long-term trend and decadal variability of the southward penetration of the East Australian Current. Geophys Res Lett 34. https://doi.org/10.1029/2007GL030393
Robinson S, Gales R, Terauds A et al (2008) Movements of fur seals following relocation from fish farms. Aquat Conserv Mar Freshwat Ecosyst 18:1189–1199
Russell L (2007) "Dirty domestics and worse cooks": aboriginal women's agency and domestic frontiers, Southern Australia, 1800–1850. Front J Women Stud 28:18–46
Salton M, Kirkwood R, Slip DJ et al (2019) Mechanisms for sex-based segregation in foraging behavior by a polygynous marine carnivore. Mar Ecol Prog Ser 624:213–226
Shaughnessy PD, Kirkwood RJ, Warneke RM (2002) Australian fur seals, *Arctocephalus pusillus doriferus*: pup numbers at Lady Julia Percy Island, Victoria, and a synthesis of the species' population status. Wildl Res 29:185–192
Shaughnessy PD, McKenzie J, Lancaster ML et al (2010) Australian fur seals establish haulout sites and a breeding colony in South Australia. Aust J Zool 58:94–103
Spence-Bailey LM, Verrier D, Arnould JPY (2007) The physiological and behavioral development of diving in Australian fur seal (*Arctocephalus pusillus doriferus*) pups. J Comp Physiol B 177:483–494
Stewardson CL, Brett M (2000) Aggressive behavior of an adult male Cape fur seal (*Arctocephalus pusillus pusillus*) towards a great white shark (*Carcharodon carcharias*). Afr Zool 35:147–150. https://doi.org/10.1080/15627020.2000.11407201

Stewardson CL, Knuckey IA (2005) South east trawl fishery: seal bycatch–guidelines for reporting and data collection. Canberra

Stewardson CL, Bester MN, Oosthuizen WH (1998) Reproduction in the male Cape fur seal *Arctocephalus pusillus pusillus*: age at puberty and annual cycle of the testis. J Zool 246:63–74

Stirling I, Warneke RM (1971) Implications of a comparison of the airborne vocalizations and some aspects of the behavior of the two Australian fur seals, *Arctocephalus* spp., on the evolution and present taxonomy of the genus. Aust J Zool 19:227–241

Stockton J (1982) Seals in Tasmanian prehistory. Proc R Soc Vict 94:53–60

Taylor S et al (2018) Utility of fur as a biomarker for persistent organic pollutants in Australian fur seals (*Arctocephalus pusillus doriferus*). Sci Total Environ 610–611:1310–1320. https://doi.org/10.1016/j.scitotenv.2017.08.204

Trillmich F (1986) Maternal investment and sex-allocation in the Galapagos fur seal, *Arctocephalus galapagoensis*. Behav Ecol Sociobiol 19:157–164

Tripovich JS, Rogers TL, Arnould JPY (2005) Species-specific characteristics and individual variation of the bark call produced by male Australian fur seals, *Arctocephalus pusillus doriferus*. Bioacoustics 15:79–96

Tripovich JS, Hall-Aspland S, Charrier I et al (2012) The behavioral response of Australian fur seals to motor boat noise. PLoS One 7:e37228. https://doi.org/10.1371/journal.pone.0037228

Troy S (1997) Territorial behavior and mating success in male New Zealand fur seals *A. forsteri*. PhD thesis, University of Melbourne

Warneke RM (1975) Dispersal and mortality of juvenile fur seals *Arctocephalus pusillus doriferus* in Bass Strait, southeastern Australia. Rapports et Proces-verbaux des Reunions Conseil International pour l'Exploration de la Mer 169:296–302

Warneke RM, Shaughnessy PD (1985) *Arctocephalus pusillus*, the South African and Australian fur seal: taxonomy, evolution, biogeography, and life history. In: Ling JK, Bryden MM (eds) Studies of sea mammals in south latitudes. South Australian Museum, Adelaide, pp 53–77

Wartzok D, Ketten DR (1999) Marine mammal sensory systems. In: Reynolds JE, Rommel SA (eds) Biology of marine mammals. Melbourne University Press, Melbourne

Woods R, Cousins DV, Kirkwood R et al (1995) Tuberculosis in a wild Australian fur seal (*Arctocephalus pusillus doriferus*) from Tasmania. J Wildl Dis 31:83–86

Chapter 28
Guadalupe Fur Seals and California Sea Lions: Two Sympatric Otariids from the California Current Ecosystem

Fernando R. Elorriaga-Verplancken, Karina Acevedo-Whitehouse, Tenaya Norris, Alexander Gutiérrez-Gutiérrez, María José Amador-Capitanachi, Ariadna Juárez-Ruiz, Julieta Sandoval-Sierra, Casandra Gálvez, and Xchel G. Moreno-Sánchez

Abstract Guadalupe fur seals (*Arctocephalus townsendi*) and California sea lions (*Zalophus californianus*) are the only two otariids that inhabit Mexican islands. California sea lions are widespread along the west coast of the Baja California Peninsula and within the Gulf of California, whereas Guadalupe fur seals are only found on Guadalupe Island and San Benito Archipelago, in the Pacific. At San Benito Archipelago, the two species are sympatric. By 2014, there were approximately 7000 California sea lions and 3700 Guadalupe fur seals in the archipelago. Stable isotope analysis from juveniles and adults of both species show that Guadalupe fur seals forage offshore with widely distributed foraging grounds, which we attribute to their squid dominated diet. California sea lions are more coastal and feed primarily on fish from higher trophic levels than the fur seals' diet. There is also a latitudinal effect in the isotopic signature with a greater proportion of Guadalupe fur seals presumed to forage at higher latitudes with lower $\delta^{15}N$ baseline values than California sea lions. Guadalupe fur seals were considered extinct 100 years ago, and the population is still recovering. Understanding its trophic ecology in relation to

F. R. Elorriaga-Verplancken (✉) · M. J. Amador-Capitanachi · A. Juárez-Ruiz · J. Sandoval-Sierra · C. Gálvez · X. G. Moreno-Sánchez
Instituto Politécnico Nacional, Centro Interdisciplinario de Ciencias Marinas (CICIMAR-IPN), Departamento de Pesquerías y Biología Marina, La Paz, Baja California Sur, Mexico
e-mail: felorriaga@ipn.mx

K. Acevedo-Whitehouse
Unit for Basic and Applied Microbiology, School of Natural Sciences, Autonomous University of Queretaro, Queretaro, Mexico

T. Norris
The Marine Mammal Center, Sausalito, CA, USA

A. Gutiérrez-Gutiérrez
Departamento Académico de Biología Marina, Universidad Autónoma de Baja California Sur (UABCS), La Paz, Baja California Sur, Mexico

C. Campagna, R. Harcourt (eds.), *Ethology and Behavioral Ecology of Otariids and the Odobenid*, Ethology and Behavioral Ecology of Marine Mammals,
https://doi.org/10.1007/978-3-030-59184-7_28

potential competitors, such as the California sea lion, is therefore of great importance especially in a region where there have been recent large-scale impacts from oceanographic warming anomalies.

Keywords Guadalupe fur seal · California sea lion · Sympatric · Stable isotope analysis · Trophic ecology · Diet · Competition · Foraging ecology · Population recovery · Warm oceanographic anomalies

28.1 Introduction

Two otariid species occur on Mexican islands: the Guadalupe fur seal (*Arctocephalus townsendi*) and the California sea lion (*Zalophus californianus*). The Guadalupe fur seal is one of only two fur seals in the Northern Hemisphere, and the only member of the genus *Arctocephalus* found north of the equator. This fur seal was hunted intensively for its pelt, reaching commercial extinction by 1894 (Townsend 1931). Recovery originated from a few individuals from Guadalupe Island that escaped the slaughter (Hubbs 1956). The current population is 34,000–44,000 individuals (García-Aguilar et al. 2018). By the end of the 1990s, the species had re-colonized the small San Benito Archipelago (Maravilla-Chavez and Lowry 1999). Guadalupe fur seals breed annually in July and raise one pup for about 9 months (Gallo-Reynoso and Figueroa-Garranza 2010). The peak abundance at San Benito Archipelago is in July and lowest in February and immature individuals make up the majority of the colony, having migrated from Guadalupe Island (Elorriaga-Verplancken et al. 2016a). These fur seals mainly prey on squids of different species (e.g. *Dosidicus gigas*, *Onychoteuthis* spp, *Ommastrephes bartramii*, and *Doryteuthis opalescens*) and myctophid fishes (Amador-Capitanachi et al. 2017; Juárez-Ruiz et al. 2018; Amador-Capitanachi et al. 2020).

The California sea lion is more widely distributed and abundant than the Guadalupe fur seal, with a population of around 387,000 distributed along several rookeries located on islands off the coast of California, to Baja California and into the Gulf of California (Szteren et al. 2006; Milanés-Salinas 2012; Carretta et al. 2014). This sea lion is resident year-round at a breeding colony on San Benito Archipelago, with around 7000 individuals (Elorriaga-Verplancken et al. 2015). The California sea lion breeding season takes place from late May through August. Females give birth to one pup and nurse it for about 12 months in Mexican colonies (García-Aguilar and Aurioles-Gamboa 2003). Adult males migrate north after breeding, wintering from central California to Washington State, in the USA. A smaller number of animals migrate as far north as British Columbia, the northern Gulf of Alaska, the Alaska Peninsula, and the eastern Aleutian Islands (Maniscalco et al. 2004). Other sea lions may remain in the Gulf of California throughout the year (Peterson and Bartholomew 1967; Le Boeuf et al. 1983). Male dispersal may diminish potential competition with adult females (Elorriaga-Verplancken et al. 2013), which restrict their foraging trips to areas close (< 370 km) to their breeding sites throughout the year (Melin et al. 2008; Kuhn and Costa 2014). These trips last around three days for Pacific rookeries (Kuhn and Costa 2014), and 1.3–2 days for the Gulf of California (García-Aguilar and Aurioles-Gamboa 2003). California sea lion foraging trips occur closer to the coast, whereas Guadalupe fur seals travel farther offshore. In general, the California sea lion's diet includes more fish than squid (Lowry et al. 1991; Porras-Peters et al. 2008). On San Benito, they prey on Pacific argentine (*Argentina sialis*), Panama hake (*Merluccius angustimanus*), opalescent squid (*Doryteuthis opalescens*), sand rockfishes (*Sebastes* spp.), among others (Pablo-Rodríguez et al. 2016; Aurioles-Gamboa and Camacho-Ríos 2007).

On Guadalupe Island, the Guadalupe fur seal population is large and the California sea lion colony is small (~300 individuals). In contrast, on San Benito Archipelago, an estimated of around 3700 Guadalupe fur seals and 7000 California sea lions were present during the summer of 2014 (Elorriaga-Verplancken et al. 2016a). This sympatry provides a unique opportunity to compare the trophic ecology of both species using stable isotope analysis, an effective tool for assessing the foraging habits of top predators like pinnipeds (Newsome et al. 2010).

28.2 The Use of Stable Isotopes in Trophic Ecology

Carbon ($\delta^{13}C$) and nitrogen ($\delta^{15}N$) are the stable isotopes most frequently used in trophic ecology. The analysis does not identify the identity of consumed prey; however, values of $\delta^{13}C$ can be used to infer habitat use based on physicochemical and biological factors (see also Franco Trecu et al. Chap. 29). These factors include the isotopic composition and concentration of dissolved CO_2 available to primary producers (Goericke and Fry 1994), the taxonomic composition and growth rate of phytoplankton (Fry and Wainright 1991), and the influence of benthic macroalgae from coastal areas, that are ^{13}C-enriched relative to phytoplankton from the offshore ecosystem, sometimes by up to 5.0–6.0‰ (Dunton and Schell 1987; Michener and Schell 1994; France 1995). As a result it is traditionally assumed that consumers close to shore have higher $\delta^{13}C$ values relative to offshore consumers.

In turn, $\delta^{15}N$ can be used to evaluate the trophic level and niche breadth because of the heavy isotope accumulation that occurs in regular increments between prey and predator as they move up the trophic chain (Owens 1987; Minagawa and Wada 1984; Hobson et al. 1997). This means that consumers that feed close to the trophic web base (e.g. zooplankton) show lower $\delta^{15}N$ values than consumers that feed close to the top of the trophic web (e.g. large fishes, marine mammals, etc.).

Importantly, spatial variation in $\delta^{15}N$ and $\delta^{13}C$ values at the base of the food web is negatively correlated with latitude in the eastern North Pacific Ocean (Altabet et al. 1999; Burton and Koch 1999). One consequence is that consumers that feed close to polar areas present lower $\delta^{15}N$ and $\delta^{13}C$ values than those that feed in tropical –temperate regions, even if they have similar trophic positions. The base values are important if large geographic differences are involved (Elorriaga-Verplancken et al. 2018).

These $\delta^{15}N$ baseline differences relate to the influence of oxygen minimum zones at intermediate depths and their corresponding denitrification by bacterial activity, which increases the baseline $\delta^{15}N$ values (Wada and Hattori 1991; Voss et al. 2001). For $\delta^{13}C$, latitudinal variation relates to the solubility of CO_2 (^{12}C-enriched) and its relationship with water temperature and so decreases at middle relative to higher latitudes (Goericke and Fry 1994).

Isotopic data provide information at distinct time scales that are tissue-specific, ranging from a few days in blood plasma (Sinisalo et al. 2008) to years in dental collagen (Elorriaga-Verplancken et al. 2013).

Many trophic ecology studies have been carried out on pinnipeds using stable isotopes; these include inter- or intra-specific differences in trophic position (Burton and Koch 1999; Velázquez-Castillo and Elorriaga-Verplancken 2017) and habitat use (Lewis et al. 2006; Elorriaga-Verplancken et al. 2018), latitudinal segregation (Aurioles-Gamboa et al. 2006; Cherel et al. 2009; Amador-Capitanachi et al. 2017) and dietary diversification (Páez-Rosas and Aurioles-Gamboa 2010; Juárez-Ruiz et al. 2018). Previously, comparative diet studies of Guadalupe fur seals and California sea lions have been performed using scat analysis (Aurioles-Gamboa and Camacho-Ríos 2007) or stable isotopes mostly from pup fur, which is an indirect proxy for maternal foraging (Elorriaga-Verplancken et al. 2016a); or using small sample sizes by species (Pablo-Rodríguez et al. 2016). Most of these studies have focused on relatively short time frames over which the isotopic signatures were integrated (days and three to four months). Here we directly evaluate the trophic ecology of male juvenile Guadalupe fur seals and adult female California sea lions at San Benito Archipelago, using fur samples collected over several years.

28.3 How Was This Work Done?

During the 2013–2018 breeding seasons for both species (one week in mid July), fur samples from 31 Guadalupe fur seals and 20 California sea lions were collected on San Benito Archipelago (28°18.482′N, 115°33.737′W), a group of three islands off the Pacific coast of the Baja California Peninsula in Mexico (Fig. 28.1). The individuals sampled were male juveniles (Guadalupe fur seal, Fig. 28.2) and adult females (California sea lion, Fig. 28.3) which represent the most abundant class (besides pups) for each species on the San Benito Archipelago (Elorriaga-Verplancken et al. 2015, 2016a). Animals were immobilized using hoop nets and fur samples were taken from the dorsal area of each individual. These samples were processed and analyzed in a mass spectrometer.

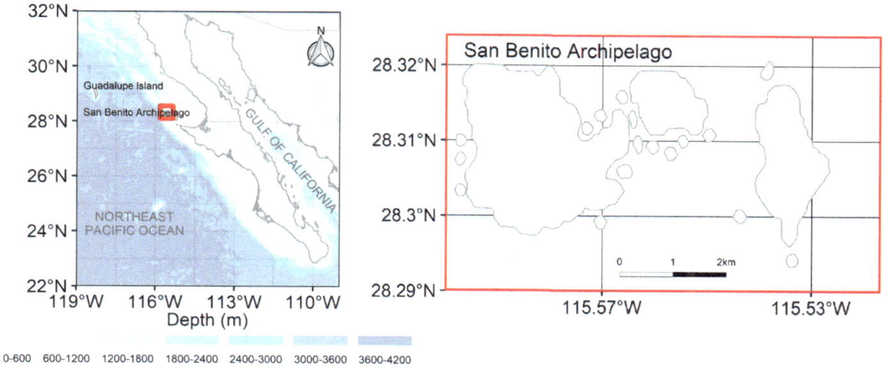

Fig. 28.1 Study area. San Benito Archipelago in Baja California, Mexico

Fig. 28.2 Male juvenile Guadalupe fur seal from San Benito Archipelago, Mexico

28.4 Our Findings

There were significant differences (U-MW, $p < 0.05$) between the two species for both isotopic ratios. California sea lions had consistently higher values ($17.9 \pm 0.4‰$ for $\delta^{15}N$ and $-15.9 \pm 0.4‰$ for $\delta^{13}C$) relative to Guadalupe fur seal values ($17.0 \pm 0.9‰$ for $\delta^{15}N$ and $-17.2 \pm 0.5‰$ for $\delta^{13}C$). The isotopic niches were greatest for Guadalupe fur seals, considering all values ($4.9‰^2$) and also the most significant within their ellipse ($1.2‰^2$), compared to California sea lions, which had niche areas of $2.3‰^2$ and $0.7‰^2$, respectively (Fig. 28.4).

Fig. 28.3 Adult Female California sea lion from San Benito Archipelago, Mexico

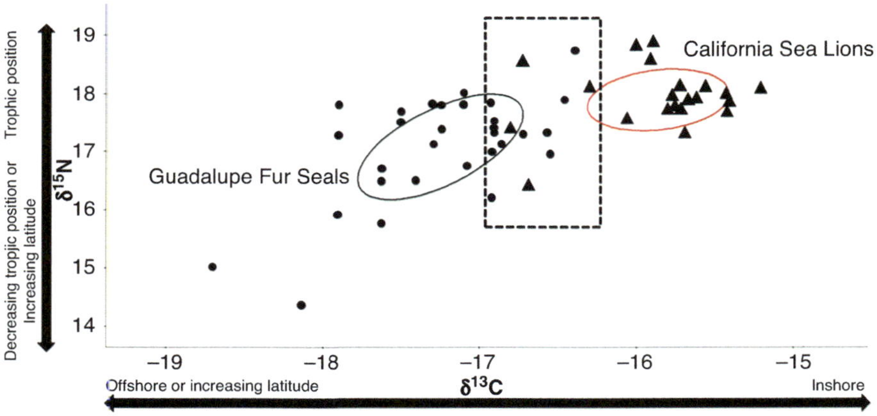

Fig. 28.4 Isotopic niches of Guadalupe fur seals (circles) and California sea lions (triangles) from the San Benito Archipelago, Mexico

These findings suggest foraging segregation exists between the two otariid species inhabiting the San Benito Archipelago; this pattern reflects niche partitioning (prey and/or habitat). The higher $\delta^{15}N$ and $\delta^{13}C$ values for adult female California sea lions relative to male juvenile Guadalupe fur seals may indicate that sea lions occupy a higher trophic position (discussed in more detail below) and use habitats closer to the coast than fur seals. For California sea lions, stomach content analyses indicate coastal foraging with a predominantly fish diet (Aurioles-Gamboa and Camacho Ríos 2007) and relatively short feeding trips (approximately three days), while satellite tracking has confirmed movements <370 km to their breeding sites throughout the year (Melin et al. 2008; Kuhn and Costa 2014).

By contrast, on both Guadalupe Island and San Benito Archipelago, Guadalupe fur seals consume a diet composed of squid and deep-sea fish like myctophids (Amador-Capitanachi et al. 2017). Moreover, sightings of juvenile Guadalupe fur seals in Mexico and the U.S. and satellite tracking of tagged juvenile Guadalupe fur seals at Guadalupe Island reveal offshore foraging (i.e., offshore of the continental shelf) with foraging excursions as far north as Washington State and as far south the central Mexican Pacific Ocean, and into the Gulf of California (Gallo-Reynoso et al. 2008; Elorriaga-Verplancken et al. 2016b; Amador-Capitanachi 2018; Ortega-Ortiz et al. 2019; NOAA Fisheries 2019). These much larger displacements by juvenile Guadalupe fur seals, would explain why the isotopic niche (sum of both isotopic ratios) is larger for this species than it is for the California sea lion. That is, the wider the foraging area, the more isotopic variation associated with different habitats can be found at the basal level (Altabet et al. 1999; Burton and Koch 1999; Lewis et al. 2006; Elorriaga-Verplancken et al. 2016c).

Traditionally, the closer the consumers are to the trophic web base, the lower the $\delta^{15}N$ values will be, and vice-versa. The trophic level determined using scat analysis was similar for both species, with values between 4 and 4.5 (Aurioles-Gamboa and Camacho Ríos 2007; Pablo-Rodríguez et al. 2016; Amador-Capitanachi et al. 2017).

However, our $\delta^{15}N$ values were higher in female California sea lions relative to juvenile Guadalupe fur seals at San Benito Archipelago, suggesting they feed at a higher trophic level. This may be a consequence of prey size, as ~100 kg female California sea lion are able to consume larger, higher trophic level prey, including mostly fishes (Aurioles-Gamboa and Camacho Ríos 2007), in contrast to juvenile Guadalupe fur seals, typical from the San Benito location, which at around half the body size of California sea lion (Fig. 28.5) may need to consume smaller lower trophic level prey, mostly small squid or myctophids (Juárez-Ruiz et al. 2018).

The $\delta^{15}N$ variation between the two species may also reflect the higher latitudes reached by Guadalupe fur seals given the inverse relationship between latitude and basal $\delta^{15}N$ and $\delta^{13}C$ values (Altabet et al. 1999). Animals foraging at higher latitudes tend to have lower isotopic values regardless of trophic level (Burton and Koch 1999; Elorriaga-Verplancken et al. 2016c; Amador-Capitanachi et al. 2017). Even though some Guadalupe fur seals also travel south, most trips are northward into the California Current, sometimes as long as 1500 km (Amador-Capitanachi 2018), assimilating (mostly) the depleted isotopic signature of those cold water latitudes. This relationship with latitude may be the reason why the correlation between the two isotopic ratios showed a significant relationship ($r = 0.609$, $p = 0.0003$) for Guadalupe fur seals but not for California sea lions ($r = 0.202$, $p = 0.383$; Fig. 28.6).

A similar latitudinal isotopic pattern has been reported for northern elephant seals (*Mirounga angustirostris*) that forage in the North Pacific, as they show lower isotopic ratios than resident harbor seals (*Phoca vitulina*) that forage off the coast

Fig. 28.5 Size perspective between a Guadalupe fur seal (right) and a California sea lion (left)

of the Baja California Peninsula (Elorriaga-Verplancken et al. 2016c). Moreover, a relationship similar to that described here, between Guadalupe fur seal and California sea lions, is also clear when comparing the two major rookeries of Guadalupe fur seals. Amador-Capitanachi et al. (2017) reported significantly lower $\delta^{15}N$ values from Guadalupe fur seals on Guadalupe Island. Scat analysis suggested a higher trophic level diet composed predominantly of jumbo squid (*Dosidicus gigas*), relative to the Guadalupe fur seal on the San Benito Archipelago, feeding mostly on lower trophic level opalescent squid (*Doryteuthis opalescens*). The $\delta^{15}N$ disparity may reflect the more northerly foraging habitats of female adult Guadalupe fur seals from Guadalupe Island, which show a lower basal signal relative to foraging areas likely to be exploited by their younger counterparts from the San Benito Archipelago.

There is an area in the center of the isotopic space where both species may converge (see the rectangle in Fig. 28.4), particularly where a few of the California sea lions (15% of the total individuals analyzed) may have foraged in oceanic waters or displaced northwards, mirroring the Guadalupe fur seal. Given the persistent occurrence of warm anomalies (e.g. The Blob) in the northeastern Pacific in recent years (Di Lorenzo and Mantua 2016), which has impacted the California sea lion in western Baja California, resulting in decreased average body mass and increased foraging effort with prolonged trips farther from the coast (Elorriaga-Verplancken et al. 2016a). Satellite tracking in California has also shown California sea lion foraging trips are extending further from the coast during these warm climatic anomalies (Weise et al. 2006; Kuhn and Costa 2014). These warming anomalies have had widespread impacts on marine species (e.g. Cavole et al. 2016), including increased foraging effort by Guadalupe fur seals (Elorriaga-Verplancken et al. 2016a), as well as interannual prey preferences (Amador-Capitanachi et al. 2020), and a negative impact on pups body condition (Gálvez et al. 2020). If these

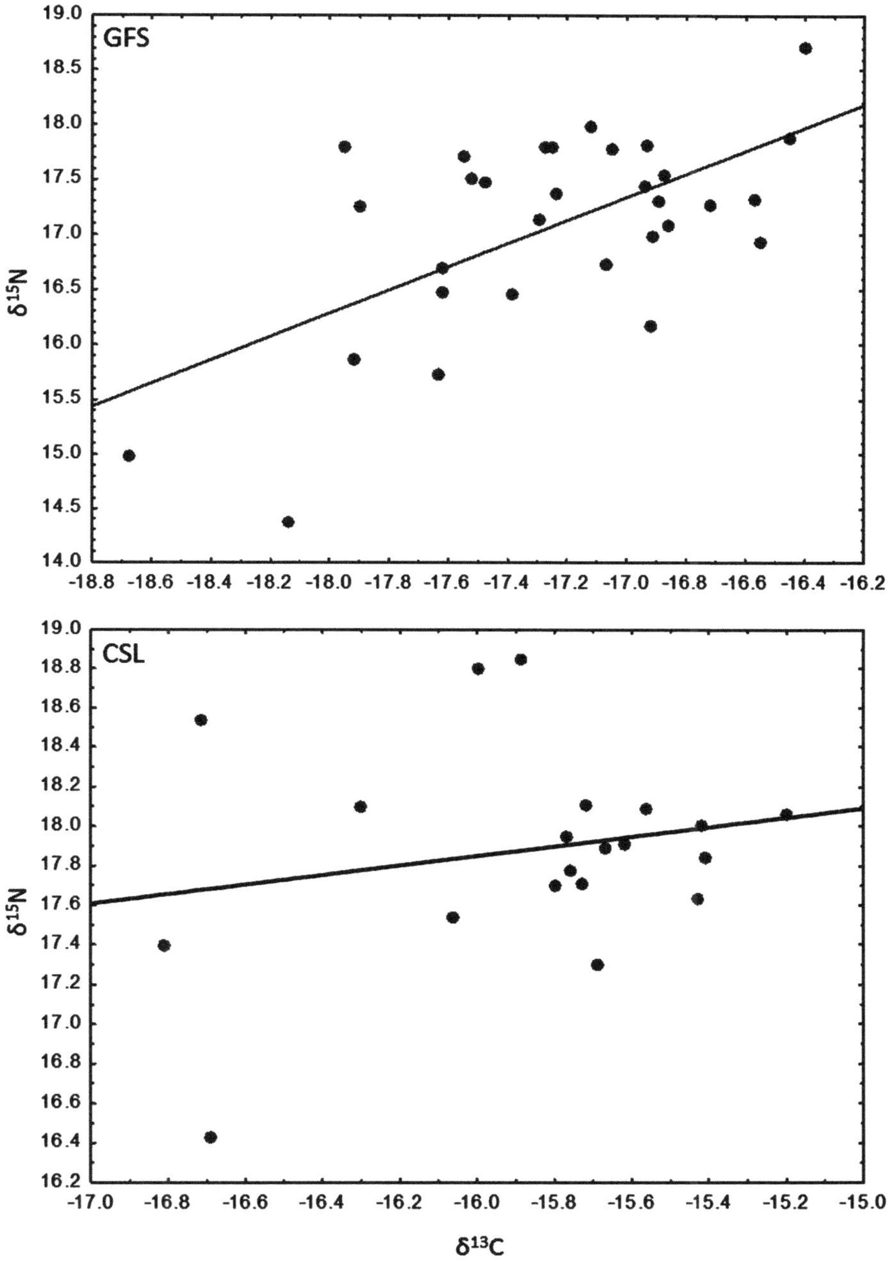

Fig. 28.6 Correlation between isotopic ratios of Guadalupe fur seals (GFS) and California sea lions (CSL) from the San Benito Archipelago, Mexico

abnormal oceanographic conditions continue in this region of the Pacific in the near future, there could be further changes to California sea lion foraging habits. California sea lions tend to be more generalist and display deeper dives than Guadalupe fur seal (García-Rodríguez and Aurioles-Gamboa 2004; Gallo-Reynoso et al. 2008; Melin et al. 2008; Amador-Capitanachi et al. 2017); however, under persistent warming anomalies, California sea lion feeding trips and habitat use may become increasingly similar (longer and offshore) to those of Guadalupe fur seals, especially those that stay relatively closer, in oceanic waters adjacent to San Benito. In addition to the impact that foraging shifts could have on the health and immune system of these otariids (Banuet-Martínez et al. 2017; Flores-Morán et al. 2017), these changes would potentially increase competition between these two species, potentially affecting both, particularly the Guadalupe fur seal, which is still undergoing recovery and recolonization of the San Benito Archipelago (Elorriaga-Verplancken et al. 2016a) and which was seriously impacted by environmental conditions that led to unusual mortality events in California (NOAA Fisheries 2019).

More studies are necessary to further understand long-term impacts of climate change on trophic relationships between these two sympatric otariid species. Isotopic assessments should be coupled, inter-annually, with scat analyses, to determine the taxonomic prey spectra for each otariid and evaluate changes over time. Such trophic ecology studies are important, but especially when they are related to population estimates and variations over the same time frames. Given that the Guadalupe fur seal was considered extinct only 70–100 years ago, enhanced knowledge of how they adapt to a changing world becomes ever more important.

28.5 Final Remarks

Guadalupe fur seals and California sea lions are the only two otariids inhabiting Mexican islands. The San Benito Archipelago is the location in Mexico where both species present the strongest sympatry. Despite this coexistence on land, there is marked variation in trophic niche, reflecting feeding segregation that may reduce potential competition for food resources. They do not only consume mostly different prey (fish vs. squid, mostly), but they do it in different locations (mostly shore vs. offshore). Low isotopic ratios in Guadalupe fur seals indicate offshore foraging ($\delta^{13}C$), but may also reflect their foraging at higher latitudes, given the inverse relationship between latitude and basal $\delta^{15}N$ and $\delta^{13}C$ values in regions like the Northeastern Pacific. More studies are critical to better understand long-term impacts of climate change on trophic relationship between these two sympatric otariid species, especially because of persistent oceanographic anomalies in the Northeastern Pacific and the fact that the Guadalupe fur seal is still under recovery.

References

Altabet MA, Pilskaln C, Thunell R et al (1999) The nitrogen isotope biogeochemistry of sinking particles from the margin of the Eastern North Pacific. Deep-Sea Res 46:655–679. https://doi.org/10.1016/S0967-0637(98)00084-3

Amador-Capitanachi MJ, Moreno-Sánchez XG, Juárez-Ruiz A et al (2017) Trophic variation between the two existing Guadalupe fur seal colonies on Guadalupe Island and the San Benito Archipelago, México. Aquat Mamm 43(1):14–25. https://doi.org/10.1578/AM.43.1.2017.14

Amador-Capitanachi MJ (2018) Variabilidad en el nicho isotópico (2013–2016) y análisis del desplazamiento mediante telemetría en el lobo fino de Guadalupe (*Arctocephalus philippii townsendi*). Dissertation. CICIMAR-IPN

Amador-Capitanachi MJ, Moreno-Sánchez XG, Ventura-Domínguez PD et al (2020) Ecological implications of unprecedented warm water anomalies on interannual prey preferences and foraging areas of Guadalupe fur seals. Mar Mamm Sci 36(4):1254–1270. https://doi.org/10.1111/mms.12718

Aurioles-Gamboa D, Camacho-Ríos FJ (2007) Diet and feeding overlap of two otariids, *Zalophus californianus* and *Arctocephalus townsendi*: implications to survive environmental uncertainty. Aquat Mamm 33:315–326. https://doi.org/10.1578/AM.33.3.2007.315

Aurioles-Gamboa D, Koch P, Le Boeuf BJ (2006) Differences in foraging location of Mexican and California elephant seals: evidence from stable isotope in pups. Mar Mamm Sci 22(2):326–338. https://doi.org/10.1111/j.1748-7692.2006.00023.x

Banuet-Martínez M, Espinoza-de-Aquino W, Elorriaga-Verplancken FR et al (2017) Climatic anomaly affects the immune competence of California sea lions. PLoS One 12(6):e0179359. https://doi.org/10.1371/journal.pone.0179359

Burton RK, Koch PL (1999) Isotopic tracking of foraging and long-distance migration in northeastern Pacific pinnipeds. Oecologia 119:578–585. https://doi.org/10.1007/s004420050822

Carretta JV, Oleson E, Weller DW et al (2014) U.S. Pacific Marine Mammal Stock Assessments 2013. National Oceanic and Atmospheric Administration Technical Memorandum NMFS-SWFSC-532

Cavole LM, Demko AM, Diner RE et al (2016) Biological impacts of the 2013–2015 warm-water anomaly in the Northeast Pacific: winners, losers, and the future. Oceanography 29:273–285. https://doi.org/10.5670/oceanog.2016.32

Cherel Y, Kernaleguen L, Richard P et al (2009) Whisker isotopic signature depicts migration patterns and multi-year intra- and inter-individual foraging strategies in fur seals. Biol Lett 5 (6):830–832. https://doi.org/10.1098/rsbl.2009.0552

Di Lorenzo E, Mantua N (2016) Multi-year persistence of the 2014/15 North Pacific marine heatwave. Nat Clim Chang 6(11):1042–1047. https://doi.org/10.1038/nclimate3082

Dunton K, Schell DM (1987) Dependence of consumers on macroalgal (*Laminaria solidungula*) carbon in an Arctic kelp community: $\delta^{13}C$ evidence. Mar Biol 93(4):615–625. https://doi.org/10.1007/BF00392799

Goericke R, Fry B (1994) Variations of marine ^{13}C with latitude, temperature, and dissolved CO_2 in the world ocean. Global Biogeochem Cycles 8:85–90. https://doi.org/10.1029/93GB03272

Elorriaga-Verplancken FR, Aurioles-Gamboa D, Newsome SD et al (2013) $\delta^{15}N$ and $\delta^{13}C$ values in dental collagen as a proxy for sex and age variation in foraging strategies of California sea lions. Mar Biol 160(3):641–652. https://doi.org/10.1007/s00227-012-2119-y

Elorriaga-Verplancken FR, Ferretto G, Angell O (2015) Current status of the California sea lion (*Zalophus californianus*) and the northern elephant seal (*Mirounga angustirostris*) at the San Benito Archipelago, Mexico. Cienc Mar 41(4):269–281. https://doi.org/10.7773/cm.v41i4.2545

Elorriaga-Verplancken FR, Sierra-Rodríguez GE, Rosales-Nanduca H et al (2016a) Impact of the 2015 El Niño-Southern Oscillation on the abundance and foraging habits of Guadalupe fur seals and California sea lions from the San Benito Archipelago, Mexico. PLoS One 11(5):e0155034. https://doi.org/10.1371/journal.pone.0155034

Elorriaga-Verplancken FR, Rosales-Nanduca H, Robles R (2016b) Unprecedented records of Guadalupe fur seals in La Paz Bay, Southern Gulf of California, Mexico, as a possible result of warm conditions in the northeastern Pacific. Aquat Mamm 42(3):261–267. https://doi.org/10.1578/AM.42.3.2016.261

Elorriaga-Verplancken FR, Morales-Luna L, Heckel G et al (2016c) Foraging ecology of harbor seals (*Phoca vitulina*) and northern elephant seals (*Mirounga angustirostris*) from Baja California, Mexico: inferences from stable isotopes in pups. J Mar Biol Assoc UK 96(1):903–908. https://doi.org/10.1017/S0025315415002143

Elorriaga-Verplancken FR, Sandoval-Sierra J, Paniagua-Mendoza A et al (2018) Seasonality and potential foraging grounds of migratory California sea lions from La Paz Bay, Southern Gulf of California, Mexico. Aquat Mamm 44(1):56–61. https://doi.org/10.1578/AM.44.1.2018.56

Flores-Morán A, Banuet-Martínez M, Elorriaga-Verplancken FR et al (2017) Atypical red blood cells are prevalent in California sea lion pups born during anomalous sea surface temperature events. Physiol Biochem Zool 90(5):564–574. https://doi.org/10.1086/692919

France RL (1995) Carbon-13 enrichment in benthic compared to planktonic algae: Foodweb implications. Mar Ecol Prog Ser 124:307–312

Fry B, Wainright SC (1991) Diatom sources of ^{13}C-rich carbon in marine food webs. Mar Ecol Prog Ser 76:149–157. https://doi.org/10.3354/meps076149

Gallo-Reynoso JP, Figueroa-Carranza AL (2010) Pup growth of the Guadalupe fur seal, *Arctocephalus townsendi*. Therya 1(1):75–90

Gallo-Reynoso JP, Figueroa-Carranza AL, Le Boeuf BJ (2008) Foraging behavior of lactating Guadalupe fur seal females. In: Lorenzo C, Espinoza E, Ortega J (eds) Avances en el estudio de los mamíferos de México, Publicaciones Especiales, vol II. Asociación Mexicana de Mastozcologia AC, Mexico City, pp 505–614

Gálvez C, Pardo M, Elorriaga-Verplancken FR (2020) Impacts of extreme ocean warming on the early development of a marine top predator: The Guadalupe fur seal. Prog Oceanogr 180:102220. https://doi.org/10.1016/j.pocean.2019.102220

García-Aguilar MC, Aurioles-Gamboa D (2003) Breeding season of the California sea lion (*Zalophus californianus*) in the Gulf of California. Aquat Mamm 29(1):67–76. https://doi.org/10.1578/016754203101024086

García-Aguilar MC, Elorriaga-Verplancken FR, Rosales-Nanduca H et al (2018) Population status of the Guadalupe fur seal (*Arctocephalus townsendi*). J Mammal 99(6):1522–1528. https://doi.org/10.1093/jmammal/gyy132

García-Rodríguez F, Aurioles-Gamboa D (2004) Spatial and temporal variations in the diet of the California sea lion (*Zalophus californianus*) in the Gulf of California, México. Fish Bull 102 (1):47–62

Hobson KA, Sease JL, Merrick RL et al (1997) Investigating trophic relationships of pinnipeds in Alaska and Washington using stable isotope ratios of nitrogen and carbon. Mar Mamm Sci 13:114–132

Hubbs CL (1956) Back from oblivion, Guadalupe fur seal: still a living species. Pac Discov 9:14–21

Kuhn CE, Costa DP (2014) Interannual variation in the at-sea behavior of California sea lions (*Zalophus californianus*). Mar Mamm Sci 30(4):1297–1319. https://doi.org/10.1111/mms.12110

Juárez-Ruiz A, Elorriaga-Verplancken FR, Moreno-Sánchez XG et al (2018) Diversification of foraging habits among Guadalupe fur seals from their only well-established breeding colony, Guadalupe Island, Mexico. Mar Biol 165:86. https://doi.org/10.1007/s00227-018-3339-6

Le Boeuf BJ, Aurioles-Gamboa D, Condit R et al (1983) Size and distribution of the California sea lion population in Mexico. Proc Calif Acad Sci 43(4):77–85

Lewis R, Connell TCO, Lewis M et al (2006) Sex-specific foraging strategies and resource partitioning in the southern elephant seal (*Mirounga leonina*). Proc R Soc 273 (1603):2901–2907. https://doi.org/10.1098/rspb.2006.3642

Lowry MS, Stewart BS, Heath CB et al (1991) Seasonal and annual variability in the diet of California sea lions *Zalophus californianus* at San Nicolas Island, California, 1981-1986. Fish Bull 89(2):331–336

Maniscalco JM, Wynne K, Pitcher KW et al (2004) The occurrence of California Sea Lions (*Zalophus californianus*) in Alaska. Aquat Mamm 30(3):427–433. https://doi.org/10.1578/AM.30.3.2004.427

Maravilla-Chavez MO, Lowry MS (1999) Incipient breeding colony of Guadalupe fur seals at Isla Benito del Este, Baja California, Mexico. Mar Mamm Sci 15(1):239–241. https://doi.org/10.1111/j.1748-7692.1999.tb00796.x

Melin SR, DeLong RL, Siniff DB (2008) The effects of El Niño on the foraging behavior of lactating California sea lions (*Zalophus californianus californianus*) during the nonbreeding season. Can J Zool 86(3):192–206. https://doi.org/10.1139/Z07-132

Michener RH, Schell DM (1994) Stable isotope ratios as tracers in marine aquatic food webs. In: Michener RH, Lajtha K (eds) Stable isotopes in ecology and environmental science, 2nd edn. Blackwell Scientific, Boston, pp 138–157

Milanés-Salinas MA (2012) Abundancia y distribución del lobo marino de California (*Zalophus californianus*) en el Pacífico Mexicano. Dissertation. UABC

Minagawa M, Wada E (1984) Stepwise $\delta^{15}N$ enrichment along food chains. Further evidence and the relation between $\delta^{15}N$ and animal age. Geochim Cosmochim Acta 48:1135–1140. https://doi.org/10.1016/0016-7037(84)90204-7

Newsome SD, Clementz MT, Koch PL (2010) Using stable isotope biogeochemistry to study marine mammal ecology. Mar Mamm Sci 26(3):509–572. https://doi.org/10.1111/j.1748-7692.2009.00354.x

NOAA Fisheries (2019) Guadalupe fur seal unusual mortality event in California. https://www.fisheries.noaa.gov/national/marine-life-distress/2015-2019-guadalupe-fur-seal-unusual-mortality-event-california. Accessed 25 Apr 2018

Ortega-Ortiz CD, Vargas-Bravo MH, Olivos-Ortiz A et al (2019) Guadalupe fur seal encounters in the Mexican Central Pacific during 2010–2015: a dispersion pattern related to the species recovery? Aquat Mamm 45(2):244–252. https://doi.org/10.1578/AM.45.2.2019.246

Owens PJN (1987) Natural variations in ^{15}N in the marine environment. Adv Mar Biol 24:389–451

Pablo-Rodríguez N, Aurioles-Gamboa D, Montero-Muñoz J (2016) Niche overlap and habitat use at distinct temporal scales among the California sea lions (*Zalophus californianus*) and Guadalupe fur seals (*Arctocephalus philippii townsendi*). Mar Mamm Sci 2(2):466–489. https://doi.org/10.1111/mms.12274

Páez-Rosas D, Aurioles-Gamboa D (2010) Alimentary niche partitioning in the Galapagos sea lion, *Zalophus wollebaeki*. Mar Biol 157:2769–2781. https://doi.org/10.1007/s00227-010-1535-0

Peterson R, Bartholomew G (1967) The natural history and behavior of the California sea lion. Am Soc Mamm Spec Special Publ 1:79

Porras-Peters H, Aurioles-Gamboa D, Cruz-Escalona V et al (2008) Trophic level and overlap of sea lion (*Zalophus californianus*) in the Gulf of California, Mexico. Mar Mamm Sci 24(3):554–576. https://doi.org/10.1111/j.1748-7692.2008.00197.x

Sinisalo T, Jones RI, Helle E et al (2008) Changes in diets of individual Baltic ringed seals (*Phoca hispida botnica*) during their breeding season inferred from stable isotope analysis of multiple tissues. Mar Mamm Sci 24(1):159–170. https://doi.org/10.1111/j.1748-7692.2007.00170.x

Szteren D, Aurioles-Gamboa D, Gerber LR (2006) Population status and trends of the California sea lion (*Zalophus californianus californianus*) in the Gulf of California, Mexico. Alaska Sea Grant College Program AK-SG-06:1–17

Townsend CH (1931) The fur seal of the California islands with new descriptive and historical matter. Fortschr Zool 9:442–457

Velázquez-Castillo M, Elorriaga-Verplancken FR (2017) Isotopic evidence for intersexual foraging variation in northern elephant seals from Baja California, Mexico. Mar Biol 164(8):168. https://doi.org/10.1007/s00227-017-3192-z

Voss M, Dippner JW, Montoya JP (2001) Nitrogen isotope patterns in the oxygen-deficient waters of the Eastern Tropical North Pacific Ocean. Deep-Sea Res I 48:1905–1921. https://doi.org/10.1016/S0967-0637(00)00110-2

Wada E, Hattori A (1991) Nitrogen in the sea: forms, abundances and rate processes. CSC Press, Boca Raton

Weise MJ, Costa DP, Kudela RM (2006) Movement and diving behavior of male California sea lion (*Zalophus californianus*) during anomalous oceanographic conditions of 2005 compared to those of 2004. Geophys Res Lett 33:L22S10. https://doi.org/10.1029/2006GL027113

Chapter 29
Fur Seal Trophic Ecology in Uruguay

Valentina Franco-Trecu, Daniel E. Naya, and Pablo Inchausti

Abstract We describe the foraging ecology of South American fur seals (*Arctocephalus australis*) at Isla de Lobos, Uruguay, the northernmost colony of the species in the Atlantic. We use multiple approaches to investigate diet and behavior at sea. South American fur seal females are epi- or meso-pelagic divers that feed on vertically migrating prey of high-energy content (e.g., Argentine short-fin squid—*Illex argentinus*, anchovy—*Engraulis anchoita* and hake—*Merluccius hubbsi*). Fur seal females travel far to forage, focusing their efforts over the slope of the continental shelf. Different degrees of individual trophic specialization and foraging away from the coast reduce the overlap with coastal fisheries and may explain the growth of this population during the past 60 years.

Keywords *Arctocephalus australis* · South American fur seal · Diet · Scat · Stable isotopes · Diving behavior · Satellite tracking · Individual specialization · Foraging behavior

Mammalian females can meet energetic demands of gestation and lactation by increasing the amount and/or quality of trophic resources (Thompson et al. 2012). Income breeders, i.e. those that replenish energy stores to successfully complete their reproductive cycle (Boyd 2000), such as fur seals and sea lions, are good models to test this idea, as offspring dependence is prolonged (Sepulveda and Harcourt Chap. 3). Pup survival depends not only on the ability of the mother to replenish her energy stores, but also on her body condition at the time of pupping and the

V. Franco-Trecu (✉) · D. E. Naya
Departamento de Ecología y Evolución, Facultad de Ciencias, Universidad de la República, Montevideo, Uruguay
e-mail: vfranco-trecu@fcien.edu.uy

P. Inchausti
Departamento de Ecología, Centro Universitario Regional Este, Universidad de la República, Montevideo, Uruguay

C. Campagna, R. Harcourt (eds.), *Ethology and Behavioral Ecology of Otariids and the Odobenid*, Ethology and Behavioral Ecology of Marine Mammals,
https://doi.org/10.1007/978-3-030-59184-7_29

fasting tolerance of the pup that remains at the rookery while mother is away at sea (Costa and Toro Chap. 2). Female foraging strategies are thus affected by maternal investment demands (Sepulveda and Harcourt Chap. 3). Understanding these strategies is relevant also for conservation efforts in the face of climate change, as climate change may drive potential shifts in prey distribution and abundance.

29.1 South American Fur Seals in Uruguay

South American fur seals (*Arctocephalus australis*) are a mid-size and dimorphic otariid, with adult males weighing about 200 kg and adult females about 45 kg. They breed in large aggregations, on islands and the mainland along the Atlantic and Pacific coasts of South America. Uruguayan rookeries represent the northernmost Atlantic breeding colonies. The Uruguayan population was exploited between 1873 and 1949, with an estimated 527,000 individuals harvested. From 1950 until 1991, the Uruguayan Government managed the harvest, killing *ca.* 300,000 fur seals, mostly adult and subadult males (Ponce de León 2000). The Uruguayan population of fur seals has increased at an annual rate of 1.5% for the past 60 years (Franco-Trecu et al. 2019). Dispersion from this population may explain the rapid growth of South American fur seal colonies in Argentina and southern Chile (Crespo et al. 2015).

Fig. 29.1 A birth of a South American fur seal at Isla de Lobos, Uruguay. Newborn pups weigh about 4 kg. The bond with the mother lasts almost 1 year

Females give birth to a single pup between late November and early January (Fig. 29.1), and enter into oestrous and mate during December and January (Franco-Trecu et al. 2014b). Approximately ten days after parturition, females leave the colony for their first foraging trip. After that, and for about 11 months, mothers alternate periods on land feeding their offspring, and trips to sea to forage, while their pups wait at the colony. Accordingly, foraging behaviour of mothers is central to pup survival, and this is reflected in population success.

This chapter compiles data on diet composition and foraging behaviour of females, using scat and stable isotope analysis, satellite tracking and dive recorders. Such studies illustrate fur seal biology at sea and on land, at individual and population levels.

29.2 Diet

Diet composition can be inferred from analyses of hard items found in scats, such as fish otoliths and squid beaks (Naya et al. 2002). The same items may be found in regurgitations (Allum and Maddigan 2012) and stomach contents of dead individuals (Koen Alonso et al. 2000). These approaches yield high-resolution information on species composition and body sizes of prey (Santos et al. 2012); but are vulnerable to biases due to differential digestion, retention and recovery of prey remains (Arim and Naya 2003; Klare et al. 2011). Scat analysis is the least invasive of the traditional indirect methods to determine diet composition, and yields the largest sample size for lowest effort and cost (Trites and Joy 2005).

The first systematic study of the diet of South American fur seals in Uruguay was based on scats, conducted in 1995–1998 at Isla de Lobos (Fig. 29.2). Four prey were the most important, comprising more than 80% of the relative frequency of occurrence: striped weakfish (*Cynoscion guatucupa*), cutlassfish (*Trichiurus lepturus*), Argentine anchovy (*Engraulis anchoita*), and cephalopods, mainly Argentine shortfin squid (*Illex argentinus*) (Naya et al. 2002). A decade later, a follow up study found that the original target species were still important but suggested that Argentine hake (*Merluccius hubbsi*) were now also a major prey item, with a reduction in cutlassfish (Franco-Trecu et al. 2013).

Stable isotope signatures (mainly of $\partial^{13}C$ and $\partial^{15}N$) present in predator and prey tissues (Franco-Trecu et al. 2012, 2013, 2014a) avoid biases of scat sampling for diet composition. Nitrogen isotopic values estimate the trophic position of a population based on the stepwise enrichment of $\partial^{15}N$ between prey and predator of *ca.* 3.4‰ per trophic level (Post 2002). Carbon isotopic values, conversely, indicate the origin of energy sources (DeNiro and Epstein 1978). In our studies, isotopic values from skin samples confirmed that South American fur seals foraged pelagically and at low trophic levels, targeting Argentine hake, anchovy and short-fin squid (Franco-Trecu et al. 2012, 2013, 2014a).

Fig. 29.2 Isla de Lobos, off the coast of Uruguay, hosts the largest population of South American fur seals in the Atlantic

29.3 Individuals Prefer a Broad Diet

Trophic specialization occurs when individuals use a narrower subset of trophic resources than those used by their population as a whole (Bolnick et al. 2003). Specialization may be estimated using stable isotopes by sampling (i) the same tissues in the same individuals at different times (e.g., Bearhop et al. 2004); (ii) different portions of metabolically inert tissues with continuous growth (e.g., whiskers, teeth), which work as sequential archives over long periods of time (e.g., Franco-Trecu et al. 2014a); and (iii) tissues with different turnover rates (Franco-Trecu et al. 2012), representing different temporal windows (i.e. serum for weeks, red cells for months). Stable isotope values from the sequential sampling of whiskers indicated that some individuals are relatively specialized feeders, while others are generalists (Lima et al. 2019). Stable isotope analysis of two tissues with different turnover rates—reflecting diet over different temporal windows—suggested dietary changes between pre- and post-partum periods, with post-partum females consuming a less diverse diet (Franco-Trecu et al. 2012).

29.4 Females Travel Far and a Long Time; Pups Wait and Fast

Satellite telemetry revealed that females foraged relatively close to the rookery (within *ca.* 60 km) during the first trip after giving birth, reducing time away from small young pups. These same females travelled farther in subsequent trips, reaching as far as the continental shelf edge, about 500 km from the colony (Fig. 29.3). Distant trips were longer: about seven days for the first trip versus ~13 days for subsequent foraging incursions. Individual female foraging areas covered a surface that was almost as large as that of the feeding ground of the overall population. Foraging areas extensively overlapped among females, suggesting little individual differences in foraging grounds (Franco-Trecu 2015).

29.5 Short, Shallow and Nocturnal Divers

Females were pelagic foragers that dove mostly at night (Fig. 29.4). Dives were short: 80% lasted less than one minute, and 40% less than 30 s. The longest dive was of 7 min. Most dives (~55%) were shallow (<20 m); about 5% were deeper than 60 m (max: 207 m). Lactating females dove deeper during the day only in the first postpartum foraging trip. Overall, nocturnal dives were shorter, with less time at the bottom than diurnal ones (Franco-Trecu 2015).

29.6 Comparative Remarks

In Uruguay, the South American fur seal is sympatric with the southern sea lion (*Otaria flavescens*), forming mixed and dense colonies. Comparison of diet composition based on scats and stable isotope analyses revealed that these two otariids share prey species (Franco-Trecu et al. 2013), despite almost no overlap in isotopic niche space (Franco-Trecu et al. 2012, 2013, 2014a). Sea lions feed at a higher trophic level, preferring near-shore resources, and with a higher degree of individual trophic specialization (Franco-Trecu et al. 2014a). Foraging areas of sea lions overlap with coastal fisheries (Riet Sapriza et al. 2013), which has led to conflicts with artisanal fishermen (Riet Sapriza et al. 2013; see also Crespo Chap. 24).

Fur seals (Arctocephalinae) are predominantly pelagic feeders (Costa and Gales 2003; Costa et al. 2001); while sea lions (Otariinae) tend to be benthic feeders that frequently forage near their aerobic dive limit (Hückstadt et al. 2016), and hence spend more energy gathering food than fur seals (Costa and Gales 2003; Costa et al. 2001; Costa and Toro Chap. 2). Therefore, sea lions have a smaller margin, compared to fur seals, to increase search effort if food resources become scarce due to fishing pressure or climatic variations (Costa et al. 2001). These almost

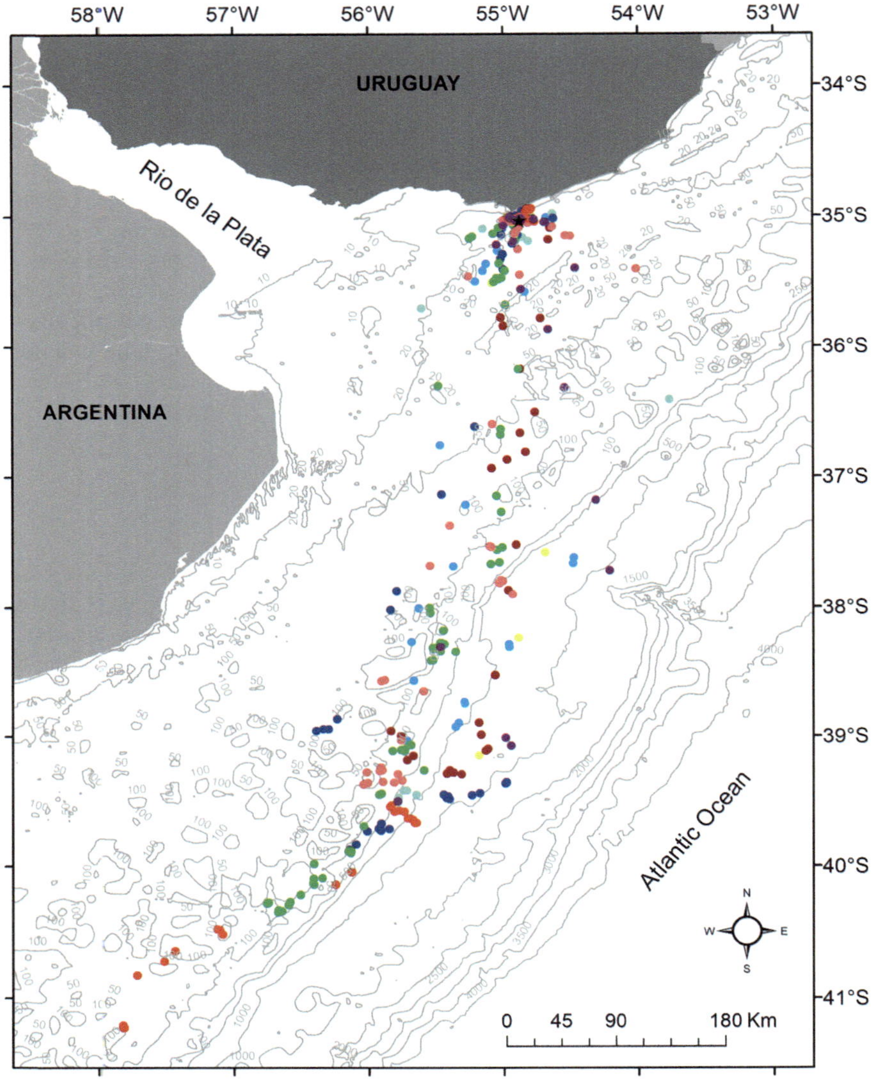

Fig. 29.3 Distribution at sea of adult South American fur seal females that departed from Isla de Lobos (asterisk). Colors code for different individuals. Animals travel to the edge of the shelf, that is, up to 500 km from the colony. Map by Caterina Dimitriadis

universal differences in foraging strategy may explain the different responses exhibited by their populations after commercial harvesting, with fur seal populations usually showing higher abundances and faster growth rates than those of sea lions (Costa et al. 2006; Costa and Toro Chap. 2).

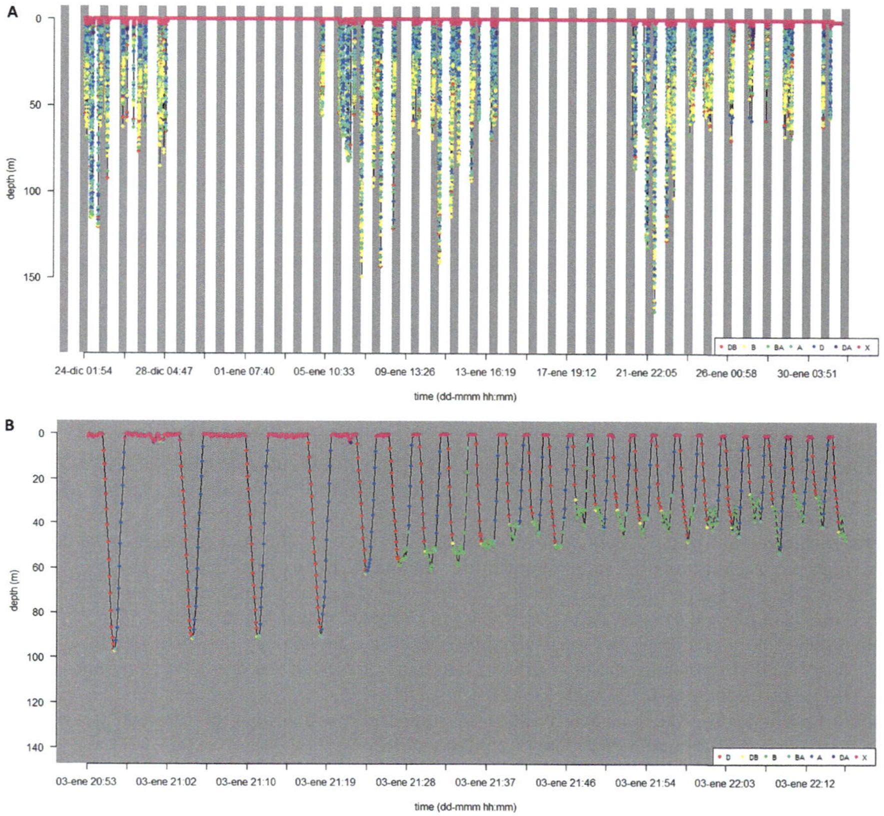

Fig. 29.4 (**a**) A typical dive profile from a lactating South American fur seal female during austral summer at Isla de Lobos, Uruguay. Complete dive profile for three foraging trips. The horizontal axis shows dates, while vertical axis shows depth. Gray bars designate night hours and white bars day hours. (**b**) A close up of a dive profile shows typical pattern of epi-pelagic divers (a V shaped dive pattern)

References

Allum LL, Maddigan FW (2012) Unusual stability of diet of the New Zealand fur seal (*Arctocephalus forsteri*) at Banks Peninsula. N Z Mar Freshw J 47:91–96

Arim M, Naya DE (2003) Pinniped diets inferred from scats: analysis of biases in prey occurrence. Can J Zool 81:67–73

Bearhop S, Adams CE, Waldrons S et al (2004) Determining trophic niche width: a novel approach using stable isotope analysis. J Anim Ecol 73:1007–1012

Bolnick DI, Svanbäck R, Fordyce JA et al (2003) The ecology of individuals: incidence and implications of individual specialization. Am Nat 161:1–28

Boyd IL (2000) State-dependent fertility in pinnipeds: contrasting capital and income breeders. Funct Ecol 14(5):623–630

Costa DP, Gales NJ (2003) Energetics of a benthic diver: seasonal foraging ecology of the Australian sea lion, *Neophoca cinerea*. Ecol Monogr 73:27–43

Costa DP, Gales NJ, Goebel ME (2001) Aerobic dive limit: how often does it occur in nature? Comp Biochem Physiol A Mol Integr Physiol 129:771–783

Costa DP, Weise MJ, Arnould JPY (2006) Potential influences of whaling on the status and trends of pinniped populations. In: Estes JA, Williams TM, Doak D, DeMaster D (eds) Whales, whaling and ocean ecosystems. University of California Press, Berkeley, CA, pp 344–359

Crespo EA et al (2015) Status, population trend and genetic structure of South American fur seals *Arctocephalus australis* in southwestern Atlantic waters. Mar Mamm Sci 31:866–890

DeNiro MJ, Epstein S (1978) Influence of diet on the distribution of carbon isotopes in animals. Geochim Cosmochim Acta 42:495–506

Franco-Trecu V (2015) Tácticas comportamentales de forrajeo y apareamiento y dinámica poblacional de dos especies de otáridos simpátricas con tendencias poblacionales contrastantes. PEDECIBA, Universidad de la República

Franco-Trecu V, Aurioles-Gamboa D, Arim M, Lima M (2012) Prepartum and postpartum trophic segregation between sympatrically breeding female *Arctocephalus australis* and *Otaria flavescens*. J Mammal 93(2):514–521

Franco-Trecu V, Drago M, Riet-Sapriza FG et al (2013) Bias in diet determination: incorporating traditional methods in Bayesian mixing models PLoS One. https://doi.org/10.1371/journal.pone.0080019

Franco-Trecu V, Aurioles-Gamboa D, Inchausti P (2014a) Individual trophic specialisation and niche segregation explain the contrasting population trends of two sympatric otariids. Mar Biol 161:609–618

Franco-Trecu V, Costa P, Schramm Y et al (2014b) Sex on the rocks: breeding tactics and reproductive success of. South American fur seal males. Behav Ecol 25(6):1513–1523

Franco-Trecu V, Drago M, Grandi MF et al (2019) Abundance and population trends of the South American fur seal (*Arctocephalus australis*) in Uruguay. Aquat Mamm 45:48–55

Hückstadt LA et al (2016) Regional variability in diving physiology and behavior in a widely distributed air-breathing marine predator, the South American sea lion, *Otaria byronia*. J Exp Biol. https://doi.org/10.1242/jeb.138677

Klare U, Kamler JF, Macdonald DW (2011) A comparison and critique of different scat-analysis methods for determining carnivore diet. Mammal Rev 41:294–312

Koen Alonso M, Crespo EA, Pedraza SN et al (2000) Food habits of the South American sea lion, *Otaria flavescens,* of Patagonia, Argentina. Fish Bull 98:250–263

Lima R, Franco-Trecu V, Vales D, Inchausti P, Secchi E et al (2019) Individual foraging specialization and sexual niche segregation in South American fur seals. Mar Biol 116:32

Naya DE, Arim M, Vargas R (2002) Diet of South American fur seals (*Arctocephalus australis*) in Isla de Lobos, Uruguay. Mar Mamm Sci 18:734–745

Ponce de León, A. 2000. Taxonomía, sistemática y sinopsis de la biología y ecología de los pinipedios de Uruguay. En: Rey, M. y F. Amestoy (Ed.). Sinopsis de la biología y ecología de las poblaciones de lobos finos y leones marinos de Uruguay. Pautas para su manejo y Administración. Proyecto URU/92/003. Instituto Nacional de Pesca – Programa de las Naciones Unidas para el Desarrollo. 117 pp.

Post DM (2002) Using stable isotopes to estimate trophic position: models, methods, and assumptions. Ecology 83(3):703–718

Riet Sapriza FG, Costa DP, Franco-Trecu V et al (2013) Foraging behavior of lactating South American sea lions, *Otaria flavescens* and spatial-resource overlap with the Uruguayan fisheries. Deep-Sea Res II 88-89:106–119

Santos MM, Juares MA, Rombola EF et al (2012) Over-representation of bird prey in pellets of South Polar Skuas. J Ornithol 153:979–983. https://doi.org/10.1007/s10336-012-0840-4

Thompson ME, Muller MN, Wrangham RW (2012) The energetics of lactation and the return to fecundity in wild chimpanzees. Behav Ecol 23:1234–1241. https://doi.org/10.1093/beheco/ars107

Trites AW, Joy R (2005) Dietary analysis from fecal samples: how many scats are enough? J Mammal 84:704–712

Chapter 30
Galapagos Sea Lions and Fur Seals, Adapted to a Variable World

Marjorie Riofrío-Lazo and Diego Páez-Rosas

Abstract Galapagos otariids (Galapagos sea lions *Zalophus wollebaeki*, and Galapagos fur seal *Arctocephalus galapagoensis*) have developed a number of behavioral adaptations to living in a tropical ecosystem exemplified by high environmental variability and subsequent oceanographic disturbances that lead to unpredictable marine productivity. Both species have been subject to selection pressures to reduce overall body size and energy requirements compared to otariids living in more predictably productive systems. The environmental stochasticity also appears to have modified both reproductive and foraging behavior. Galapagos otariids respond to limited food supplies with prolonged lactation, thus enhancing the chances of pup survival, and have developed trophic specialization with a high level of flexibility that reduces interspecific competition, combined with prey-switching during El Niño events. Galapagos fur seal mothers adjust the length of foraging trips and the duration of visits to attend pups in response to fluctuations in food. Galapagos sea lions reduce competition among individuals by displaying diverse foraging strategies in relation to prey availability. These mechanisms make life viable in the Galapagos Islands. Although the populations of these otariids have slightly increased in the last 5 years, they may not have recovered yet from the drastic reductions caused by El Niño events during the years 1982/83 and 1997/98. This shows the high vulnerability of these otariids to temporal changes in environmental conditions of the region and how the natural climatic variation of this ecosystem is an important stressor factor of their populations.

M. Riofrío-Lazo (✉)
Universidad San Francisco de Quito, Galapagos Science Center, Isla San Cristóbal, Islas Galápagos, Ecuador
e-mail: mriofriol@usfq.edu.ec

D. Páez-Rosas
Universidad San Francisco de Quito, Galapagos Science Center, Isla San Cristóbal, Islas Galápagos, Ecuador

Dirección Parque Nacional Galápagos, Unidad Técnica Operativa San Cristóbal, Isla San Cristóbal, Islas Galápagos, Ecuador

C. Campagna, R. Harcourt (eds.), *Ethology and Behavioral Ecology of Otariids and the Odobenid*, Ethology and Behavioral Ecology of Marine Mammals,
https://doi.org/10.1007/978-3-030-59184-7_30

Keywords Environmental stochasticity · Foraging patterns · Trophic specialization · Population trends · Behavioral adaptations · Galapagos · Fur seals · Sea lions · El Niño

30.1 Introduction

The Galapagos sea lion (*Zalophus wollebaeki*) and Galapagos fur seal (*Arctocephalus galapagoensis*) are otariids endemic to the Galapagos Archipelago and adapted to this hotspot of local productivity in the midst of a tropical environment in the equatorial Pacific (Churchill et al. 2014). The Galapagos Archipelago is a region where levels of marine productivity are strongly influenced by seasonal oceanographic currents and by a pattern of upwelling that turn this area into a biodiversity hotspot for marine predators (Palacios et al. 2006). This unique oceanographic setting has contributed to sporadic colonization of the islands leading to the evolution and presence of the divergent species such as the Galapagos penguin (*Spheniscus mendiculus*) and the Galapagos otariids with ecological characteristics that are not otherwise typical of a low-latitude ecosystem (Okey et al. 2004; Nims et al. 2008).

Although the oceanic waters around the Galapagos Islands are considered to be an upwelling system in an equatorial region (Palacios et al. 2006; Schaeffer et al. 2008), they are less productive and more unpredictable than higher latitude upwelling systems (Martin et al. 1994; Sakamoto et al. 1998). This creates an ecological challenge for the otariids living in the Galapagos, as they depend on an ocean with poorer conditions than other otariids living at higher latitudes (Villegas-Amtmann et al. 2017).

Moreover, oceanographic-atmospheric disturbances such as the El Niño and La Niña events, which are opposite effects of the ENSO (El Niño Southern Oscillation), occur periodically in Galapagos, causing fluctuations in marine productivity with resulting demographic impacts on top predators (Trillmich and Dellinger 1991; Riofrío-Lazo et al. 2017). The intensity, length and frequency of ENSO events vary with time (Pennington et al. 2006) but appear to be increasing, probably influenced by global warming and/or larger oscillations over 25-year periods (Capotondi et al. 2015). The environmental variability and unpredictability in the productivity of the Galapagos Archipelago have dramatically affected the abundance and life history strategies of Galapagos otariids (Páez-Rosas et al. 2012).

Otariids respond quickly to changes in environmental conditions (Moore 2008). In the Galapagos, these species can be used as indicators of the health of the system in the areas that they inhabit (Drago et al. 2016; Páez-Rosas and Guevara 2017). Fur seals and sea lions depend on terrestrial habitat for breeding and rearing offspring, and the marine habitat for feeding and moving around; therefore, they are exposed to a greater number of environmental drivers affecting survival (Montero-Serra et al. 2014).

Currently, the two species of Galapagos otariids are among the most important species listed as targets of conservation efforts in the Archipelago (Páez-Rosas and Guevara 2017). Both species are red listed as Endangered by the International Union for Conservation of Nature (IUCN) as their populations have undergone drastic declines (50%) since 1978 (Trillmich 2015a, b). This is in part due to episodes of El Niño events combined with anthropogenic stressors that contribute to the deterioration of their habitats (Trillmich and Dellinger 1991; Alava and Salazar 2006; Riofrío-Lazo et al. 2017).

In the face of these fluctuating environmental conditions producing intense selective pressures on survival and reproduction, Galapagos otariids have evolved mechanisms to cope with the uncertainty of their environment. We provide an overview of their biology focusing on the behavioral adaptations on which they rely for continuity.

30.2 Galapagos Fur Seals

This, the smallest of all otariids, breeds on eight rookeries in the western and northern islands (Fig. 30.1). Most (95%) of its population is in Fernandina (Cabo Hammond and Cabo Douglas) and north of Isabela Island (Punta Vicente Roca and Cabo Marshall). It is also found in small numbers on other islands such as Genovesa and Rábida islands. The species is sympatric with the Galapagos sea lion on Fernandina, Isabela, Santiago, Rábida, Pinta, Marchena and Genovesa (Páez-Rosas et al. 2012). The western and northern islands are situated in the region of strongest upwelling in the Galapagos Archipelago with cool waters, average Sea Surface Temperature (SST) around 18 °C and high productivity (Palacios et al. 2006).

The fur seal of the Galapagos is closely related to the South American fur seal (*A. australis*). Indeed, it was considered a subspecies of this otariid based on morphometric cranial studies (Brunner 2003). Subsequent genetic studies suggested a relatively long divergence time between these two taxa (0.7 million years; Higdon et al. 2007), which provided support to classify them as different species (Oliveira and Brownell 2014). Further studies are required for better understanding the taxonomic status of Galapagos fur seal (Berta and Churchill 2012).

Galapagos fur seals are typical fur seals, being polygynous and sexually dimorphic (Fig. 30.2, Table 30.1). They live in small rookeries throughout the year and exhibit high fidelity to their breeding rookeries (Páez-Rosas 2011). They do not migrate, although vagrants have been sighted along the coast of Ecuador, Peru, Colombia, Costa Rica, El Salvador and Mexico (Capella et al. 2002; Aurioles-Gamboa et al. 2004; Félix et al. 2007; Montero-Cordero et al. 2010; Ibarra et al. 2016; Páez-Rosas et al. 2017a). They rest, thermoregulate and breed on rocky shores, near deep productive waters, and use crevices, caves, large boulders and rock ledges for shade and shelter (Trillmich 1987).

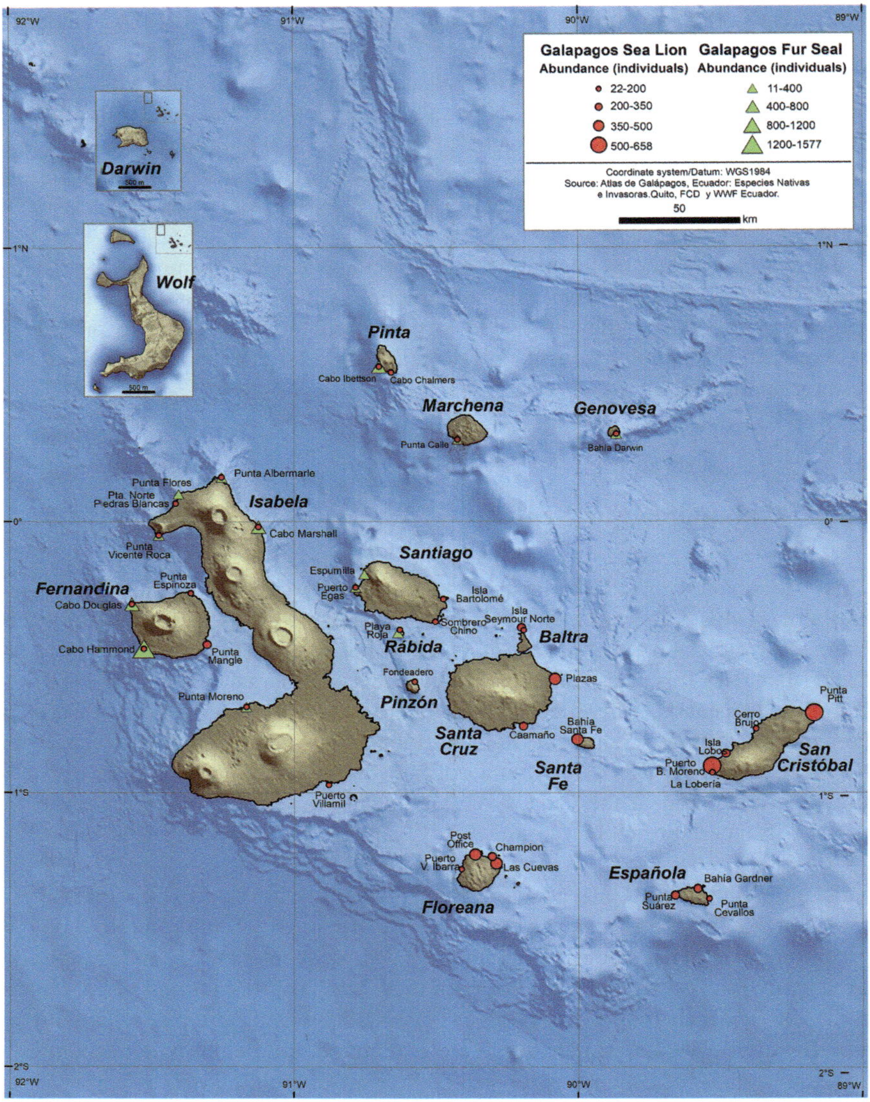

Fig. 30.1 Map of the Galapagos Islands showing all breeding rookeries of Galapagos sea lions (*Zalophus wollebaeki*, red symbols) and Galapagos fur seals (*Arctocephalus galapagoensis*, green symbols) in the archipelago. The symbol size is according to the individuals' abundance in each rookery (Map courtesy of Byron X. Delgado, GIS Research/Knowledge Management, CDF)

Male territories are small, about 20m^2 and must have access to water (Bonner 1984). Territorial males generally return to their previous territory for up to 3 subsequent years, while subadult males roam throughout the north and western of archipelago until they are physically mature and large enough to compete for a

Fig. 30.2 Galapagos fur seals (*Arctocephalus galapagoensis*) have strong sexual dimorphism and form small groups associated with rocky slopes. Left: adult male *A. galapagoensis* arriving at the Cabo Marshall rookery (Isabela Island) after its feeding trip. Right: a smaller adult female vocalizing while taking care of its territory on Fernandina Island (Photos courtesy of Diego Páez-Rosas, Galapagos Pinnipeds Project, USFQ-DPNG)

Table 30.1 Descriptive data (mean values) for adult Galapagos fur seals

Sex category	Weight (kg)	Length (cm)	Age of sexual maturity (year)	Age of reproductive individuals (year)	Longevity (year)
Males	70	150	6	10 ± 2	20
Females	40	120	4	10 ± 2	20

Weight and length values taken from Trillmich (1986a). Age of sexual maturity, age of reproductive individuals and longevity taken from Trillmich (2015b)

territory (Horning and Trillmich 1999). Males do not maintain the territory for the whole reproductive season but may extend their stay by making short foraging trips at night. The period of territory tenure decreases with the following reproductive season from 27 to 15 days in median (Aurioles-Gamboa and Trillmich 2018).

Females give birth to one pup per season between August and November (Trillmich 1987), after 11 months of gestation (Trillmich 1987). Mother-pup recognition is established by an intense calling activity within the first days after parturition (Trillmich 1986a). Mother often makes a final olfactory check before allowing a pup to suckle and exclusively suckles her pup, rejecting non-fillial young, often aggressively (Trillmich 1986a, Tyack 2001). The maternal care strategy involves cycles of nursing interspersed with periods at sea. This "feeding cycle" consists of a perinatal period of 7 ± 1.2 days during which the mother stays with the pup, followed by very short feeding trips by the mother (17.9 ± 10.6 h) with returns to feed the pup (Trillmich 1986a; Villegas Amtmann et al. 2013). Pups depend on their mothers longer than any other fur seal species. Maternal investment increases with the age of the young and is higher among mothers rearing sons (Trillmich and Wolf 2008). Weaning occurs 18–36 months after birth, depending on environmental conditions (Trillmich 1987). This long investment of mothers in the pups has been hypothesized to be an adaptation to cope with a seasonally reduced food availability in the Galapagos that makes it difficult for young animals to find food on their own. It may also be a strategy to reduce risk of mortality by predation of young by sharks

(Trillmich and Wolf 2008), by reducing the time that naive young spend foraging in the sea.

Mothers often nurse older young and newborn simultaneously. Many of these newborns die due to competition for milk with yearlings (Trillmich and Wolf 2008). Thus, females reproduce successfully only on average every other year. Being born in the presence of an older sibling is reflected in decreased growth rates during early life and reduced survival prospects of the younger sibling, but represents insurance against loss of older pup or extra reproductive value (Trillmich and Wolf 2008). The effects on the newborn's body mass are much stronger during warmer El Niño years and if the older sibling is male. Mothers aggressively defend the younger sibling during the conflict between siblings of different cohorts. Maternal aggression forcing the older sibling to wean only occurs if it reaches a size big enough to feed by itself (Trillmich and Wolf 2008).

30.2.1 Foraging Ecology

Before giving birth, females spend about ten days at sea feeding, a relatively long period compared with the usual duration of foraging trips. Time at sea has a clear relationship with the lunar cycle. The shortest trips (10–20 h) occur around full moon, and the longest (40–70 h) between the decreasing half-moon and new moon (Trillmich 1986a). The proximity of resources to the rookeries enables females to spend a shorter time at sea and return more frequently to land to attend the offspring. In years of unusually good feeding conditions (for example in 1981 when yearlings of both sexes were almost as heavy as 2-year-olds in previous years), females apparently regulated feeding efforts by staying ashore longer rather than by shortening their foraging trips (Trillmich 1986a). By contrast, during anomalous conditions (El Niño years) females extend their trips to sea (Trillmich and Limberger 1985).

Foraging fur seals tend to return to the same foraging grounds in part due to the fact that the areas of greatest productivity are located mainly off the western side of the Archipelago (Páez-Rosas et al. 2012, 2014a, b). Females and juveniles restrict their foraging trips to areas located less than 70 km from the coast (Jeglinski et al. 2013; Villegas-Amtmann et al. 2013), while males undertake longer trips, particularly during the months leading up to the breeding season (Trillmich 1987).

Galapagos fur seal dives are shallow (usually less than 30 m) and at night (Trillmich and Limberger 1985). They feed on organisms from the deep scattering layer, mainly small cephalopods (*Onychoteuthis banksii, Dosidicus gigas*) and myctophids (Dellinger and Trillmich 1999; Páez-Rosas et al. 2012). Juveniles and adults feed on a wide range of prey items at low trophic levels (Clarke and Trillmich 1980; Dellinger and Trillmich 1999).

In rookeries where Galapagos fur seals live in sympatry with sea lions, they partition feeding habitats (Páez-Rosas et al. 2012), display different foraging strategies and target different prey (Villegas Amtmann et al. 2013). This habitat specialization is already present during juvenile stages in the Galapagos fur seal (Jeglinski et al. 2013). During strong El Niño or La Niña years, some overlap may occur in pelagic foraging zones but not in prey items as prey specialization persists (Páez-Rosas et al. 2012).

30.2.2 Demographics

Fur seals were intensively exploited during the nineteenth century. The population was drastically reduced but recovered later during the twentieth century (Alava et al. 2017). This demographic decline could, however, have reduced genetic variability, exposing them to the risks of consanguinity and genetic drift (Lopes et al. 2015). The Galapagos fur seal population shows strong matrilineal structure where close rookeries display isolated mitochondrial haplotypes and it appears that it is the males that ensure genetic flow in the population by their movements among the rookeries (Lopes et al. 2015).

The total population of the Galapagos fur seal based on counts in the last 5 years (2014–2018) is 2317 ± 455 individuals (Páez-Rosas et al. unpublished data). The total population in the archipelago has always been low in comparison to Galapagos sea lions even though the casual observer might be misled as in sympatric rookeries (Cabo Hammond and Cabo Douglas), the number of Galapagos fur seals is approximately five times that of Galapagos sea lions (Páez-Rosas et al. unpublished data).

Galapagos fur seal population abundance declined 74–80% over a period of 23 years (2.3 generations; Trillmich 2015a, Table 30.2). Population monitoring 2002 to 2005 did indicate a reduction of more than 60% in the last 37 years relative to counts from 1978 (Alava et al. 2017) however, these results are not the best estimate of Galapagos fur seal abundance (Trillmich 2015a).

Since 2014 to the present, the method of counting fur seals and sea lions has been standardized to direct counts in the rookeries during annual censuses of the entire Archipelago. The fur seal population seems to have remained stable over the past 5 years. The average count for 2014 to 2018 is close to the count in 2001, suggesting few changes in abundance over the last 17 years. This still represents a reduction of

Table 30.2 Galapagos fur seal population abundance recorded during different censuses

Census year	Population counted	Population estimated	Reference
1978	9785	30,000	Trillmich (1987)
2001	2733	6000–8000	Alava and Salazar (2006)
2002–2005	–	15,000	Salazar and Denkinger (2010)
2014–2018	2317	–	Páez-Rosas et al. unpublished data

Population counted is the total of animals counted in the rookeries and/or at sea
Population estimated is the total population size based on correction factors applied to the counts in 1978 and 2001, and based on extrapolation from 2002–5 main rookery surveys. The methods for population estimates are different by authors and it is not clear how they made the calculus. For 2014 to 2018, the population counted is the average value during this period

more than 50% of the population over the past four generations (1978–2018) and so the IUCN conservation status of the species remains Endangered.

The population reduction may in part be explained by the impacts of strong El Niño events in 1982/83 and 1997/98, which caused acute shortages of prey and high rates of mortality (Trillmich and Limberger 1985; Trillmich and Dellinger 1991; Alava and Salazar 2006). Since then, El Niño events of variable intensity have been recorded every 4–5 years, which may explain a lack of further recovery (Riofrío-Lazo et al. 2017).

30.3 Galapagos Sea Lion

30.3.1 Species Description and Reproduction

The Galapagos sea lion is the most abundant of the two otariids of the Galapagos Islands. It is distributed throughout the archipelago (Fig. 30.1). Its larger rookeries are located in islands of the southeastern region (San Cristóbal, Floreana, Española and Santa Fe) (Riofrío-Lazo et al. 2017), under the influence of the Humboldt Current. Sea Surface temperatures are cool for a tropical region (average 23 °C), albeit not as cool as those of the western region (Palacios et al. 2006).

The Galapagos sea lion was in the past considered a subspecies of the California sea lion (*Zalophus californianus*), but new genetic evidence suggests a separation of 2.5 ± 0.5 million years among the three species of *Zalophus* genus (*Z. californianus*, *Z. wollebaeki* and *Z. japonicas*, extinct; Wolf et al. 2007a). Moreover, recent studies suggest a degree of genetic and morphological divergence between Galapagos sea lion populations from the western region (Fernandina and Isabela) and the central and southeastern regions of the archipelago (Wolf et al. 2008; Jeglinski et al. 2015). The difference in the genetic structure among Galapagos sea lion populations is the result of intra- and inter-specific niche segregation rather than geographic barriers (Wolf et al. 2008).

Like most pinnipeds, the sea lion of the Galapagos shows marked sexual size dimorphism (Fig. 30.3, Table 30.3) already evident at birth, when females weigh on average 5.8 kg and males 6.7 kg (Kraus et al. 2013). Sea lions are highly gregarious, settling on islands, islets and sites protected from predators, forming rookeries maintained throughout the year. Their territories include aquatic habitat for thermo-regulation and copulation, and terrestrial habitat for resting, breeding and nursing pups (Montero-Serra et al. 2014). They are a non-migratory otariid, with vagrants recorded along the coasts of Mexico, Colombia and Ecuador (Ceballos et al. 2010).

Galapagos sea lions are polygynous, and show high fidelity to breeding rookeries (Páez-Rosas 2011). A typical rookery has three zones: one of juveniles, one of bachelors, and the breeding area (Páez-Rosas 2011). Territorial adult males arrive at the rookery early in the reproductive period, and force adult and sub-adult males to occupy marginal areas (bachelors' zone; Kunc and Wolf 2008). This competition among males, mainly during the reproductive periods, leads to sexual segregation in

Fig. 30.3 Galapagos sea lions (*Zalophus wollebaeki*) have large sexual dimorphism and form dense aggregations along the shoreline. Left: adult male *Z. wollebaeki* taking care of its territory at the Punta Pitt rookery (San Cristóbal Island). Right: a smaller adult female enters its rookery on Española island (Photos courtesy of Diego Páez-Rosas, Galapagos Pinnipeds Project, USFQ-DPNG)

Table 30.3 Descriptive data (range values) for adult Galapagos sea lions

Sex category	Weight (kg)	Length (cm)	Age of sexual maturity (year)	Age of physical maturity (year)	Longevity (year)
Males	150–200	170–210	4–6	8–9	17
Females	60–95	156–176	4–6	8–9	20–22

Weight and length values were taken from Trillmich et al. (2014). Age of sexual and physical maturity were taken from Peterson and Bartholomew (1967). Longevity was taken from Riedman (1990)

the rookeries (Trillmich et al. 2014). Females are distributed irregularly in clusters or aggregations along the beaches. Juveniles gather away from territories (Wolf et al. 2005).

Male territories occur where female densities are highest. In contrast to some other otariid species, male mating success may depend on being chosen by females instead of copulating with each receptive female that goes into oestrus on their territories (Trillmich et al. 2014). The consequence of this reproductive system is that non-territorial males sire more than 50% of the annual number of pups (Pörschmann et al. 2010). This is related to the longer time spent by non-territorial males compared to territorial males in the rookery. The long reproductive season of about 5 months reduces the opportunity for individual males to monopolize large numbers of females and successfully sire offspring (Pörschmann et al. 2010). Territorial males tend to return each season to the territories where they had previously obtained matings (Fig. 30.4a), whereas non-territorial males depend upon their ability to sneak into the neighboring territories (Fig. 30.4b; Wolf et al. 2005; Meise et al. 2013).

Areas occupied by aggregations of females and their descendants are independent of the established territory boundaries and are explained by the similar space use of the aggregation members (Wolf et al. 2007b). Females from the same aggregation are more related than individuals from different groups. This is because the

Fig. 30.4 (**a**) Galapagos sea lion rookeries have territorial males (*) surrounded by females and young in inshore areas. These animals remain active and vocalizing to prevent the presence of other males. (**b**) Areas outside the rookeries where non-territorial males (*) are generally resting. These animals are waiting to be able to enter the rookeries and copulate with available females (Photos courtesy of Diego Páez-Rosas, Galapagos Pinnipeds Project, USFQ-DPNG)

aggregations are made up of mother-offspring pairs and a high number of siblings and half siblings (Wolf and Trillmich 2008). Adult males are less likely to meet in groups of close relatives as they move away from their home ranges (Trillmich et al. 2014). However, the spatial habitat preferences, the limited availability of shady places for thermoregulation and the high overlap of their home ranges within the rockery provide the opportunity for long-lasting social relationships among males (Meise et al. 2013).

Females reproduce annually, giving birth to one pup, after 11 months of gestation. The breeding season (August–January) varies slightly depending on the island (Trillmich et al. 2014). Births tend to begin earlier in the western and later in the southeastern part of the Galapagos (Trillmich 1986b). Most of the births occur in September to November but may also extend into January, as has been observed on San Cristobal Island in the east (Drago et al. 2016) and Caamaño Islet in the center (Pörschmann et al. 2010). Yearlings compete with newborns for mother's milk, and that often results in high pup mortality (Trillmich and Wolf 2008). Mothers may thus lose the pup of the year, reproducing successfully only every two to three seasons. On rare occasions a trio is formed (Trillmich and Wolf 2008), in which a female can successfully breastfeed the newborn and the older offspring.

After birth, the females' behavior is highly aggressive defending the area around the newborn and calling frequently to it (Trillmich et al. 2014). Females stay on land with the newborn pups for a 4–7 day perinatal period. This period helps to establish an exclusive, mutual bond between mother and pup (Trillmich 1981). Females are income breeders; they show a feeding cycle that alternates foraging trips that last from few hours to 4 days, and that increase with pup's age, with suckling bouts on land lasting 5 hours on average (Trillmich 1986b; Villegas-Amtmann et al. 2008; Jeglinski et al. 2012: Piedrahita 2015).

Lactation generally lasts for more than 1 year and may continue up to 3 years (Trillmich and Wolf 2008). That means females are nursing all their reproductive lives as long as the pup or juvenile offspring survives (Trillmich et al. 2014). Females respond to the energy demand of offspring growth by increasing their time at sea. There are no differences in the attendance behavior between mothers rearing sons and daughters (Piedrahita et al. 2014). The higher energy requirements of sons compared to the daughters are covered by the mother increasing her efficiency in the use of time outside the rookery, resting less on land and diving more intensely when feeding. Sons are more efficient than daughters in the use of milk transferred by the mother through a more efficient digestion or a lower metabolic rate (Piedrahita et al. 2014).

Yearlings offspring compensate the reduction in maternal milk intake by increasing solid food consumption through independent foraging dives (Jeglinski et al. 2012). Their time budget differs according to sex; pup males contribute much less to their own maintenance requirements than pup females as they show less independent foraging during the latter part of lactation (Piedrahita et al. 2014). Yearling diving behavior is different when the mother is present than when she is absent. Juveniles most often stay ashore after the mother leaves the rookery, so they do not develop their foraging skills through observational learning of maternal foraging behavior (Piedrahita et al. 2014).

30.3.2 Foraging Behavior and Feeding

The Galapagos sea lion is a highly philopatric species but with great mobility to exploit different environments during its foraging trips (Villegas-Amtmann et al. 2008; Jeglinski et al. 2013). Foraging trips are relatively short compared to other sea lions, with an average duration of 44 h at sea alternated with stretches on land between 7 and 48 h (Páez-Rosas et al. 2017b). Juveniles and adults may haul out at numerous sites, including other sea lion rookeries (Villegas-Amtmann et al. 2008) but return to their home rookeries at the beginning of the breeding season. The maximum distance recorded from the rookery for adult females ranged from 50 to 97 km, while juveniles travelled maximum distances of less than 15 km (Villegas-Amtmann et al. 2008; Jeglinski et al. 2013; Páez-Rosas et al. 2017b).

Galapagos sea lions forage in pelagic and benthic zones, on the shelf of the archipelago and along its edge in the western part (Fig. 30.5; Villegas-Amtmann et al. 2008, 2013; Jeglinski et al. 2012, 2013; Páez-Rosas et al. 2017b). Adult females forage day and night but predominantly at night; and juveniles forage to shallow depths (Jeglinski et al. 2012; Páez-Rosas et al. 2017b). Despite being the smallest of the sea lions, they are among the deepest divers (Villegas-Amtmann et al. 2017): yearlings dive to a maximum depth of 367 m, while adult females reach ~600 m (Jeglinski et al. 2012; Páez-Rosas et al. 2017b).

Adult females differ in dive behavior among rookeries. Females display at least three foraging strategies based on the dive depth: epipelagic, mesopelagic, and benthic (Villegas-Amtmann et al. 2008). This suggests that individuals may have different diets within the same population (Páez-Rosas and Aurioles-Gamboa 2010; Páez-Rosas et al. 2017b). Diving behavior varies with rookery and region of the archipelago, and is likely related to prey availability, oceanographic characteristics

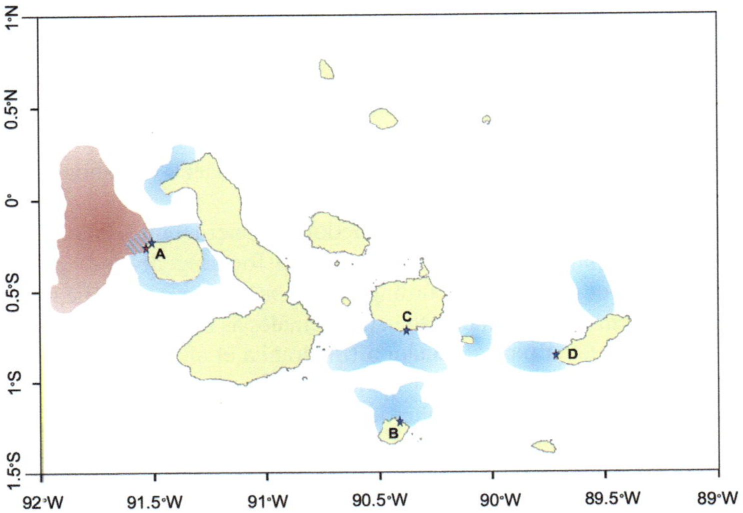

Fig. 30.5 Map of the foraging ranges of Galapagos sea lions (blue) and Galapagos fur seals (red) based on 95% contour kernel analysis. The stars represent the more important rookeries in the archipelago: (**a**) Cabo Douglas—Fernandina, (**b**) Post Office—Floreana, (**c**) Caamaño—Santa Cruz, and (**d**) El Malecón—San Cristóbal Island. The area of overlap between both species is represented by a mixed texture near Fernandina Island. The probability of overlap gradually decreases due to offshore/pelagic strategy of Galapagos fur seals (Data shown in this image derive from Villegas-Amtmann et al. 2008, 2013; Jeglinski et al. 2012, 2013; Páez-Rosas et al. 2017b)

and the topography of the area (Páez-Rosas et al. 2014a, 2017b). In this manner, niches are partitioned, and this may reduce competition, particularly among females, intensified by the extended breeding period and pup care that prevails in this species (Páez-Rosas and Aurioles-Gamboa 2010).

The Galapagos sea lion diet is composed of a large variety of small benthic and pelagic fish belonging to the Clupeidae, Myctophidae, Carangidae, Engraulidae and Serranidae (Dellinger and Trillmich 1999; Páez-Rosas and Aurioles-Gamboa 2010, 2014). Despite this wide trophic spectrum, a degree of trophic specialization is identified in the diet of individuals as a result of the high frequency of consumption of a limited number of prey items (Páez-Rosas and Aurioles-Gamboa 2010; Páez-Rosas et al. 2017b). At the center of the archipelago, the diet is different than that of animals from the western and southeastern sites. In the central region, sea lion females display an inshore foraging strategy, mainly feeding on sardines (*Sardinops sagax*) and myctophids (Dellinger and Trillmich 1999; Páez-Rosas and Aurioles-Gamboa 2014). In the southeast, female sea lions feed on prey from different environments, including benthic (*Semicossyphus darwini*) and pelagic (*S. sagax*) fish. In the west, they feed primarily on benthic fish (Páez-Rosas et al. 2014b). These ecological differences among Galapagos sea lion populations of these regions concord with the genetic differentiation found in this species (Wolf et al. 2008).

In rookeries in the western region, where the sea lions live in sympatry with fur seals, the two species specialize within different foraging niches (Dellinger and Trillmich 1999; Páez-Rosas et al. 2012). This may decrease interspecific competition for food. However, the diet overlaps during anomalous El Niño years, with the sea lions feeding on a variety of habitats and species at similar trophic levels to those consumed by the fur seals (Páez-Rosas et al. 2012). This suggests the adaptive capacity of sea lions to life in a variable ecosystem (Páez-Rosas et al. 2017b).

30.3.3 Population Trends

The Galapagos sea lion breeds on almost all the islands of the archipelago. Approximately 56% of the population inhabits the southeastern region (Riofrío-Lazo et al. 2017). Population estimates differ from the first census in 1978 until more recent counts, probably reflecting differences in the methods used.

In 1978, a direct count of 8000 animals suggested an estimated population of 30,000–40,000 individuals (Trillmich 1979). In 2001, the direct count was 4937 sea lions, resulting in an estimated 14,000–16,000 individuals (Alava and Salazar 2006). As the rationale behind the correction factors was not provided, the estimates may not be compared. Even with differences in collection methods the data suggest that the population had declined substantially (Trillmich 2015b).

The El Malecón rookery, at San Cristobal Island, is the largest in the archipelago. Population monitoring has been conducted annually, since 2005 and there has been an increase of 2% per year, from 2005 to 2015 (Riofrío-Lazo et al. 2017). Despite this small increase, the rookery may still be affected by anthropogenic disturbances. It is near the town, and is also exposed to high environmental variability (Páez-Rosas and Guevara 2017).

Since 2014, counts have followed a standardized method. In 2015, 2300 sea lions were recorded in the southeastern region (Riofrío-Lazo et al. 2017). The current population size in the entire archipelago based on counts over the past five years (2014–2018) is 4500 ± 427 individuals (Páez-Rosas et al. unpublished data), an increase in total population of 3% per year, from 2015 to 2018. As counts for 2014 to 2018 remain close to that recorded in 2001, the population may have been relatively stable in the last 17 years. If so, this suggests a reduction of more than 50% of the population over the last four generations (1978–2018). Thus, the Galapagos sea lion remains an IUCN Endangered species and like the Galapagos fur seal, there may be residual impacts of the catastrophic 1982/83 and 1997/98 strong El Niño events (Trillmich and Dellinger 1991; Alava and Salazar 2006; Páez-Rosas et al. 2012; Trillmich and Limberger 1985). Like the Galapagos fur seal, the recurrence of El Niño events every 4–5 years may have inhibited the ability to recover from the two strong events of 1982/83 and 1997/98.

30.4 Adaptations of Galapagos Otariids to Ecosystem Uncertainties

The Galapagos fur seal and sea lions are the smallest of all fur seal and sea lion species. This may be an adaptation to life in a tropical, less productive environment than other species (Trillmich et al. 2014). The energetic costs of living in a warm equatorial climate, with lower resource availability, may have selected for a reduction in metabolic rate in relation to other otariids (Trillmich and Kooyman 2001). This highlights the fundamental role of the physical environment in shaping the physiology of these species (Trillmich and Kooyman 2001; Villegas-Amtmann et al. 2017). Lower energy requirements may be an adaptation to reduced prey availability.

Reduced prey availability causes females to increase their foraging effort, and spend more time at sea to obtain the same amount of prey than other species in less time. If a lactating female spends more time to deliver the same amount of energy, the pup receives less overall energy. As more of the pup's energy is spent on maintenance, pup growth is slow and pup survival high (Costa 2009). During strong El Niño years, female Galapagos fur seals prolong the duration of foraging trips up to three times longer than in normal years, and this causes rapid undernourishment of their pups and even subsequent mortality (Trillmich and Limberger 1985).

Female Galapagos fur seals also show behavioral adaptations to fluctuations in food abundance over the short- and long-term (days and weeks). They apparently respond to short-term fluctuations by adjusting the length of foraging trips. They respond to long-term changes, either energy demands of the young or in food abundance, by changing the duration of visits to the rookery. In this way, it is probable that fur seals spend less time ashore and make more foraging trips during the warm season (January to May) when the food supply is lower (Trillmich 1986a).

El Niño effects include drastic reductions in population (mainly the younger age classes, which may disappear completely), severe malnutrition accompanied by a reduction in the growth rate of pups, and changes in the use of their foraging areas and diet (Trillmich and Dellinger 1991; Páez-Rosas et al. 2012, 2017b). Related to the growth of pups, there is a close relationship between SST and the early development of Galapagos sea lion pups, with a decrease in mass and growth rate during the first 60 days of life with an increase of SST (Kraus et al. 2013; Mueller et al. 2011). Thus, during warm conditions we find lower weight and smaller pups.

Longer lactation is energetically more expensive overall, but may counteract the seasonal availability of food resources in the Galapagos. Both Galapagos fur seals and the Galapagos sea lion are among the otariids with the longest lactation periods (2–3 years; Trillmich 1987). Galapagos sea lion 1- and 2-year-old individuals often continue to suckle and weaning occurs at a time when they are independently foraging (Trillmich and Wolf 2008). The extension of maternal investment in these otariids is a response mechanism to increase the chances of pup survival when animals are exposed to food stress due to drastic changes in the availability of prey in their habitats (Gentry and Kooyman 1987; Trillmich and Wolf 2008). The diversification of feeding strategies in the Galapagos sea lion is a mechanism to reduce intra-specific competition even between individuals of the same rookery, and may be an adaptation to prey availability and topography of their feeding grounds (Páez-Rosas et al. 2017b).

Trophic specialization of Galapagos sea lions decreases the level of inter-specific food competition with Galapagos fur seals. These species exploit distinct foraging niches and their diets do not overlap except during anomalous years (Páez-Rosas et al. 2012). This occurs when El Niño events drive changes in abundance and distribution of their main prey species to force both otariids to feed on prey at a similar trophic level and in a similar region (Dellinger and Trillmich 1999; Páez-Rosas et al. 2012). This points to the high level of foraging flexibility that Galapagos sea lion has developed to incorporate ecological equivalents to its diet and counteract the lack of food in its habitat, and thus increase its survival in a highly demanding ecosystem in terms of prey availability (Páez-Rosas et al. 2017b).

These adaptation mechanisms show how Galapagos otariids have been able to persist in the archipelago even with limited resources and under variable climatic conditions. Hence, they can be considered bio-indicators of changes in the trophic web of this ecosystem influenced by environmental variability and short-term warming events (Páez-Rosas et al. 2012, 2014a). Continued monitoring of these species will help inform our understanding of localized strong ecological effects arising from global climate change.

Acknowledgements We thank the Galapagos National Park for granting us research permits to collect the data exposed in this chapter, and for help provided during the logistical phase. We also thank the Universidad San Francisco de Quito for financial and logistical support during the preparation of this manuscript. We are grateful to Jorge Torres and the GNP rangers for their help in collecting and processing the data exposed in our work, as well as Byron X. Delgado and the Knowledge Management Department of the Charles Darwin Foundation for help in the preparation of Fig. 30.2. Finally, we thank the Galapagos Science Center for providing the facilities for information processing and analysis.

References

Alava JJ, Salazar S (2006) Status and conservation of otariids in Ecuador and the Galapagos Islands. In: Trites AW, Atkinson SK, Demaster DP et al (eds) Sea lions of the world. Fairbanks, Alaska Sea Grant College Program, p 495–519

Alava JJ, Denkinger J, Jiménez PJ et al (2017) Population status, anthropogenic stressors and conservation of the Galapagos fur seal (*Arctocephalus galapagoensis*): an overview. In: Alava JJ (ed) Tropical Pinnipeds: bio-ecology, threats and conservation, 1st edn. CRC Press/Taylor & Francis Group, Abingdon, pp 120–131

Aurioles-Gamboa D, Trillmich F (2018). Guadalupe, Juan Fernández, and Galapagos fur seals: *Arctocephalus townsendi*, *A. philippii*, *A. galapagoensis*. In: Würsig B, Thewissen JGM, Kovacs KM (eds) Encyclopedia of Marine Mammals, 3rd edn. Academic Press, p 435–440

Aurioles-Gamboa D, Schramm Y, Mesnick S (2004) Galapagos fur seals, *Arctocephalus galapagoensis*, in México. Lat Am J Aquat Mamm 3(1):77–80

Berta A, Churchill M (2012) Pinniped taxonomy: review of currently recognized species and subspecies, and evidence used for their description. Mammal Rev 42(3):207–234
Bonner N (1984) Seals of the Galapagos Islands. Biol J Linn Soc 21:177–184
Brunner S (2003) Fur seals and sea lions (Otariidae): identification of species and taxonomic review. Syst Biodivers 1(3):339–439
Capella J, Flórez-González L, Falk P et al (2002) Regular appearance of otariid pinnipeds along the Colombian Pacific coast. Aquat Mamm 28(1):67–72
Capotondi A, Wittenberg A, Newman M et al (2015) Understanding ENSO diversity. Bull Am Meteorol Soc 96:921–938
Ceballos G, Pompa S, Espinoza E et al (2010) Extralimital distribution of Galapagos (*Zalophus wollebaeki*) and northern (*Eumetopias jubatus*) sea lions in Mexico. Aquat Mammal 36 (2):188–194
Churchill M, Boessenecker R, Clementz M (2014) Colonization of the Southern Hemisphere by fur seals and sea lions (Carnivora: otariidae), revealed by combined evidence phylogenetic and Bayesian biogeographic analysis. Zool J Linnean Soc 172:200–225
Clarke MR, Trillmich F (1980) Cephalopods in the diet of fur seals of the Galapagos Islands. J Zool 190:211–215
Costa DP (2009) Energetics. In: Perrin WF, Wursig B, Thewissen JGM (eds) Encyclopedia of marine mammals. Academic, San Diego, pp 383–391
Dellinger T, Trillmich F (1999) Fish prey of the sympatric Galapagos fur seals and sea lions: seasonal variation and niche separation. Can J Zool 77(8):1204–1216
Drago M, Franco-Trecu V, Cardona L et al (2016) Stable isotopes reveal long-term fidelity to foraging grounds in the galapagos sea lion (*Zalophus wollebaeki*). PLoS One 11:e0147857. https://doi.org/10.1371/journal.pone.0147857
Félix F, Jiménez P, Falconí F et al (2007) New cases and first births of the Galapagos fur seal *Arctocephalus galapagoensis* (Heller, 1904) from the mainland coast of Ecuador. Rev Biol Mar Oceanogr 42(1):77–82. https://doi.org/10.4067/S0718-19572007000100009
Gentry RL, Kooyman GL (1987) Fur seals: Maternal Strategies on Land and at Sea. Princeton University Press
Higdon JW, Bininda-Emonds ORF, Beck RMD et al (2007) Phylogeny and divergence of the pinnipeds (Carnivora: Mammalia) assessed using a multigene dataset. BMC Evol Biol 7:216. https://doi.org/10.1186/1471-2148-7-216
Horning M, Trillmich F (1999) Lunar cycles in diel prey migrations exert a stronger effect on the diving of juveniles than adult Galapagos fur seals. Proc R Soc Lond B 266:1127–1132
Ibarra R, Herrera N, Paniagua W et al (2016) Presencia de leones marinos (Carnivora, Pinnipedia, Otariidae) en El Salvador [Presence of sea lions (Carnivora, Pinnipedia, Otariidae) in El Salvador]. Revista Comunicaciones Científicas y Tecnológicas 2(1):82–90
Jeglinski JWE, Werner C, Robinson PW et al (2012) Age, body mass and environmental variation shape the foraging ontogeny of Galapagos sea lions. Mar Ecol Prog Ser 453:279–296
Jeglinski JWE, Goetz KT, Werner C, Costa DP, Trillmich F (2013) Same size–same niche? Foraging niche separation between sympatric juvenile Galapagos sea lions and adult Galapagos fur seals. J Anim Ecol 82:694–706. https://doi.org/10.1111/1365-2656.12019
Jeglinski JW, Wolf JB, Werner C et al (2015) Differences in foraging ecology align with genetically divergent ecotypes of a highly mobile marine top predator. Oecologia 179(4):1041–1052
Kraus C, Mueller B, Meise K et al (2013) Mama's boy: sex differences in juvenile survival in a highly dimorphic large mammal, the Galapagos sea lion. Oecologia 17(4):893–903
Kunc HP, Wolf JB (2008) Seasonal changes of vocal rates and their relation to territorial status in male Galápagos sea lions (*Zalophus wollebaeki*). Ethology 114(4):381–388
Lopes F, Hoffman J, Valiati V et al (2015) Fine-scale matrilineal population structure in the Galapagos fur seal and its implications for conservation management. Conserv Genet 16 (5):1099–1113. https://doi.org/10.1007/s10592-015-0725-1
Martin JH, Coale KH, Johnson KS et al (1994) Testing the iron hypothesis in ecosystems of the equatorial Pacific-Ocean. Nature 371(6493):123–129

Meise K, Krüger O, Piedrahita P et al (2013) Site fidelity of male Galapagos sea lions: a lifetime perspective. Behav Ecol Sociobiol 67(6):1001–1011

Montero-Cordero A, Martínez-Fernández D, Hernández-Mora G (2010) Mammalia, Carnivora, Otariidae, *Arctocephalus galapagoensis* Heller, 1904: first continental record for Costa Rica. Checkl J 6(4):630–632

Montero-Serra I, Páez-Rosas D, Murillo JC et al (2014) Environment-driven changes in terrestrial habitat use and distribution of the Galapagos sea lion. Endanger Species Res 24(1):9–19

Moore S (2008) Marine mammals as ecosystem sentinels. J Mammal 89(3):534–540

Mueller B, Pörschmann U, Wolf JB (2011) Growth under uncertainty: the influence of marine variability on early development of Galapagos sea lions. Mar Mamm Sci 27(2):350–365

Nims BD, Vargas FH, Merkel J (2008) Low genetic diversity and lack of population structure in the endangered Galapagos penguin (*Spheniscus mendiculus*). Conserv Genet 9(6):1413–1420

Okey TA, Banks S, Born AF et al (2004) A trophic model of a Galápagos subtidal rocky reef for evaluating fisheries and conservation strategies. Ecol Model 172(2–4):383–401

Oliveira LA, Brownell RL (2014) Taxonomic status of two subspecies of South American fur seals: *Arctocephalus australis australis* vs. *A a gracilis*. Mar Mamm Sci 30(3):1258–1263

Páez-Rosas D (2011) Ecología trófica de los pinnípedos de las Islas Galápagos: análisis temporal y espacial. PhD Dissertation, Instituto Politécnico Nacional, Mexico

Páez-Rosas D, Aurioles-Gamboa D (2010) Alimentary niche partitioning in the Galapagos sea lion, *Zalophus wollebaeki*. Mar Biol 157(12):2769–2781

Páez-Rosas D, Aurioles-Gamboa D (2014) Spatial variation in the foraging behaviour of the Galapagos sea lions (*Zalophus wollebaeki*) assessed using scat collections and stable isotope analysis. J Mar Biol Assoc UK 94:1099–1107

Páez-Rosas D, Guevara N (2017) Management strategies and conservation status of Galapagos sea lion populations at San Cristóbal Island, Galapagos, Ecuador. In: Alava JJ (ed) Tropical Pinnipeds: bio-ecology, threats and conservation, 1st edn. CRC Press/Taylor & Francis Group, Abingdon, pp 159–175

Páez-Rosas D, Aurioles-Gamboa D, Alava JJ et al (2012) Stable isotopes indicate differing foraging strategies in two sympatric otariids of the Galapagos Islands. J Exp Mar Biol Ecol 424:44–52

Páez-Rosas D, Riofrío-Lazo M, Aurioles-Gamboa D (2014a) Flexibility in the foraging strategies of the Galapagos sea lion inferred from a multiple approach analysis. In: Vinueza L, Denkinger J (eds) The Galapagos marine reserve: a dynamic socio-ecological system. Springer, New York, pp 71–80

Páez-Rosas D, Rodríguez-Pérez M, Riofrío-Lazo M (2014b) Competition influence in the segregation of the trophic niche of otariids: a case study using isotopic bayesian mixing models in Galapagos pinnipeds. Rapid Commun Mass Spectrom 28(23):2550–2558

Páez-Rosas D, Valdovinos LA, Elorriaga-Verplancken FR (2017a) Northernmost record of the Galapagos fur seal (*Arctocephalus galapagoensis*): a consequence of anomalous warm conditions around the Galapagos archipelago. Aquat Mamm 43(6):629–635

Páez-Rosas D, Villegas-Amtmann S, Costa D (2017b) Intraspecific variation infeeding strategies of Galapagos sea lions: a case of trophic specialization. PLoS One 12(10):e0185165

Palacios DM, Bograd SJ, Foley DG et al (2006) Oceanographic characteristics of biological hot spots in the North Pacific: a remote sensing perspective. Deep Sea Res Part 2 Top Stud Oceanogr 53(3–4):250–269

Pennington JT, Mahoney KL, Kuwahara VS et al (2006) Primary production in the eastern tropical Pacific: a review. Progress Ocean 69(2–4):285–317

Peterson RS, Bartholomew GA (1967) The natural history and behavior of the California sea lion. Am Soc Mammal, Spec Pub No. 1, p 79

Piedrahita PM (2015) Mother-offspring relations and transition to independence in the Galapagos sea lion (*Zalophus Wollebaeki*). PhD Dissertation, University of Bielefeld, Germany

Piedrahita P, Meise K, Werner C et al (2014) Lazy sons, self-sufficient daughters: are sons more demanding? Anim Behav 98:69–78

Pörschmann U, Trillmich F, Mueller B et al (2010) Male reproductive success and its behavioural correlates in a polygynous mammal, the Galapagos sea lion (*Zalophus wollebaeki*). Mol Ecol 19 (12):2574–2586

Riedman M (1990) The Pinnipeds: seals, sea lions and walruses. University of California Press, Oxford

Riofrío-Lazo M, Arreguín-Sánchez F, Páez-Rosas (2017) Population abundance of the endangered Galapagos sea lion *Zalophus wollebaeki* in the southeastern Galapagos Archipelago. PLoS One 12(1):e0168829

Sakamoto CM, Millero FJ, Yao WS et al (1998) Surface seawater distributions of inorganic carbon and nutrients around the Galapagos Islands: results from the PlumEx experiment using automated chemical mapping. Deep Sea Res Part 2 Top Stud Oceanogr 45(6):1055–1071

Salazar S, Denkinger J (2010) Possible effects of climate change on the populations of Galapagos pinnipeds. Noticias de Galápagos 67:45–49

Schaeffer BA, Morrison JM, Kamykowski D et al (2008) Phytoplankton biomass distribution and identification of productive habitats within the Galapagos Marine Reserve by MODIS, a surface acquisition system, and in-situ measurements. Remote Sens Environ 112(6):3044–3054

Trillmich F (1979) Galapagos sea lions and fur seals. Noticias de Galápagos 29:8–14

Trillmich F (1981) Mutual mother-pup recognition in Galapagos fur seals and sea lions: cues used and functional significance. Behaviour 8(1):21–42

Trillmich F (1986a) Maternal investment and sex-allocation in the Galapagos fur seal, *Arctocephalus galapagoensis*. Behav Ecol Sociobiol 19(3):157–164

Trillmich F (1986b) Attendance behavior of Galapagos sea lions. In: Gentry RL, Kooyman GL (eds) Fur seals: maternal strategies on land and at sea. Princeton University Press, Princeton, pp 196–208

Trillmich F (1987) Galapagos fur seal, *Arctocephalus galapagoensis*. In: Croxall JP, Gentry RL (eds) Status, biology, and ecology of fur seals. NOAA Technical Report NMFS, vol 51, pp 23–27

Trillmich F (2015a) *Arctocephalus galapagoensis*. The IUCN red list of threatened species 2015:e. T2057A45223722. https://doi.org/10.2305/IUCN.UK.2015-2.RLTS.T2057A45223722.en. Accessed 10 May 2019

Trillmich F (2015b) *Zalophus wollebaeki*. The IUCN red list of threatened species 2015:e. T41668A45230540. https://doi.org/10.2305/IUCN.UK.2015-2.RLTS.T41668A45230540.en. Accessed 10 May 2019

Trillmich F, Dellinger T (1991) The effects of El Niño on Galapagos Pinnipeds. In: Trillmich F, Ono KA (eds) Pinnipeds and El Niño. Ecological studies (analysis and synthesis), vol 88. Springer, Heidelberg, pp 66–74

Trillmich F, Kooyman GL (2001) Field metabolic rate of lactating female Galapagos fur seals (*Arctocephalus galapagoensis*): the influence of offspring age and environment. Comp Biochem Physiol A 129(4):741–749

Trillmich F, Limberger D (1985) Drastic effects of El Niño on Galapagos pinnipeds. Oecologia 67 (1):19–22

Trillmich F, Wolf JB (2008) Parent-offspring and sibling conflict in Galapagos fur seals and sea lions. Behav Ecol Sociobiol 62(3):363–375

Trillmich F, Jeglinski JWE, Meise K et al (2014) The Galapagos sea lion: adaptation to spatial and temporal diversity of marine resources within the Archipelago. In: Denkinger J, Vinueza L (eds) The Galapagos marine reserve. Springer, New York, pp 61–70

Tyack PL (2001) Marine mammal social organization and communication. In: Steele JH (ed) Encyclopedia of ocean sciences, 2nd edn. Academic Press, New York, pp 615–621

Villegas-Amtmann S, Costa DP, Tremblay Y et al (2008) Multiple foraging strategies in a marine apex predator, the Galapagos sea lion *Zalophus wollebaeki*. Mar Ecol Prog Ser 363:299–309

Villegas-Amtmann S, Jeglinski JW, Costa DP et al (2013) Individual foraging strategies reveal niche overlap between endangered Galapagos pinnipeds. PLoS One 8(8):e70748

Villegas-Amtmann S, McDonald BI, Páez-Rosas D et al (2017) Adapted to change: low energy requirements in a low and unpredictable productivity environment, the case of the Galapagos sea lion. Deep Sea Res Part 2 Top Stud Oceanogr 140:94–104

Wolf JBW, Kauermann G, Trillmich F (2005) Males in the shade: habitat use and sexual segregation in the Galápagos sea lion (*Zalophus californianus wollebaeki*). Behav Ecol Sociobiol 59 (2):293–302

Wolf JBW, Tautz D, Trillmich F (2007a) Galápagos and Californian sea lions are separate species: genetic analysis of the genus Zalophus and its implications for conservation management. Front Zool 13:1–13. https://doi.org/10.1186/1742-9994-4-20

Wolf JBW, Mawdsley D, Trillmich F, James R (2007b) Social structure in a colonial mammal: unravelling hidden structural layers and their foundations by network analysis. Anim Behav 74:1293–1302

Wolf JBW, Trillmich F (2008) Kin in space: social viscosity in a spatially and genetically substructured network. Proc R Soc Lond B 275:2063–2069

Wolf JBW, Harrod C, Brunner S et al (2008) Tracing early stages of species differentiation: ecological, morphological and genetic divergence of Galápagos sea lion populations. BMC Evol Biol 8:150. https://doi.org/10.1186/1471-2148-8-150

Index

C. Campagna, R. Harcourt (eds.), *Ethology and Behavioral Ecology of Otariids and the Odobenid*, Ethology and Behavioral Ecology of Marine Mammals, https://doi.org/10.1007/978-3-030-59184-7

C

N

O